D0848573

DISEASES OF THE MOTOR
SYSTEM

HANDBOOK OF
CLINICAL NEUROLOGY

Editors

PIERRE J. VINKEN   GEORGE W. BRUYN
HAROLD L. KLAWANS

Editorial Advisory Board

R. D. ADAMS, S. H. APPEL, E. P. BHARUCHA, M. CRITCHLEY
C. D. MARSDEN, H. NARABAYASHI, A. RASCOL
S. REFSUM, L. P. ROWLAND, F. SEITELBERGER

VOLUME 59

ELSEVIER SCIENCE PUBLISHERS · AMSTERDAM
ELSEVIER SCIENCE PUBLISHING CO., INC. · NEW YORK

DISEASES OF THE MOTOR SYSTEM

Editors

PIERRE J. VINKEN   GEORGE W. BRUYN
HAROLD L. KLAWANS

This volume has been co-edited by

J. M. B. VIANNEY ᴅᴇ JONG

REVISED SERIES 15

ELSEVIER SCIENCE PUBLISHERS · AMSTERDAM
ELSEVIER SCIENCE PUBLISHING CO., INC. · NEW YORK

Library of Congress Cataloging-in-Publication Data

Diseases of the motor system | editors,
Pierre J. Vinken, George W. Bruyn, Harold L. Klawans;
this volume has been co-edited by J.M.B.V. De Jong.
p. cm. — (Handbook of clinical neurology; rev. ser., 15 = v. 59)
Includes bibliographical references and index.
ISBN 0-444-81278-4
1. Movement disorders, 2. Motor neurons — Diseases. I. Vinken, P. J.
II. Bruyn, G. W. III. Klawans, Harold L. IV. Jong, J. M. B.
V. de. V. Series: Handbook of clinical neurology; v. 59.
[DNLM: 1. Movement Disorders. 2. Muscular Atrophy, Spinal.
3. Neuromuscular Diseases. WL 100 H236 v. 59]
RC332.H3 vol. 59 [RC376.5]
616.8 s—dc20 [616.7'4]
DNLM/DLC
for Library of Congress 91-16412
CIP

ISBN for the series: 0 444 90404 2
ISBN for this volume: 0 444 81278 4

212 illustrations, 53 tables

PUBLISHED BY
ELSEVIER SCIENCE PUBLISHERS B.V.
P.O. BOX 1527
1000 BM AMSTERDAM

SOLE DISTRIBUTORS FOR THE USA AND CANADA:
ELSEVIER SCIENCE PUBLISHING CO. INC.
655 AVENUE OF THE AMERICAS
NEW YORK. NY 10010

PRINTED IN THE NETHERLANDS

Foreword

The considerable acceleration in research devoted to the motor system disorders, which we have witnessed over the last decade, has repeatedly delayed the realization of the present volume. Since the original conception as to contents and format, chapters and the mosaic of their arrangement had to be revised several times because of the incessant stream of new data arriving at both the auctorial and editorial benches. However, there comes a moment when the volume has to appear. Undoubtedly, molecular-biological discoveries will have their effect on the contents of the present volume within a few years.

Today's elucidation of causative gene sites within the genome implies a fundamental stride on the path towards final understanding of diseases, which for far too long have been referred to as (for want of a better term) 'heredodegenerations'. Many contemporary readers of this volume will see the day when treatment of these disorders will no longer be pathetically restricted to alleviation or palliation, but will be preventive.

As long as principal clarifications have not been achieved, the Editors have retained (as the reader will notice) the attitude of reluctant splitters rather than of lumpers. Traditional ways of thinking, under the pressure of modern research data, will have to yield to drastic modification or complete replacement. Still, it would be worthwhile, for any younger reader of this volume on 'system' disorders, to obtain the background for daily medical practice: the thoughts of Schaffer, of Spielmeyer and of Spatz in the development of the concept of 'system-atrophies' were lucidly reviewed by van Bogaert (1948). Noble, perfectionistic Spielmeyer, in fact, denied the notion of purely elective maladies nearly three-quarters of a century ago.

The abscissa and ordinate of this volume are topistic (from peripheral to central motor disease) and medical (from diagnosis to treatment). It was felt that it would be of service to both clinically and scientifically active neurologists to provide an updated and corrected WFN classification of motor neurone disorders as first chapter. Having been made possible through the cooperation of the WFN Research Committee, this chapter constitutes a quick orientation as a new feature. Where possible (another new feature), disorders have been provided with the 6-digit identification numbers as listed in McKusick's (1990) catalogue. The

v

chapters on differential diagnosis cannot claim exhaustiveness, but they do cover most of the field.

Although the World Federation of Neurology is taking pains to establish an international standard nomenclature in the near future, our present readers must not forget that the term 'Motor Neurone Disease', mostly used as a collective name for all diseases of the motor system, is a synonym for amyotrophic lateral sclerosis in the United Kingdom. Also, 'demyelination' usually signifies loss of myelin with virtual sparing of axons, except in the United States where the term connotes tract degeneration (cf. Chapter 11 by Tyler and Shefner, pages 169–216).

During the final stages in the preparation of this volume, we were greatly saddened by the loss of two members of our Editorial Advisory Board. On 20th August, 1990, Professor Russell N. DeJong died in Mt. Pleasant, Michigan, and some seven weeks later, on 2nd October, Professor J. Oscar Trelles died in Lima. Russell DeJong joined the Editorial Advisory Board at the inception of the original series of the Handbook in the mid-1960s. He combined the talents of a conscientious and thoroughly experienced clinician and a natural editor in his quiet way of handling matters and was often decisive in the selection and arrangement of specific chapters over the past 60 volumes. We mourn his passing. Oscar Trelles, who joined the Board in 1974, was trained in the French School of neurology in the early 1930s. Throughout the years he was unstinting in sharing with us his expertise and long experience. He has left his mark on all volumes, and 'Viral Diseases', published in 1989, includes one of his last published works. We salute his memory. DeJong and Trelles were grand old men of neurology.

September 1991

P.J.V.
G.W.B.
H.L.K.
J.M.B.V. de J.

REFERENCES

McKUSICK, V. A., C. A. FRANCOMANO and S. E. ANTONARAKIS: Mendelian Inheritance in Man, 9th ed. Baltimore, Johns Hopkins (1990).

VAN BOGAERT, L.: Maladies nerveuses systématisées et problèmes de l'hérédité. Acta Neurol. Psychiat. Belg. 48 (1948) 308–329.

List of contributors

O. Appenzeller
Lovelace Medical Foundation, Research Department, 2425 Ridgecrest Drive SE, Albuquerque, NM 87108, U.S.A.   133

P. G. Barth
Department of Neurology, Academic Medical Centre, University of Amsterdam, Meibergdreef 9, 1105 AZ Amsterdam, The Netherlands   367

A. Bergin
Department of Neurogenetics, Room 503, The Kennedy Institute, 707 N. Broadway, Baltimore, MD 21205, U.S.A.   351

P. A. Bolhuis
Department of Neurology, Academic Medical Centre, University of Amsterdam, Meibergdreef 9, 1105 AZ Amsterdam, The Netherlands   367

G. W. Bruyn
Department of Neurology, Academic Hospital, State University Leiden, Rijnsburgerweg 10, 2333 AA Leiden, The Netherlands   217, 319, 425, 475

R. P. M. Bruyn
Department of Neurology, Oudenrijn Hospital, Van Heuven Goedhartlaan 1, 3527 CE Utrecht, The Netherlands   301, 319, 475

J. M. B. V. De Jong
Department of Neurology, Academic Medical Centre, University of Amsterdam, Meibergdreef 9, 1105 AZ Amsterdam, The Netherlands   1, 383

M. De Visser
Department of Neurology, Academic Medical Centre, University of Amsterdam, Meibergdreef 9, 1105 AZ Amsterdam, The Netherlands   367

R. M. Garruto
Laboratory of Central Nervous System Studies, National Institute of Neurological Disorders and Stroke, National Institutes of Health, Building 36, Room 5B-21, Bethesda, MD 20892, U.S.A.   253

M. Osame
*Third Department of Internal Medicine, Kagoshima University School of
Medicine, Kagoshima, Japan*   447

M. Osawa
*Department of Pediatrics, Tokyo Women's Medical College, 8–1 Kawada-
cho, Shinjuku-ku, Tokyo 162, Japan*   51

G. W. Padberg
*Department of Neurology, University Hospital, Rijnsburgerweg 10, 2333
AA Leiden, The Netherlands*   41

G. C. Román
*Neuroepidemiology Branch, National Institute of Neurological Disorders
and Stroke, National Institutes of Health, Bethesda, MD 20892, U.S.A.* 447

Ph. Scheltens
*Department of Neurology, Free University Hospital, PO Box 7057, 1007
MB Amsterdam, The Netherlands*   301

J. M. Shefner
*Division of Neurology, Brigham and Women's Hospital, Neurology and
Spinal Cord Injury Service, Veteran's Administration Medical Center,
West Roxbury; and Department of Neurology, Harvard Medical School,
75 Francis Street, Boston, MA 02115, U.S.A.*   169

H. Shiraki
*Shiraki Institute of Neuropathology, 6–15–16–301 Honkomagome,
Bunkyo-ku, Tokyo 113, Japan*   273

K. Shishikura
*Department of Pediatrics, Tokyo Women's Medical College, 8–1 Kawa-
dacho, Shinjuku-ku, Tokyo 162, Japan*   51

P. A. E. Sillevis Smitt
*Department of Neurology, Academic Medical Centre, University of
Amsterdam, Meibergdreef 9, 1105 AZ Amsterdam, The Netherlands*   383

C. H. Smith
*Departments of Ophthalmology, Medicine and Neurology, University of
Washington, 1229 Madison, Suite 1490, Seattle, WA 98104, U.S.A.*   325

R. A. Smith
*Center for Neurologic Study, 11211 Sorrento Valley Road, San Diego,
CA 92121, U.S.A.*   459

J. Troost
*Department of Neurology, Medisch Spectrum Twente, Enschede Hospital,
Haaksbergerstraat 55, 7513 ER Enschede, The Netherlands*   97

Contents

Handbook of Clinical Neurology, Vol. 15 (59): Diseases of the Motor System
J.M.B.V. de Jong, editor
© Elsevier Science Publishers B.V., 1991

The World Federation of Neurology classification of spinal muscular atrophies and other disorders of motor neurons*

J. M. B. VIANNEY DE JONG

Department of Neurology, Academic Medical Center, Amsterdam, The Netherlands

In 1988 the World Federation of Neurology (WFN) Research Committee Research Group on Neuromuscular Diseases produced a revised update of its comprehensive classification of the neuromuscular diseases, first published in 1968. After Professor L. P. Rowland of New York and Professor J. G. McLeod of Sydney had completed the revision and the members of the Executive Committee had provided their comments, the classification was published (WFN Research Committee 1988). By permission of the Committee, the first section of the revised classification is reproduced here. Selected references have been given to some conditions that are not described in detail in Walton's (1988) 'Disorders of Voluntary Muscle'. The Committee has used the term amyotrophy not in the literal sense of 'muscular atrophy' but as the equivalent of 'neurogenic atrophy' as opposed to myopathy or primary disease of muscle; it has opted for a genetic classification of the spinal muscular atrophies and other motor disorders.

The catalogue numbers for heritable conditions listed by McKusick et al. (1990) are indicated in parentheses for each disorder. The editors of the present volume have been greatly assisted by the invaluable work of the WFN Committee and

gladly acknowledge their indebtedness. Some errors in the references have been corrected.

SPINAL MUSCULAR ATROPHIES AND OTHER DISORDERS OF MOTOR NEURONS

A. Heritable

A–1 Autosomal recessive: biochemical abnormality unknown

1. Spinal muscular atrophy (SMA) type 1. Infantile spinal muscular atrophy, proximal (Werdnig-Hoffmann) (253300).
2. Spinal muscular atrophy type 2. Infantile spinal muscular atrophy, arrested; spinal muscular atrophy, intermediate type, chronic infantile form (253550).
3. Spinal muscular atrophy type 3. Proximal spinal muscular atrophy, juvenile (Kugelberg-Welander) (253400).
4. Proximal spinal muscular atrophy of adults (271150), indistinguishable from recessive and X-linked forms.
5. Scapuloperoneal muscular atrophy (271220).
6. Distal spinal muscular atrophy of childhood (spinal form of Charcot-Marie-Tooth disease) (271120).
7. Distal spinal muscular atrophy of adults (271120) (Meadows and Marsden 1969;

---

*The classification was first published in Journal of the Neurological Sciences 86 (1988) 333–360.

Harding and Thomas 1980); see also autosomal dominant form.

8. Spinal muscular atrophy with microcephaly and mental retardation (271110) (Spiro et al. 1967).

9. Spinal muscular atrophy, Ryukyuan type (271200) (Kondo et al. 1970).

10. Muscular atrophy, progressive, with chorea and optic atrophy (253500).

11. Arthrogryposis multiplex congenita, neurogenic type (208100) (Drachman and Banker 1961; Krugliak et al. 1978; Hageman et al. 1984).

12. Arthrogryposis multiplex congenita with pulmonary hypoplasia (Pena-Shokeir syndrome type I) (208150) (Moerman et al. 1983).

13. Multiple contracture syndrome, Finnish type (253310); same as Pena-Shokeir syndrome type I, except fatal in neonatal period (Herva et al. 1985).

14. Arthrogryposis-like disorder (Kuskokwim disease (208200), a disorder of Eskimo people, classification uncertain.

15. Charcot-Marie-Tooth disease, progressive ataxia and tremor (214380), consistent with 'neuronal form of Charcot-Marie-Tooth disease' or hereditary motor and sensory neuropathy type II (11821), (see III.A.) 1.b (Harding 1980; Bouchard et al. 1984).

16. Charcot-Marie-Tooth disease (CMT4) (214400), probably same as 15.

17. Bulbar palsy, progressive, of childhood (Fazio-Londe disease) (211500) (Gomez et al. 1962; Albers et al. 1983).

18. Bulbar palsy, progressive, with deafness (pontobulbar palsy with deafness; Brown-Vialetto-van Laere syndrome) (211530) (Boudin et al. 1971; Gallai et al. 1981; Brucher et al. 1981).

19. Bulbar palsy, with deafness and retinitis pigmentosa (Alberca et al. 1980).

20. Bulbar palsy with olivopontocerebellar atrophy (Lapresle and Annabi 1979).

21. Spinal muscular atrophy with amyotrophic lateral sclerosis, spinocerebellar ataxia and deafness (Gemignani 1986).

22. Spinal muscular atrophy with optic atrophy and deafness (Rosenberg et al. 1967; Iwas-

hita et al. 1970; Chalmers and Mitchell 1987).

23. Spinal muscular atrophy with deafness (Rosemberg et al. 1982).

24. Spinal muscular atrophy with ophthalmoplegia (Matsunaga et al. 1973; Dubrovsky et al. 1981).

25. Spinal muscular atrophy with retinitis pigmentosa (Pearn et al. 1978; Furukawa et al. 1968).

26. Spinal muscular atrophy with retinitis pigmentosa and hereditary spastic paraplegia (Kjellin syndrome) (Kjellin 1959; Harding 1984).

27. Troyer syndrome (spastic paraparesis, childhood onset, with distal muscle wasting) (275900) (Cross and McKusick 1967; Neuhäuser et al. 1976).

28. Spinal muscular atrophy and mental retardation (Staal et al. 1975).

29. Spinal muscular atrophy, mental retardation, seizures and orofacial dysplasia (Landau et al. 1976).

30. Spinal muscular atrophy, restricted forms, e.g. one hand (Sobue et al. 1978; Harding et al. 1983), both hands (Serratrice 1984a), quadriceps (Furukawa et al. 1977; Serratrice et al. 1985) (See also G.1 and A.3).

31. Spinal muscular atrophy with recessive spinocerebellar degeneration or Friedreich ataxia (229300) (Singh and Sham 1964; Boudouresques et al. 1971).

32. Spinal muscular atrophy with spastic paraplegia, mental retardation and ichthyosis (Sjögren-Larsson syndrome) (270200) (Sjögren and Larsson 1957; McNamara et al. 1975). See also V.A-2.j for congenital myopathy with Sjögren-Larsson syndrome.

33. Neuronal intranuclear hyaline inclusion disease (Sung et al. 1980); single sporadic case, manifest by amyotrophy, dementia, choreoathetosis, seizures, sphincter disorder.

34. Spinal muscular atrophy or axonal neuropathy in xeroderma pigmentosum (278800) (Thrush et al. 1974); spinal cord may be normal at autopsy (Röytta and Anttinen 1986).

35. Spinal muscular atrophy in infantile neu-

roaxonal dystrophy (INAD; Seitelberger disease) (256600) (Huttenlocher 1967; Ule 1972).

36. Spinal muscular atrophy in Hallervorden-Spatz disease (neuroaxonal dystrophy, late infantile) (234200) (Bots and Gilles 1973).

37. Spinal muscular atrophy in amyotrophic choreo-acanthocytosis (100500); inheritance patterns may seem either recessive or dominant (Serra et al. 1986).

A–2 Autosomal recessive, biochemical abnormality known

1. Spinal muscular atrophy with hexosaminidase deficiency (272800).

2. Spinal muscular atrophy with lysosomal enzyme deficiencies (Goto et al. 1983).

3. Spinal muscular atrophy with phenylketonuria (261600) (Meier et al. 1975).

4. Spinal muscular atrophy with hydroxyisovaleric aciduria (Eldjarn et al. 1970) (210200).

5. Spinal muscular atrophy or amyotrophic lateral sclerosis with ceroid lipofuscinosis (Iseki et al. 1987; Ashwal et al. 1984).

A–3 Autosomal dominant, biochemical abnormality unknown

1. Muscular atrophy, ataxia, retinitis pigmentosa and diabetes mellitus (158500) (Furukawa et al. 1968b).

2. Muscular atrophy, juvenile spinal (Kugelberg-Welander) (158600) proximal type, usually autosomal recessive, but dominant forms reported (Tsukagoshi et al. 1966; Pearn 1978). Variants include the monomelic form which is usually sporadic (see also G.1 and A-1.30) (Sobue et al. 1978) and SMA of facioscapulohumeral distribution (Fenichel et al. 1967; Furukawa and Toyokura 1976); problems of classification were discussed (Hausmanowa-Petrusewicz et al. 1985; Zerres et al. 1983).

3. Muscular atrophy, progressive, malignant (158650), fatal within one year, but not clearly different from autosomal dominant amyotrophic lateral sclerosis (Zatz et al. 1971; Manta et al. 1983).

4. Muscular atrophy, progressive, with amyotrophic lateral sclerosis (158700); also clearly different from ALS (105400 below A-3.6 or 'peroneal muscular atrophy with pyramidal features' (Harding and Thomas 1984).

5. Amyotrophic dystonic paraplegia (105300) (Gilman and Horenstein 1964).

6. Amyotrophic lateral sclerosis (105400), includes typical forms (Mulder et al. 1986), as well as those with degeneration of posterior columns and other tracts in spinal cord (Engel et al. 1959), and forms associated with dementia, parkinsonism or tics (Hudson 1981; Spitz et al. 1985). Inheritance of Madras type uncertain, with predilection for bulbar palsy and hearing loss (Arjundas 1977).

7. Amyotrophic lateral sclerosis-parkinsonism-dementia complex of Guam (105500), probably not heritable (Garruto et al. 1985; Spencer et al. 1987).

8. Spinal muscular atrophy with olivopontocerebellar atrophy type IV (OPCA IV; Schut-Haymaker-type OPCA) (164600) (Landis et al. 1974).

9. Amyotrophic lateral sclerosis with dementia (105550): said to differ from other forms (Pinsky et al. 1975).

10. Distal spinal muscular atrophy with vocal cord paralysis (Young and Harper 1980; Serratrice et al. 1984b).

11. Spinal muscular atrophy with bulbar palsy (Dobkin and Verity 1976).

12. Arthrogryposis multiplex congenita, distal, type 1 (108120); uncertain whether this is neurogenic or myopathic (Hall et al. 1982a; McCormack et al. 1980; Fleury and Hageman 1985).

13. Charcot-Marie-Tooth disease, neuronal type (CMT3) (118210), includes forms linked to Duffy blood group on chromosome 1, and those that are not linked (Guiloff et al. 1982); related to spinal muscular atrophy, distal type?

14. Scapuloperoneal atrophy with cardiopathy (Emery-Dreifuss muscular dystrophy, autosomal dominant type) (181350; also 158700); neurogenic status uncertain (Cha-

krabati et al. 1981; Fenichel et al. 1982;
Miller et al. 1985).

15. Scapuloperoneal amyotrophy (Kaeser syndrome) (scapuloperoneal syndrome, neurogenic type) (181400).

16. Spastic paraplegia with amyotrophy of
hands (Silver disease) (182700) (Silver 1966;
Van Gent et al. 1985).

17. Spinal muscular atrophy, distal (182960);
similar to recessive form (Harding and
Thomas 1980); may also be sporadic and
confined to the hands (O'Sullivan and
McLeod 1978).

18. Spinal muscular atrophy, facioscapulohumeral type (182970) (Fenichel et al. 1967).

19. Spinal muscular atrophy, proximal, adult
type (Finkel late-onset type SMA included)
(182980) (Richieri-Costa et al. 1981). Clinically not different from recessive and X-
linked forms.

20. Spinocerebellar ataxia with rigidity and
peripheral neuropathy (183050) (Ziegler et
al. 1972); includes fasciculations and parkinsonism as well as signs of peripheral
neuropathy.

21. Spinal muscular atrophy with myoclonus
(159950) (Jankovic and Rivera 1979).

22. Spinal muscular atrophy with Joseph disease (109150) (Also Azorean neurologic
disease, Machado-Joseph disease, spino-
pontine atrophy, nigrospinodentatal degeneration) (Rosenberg and Fowler 1981).

23. Scapuloperoneal atrophy with cardiopathy
and inflammatory myopathy (Jennekens et
al. 1975); neurogenic status not certain.

24. Amyotrophic lateral sclerosis with Pick
disease of brain (dementia with lobar atrophy and neuronal cytoplasmic inclusions)
(172700) (De Morsier 1967).

25. Amyotrophic lateral sclerosis with autosomal dominant familial parkinsonism
(168600) (Brait et al. 1973).

26. Spinal muscular atrophy with Huntington
disease (143100) (Fotopulos and Schulz
1966; Serratrice et al. 1984a).

27. Amyotrophic lateral sclerosis with Luyso-
pallidal-nigral atrophy (Gray et al. 1985).

28. Spinal muscular atrophy with pallidonigral
degeneration (Serratrice et al. 1983).

*A–4   X-linked recessive, biochemical disorder
unknown*

1. Spinal and bulbar muscular atrophy (Kennedy disease); bulbospinal muscular atrophy; X-linked muscular atrophy, benign,
with hypertrophy of calves (313200). Kennedy disease localised to Xq28. Juvenile
form, with calf enlargement, probably
different condition (Bouwsma and Van
Wijngaarden 1980); may be associated with
hyperlipidaemia (Quarfordt et al. 1970).

2. Spinal muscular atrophy, juvenile, proximal; similar to Kugelberg-Welander syndrome but X-linked, (313200); (Tsukagoshi
et al. 1970); possibly same as Kennedy
syndrome.

3. Scapuloperoneal muscular atrophy with
cardiopathy, X-linked (310300); (Mawatari
et al. 1973); probably Emery-Dreifuss muscular dystrophy (Rowland et al. 1979).

4. Charcot-Marie-Tooth peroneal muscular
atrophy and Friedreich ataxia, combined
(302900); uncertain whether anterior horn
cell disease or peripheral neuropathy, based
on single report (Van Bogaert and Moreau
1939–1941).

5. Spinal muscular atrophy, X-linked, facioscapulohumeral distribution (Skre et al.
1978).

6. Arthrogryposis multiplex congenita with
renal and hepatic abnormality (301820);
includes rarefaction of anterior horns, reported in only one family (Nezelof et al.
1979).

7. Arthrogryposis multiplex congenita, distal
(301830), X-linked; includes one type with
anterior horn cell disease; others attributed
to non-progressive intrauterine myopathy
or connective tissue disease (Hall et al.
1982b).

A–5   X-linked dominant, lethal in males

1. Infantile spinal muscular atrophy in incontinentia pigmenti (Bloch-Sulzberger syndrome) (308300) (Larsen et al. 1987).

B. Congenital and developmental abnormalities

1. Möbius syndrome (agenesis of cranial nerve
nuclei); Möbius syndrome with peripheral

neuropathy and hypogonadism (Abid et al. 1978), Möbius syndrome with absence of pectoral muscle.

2. Congenital absence of muscles (e.g., pectorals, abdominals), might better be classed with disorders of muscle, but it is not known whether total motor units (including anterior horn cells) are absent. Congenital absence of pectoral muscle with syndactyly (David 1972).

3. Amyotrophy with developmental anomalies of the spinal cord or nerve roots: hydromyelia, syringomyelia or syringobulbia (often with Chiari malformation); spinal dysraphism; meningomyelocele; aplasia of spinal cord (amyelia).

4. Spinal muscular atrophy with pontocerebellar hypoplasia (Goutières et al. 1977).

5. Arthrogryposis multiplex congenita of nonneural, non-myopathic origin, sporadic, presumably not heritable and presumably intrauterine or developmental disorder of joints; accounts for most cases.

C. Disorders of motor neurons attributed to physical causes

1. Trauma: direct injury of spinal cord (birth injury; spinal fracture); traumatic haematomyelia.

2. Amyotrophy due to destruction or compression or compressive ischemia of anterior horn cells: intramedullary or extramedullary spinal cord tumours; infectious mass lesions (tuberculoma, gumma, parasitic cysts); spontaneous hydromyelia.

3. Amyotrophy due to ischaemia of anterior horns:
a. Clamping aorta for abdominal vascular surgery.
b. Occlusion or stenosis of anterior spinal artery.
c. Progressive vascular myelopathy (Jellinger et al. 1962).

4. Amyotrophy after electrical injury, mechanism uncertain (Farrell and Neumayer 1968; Holbrook et al. 1970).

5. Amyotrophy after radiotherapy (Lagueny and Starr 1985).

D. Disorders of motor neurons attributed to toxins, chemicals or heavy metals

1. Tetanus toxin.
2. Strychnine.
3. Botulinum toxin (adult and infantile forms) (Clay et al. 1977; Arnon, 1980).
4. Lead (Campbell et al. 1970; Boothby et al. 1974).
5. Mercury (Kantarjian 1961).
6. Organophosphates (tri-*ortho*-cresyl phosphate, ginger jake paralysis).
7. Saxitoxin and related marine toxins.
8. Dapsone motor neuropathy (not clear whether this is an axonal peripheral neuropathy or motor neuron disorder) (Homeida et al. 1980).
9. Phenytoin motor neuropathy (Direkze and Fernando 1977).

E. Disorders of motor neurons attributed to viral infection

E–1 Acute disorders
1. Paralytic acute anterior poliomyelitis:
a. Due to poliomyelitis virus.
b. Due to other enteroviruses (such as coxsackievirus).
2. Amyotrophy in Russian spring-summer encephalitis.
3. Herpes zoster (Thomas and Howard 1972).
4. Amyotrophy with acute haemorrhagic conjunctivitis (Bharucha et al. 1972; Kono et al. 1974).
5. Amyotrophy with asthma (Beede et al. 1980).
6. Amyotrophy in acute transverse myelitis, cause undetermined.

E–2 Subacute or chronic disorders
1. Amyotrophy in Creutzfeldt-Jakob disease (Allen et al. 1971) (123400).
2. Amyotrophy or amyotrophic lateral sclerosis due to human immunodeficiency virus (Horoupian et al. 1984; Hoffman et al. 1985).
3. Persistent infection by poliovirus in agammaglobulinaemia.
4. Late post-poliomyelitis muscular atrophy

(post-polio syndrome) (Dalakas et al. 1984; Cashman et al. 1987).

5. Syndrome of benign fasciculation and cramps in survivors of paralytic poliomyelitis or other forms of acute myelitis (Fetell et al. 1982).

6. Amyotrophy after encephalitis lethargica (Greenfield and Matthews 1954).

F. Disorders of motor neurons with immunological abnormality

1. Motor neuron diseases with monoclonal paraproteinaemia (including Waldenström macroglobulinaemia, multiple myeloma, chronic lymphatic leukaemia) (Shy et al. 1986).

2. Amyotrophy with Hodgkin's disease (Rowland and Schneck 1963; Walton et al. 1968; Schold et al. 1979).

3. Carcinomatous motor neuron disease (Brain et al. 1965) not currently regarded as a specific syndrome, but amyotrophy may be seen in paraneoplastic encephalomyelopathy, or may improve after removal of tumour (Mitchell and Olczak 1979).

G. Disorders of motor neurons of undetermined aetiology

1. Motor neuron diseases of adults (sporadic): amyotrophic lateral sclerosis, progressive spinal muscular atrophy, progressive bulbar amyotrophic lateral sclerosis, or mixed forms.
   Includes sporadic cases with atypical features: e.g. ophthalmoplegia (Harvey et al. 1979); nystagmus (Kushner et al. 1984); sporadic multisystem disorders with amyotrophy, cerebellar disorder and ophthalmoplegia (Hayashi et al. 1986); monomelic and other restricted forms (Serratrice 1983; Hirayama et al. 1987).

2. Juvenile motor neuron diseases (sporadic): spinal muscular atrophy or amyotrophic lateral sclerosis (Van Bogaert 1925; Nelson and Prensky 1972).

3. Amyotrophy in Shy-Drager syndrome (progressive multisystem degeneration, progressive autonomic failure of central origin), with orthostatic hypotension, sphincter disorders, parkinsonism or cerebellar disorder.

4. Amyotrophy in sporadic Pick disease (Minauf and Jellinger 1969).

5. Chronic neurogenic atrophy of the quadriceps (sporadic) (Furukawa et al. 1977; Serratrice et al. 1985).

H. Disorders of motor neurons in endocrine disorders

1. Tetany (hypocalcaemia; hypomagnesaemia; alkalosis).

2. Amyotrophy in hypoglycaemic hyperinsulinism (Harrison 1976).

3. Amyotrophy, fasciculation and upper motor neuron signs (amyotrophic lateral sclerosis) in hyperthyroidism (Mottier et al. 1981; Fisher et al. 1985).

4. Amyotrophy in hyperparathyroidism (Patten and Pages 1984).

I. Disorders of motor neurons manifest by hyperactivity

(Note): There is debate about the pathogenesis of several syndromes of hyperactivity; many believe that only peripheral nerves are involved but others implicate the perikaryon, alone or in addition to peripheral nerve disorder; this is discussed in Rowland (1985). (Some may be forms of dystonia or may originate in upper motor neurons.)

1. Ordinary muscle cramps.

2. Benign fasciculation-cramp syndrome (syndrome of Foley and Denny-Brown).

3. Occupational cramps and writers' cramp (focal action-induced dystonia).

4. Isaacs syndrome (neuromyotonia; Isaacs-Mertens syndrome), manifest by myokymia, impaired muscle relaxation (pseudomyotonia) and abnormal postures of the limbs. Most cases are sporadic; a few are paraneoplastic (Walsh 1976) and some are heritable (McGuire et al. 1984).

5. Tetanus.

6. Strychnine intoxication.

7. Stiff man syndrome (Moersch-Woltman syndrome).

8. Satoyoshi syndrome (Satoyoshi et al. 1967). See also V.D.–3.7, original WFN document.

9. Myelopathy with rigidity, spasm or continuous motor unit activity (Howell et al. 1979; Whiteley et al. 1976).

10. Myokymia-hyperhidrosis-impaired muscle relaxation (Gamstorp et al. 1959); probably the same as the Isaacs syndrome.

11. Black widow spider bite.

12. Tetany.

13. Spinal myoclonus.

14. Facial myokymia.

15. Hemifacial spasm (see also III C-1 3a, original WFN document).

16. ?'Painful legs and moving toes' (Spillane et al. 1971).

17. ?Ekbom syndrome ('restless legs') (Harriman et al. 1970).

REFERENCES

ABID, F., R. HALL, P. HUDGSON and R. WEISER: Moebius syndrome, peripheral neuropathy and hypogonadotrophic hypogonadism. J. Neurol. Sci. 35 (1978) 309–315.

ALBERCA, R., C. MONTERO, A. IBAÑEZ, D. I. SEGURA and G. MIRANDA-NIEVES: Progressive bulbar paralysis associated with neural deafness. Arch. Neurol. (Chic.) 37 (1980) 214–216.

ALBERS, J. W., S. ZIMNOWODZKI, C. M. LOWREY and B. MILLER: Juvenile progressive bulbar palsy. Arch. Neurol. (Chic.) 40 (1983) 351–353.

ALLEN, I. V., E. DERMOTT, J. H. CONNOLLY and L. J. HURWITZ: A study of a patient with the amyotrophic form of Creutzfeldt-Jakob disease. Brain 94 (1971) 715–724.

ARJUNDAS, G.: Experiences with neuromuscular disorders in South India. Neurology (India) 25 (1977) 1–18.

ARNON, S. S.: Infant Botulism. Annu. Rev. Med. 31 (1980) 541–560.

ASHWAL, S., TH. V. THRASHER, D. R. RICE and D. A. WENGER: A new form of sea-blue histiocytosis associated with progressive anterior horn cell and axonal degeneration. Ann. Neurol. 16 (1984) 184–192.

BEEDE, H. E. and R. W. NEWCOMB: Lower motor neuron paralysis in association with asthma. Johns Hopkins Med. J. 147 (1980) 186–187.

BHARUCHA, E. P., V. P. MONDKAR, N. H. WADIA, P. F. IRANI and S. M. KATRAK: Neurological complications of a new conjunctivitis. Lancet 2 (1972) 970–971.

BOOTHBY, J. A., P. V. DE JESUS and L. P. ROWLAND: Reversible forms of motor neuron disease. Arch. Neurol. (Chic.) 31 (1974) 18–23.

BOTS, G. TH. A. M. and A. STAAL: Amyotrophic lateral sclerosis-dementia complex, neuroaxonal dystrophy, and Hallervorden-Spatz disease. Neurology 23 (1973) 35–39.

BOUDIN, G., B. PEPIN, J. C. VERNANT, B. GAUTIER and H. GOUEROU: Cas familial de paralysie bulbo-pontine chronique progressive avec surdité. Rev. Neurol. 124 (1971) 90–92.

BOUDOURESQUES, J., M. TOGA, R. KHALIL, A. GOSSET, R. A. VIGOUROUX and J. F. PELLISSIER: Forme amyotrophique d'une dégénérescence spino-cérébelleuse. Rev. Neurol. (Paris) 125 (1971) 25–38.

BOUWSMA, G. and G. K. VAN WIJNGAARDEN: Spinal muscular atrophy and hypertrophy of the calves. J. Neurol. Sci. 44 (1980) 275–279.

BRAIN, R., P. B. CROFT and M. WILKINSON: Motor neurone disease as a manifestation of neoplasm. Brain 88 (1965) 479–500.

BRAIT, K., S. FAHN and G. A. SCHWARZ: Sporadic and familial parkinsonism and motor neuron disease. Neurology 23 (1973) 990–1002.

BRUCHER, J. M., R. DOM, A. LOMBAERT and H. CARTON: Progressive bulbar palsy with deafness. Clinical and pathological study of two cases. Arch. Neurol. (Chic.) 38 (1981) 186–190.

CAMPBELL, A. M. G., E. R. WILLIAMS and D. BARLTROP: Motor neurone disease and exposure to lead. J. Neurol. Neurosurg. Psychiatry 33 (1970) 877–885.

CASHMAN, N. R., R. MASELLI, R. J. WOLLMANN, R. ROOS, R. SIMON and J. P. ANTEL: Late denervation in patients with antecedent paralytic poliomyelitis. N. Engl. J. Med. 317 (1987) 7–12.

CHAKRABATI, A. and J. M. S. PEARCE: Scapuloperoneal syndrome with cardiomyopathy: report of a family with autosomal dominant inheritance and unusual features. J. Neurol. Neurosurg. Psychiatry 44 (1981) 1146–1152.

CHALMERS, D. and J. D. MITCHELL: Optico-acoustic atrophy in distal spinal muscular atrophy. J. Neurol. Neurosurg. Psychiatry 50 (1987) 238–239.

CLAY, S. A., J. C. RAMSEYER, L. S. FISHMAN and R. P. SEDGWICK: Acute infantile motor unit disorder. Arch. Neurol. (Chic.) 34 (1987) 246–249.

CORNELL, J., S. SELLARS and P. BEIGHTON: Autosomal recessive inheritance of Charcot-Marie-Tooth disease associated with sensorineural deafness. Clin. Genet. 25 (1984) 163–165.

CROSS, H. E. and V. A. MCKUSICK: The Troyer syndrome: a recessively inherited form of spastic paraplegia with distal muscle wasting. Arch. Neurol. (Chic.) 16 (1967) 473–485.

DALAKAS, M. C., J. L. SEVER, D. L. MADDEN, N. M. PAPADOPOULOS, I. C. SHEKARCHI, P. ALBRECHT and A. KREZLEWICZ: Late postpoliomyelitis muscular atrophy: clinical, virologic, and immunologic studies. Rev. Inf. Dis. 6 (1984) Suppl. 2: S 562-S 567.

DAVID, T. J.: Nature and etiology of the Poland anomaly. N. Engl. J. Med 287 (1972) 487–489.

DE MORSIER, G.: Un cas de maladie de Pick avec sclérose latérale amyotrophique terminale. Contribution à la sémiologie temporale. Rev. Neurol. (Paris) 116 (1967) 373–382.

DIREKZE, M. and P. S. L. FERNANDO: Transient anterior horn cell dysfunction in diphenylhydantoin therapy. Eur. Neurol. 15 (1977) 131–134.

DOBKIN, B. H. and M. A. VERITY: Familial progressive bulbar and spinal muscular atrophy. Neurology 26 (1976) 754–763.

DRACHMAN, D. B. and B. Q. BANKER: Arthrogryposis multiplex congenita. Arch. Neurol. (Chic.) 5 (1961) 77–93.

DUBROVSKY, A., L. TARATUTO and R. MARTINO: Distal spinal muscular atrophy and ophthalmoparesis. A case with selective type 2 fiber hypotrophy. Arch. Neurol. 38 (1981) 594–596.

ELDJARN, L., E. JELLUM, O. STOKKE, H. PANDE and P. E. WARLER: β-Hydroxyisovaleric aciduria and β-methylcrotonylglycinuria: a new inborn error of metabolism. Lancet ii (1970) 521–522.

ENGEL, W. K., L. T. KURLAND and I. KLATZO: An inherited disease similar to amyotrophic lateral sclerosis with a pattern of posterior column involvement. An intermediate form? Brain 82 (1959) 203–220.

FARRELL, D. F. and A. STARR: Delayed neurological sequelae of electrical injuries. Neurology 18 (1969) 601–606.

FENICHEL, G. M., E. S. EMERY and P. HUNT: Neurogenic atrophy simulating facioscapulohumeral dystrophy. Arch. Neurol. (Chic.) 17 (1967) 257–260.

FENICHEL, G. M., Y. CHU SUL, A. W. KILROY and R. BLOUIN: An autosomal-dominant dystrophy with humeropelvic distribution and cardiomyopathy. Neurology (NY) 32 (1982) 1399–1401.

FETELL, M. R., G. SMALLBERG, L. D. LEWIS, R. E. LOVELACE, A. P. HAYS and L. P. ROWLAND: A benign motor neuron disorder: delayed cramps and fasciculation after poliomyelitis or myelitis. Ann. Neurol. 11 (1982) 423–427.

FISHER, J., J. E. MATEER, I. ULRICH and J. A. GUTRECHT: Pyramidal tract deficits and polyneuropathy in hyperthyroidism. Am. J. Med. 79 (1985) 1041–1044.

FLEURY, P. and G. HAGEMAN: A dominantly inherited lower motor neuron disorder presenting at birth with associated arthrogryposis. J. Neurol. Neurosurg. Psychiat. 48 (1985) 1037–1048.

FOTOPULOS, D.: Huntington Chorea und chronisch-progressive spinale Muskelatrophie. Psychiat. Neurol. Med. Psychol. (Leipzig) 18 (1966) 63–70.

FURUKAWA, T. and Y. TOYOKURA: Chronic spinal muscular atrophy of facioscapulohumeral type. J. Med. Genet. 13 (1976) 285–289.

FURUKAWA, T., A. TAKAGI, K. NAKAO, H. SUGITA, H. TSUKAGOSHI and T. TSUBAKI: Hereditary muscular atrophy with ataxia, retinitis pigmentosa, and diabetes mellitus. Neurology 18 (1968) 942–947.

FURUKAWA, T., N. AKAGAMI and S. MARUYAMA: Chronic neurogenic quadriceps amyotrophy. Ann. Neurol. 2 (1977) 528–530.

GALLAI, V., J. M. HOCKADAY, J. T. HUGHES, D. J. LANE, D. R. OPPENHEIMER and G. RUSHWORTH: Ponto-bulbar palsy with deafness (Brown-Vialetto-van Laere syndrome). J. Neurol. Sci. 50 (1981) 259–275.

GARRUTO, R. M., R. YANAGIHARA and D. C. GAJDUSEK: Disappearance of high-incidence amyotrophic lateral sclerosis and parkinsonism-dementia of Guam. Neurology 35 (1985) 193–198.

GEMIGNANI, F.: Spinocerebellar ataxia with localized amyotrophy of the hands, sensorineural deafness and spastic paraparesis in two brothers. J. Neurogenet. 3 (1986) 125–133.

GILMAN, S. and S. HORENSTEIN: Familial amyotrophic dystonic paraplegia. Brain 87 (1964) 51–66.

GOMEZ, M. R., V. CLERMONT and J. BERNSTEIN: Progressive bulbar paralysis in childhood (Fazio-Londe's disease). Arch. Neurol. (Chic.) 6 (1962) 77–83.

GOTO, I., H. NAKAI, T. TABIRA, Y. TANAKA, H. SHIBASAKI and Y. KUROIWA: Juvenile neurogenic muscle atrophy with lysosomal enzyme deficiencies: new disease variant of mucopolysaccharidosis. J. Neurol. 229 (1983) 45–54.

GOUTIÈRES, F., J. AICARDI and E. FARKAS: Anterior horn cell disease associated with pontocerebellar hypoplasia in infants. J. Neurol. Neurosurg. Psychiatry 40 (1977) 370–378.

GRAY, F., J. F. EIZENBAUM, J. D. DEGOS and J. POIRIER: Luyso-pallidal-nigral atrophy and amyotrophic lateral sclerosis. Acta Neuropathol. (Berl.) 66 (1985) 78–82.

GREENFIELD, J. G. and W. B. MATTHEWS: Post-encephalitic Parkinsonism with amyotrophy. J. Neurol. Neurosurg. Psychiat. 17 (1954) 50–56.

GUILOFF, R. J., P. K. THOMAS, M. CONTRERAS, S. ARMITAGE, G. SCHWARZ and E. M. SEDGWICK: Evidence for linkage of type I hereditary motor and sensory neuropathy to the Duffy locus on chromosome 1. Ann. Hum. Genet. 46 (1982) 25–27.

HAGEMAN, G., F. G. I. JENNEKENS, J. K. VETTE and J. WILLEMSE: The heterogeneity of distal arthrogryposis. Brain Dev. Neurol. 6 (1984) 273–283.

HALL, J. G., S. D. REED and G. GREENE: The distal arthrogryposes: delineation of new entities — review and nosologic discussion. Am. J. Med. Genet. 11 (1982) 185–239.

HALL, J. G., S. D. REED, C. I. SCOTT, J. G. ROGERS, K. L. JONES and A. CAMARANO: Three distinct types of X-linked arthrogryposis seen in 6 families. Clin. Genet. 21 (1982) 81–97.

HARDING, A. E.: The Hereditary Ataxias and Related Disorders. Edinburgh, Churchill-Livingstone (1984) 266 pp.

HARDING, E. A. and P. K. THOMAS: Hereditary distal spinal muscular atrophy. J. Neurol. Sci. 45 (1980) 337–348.

HARDING, E. A. and P. K. THOMAS: Peroneal muscular atrophy with pyramidal features. J. Neurol. Neurosurg. Psychiatry 47 (1984) 168–172.

HARDING, E. A., P. G. BRADBURY and N. M. F. MURRAY: Chronic asymmetrical spinal muscular atrophy. J. Neurol. Sci. 59 (1983) 69–83.

HARRIMAN, D. G. F., D. TAVERNER and A. I. WOOLF: Ekbom's syndrome and burning paraesthesiae. Brain 93 (1970) 393–406.

HARRISON, M. J. G.: Muscle wasting after prolonged hypoglycaemic coma: case report with electrophysiological data. J. Neurol. Neurosurg. Psychiatry 39 (1976) 465–470.

HARVEY, D. G., R. M. TORACK and H. E. ROSENBAUM: Amyotrophic lateral sclerosis with ophthalmoplegia. Arch. Neurol. (Chic.) 36 (1979) 615–617.

HAUSMANOWA-PETRUSEWICZ, I., J. ZAREMBA and J. BORKOWSKA: Chronic spinal muscular atrophy of childhood and adolescence: problems of classification and genetic counselling. J. Med. Genet. 22 (1985) 350–353.

HAYASHI, Y., K. NAGASHIMA, Y. URANO and M. IWATA: Spinocerebellar degeneration with prominent involvement of the motor neuron system: autopsy report of a sporadic case. Acta Neuropathol. (Berl.) 70 (1986) 82–85.

HERVA, R., J. LEISTI, P. KIRKINEN and U. SEPPANEN: A lethal autosomal recessive syndrome of multiple congenital contractures. Am. J. Med. Genet. 20 (1985) 431–439.

HIRAYAMA, K., M. TOMONAGA, K. KITANO, T. YAMADA, S. KOJIMA and K. ARAI: Focal cervical poliopathy causing juvenile muscular atrophy of distal upper extremity: a pathological study. J. Neurol. Neurosurg. Psychiatry 50 (1987) 285–290.

HOFFMAN, P. M., B. W. FESTOFF, L. T. GIRON, L. C. HOLLENBECK, R. M. GARRUTO and F. W. RUSCETTI: Isolation of LAV/HTLV-III from a patient with amyotrophic lateral sclerosis. N. Engl. J. Med. 313 (1985) 324–325.

HOLBROOK, L. A., F. X. M. BEACH and J. R. SILVER: Delayed myelopathy: a rare complication of severe electric burns. Br. Med. J. 4 (1970) 659–660.

HOMEIDA, M., A. BABIKR and T. K. DANESHMEND: Dapsone induced optic atrophy and motor neuropathy. Br. Med. J. 281 (1980) 1180.

HOROUPIAN, D. S., P. PICK, I. SPIGLAND, P. SMITH, R. PORTENOY, R. KATZMAN and S. CHO: Acquired immune deficiency syndrome and multiple tract degeneration in a homosexual man. Ann. Neurol. 15 (1984) 502–505.

HOWELL, D. A., A. J. LEE and P. J. TOGHILL: Spinal internuncial neurones in progressive encephalomyelitis with rigidity. J. Neurol. Neurosurg. Psychiat. 42 (1979) 773–785.

HUDSON, A. J.: Amyotrophic lateral sclerosis and its association with dementia. Brain 104 (1981) 217–247.

HUTTENLOCHER, P. R. and F. H. GILLES: Infantile neuroaxonal dystrophy. Neurology 17 (1967) 1174–1184.

ISEKI, E., N. AMANO, S. YOKOI, Y. YAMADA, K. SUZUKI and M. YAZAKI: A case of adult neuronal ceroid-lipofuscinosis with the appearance of membranous cytoplasmic bodies localized in the spinal anterior horn. Acta Neuropathol. (Berl.) 72 (1987) 362–368.

IWASHITA, H., N. INOUE, S. ARAKI and Y. KUROIWA: Optic atrophy, neural deafness, and distal neurogenic amyotrophy. Arch. Neurol. (Chic.) 22 (1970) 357–364.

JANKOVIC, J. J. and V. M. RIVERA: Hereditary myoclonus and progressive distal muscular atrophy. Ann. Neurol. 6 (1979) 227–231.

JELLINGER, K. and E. NEUMAYER: Myélopathie progressive d'origine vasculaire. Acta Neurol. Psychiat. Belg. 62 (1962) 944–956.

JENNEKENS, F. G. I., H. F. M. BUSCH, N. M. VAN HEMEL and R. A. HOOGLAND: Inflammatory myopathy in scapulo-ilio-peroneal atrophy with cardiopathy. Brain 98 (1975) 709–722.

KANTARJIAN, A. D.: A syndrome clinically resembling amyotrophic lateral sclerosis following chronic mercurialism. Neurology 11 (1961) 639–644.

KJELLIN, K.: Familial spastic paraplegia with amyotrophy, oligophrenia, and central retinal degeneration. Arch. Neurol. (Chic.) 1 (1959) 133–140.

KONDO, K., T. TSUBAKI and F. SAKAMOTO: The Ryukyan muscular atrophy. An obscure heritable neuromuscular disease found in the islands of Southern Japan. J. Neurol. Sci. 11 (1970) 359–382.

KONO, R., K. MIYAMURA, E. TAJIRI, S. SHIGA, A. SASAGAWA, P. F. IRANI, S. M. KATRAK and N. H. WADIA: Neurologic complications associated with acute hemorrhagic conjunctivitis virus infection and its serologic confirmation. J. Infect. Dis. 129 (1974) 590–593.

KRUGLIAK, L., N. GADOTH and A. J. BEHAR: Neuropathic form of arthrogryposis multiplex congenita: report of 3 cases with complete necropsy, including the first reported case of agenesis of muscle spindles. J. Neurol. Sci. 37 (1978) 179–185.

KUSHNER, M. J., M. PARISH, A. BURKE, M. BEHRENS, A. P. HAYS, B. FRAME and L. P. ROWLAND: Nystagmus in motor neuron disease. Clinicopathological study of two cases. Ann. Neurol. 16 (1984) 71–77.

LAGUENY, A., M. AUPY, P. AUPY, X. FERRER, P. HENRY and J. JULIEN: Syndrome de la corne antérieure post-radiothérapeutique. Rev. Neurol. (Paris) 141 (1985) 222–227.

LANDAU, W. M., TORACK, R. M. and M. A. GUGGENHEIM: Congenital retardation and central motor defect with later evolution of seizure disorder, orofacial dysplasia, and amyotrophy. Neurology 26 (1976) 869–873.

LANDIS, D. N. M., R. N. ROSENBERG, S. C. LANDIS, L. SCHUT and W. L. NYHAN: Spinal muscular atrophy with olivopontocerebellar atrophy type IV. Arch. Neurol. (Chic.) 31 (1974) 295–307.

LAPRESLE, J. and A. ANNABI: Olivopontocerebellar atrophy with velopharyngolaryngeal paralysis: a contribution to the somatotopy of the nucleus ambiguus. J. Neuropathol. Exp. Neurol. 138 (1979) 401–406.

LARSEN, R., S. ASHWAL and N. PECKHAM: Incontinentia pigmenti: association with anterior horn degeneration. Neurology 37 (1987) 446–450.

MANTA, P., N. KALFAKIS, D. VASILOPOULOS and M. SPENGOS: An unusual case of lower motor neuron disease. J. Neurol. 230 (1983) 141–142.

MATSUNAGA, J., T. INOKUCHI, A. OHNISHI and Y. KUROIWA: Oculopharyngeal involvement in familial neurogenic muscular atrophy. J. Neurol. Neurosurg. Psychiatry 36 (1973) 104–111.

MAWATARI, S. and K. KATAYAMA: Scapuloperoneal muscular atrophy with cardiopathy. An X-linked recessive trait. Arch. Neurol. (Chic.) 28 (1973) 55–59.

MCCORMACK, M. K., P. J. COPPOLA-MCCORMACK and M.-L. LEE: Autosomal-dominant inheritance of distal arthrogryposis. Am. J. Med. Genet. 6 (1980) 163–169.

MCGUIRE, S. A., J. J. TOMASOVIC and N. ACKERMAN: Hereditary continuous muscle fiber activity. Arch. Neurol. (Chic.) 41 (1984) 395–396.

MCKUSICK, V. A., C. A. FRANCOMANO and I. E. ANTONUOR-AKIS: Mendelian Inheritance in Man, 10th ed. Baltimore, Johns Hopkins University Press (1990).

MCNAMARA, J. O., J. R. CURRAN and H. H. ITABASHI: Congenital ichthyosis with spastic paraplegia of adult onset. Arch. Neurol. (Chic.) 32 (1975) 699–701.

MEADOWS, J. C. and C. D. MARSDEN: A distal form of chronic spinal muscular atrophy. Neurology 19 (1969) 53–58.

MEIER, C., J. LÜTSCHG, F. VASSELLA and A. BISCHOFF: Progressive neural muscular atrophy in a case of phenylketonuria. Dev. Med. Child. Neurol. 17 (1975) 625–630.

MILLER, R. G., R. B. LAYZER, M. A. MELLENTHIN, M. GOLABI, R. A. FRANCOZ and J. C. MALL: Emery-Dreifuss muscular dystrophy with autosomal dominant transmission. Neurology 35 (1985) 1230–1233.

MITCHELL, D. M. and S. A. OLCZAK: Remission of a syndrome indistinguishable from motor neurone disease after resection of bronchial carcinoma. Br. Med. J. 2 (1979) 176–177.

MOERMAN, PH., J. P. FRYNS, P. GODDURIS and J. M. LAUWER-YNS: Multiple ankylosis, facial anomalies, and pulmonary hypoplasia associated with severe antenatal spinal muscular atrophy. J. Pediatr. 103 (1983) 238–241.

MOTTIER, D., G. BERGERET, H. F. PERREAUT, A. MISSOUM, J. BASTARD and D. MABIN: Myopathie thyroïdienne chronique simulant une sclérose latérale amyotrophique. Nouv. Presse Méd. 10 (1981) 1655.

MULDER, D. W., L. T. KURLAND, K. P. OFFORD and C. M. BEARD: Familial adult motor neuron disease. Neurology 36 (1986) 511–517.

NELSON, J. S. and A. L. PRENSKY: Sporadic juvenile amyotrophic lateral sclerosis. Arch. Neurol. (Chic.) 27 (1972) 300–306.

NEUHÄUSER, G., C. WIFFLER and J. M. OPITZ: Familial spastic paraplegia with distal muscle wasting in the Old Order Amish; atypical Troyer syndrome or 'new' syndrome. Clin. Genet. 9 (1976) 315–323.

NEZELOF, C., M. C. DUPART, F. JAUBERT and E. ELIACHAR: A lethal familial syndrome associating arthrogryposis multiplex congenita, renal dysfunction, and a cholestatic and pigmentary liver disease. J. Pediatr. 94 (1979) 258–260.

O'SULLIVAN, P. J. and J. G. MCLEOD: Distal chronic spinal muscular atrophy involving the hands. J. Neurol. Neurosurg. Psychiatry 41 (1978) 653–658.

PATTEN, B. M. and M. PAGES: Severe neurological disease associated with hyperparathyroidism. Ann. Neurol. 15 (1984) 453–456.

PEARN, J.: Autosomal dominant spinal muscular atrophy. J. Neurol. Sci. 38 (1978) 263–275.

PEARN, J. H., P. HUDGSON and J. N. WALTON: A clinical and genetic study of spinal muscular atrophy of adult onset. Brain 101 (1978) 591–606.

PINSKY, L., M. H. FINLAYSON, I. LIBMAN and B. H. SCOTT: Familial amyotrophic lateral sclerosis with dementia: a second Canadian family. Clin. Genet. 7 (1975) 186–191.

QUARFORDT, S. H., D. C. DEVIVO, W. K. ENGEL, R. I. LEVY and D. S. FREDRICKSON: Familial adult-onset proximal spinal muscular atrophy. Report of a family with type II hyperlipoproteinemia. Arch. Neurol. (Chic.) 22 (1970) 541–549.

RICHIERI-COSTA, A., A. ROGATKO, R. LEVISKY, N. FINKEL and O. FROTA-PESSOA: Autosomal dominant late adult spinal muscular atrophy, type Finkel. Am. J. Med. Genet. 9 (1981) 119–128.

ROSEMBERG, S., C. L. P. LANCELLOTTI, F. ARITA, C. CAMPOS and N. P. DE CASTRO: Progressive bulbar paralysis of childhood (Fazio-Londe disease) with deafness. Eur. Neurol. 21 (1982) 84–89.

ROSENBERG, R. N. and A. CHUTORIAN: Familial optico-acoustic nerve degeneration and polyneuropathy. Neurology 17 (1967) 827–832.

ROSENBERG, R. N. and H. L. FOWLER: Autosomal dominant motor system disease of the Portuguese: a review. Neurology (NY) 31 (1981) 1124–1126.

ROWLAND, L. P.: Cramps, spasms and muscle stiffness. Rev. Neurol. (Paris) 141 (1985) 261–273.

ROWLAND, L. P. and S. A. SCHNECK: Neuromuscular disorders associated with malignant neoplastic disease. J. Chronic. Dis. 16 (1963) 777–795.

ROWLAND, C. P., M. FERELL, M. OLARTE, A. HAYS, N. SINGH and F. E. WANAT: Emery-Dreifuss muscular dystrophy. Ann. Neurol. 5 (1979) 111–117.

RÖYTTÄ, M. and A. ANTTINEN: Xeroderma pigmentosum with neurological abnormalities. A clinical and neuropathological study. Acta Neurol. Scand. 73 (1986) 191–199.

SATOYOSHI, E. and K. YAMADA: Recurrent muscle spasm of central origin. Arch. Neurol. (Chic.) 16 (1967) 254–264.

SCHOLD, S. C., E. S. CHO, M. SOMASUNDARAM and J. B. POSNER: Subacute motor neuropathy: a remote effect of lymphoma. Ann. Neurol. 5 (1979) 271–287.

SERRA, S., A. XERRA and A. ARENA: Amyotrophic choreo-acanthocytosis: a new observation in Southern Europe. Acta Neurol. Scand 73 (1986) 481–486.

SERRATRICE, G.: Classification of adult chronic spinal muscular amyotrophies. Cardiomyology 2 (1983) 255–273.

SERRATRICE, G.: Amyotrophie spinale distale chronique localisée aux deux membres supérieurs (type

O'Sullivan et MacLeod). Rev. Neurol. (Paris) 140 (1984a) 368–369.

SERRATRICE, G. T., M. TOGA and J. F. PELLISSIER: Chronic spinal muscular atrophy and pallidonigral degeneration: report of a case. Neurology 33 (1983) 306–310.

SERRATRICE, G., G. CREMIEUX, J. F. PELLISSIER and J. POUGET: Deux cas de syndrome de Fotopulos (amyotrophie spinale chronique de la ceinture scapulaire et chorée chronique). Nouv. Presse Méd. 13 (1984a) 1274.

SERRATRICE, G., J. F. PELLISSIER, J. L. GASTAUT and C. DESNUELLE: Amyotrophie spinale chronique avec paralysie des cordes vocales: syndrome de Young et Harper. Rev. Neurol. (Paris) 140 (1984b) 657–658.

SERRATRICE, G., A. POU-SERRADEL, J. F. PELLISSIER, H. ROUX, J. LAMARCO-CIVRO and J. POUGET: Chronic neurogenic quadriceps myopathies. J. Neurol. 232 (1985) 150–153.

SHY, M. E., L. P. ROWLAND, T. SMITH, W. TROJABORG, N. LATOV, W. SHERMAN, M. A. PESCE, R. E. LOVELACE and E. F. OSSERMAN: Motor neuron disease and plasma cell dyscrasia. Neurology 36 (1986) 1429–1436.

SILVER, J. R.: Familiar spastic paraplegia with amyotrophy of the hands. Ann. Hum. Genet. 30 (1966) 69–75.

SINGH, H. and R. SHAM: Heredofamilial ataxia with muscle fasciculations. A report of 2 cases in brothers. Br. J. Clin. Pract. 18 (1964) 91–92.

SJÖGREN, T. and T. LARSSON: Oligophrenia in combination with congenital ichthyosis and spastic disorders: a clinical and genetic study. Acta Psychiatr. Scand. 32 (1957) Suppl. 113, 1–112.

SKRE, H., S. I. MELLGREN, P. BERGSHOLM and J. E. SLAGSVOLD: Unusual type of neural muscular atrophy with a possible X-chromosomal inheritance pattern. Acta Neurol. Scand. 58 (1978) 249–260.

SOBUE, I., N. SAITO, M. IIDA and K. ANDO: Juvenile type of distal and segmental muscular atrophy of upper extremities. Ann. Neurol. 3 (1978) 429–432.

SPENCER, P. S., P. B. NUNU, J. HUGON, A. C. LUDOLPH, S. M. ROSS, D. N. ROY and R. C. ROBERTSON: Guam amyotrophic lateral sclerosis-Parkinsonism-dementia linked to a plant excitant neurotoxin. Science 237 (1987) 517–522.

SPILLANE, J. D., P. W. NATHAN, R. E. KELLY and C. D. MARSDEN: Painful legs and moving toes. Brain 94 (1971) 541–556.

SPIRO, A. J., M. H. FOGELSON and A. C. GOLDBERG: Microcephaly and mental subnormality in chronic progressive spinal muscular atrophy of childhood. Dev. Med. Child. Neurol. 9 (1967) 594–601.

SPITZ, M. C., J. JANKOVIC and J. M. KILLIAN: Familial tic disorder, parkinsonism, motor neuron disease, and acanthocytosis: a new syndrome. Neurology 35 (1985) 366–370.

STAAL, A., L. N. WENT and H. F. M. BUSCH: An unusual form of spinal muscular atrophy with mental retardation occurring in an inbred population. J. Neurol. Sci. 25 (1975) 57–64.

SUNG, J. H., M. RAMIREZ-LASSEPAS, A. R. MASTRI and S. M. LARKIN: An unusual degenerative disorder of neurons associated with a novel intranuclear hyaline inclusion (neuronal intranuclear inclusion disease). J. Neuropathol. Exp. Neurol. 39 (1980) 107–130.

THOMAS, J. E. and F. M. HOWARD: Segmental zoster paresis — a disease profile. Neurology 22 (1972) 459–466.

TRUSH, D. C., G. HOLTI, W. G. BRADLEY, M. J. CAMPBELL and J. WALTON: Neurological manifestations of xeroderma pigmentosum in two siblings. J. Neurol. Sci. 22 (1974) 91–104.

TSUKAGOSHI, H., H. SUGITA, T. FURUKAWA, T. TSUBAKI and E. ONO: Kugelberg-Welander syndrome with dominant inheritance. Arch. Neurol. (Chic.) 14 (1966) 378–381.

TSUKAGOSHI, H., H. SHOJI and T. FURUKAWA: Proximal neurogenic muscular atrophy in adolescence and adulthood with X-linked recessive inheritance. Neurology 20 (1970) 1188–1193.

ULE, G.: Progressive neurogene Muskelatrophie bei neuroaxonaler Dystrophie mit Rosenthalschen Fasern. Acta Neuropathol. (Berl.) 21 (1972) 332–339.

VAN BOGAERT, L.: La sclérose latérale amyotrophique et la paralysie bulbaire progressive chez l'enfant. Rev. Neurol. (Paris) I (1925) 180–192.

VAN BOGAERT, L. and M. MOREAU: Combinaison de l'amyotrophie de Charcot-Marie-Tooth et de la maladie de Friedreich chez plusieurs membres d'une même famille. Encéphale 34 (1939–1941) 312–322.

VAN GENT, E. M., R. A. HOOGLAND and F. G. I. JENNEKENS: Distal amyotrophy of predominantly the upper limbs with pyramidal features in a large kinship. J. Neurol. Neurosurg. Psychiatry 48 (1985) 266–269.

WALSH, J. C.: Neuromyotonia: an unusual presentation of intrathoracic malignancy. J. Neurol. Neurosurg. Psychiatry 39 (1976) 1086–1091.

WALTON, J.: Diseases of voluntary muscle, 5th ed. Edinburgh, Churchill-Livingstone (1988).

WALTON, J. N., B. E. TOMLINSON and G. W. PEARCE: Subacute 'poliomyelitis' and Hodgkin's disease. J. Neurol. Sci. 6 (1968) 435–445.

WHITELEY, A. M., M. SWASH and H. URICH: Progressive encephalomyelitis with rigidity. Its relation to subacute myoclonic spinal neuronitis and to the 'stiff man syndrome'. Brain 99 (1976) 27–42.

WORLD FEDERATION OF NEUROLOGY RESEARCH COMMITTEE RESEARCH GROUP ON NEUROMUSCULAR DISEASES. J. Neurol. Sci. 86 (1988) 333–360.

YOUNG, I. D. and P. S. HARPER: Hereditary distal spinal muscular atrophy with vocal cord paralysis. J. Neurol. Neurosurg. Psychiatry 43 (1980) 413–418.

ZATZ, M., C. PENHA-SERRANO, D. FROTA-PESSOA and D. KLEIN: A malignant form of neurogenic muscular atrophy in adults, with dominant inheritance. J. Genet. Hum. 19 (1971) 337–354.

ZERRES, K. and T. GRIMM: Genetic counselling in families with spinal muscular atrophy type Kugelberg-Welander. Hum. Genet. 65 (1983–1984) 74–75.

ZIEGLER, D. W., R. N. SCHIMKE, J. J. KEPES, D. L. ROSE and G. KLINKERFUSS: Late onset ataxia, rigidity, and peripheral neuropathy. Arch. Neurol. (Chic.) 27 (1972) 52–66.

POSTSCRIPT

In this classification one misses:

Congenital suprabulbar paresis with mild mental retardation (185480) of Klippel and Pierre-Weil (1909).

Refs: Klippel and Pierre-Weil: Syndrome labio-glosso-laryngé pseudo-bulbaire héréditaire et familial. Rev. Neurol. (Paris) 17 (1909) 102–103.

Worster-Drought, C: Congenital suprabulbar pareses. J. Laryng. Otol. 70 (1956) 453–463.

Multiple system atrophies with mental retardation and infantile osteoporosis.

Refs: Neimann, N., M. Vidailhet, J. J. Martin, M. Andre, J. Floquet and J. Grignon: Fragilité osseuse, amyotrophie, arriération et lésions dégénératives du système nerveux central. Arch. Fr. Pédiatr. 30 (1973) 899–913.

Neimann, N., J. J. Martin, M. Vidailhet, J. Floquet, M. Pierson and A. Bajolle: Atrophies systematisées multiples, arriération mentale, amyotrophie neurogène et fragilité osseuse congénitale. J. Neurol. Sci. 30 (1976) 287–297.

Amyotrophic lateral sclerosis with polyglucosan bodies (20250).

Refs: Orthner, H. H., P. E. Becker and D. Müller: Recessiv erbliche amyotrophische Lateralsklerose mit 'Lafora-Körpern'. Arch. Psychiatr. Nervenkr. 217 (1973) 387–412.

Barth, H., C. Kemmer, D. Kunze, and B. Sachs: Amyotrophe Lateralsklerose mit Myoklonuskörpern. Zentralbl. Allg. Pathol. 120 (1976) 333-342.

Handbook of Clinical Neurology, Vol. 15 (59): Diseases of the Motor System
J.M.B.V. de Jong, editor
© Elsevier Science Publishers B.V., 1991

CHAPTER 2

Adult progressive muscular atrophy and hereditary spinal muscular atrophies*

FORBES H. NORRIS

ALS and Neuromuscular Research Foundation and Department of Neurology, Pacific Presbyterian Medical Center, San Francisco, CA, USA

Perhaps the most critical adult lower motor neuron syndrome is progressive muscular atrophy (PMA) because it is, with infantile Werdnig-Hoffmann disease, the most pure of the clinical diseases of the lower motor neuron. Unlike Werdnig-Hoffmann disease, where the genetic background clearly points toward a failure of production/delivery/action of a specific (though at this time hypothetical) motor neuron growth factor, PMA is far more mysterious. It strikes an otherwise healthy young or middle-aged adult, then over many years or decades brings him to bed from paralysis.

In some classifications, PMA is denoted progressive spinal muscular atrophy (Walton and Thomas 1988, p. 339) but in the literature *spinal muscular atrophy*, as will be seen below, has become closely associated with the early-onset, genetic diseases. Many authors have included PMA with the adult, idiopathic motor neuron diseases (MND) of which amyotrophic lateral sclerosis (ALS) is the most common, but the abbreviation PMA/MND is cumbersome, so PMA will be used in this discussion.

The onset in PMA is usually in the hand and distal arm muscles, and usually atrophy is remarked before weakness brings the patient to a

physician. Fasciculation is usually noticed after the physician sees it. The onset is asymmetrical but there is usually bilateral involvement when the neurologist sees the patient. The muscle stretch reflexes are depressed in the arms though at an early stage the biceps and pectoralis reflexes may be intact. Some patients complain of strange sensations in the fingers and hands, though sensory testing is normal, as is the remainder of the neurological examination. Cramps may be a frequent complaint but further weakness with cold exposure (chilling) is noted more often (Müller 1952), and in some patients is the first symptom. Useful clinical clues to MND are the global involvement of the hand muscles and atrophy (or reduction) of the overlying subcutaneous fat (Mulder 1984).

PMA progresses only very slowly, over decades. The average case presenting as PMA is actually the lower motor onset of amyotrophic lateral sclerosis (ALS), in which rapid progression and appearance of upper motor signs soon indicate that diagnosis. Those few cases which do continue with very slow progression of purely lower motor involvement comprise PMA. Occasional cases of rapid progression have been reported as PMA but in some of these there are other causes of neural injury, such as diabetic neuropathy (Scully et al. 1987) or gammopathy (Rao et al. 1986).

---

*Supported by the ALS and Neuromuscular Research Foundation and the Hedco Foundation.

Distal leg and foot muscle involvement usually follow, with spread to the proximal arm and shoulder girdle muscles apparent in the examination at that stage. Many years later, proximal leg and pelvic girdle muscle weakness will necessitate a wheelchair for ambulation, and by this stage there may also be difficulty in sitting, both in rising and in sustaining sitting posture more than an hour or two. In other words, the segmental involvement usually commences (at least clinically) in the lower cervical segments, then skips to the lower lumbar and upper sacral segments, then slowly ascends. Last involved are the muscles of respiration and, rarely, the lower bulbar motor neurons. Before that stage, however, loss of all the muscle stretch reflexes has indicated even more diffuse disease.

In some cases of apparent PMA, the onset is in the proximal arm muscles rather than the forearms and hands. Perhaps these cases, even though sporadic (non-familial) and idiopathic, represent significant overlap with other syndromes of SMA (see below), but one must question whether a respiratory muscle onset (Miller et al. 1957) is not really the onset of ALS.

F.-A. Aran and Duchenne de Boulogne are usually credited with first describing PMA in the eponym 'Aran-Duchenne disease'. Because Duchenne was senior and subsequently claimed the first description, some authors have awarded him the priority in the term 'Duchenne-Aran disease', but careful review of the original reports and the relevant records from that time fail to support Duchenne's claims (Norris 1975, 1991; Bonduelle 1990). Aran (1850) actually credited Professor Rayer with discovery of his index case while on rounds at la Charité, and he found somewhat similar cases reported earlier by Bell and Darwall, possibly even as early as 1795 by Graves (Aran 1850).

Cruveilhier (1853) correctly identified spinal cord and ventral root disease as the cause of the muscular atrophy in Aran's (1850) case 8, but it remained for Charcot (1873; and Joffroy 1869) to focus attention on the ventral horn cells and confirm Cruveilhier. By 1882, Bramhall could write that 'the paralysis presents all the characteristic features which result from a lesion of the anterior horn'.

The next major development was presented in Gowers' (1902) lecture on abiotrophy. The concept is simple, and based on probably thousands of years of folklore. He theorized that every cell is endowed at birth with a certain 'life energy', to be carefully husbanded throughout life to permit good function into old age. Overuse of this life energy would cause the cell to expire prematurely. Thus, the cause of PMA was overactivity in earlier years; the corollary (still recommended today by many physicians) is that the patient should go home, take to bed, and 'conserve the remaining strength'.

The incidence of PMA is not established. Kurland et al. (1969) found that a rising rate for ALS corresponded to a falling incidence of PMA, which suggests that more physicians are recognizing the PMA-like lower motor neuron onset of ALS as ALS rather than PMA. In a prospective population-based study in northern California (Norris et al. 1989, 1991) there were only 17 PMA cases among 761 idiopathic MND cases. The ALS incidence in this defined population was about one new case/100 000 general population/year, suggesting an incidence for PMA of about 0.02/100 000/year.

Risk factors and familial neurologic disorders for this PMA population are shown in Table 1, which shows very little difference from a normal adult population, though the PMA onset at 47 years was significantly less than that for sporadic ALS (Table 1). The PMA male:female ratio was 7.5 in this population compared to 1.3 for sporadic ALS and 1.2 for familial ALS.

Likewise, the pathology in PMA is only described in occasional reports, most in connection with ALS, again reflecting the current opinion that most cases of PMA are the lower motor onset of ALS. After decades with little or no bulbar involvement and no upper motor signs, nor any in the necropsy, surely cases of PMA should be given special attention by neuropathologists, especially in comparison with ALS.

In past studies the pathology found in PMA was that seen in the ventral horns in ALS, namely loss of neurons, shrinkage of the remaining neurons and an apparent increase of lipofuscin due to this shrinkage (McHolm et al. 1984), and disorganization of the Nissl substance

TABLE 1

Risk factors in PMA compared to ALS (from Norris et al. 1989, 1991).

Comparison	PMA	Sporadic ALS	Familial ALS
Onset age*	46±4	59±1	51±2
Male:Female	7.5	1.3	1.2
Survival months*			
(to 31 Dec. 1990)	200±30	44±1	60±8
Total cases	17	627	56
Cases with:			
Neoplasm	1	21	0
Trauma**	1	51	4
Polio†	0	6	1
Exposure to:			
Metal	0	18	3
Other	0	1	0
Family history			
MND	0	0	56
Paralysis,			
unclassifiable	1	3	3
Parkinsonism	2	33	3
Alzheimer disease	0	9	1

*Mean ± standard error.
**Fracture, concussion, major joint dislocation within 5 years before onset.
†Paralytic polio-like illness in past.

with clearing of the peripheral cytoplasm (peripheral chromatolysis). Neuronophagia was extremely rare, vascular cuffing and meningeal infiltrates virtually non-existent, in contrast to their occasional presence in ALS. The ventral roots were atrophied but, no matter how severe the neuronal loss, the gross appearance of the cord and ventral horn was usually normal (Norris 1975, pp. 14–19). In cases including ALS, Swash et al. (1986) found no pattern in the sites of anterior horn cell involvement in the cervical cord at the C-8 level. By their technique, however, they were unable to confirm the presence of the anatomical columns long accepted on the basis of previous work by several investigators.

Kato et al. (1988) have reported a case of lower motor neuronopathy in which Lewy and Bunina bodies were found in addition to 'cord-like thickenings of cell processes.' They noted one and perhaps two other cases in the literature and suggested that this unusual pathology indicates a special subtype of PMA. Very short courses (9 months), however, raise the possibility of ALS with death from respiratory failure before there was upper motor involvement.

Unique neuronal 'conglomerates' were found in a case of PMA studied by Schochet et al. (1969). Neuronal loss was minimal; most of the ventral horn cells possessed large cytoplasmic hyalin bodies. The latter proved on electron microscopy to be tangled masses of neurofibrils. Further analysis including results from other cases indicate that such spheroids correlate roughly with the relatively recent onset of motor ganglion degeneration. The neuronal masses in these cases are clearly different from the intra-nuclear hyalin inclusions found by Sung et al. (1980) in a multiple system atrophy which became symptomatic in childhood.

A frequent PMA alteration is gliosis of the affected ventral horns, to be found also in the ventral horns of segments which seem to possess normal numbers of neurons. Conversely, occasional ventral horns show marked loss or degeneration of neurons with little glial change.

The muscular atrophy consists mainly of a decrease in diameter of the denervated muscle fibers without major architectural changes. The practical value of the muscle biopsy is to demonstrate a mainly neurogenic disorder and to exclude a different disorder such as polymyositis. Scattered degenerating muscle fibers may be seen and other alterations in the muscle fibers can be detected by electron microscopic study (Norris 1975, p. 18). It is not clear how any of these myopathic changes relate to what is generally accepted as a neurogenic process. Possible explanations include overwork stress on weakened muscles with subsequent degeneration in some of the muscle fibers (Peach 1990). Long-standing, severe denervation results in fibrosis and muscle fiber breakdown which is ultimately a myopathic process (Pearce and Harriman 1966). It should be recalled that experimental sciatic nerve transection produces myopathic alterations in the denervated muscles (Adams et al. 1962).

In summary, the main findings in PMA are also found in ALS, aside from the cerebral and pyramidal lesions in ALS. The present series includes no case studied by the most modern methods, and the literature provides no clue to any significant difference in PMA from ALS. Accepting that some cases of SMA overlap PMA, it is of interest that Ono et al. (1989) found a skin abnormality in ALS that was not present in their SMA cases. Mapelli (1990) was not able to confirm such a skin disorder, but a secondary skin alteration in both PMA and ALS would not be surprising in view of the loss of subcutaneous fat (Mulder 1984) which can progress to severe cachexia (Norris et al. 1978).

Diagnostic studies include electromyography (EMG) and nerve conduction velocity measurements with tests of neuromuscular transmission. The latter is in search of myasthenia gravis early in the illness, before the reflexes diminish or disappear, or a myasthenic state indicating a trial of anti-cholinergic medication either then or later in the illness (Denys and Norris 1979; Mulder 1984). Both the motor and sensory nerve conduction velocities are normal early in PMA, and usually throughout the illness though sometimes the motor nerve conduction is reduced later in very atrophic muscles as in ALS (Lambert 1969). Early or disproportionate reduction in NCV points toward a primarily neuropathic process rather than MND (Auer et al. 1989). As indicated above, it would be highly desirable to differentiate PMA from early ALS, and we can look to central conduction studies to help detect early corticospinal tract disease. Ingram and Swash (1987) found slowing of central motor conduction only when there was clinical evidence of upper motor involvement (i.e. ALS) but Hugon et al. (1987) could not confirm this. Another electrographic technology is electronystagmography, which could be predicted to show abnormalities in bulbar syndromes and ALS not to be found in PMA, but that was not the case: all the MND had frequent abnormalities (Lebo et al. 1983).

It is important that the EMG study include proximal as well as distal muscles, even though the proximal muscles may not seem to be affected, in order to exclude polymyositis. The pathophysiology in polymyositis can predominate in the neurovascular bundles and cause both neurogenic lesions of the muscle fibers supplied by such a bundle, as well as direct (myogenic) lesions in the contiguous muscle fibers. Such a process will produce a mixture of neurogenic and myogenic

(or neuromyopathic) abnormalities in the EMG and the muscle biopsy. The latter may not show the inflammation, which is often segmental along the length of a neurovascular bundle, and so missed in random sections. An instructive case was reported where polymyositis was misdiagnosed as PMA through such an occurrence (Norris 1975, case 3, p. 26). This patient was not cured, but his hopeless diagnosis was changed to a more optimistic one and he did improve with appropriate therapy. The first step, therefore, in a new patient suspected of having PMA (or any other of the idiopathic, incurable MNDs) is great care to exclude carefully the other, potentially treatable causes of the syndrome.

DIFFERENTIAL DIAGNOSIS

It is wise to consider muscle biopsy and perhaps nerve biopsy in all exclusively lower motor neuron syndromes, to aid in detection of atypical polymyositis, some rare myopathies such as debrancher deficiency (Cornelio et al. 1984), and dysimmune polyneuritis. There is great current interest in multifocal nerve conduction blocks causing a PMA-like syndrome (Pestronk et al. 1989, 1990; Sadiq et al. 1990). Specific immunoglobulin dyscrasias, especially to GM_1 gangliosides, have been detected in many of these cases and some have stabilized or improved with vigorous immunosuppression (Pestronk et al. 1988; Shy et al. 1990). Pestronk et al. (1990) have identified two other groups of lower motor neuron patients, both with slow motor nerve conduction but no focal conduction blocks: one group with elevated antibodies to GM_1 ganglioside, the other with antibodies to asialo-GM_1 ganglioside and Gal(β1-3)GalNAc disaccharide. The resulting interest in immunosuppression has been heightened by Engelhardt et al. (1989) describing an experimental lower MND induced by autoimmunity.

In ALS renal biopsy tissue, Oldstone et al. (1973) found immune complex in glomeruli and mesangia of patients having the most rapid courses. We immediately began efforts at immunosuppression, including plasma exchange. Some of the patients seemed to improve early in the trial, but a placebo effect now seems likely

Contents: Chapter 1. The World Federation of Neurology classification of spinal muscular Datrophies and other disorders of motor neurons *(J.M.B.V. de Jong)*. Chapter 2. Adult progressive muscular atrophy and hereditary spinal muscular atrophies *(F.H. Norris)*. Chapter 3. The postpolio syndrome *(D.W. Mulder)*. Chapter 4. Special forms of spinal muscular atrophy *(G.W. Padberg)*. Chapter 5. Werdnig-Hoffmann disease and variations *(M. Osawa and K. Shishikura)*. Chapter 6. Wohlfart-Kugelberg-Welander disease *(S. Zierz and K. Zerres)*. Chapter 7. Spinal muscular atrophy of infantile and juvenile onset, due to metabolic derangement *(J. Troost)*. Chapter 8. Non-progressive juvenile atrophy of the distal upper limb *(Hirayama's disease) (K. Hirayama)*. Chapter 9. Progressive bulbar paralysis of childhood *(M.R. Gomez)*. Chapter 10. Progressive dysautonomias *(O. Appenzeller)*. Chapter 11. Amyotrophic lateral sclerosis *(H.R. Tyler and J. Shefner)*. Chapter 12. Progressive bulbar palsy in adults *(G.W. Bruyn)*. Chapter 13. Dementia and parkinsonism in amyotrophic lateral sclerosis *(A.J. Hudson)*. Chapter 14. Familial amyotrophic lateral sclerosis *(D.B. Williams)*. Chapter 15. Amyotrophic lateral sclerosis in the Mariana Islands *(R.M. Garruto and R. Yanagihara)*. Chapter 16. Amyotrophic lateral sclerosis and Parkinsonism-dementia in the Kii Peninsula - comparison with the same disorders in Guam and with Alzheimer's disease *(H. Shiraki and Y. Yase)*. Chapter 17. Hereditary spastic paraparesis *(Strümpel-Lorrain) (R.P.M. Bruyn and Ph. Scheltens)*. Chapter 18. Ferguson-Critchley syndrome *(R.P.M. Bruyn and G.W. Bruyn)*. Chapter 19. Hereditary spastic paraplegia with retinal disease *(C.G. Wells, W.T. Longstretch Jr. and C.H. Smith)*. Chapter 20. Hereditary secondary dystonias *(E.J. Novotny)*. Chapter 21. Spastic paraparesis due to metabolic disorders *(H.W. Moser, A. Bergin and S. Naidu)*. 22. Differential diagnosis of spinal muscular atrophies and other disorders of motor neurons with infantile or juvenile onset *(M. de Visser, P.A. Bolhuys and P.G. Barth)*. Chapter 23. Differential diagnosis of sporadic amyotrophic lateral sclerosis, progressive spinal muscular atrophy and progressive bulbar palsy in adults *(E.S. Louwerse, P.A.E. Sillevis Smitt and J.M.B.V de Jong)*. Chapter 24. Differential diagnostic work-up of spastic para*(tetra)*plegia *(G.W. Bruyn)*. Chapter 25. HTLV-1-Associated motor neuron disease - G.C. Roman, J.-C. Vernant and M. Osame). Chapter 26. Palliative treatment of motor neuron disease *(R.A. Smith, E. Gillie and J. Licht)*. Chapter 27. Hemiatrophies and hemihypertrophies *(G.W. Bruyn and R.P.M. Bruyn)*.

Send your order to your bookseller or
ELSEVIER SCIENCE PUBLISHERS
P.O. Box 211, 1000 AE Amsterdam, The Netherlands

Distributor in the U.S.A. and Canada:
ELSEVIER SCIENCE PUBLISHING CO., INC.
P.O. Box 882, Madison Square Station, New York, NY 10159

Continuation orders for series are accepted.

Orders from individuals must be accompanied by a remittance, following which books will be supplied postfree.

The Dutch guilder price is definitive. US $ prices are subject to exchange rate fluctuations. Prijzen zijn excl. B.T.W.

Neurology (clinical), Neurogenetics

Handbook of Clinical Neurology

edited by **P.J. Vinken, G.W. Bruyn**, *and* **H.L. Klawans**

Diseases of the Motor System

co-edited by **J.M.B.V. de Jong**

Handbook of Clinical Neurology Volume 59 (15)

1991 564 pages
Price: US $ 200.00 / Dfl. 390.00
Subscription price: US $ 170.00 / Dfl. 332.00
ISBN 0-444-81278-4
PUBLICATION: 4TH QUARTER 1991

The incessant stream of new data generated by the considerable acceleration in research devoted to the motor system disorders in the last decade is decidedly reflected in this book.

The ordinate angle forming the abscissa of the volume is not a narrow one, expounding on topics from peripheral to central motor disease, and on the medical side from diagnosis to treatment.

New features of the book include an updated WFN classification of motor neurone disorders. Where possible, disorders have been given 6-digit identification numbers as listed in McKusick's (1990) catalogue.

This volume further equals the excellence for which the *Handbook of Clinical Neurology* has for years been renowned, and divergence from the previous volume is a direct reflection of how new volumes in this series stay in the forefront of neurological developments, pointing out the day when treatment of these disorders will not be alleviative or palliative, but preventive.

ELSEVIER

Amsterdam

ELSEVIER SCIENCE PUBLISHERS

because further vigorous treatment was only associated with typical MND deterioration. Thus, it does not seem reasonable to recommend more intensive trial of such treatment in PMA. This consideration remains open in the malignant disease ALS, where Drachman and Kuncl (1989) propose that the pathology might include an unconventional dysimmune process, resistant to conventional treatment, thus opening a potential therapeutic door for patients willing to undergo the risks of even more intensive immunosuppression. Lampson et al. (1990) found evidence that in MND patients the antibody can be produced locally in the tissue as a secondary or epiphenomenon, and Salazar-Grueso et al. (1990) found about as many blood antibody elevations in other neurologic (disease control) patients, though in one MND case there was improvement during prednisone treatment.

Chronic inflammatory polyradiculopathy

Dyck et al. (1975) described a series of patients in whom inflammatory disease had greatest impact on the spinal nerve roots and the peripheral nerves. Administration of corticosteroids appeared to benefit some of these cases, linking them to the cases just described, but many were readily distinguishable from idiopathic MND because of sensory loss, impaired bowel or bladder control, etc. Among MND patients referred here, fewer than 0.1% have displayed spinal fluid findings consistent with inflammatory disease.

While this series so far contains no definite example of luetic amyotrophy (El Alaoui-Faris et al. 1990), the possibility of its occurrence mandates that the spinal fluid be examined at least once in every patient diagnosed as having PMA. Spinal meningovascular lues can produce a rather pure motor neuron picture for the clinician and be missed in cases where the blood serology is negative and the CSF is not examined. A possible early case (1795) was mentioned by Aran (1850). The CSF serology is positive although in some cases the blood reaction might be negative (Escobar et al. 1970). The blood fluorescent treponemal antibody absorption (FTA-abs) test and spinal fluid Venereal Disease

Research Laboratory (VDRL) test should be obtained in every new MND patient who lacks a clearly positive family history.

Amyotrophic spondylotic myeloradiculopathy

The typical case of symptomatic cervical/lumbar spondylosis can be distinguished from primary MND by the presence of sensory symptoms or findings and by absence on examination of extensive motor findings usually present in MND, particularly the widespread EMG abnormalities; the patient with MND usually has more extensive motor findings than the symptoms may indicate and, with rare exceptions, these are entirely motor. Thacker et al. (1988) found that EMG and nerve conduction studies were more helpful when F-wave latency and conduction were added. The problem best known in this regard is cervical spondylosis, but lumbar spondylosis can also be a problem, and sometimes both conditions afflict the same patient and are more apt to depress all the muscle stretch reflexes, as in PMA, whereas cervical spondylosis alone is more apt to cause amyotrophic arms but spastic legs, thus mimicking ALS.

The most satisfactory diagnostic test for amyotrophic spondylotic myeloradiculopathy, after the EMG, is the myelogram (Dorsen and Ehni 1979). In one case, even though the patient refused lumbar puncture, a computerized tomogram of the cervical canal demonstrated significant narrowing in the anteroposterior diameter at several levels; at one point in the sagittal plane, the diameter was only 5 mm. There were extensive changes in the plain X-rays, suggesting a process of long duration with heavy calcification. The computerized tomogram may be less reliable than the myelogram, however, for patients with softer lesions and shorter courses. While seeing increasing and valuable application to this problem, the magnetic resonance image of calcified lesions can also be misleadingly benign. Thus far, this Center has had no autopsies in cases of this type; the lesions seen at surgery were those described previously by other authors (Wilkinson 1971) and provided no clue to the clinical sparing of sensory functions.

Heavy-metal toxicity

Since the time of Aran (1850), occasional cases
of exposure to heavy metals have been reported
in patients afflicted with MND, with lead best
known in this regard. Mercury in various organic
forms and selenium have also been associated
with MND (Conradi et al. 1982).

For MND patients with clear histories of
exposure to a heavy metal, we recommend an
open bone biopsy to permit mass spectrographic
quantitation of the particular element in a dehy-
drated specimen of cortical bone (Conradi et al.
1982). Patients with elevated levels of metal in
the cortical bone should be treated with the
standard chelation therapy, intravenous calcium
disodium edetate (EDTA) during 24-h urine
collections for measurement of other heavy metals
(as controls) besides the metal in question. The
urine should be collected at least once before
EDTA is administered and then every 24 h during
the treatment. Excretion of the suspect metal
more than threefold above the control level on 3
or more days during the 5 days of intravenous
treatment may indicate significant mobilization
of the metal. Review suggests, however, that the
optimal schedule for administration and dosage
of EDTA, and the criteria for abnormality of
metal excretion in the urine, have never been
fully determined (Conradi et al. 1982). Some
degree of sustained recovery may be viewed as
further evidence that the amount of metal in the
body was elevated and that the administration
of EDTA was of benefit.

Despite the rarity of exposure to heavy metals
and the difficulty of establishing the diagnosis,
patients should be interviewed routinely for this
condition. Screening of urine for its heavy metals
should perhaps be abandoned in favor of the
bone biopsy when the history clearly reveals
undue exposure of a heavy metal.

Fluorotic amyotrophy is seen only in areas
having very high fluoride levels in drinking water.
The resulting bone disorder can cause extensive
encroachment on the spinal cord and its roots,
thus leading to severe muscular atrophy and
eventually tetraplegia (Siddiqui 1973).

Other toxins. The prevalence of MND on Guam
raised early the possibility of a toxic agent there.

The Chamorran natives of Guam formerly ingested a nut, the cycad, obtained from the plant
Cycas circinalis, which has potential for carcino-
genesis and neural damage. The use of cycad
declined, however, in the face of a continued
high incidence of motor neuron disease until
recently, and the necessary dosage of the toxin,
BMAA, seems higher than the Chamorrans could
ingest. This problem is addressed in a recent
symposium volume (Rose and Norris 1990).

The high rate of juvenile-onset ALS with
deafness in men in southern India (Jagannathan
1973) is also suggestive of an environmental
toxin, as is the high incidence of MND in New
Guinea (Gajdusek 1963).

Some other actual or putative motor neuron
toxins are reviewed elsewhere (Norris 1975,
pp. 26–28; Spencer and Schaumburg 1980; Conradi et al. 1982; Rose and Norris 1990). Even
the familiar drug, phenytoin, may be harmful to
motor neurons in some cases (Direkze and
Fernando 1977). Specific excitotoxins are presently attracting the greatest research interest
(Teitelbaum et al. 1990; Rose and Norris 1990).

Posttraumatic amyotrophy

Many cases of PMA and ALS have been reported
because of the onset shortly after a significant
trauma (Norris 1975, pp. 28–29; Riggs 1985).
These are usually isolated cases and the question
of a coincidence remains unresolved. Gresham et
al. (1987) found no relationship to trauma in
their clinic. While also agreeing with a coincidental relationship, Milanese and Martin (1970)
suggested that the role of trauma is to modify
the expression of any subsequent MND. Other
individuals have worse trauma and never develop
MND. In the present series, just one of the 17
PMA patients reported major trauma within 5
years of the onset of paralysis (Table 1).

Electrical injury. This trauma is mentioned
separately because there does seem to be an
association between the spinal passage of large
electric currents and segmental non-progressive
amyotrophy in survivors (Panse 1970; Holbrook
et al. 1970). A less clear association is the
development of further generalized, progressive

amyotrophy in a minority of these survivors, totalling about a dozen patients in the world literature (Sirdofsky and Hawley 1990). The present northern California series (Norris et al. 1989, 1991) has only one patient with MND following such an injury. If electric currents cause MND, more cases should have been seen.

Metabolic and endocrine disorders

Two cases of severe MND apparently caused by hyperparathyroidism have been seen in the northern California series; both patients are improving slowly after parathyroidectomy, one from ventilator-dependence. A notable feature in both cases was a normal serum calcium measurement in one of 3 tests in one patient and 2 of 4 tests in the other. Neither patient had marked hypercalcemia at any time, as noted by Patten and Pages (1984). Possibly an increased parathyroid hormone level has some unappreciated neurotoxic effect on motor neurons. Ventral horn cell disease was verified by necropsies in one case of Patten and Pages (1984) and in the patient studied by Dubas et al. (1989).

Hypophosphatemic osteomalacia may also mimic MND except for reversibility on administration of neutral phosphate (Mallette and Pattern 1977). An occasional case of hypermagnesemia may also mimic MND (Castelbaum et al. 1989). Wohlfart-Kugelberg-Welander disease was associated with hyperlipoproteinemia type II in one affected family (Quarfordt et al. 1970). Aside from the obvious possibilities of thyrotoxicosis and diabetic amyotrophy (Chokroverty et al. 1977) mimicking MND, there has been no other consistently detectable metabolic/endocrine disorder in MND (Norris 1975, pp. 29–31).

Arteriosclerotic myelopathy

Only brief attention will be given here to arteriosclerotic myelopathy, chiefly because of its rarity as a progressive condition at this Center. Jellinger and Neumayer (1962), however, collected 21 cases of progressive vascular myelopathy within 10 years. In an earlier series (Norris 1969), one PMA patient had many arteriosclerotic occlusions in radicular and leptomeningeal arteries,

which presumably caused the circumferential myelin pallor found in the spinal cord. Hemiplegic amyotrophy (Chokroverty et al. 1976) may result in part from transsynaptic disturbance of innervation and a similar dysfunction may be involved in the atrophy caused by some parietal lobe lesions (Gastaut and Benaim 1988). The hemi- or monoplegic character of these poststroke syndromes, and eventual stabilization, should readily differentiate them from MND.

Postpoliomyelitic amyotrophy (Postpolio syndrome, progressive postpolio muscular atrophy)

Some persons recover from polio and later develop typical ALS, including rapid deterioration with a fatal outcome (Norris et al. 1990). Fortunately, most who develop late deterioration have the less malignant postpolio syndrome. Campbell et al. (1969), then Mulder et al. (1972) drew our attention to this subgroup of patients with MND: focal paralysis resulted from a febrile illness (presumably poliomyelitis) many years before; relative stability followed recovery from the illness; in later years, further weakness develops, with a corresponding increased atrophy in the part previously paretic; there is subsequent extension to other parts of the body but a much slower course than ALS, and no corticospinal signs, i.e. the clinical picture of PMA. The patients with greatest paralysis in the acute illness and the greatest recovery in the convalescence are most at risk (Klingman et al. 1988). Special EMG studies point to disintegration near the synapse of axonal sprouts to re-innervated muscle fibers as the cause of this syndrome (Weichers and Hubbell 1981; Cashman et al. 1987). Pezeshkpour and Dalakas (1988) were able to examine 3 spinal cords and found continuing inflammatory change. Brahic et al. (1985) found evidence for polio or Theiler virus genome in only one of the 15 ALS cords they studied; one of 6 controls was also positive.

Even more common is the benign syndrome of cramps and fasciculation after recovery of strength from polio (Fetell et al. 1982). In an earlier series, the author saw such a patient 60 years after the acute illness.

Other viral infections can also cause amyotrophy, generally as a relatively acute manifestation of the febrile illness. These cases should cause no confusion with PMA unless the muscle denervation is masked by some other feature of the illness. For example, in zoster paresis (Thomas and Howard 1972) the painful skin eruption may outweigh any motor symptoms or prevent neuromuscular examination during the acute phase, so that weakness may appear to develop later. Hemorrhagic conjunctivitis virus (Bharucha and Mondkar 1972; Wadia et al. 1972) may have caused permanent (though not progressive) paralysis in one case. An unknown agent causes the rare syndrome of progressive rigid encephalomyelitis (Whiteley et al. 1976).

The natural viruses to suspect in this regard are those which are prominent causes of acute encephalitis and poliomyelitis though searches for elevated serum antibody titers and tissue cultures have been largely negative in MND (Castalano 1972; Cremer et al. 1973; Oldstone et al. 1973). Roos et al. (1980) found no polio genetic material in tissues from a patient who died of ALS many years after suffering from probable polio; Weiner et al. (1980) were unable to detect virus in 5 ALS autopsies. The one positive ALS case and the positive control found by Brahic et al. (1985) have been noted above.

The human immunodeficiency virus (HIV) has been found in one ALS patient (Hoffmann et al. 1985). In a second case (Horoupian et al. 1984), clear upper motor lesions were produced but alterations of the lower motor neurons could have been secondary. In any event, the presence of an adenovirus in the spinal fluid prevents conclusions about the actual role of the HIV agent in the neuropathology. In the absence of further reports and with a rising rate of this infection, the HIV agent does not seem very likely to cause lower motor MND.

Postencephalitic amyotrophy. Other authors had mentioned sporadic cases of amyotrophy after encephalitis but Wimmer and Neel (1928) drew attention to this association (Norris 1975, pp. 21–24). In 2 of their 3 cases with postmortem studies, there were spinal cord foci of perivascular infiltration by lymphocytes or monocytes. Norris et al. (1969) found continued inflammatory

changes in another case, and the postpolio inflammatory changes found by Pezeshkpour and Dalakas (1988) have already been noted. Such cases raise the possibility that chronic viral infection can follow acute encephalitis (or poliomyelitis) and later cause an amyotrophic disorder. The models for such chronic or slow virus infection are kuru, progressive multifocal leukoencephalopathy, subacute sclerosing panencephalitis (Dawson's inclusion body encephalitis) and Creutzfeldt-Jakob disease, the last of greatest interest because of its tendency to cause amyotrophy (Allen et al. 1971). For detailed discussions, there are relevant chapters in other volumes of this Handbook.

MISCELLANEOUS ASSOCIATIONS

Myasthenic states. The electrographic study (EMG and nerve conduction measurements) should include assessment of neuromuscular transmission in several nerve-muscle pairs. This caution is not so much for myasthenia gravis (which is unlikely with the typical atrophy and reflex changes usually seen in PMA) but for the Lambert-Eaton myasthenic myopathy, which was actually found in one patient after this author had examined him and agreed with the referral diagnosis of PMA. It was several months after detection of the Lambert-Eaton syndrome that a rapidly enlarging axillary lymph node showed small-cell (oat-cell) carcinoma. After mediastinal and axillary radiation in 1975, the myasthenic syndrome remitted and he continues in his profession of school administration, 15 years later.

Radiation myelopathy may have a vascular basis (Jellinger and Sturm 1971) and some cases present with a progressive amyotrophy for several years (Maier et al. 1969), which in some cases has been so pure clinically as to suggest ventral horn cell susceptibility to the radiation (Sadowski et al. 1976; Lagueny et al. 1985). Since this has been reported in patients receiving therapy for neoplasms, these cases might also represent instances of carcinomatous amyotrophy (see below).

Carcinomatous neuromyopathy. Some patients with neoplasms develop neurological syndromes which are not due to metastases, the effect of anti-cancer treatments or evident metabolic derangements. MND has been noted (Norris and Engel 1965) in cancer patients tending to be older than usual and with more slowly evolving neurologic symptoms (Brain et al. 1965). With further experience, the lymphomas seem most likely to cause such a remote effect (Schold et al. 1979). The ventral horn pathology was also unusual for evidence of inflammation in some fatal cases (Walton et al. 1968; Norris et al. 1969; Adams et al. 1970); other patients improved after the malignancy was treated (Peters and Clatanoff 1968; Buchanan and Malamud 1973; Mitchell and Olczak 1979; Evans et al. 1990). This subgroup of MND probably overlaps that described above, in which lymphoma and plasma cell dyscrasias produced neural antibodies (Shy et al. 1986), also the POEMS syndrome (Bardwick et al. 1980; Dalakas and Engel 1981).

Benign monomelic amyotrophy is a consideration very early in PMA, especially in Asians (Sobue et al. 1978; Peiris et al. 1989) in whom the monomelic disorder was first described (Hirayama et al. 1959). The clinical picture is rather simple: a healthy older boy or young man, with no family history of neurologic disease, develops amyotrophy in one hand, extending up the forearm over several years, after which the disorder stabilizes. Cold sensitivity may be marked. The muscle stretch reflexes are usually depressed, but may be normal or even hyperactive. Sobue et al. (1978) described their findings in 47 Japanese patients. In a series of 12 Chinese patients, tremor was prominent in 6, though whether simply due to contraction fasciculation or another dysfunction cannot be determined from the report by Loong et al. (1975). Focal cord atrophy was found in one case by means of metrizamide computed tomography (Metcalf et al. 1987). Muscle and nerve electrographic and biopsy studies pointed to focal anterior horn cell disease (Sobue et al. 1978; Gourie-Devi et al. 1984; Chaine et al. 1988) which was confirmed in an autopsy by Hirayama et al. (1987). Similarly benign monomelic amyotrophy has been reported in the leg of one patient (Riggs et al. 1984).

The importance to the patient of diagnosing this disorder is evident from its greater benignity than PMA. There seems to be no way at this time of making an early, firm differentiation. Denervation in the other hand and arm muscles can be detected early by EMG study, but in some of the 'monomelic' cases there has been progression to the other limb. Sobue et al. (1978) had 24 bilateral among 71 total cases; Loong et al. (1975) had 2 bilateral among 14 cases. The Indian cases (Singh et al. 1980) tend to follow a febrile illness (polio??) and show disproportionate tremor. Serratrice (1984) regards bilateral cases as sufficiently different to merit the eponym O'Sullivan and McLeod syndrome (1978). In 2 of 5 patients with the bilateral syndrome, Gaio et al. (1989) found EMG evidence of denervation also in distal leg muscles. Their magnetic resonance images of the cervical cord seemed flattened from C-5 to T-1, consistent with segmental atrophy. The strong male and Asian preponderance point toward a recessive, sex-linked genetic origin and recently Tandan et al. (1990a) reported this segmental atrophy in twins.

The very rare cases of bilateral quadriceps atrophy (Furukawa et al. 1977; Serratrice et al. 1985) presumably belong in this group, though Gourie-Devi et al. (1984) are inclined to classify them as atypically limited Wohlfart-Kugelberg-Welander SMA (see below).

Lyme disease

Halperin et al. (1990) found evidence of Lyme disease in 3 ALS patients who resided in a hyperendemic area and manifested mainly lower motor involvement; antibiotic treatment was followed by improvement. Mandell et al. (1989) showed no difference in serum Lyme titers of ALS patients compared to controls. Since the fluorescent treponemal antibody test also reacts with Lyme antibodies, the present series contains over 500 negatives in PMA and ALS. One ALS patient was positive, but continued to progress during antibiotic treatment. Long term follow-ups of Halperin's (1990) cases will be necessary to answer the question whether Lyme infection causes an ALS/PMA syndrome, contributes to or accelerates the MND process, or occurs coincidentally and simply adds to the MND symptoms.

Madras MND (Jagannathan 1973) is found in boys and young men in Southern India and early on may have some resemblance to the above syndrome of bilateral 'monomelic' paralysis, but there is spastic weakness and Babinski signs, so juvenile ALS would be the usual differentiation. Moreover, 60% have cranial nerve palsies, including nerve deafness in 33% (Gourie-Devi et al. 1984). With such a geographic localization, natural concerns are about an environmental toxin and unrecorded in-breeding. There may be overlap between the Madras cases, progressive bulbospinal muscular atrophy (see below) and progressive pontobulbar palsy with deafness (Alberca et al. 1980; Brucher et al. 1981).

Asthma has been associated with lower motor neuron disease resembling polio (Manson and Thong 1980; Beede and Newcomb 1980) in children with immunologic abnormalities. This syndrome has never been encountered by this reviewer in experience with over 3500 cases of adult MND.

The *Shy-Drager syndrome* is of course the prototype for motor atrophy in multiple systems. The very rare cases seen personally had relatively minor amyotrophy in the face of major symptoms from hypotension and rigidity, though the first case referred to Dr. Shy had been diagnosed as ALS.

HEREDITARY SPINAL MUSCULAR ATROPHIES (SMAs) (010100)

Another way to seek understanding of PMA is through comparison with the various SMAs, though genetic backgrounds and earlier ages of onset obviously provide immediate differentiation from PMA. On the other hand, the strong male predominance in PMA may indicate a sex-linked genetic disposition, which has very low penetrance so that the family histories seem to be negative.

Distal chronic spinal muscular atrophy (271120) (Meadows and Marsden 1969; McLeod and Prineas 1971) is perhaps the closest SMA to PMA. The onset is in the hand muscles in childhood, sometimes infancy as in case 2 of

Meadows and Marsden (1969), with autosomal dominant inheritance and very slow progression. As in PMA the first appearance of weakness may be brought out by chilling. Iwashita et al. (1970) reported two siblings who also suffered from optic atrophy and nerve deafness. Chalmers and Mitchell (1987) had a similar case, but there was no family history and the reflexes were normal.

It is not clear whether this SMA differs substantially from that (182960) noted by Nelson and Amick (1966) then described in a large series by Harding and Thomas (1980a), who emphasized a peroneal muscular atrophy-appearance with preservation of the muscle stretch reflexes; the mean onset age was 25 years. Of 34 patients, 18 gave no family history of similar disease, whereas 5 families had an autosomal dominant pattern. Perhaps this is a 'mixed bag' with some of the sporadic adult cases overlapping PMA; the present series has one such patient. Jankovic and Rivera (1979) studied a family with stimulus-sensitive myoclonus in addition. Another family described by Furukawa et al. (1968a) also had ataxia, retinitis pigmentosa and diabetes mellitus, and 2 of the patients of Harding and Thomas (1980a) had mild upper extremity incoordination. At the other extreme, Nedelec et al. (1987) described half-brothers with only one leg involved, perhaps a condition more related to benign monomelic amyotrophy (see above).

It is hard to fit into either group the family with very rapidly fatal disease reported by Zatz et al. (1971). The clinical homogeneity in that family is consistent with familial ALS.

Neuronal Charcot-Marie-Tooth disease (118210) must be considered when PMA-like symptoms commence in the legs and the family history is not known. The peripheral motor nerve conductions tend to be normal or nearly so, and onion-bulbs are not present in nerve biopsies, which instead show axonal lesions. Dyck and Lambert (1968) first separated this form from the more common hypertrophic, onion-bulb peripheral neuropathy. Clinical clues to the differentiation that had been present for many years were the later onset of weakness in the neuronal form, and the predominance of motor over sensory signs and symptoms (Harding and Thomas

1980b). This Center continues to follow several patients from two families whose sensory examinations are only notable for loss of vibratory perception in the feet, and reduction but not absence of the perception at the knees, who are now in the 6th and 7th decades of life, though the paralysis has led to wheelchair-dependence. Tandan et al. (1990b) reported a family with ptosis, Parkinsonism and dementia. Survival past the 7th decade was common and l-dopa treatment was beneficial. The ventral horn cell lesion was confirmed in the necropsies of two cases, which also showed that the Parkinsonism was caused by nigral gliosis with loss of pigment. Hyperreflexia in life indicated corticospinal pathway involvement, which may provide a link to the family with SMA and cerebellar ataxia described by Hopf et al. (1971), and the family with ataxia and tremor (214380) investigated by Bouchard et al. (1980), and possibly older cases such as that of Van Bogaert and Moreau (1939). Cornell et al. (1984) had a family with paralysis and deafness (118300), Rosenberg and Chutorian (1967) one with optic atrophy in addition to deafness; these cases may indicate a link to progressive pontobulbar palsy (Alberca et al. (1980; Brucher et al. 1981).

Progressive bulbospinal muscular atrophy (313200) was described by Magee (1960) and probably Tsukagoshi et al. (1965). Kennedy et al. (1968) proved anterior horn cell disease in an autopsied case. In a biopsied case, Penisson-Besnier et al. (1989) found only axon disease in the superficial peroneal nerve. The condition is usually sex-linked and presents with very alarming but mild and only slowly progressive bulbar palsy accompanied by shoulder and pelvic girdle amyotrophy plus gynecomastia. The distal extremity muscles can also be involved over a course of decades. The 4 cases seen at this Center (one from outside the study population) all had the most notable fasciculation in the tongue and face muscles. Two are otherwise very functional in the 7th and 9th decades. Only one has a known affected relative, his twin brother. Barkhaus et al. (1982) and Harding et al. (1982) found an association with diabetes mellitus. Schanen et al. (1984) also found hypogonadism and sexual dysfunction. Both hyper- and hypolipoproteinemia have been reported in some families, probably as an epiphenomenon or even coincidentally (Warner et al. 1990). Presumably from chronic denervation, as discussed above, Ringel et al. (1978) found a mixed neurogenic and myogenic pattern in limb muscle biopsies. Low sensory potential amplitudes and loss of large myelinated sural nerve fibers in biopsies suggest a subclinical sensory lesion (Pouget et al. 1986), which was confirmed by Wilde et al. (1987) and Sobue et al. (1989).

A related or variant condition with juvenile onset may have autosomal dominant (Dobkin and Verity 1976) or recessive transmission (Summers et al. 1987). In the former, progression may be more rapid and accompanied by upper facial involvement (ptosis), distal rather than proximal arm weakness, a cardiac conduction defect and a muscle mitochondrial abnormality ('ragged red' fibers), clearly more than the syndrome under discussion, and without further notice in the literature available for this review. The autosomal recessive form of bulbospinal muscular atrophy (211530) is also difficult to evaluate because of its rarity. Summers et al. (1987) associate it with nerve deafness and credit Vialetto (1936) and Van Laere (1966) with recognition of the condition. Another autosomal dominant form has adult onset and laryngeal palsy as the main bulbar manifestation (Young and Harper 1980; Serratrice et al. 1984b). If the family with oculopharyngeal involvement (Matsunaga et al. 1973) is classified in this group, it is of interest that the bulbar symptoms could develop without any evidence of spinal cord disease.

It may not be wise to subdivide such cases too finely, but certainly they should continue to be sought for their potential value in providing material for genetic analysis. Perhaps these patients have one gene abnormality in common with familial ALS, thus causing overlap of signs and symptoms, and a second gene alteration which provides partial protection, so that the lesions develop less extensively and much more slowly than in ALS.

This is a good place to emphasize again the importance of a thorough investigation in every case of MND, so as to come to the best possible diagnosis even though it may not lead to thera-

peutic opportunity. In each of the 4 cases of
bulbospinal atrophy seen at this Center, the
previous neurologists had diagnosed ALS and
forecast death within a year because of such
marked bulbar symptoms. Two of the 4 were
able to adjust to such devastating news and were
able to continue their lives with equanimity. The
third patient, however, went into an agitated
depression from which he never recovered; his
life was ruined. The fourth patient seemed calm
on the surface, but had silently developed a large
peptic ulcer, which hemorrhaged fatally just a
month after the workup and correct neurologic
diagnosis here.

Wohlfart-Kugelberg-Welander disease (253400)
is an autosomal recessive (but occasionally dominant) proximal SMA having onset usually in
childhood, averaging 9 years of age in the original
series of Kugelberg and Welander (1956). They
also pointed out that the atrophic weakness in
the forearms and hands affected the flexor
muscles more than the extensors, the reverse of
what we commonly see in adult MND. Cases
have been reported with adult onset of paralysis
and recessive inheritance (Bundey and Lovelace
1975; Pearn et al. 1978), also adult onset with
dominant heredity (182980) (Finkel 1962). Quarfordt et al. (1970) and Tsukagoshi et al. (1970)
reported adult-onset with an X-linked, recessive
genetic pattern; the former cases were also
atypical because, of bulbar involvement and
association with hyperlipoproteinemia type II
(313200). Pearn et al. (1978) emphasized that
their adult onset, autosomal recessive cases had
good athletic records as youngsters; as in so
many amyotrophies, cold sensitivity was a prominent complaint. Dubowitz (1964) and Furukawa
et al. (1968b) found that even with dominant
inheritance, males were more severely affected,
suggesting an additional X-linked factor. Bouwsma and Van Wijngaarden (1980) found calf
hypertrophy and serum CK elevations only in 23
of 59 males in a series of 100 patients. Spira's
(1963) cases of autosomal recessive disease were
notable for calf muscle hypertrophy and an
apparent increase of fasciculation with aging. In
Melanesian families, Scrimgeour and Mastaglia
(1984) noted rapid progression and early death,
possibly as early as age 14. Amick et al. (1966)

reported a family with an autosomal dominant
pattern and members with the infantile Werdnig-
Hoffmann (one case symptomatic at age 22
months) and juvenile Wohlfart-Kugelberg-Welander disease. These disorders must therefore
represent a clinical spectrum of the same SMA.
Meadows et al. (1969) also encountered Werdnig-
Hoffman disease in sibs as well as a range of
other findings from bulbar involvement to pyramidal signs, tremor and incontinence; they commented that 'the concept of the Kugelberg-
Welander syndrome has widened since it was
initially described'. Serratrice et al. (1983) reported an autopsy-proven sporadic case with
pallidonigral degeneration in addition.

Friedreich's ataxia (229300) is frequently mentioned as a cause of heritable amyotrophy (Boudouresques et al. 1971) but the latter is nearly
always a minor part of the neurologic picture
and not clearly anterior horn cell in origin
though McLeod (1970) found little evidence of
peripheral motor neuropathy and Singh and
Sham (1964) saw tongue fibrillation in one case.
In 17 autopsies of English cases, Hewer (1968)
found no ventral horn disease. Sigwald et al.
(1964) found spinal motor neuron disease in
familial ataxia but the clinical syndrome was
more extensive, and nigral lesions were also
found. Histiocytes were not reported in this
family. Progressive lower motor neuronopathy
(plus spasticity, choreo-athetosis, seizures and
mental retardation) (164600) was documented by
Landis et al. (1974) also in sea-blue histiocytosis
disease (269600) by Ashwal et al. (1984). This
general group may also include the family with
ataxia, rigidity, fasciculation, motor neuropathy
and Parkinsonism (183050) described by Ziegler
et al. (1972). The family studied by Eto et al.
(1990) would seem to belong in this group despite
certain resemblances to Joseph (Machado) disease. Motor neuronopathy was well established
in two of their cases and in a similar case
investigated by Boller and Segarro (1976).

Scapulohumeral muscular atrophy (181400)
(Kaeser 1965), or neurogenic scapuloperoneal
syndrome, was only differentiated from a somewhat similar picture caused by muscular dystrophy through careful investigation of the Swiss
family F. Having onset between ages 30 and 50,

with slow progression for up to 30 years, this condition is inherited as an autosomal dominant character. Weakness begins in the lower legs (as in peroneal muscular atrophy), ascends to the pelvic girdle muscles, then over many years involves the shoulder girdle and upper arm muscles, and even facial and swallowing functions. In study of tissues from the first autopsied case, Dr. G. M. Shy found neurogenic muscular atrophy, but only central chromatolysis and vacuolar changes in neurons of the cranial VII, IX, and X nuclei (Kaeser 1965). In a follow-up with the first complete autopsy, Probst et al. (1977) confirmed normal numbers of neurons amidst a neuropil abnormality caused by axonal swelling and intraaxonal masses.

Similar cases develop early facial paresis (182970) (Fenichel et al. 1967), and in others there is cardiopathy and prominent myopathic changes in the skeletal muscles (181350) (Mawatori and Katayama 1973; Takahashi et al. 1975; Fenichel et al. 1982). Pearn (1980) only differentiated by the genetic pattern, i.e. autosomal dominant vs. recessive vs. sex-linked. In one family with dominant transmission, the disease developed in middle age but progressed rapidly to fatal respiratory muscle weakness within 3 years (Jansen et al. 1986). In another family with sex-linked transmission, one member apparently died of Werdnig-Hoffmann disease; 2 siblings with this scapuloperoneal syndrome also had Babinski signs when examined (Skre et al. 1978). In still another family, Fotopulos (1966) found both the scapuloperoneal syndrome and Huntington's chorea in a father and daughter (143100). Serratrice et al. (1984a) reported two further cases in which other affected family members could not be identified.

Familial muscle cramps (158400), with or without frequent fasciculation or myokymia, raise concern about the onset of MND (Lanska et al. 1986). The association of cramps and the type of fasciculation termed myokymia was made many years ago by Denny-Brown and Foley (1948). Norris et al. (1957) showed probable central origin of cramps. Van den Berghe et al. (1980) found that familial cramps tend to diminish after about age 25, whereas Ricker and Moxley (1990) noted the age of onset to be 15–30 years. Van

den Berghe et al. (1980) found EMG and muscle biopsy evidence of a neurogenic disorder, which was confirmed by Lazaro et al. (1981) if 'an occasional angulated fiber' in the muscle biopsy is truly beyond the normal (Bjornskov et al. [1986] found up to 11% small angulated fibers in each of 6 control patients). Lazaro et al. (1981) also found slow peripheral nerve motor conduction in their propositus but a glucose tolerance test was not done, nor a nerve biopsy. Frankel et al. (1974) believed that their patient's 'restless legs' were caused by an occult motor neuron disorder.

Hexosaminidase deficiency (272800) (Johnson 1981) can cause SMA of adult onset, nearly always in Ashkenazi Jewish people. A man with the onset at age 15 (probably earlier) developed a proximal-SMA picture like Wohlfart-Kugelberg-Welander disease; he was correctly diagnosed by means of blood measurement for hexosaminidase A and rectal biopsy for ganglion cells (Johnson et al. 1982). An older patient had a slowly progressing ALS syndrome (Yaffe et al. 1979). The clinical spectrum may be broader and rectal biopsy can aid in diagnosing suspect cases (Thomas et al. 1989). In survey of 102 ALS patients, however, Gudesblatt et al. (1988) found no such cases, even though 52 of their patients were considered to be atypical because of onsets under age 35, prolonged courses and family histories of MND.

Neuroacanthocytosis (100500) has normal serum lipoproteins but severe neurologic disease including dystonias or chorea, and areflexic proximal muscle weakness (Aminoff 1972). The acanthocytosis may occur without neurologic involvement (Levine et al. 1968). Both autosomal dominant and recessive inheritance are reported. In the recessive family of Spitz et al. (1985), the wasting was distal, Tourette-like tics prominent, and there was also supranuclear ophthalmoplegia in one member.

Familial spherocytosis (Lagreze et al. 1987) is another disorder causing characteristic erythrocyte abnormality and adult-onset amyotrophy plus chorea and dementia. In one autopsy, the greatest neuronal loss was in the striatum and the spinal ventral horns. Of possible significance for consideration of auto-immunity in PMA (see above) was the presence of CSF oligoclonal IgG production, presumably a secondary phenomenon.

Congenital ichthyosis with spastic paraplegia and SMA (Sjögren-Larsson syndrome) (270200) can lead to neurologic manifestations in adult life, though in these cases the amyotrophy is a minor part of the clinical picture (McNamara et al. 1975).

SMA with Joseph disease (109150) type III (Machado disease) is referred to by Rosenberg and Fowler (1981) as 'ataxia-ALS syndrome' because of widespread amyotrophy, presumably from motor neuron lesions though the late sensory impairment suggests a peripheral neuropathic component. In one autopsy in type I, Woods and Schaumberg (1972) found anterior horn cell as well as other lesions. Motor neuronopathy was confirmed by Romanul et al. (1977), but in an atypical family. Eto et al. (1990) note marked similarities to this syndrome in their family with spinocerebellar degeneration. They very reasonably conclude that exact differentiation awaits the relevant advances in genetic analysis.

Acid maltase deficiency produced adult-onset Wohlfart-Kugelberg-Welander disease-like paralysis in the patient of Pongratz et al. (1984).

Familial spastic paraplegia turns out to be another 'mixed bag'. The Troyer syndrome (Cross and McKusick 1967; Neuhauser et al. 1976) is an autosomal recessive (275900); clinically there are variants with dystonia but autosomal dominant inheritance (105300), and without dystonia but with amyotrophy (182600) as reported by Silver (1966).

Miscellaneous conditions include neurofibromatosis, which in the present series led to motor neuropathy mimicking PMA in 2 patients. Thomas et al. (1990) describe 3 cases of distal sensorimotor neurofibromatous neuropathy. In view of the heritability of the basic immunologic responses, it is not surprising that sibs have developed neuropathic gammopathy (Jensen et al. 1988).

CONCLUDING REMARKS

Clearly these variably similar but different heritable syndromes indicate the potential for abnor-

mality in any one of multiple gene sites to act directly or indirectly against the upper or lower motor neuron, or both. The complexity is increased by the possibility that one genetic abnormality may protect against the disorder caused by another gene defect. Another possibility is more complex: genetic sensitivity to an environmental toxin (or virus, despite the negative studies to date). Even more potential linkages and etiologies for MND have been listed by the Research Group on Neuromuscular Disease (Walton and Thomas 1988).

We have noted that the male predominance in PMA might indicate a sex-linked condition. As stated at the outset, the approach to Werdnig-Hoffmann disease that now seems most likely is search for a specific motor neuron growth factor (there may be several), by analogy with experimental auto-immune sympathectomy. An important alternative is search for the genetic code that programs the death of the surplus motor neurons in the embryo (Sarnat et al. 1989). The very direction in which to study PMA, however, remains a great challenge. PMA seems more closely related to ALS than to the SMAs, particularly since the major SMAs are linked to chromosome 5 (Munsat et al. 1990), familial ALS to chromosome 21 (Siddique et al. 1991).

Meantime, practical physicians must care for the patients suffering these diseases. Much can be done to alleviate the various problems (Norris et al. 1985) and Chapter 26 by R. A. Smith deals with such symptomatic treatments. Modes of hereditary transmission, a vital requirement for genetic counseling in the SMAs and familial ALS, is discussed in Chapters 4–9 and 14.

REFERENCES

ADAMS, R. D. ET AL.: Clinicopathologic conference. New Engl. J. Med. 283 (1970) 806–814.
ADAMS, R. D., D. DENNY-BROWN and C. M. PEARSON: Diseases of Muscle, 2nd ed. New York, Hoeber-Harper (1962) 138–157.
ALBERCA, R., C. MONTERO, A. IBANEZ, D. I. SEGURA and G. MIRANDA-NIEVES: Progressive bulbar paralysis associated with neural deafness, a nosological entity. Arch. Neurol. (Chic.) 37 (1980) 214–216.
ALLEN, I. V., E. DERMOTT, J. H. CONNOLLY and L. J. HURWITZ: A study of a patient with the amyotrophic

form of Creutzfeldt-Jakob disease. Brain 94 (1971) 715–724.

AMICK, L. D., H. L. SMITH and W. W. JOHNSON: An unusual spectrum of progressive spinal muscular atrophy. Acta Neurol. Scand. 42 (1966) 275–295.

AMINOFF, M. J.: Acanthocytosis and neurological disease. Brain 95 (1972) 749–760.

ASHWALL, S., T. V. TRASHER, D. R. RICE and D. A. WENGER: A new form of sea-blue histiocytosis associated with progressive anterior horn cell and axonal degeneration. Neurology (Minneap.) 16 (1984) 184–192.

AUER, R. N., R. B. BELL and M. A. LEE: Neuropathy with onion bulb formations and pure motor manifestations. Can. J. Neurol. Sci. (1989) 16: 194–197.

BARDWICK, P. A., N. J. AVAIFLER, G. N. GILL ET AL.: Plasma cell dyscrasia with polyneuropathy, organomegaly, endocrinopathy, M protein, and skin changes: the POEMS syndrome. Medicine 59 (1980) 311–322.

BARKHAUS, P. E., W. R. KENNEDY, L. Z. STERN and R. B. HARRINGTON: Hereditary proximal spinal and bulbar motor neuron disease of late onset: a report of six cases. Arch. Neurol. (Chic.) 39 (1982) 112–116.

BEEDE, H. E. and R. W. NEWCOMB: Lower motor neuron paralysis in association with asthma. Johns Hopkins Med. J. 147 (1980) 186–187.

BJORNSKOV, E. K., F. H. NORRIS and J. M. MOWER-KUBY: Quantitative axon terminal and end-plate morphology in amyotrophic lateral sclerosis. Arch. Neurol. (Chic.) 41 (1984) 527–530.

BHARUCHA, E. P. and V. P. MONDKAR: Neurological complications of a new conjunctivitis. Lancet 3 (1972) 970 (letter).

BOLLER, F. and J. M. SEGARRA: Spino-pontine degeneration. In: P. J. Vinken and G. W. Bruyn (Eds): Handbook of Clinical Neurology, Vol. 21. Amsterdam, North-Holland, (1976) 389–402.

BONDUELLE, M.: Histoire de la neurologie: Aran-Duchenne? Duchenne-Aran? La querelle de l'atrophie musculaire progressive. Rev. Neurol. (Paris) 146 (1990) 97–106.

BOUCHARD, J., P. BEDARD and R. BOUCHARD: Study of a family with progressive ataxia and severe distal amyotrophy. Can. J. Neurol. Sci. 7 (1980) 345–349.

BOUDOURESQUES, J., M. TOGA, R. KHALIL, A. GOSSET, R. A. VIGOUROUX and J. F. PELLISSIER: Forme amyotrophique d'une dégénérescence spino-cérébelleuse. Rev. Neurol. (Paris) 125 (1971) 25–38.

BOUWSMA, G. and G. K. VAN WIJNGAARDEN: Spinal muscular atrophy and hypertrophy of the calves. J. Neurol. Sci. 44 (1980) 275–279.

BRAHIC, M., R. A. SMITH, C. J. GIBBS, R. M. GARRUTO, W. W. TOURTELLOTTE and E. CASH: Detection of picornavirus sequences in nervous tissue of amyotrophic lateral sclerosis and control patients. Ann. Neurol. 18 (1985) 337–343.

BRAIN, L., P. B. CROFT and M. WILKINSON: Motor neurone disease as a manifestation of neoplasm. Brain 88 (1965) 479–500.

BRAMWELL, B.: The Diseases of the Spinal Cord. London, Simpkin & Marshall (1882) pp. 183–189.

BRUCHER, J. M., R. DOM, A. LOMBAERT and H. CARTON: Progressive pontobulbar palsy with deafness. Arch. Neurol. (Chic.) 38 (1981) 186–190.

BUCHANAN, D. S. and N. MALAMUD: Motor neuron disease with renal cell carcinoma and postoperative neurologic remission. Neurology (Minneap.) 23 (1973) 891–894.

BUNDEY, S. and R. E. LOVELACE: A clinical and genetic study of chronic proximal spinal muscular atrophy. Brain 98 (1975) 455–472.

CAMPBELL, A. M. G., E. R. WILLIAMS and D. BARLTROP: Late motor neuron degeneration following poliomyelitis. Neurology (Minneap.) 19 (1969) 1101–1106.

CASHMAN, N. R., R. MASELLI, R. L. WOLLMANN, R. ROOS, R. SIMON and J. P. ANTEL: Late denervation in patients with antecedent paralytic poliomyelitis. N. Engl. J. Med. 317 (1987) 7–12.

CASTELBAUM, A. R., P. D. DONOFRIO, F. O. WALKER and B. T. TROOST: Laxative abuse causing hypermagnesemia, quadriparesis, and neuromuscular junction defect. Neurology (Minneap.) 39 (1989) 746–747.

CASTALANO, L. W.: Herpes virus hominis antibody in multiple sclerosis and amyotrophic lateral sclerosis. Neurology (Minneap.) 22 (1972) 473–478.

CHAINE, P., P. BOUCHE, J. M. LEGER, D. DORMONT and H. P. CATHALA: Atrophie musculaire progressive localisée à la main. Rev. Neurol. (Paris) 144 (1988) 759–763.

CHALMERS, N. and J. D. MITCHELL: Optico-acoustic atrophy in distal spinal muscular atrophy. J. Neurol. Neurosurg. Psychiatry 50 (1987) 238–239.

CHARCOT, J.-M.: Leçons sur les maladies du système nerveux. IInd Series, collected by Bourneville. Paris, Delahaye (1873) pp. 192–242. (English translation by G. Sigerson. London, New Sydenham Society (1881), reprinted by N. Y. Acad. Med., New York, Hafner (1962) pp. 163–204.)

CHARCOT, J.-M. and A. JOFFROY: Deux cas d'atrophie musculaire progressive. Arch. Physiol. 2 (1869) 354–367, 745–760.

CHOKROVERTY, S., M. G. REYES, F. A. RUBINO and K. D. BARRON: Hemiplegic amyotrophy. Arch. Neurol. (Chic.) 33 (1976) 104–110.

CHOKROVERTY, S., M. G. REYES, F. A. RUBINO and H. TONAKI: The syndrome of diabetic amyotrophy. Ann. Neurol. 2 (1977) 181–194.

CONRADI, S., L. O. RONNEVI and F. H. NORRIS: Motor neuron disease and toxic metals. In: L. P. Rowland (Ed.), Human Motor Neuron Diseases. New York, Raven Press (1982) 201–229.

CORNELIO, F., N. BRESOLIN, P. A. SINGER, S. DIMAURO and L. P. ROWLAND: Clinical varieties of neuromuscular disease in debrancher deficiency. Arch. Neurol. (Chic.) 41 (1984) 1027–1032.

CORNELL, J., S. SELLARS and P. BEIGHTON: Autosomal recessive inheritance of Charcot-Marie-Tooth disease associated with sensorineural deafness. Clin. Genet. 25 (1984) 163–165.

CREMER, N. E., L. OSHIRO, F. H. NORRIS and E. H. LENNETTE: Culture of tissues from patients with amyotrophic lateral sclerosis. Arch. Neurol. (Chic.) 29 (1973) 331–333.

CROSS, H. E. and V. A. MCKUSICK: The Troyer syndrome. Arch. Neurol. (Chic.) 16 (1967) 473–485.

CRUVEILHIER, J.: Sur la paralysie musculaire progressive atrophique. Arch. Gén. Méd. 91 (1853) 561–603.

DALAKAS, M. C. and W. K. ENGEL: Polyneuropathy with monoclonal gammopathy: studies of 11 patients. Ann. Neurol. 10 (1981) 45–52.

DALAKAS, M. C., G. ELDER, M. HALLETT, J. RAVITS, M. BAKER, N. PAPADOPOULOS, P. ALBRECTS and J. SEVER: A long-term follow-up study of patients with post-poliomyelitis neuromuscular symptoms. N. Engl. J. Med. 314 (1986) 959–963.

DENNY-BROWN, D. and J. M. FOLEY: Myokymia and the benign fasciculation of muscular cramps. Trans. Assoc. Am. Phys. 61 (1948) 88–96.

DENYS, E. H. and F. H. NORRIS: Amyotrophic lateral sclerosis: impairment of neuromuscular transmission. Arch. Neurol. (Chic.) 36 (1979) 74–80.

DIREKZE, M. and P. S. L. FERNANDO: Transient anterior horn cell dysfunction in diphenylhydantoin therapy. Eur. Neurol. 15 (1977) 131–134.

DOBKIN, B. H. and M. A. VERITY: Familial progressive bulbar and spinal muscular atrophy: juvenile onset and late morbidity with ragged-red fibers. Neurol. (Minneap.) 26 (1976) 754–763.

DORSEN, M. and G. EHNI: Cervical spondylotic radiculopathy producing manifestations mimicking primary muscular atrophy. Neurosurgery 5 (1979) 427–431.

DRACHMAN, D. B. and R. W. KUNCL: Amyotrophic lateral sclerosis: An unconventional autoimmune disease? Ann. Neurol. 26 (1989) 269–274.

DUBAS, F., P. BERTRAND and J. ÉMILE: Amyotrophie spinale progressive et adénome parathyroidien. Rev. Neurol. (Paris) 145 (1989) 65–68.

DUBOWITZ, V.: Infantile muscular atrophy: a prospective study with reference to a slowly progressive variety. Brain 87 (1964) 707–718.

DYCK, P. J. and E. H. LAMBERT: Lower motor and primary sensory neuron diseases with peroneal muscular atrophy, II. Arch. Neurol. (Chic.) 18 (1968) 619–625.

DYCK, P. J., A. C. LAIS, M. OHTAS ET AL.: Chronic inflammatory polyradiculoneuropathy. Mayo Clin. Proc. 50 (1975) 621–637.

EL ALAOUI-FARIS, M., A. MEDEJEI, K. AL ZEMMOURI, M. YAHYAOUI and T. CHKILI: Le syndrome de sclérose latérale le amyotrophique d'origine syphilitique. Rev. Neurol. (Paris) 146 (1990) 41–44.

ENGELHARDT, J. I., S. H. APPEL and J. M. KILLIAN: Experimental autoimmune motoneuron disease. Ann. Neurol. 26 (1989) 368–376.

ESCOBAR, M. R., H. P. DALTON and M. J. ALLISON: Fluorescent antibody tests for syphilis using cerebrospinal fluid. Am. J. Clin. Pathol. 53 (1970) 886–890.

ETO, E., S. M. SUMI, T. D. BIRD, T. MCEVOY-BUSH, M. BOEHNKE and G. SCHELLENBERG: Family with dominantly inherited ataxia, amyotrophy, and peripheral sensory loss. Arch. Neurol. (Chic.) 47 (1990) 968–974.

EVANS, B. K., C. FAGAN, T. ARNOLD, E. J. DROPCHO and

S. J. OH: Paraneoplastic motor neuron disease and renal cell carcinoma. Neurology (Minneap.) 40 (1990) 960–962.

FENICHEL, G. M., E. S. EMERY and P. HUNT: Neurogenic atrophy simulating facioscapulohumeral dystrophy. Arch. Neurol. (Chic.) 17 (1967) 257–260.

FENICHEL, G. M., Y. C. SUL, A. W. KILROY and R. BLOVIN: An autosomal dominant dystrophy with humeropelvic distribution and cardiomyopathy. Neurology (Minneap.) 32 (1982) 1399–1401.

FETELL, M. R., G. SMALLBERG, L. D. LEWIS, R. E. LOVELACE, A. P. HAYS and L. P. ROWLAND: A benign motor neuron disorder: delayed cramps and fasciculation after poliomyelitis or myelitis. Ann. Neurol. 11 (1982) 423–427.

FINKEL, N.: A forma pseudomiopatica tardia da atrofia muscular progressiva heredo-familial. Arq. Neuro-Psiquiatr. 20 (1962) 307–322 (from McKusick 1990, 18298).

FOTOPULOS, D.: Huntington-Chorea und chronisch-progressive spinale Muskelatrophie. Psychiat. Neurol. Psychol. 18 (1966) 63–70.

FRANKEL, B. L., B. M. PATTEN and J. C. GILLIN: Restless legs syndrome. J. Am. Med. Assoc. 230 (1974) 1302–1303.

FURUKAWA, T. and Y. TOYOKURA: Chronic spinal muscular atrophy of facioscapulohumeral type. J. Med. Gen. 13 (1976) 285–289.

FURUKAWA, T., A. TAKAGI, K. NAKAO, H. SUGITA, H. TSUKAGOSHI and T. TSUBAKI: Hereditary muscular atrophy with ataxia, retinitis pigmentosa, and diabetes mellitus. Neurology (Minneap.) 18 (1968a) 942–947.

FURUKAWA, T., K. NAKAO, H. SUGITA and H. TSUKAGOSHI: Kugelberg-Welander disease with particular reference to sex-influenced manifestations. Arch. Neurol. (Chic.) 19 (1968b) 156–162.

FURUKAWA, T., N. AKAGAMI and S. MARUYAMA: Chronic neurogenic quadriceps amyotrophy. Ann. Neurol. 2 (1977) 528–530.

GAIO, J. M., B. LECHEVALIER, M. HOMMEL, F. VIADER, F. CHAPON and J. PERRET: Amyotrophie spinale chronique des membres superieurs de l'adulte jeune (syndrome de O'Sullivan et McLeod). Rev. Neurol. (Paris) 145 (1989) 163–168.

GAJDUSEK, D. C.: Motor-neuron disease in natives of New Guinea. N. Engl. J. Med. 268 (1963) 474–476.

GASTAUT, J. L. and L. J. BENAIM: L'amyotrophie d'origine pariétale: Rev. Neurol. (Paris) 144 (1988) 301–305.

GOURIE-DEVI, M., T. G. SURESH and S. K. SHANKAR: Monomelic amyotrophy. Arch. Neurol. (Chic.) 41 (1984) 388–394.

GOWERS, W. R.: A lecture on abiotrophy. Lancet 2 (1902) 1003–1007.

GRESHAM, L. S., C. A. MOLGAARD, A. L. GOLBECK and R. SMITH: Amyotrophic lateral sclerosis and history of skeletal fracture: a case control study. Neurology (Minneap.) 37 (1987) 717–719.

GUDESBLATT, M., M. D. LUDMAN, J. A. COHEN, R. J. DESNICK,

S. CHESTER, G. A. GRABOWSKI and J. T. CAROSCIO: Hexosaminidase A activity and amyotrophic lateral sclerosis. Muscle Nerve 11 (1988) 227–230.

HALPERIN, J. J., G. P. KAPLAN, S. BRAZINSKY, T. F. TSAI, T. CHENG, A. IRONSIDE, P. WU, J. DELFINER, M. GOLIGHTLY, R. H. BROWN, R. J. DATTWYLER and B. J. LUFT: Immunologic reactivity against *Borrelia burgdorferi* in patients with motor neuron disease. Arch. Neurol. (Chic.) 47 (1990): 586–594.

HARDING, A. E. and P. K. THOMAS: Hereditary distal spinal muscular atrophy. J. Neurol. Sci. 45 (1980a) 337–348.

HARDING, A. E. and P. K. THOMAS: The clinical features of hereditary motor and sensory neuropathy types I and II. Brain 103 (1980b) 259–280.

HARDING, A. E., P. K. THOMAS, M. BARAITSER, P. G. BRADBURY, J. A. MORGAN-HUGHES and J. R. PONSFORD: X-linked recessive bulbospinal neuronopathy: a report of ten cases. J. Neurol. Neurosurg. Psychiatry 45 (1982) 1012–1019.

HEWER, R. L.: Study of fatal cases of Friedreich's ataxia. Br. Med. J. 3 (1968) 649–652.

HIRAYAMA, K., Y. TOYOKURA and T. TSUBAKI: Juvenile muscular atrophy of unilateral upper extremity. Psychiat. Neurol. Jpn. 61 (1959) 2190–2197 (from Gourie-Devi, Suresh and Shankar 1984).

HIRAYAMA, K., M. TOMONAGA, K. KITANO, T. YAMADA, S. KOJIMA and K. ARAI: Focal cervical poliopathy causing juvenile muscular atrophy of distal upper extremity: a pathological study. J. Neurol. Neurosurg. Psychiatry 50 (1987) 285–290.

HOFFMAN, P., B. FESTOFF, L. T. GIRON, L. C. HOLLENBECK, R. M. GARRUTO and F. W. RUSCETTI: Isolation of LAV/HTLV-III from a patient with amyotrophic lateral sclerosis. N. Engl. J. Med. 313 (1985) 324–325.

HOLBROOK, L. A., F. X. M. BENCH and J. R. SILVER: Delayed myelopathy: a rare complication of severe electrical burns. Br. Med. J. 4 (1970) 659–660.

HOPF, H. C., F. DUENSING, K. LOWITZSCH and R. KRÖNKE: Hereditäre cerebellare Ataxie mit spinaler Muskelatrophie. Z. Neurol. 199 (1971) 344–352.

HOROUPIAN, D. S., P. PICK, I. SPIGLAND, P. SMITH, R. PORTENOY, R. KATZMAN and S. CHO.: Acquired immune deficiency syndrome and multiple tract degeneration in a homosexual man. Ann. Neurol. 15 (1984) 502–505.

HUGON, J., M. LUBEAU, F. TABARAUD, F. CHAZOT, J. M. VALLAT and M. DUMAS: Central motor conduction in motor neuron disease. Ann. Neurol. 22 (1987) 544–546.

INGRAM, D. A. and M. SWASH: Central motor conduction is abnormal in motor neuron disease. J. Neurol. Neurosurg. Psychiatry 50 (1987) 159–166.

IWASHITA, H., N. INOUE, S. ARAKI and Y. KUROIWA: Optic atrophy, neural deafness, and distal neurogenic amyotrophy. Arch. Neurol. (Chic.) 22 (1970) 357–364.

JAGANNATHAN, K.: Juvenile motor neurone disease. In: V. D. Spillane (Ed.), Tropical Neurology. London, Oxford University Press (1973) 127–130.

JANKOVIC, J. and V. M. RIVERA: Hereditary myoclonus and progressive distal muscular atrophy. Ann. Neurol. 6 (1979) 227–231.

JANSEN, P. H. P., E. M. G. JOOSTEN, H. H. J. JASPAR and H. M. VINGERHOETS: A rapidly progressive autosomal dominant scapulohumeral form of spinal muscular atrophy. Ann. Neurol. 20 (1986) 538–540.

JELLINGER, K. and E. NEUMAYER: Myélopathie progressive d'origine vasculaire. Acta Neurol. Psychiatr. Belg. 62 (1962) 944–956.

JELLINGER, K. and K. W. STURM: Delayed radiation myelopathy in man. J. Neurol. Sci. 14 (1971) 389–408.

JENSEN, T. S., H. D. SCHRODER, V. JONSSON, J. ERNERUDH, B. STIGSBY, Z. KAMIENIECKA, E. HIPPE and W. TROJABORG: IgM monogammopathy and neuropathy in two siblings. J. Neurol. Neurosurg. Psychiatry 51 (1988) 1308–1315.

JOHNSON, W. G.: The clinical spectrum of hexosaminidase deficiency diseases. Neurology (Minneap.) 31 (1981) 1453–1456.

JOHNSON, W. G., H. J. WIGGER, H. R. KARP, L. M. GLAUBIGER and L. P. ROWLAND: Juvenile spinal muscular atrophy: a new hexosaminidase deficiency phenotype. Ann. Neurol. 11 (1982) 11–16.

KAESER, H. E.: Scapuloperoneal muscular atrophy. Brain 88 (1965) 407–418.

KATO, T., T. KATAGIRI, A. HIRANO, H. SASAKI and S. ARAI: Sporadic lower motor neuron disease with Lewy body-like inclusions: a new subgroup? Acta Neuropathol. 76 (1988) 208–211.

KENNEDY, W. R., M. ALTER and J. H. SUNG: Progressive proximal spinal and bulbar muscular atrophy of late onset. Neurol. (Minneap.) 18 (1968) 671–680.

KLINGMAN, J., H. CHUI, M. GORGIAT and J. PERRY: Functional recovery: a major risk factor for the development of postpoliomyelitis muscular atrophy. Arch. Neurol. (Chic.) 45 (1988) 645–647.

KUGELBERG, E. and L. WELANDER: Heredofamilial juvenile muscular atrophy simulating muscular dystrophy. Arch. Neurol. Psychiatry 75 (1956) 500–509.

KURLAND, L. T., N. W. CHOI and G. P. SAYRE: Implications of incidence and geographic patterns on the classification of amyotrophic lateral sclerosis. In: F. H. Norris and L. T. Kurland (Eds.), Motor Neuron Diseases. New York-London, Grune and Stratton (1969) 28–50.

LAGREZE, H. L., B. R. BROOKS, S. A. MOORE, S. E. SPENCER, S. E. KORNGUTH, A. J. BRIDGES, R. L. LEVINE and S. B. PERLMAN: Familial amyotrophy, neuropathy, chorea, and dementia with spherocytosis/elliptocytosis: a new syndrome. Neurology (Minneap.) 37 (1987) 139 (abstract).

LAGUENY, A., M. AUPY, P. AUPY, X. FERRER, P. HENRY and J. JULIEN: Syndrome de la corne antérieure post-radiothérapique. Rev. Neurol. (Paris) 141 (1985) 222–227.

LAMBERT, E. H.: Electromyography in amyotrophic lateral sclerosis. In: F. H. Norris and L. T. Kurland (Eds.), Motor Neuron Diseases. New York-London, Grune and Stratton (1965) 135–153.

LAMPSON, L. A., P. D. KUSHNER and R. A. SOBEL: Major

histocompatibility complex antigen expression in the affected tissues in amyotrophic lateral sclerosis. Ann. Neurol. 28 (1990) 365–372.

LANDIS, D. M., R. N. ROSENBERG, S. C. LANDIS, L. SCHUT and W. L. NYHAN: Olivopontocerebellar degeneration. Arch. Neurol. (Chic.) 31 (1974) 295–307.

LANSKA, D. J., R. L. RUFF, S. D. WHEELER, R. ZEILER and P. GALLOWAY: Myokymia in Isaac's syndrome and motor neuron disease. Neurology (Minneap.) 36 (1986) 239 (abstract).

LAZARO, R. P., R. D. ROLLINSON and G. M. FENICHEL: Familial cramps and muscle pain. Arch. Neurol. (Chic.) 38 (1981) 22–24.

LEBO, C. P., F. H. NORRIS and E. F. STEINMETZ: Electronystagmography in amyotrophic lateral sclerosis. Arch. Neurol. (Chic.) 40 (1983) 525–526 (letter).

LEVINE, I. M., J. W. ESTES and J. M. LOONEY: Hereditary neurological disease with acanthocytosis: a new syndrome. Arch. Neurol. (Chic.) 19 (1968) 403–409.

LOONG, S. C., M. H. L. YAP and I. P. NEI: An unusual form of motor neuron disease. Proc. Singapore Med. Soc. 10 (1975) 21–24.

MAGEE, K. R.: Familial progressive bulbar-spinal muscular atrophy. Neurology (Minneap.) 10 (1960) 295–305.

MAIER, J. G., R. H. PERRY, W. SAYLOR and M. H. SULAK: Radiation myelitis of the dorsolumbar spinal cord. Radiology 93 (1969) 153–160.

MALLETTE, L. E. and B. M. PATTEN: Neurogenic muscle atrophy and osteomalacia in adult Fanconi syndrome. Ann. Neurol. 1 (1977) 131–137.

MANDELL, H., A. C. STEERE, B. N. REINHARDT ET AL.: Lack of antibodies to Borrelia burgdorferi in patients with amyotrophic lateral sclerosis. N. Engl. J. Med. 320 (1989) 255–256.

MANSON, J. I. and Y. H. THONG: Immunological abnormalities in the syndrome of poliomyelitis-like illness associated with acute bronchial asthma. Arch. Dis. Child. 55 (1980) 26–32.

MAPELLI, G.: Is there a characteristic dermal involvement in amyotrophic lateral sclerosis? J. Neurol. Sci. 95 (1990) 231–232 (letter).

MATSUNAGA, M., T. INOKUCHI, A. OHNISHI and Y. KUROIWA: Oculopharyngeal involvement in familial neurogenic muscular atrophy. J. Neurol. Neurosurg. Psychiatry 36 (1973) 104–111.

MAWATORI, S. and K. KATAYAMA.: Scapuloperoneal muscular atrophy with cardiopathy, an X-linked recessive trait. Arch. Neurol. (Chic.) 28 (1973) 55–59.

MCHOLM, G. B., M. J. AGUILAR and F. H. NORRIS: Lipofuscin in amyotrophic lateral sclerosis. Arch. Neurol. (Chic.) 41 (1984) 1187–1188.

MCKUSICK, V. A.: Mendelian Inheritance in Man, 8th ed. Baltimore, Johns Hopkins University Press (1990).

MCLEOD, J. G.: An electrophysiological and pathological study of peripheral nerves in Friedreich's ataxia. J. Neurol. Sci. 12 (1971) 333–349.

MCLEOD, J. G. and J. W. PRINEAS: Distal type of chronic spinal muscular atrophy. Brain 94 (1971) 703–714.

MCNAMARA, J. O., J. R. CURRAN and H. H. ITABASHI: Congenital ichthyosis with spastic paraplegia of adult onset. Arch. Neurol. (Chic.) 32 (1975) 699–701.

MEADOWS, J. C. and C. D. MARSDEN: A distal form of chronic spinal muscular atrophy. Neurology (Minneap.) 19 (1969) 53–58.

MEADOWS, J. C., C. D. MARSDEN and D. G. F. HARRIMAN: Chronic spinal muscular atrophy in adults; Part 1: the Kugelberg-Welander syndrome. J. Neurol. Sci. 9 (1969) 527–550.

METCALF, J. C., J. B. WOOD and T. E. BERTORINI: Benign focal amyotrophy: metrizamide CT evidence of cord atrophy. Muscle Nerve 10 (1987) 338–345.

MILANESE, C. and L. MARTIN: Amyotrophic lateral sclerosis and trauma. Acta Neurol. Belg. 70 (1970) 482–491.

MILLER, R. D., D. W. MULDER, W. S. FOWLER and A. M. OLSEN: Exertional dyspnea: a primary complaint in progressive muscular atrophy and amyotrophic lateral sclerosis. Ann. Intern. Med. 46 (1957) 119–126.

MITCHELL, D. M. and S. A. OLCZAK: Remission of a syndrome indistinguishable from motor neurone disease after resection of bronchial carcinoma. Br. Med. J. 3 (1979) 176–177.

MULDER, D. M.: Motor neuron disease. In: P. J. Dyck, P. K. Thomas, E. H. Lambert and R. Bunge (Eds.), Peripheral neuropathy, ed. II., Vol. II. Philadelphia, Saunders (1984) 1525–1536.

MULDER, D. W., R. A. ROSENBAUM and D. D. LAYTON: Late progression of poliomyelitis or 'forme fruste' amyotrophic lateral sclerosis? Mayo Clin. Proc. 47 (1972) 756–761.

MÜLLER, R.: Progressive motor neuron disease in adults. Acta Psychiatr. Neurol. 27 (1952) 137–156.

MUNSAT, T. L., L. SKERRY, B. KORF, B. POBER, Y. SCHAPIRA, G. G. GASCON, S. M. AL-RAJEH, V. DUBOWITZ, K. DAVIES, L. M. BRZUSTOWICZ, G. K. PENCHASZADEH and T. C. GILLIAM: Phenotypic heterogeneity of spinal muscular atrophy mapping to chromosome 5q11.2-13.3 (SMA 5q). Neurology (Minneap.) 40 (1990) 1831–1836.

NEDELEC, C., F. DUBAS, J. L. TRUELLE, F. POUPLARD, F. DELESTRE and I. PENISSON-BESNIER: Amyotrophie spinale progressive distale et asymétrique des membres inférieurs à caractère familial. Rev. Neurol. (Paris) 143 (1987) 765–767.

NELSON, J. W. and L. D. AMICK: Heredofamilial progressive spinal muscular atrophy. Neurology (Minneap.) 16 (1966) 306 (Abstract).

NEUHAUSER, G., C. WIFFLER and J. M. OPITZ: Familial spastic paraplegia with distal muscle wasting in the Old Order Amish; atypical Troyer syndrome or 'new' syndrome. Clin. Gen. 9 (1976) 315–323.

NORRIS, F. H.: Adult spinal motor neuron disease: Progressive muscular atrophy (Aran's disease) in relation to amyotrophic lateral sclerosis. In: P. J. Vinken and G. W. Bruyn (Eds.), Handbook of Clinical Neurology, System Disorders and Atrophies, Vol. 22, Part II. Amsterdam, North-Holland Publ Co (1975) 1–56.

NORRIS, F. H.: ALS: The clinical disorder. In: R. A.

Smith (Ed.), Handbook of Amyotrophic Lateral Sclerosis. New York, Marcel Dekker (1991) in press.

NORRIS, F. H. and W. K. ENGEL: Carcinomatous amyotrophic lateral sclerosis. In: L. Brain and F. H. Norris (Eds.), The Remote Effects of Cancer on the Nervous System. New York-London, Grune and Stratton (1965) 24–34.

NORRIS, F. H., E. L. GASTEIGER and P. O. CHATFIELD: An electromyographic study of induced and spontaneous muscle cramps. Electroenceph. Clin. Neurophysiol. 9 (1957) 139–147.

NORRIS, F. H., W. H. MCMENEMEY and R. O. BARNARD: Anterior horn pathology in carcinomatous neuromyopathy compared to other forms of motor neuron disease. In: F. H. Norris and L. T. Kurland (Eds.), Motor Neuron Diseases. New York, Grune and Stratton (1969) 100–111.

NORRIS, F. H., E. H. DENYS and K. S. Ü: Old and new clinical problems in amyotrophic lateral sclerosis. In: T. Tsubaki and Y. Toyokura (Eds.), Amyotrophic Lateral Sclerosis. Tokyo, Tokyo University Press (1978) 3–26.

NORRIS, F. H., R. A. SMITH and E. H. DENYS: Motor neuron disease: towards better care. Br. Med. J. 291 (1985) 259–262.

NORRIS, F. H., E. H. DENYS, K. S. Ü and E. MUKAI: Population study of amyotrophic lateral sclerosis. Ann. Neurol. 26 (1989) 139–140 (abstract).

NORRIS, F. H., E. H. DENYS and K. S. Ü: Polio and amyotrophic lateral sclerosis. Neurology (Minneap.) 40 (1990) 1150 (letter).

NORRIS, F. H., E. H. DENYS, K. S. Ü, E. MUKAI, F. H. NORRIS III, D. HOLDEN and L. ELIAS: The natural history of amyotrophic lateral sclerosis in a defined population. in preparation.

OLDSTONE, M. A. B., C. B. WILSON, D. DALESSIO, F. H. NORRIS and E. NELSON: Immunoglobulin deposits in amyotrophic lateral sclerosis. Trans. Am. Neurol. Assoc. 98 (1973) 31–32.

ONO, S., T. MANNEN and Y. TOYOKURA: Differential diagnosis between amyotrophic lateral sclerosis and spinal muscular atrophy by skin involvement. J. Neurol. Sci. 91 (1989) 301–310.

O'SULLIVAN, D. J. and J. G. MCLEOD: Distal chronic spinal muscular atrophy involving the hands. J. Neurol. Neurosurg. Psychiatry 41 (1978) 653–658.

PANSE, F.: Electrical lesions of the nervous system. In: P. J. Vinken and G. W. Bruyn (Eds.), Handbook of Clinical Neurology, Vol. 7. Amsterdam, North-Holland Publ Co; New York, American Elsevier (1970) 344–387.

PATTEN, B. M. and M. PAGES: Severe neurological disease associated with hyperparathyroidism. Ann. Neurol. 15 (1984) 453–456.

PEACH, P. E.: Overwork weakness with evidence of muscle damage in a patient with residual paralysis from polio. Arch. Phys. Med. Rehabil. 71 (1990) 248–250.

PEARCE, J. and D. G. F. HARRIMAN: Chronic spinal muscular atrophy. J. Neurol. Neurosurg. Psychiatry 29 (1966) 509–520.

PEARN, J.: Classification of spinal muscular atrophies. Lancet 2 (1980) 919–921.

PEARN, J. H., P. HUDGSON and J. N. WALTON: A clinical and genetic study of spinal muscular atrophy of adult onset: the autosomal recessive form as a discrete disease entity. Brain 101 (1978) 591–606.

PEIRIS, J. B., K. N. SENEVIRATNE, H. R. WICKREMASINGHE, S. B. GUNATILAKE and R. GAMAGE: Non-familial juvenile distal spinal muscular atrophy of upper extremity. J. Neurol. Neurosurg. Psychiatry 52 (1989) 314–319.

PENISSON-BESNIER, I., F. DUBAS, F. DELESTRE and J. EMILE: Atteinte du nerf sensitif dans l'amyotrophie bulbo-spinale liée a l'X (syndrome de Kennedy). Neurophysiol. Clin. 19 (1989) 163–170.

PESTRONK, A., D. R. CORNBLATH, A. A. ILYAS, H. BABA, R. H. QUARLES, J. W. GRIFFIN, K. ALDERSON and R. N. ADAMS: A treatable multifocal motor neuropathy with antibodies to GM_1 ganglioside. Ann. Neurol. 24 (1988) 73–78.

PESTRONK, A., R. N. ADAMS, D. CORNBLATH, R. W. KUNCL, D. B. DRACHMAN and L. CLAWSON: Patterns of serum IgM antibodies to GM_1 and GD_{1a} gangliosides in amyotrophic lateral sclerosis. Ann. Neurol. 25 (1989) 98–102.

PESTRONK, A., V. CHAUDHRY, E. L. FELDMAN, J. W. GRIFFIN, D. R. CORNBLATH, E. H. DENYS, M. GLASBERG, R. W. KUNCL, R. K. OLNEY and W. C. YEE: Lower motor neuron syndromes defined by patterns of weakness, nerve conduction abnormalities, and high titers of antiglycolipid antibodies. Ann. Neurol. 27 (1990) 316–326.

PETERS, H. A. and D. V. CLATANOFF: Spinal muscular atrophy secondary to macroglobulinemia. Neurology (Minneap.) 18 (1968) 101–108.

PEZESHKPOUR, G. H. and M. C. DALAKAS: Long-term changes in the spinal cords of patients with old poliomyelitis. Arch. Neurol. (Chic.) 45 (1988) 505–508.

PONGRATZ, D., H. KÖTZNER, G. HÜBNER, T. DURFEL and O. H. WIELAND: Adulte Form des Mangels an säurer Maltase unter dem Bild einer progressiven spinalen Muskelatrophie. Dtsch. Med. Wochenschr. 109 (1984) 537–541.

POUGET, J., J. F. PELLISSIER, J. C. SAINT-JEAN and G. SERRATRICE: Peripheral neuropathy in X-linked recessive bulbo-spinal muscular atrophy. Muscle Nerve 9, Suppl. 5 (1986) 119 (abstract).

PROBST, A., J. ULRICH, H. E. KAESER and P. HEITZ: Scapuloperoneal muscular atrophy: full autopsy report. Exp. Neurol. 16 (1977) 181–196.

QUARFORDT, S. H., D. C. DEVIVO, W. K. ENGEL, R. I. LEVY and D. S. FREDRICKSON.: Familial adult-onset proximal spinal muscular atrophy. Arch. Neurol. (Chic.) 22 (1970) 541–549.

RAO, K., M. GRUNNET, M. BARWICK and M. DENAYER: Rapidly progressive fatal adult spinal muscular atrophy with monoclonal gammopathy. Muscle Nerve 9, Suppl. 5 (1986) 120 (abstract).

RICKER, K. and R. T. MOXLEY III: Autosomal dominant cramping disease. Arch. Neurol. (Chic.) 47 (1990) 810–812.

RIGGS, J. E.: Trauma and amyotrophic lateral sclerosis. Arch. Neurol. (Chic.) 42 (1985) 205 (letter).

RIGGS, J. E., S. S. SCHOCHET and L. GUTMANN: Benign focal amyotrophy: variant of chronic spinal muscular atrophy. Arch. Neurol. (Chic.) 41 (1984) 678–679.

RINGEL, S. P., N. S. LAVA, M. M. TREIHAFT, M. L. LUBS and H. A. LUBS: Late-onset X-linked recessive spinal and bulbar muscular atrophy. Muscle Nerve 1 (1978) 297–307.

ROMANUL, F. C. A., H. L. FOWLER, J. RADVANY, R. G. FELDMAN and M. FEINGOLD: Azorean disease of the nervous system. N. Engl. J. Med. 296 (1977) 1505–1508.

ROOS, R. P., M. V. VIOLA, R. WOLLMANN, M. H. HATCH and J. P. ANTEL: Amyotrophic lateral sclerosis with antecedent poliomyelitis. Arch. Neurol. (Chic.) 37 (1980) 312–313.

ROSE, F. C. and F. H. NORRIS (Eds.): Amyotrophic Lateral Sclerosis: Recent Advances in Neurotoxicology and Neuroepidemiology. London, Smith-Gordon (1990).

ROSENBERG, R. N. and A. CHUTORIAN: Familial opticoacoustic degeneration and polyneuropathy. Neurology (Minneap.) 17 (1967) 827–832.

ROSENBERG, R. N. and H. L. FOWLER: Autosomal dominant motor system disease of the Portuguese: a review. Neurology (Minneap.) 31 (1981) 1124–1126.

SADIQ, S. A., F. P. THOMAS, K. KILIDIREAS, S. PROTOPSALTIS, A. P. HAYS, K.-W. LEE, S. N. ROMAS, N. KUMAR, L. VANDENBERG, M. SANTORO, D. J. LANGE, D. S. YOUNGER, R. E. LOVELACE, W. TROJABORG, W. H. SHERMAN, J. R. MILLER, J. MINUK, M. A. FEHR, R. I. ROELOFS, D. HOLLANDER, F. T. NICHOLS III, H. MITSUMOTO, J. J. KELLEY, JR., T. R. SWIFT, T. L. MUNSAT and N. LATOV: The spectrum of neurologic disease associated with anti-GM$_1$ antibodies. Neurology (Minneap.) 40 (1990) 1067–1072.

SADOWSKY, C. H., E. SACHS and J. OCHOA: Postradiation motor neuron syndrome. Arch. Neurol. (Chic.) 33 (1976) 786–787.

SAFDARI, H. and E. P. RICHARDSON: Subacute necrotizing polioencephalopathy. Arch. Neurol. (Chic.) 36 (1979) 638–642.

SALAZAR-GRUESO, F., M. J. ROUTBORT, J. MARTIN, G. DAWSON and R. P. ROOS: Polyclonal IgM anti-GM$_1$ ganglioside antibody in patients with motor neuron disease and variants. Ann. Neurol. 27 (1990) 558–563.

SARNAT, H. B., P. JACOB and C. JIMINÉZ: Atrophie spinale musculaire: l'évanouissement de la fluorescence à l'arn des neurones moteurs en dégénérescence. Rev. Neurol. (Paris) 145 (1989) 305–311.

SCHANEN, A., J. MIKOL, C. GUIZIOU, C. VITAL, M. COQUET, A. LAGUENY, J. JULIEN and M. HAGUENAU: Forme familiale liée au sexe d'amyotrophie spinale progressive de l'adulte. Rev. Neurol. (Paris) 140 (1984) 720–727.

SCHOCHET, S. S., J. M. HARDMAN, P. P. LADEWIG and K. M. EARLE: Intraneuronal conglomerates in sporadic motor neuron disease. Arch. Neurol. (Chic.) 20 (1969) 548–553.

SCHOLD, S. C., E. S. CHO, M. SOMASUNDARAM and J. B. POSNER: Subacute motor neuronopathy: a remote effect of lymphoma. Ann. Neurol. 5 (1979) 271–287.

SCRIMGEOUR, E. M. and F. L. MASTAGLIA: Late-childhood-onset spinal muscular atrophy in three Melanesian families in Papua New Guinea. Am. J. Med. Gen. 19 (1984) 769–777.

SCULLY, R. E., E. J. MARK, W. F. MCNEELY and B. U. MCNEELY: Case Records of the Massachusetts General Hospital. N. Engl. J. Med. 316 (1987) 1326–1335.

SERRATRICE, G.: Amyotrophie spinale distale chronique localisée aux deux membres supérieurs (type O'Sullivan et McLeod). Rev. Neurol. (Paris) 140 (1984) 368–369.

SERRATRICE, G., G. CREMIEUX, J. F. PELLISSIER and J. POUGET: Deux cas de syndrome de Fotopullos (amyotrophie spinale chronique de la ceinture scapulaire et chorée chronique). Presse Med. 13 (1984a) 1274.

SERRATRICE, G., J. F. PELLISSIER, J. L. GASTAUT and C. DESNUELLE: Amyotrophie spinale chronique avec paralysie des cordes vocales: syndrome de Young et Harper. Rev. Neurol. (Paris) 11 (1984b) 657–658.

SERRATRICE, G., A. POU-SERRADEL, J. F. PELLISSIER, H. ROUX, J. LAMARCO-CIVRO and J. POUGET: Chronic neurogenic quadriceps amyotrophies. J. Neurol. 232 (1985) 150–153.

SERRATRICE, G. T., M. TOGA and J. F. PELLISSIER: Chronic spinal muscular atrophy and pallidonigral degeneration: report of a case. Neurology (Minneap.) 33 (1983) 306–310.

SHY, M. E., L. P. ROWLAND, T. SMITH, W. TROJABORG, N. LATOV, W. SHERMAN, M. A. PESCE, R. E. LOVELACE and E. F. OSSERMAN: Motor neuron disease and plasma cell dyscrasia. Neurology (Minneap.) 36 (1986) 1429–1436.

SHY, M. E., T. HEIMAN-PATTERSON, G. J. PARRY, A. TAHMOUSH, V. A. EVANS and P. K. SCHICK: Lower motor neuron disease in a patient with auto-antibodies against Gal (β1-3)GalNAc in gangliosides GM$_1$ and GD$_{1b}$: improvement following immunotherapy. Neurology (Minneap.) 40 (1990) 842–844.

SIDDIQUE, T., D. A. FIGLEWICZ, M. A. PERICAK-VANCE ET AL.: Linkage of a gene causing familial amyotrophic lateral sclerosis to chromosome 21 and evidence of genetic locus heterogeneity. N. Engl. J. Med. 324 (1991) 1381–1384.

SIDDIQUI, A. H.: Endemic fluorosis in India. In: J. D. Spillane (Ed.), Tropical Neurology. London, Oxford Press (1973) 124–126.

SIGWALD, J., J. LAPRESLE, P. RAVERDY, P. and J. RECONDO: Atrophie cérébelleuse familiale avec association de lésions nigériennes et spinales. Presse Méd. 72 (1964) 557–562.

SILVER, J. R.: Familial spastic paraplegia with amyotrophy of the hands. J. Neurol. Neurosurg. Psychiatry 29 (1966) 135–144.

SINGH, H. and R. SHAM: Heredofamilial ataxia with muscle fasciculations. Br. J. Clin. Pract. 18 (1964) 91–92.

SINGH, N., K. K. SACHDEV and A. K. SUSHEELA: Juvenile muscular atrophy localized to arms. Arch. Neurol. (Chic.) 37 (1980) 297–299.

SIRDOFSKY, M. D. and R. J. HAWLEY: Progressive motor neuron disease associated with electrical injury. Muscle Nerve 14 (1991) in press.

SKRE, H., S. I. MELLGREN, P. BERGSHOLM and J. E. SLAGSVOLD: Unusual type of neural muscular atrophy with a possible X-chromosomal inheritance pattern. Acta Neurol. Scand. 58 (1978) 249–260.

SOBUE, G., Y. HASHIZUME, E. MUKAI, M. HIRAYAMA, T. MITSUMA and A. TAKAHASHI: X-linked recessive bulbospinal neuronopathy. Brain 112 (1989) 209–232.

SOBUE, I., N. SAITO, M. IIDA and K. ANDO: Juvenile type of distal and segmental muscular atrophy of upper extremities. Ann. Neurol. 3 (1978) 429–432.

SOMMER, C and M. SCHROEDER: Hereditary motor and sensory neuropathy with optic atrophy. Arch. Neurol. 46 (1989) 973–977.

SPENCER, P. S. and H. H. SCHAUMBURG (Eds.): Experimental and Clinical Neurotoxicology. Baltimore, Williams & Wilkins (1980).

SPIRA, R.: Neurogenic, familial, girdle type muscular atrophy. Confin. Neurol. 23 (1963) 245–255.

SPITZ, M. C., J. JANKOVIC and J. M. KILLIAN: Familial tic disorder, Parkinsonism, motor neuron disease, and acanthocytosis: a new syndrome. Neurology (Minneap.) 35 (1985) 366–370.

SUMMERS, B. A., M. SWASH, M. S. SCHWARTZ and D. A. INGRAM: Juvenile-onset bulbospinal muscular atrophy with deafness: Vialetta-van Laere syndrome or Madras-type motor neuron disease? J. Neurol. 234 (1987) 440–442.

SUNG, J. H., M. RAMIREZ-LASSEPAS, A. R. MASTRI and S. M. LARKIN: An unusual degenerative disorder of neurons associated with a novel intranuclear hyaline inclusion (neuronal intranuclear hyaline inclusion disease). Neuropathol. Exp. Neurol. 39 (1980) 107–115.

SWASH, M., M. LEADER, A. BROWN and K. W. SWETTENHAM: Focal loss of anterior horn cells in the cervical cord in motor neuron disease. Brain 109 (1986) 939–952.

TAKAHASHI, K., H. NAKAMURA, R. NAKASHIMA and O. DAIMARU: A pathologic study of scapuloperoneal muscular atrophy. Adv. Neurol. Sci. 19 (1975) 40–42.

TANDAN, R., K. R. SHARMA, W. G. BRADLEY, H. BEVAN and P. JACOBSEN: Chronic segmental spinal muscular atrophy of upper extremities in identical twins. Neurology 40 (Minneap.) (1990a) 236–239.

TANDAN, R., R. TAYLOR, A. ADESINA, K. SHARMA, T. FRIES and W. PENDLEBURY: Benign autosomal dominant syndrome of neuronal Charcot-Marie-Tooth disease, ptosis, Parkinsonism, and dementia. Neurology (Minneap.) 40 (1990b) 773–779.

TEITELBAUM, J. S., R. J. ZATORRE, S. CARPENTER, D. GENDRON, A. C. EVANS, A. GJEDDE and N. R. CASHMAN: Neurologic sequelae of domoic acid intoxication due to the ingestion of contaminated mussels. N. Engl. J. Med. 322 (1990) 1781–1787.

THACKER, A. K., S. MISRA and B. C. KATIYAR: Nerve conduction studies in upper limbs of patients with cervical spondylosis and motor neurone disease. Acta Neurol. Scand. 78 (1988) 45–48.

THOMAS, J. E. and F. M. HOWARD: Segmental zoster paresis — a disease profile. Neurology (Minneap.) 22 (1972) 459–466.

THOMAS, P. K., R. H. M. KING, T. R. CHIANG, F. SCARAVILLI, A. K. SHARMA and A. W. DOWNIE: Neurofibromatous neuropathy. Muscle Nerve 13 (1990) 93–101.

THOMAS, P. K., E. YOUNG and R. H. M. KING: Sandhoff disease mimicking adult-onset bulbospinal neuronopathy. J. Neurol. Neurosurg. Psychiatry 52 (1989) 1103–1106.

TSUKAGOSHI, H., T. NAKANISHI, K. KONDO and T. TSUBAKI: Hereditary proximal neurogenic spinal muscular atrophy in adult. Arch. Neurol. (Chic.) 12 (1965) 597–603.

TSUKAGOSHI, H., H. SHOJI and T. FURUKAWA: Proximal neurogenic muscular atrophy in adolescence and adulthood with X-linked recessive inheritance. Neurology (Minneap.) 20 (1970) 1188–1190.

VAN BOGAERT, L. and M. MOREAU: Combinaison de l'amyotrophie de Charcot-Marie-Tooth et de la maladie de Friedreich chez plusieurs membres d'une même famille. Encéphale 34 (1939) 312–322.

VAN DEN BERGHE, P., J. A. BUICKE and R. DOM: Familial muscle cramps with autosomal dominant transmission. Eur. Neurol. 19 (1980) 207–212.

VAN LAERE, J.: Paralysie bulbo-pontine chronique progressive familiale avec surdite. Rev. Neurol. (Paris) 115 (1966) 289–295.

VIALETTO, E.: Contributo alla forma ereditaria della paralisi bulbare progressiva. Rio. sper. Freniatr. 40 (1936) 1–24.

WADIA, N. H., P. F. IRANI and S. M. KATRAK: Neurological complications of a new conjunctivitis. Lancet 2 (1972) 970–971 (letter).

WALTON, J. N. and P. K. THOMAS: World Federation of Neurology Research Committee: Research Group on Neuromuscular Diseases. J. Neurol. Sci. 86 (1988) 333–360.

WALTON, J. N., B. E. TOMLINSON and G. W. PEARCE: Subacute "poliomyelitis" and Hodgkin's disease. J. Neurol. Sci. 6 (1968) 435–445.

WARNER, C. L., S. SERVIDEI, D. J. LANGE, E. MILLER, R. E. LOVELACE and L. P. ROWLAND: X-linked spinal muscular atrophy (Kennedy's syndrome). Arch. Neurol. (Chicago) 47 (1990) 1117–1120.

WEICHERS, D. O. and S. L. HUBBELL: Late changes in the motor unit after acute poliomyelitis. Muscle Nerve 4 (1981) 524–528.

WEINER, L. P., S. A. STOHLMAN and R. L. DAVIS: Attempts to demonstrate virus in amyotrophic lateral sclerosis. Neurol. (Minneap.) 30 (1980) 1319–1322.

WHITELEY, A. M., M. SWASH and H. URICH: Progressive encephalomyelitis with rigidity. Brain 99 (1976) 27–42.

WILDE, J., T. MOSS and D. THRUSH: X-linked bulbo-spinal neuronopathy: a family study of three patients. J. Neurol. Neurosurg. Psychiatry 50 (1987) 279–284.

WILKINSON, M. (Ed.): Cervical Spondylosis, Its Early Diagnosis and Treatment. London, W. Heinemann (1971).

WIMMER, A. and A. V. NEEL: Les amyotrophies systématisées dans l'encephalite épidémique chronique. Acta Psychiat. Neurol. 3 (1928) 319–365.

WOODS, B. I. and H. H. SCHAUMBURG: Nigro-spino-dental degeneration with nuclear ophthalmoplegia. J. Neurol. Sci. 17 (1972) 149–166.

YAFFE, M. G., M. KABACK, M. GOLDBERG, J. MILES, H. ITABASHI, H. MCINTYRE and T. MOHANDAS: An amyotrophic lateral sclerosis-like syndrome with hexosaminidase-A deficiency: a new type of GM_2 gangliosidosis. Neurology (Minneap.) 29 (1979) 611 (abstract).

YOUNG, I. D. and P. S. HARPER: Neurogenic atrophy simulating facioscapulohumeral dystrophy. J. Neurol. Neurosurg. Psychiatry 43 (1980) 413–418.

ZATZ, M., C. PENHA-SERRANO, O. FROTA-PESSOA and D. KLEIN: A malignant form of neurogenic muscular atrophy in adults, with dominant inheritance. J. Genet. Hum. 19 (1971) 337–354.

ZIEGLER, D. K., R. N. SCHIMKE, J. J. KEPES, D. L. ROSE and G. KLINKERFUSS: Late onset ataxia, rigidity, and peripheral neuropathy. Arch. Neurol. (Chic.) 27 (1972) 52–56.

CHAPTER 3

The postpolio syndrome

DONALD W. MULDER

Department of Neurology, Mayo Clinic and Mayo Medical School, Rochester, MN, USA

Paralytic poliomyelitis causes spotty muscular atrophy and weakness. In the young patient the weak muscles are associated with a lack of growth in the affected limbs. The disability of patients who had paralytic poliomyelitis increases as the patient ages. This progression of muscle weakness and atrophy long after the acute infection had subsided has been reported by many observers such as Potts (1903), Campbell et al. (1969), Mulder et al. (1972), Codd et al. (1985), and Dalakas and Hallett (1988).

These sequelae of poliomyelitis have been confused with other diseases of the motor system. Cornil and Lépine (1876) described the late symptoms of poliomyelitis as a form of progressive muscular atrophy. The late sequelae of poliomyelitis have frequently been misdiagnosed as amyotrophic lateral sclerosis (ALS) (Mulder et al. 1972; Dalakas 1990). Poliomyelitis was reported by others to predispose individuals to ALS (Zilkha 1962; Poskanzer et al. 1969). Recently Armon et al. (1990) reported that ALS rarely occurs in patients who have had paralytic polio. They suggested that paralytic poliomyelitis may protect the patient against ALS. Most neurologists now agree that the postpolio syndromes can be clearly differentiated from ALS by history and observation (Mulder et al. 1972;

Dalakas 1990). The clinical similarity of the two conditions led to the hypothesis that ALS may be the result of an acute illness, perhaps viral, early in life (Mulder et al. 1972).

The disappearance of the epidemics of poliomyelitis with the discovery of effective vaccines in the 1950s (Paul 1971) resulted in a generation of physicians and patients with little awareness of these sequelae. In the past decade, physicians and patients have rediscovered this syndrome and named it the postpolio syndrome (PPS) (Dalakas and Hallett 1988). The PPS has been described as a new disorder and patients fear that it may be the result of a recrudescence of the poliomyelitis virus (Mulder 1991). This rediscovery of a previously well-known syndrome has occasioned great anxiety in the half million survivors of the polio epidemics who now fear that their hard won recovery is about to be lost, returning them, once again, paralyzed to the hospital (Mulder 1991).

POLIOMYELITIS EPIDEMICS

During the great epidemics, most patients with acute poliomyelitis had a clinically nonapparent or minor illness. In a smaller number of patients the illness was manifest by non-specific signs of an upper respiratory infection and gastroenteritis or the symptoms and findings of an aseptic meningitis, or both. The paralytic form of the disease occurred in fewer cases and was usually preceded by such minor illness. The onset of flaccid paralysis, often spotty, began several days after the onset of fever and reached its maximum in several days. The weakness might involve any of the skeletal muscles and was often associated with signs of involvement of systems other than the motor system (Paul 1971). Patients with paralytic poliomyelitis who recovered were known to have severe residual weakness and often deformity, although most were physically active and denied any disability.

The pathology of acute poliomyelitis has been extensively reported elsewhere (Bodian 1949). The significant central nervous system pathology was limited primarily to the motor cells and even in minimally affected segments of the spinal cord only 10% of the motor neurons might appear normal (Bodian 1949). Sharrard (1953, 1955) reported that pathologic study of the spinal cord of patients who had recovered from their acute infection revealed that motor cell destruction in the anterior spinal cord was always more severe than expected. In muscles in which no weakness had been demonstrated, the corresponding anterior horn cells in the spinal cord might have a loss of up to 40% of the motor cells that innervated the muscles. He reported that the number of motor neurons necessary to produce muscle activity was much lower than had been expected.

The patients who now are complaining of the residuals of paralytic poliomyelitis had their acute illness in the great epidemics between 1945 and 1955. The number of patients afflicted with acute poliomyelitis was highest in the latter part of the 1940s and in the early 1950s, peaking in 1952 when 56 000 cases were reported to the United States Public Health service (Codd and Kurland 1985). During those years, improvements in medical care, particularly improved respiratory care of patients with acute polio, enabled many patients who were severely paralyzed to survive their acute illness. Thus there now may be 200 000–300 000 patients in the USA who had poliomyelitis and survived with severe disability (Dalakas and Hallett 1988). It is now 30–40 years after their acute illness and many are more

than 60 years of age. New symptoms and signs of weakness are now developing in many of them, which they and their physicians did not expect (Codd and Kurland 1985).

DEFINITION OF POSTPOLIO SYNDROME

Patients now diagnosed as having the PPS had paralytic poliomyelitis followed by severe residual weakness and atrophy. Three kinds of late progression can be differentiated in these patients. (1) The decreasing energy level of the aging patients no longer enables them to compensate for severe residual disabilities. (2) Secondary unrelated illness such as malignancy or degenerative arthritic joint changes interfere with the patients' ability to compensate. (3) Progressive weakness and atrophy of specific muscles develop in a smaller number of patients with poliomyelitis. This condition has been called progressive postpolio muscular atrophy (PPMA) (Dalakas and Hallett 1988).

These causes of the PPS are not mutually exclusive and in one patient all three may be operative. The third group of patients are the principal subject of this chapter.

EPIDEMIOLOGY

The incidence of the PPS in the approximately 300 000 postpolio patients in the US is difficult to estimate (Codd and Kurland 1985). The estimates that are available are based on patients who had paralytic poliomyelitis, excluding the many patients who had nonparalytic poliomyelitis. It is probable that all patients with paralytic poliomyelitis will demonstrate increasing disability as they age compared with a similar cohort of persons who did not have paralytic poliomyelitis. Preliminary population-based epidemiologic studies suggest that increasing weakness (PPS) will develop in from 22% (Codd et al. 1985) to 55% (Windebank et al. 1987) of patients with paralytic polio. These observations are based on self-reporting and the authors noted that the only way to document increasing muscle weakness will be with serial observations (Windebank et al. 1987). PPMA is uncommon (Mulder et al. 1972), and its true frequency awaits the completion of population-based studies (Windebank et al. 1987).

Poliomyelitis is now uncommon in the western industrialized countries. The incidence of PPSs is expected to soon diminish in these countries. Developing, underdeveloped, and tropical regions continue to have a high incidence of acute poliomyelitis (Codd and Kurland 1985). The PPSs can be expected to remain a significant medical problem for many years in these countries.

CLINICAL SYMPTOMS AND FINDINGS

The new complaints of patients with PPMA are insidious in onset, occurring many years after the acute disease. The patients initially report a gradual decrease in functional capacity, with excessive fatigue and slowly progressive flaccid muscular weakness, which is often asymmetric. The new weakness is associated with muscle atrophy and fasciculations. The lower limbs are most severely affected and muscles that had been most severely involved after the original infection are those most commonly affected. Muscles that the patient may have thought were uninvolved in the acute illness are more rarely involved. Persons with involvement of the respiratory musculature during the acute illness often become aware of increasing difficulty with breathing (Fischer 1987). Increased discomfort is frequent about joints already compromised by muscle weakness. Sensory abnormalities are not a part of the syndrome and the deep reflexes are, for the most part, absent or normal. Extensor toe signs were reported by several authors but were difficult to evaluate because of the age of the patients and the weakness of small muscles in the foot (Zilkha 1962; Mulder et al. 1972).

In one group of 34 patients referred to a diagnostic center because of progressive weakness, many years after an illness diagnosed as poliomyelitis, the diagnosis of PPMA was based on clinical symptoms and findings (Mulder et al. 1972). These included: (1) a credible history of poliomyelitis; (2) a partial recovery of motor function; (3) a period of stabilization of the recovery of at least 10 years; and (4) the

subsequent development of muscular weakness and atrophy.

Review of the clinical records of these patients revealed that they had had severe paralytic poliomyelitis, during which they were bedfast for months. The onset of acute polio was early in life (median age at onset, 5 years). The patients continued to have severe weakness after their recovery although they were able to resume or obtain gainful employment. This remarkable adjustment to muscle weakness, which was frequently severe, was maintained in these 34 patients for an average of 37 years before the onset of new muscle weakness. The new weakness began asymmetrically and was gradual in onset. It was most pronounced in muscles that had been most severely affected in the acute illness. Occasionally a muscle was found to be weak which the patient reported had not been involved in the initial episode of acute poliomyelitis; however, careful review of the original hospitalization records revealed more widespread weakness than the patient remembered.

Although the clinical syndrome resembles ALS (Mulder et al. 1972), the prognosis and clinical course are clearly different. The rate of progression of PPMA is very slow (Dalakas and Hallett 1988) and the muscular atrophy and weakness remain in previously affected muscles, unlike ALS in which the disease diffusely affects skeletal muscles and leads to death in a few years (Yoshida et al. 1986).

LABORATORY FINDINGS

Electrophysiologic studies reveal widespread abnormalities in patients who have PPMA (Mulder et al. 1972; Wiechers 1987). These studies reveal fibrillation potentials, fasciculations, and abnormalities in the amplitude of the compound action potential and in the motor unit potential count in patients who had acute poliomyelitis and now have severe residual muscle atrophy and weakness (Windebank et al. 1987). These findings may be noted in clinically uninvolved limbs (Bromberg and Waring 1990). These studies correlate with the severity of limb involvement but do not identify the presence or likelihood of progression of weakness (Cashman et al. 1987b; Windebank et al. 1987). In our experience, conduction velocities in motor and sensory nerves of affected extremities as well as distal latencies of these nerves were normal (Mulder et al. 1972). Hodes (1948) reported defects of neuromuscular transmission in postpolio patients and these findings have been reported by others (Trojan et al. 1990). These electrodiagnostic studies do not differentiate PPMA patients from patients with ALS or other motor neuron diseases or from patients who had paralytic polio but do not have progression of weakness and atrophy (Mulder et al. 1972; Cashman et al. 1987a).

The biopsies of involved muscles in patients with PPMA revealed new and old neurogenic changes along with myogenic features similar to those seen in other chronic neurogenic diseases (Cashman et al. 1987a; Dalakas 1988). Minimal interstitial or perivascular inflammation was noted in 6 of 16 biopsy specimens from newly symptomatic muscles (Dalakas 1988). These findings also do not differentiate patients with PPMA from those with prior paralytic poliomyelitis who have no progressive symptoms or from those with other motor neuron diseases (Cashman et al. 1987a). Pezeshkpour and Dalakas (1988) reported the pathologic findings in the spinal cord of 8 patients who had poliomyelitis and died a mean of 20.7 years after their acute infection. Three of these patients had a slowly progressive weakness that could have been PPMA. In all patients there were loss or atrophy of motor neurons, severe reactive gliosis, and no involvement of the corticospinal tracts. In addition, a mild to moderate perivascular and interparenchymal inflammation was observed in the involved areas of the spinal cord. There was no difference in the pathologic findings reported in 3 of the patients who had PPMA by history and in the 5 patients who had no evidence of progression of their disease. Routine laboratory studies are usually reported as normal in patients with PPMA except for the serum creatine kinase value, which is often increased (Dalakas, 1988). Virologic studies and immunologic studies have produced no evidence to suggest a reactivation of the polio virus in patients with PPMA (Dalakas and Hallett 1988).

There are no serologic, electrodiagnostic, pathologic, or other modes of investigation that differentiate PPMA patients from those with other motor neuron diseases or from patients who had paralytic poliomyelitis but have had no progression of their symptoms. Thus the diagnosis remains dependent on an accurate history and examinations repeated at sufficiently long intervals to confirm the progression of muscular atrophy and weakness.

TREATMENT

Compassionate and effective care of these patients requires that physicians understand the extraordinary ordeal patients with acute poliomyelitis suffered and their subsequent heroic endeavors to compensate for significant and permanent handicaps. Psychologic studies confirm the emotional strength of these patients who have spent a lifetime successfully battling severe motor deficits and societal prejudices against the handicapped (Windebank et al. 1987). In these aging patients, it is imperative to search for other debilitating diseases that may have caused the progressive weakness. Such diseases may be treatable. Most of these patients have degenerative changes of their joints, many of which may be corrected or assisted (Perry 1985). In some patients merely changing life-style may be sufficient to help them to adapt. In others the use of additional assistive devices or exercise regimens carefully tailored to individual requirements may return them to their previous level of functioning (Kenny 1937; Maynard 1985).

In treating these patients the physician should recognize that the individual with a PPS has had a lifetime of experience in treating his own unique illness and that we can only offer specialized suggestions. Finally, we should reassure patients that their weakness will not lead to ALS or a similar disorder. More complete discussion of treatment may be found in the monographs edited by Halstead and Wiechers (1985, 1987) or in the recent monograph edited by Munsat (1991).

DISCUSSION

In patients who have had poliomyelitis with severe residuals of muscular atrophy and weakness, progressive symptoms of fatigue and weakness — the PPS — may later develop. In the majority of patients the new symptoms are related to aging or secondary diseases such as degenerative arthritis. These symptoms can often be improved by treatment of the secondary disease, physical therapy, or simple life-style adjustment.

A much smaller group of postpolio patients have a progressive muscular atrophy and weakness (PPMA) which begins long after their acute poliomyelitis has subsided. These patients cannot now be differentiated by any laboratory procedure, including muscle biopsy or electrodiagnostic studies. The differentiation is dependent on the history of poliomyelitis and repetitive examinations that demonstrate progressive weakness and atrophy. The cause of the syndrome of PPMA remains unknown. There is no evidence that it results from a recrudescence of the polio virus. There is no evidence that a second neurologic disease such as ALS is more likely to develop in the patient who has had poliomyelitis. We have suggested (Mulder et al. 1972) that the progressive muscular atrophy is related to the loss of neurons with normal aging (Tomlinson and Irving 1977) in a patient who has previously had severe anterior horn cell damage. Dalakas (1990) suggested that in this syndrome the few remaining motor neurons are stressed to control a larger than normal area of motor control and that these hyperfunctioning motor neurons can no longer maintain their metabolic needs. This results in a slow deterioration of the motor terminals, with progressive muscular atrophy and weakness.

Increasing weakness and fatigue develop in patients who have had paralytic poliomyelitis as they age (PPS). A progressive muscular atrophy (PPMA) develops in a smaller number of these patients. This is a slowly progressive syndrome which is benign compared to other motor neuron diseases such as ALS. This syndrome does not progress to ALS and one study suggested that it protects against ALS. Therapy for PPS and PPMA is effective and often allows patients to return to their previous levels of activity. That many patients who suffered severe polio many years ago now report no complaints is surprising.

REFERENCES

ARMON, C., J. R. DAUBE, A. J. WINDEBANK and L. T. KURLAND: How frequently does classic amyotrophic lateral sclerosis develop in survivors of poliomyelitis? Neurology 40 (1990) 172–174.

BODIAN, D.: Histopathologic basis of clinical findings in poliomyelitis. Am. J. Med. 6 (1949) 563–578.

BROMBERG, M. B. and W. P. WARING: Prior poliomyelitis: evidence for denervation in clinically uninvolved limbs. Neurology 40, Suppl. 1 (1990) 430.

CAMPBELL, A. M. G., E. R. WILLIAMS and J. PEARCE: Late motor neuron degeneration following poliomyelitis. Neurology 19 (1969) 1101–1106.

CASHMAN, N. R., R. MASELLI, R. L. WOLLMANN, R. ROOS, E. NICHOLS, F. BROWN, R. SIMON and J. P. ANTEL: Electromyography and muscle biopsy do not distinguish newly symptomatic from asymptomatic patients with prior paralytic poliomyelitis (abstract). Neurology 37, Suppl. 1 (1987a) 214.

CASHMAN, N. R., R. MASELLI, R. L. WOLLMANN, R. ROOS, R. SIMON and J. P. ANTEL: Late denervation in patients with antecedent paralytic poliomyelitis. N. Engl. J. Med. 317 (1987b) 7–12.

CODD, M. B. and L. T. KURLAND: Polio's late effects. The 1986 Medical and Health Annual. Chicago: Encyclopedia Britannica (1986) 249–252.

CODD, M. B., D. W. MULDER, L. T. KURLAND, C. M. BEARD and W. M. O'FALLON: Poliomyelitis in Rochester, Minnesota, 1935–1955: epidemiology and long-term sequelae: a preliminary report. In: L. S. Halstead and D. O. Wiechers (Eds.), Late Effects of Poliomyelitis. Miami, Symposia Foundation (1985) 121–134.

CORNIL, V. and R. LÉPINE: Sur un cas de paralysie générale spinale antérieure subaiguë, suivi d'autopsie. C. R. Soc. Biol. 1875, Par., 1876, 6. s. II, 75–82.

DALAKAS, M. C.: Morphologic changes in the muscles of patients with postpoliomyelitis neuromuscular symptoms. Neurology 38 (1988) 99–104.

DALAKAS, M. C.: Post-poliomyelitis motor neuron disease: what did we learn in reference to amyotrophic lateral sclerosis? In: A. J. Hudson (Ed.), Amyotrophic Lateral Sclerosis: Concepts in Pathogenesis and Etiology. Toronto, University of Toronto Press (1990) 326–357.

DALAKAS, M. C. and M. HALLETT: The post-polio syndrome. In: F. Plum (Ed.), Advances in Contemporary Neurology. Philadelphia, F. A. Davis Company (1988) 51–94.

FISCHER, D. A.: Sleep-disordered breathing as a late effect of poliomyelitis. Birth Defects 23, No. 4 (1987) 115–119.

HALSTEAD, L. S. and D. O. WIECHERS: Late Effects of Poliomyelitis. Miami, Symposia Foundation (1985).

HALSTEAD, L. S. and D. O. WIECHERS: Research and Clinical Aspects of the Late Effects of Poliomyelitis. Birth Defects 23, No. 4 (1987).

HODES, R.: Electromyographic study of defects of

neuromuscular transmission in human poliomyelitis. Arch. Neurol. Psychiatry 60 (1948) 457–473.

KENNY, E.: Infantile Paralysis and Cerebral Diplegia. Sydney, Australia, Angus and Robertson Limited (1937).

MAYNARD, F. M.: Post-polio sequelae–differential diagnosis and management. Orthopedics 8 (1985) 857–861.

MULDER, D. W.: Post-polio syndrome: past, present and future. In: T. L. Munsat (Ed.), Post-Polio Syndrome. Boston, Butterworth-Heinemann (1991) 1–8.

MULDER, D. W., R. A. ROSENBAUM and D. D. LAYTON, JR.: Late progression of poliomyelitis or forme fruste amyotrophic lateral sclerosis. Mayo Clin. Proc. 47 (1972) 756–761.

MUNSAT, T. L.: Post-Polio Syndrome. Boston, Butterworth-Heinemann (1991) 1–8.

PAUL, J. R.: A History of Poliomyelitis. New Haven, Yale University Press (1971).

PERRY, J.: Orthopedic management of post-polio sequelae. In: L. S. Halstead and D. O. Wiechers (Eds.), Late Effects of Poliomyelitis. Miami, Symposia Foundation (1985) 193–207.

PEZESHKPOUR, G. H. and M. C. DALAKAS: Long-term changes in the spinal cords of patients with old poliomyelitis: signs of continuous disease activity. Arch. Neurol. 45 (1988) 505–508.

POSKANZER, D. C., H. M. CANTOR and G. S. KAPLAN: The frequency of preceding poliomyelitis in amyotrophic lateral sclerosis. In: F. H. Norris, Jr. and L. T. Kurland (Eds.), Motor Neuron Diseases: Research on Amyotrophic Lateral Sclerosis and Related Disorders. New York, Grune and Stratton (1969) 286–290.

POTTS, C. S.: A case of progressive muscular atrophy occurring in a man who had had acute poliomyelitis nineteen years previously: with a review of the literature bearing upon the relations of infantile spinal paralysis to the spinal diseases of later life. Univ. Pennsylvania Med. Bull. 16 (1903) 31–37.

SHARRARD, W. J. W.: Correlation between changes in the spinal cord and muscle paralysis in poliomyelitis — a preliminary report. Proc. R. Soc. Med. 46 (1953) 346–349.

SHARRARD, W. J. W.: The distribution of the permanent paralysis in the lower limb in poliomyelitis: a clinical and pathological study. J. Bone Joint Surg. [British] 37 (1955) 540–558.

TOMLINSON, B. E. and D. IRVING: The number of limb motor neurons in the human lumbosacral cord throughout life. J. Neurol. Sci. 34 (1977) 213–219.

TROJAN, D. A., D. GENDRON and N. R. CASHMAN: Neuromuscular junction transmission in the postpoliomyelitis syndrome. Neurology 40, Suppl. 1 (1990) 429.

WIECHERS, D. O.: Reinnervation after acute poliomyelitis. Birth Defects 23, No. 4 (1987) 213–220.

WINDEBANK, A. J., J. R. DAUBE, W. J. LITCHY, M. CODD, E. Y. S. CHAO, L. T. KURLAND and R. IVERSON: Late sequelae of paralytic poliomyelitis in Olmsted County, Minnesota. Birth Defects 23, No. 4 (1987) 27–37.

YOSHIDA, S., D. W. MULDER, L. T. KURLAND, C-P CHU and H. OKAZAKI: Follow-up study on amyotrophic lateral sclerosis in Rochester, Minn., 1925 through 1984. Neuroepidemiology 5 (1986) 61–70.

ZILKHA, K. J.: Discussion on motor neuron disease. Proc. R. Soc. Med. 55 (1962) 1028–1029.

Handbook of Clinical Neurology, Vol. 15 (59): Diseases of the Motor System
J.M.B.V. de Jong, editor
© Elsevier Science Publishers B.V., 1991

CHAPTER 4

Special forms of spinal muscular atrophy

GEORGE W. PADBERG

Department of Neurology, Leiden University, Leiden, The Netherlands

Special forms of spinal muscular atrophy (SMA) have been recognized in recent years because of their clinical resemblance to diseases of a different pathogenesis, or because of the occurrence of special features, presumably not fortuitous associations. In all instances, these nosological endeavours led to more or less explicit statements saying that the disorders under study were separate genetic entities. It is a bit odd to write about the clinical features of these disorders while the fundamental question about genetic heterogeneity may be answered shortly; recently the gene for autosomal recessive childhood SMA has been found linked to the DNA markers D5S6 and D5S39 located on the long arm of chromosome 5 (Brzustowicz et al. 1990; Melki et al. 1990). Subsequently autosomal recessive infantile SMA was also found to be linked to D5S39 and other markers in the same regions (Gilliam et al. 1990). These studies strongly suggest that adolescent SMA and infantile SMA are allelic conditions. Still, two families did not show linkage indicating that spinal muscular atrophy might be genetically heterogeneous (Gilliam et al. 1990). Linkage studies of the Ryukyuan cases might be helpful to solve the claim that this autosomal recessive SMA with onset in infancy is different from the usual forms. Now the gene for facioscapulohumeral muscular dystrophy (FSHD) has been located on chromosome 4 (Padberg et al. 1990; Wijmenga

et al. 1990) the problem whether an SMA in facioscapulohumeral distribution exists, can be tackled. Similar studies will clarify the status of scapuloperoneal muscular dystrophy (SPD) and scapuloperoneal spinal muscular atrophy (SPSMA). Also, the relationship between X-linked SPSMA, X-linked bulbospinal SMA (BSMA) and X-linked proximal SMA might be studied now the gene for X-linked BSMA has been located proximally on the long arm of the X-chromosome linked to the marker DXYS1 (Fischbeck et al. 1986). Genetic studies are of particular importance since it has been demonstrated repeatedly that our clinical, electrophysiological and morphological criteria are often inadequate in making the distinction between neurogenic and myopathic disorders (Lunt et al. 1989).

FACIOSCAPULOHUMERAL SPINAL MUSCULAR ATROPHY (182970)

The adjective facioscapulohumeral (FSH) was coined by Landouzy and Dejerine to characterize the main features of the muscular dystrophy that subsequently was to bear their names (Landouzy and Dejerine 1885, 1886). The choice of terms, however, was unfortunate since upper arm weakness and atrophy is not an early sign, and usually develops some time after the shoulder fixators and foot extensor weakness (Padberg, 1982). Thus, most patients with FSHD go through the phase of a scapuloperoneal syndrome, which is most striking in those patients in whom facial weakness is very mild (Kazakov et al. 1976). Therefore, most authors doubt that autosomal dominant SPD is genetically distinct from autosomal dominant FSHD, an issue that will be settled now linkage is found in FSHD. In the SMA literature the term FSH is used in 2 ways: by comparison with the clinical picture of FSHD, or by taking the literal content of the term as description of the main features of the syndrome. The former has happened often, the latter only rarely.

The first description of FSHSMA was published by Fenichel et al. (1967). Their patient was a 17-year-old girl with progressive facial and shoulder girdle weakness and atrophy. The extraocular, jaw and tongue muscles were not involved, which renders this clinical picture similar to FSHD. A first EMG was reportedly normal, while a second EMG revealed occasional fasciculations, but no denervation potentials or abnormal conduction velocities, so that the final impression was irritable muscle with no concrete evidence of either a neuropathic or myopathic pattern. A biopsy of the left quadriceps muscle showed occasional small angulated fibres which was thought to suggest early denervation atrophy. The mother of this patient also had facial and shoulder girdle weakness and, as her daughter, no involvement of the extraocular, masseter and tongue muscles. The EMG closely resembled that of her daughter with, in addition, fibrillation potentials in the thenar muscles. On the basis of the latter, it was concluded that this patient suffered a neurogenic disorder; a muscle biopsy was not performed. Apparently the authors were not aware that angulated fibres might be present in biopsies of patients with FSHD (Dubowitz 1985). Although Mussini et al. (1988) suggested that patients with FSHD might be separated in distinct groups on the basis of angulated fibres or infiltrates in the muscle biopsies, we do not think this has any genetic implication, since we have observed patients with and without angulated fibres or infiltrates within the same family. Occasionally, even small groups of atrophic fibres

and fibre type predominance may be found. Therefore, the two patients of Fenichel et al. (1967) offer no evidence to consider them distinct from other cases of FSHD.

A similar diagnostic situation was encountered by Mares et al. (1964) in a father and his 3 children with a FSH syndrome, who all had a reduced interference pattern with action potentials of increased amplitude on EMG examination, and myopathic muscle biopsies. The father and son with a progressive FSH syndrome reported by Ricker et al. (1968), showed an identical problem; EMG in the son demonstrated myopathic and neurogenic features and a muscle biopsy revealed myopathic changes with small angulated fibres. Probably all these patients had FSHD. Also, the mother and her two children described by Furakawa and Toyokura (1976) had both myopathic and neurogenic characteristics on EMG and muscle biopsy. Although they had no bulbar weakness except for the facial muscles, they all had fasciculations on EMG examinations, which apparently weighted heavily in the decision how to name the disorder. In the abstract of Furakawa et al. (1981) about 13 patients out of 8 families, the diagnosis of FSHSMA depended to a large extent on EMG examinations, but the report does not allow firm conclusions. As long as small angulated fibres are quite common in muscle biopsies of patients with FSHD (Dubowitz 1985) and as long as neurogenic findings on EMG are not rare in FSHD (McComas 1977), it is reasonable to question the existence of autosomal dominant FSHSMA.

A few reports on sporadic cases with FSHSMA have been published. The first patient described by Furakawa et al. (1969) had facial weakness and progressive shoulder girdle weakness since he was 18 years old. The tongue and pharynx muscles were not affected. EMG showed large action potentials of long duration and a biopsy revealed large groups of small fibres. His parents and 6 sibs were said to be 'normal'. The second patient reported by the same authors had shoulder girdle and upper arm weakness only, and neurogenic characteristics on EMG examination. A biopsy of the biceps brachii was normal. Patel and Swami (1969) reported a juvenile case with shoulder girdle weakness and equivocal facial weakness, and the man studied by Krüger and Frank (1974) had facial and shoulder girdle weakness with fasciculations in the upper arm. EMG and muscle biopsy showed neurogenic changes.

The sporadic cases testify that a neurogenic atrophy in FSH distribution can be observed occasionally; still it is unclear how these rare cases relate to other forms of chronic SMA. The proband in the family of Skre et al. (1978) had FSHSMA, but other members had SPSMA, whereas the pattern of inheritance was possibly X-linked recessive, rendering this family and its proband very special and their disorder unique.

SPINAL MUSCULAR ATROPHY, RYUKYU TYPE (271200)

In 1970 Kondo et al. reported a form of autosomal recessive progressive proximal spinal muscular atrophy with onset in infancy, which they observed during a survey for neuromuscular diseases at the Ryukyu Islands in south-west Japan. The most remarkable feature of this type of SMA was that the disease progressed from early infancy till puberty and that it was stationary thereafter. The bulbar, neck and also trunk muscles remained unaffected, so that unsupported sitting was possible until advanced age. Sporadic cases of this type of SMA have been reported occasionally in other areas in Japan (Nakazato et al. 1977). The lack of follow-up studies makes it difficult to lend this entity an independent nosological place. Linkage analyses and molecular biological studies will have to prove if and how this entity differs from other forms of SMA.

SPINAL MUSCULAR ATROPHY, FINKEL TYPE (182980)

The large family with adult onset, proximal spinal muscular atrophy described by Finkel (1962), was unusual in the sense that the pattern of inheritance was not obvious from the pedigree. Finkel favoured an autosomal recessive inheritance. The pedigree however (Finkel 1962) showed a parent to child transmission in 10 sibships resulting in 26 affected and 42 non-

affected sibs. This could be compatible with autosomal dominant inheritance, possibly with an incomplete penetrance, as there was also one observation of a skipped generation, where a healthy mother had 2 affected children out of 13. At the same time another 15 affected parents in this pedigree had a total of 77 children of whom none were affected. Therefore, both autosomal dominant and autosomal recessive explanations appeared inadequate in this family. Different forms of SMA within one family have been observed, leading to the hypothesis that types I, II and III are allelic conditions (Becker 1964; Bouwsma 1986) which eventually turned out to be correct (Gilliam et al. 1990). Also, the observed ratio of affected to non-affected sibs, which is close to 1:3 in kindreds with multiple cases (Becker 1964), could be explained by the involvement of an allelic regulator or activator gene. Similarly, the observed male/female ratio of affected individuals, which was 5:3, was not well explained in Finkel's family. However, in juvenile SMA a 'female sparing' factor has been suggested (Hausmanova-Petrusewicz et al. 1985). Becker (1966) reviewed the Finkel family and suggested that the pattern of inheritance fits an autosomal dominant mode best. The study of Richieri-Costa et al. (1981) who re-examined the Finkel family, demonstrated conclusively the autosomal dominant pattern with complete penetrance and equal involvement of both sexes. Unusual findings in this family were cramps, and fits of suffocation in several sibs. The mean age of onset of this chronic progressive proximal SMA was 48 years. These observations bring the Finkel family into the more accepted categories of SMA.

BULBOSPINAL MUSCULAR ATROPHY, X-LINKED RECESSIVE (313200)

Kurland (1957) is credited for the recognition of the unusual clinical picture in 2 of his cases (Magee 1960) which subsequently was established as an independent clinical entity, consisting of adult onset, slowly progressive proximal spinal muscular atrophy with marked involvement of the bulbar musculature, and X-linked recessive inheritance (Kennedy et al. 1968). Initially, the late onset of the disease, usually in the third and fourth decade (Tsukagoshi et al. 1970), and the severe atrophy and weakness of the temporal, facial and tongue muscles attracted attention. The hereditary pattern was not clear at first (Magee 1960) but subsequent reports established the X-linked recessive transmission. EMG studies and muscle biopsies confirmed the neurogenic atrophy, which was clinically suggested by an abundance of fasciculations in the facial muscles (Harding et al. 1982). The myopathic features on EMG, which have been reported occasionally (Tsukagoshi et al. 1970; Schoenen et al. 1979) have been interpreted as secondary changes by most authors (Harding et al. 1982). Magee (1960) and Schoenen et al. (1979) noted intention tremors; Tsukagoshi et al. (1970) reported tremors as the initial manifestation of the disease. Distal involvement of muscles (Schoenen et al. 1979) occurred occasionally, as well as adolescent onset (Tsukagoshi et al. 1970). Initially, mild sensory changes were not appreciated (Kennedy et al. 1968), or attributed to the diabetes mellitus which is reported with increased frequency in this disorder. Harding et al. (1982) stressed these mild sensory findings and the often reported absence of sensory nerve action potentials. The reduced or unrecordable sensory nerve action potentials led to the term bulbo-spinal neuronopathy (Wilde et al. 1987). The autopsy case of Kennedy et al. (1968) confirmed the anterior horn cell disease and demonstrated a mild loss of nerve fibres in the peripheral nerves; a slight pallor of the fasciculus gracilis in the thoracic part of the spinal cord was interpreted as being of vascular origin. Sobue et al. (1989) reviewed the 5 autopsy cases from the literature and concluded that the sensory neurons were not studied in sufficient detail. In their own 3 cases they found a distinct sensory axonopathy, lending support to the term neuronopathy.

Most remarkably a gynaecomastia was present in many (Kennedy et al. 1968; Tsukagoshi et al. 1970; Stefanis et al. 1975; Ringel et al. 1978; Schoenen et al. 1979; Arbizu et al. 1983; Schanen et al. 1984; Sobue et al. 1989), but not in all reported cases (Barkhaus et al. 1982). Sterility (Stefanis et al. 1975) and late hypogonadism (Arbizu et al. 1983) have been described. Testicular biopsy and hormonal studies demonstrated a primary testicular failure (Arbizu et al. 1983). It was suggested that a hereditary androgen receptor failure might explain the clinical picture as androgen receptors have been found in spinal and bulbar motor neurons, and were absent in oculomotor and sacral nuclei (Weiner, 1980).

Fischbeck et al. (1986) confirmed the X-linked recessive mode of inheritance by demonstrating linkage between the putative BSMA gene and the marker DXYS1 on the proximal long arm of the X-chromosome. Additional reports (Fischbeck et al. 1987) confirmed this localisation which placed the genetic defect near the proposed location of the androgen receptor gene, which is on Xq11–12. These findings were supported by the clinical study of Mukai and Yasuma (1987). The clinical features of androgen insensitivity and the results of the linkage analysis suggest the androgen receptor as a candidate gene for this disorder. However, Fischbeck (1990), in an elegant series of studies, was unable to demonstrate abnormalities in the androgen receptor gene in BSMA. This rules out a major deletion of the gene but a point mutation in the gene cannot be excluded.

SCAPULOPERONEAL SPINAL MUSCULAR ATROPHY

The term scapuloperoneal was coined by Davidenkow (1929) to describe a syndrome of distal motor and sensory changes in the lower extremities and shoulder girdle weakness, occasionally with distal sensory abnormalities in the arms. Considered by most authors a variant of Charcot-Marie-Tooth disease, the Davidenkow syndrome, as it was called, was reported only rarely outside the East-European countries. The adjective scapuloperoneal however, started a career on its own, as it was found useful to summarize a disease state of various etiology. The term scapuloperoneal spinal muscular atrophy (SPSMA) was used for the first time by Emery et al. (1968), although the first report was by Stark in 1958. If we accept the general principle of classifying the SMAs according to age of onset (Emery 1971), it is noted that infantile SP-forms have not been reported. It is also unlikely that such a particular distribution of the muscle weakness will be detected at a very early age. Therefore, only juvenile and adult onset cases have been described. The question remains if such a classification is justified, but, so far, all cases with SPSMA belonging to one family had either juvenile or adult onset of the disease. Feigenbaum and Munsat's (1970) case 2 was the only one with onset of the disease in the shoulder girdle muscles; in all other cases the disease started in the peroneal muscles, and ran an ascending course. Another remarkable feature of SPSMA is the high incidence of foot deformities such as pes cavus or pes equinovarus. Kaeser (1965, 1970) and Mercelis et al. (1980) emphasized the often conflicting findings on EMG and muscle biopsy regarding the primary site of the lesion, and the frequent observation in the EMG of neurogenic changes in one part of the body and myopathic abnormalities in another part.

Juvenile onset

Autosomal dominant cases of juvenile onset are rare (Feigenbaum and Munsat 1970; Serratrice et al. 1976). Equally rare are the autosomal recessive cases, and in no instance could a definite conclusion on the pattern of heredity be reached. The young male reported by Negri et al. (1979) was mentally retarded. The 2 brothers reported by Mercelis et al. (1980) both presented with foot extensor weakness in the second decade. In the elder boy, EMG had shown a myopathic pattern on several occasions while the muscle biopsy findings were inconclusive. His younger brother showed neurogenic features on EMG and muscle biopsy, testifying to the taxonomical difficulties in this syndrome.

Sporadic cases and cases in which the mode of inheritance was uncertain have been reported frequently (Emery et al. 1968; Munsat 1968; Zellweger and McCormick 1968; Schuchmann 1970; Feigenbaum and Munsat 1970; Serratrice et al. 1976). Again, they are difficult to classify, as many cases showed both myopathic and neurogenic changes. Answers probably will come only when the molecular biology of the various genetic disorders is known.

Adult onset

The autosomal dominant adult onset SPSMA as described by Stark (1958) and Kaeser (1964, 1965, 1975) is clearly a particular entity. Twelve persons in 5 generations have been found affected. The disease started between 30 and 50 years with weakness and atrophy distally in the lower limbs, and ran an ascending course, affecting in the end not only the shoulder girdle muscles, but also the palatal and pharyngeal muscles. EMG and muscle biopsies showed myopathic characteristics on several occasions apart from neurogenic changes. Autopsies (Kaeser 1964) revealed neurogenic changes in muscles, and at all levels axonal swellings with lipofuscein accumulation and intra-axonal corpora amylacea in the anterior horns, and normal numbers of anterior horn cells, rendering this disease distinct from chronic SMA at the pathological level. Similar cases of Kaeser syndrome have been reported by Serratrice et al. (1976, 1982) but the cases reported by Tsukagoshi et al. (1969) had distal sensory disturbances, and should be considered examples of Davidenkow's syndrome.

The two families reported by Jennekens et al. (1975) had an autosomal dominant scapulo-ilio-peroneal syndrome, with a cardiomyopathy starting in the third or fourth decade and frequent involvement of bulbar muscles. Contractures were absent and EMG changes progressed with age from non-specific findings to abnormalities of pulse formation and conduction. Atrial paralysis, as reported in the X-linked recessive SP myopathic syndromes with cardiomyopathy (Waters et al. 1975), was absent in these families (Jennekens et al. 1975). Cardiac hypertrophy developed gradually. EMG and muscle biopsies showed both neurogenic and myopathic changes; several biopsies showed inflammatory reactions. These autosomal dominant cases resemble the picture of the 2 brothers reported by Takahashi et al. (1974). Autopsy of one of these cases showed normal numbers of anterior horn cells, which were laden with lipofuscein granules. The mode of inheritance as well as the site of the primary lesion in these cases remained unsettled; a dying back phenomenon was postulated. The 3 male patients with a neurogenic SP syndrome of juvenile onset and a progressive cardiomyopathy reported by Mawatari et al. (1973) resembled the X-linked recessive myopathic SP cases described by Waters et al. (1975); it remains to be seen if Mawatari's patients constitute separate disease entities.

The adult onset autosomal recessive neurogenic SP syndrome reported by Tandan et al. (1989) in a brother and sister is unique as similar cases have not been described before. Mild ptosis and facial weakness were present in both, whereas the brother developed dysarthria and dysphagia in the course of the disease. EMG showed neurogenic and occasional myopathic characteristics and the muscle biopsies showed small angulated fibres in both and small groups of atrophic fibres in the brother. The family with possible X-linked scapuloperoneal neurogenic atrophy reported by Skre et al. (1978) was unusual because of a variable age of onset, a variable involvement of bulbar musculature and a possible X-linked inheritance. Muscle biopsies showed mild neurogenic features in addition to myopathic characteristics, which is not unusual for an SP syndrome.

Sporadic cases of adult onset SPSMA have been reported, but are rare, and often difficult to classify as both myopathic and neuropathic characteristics are found on EMG and in muscle biopsies depending on the sites that are studied (Fotopulos and Schulz 1966; Takahashi et al. 1974; Serratrice et al. 1976).

In conclusion, it can be stated that SPSMA is a rare disorder with various ages of onset and modes of inheritance and an unsettled pathogenesis. More insight in these disorders will be gained by the advancement of the molecular biology of the more common spinal muscular atrophies.

The special forms of SMA discussed in this chapter are special indeed as they have all struggled for recognition. Some have succeeded, some have failed: it is doubtful if autosomal dominant facioscapulohumeral SMA exists at all, and if the Finkel type and the Ryukyuan type of SMA deserve an independent standing. Scapuloperoneal SMA is a rare syndrome and, most likely covers several diseases of different genetic and pathophysiological origin. The only well defined disorder among the ones discussed, is X-linked recessive bulbospinal muscular atrophy because of its particular mode of inheritance, and the genetic localisation demonstrated by linkage analysis.

Addendum

At the second workshop on FSHD on September 16, 1990 in Munich Dr. P. Lunt confirmed the report of Wijmenga et al. (1990) on the location of the FSHD-gene on chromosome 4. In addition they had tested a part of a family diagnosed as having autosomal dominant FSHSMA (Siddique et al. 1989) and found similar linkage data as for FSHD, suggesting that the 2 clinical diagnoses reflect the same genetic disorder. At the same meeting Dr. C. Sewry demonstrated that small angular fibres in FSHD express fetal myosin, which strongly suggests that they are regenerating fibres. These 2 communications offer additional arguments to the thesis in this chapter that the diagnosis of autosomal dominant FSHSMA lacks substantiation.

REFERENCES

ARBIZU, T. J., J. SANTAMARIA, J. M. GOMEZ, A. QUILEZ and J. P. SERRA: A family with adult spinal and bulbar muscular atrophy, X-linked inheritance and associated testicular failure. J. Neurol. Sci. 59 (1983) 371–382.

BARKHAUS, P. E., W. R. KENNEDY, L. Z. STERN and R. B. HARRINGTON: Hereditary proximal spinal and bulbar motor neuron disease of late onset: a report of six cases. Arch. Neurol. (Chic.) 39 (1982) 112–116.

BECKER, P. E.: Atrophia musculorum spinalis pseudomyopathica. Z. menschl. Vererb.-u. Konstitutionslehre 37 (1964) 193–220.

BECKER, P. E.: Handbook of Human Genetics, Vol. 1. Thieme, Stuttgard (1966).

BOUWSMA, G. and N. J. LESCHOT: Unusual pedigree patterns in seven families with spinal muscular atrophy; further evidence for the allelic model hypothesis. Clin. Genet. 30 (1986) 145–147.

BRZUSTOWICZ, L. M.,   T. LEHNER,   L. H. CASTILLA, G. K. PENCHASZADEH,   K. C. WILHELMSEN,   R. DANIELS, K. E. DAVIES, M. LEPPERT, F. ZITER, D. WOOD, V. DUBOWITZ, K. ZERRES, I. HAUSMANOWA-PETRUSEWICZ, J. OTT, T. L. MUNSAT and T. C. GILLIAM: Genetic mapping of chronic childhood-onset spinal muscular atrophy to chromosome 5q11.2-13.3. Nature 344 (1990) 540–541.

DAVIDENKOW, S.: Über die scapulo-peroneale Amyotrophie (Die Familie Z). Z. Gesamte Neurol. Psychiatrie 122 (1929) 628–650.

DAVIDENKOW, S.: Scapuloperoneal amyotrophy. Arch. Neurol. Psychiatry (Chic.) 41 (1939) 694–701.

DUBOWITZ, V.: Muscle Biopsy. A Practical Approach, 2nd ed. Baillière Tindall (1985).

EMERY, A. E. H.: The nosology of the spinal muscular atrophies. J. Med. Genet. 8 (1971) 481–495.

EMERY, E. S., C. M. FENICHEL and G. ENG: A spinal muscular atrophy with scapuloperoneal distribution. Arch. Neurol. (Chic.) 18 (1968) 129–133.

FEIGENBAUM, J. A. and T. L. MUNSAT: A neuromuscular syndrome of scapuloperoneal distribution. Bull. Los Angeles Neurol. Soc. 35 (1970) 47–57.

FENICHEL G. M., E. S. EMERY and P. HUNT: Neurogenic atrophy simulating facioscapulohumeral dystrophy (a dominant form). Arch. Neurol. (Chic.) 17 (1967) 257–260.

FINKEL, N.: A forma pseudomiopatica tardia da atrofia muscular progressive heredo-familial. Arq. Neuro-Psiquiatr. 20 (1962) 307–322.

FISCHBECK, K. H., V. IONASESCU, A. W. RITTER, R. IONASCESCU, K. DAVIES, S. BALL, P. BOSCH, T. BURNS, I. HAUSMANOVA-PETRUSEWICZ, J. BORKOWSKA, S. P. RINGEL and L. Z. STERN: Localisation of the gene for X-linked spinal muscular atrophy. Neurology 36 (1986) 1595–1598.

FISCHBECK, K. H., A. RITTER, Y. SHI, R. W. NUSSBAUM, J. LESKO, V. IONASESCU, P. CHANCE and A. HARDING: Linkage studies of X-linked neuropathy and spinal muscular atrophy. Cytogen. Cell. Genet. 46 (1987) 614.

FISCHBECK, K. H., D. SOUDERS and A. LA SPADA: A candidate gene for X-linked spinal muscular atrophy. Adv. Neurol. 1990 in press.

FOTOPULOS, D. and H. SCHULZ: Beitrag zur Pathogenese des scapulo-peronealen Syndroms. Neurol. Med. Psychol. (Leipzig) 18 (1966) 129–136.

FURAKAWA, T., H. TSUKAGOSHI, H. SUGITA and Y. TOYOKURA: Neurogenic muscular atrophy simulating facioscapulohumeral muscular dystrophy. J. Neurol. Sci. 9 (1969) 389–397.

FURAKAWA, T. and Y. TOYOKURA: Chronic spinal muscular atrophy of facioscapulohumeral type. J. Med. Genet. 13 (1976) 285–289.

FURAKAWA, T., H. TSUKAGOSHI and J. B. PETER: Facioscapulohumeral spinal muscular dystrophy. Clinical and genetic studies. Abstracts 12th World Congress of Neurology, 1981, Kyoto, Japan. Int. Congress Ser. 548. Excerpta Medica Amsterdam, Oxford, Princeton.

GILLIAM, T. C., L. M. BRZUSTOWICZ, L. H. CASTILLA, T. LEHNER, G. K. PENCHASZADEH, R. J. DANIELS, B. C. BYTH, J. KNOWLES, J. E. HISLOP, Y. SHAPIRA, V. DUBOWITZ, T. L. MUNSAT, J. OTT and K. E. DAVIES: Genetic homogeneity between acute and chronic forms of spinal muscular atrophy. Nature 345 (1990) 823–825.

HARDING, A. E., P. K. THOMAS, M. BARAITSER ET AL.: X-linked recessive bulbospinal neuronopathy: a report of ten cases. J. Neurol. Neurosurg. Psychiatry 45 (1982) 1012–1019.

HAUSMANOWA-PETRUSEWICZ, I., J. BOKOWSKA and Z. JANCZEWSKI: X-Linked adult form of spinal muscular atrophy. J. Neurol. 229 (1983) 175–188.

HAUSMANOWA-PETRUSEWICZ, I., J. ZAREMBA and J. BOKOWSKA: Chronic proximal spinal muscular atrophy of childhood and adolescence: problems of classification and genetic counselling. J. Med. Genet. 22 (1985) 350–353.

JENNEKENS, F. G. J., H. F. M. BUSCH, N. M. VAN HEMEL and R. A. HOOGLAND: Inflammatory myopathy in scapulo-ilio-peroneal atrophy with cardiopathy. A study of two families. Brain 98 (1975) 709–722.

KAESER, H. E.: Die familiäre scapuloperoneale Muskelatrophie. Dtsch. Z. Nervenheilk. 186 (1964) 379–394.

KAESER, H. E.: Scapuloperoneal muscular atrophy. Brain 88 (1965) 407–418.

KAESER, H. E.: Scapuloperoneal syndrome. In: P. J. Vinken and G. W. Bruyn (Eds.), Handbook of Clinical Neurology, Vol. 22, chapter 2. North-Holland Publishing Co. 1975.

KAZAKOV, V. M., D. K. BOGORODINSKY and A. A. SKOROMETZ: Myogenic scapuloperoneal dystrophy-muscular dystrophy in the K-kindred. Eur. Neurol. 13 (1975) 350–359.

KENNEDY, W. R., M. ALTER and J. H. SUNG: Progressive proximal spinal and bulbar muscular atrophy of late onset: a sex-linked trait. Neurology (Minneap.) 18 (1968) 671–680.

KONDO, K., T. TSUBAKI and F. SAKAMOTO: The Ryukyuan Muscular Atrophy. An obscure heritable neuromuscular disease found in the Islands of Southern Japan. J. Neurol. Sci. 11 (1970) 359–382.

KRÜGER, H. and M. FRANKE: Facioscapulohumeral form der neurogene Muskelatrophie. Psychiatr. Neurol. Med. Psychol. (Leipzig) 26 (1974) 295–301.

KURLAND, L. T.: Epidemiologic investigations of amyotrophic lateral sclerosis, Part 3 (A genetic interpretation of incidence and geographic distribution). Proc. Mayo Clin. 32 (1957) 449–462.

LANDOUZY, L. and J. DEJERINE: De la myopathie atrophique progressive. Rev. Méd. 5 (1885) 81–117.

LANDOUZY, L. and J. DEJERINE: De la myopathie atrophique progressive. Rev. Méd. 5 (1885) 253–366.

LANDOUZY, L. and J. DEJERINE: De la myopathie atrophique progressive. Rev. Méd. 6 (1886) 977–1027.

LOVELACE, R. E. and M. MENKEN: The scapuloperoneal syndrome: dystrophic type. Excerpta Med. Int. Congr. Ser. 193 (1969) 247–248.

LUNT, P. W.: A workshop on facioscapulohumeral (Landouzy-Dejerine) disease, Manchester 16 to 17 November 1988. J. Med. Genet. 26 (1989) 535–537.

MAGEE, K. R. and R. N. DE JONG: Neurogenic muscular atrophy simulating muscular dystrophy. Neurology (Chic.) 2 (1960) 677–682.

MAGEE, K. R.: Familial progressive bulbar-spinal muscular atrophy. Neurology (Minneap.) 10 (1960) 295–305.

MARES, A., M. CONSTANTINESCO, C. VASILESCO and M. IONESCO: Atrophie musculaire hérédo-familiale à transmission dominante. Etude clinique, électromyografique et bioptique. Rev. Roum. Neurol. 1 (1964) 295–307.

MAWATARI, S. and K. KATAYAMA: Scapuloperoneal muscular atrophy with cardiopathy. An X-linked recessive trait. Arch. Neurol. 28 (1973) 55–59.

MC COMAS, A. J.: Neuromuscular Function and Disorders. Butterworth, London (1977).

MELKI, J., S. ABDELHAK, P. SHETH, M. F. BACHELOT, P. BURLET, A. MARCADET, J. AICARDI, A. BAROIS, J. P. CARRIERE, M. FARDEAU, D. FONTAN, G. PONSOT, T. BILLETTE, C. ANGELINI, C. BARBOSA, G. FERRIERE, G. LANZI, A. OTTOLINI, M. C. BABRON, D. COHEN, A. HANAUER, F. CLERGET-DARPOUX, M. LATHROP, A. MUNNICH and J. FREZAL: Gene for chronic proximal spinal muscular atrophies maps to chromosome 5q. Nature 344 (1990) 767–768.

MERCELIS, R., J. DE MEESTER and J. J. MARTIN: Neurogenic scapuloperoneal syndrome in childhood. J. Neurol. Neurosurg. Psychiatry 43 (1980) 888–896.

MEADOWS, J. C. and C. D. MARSDEN: Scapuloperoneal amyotrophy. Arch. Neurol. 20 (1969) 9–12.

MUKAI, E. and T. YASUMA: A pedigree with protanopia and bulbospinal muscular atrophy. Neurology 37 (1987) 1019–1021.

MUNSAT, TH. L.: Infantile scapuloperoneal muscular atrophy. Neurology (Minneap.) 18 (1968) 285.

MUSSINI, J. M., H. LE NOAN, J. MUSSINI-MONTPELLIER and F. JANSSEN: Histological findings in facioscapulohumeral myopathy: is there biological heterogeneousness? Abstract. 9th International Meeting on Neuromuscular Disease. Marseilles 15–17 September (1988).

NAKAZATO, H., M. KINOSHITA and E. SATOYOSHI: A case of Ryukyuan muscular atrophy. Clin. Neurol. (Tokyo) 17 (1977) 353–356.

NEGRI, S., T. CARACENI and F. CORNELIO: A case of scapulo-tibio-peroneal syndrome: electromyographic and histoenzymologic consideration. Eur. Neurol. (Basel) 10 (1973) 31–40.

PADBERG, G. W.: Facioscapulohumeral Disease. Thesis, Leiden University (1982).

PADBERG, G. W., C. WIJMENGA, O. F. BROUWER, P. MOERER, J. L. WEBER and R. R. FRANTS: Facioscapulohumeral Dystrophy gene maps to chromosome 4. J. Neurol. Sci. 98S (1990) 550.

PATEL, A. N. and R. K. SWAMI: Muscle percussion and neostigmine test in the clinical evaluation of neuromuscular disorders. N. Engl. J. Med. 281 (1969) 523–526.

RICHIERI-COSTA, A., A. ROGATKO, R. LEVISKI, N. FINKEL and O. FROTA-PESSOA: Autosomal dominant late adult spinal muscular atrophy, type Finkel. Am. J. Med. Genet. 9 (1989) 119–128.

RICKER, K., H. G. MERTENS and K. SCHIMRIGK: The neurogenic scapulo-peroneal syndrome. Eur. Neurol. 1 (1968) 257–274.

RINGEL, S. P., N. S. LAVA, M. M. TREIHAFT, M. L. LUBS and H. A. LUBS: Late onset X-linked recessive spinal and bulbar muscular atrophy. Muscle Nerve 1 (1979) 297–307.

SCHANEN, A., J. MIKOL, C. GUIZIOU, V. VITAL, M. COQUET, A. LAGUENY, J. JULIEN and M. HAGUENAU: Forme familiale liée au sexe d'amyotrophie spinale progressive de l'adulte. Rev. Neurol. (Paris) 140 (1984) 720–727.

SCHOENEN, J., P. J. DELWAIDE, J. J. LEGROS and P. FRANCHIMONT: Motoneuropathie héréditaire – La forme proximale de l'adulte liée au sexe (ou maladie de Kennedy). J. Neurol. Sci. 49 (1979) 343–357.

SCHUCHMANN, L.: Spinal muscular atrophy of the scapulo-peroneal type. Z. Kinderheilkd. 109 (1970) 118–123.

SERRATRICE, G., J. L. GASTAUT, J. F. PELLISSIER and J. POUGET: Amyotrophies scapulo-péronières chroniques de type Stark-Kaeser. Rev. Neurol. (Paris) 132 (1976) 823–832.

SERRATRICE, G., J. F. PELLISSIER, J. POUGET, J. L. GASTAUT and D. CROS: Les Syndromes Scapulo-péroniens. Rev. Neurol. (Paris) 138 (1982) 691–711.

SERRATRICE, G., J. F. PELLISSIER, J. POUGET and J. C. SAINT-JEAN: Un syndrome neuro-endocrien: l'amyotrophie spinale pseudo-myopathique avec gynécomastie liée au chromosome X. Presse Méd. 16 (1987) 299–302.

SERRATRICE, G., J. F. PELLISSIER and J. POUGET: Neuronopathie bulbo-spinale liée à l'X: syndrome de Kennedy. Rev. Neurol. (Paris) 144 (1988) 756–758.

SIDDIQUE, T., H. ROPER, M. PERICAK-VANCE, J. SHAW, K. L. WARNER, W. Y. HUNG, K. L. PHILLIPS, P. LUNT, W. J. K. CUMMING and A. D. ROSES: Linkage analysis in the spinal muscular atrophy type of facioscapulohumeral disease. J. Med. Genet. 26 (1989) 487–489.

SKRE, H., S. J. MELLGREN, P. BERGSHOLM and J. E. SLAGSVOLD: Unusual type of neurol muscular atrophy with a possible X-chromosomal inheritance pattern. Acta Neurol. Scan. 58 (1978) 249–260.

SOBUE, G., Y. HASHIZUME, E. MUKAI, M. HIRAYAMA, T. MITSUMA and A. TAKAHASHI: X-Linked recessive bulbospinal neuromyopathy. Brain 112 (1989) 209–232.

STARK, P.: Etude clinique et génétique d'une famille atteinte d'atrophie musculaire progressive neurale (amyotrophie de Charcot-Marie). J. Génét. Hum. 7 (1959) 1–32.

STEFANIS, C., TH. PAPAPETROPOULOS, S. SCARPALEZOS, G. LYGIDAKIS and C. P. PANAYIOTOPOULOS: X-Linked spinal and bulbar muscular atrophy of late onset – A separate type of motor neuron disease? J. Neurol. Sci. 24 (1975) 493–503.

RICKER, K., H. G. MERTENS and K. SCHIMRIGK: The neurogenic scapuloperoneal syndrome. Eur. Neurol. 1 (1968) 257–274.

TAKAHASKI, K., H. NAKAMURA, K. TANAKA and T. MUTSUMURA: Neurogenic scapuloperoneal amyotrophy associated with dystrophic changes. Clin. Neurol. (Tokyo) 11 (1971) 650–658.

TAKAHASKI, K., H. NAKAMURA, R. NAKASHIMA: Scapuloperoneal dystrophy associated with neurogenic changes. J. Neurol. Sci. 23 (1974) 575–583.

TANDAN, R., A. VERMA and M. MOHIRE: Adult-onset autosomal recessive neurogenic scapuloperoneal syndrome. J. Neurol. Sci. 94 (1989) 201–209.

TOGHI, H., H. TSUKAGOSHI and Y. TOYOKURA: Neurogenic scapuloperoneal syndrome with autosomal recessive inheritance. Clin. Neurol. (Tokyo) 11 (1971) 215–220.

TSUKAGOSHI, H., T. TAKASU, M. YOSHIDA and Y. TOYOKURA: A family with scapuloperoneal muscular atrophy. Clin. Neurol. (Tokyo) 9 (1969) 511–517.

TSUKAGOSHI, H., H. SHOJI and T. FURUKAWA: Proximal neurogenic atrophy in adolescence and adulthood with X-linked recessive inheritance. Neurology (Minneap.) 20 (1970) 1188–1193.

WATERS, D. D., D. O. NUTTER, L. C. HOPKINS and E. R. DORNEY: Cardiac features of an unusual X-linked humeroperoneal neuromuscular disease. N. Engl. J. Med. 293 (1975) 1017–1022.

WEINER, L. P.: Possible role of androgen receptors in amyotrophic lateral sclerosis — A hypothesis. Arch. Neurol. 37 (1980) 129–131.

WILDE, J., T. MOSS and D. THRUSH: X-linked bulbo-spinal neuronopathy: a family study of three patients. J. Neurol. Neurosurg. Psychiatry 50 (1987) 279–284.

WIJMENGA, C., R. R. FRANTS, O. F. BROUWER, P. MOERER, J. L. WEBER and G. W. PADBERG: Location of Facio-scapulohumeral Muscular Dystrophy gene on to chromosome 4. Lancet 336 (1990) 651–653.

ZELLWEGER, H. and W. F. MCCORMICK: Scapulo-peroneal dystrophy and scapulo-peroneal atrophy. Helvet. Paediatr. Acta 23 (1968) 643–649.

Handbook of Clinical Neurology, Vol. 15 (59): Diseases of the Motor System
J.M.B.V. de Jong, editor
© Elsevier Science Publishers B.V., 1991

CHAPTER 5

Werdnig-Hoffmann disease and variants

MAKIKO OSAWA and KEIKO SHISHIKURA

Department of Pediatrics, Tokyo Women's Medical College, Tokyo, Japan

Werdnig-Hoffmann disease (WHD) (253300) is representative of the infantile proximal spinal muscular atrophies (SMA) which are characterized by anterior horn cell degeneration. The historical and clinical aspects of WHD were reviewed by V. Dubowitz in Chapter 4 of Volume 22 of the original series of the Handbook. Therefore we elaborate on some of the clinical features which have since been clarified, consider the relationship between clinical features and pathological changes, and briefly discuss some variants and other diseases in which spinal anterior horn cells may be affected.

CLINICAL CLASSIFICATION

Though recent advances in molecular genetics (to be described later) seem to have curtailed discussion of the nosological classification of SMAs, the subclassification system is still very useful for clinicians. In spite of some borderline cases, the identification of a particular patient's disease as belonging to one of the 3 types of SMA yields invaluable prognostic information and serves as a guide to management (Figs. 1, 2 and 3).

More than one classification system has been proposed. Byers and Banker (1961) devised a system based on age of onset. The most severe form had the earliest onset of weakness; in utero

or before 2 months of age. In the second group, its onset was between 2 and 12 months of age. The least severe form had its onset in the second year of life. These patients had only localized weakness and much longer survival.

The classification scheme of Fried and Emery (1971) divides SMA into 3 types. Type I, infantile WHD, has its onset before 9 months, usually before 3 months and death generally occurs by age 1 (4 at the latest). In type II the age of onset is between 3 and 15 months, 6–12 months most commonly, and death generally occurs between 10 and 25 years of age. The onset of type III is not earlier than 24 months, after 36 months in most cases, and the majority of patients live beyond age 20.

The most widely used system is that of Dubowitz (1978): SMA I, acute WHD, patients are never able to sit unsupported; SMA II, intermediate, able to sit unaided but never become ambulant; and III, juvenile or Kugelberg-Welander disease (KW) (253400), ambulant.

Hausmanowa-Petrusewicz (1970, 1984) defined SMA I as onset before 1 year, mostly before or around birth, and no ambulation, then subclassified Type I into subgroups a and b. Type Ia is comparable to acute WHD, which is fatal before age 4, and Ib, to chronic WHD, the progression of which can be slow with survival up to age 30. Type II, intermediate, was defined by onset be-

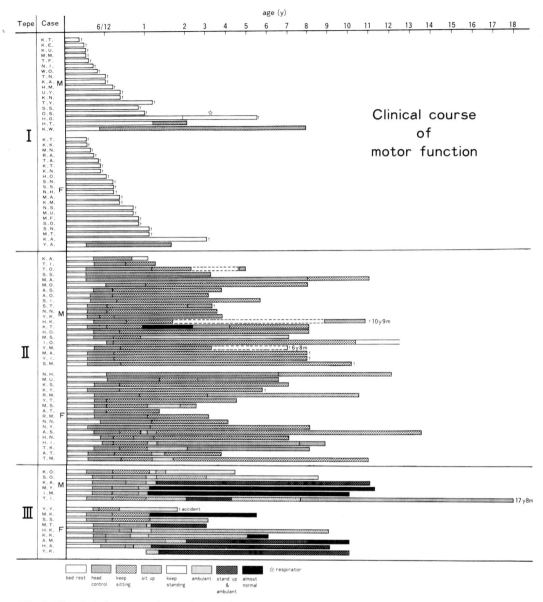

Fig. 1. The clinical courses of several outpatients followed by the Pediatric Department of Tokyo Women's Medical College are subdivided by age of onset. A minority of Type I patients, probably comparable to the Hausmanova-Petrusewicz Type Ib, develop head control. Type II patients can attain head control in infancy and maintain a sitting posture. A few patients, mainly girls, are even able to maintain a standing posture for brief periods during their clinical course. Symptom progression is much slower in Type III, with some patients actually experiencing clinical improvement over the course of time.

tween 2 and 6 years and possible early ambulation with immobility occurring by age 10–14. Type III, juvenile or KW has its onset between 1 and 15 years and patients remain mobile.

Fried and Emery highlighted the ages at onset and death, Dubowitz emphasized motor ability and Hausmanowa-Petrusewicz divided patients according to age of onset, motor ability or course, and death. Every classification system is an arbitrary subdivision of a flow chart, highlight-

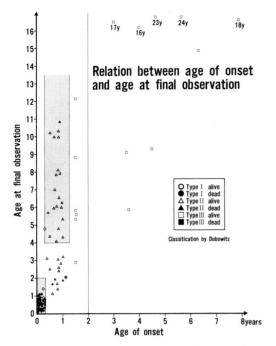

Fig. 2. The relationship between age of onset and age at death or the age at the final observation.

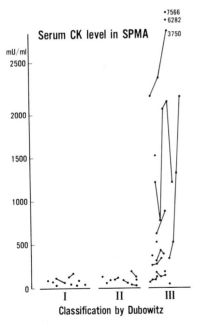

Fig. 3. Serum CK levels grouped according to Dubowitz. Note the 2 clusters of Type III patients, with high and low CK levels. Those with higher levels tend to have better prognoses.

ing either onset, which may be difficult to reconstruct and depend on parental awareness, and/or the course of the disease, which varies with the aggressiveness of treatment. The classification systems of Dubowitz and Hausmanowa-Petrusewicz seem to have been widely accepted and we should be aware of the relationship between the two.

SMA I by Dubowitz, Ia by Hausmanowa-Petrusewicz or acute WHD, has the most severe course, earliest onset and death, generally from respiratory failure, occurs around age 1, although there are a few patients whose clinical course is relatively benign and static who may survive to nearly age 4.

SMA II by Dubowitz patients have a less severe course, attain the ability to sit unsupported, may live beyond age 4, and are comparable to Type Ib of the Hausmanowa-Petrusewicz classification. The original cases described by Werdnig and Hoffmann had their onsets between 5 and 9 months, were able to sit unaided, and belong to this group.

SMA III by Dubowitz has the mildest course, patients are ambulatory and can expect to live into adulthood. According to Hausmanowa-Petrusewicz, cases who become ambulant are divided into 2 types; SMA II (intermediate), in which ambulation is transient, and SMA III (KW), with a better course.

Hereafter, we will use the subclassification system of Dubowitz unless otherwise specified.

INCIDENCE

Pearn et al. (1973) carried out a total population survey in northeastern England and estimated the gene frequency of SMA I to be 6.25×10^{-3}, a carrier rate of 1 in 80 and the incidence of SMA to be 1 per 20 000 live births. The chronic forms affect a similar number, 1 in 24 000 (Pearn et al. 1973).

An unusually high frequency of SMA I, 4 out of a total of 1600 infants, was reported in the Egyptian Kararite community in Israel (Fried and Mundel 1977). The Kararites lived as a closed, reproductively isolated, religious community for more than 10 centuries. It was assumed that the gene frequency of 0.05, a heterozygous carrier rate of 1 in 10 persons, was due to genetic drift. Pascalet-Guidon et al. (1984) also reported a remarkably high incidence of SMA I in a limited area of Reunion Island in the Indian Ocean. All 19 affected cases, who belonged to 13 sibships, descended from a common ancestor who had lived in the 17th century.

GENETICS

Although numerous pedigree analyses have confirmed that SMA is inherited as an autosomal recessive (AR) in the great majority of patients, the existence of borderline cases and of more than one type of SMA in a single family (Hausmanowa-Petrusewicz et al. 1966; Hausmanowa-Petrusewicz 1970; Bouwsma and Leshot 1986) suggests that this classification system is somewhat arbitrary.

Therefore, research has been focused on determining whether SMA in childhood is a single disease spectrum or a group of phenotypically similar, but genetically distinct, entities. Recent breakthroughs in genetics research may finally have resolved this long-standing controversy. Brzustowicz et al. (1990), and Melki et al. (1990a) performing DNA analysis on patients and their families, mapped the gene responsible for SMA II and III to chromosome 5q. Working independently, these groups narrowed the location to regions 5q 11.2–13.2 (Brzustowicz et al. 1990)

and 5q 12–14 (Melki et al. 1990a). The gene for chronic SMA is therefore located in region 5q 12–13. SMA Type I was subsequently mapped to the same area (Gilliam et al. 1990). The nearest marker locus thus far identified is D5S39 (Melki et al. 1990b).

Although some questions of heterogeneity remain (Gilliam et al. 1990), it appears that most clinical phenotypes are due to an allelic series of mutations at the same locus. Identification of closely linked marker genes will pave the way for prenatal diagnosis and carrier detection in the near future. The gene and its products will eventually be identified, making characterization possible of the actual biochemical defect, the first step in developing therapeutic interventions.

SOME CLINICAL FEATURES

The thorax often becomes progressively and disproportionately narrow with time (Fig. 4). Pascual Castroviejo (1984) noted thoracic asymmetry in 20% of his 130 SMA cases.

Atrophy of tonsils and adenoids was observed in more than 50% of affected infants and their complete absence in 18% (Hausmanowa-Petrusewicz and Fidzianska-Dolot 1984). On the other hand, Pascual Castroviejo (1984) did not note this finding in his series.

Characteristic dental features include thin

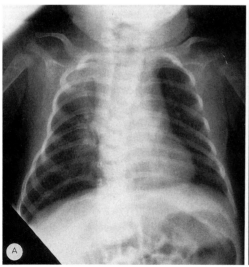
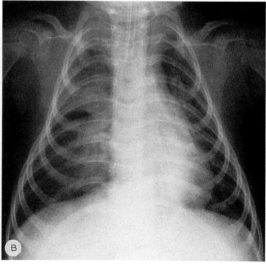

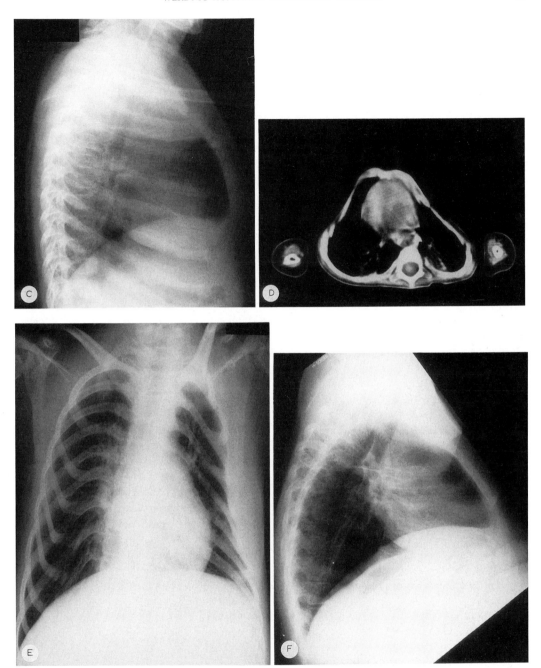

Fig. 4. Chest X-rays of patients with acute WHD at 3 months (A), 9 months (B, C) and intermediate (SMA II) at 12 years (E, F) and CT scan of acute WHD at 11 months (D). Note flaring of the lower thoracic cage (A) which becomes more pronounced as the bell shaped deformity develops (B, C). Note the unchanged (adequate) anteroposterior diameter (C, D, F) and marked lateral recession of the rib cage.

enamel over the primary molars and abnormal root formation (Farrington and Corcoran 1975).

Fragility of the skin, dermatographia and increased diaphoresis have also been reported and may be related to disturbance of sympathetic circuits (Pascual Castroviejo 1984).

LABORATORY FINDINGS

Immunological abnormalities

Ryniewicz and Pawinska (1978) reported that the Mantoux reaction is negative in 97% of children with SMA Ia, in contrast to an 85% positive rate in normal Polish children.

Metabolic abnormality

Increased urinary excretion of long-chain dicarboxylic acids in WHD (Kelley and Sladky 1986) and low carnitine levels (Carrier and Berthillier 1980), suggest that a defect in fatty acid metabolism may play a role in motor neuron disease.

Pulmonary function

Cunningham (1978) measured thoracic gas volume (TGV), resting lung volume at end expiration, by the plethysmographic technique in 9 infants with WHD. Those with intrauterine onset of disease had a significantly reduced TGV, whereas those with postnatal onset had normal lung volumes. They suspected diminished fetal breathing movements, resulting from weakness of the respiratory musculature in utero, could be responsible for the reduction in lung volume found in those infants with intrauterine onset of the disease.

Electromyography

As noted by Dubowitz (1975), EMG is a useful means of demonstrating the neurogenic atrophy associated with anterior horn cell degeneration. Typical EMG findings include spontaneous discharge activity in resting muscle, increased mean amplitude, increased amplitude during voluntary effort, prolonged duration of individual motor unit potentials (MUPs), increased maximum amplitude in the territory of motor units and, in

older patients, evidence of denervation such as pseudomyotonic bursts.

Though nerve conduction is generally considered to be normal, some decreases in velocity have been demonstrated in severe cases.

Hausmanowa-Petrusewicz and Karwanska (1986), reporting on the diagnostic and prognostic value of EMG, found different spontaneous activity patterns among the different types of SMA and that EMG patterns changed over time. Spontaneous rhythmic firing of motor units is found only in young patients with SMA Ia, b and II. Pseudomyotonic volleys eventually appeared in some long-surviving cases. MUPs in SMA Ia were characterized by some short potentials of low amplitude, as well as long potentials of high amplitude, and bimodal amplitude and duration histograms. With longer survival, amplitudes became higher and durations longer. The EMGs of milder forms, type Ib, II and particularly III, showed more features of chronic anterior horn cell involvement such as patterns of reinnervation and hypertrophy of muscle fibers.

With single fiber EMG, study of motor unit microphysiology has become possible (Stalbert and Fawcett 1984). It can be used to assess the denervation-reinnervation process which follows anterior horn cell degeneration in SMA. Reorganization of the motor unit and local concentration of component fibers is reflected in an increased fiber density (FD), a sensitive parameter of reinnervation. Jitter indicates variable time intervals between action potentials in consecutive discharges, a sign of early reinnervation. Increased FD are seen in chronic forms of SMA. The ability of surviving neurons to sustain an extra muscle fiber load, as well as the severity, chronicity and rate of the denervation-reinnervation process influence the degrees of jitter and FD changes.

Ultrasound imaging of muscle

Over the past decade, ultrasound (US) imaging has been introduced as a diagnostic tool for neuromuscular disorders (Heckmatt et al. 1980; Heckmatt and Dubowitz 1984). US is a noninvasive means of assessing muscle atrophy which cannot be detected clinically because of increased subcutaneous tissue. In SMA, muscles show an increased echo intensity while that of bone is correspondingly decreased. Very young SMA patients may have little or no muscle atrophy, exhibiting what has been termed the 'pre-pathological' state sometimes seen in early WHD. Though this finding limits the usefulness of US scanning in infants under 6 months, they reported that the results in older children correlated reasonably well with those of muscle biopsy.

The authors suggested a four-point classification system to compare the degree of change in muscle architecture with changes in echo intensity: Grade 1, normal echo, minimal histological change; Grade 2, mild to moderate increase in muscle echo, mild to moderate pathology; Grade 3, diffuse, marked increase in muscle echo with reduced bone echo, extensive pathological changes; and Grade 4, very large increase in muscle echo with faint or absent bone echo, more than 50% of muscle has been replaced by connective tissue and fat. As equipment and the interpretive skills of clinicians improve, US is likely to become an increasingly valuable means of non-invasively assessing the topographical distribution of muscle atrophy.

Computed tomography (CT) of skeletal muscle (Figs. 5, 6 and 7)

Bulcke (1984) reviewed the use of CT, which yields more precise anatomical information than US, i.e. selective visualization of areas of atrophy and hypertrophy, in evaluating neuromuscular disease. All but the earliest stages of SMA can be differentiated from most myopathies by means of CT. Based on CT studies of two KW patients, SMA findings were described as a progressively more ragged muscle outline, retraction of muscles away from the fascia toward bony insertions, visualized as reduced muscle volume on the CT scan, and ultimately as small pockets of muscular tissue submerged in a large mass of adipose tissue. De Visser and Verbeeten (1985) reported the muscle CT findings of 12 cases of benign infantile SMA, from 12 to 41 years old, including 3 cases whose onsets were before age 1.

Horikawa et al. (1986) evaluated thigh muscle CT scans taken at the upper quarter level in 5 patients with SMA II, aged 6, 7, 20, 21 and 22. The muscle CTs of the cases aged 6 and 7, revealed severely decreased cross-sectional areas of muscle, surrounded by large areas of low density, without a significant intramuscular decrease in density. According to their observations, the hamstring and adductor muscles, especially the adductor longus muscle (ALM), were less affected than the quadriceps femoris. Spotty, moth-eaten, low density areas were observed predominantly in the more severely affected muscles. In all cases in their twenties all muscles except the ALM were unidentifiable because of severe atrophy with extremely low density. These

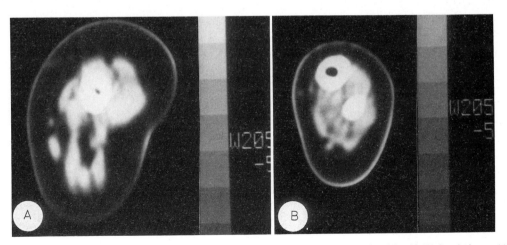

Fig. 5. Computerized tomography of skeletal muscle at left thigh (A) and left mid-calf (B) levels in an 11-month-old patient with acute WHD. Note the remarkable muscle atrophy surrounded by an extensive low density area with only slight intramuscular decrease in density.

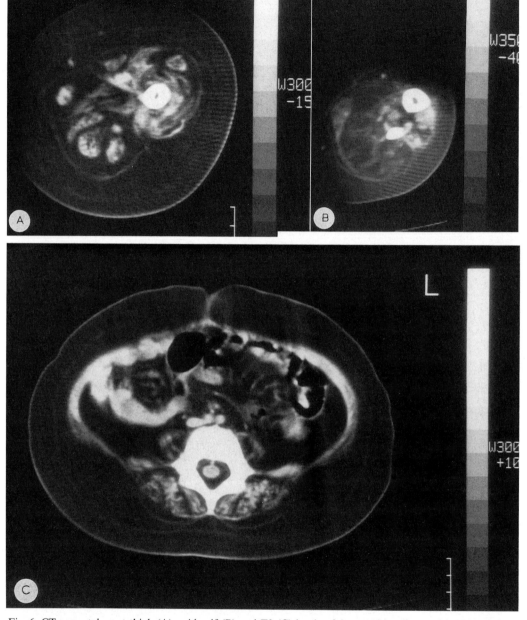

Fig. 6. CT scans taken at thigh (A), mid-calf (B) and T3 (C) levels of 6-year-old patients with SMA II. The atrophic muscles are surrounded by an extensive low density area. Spotty, moth-eaten low density areas are seen predominantly in the most severely affected muscles.

data suggest that the muscular wasting seen in SMA II starts with muscle fiber atrophy due to denervation, followed by eventual replacement of muscle with fatty tissue. Preservation of the ALM indicates that the loss of anterior horn cells does not always occur homogeneously.

Recent advances in muscle pathology

Fidzianska (1974) carried out an ultrastructural study in 7 children with WHD and classified the muscle fibers into 3 categories: (1) muscle cells of normal diameter with no distinct changes or

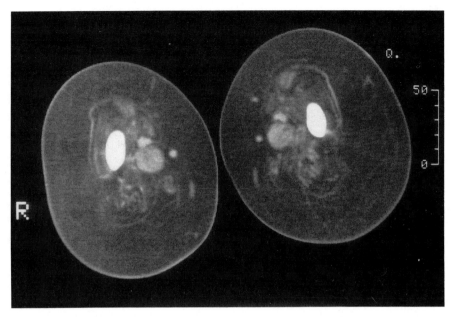

Fig. 7. Thigh CT image of a 21-year-old female with SMA II (courtesy of Dr. Horikawa). Only the adductor longus muscles, the calculated CT numbers of which were 4.3 (right) and 9.8 (left), can be identified. Other muscles are unidentifiable because of severe atrophy with extremely low density. Relative preservation of the adductor longus muscle may explain the patient's ability to maintain lower extremities in adducted posture in contrast to the frog leg posture of muscular dystrophy.

only slight changes; (2) small muscle cells with large central nuclei with the appearance of myoblasts; and (3) cells resembling myotubes with 2–4 of these cells, at different stages of maturation, sharing a common basement membrane. She concluded that myoblast and myotube-like cells represented an arrest in maturation rather than cellular atrophy. Subsequently, she compared the small atrophic fibers in amyotrophic lateral sclerosis (ALS) and those in WHD (Fidzianska 1976). Atrophic muscle fibers in ALS showed irregularity in shape and size with degenerative changes of myofibrils and accumulated nuclei, while the small muscle cells in WHD were uniform in diameter with preserved architecture and single nuclei and were scattered among numerous myotube-like cells. Special attention has been given to whether the small fibers are destined to atrophy or undergo maturational arrest. Saito (1985) observed that the satellite cells in WHD were significantly increased in number in comparison with those of control infants. The satellite cells appeared to be quiescent, displaying no morphological changes. Based

on the observation that experimental denervation of adult skeletal muscle led to an increase in the number of satellite cells in the active phase (Snow 1983), Okada et al. (1984) suggested that neural factors had not influenced satellite cell activation. They speculated that the increase in the number of satellite cells in the quiescent state had resulted from failure of the satellite cells to induce growth of muscle fibers, resulting in muscle fiber immaturity.

Walsh et al. (1987) performed immunocytochemical analysis, utilizing an antibody to neural cell adhesion molecule (N-CAM), on muscle biopsies from patients with WHD and KW. All myofibers, both atrophied and normal, were positive for N-CAM expression profiles, although only the atrophic fibers were positive in KW. They concluded that the positive N-CAM reactivity reflected unstable innervation of myofibers that had previously been innervated.

Sawchak et al. (1990) compared the immunocytochemical reactivity of WHD muscle and that of 20-week gestation muscle, using myosin-isoform-specific monoclonal antibodies (McAbs).

Only a few WHD fibers expressed prenatal myosin heavy isoform chains (MHC) as detected by reactivity with McAb, ALD 180 (specific for a prenatal MHC), whereas virtually all of the fibers from 20-week gestation muscle were strongly reactive with ALD 180. This study suggested that there was no universal arrest in maturation with respect to MHC expression and that only a small subset of 'fetal-like' (probably regenerating) fibers were reactive with prenatal-MHC isoform specific McAb, ALD 180.

The roles of nerve growth factor (NGF) and neuronal cell death during the fetal period have attracted considerable interest. Henderson and Fardeau (1988) injected neonatal rats with antibody to NGF and found almost complete selective destruction of the sympathetic nervous system. NGF plays a role in the regulation of neuronal cell population size and development in different parts of the nervous system. Based on several in vivo and in vitro results suggesting that the muscle itself may, at certain stages, produce motoneuron growth factors, they speculated that the motoneuron death observed in SMA might be accelerated by a malfunction in this trophic support system. Schmalbruch (1988) observed loss of almost all motoneurons and two thirds of the sensory neurons in newborn rats after sciatic nerve section. He concluded that lack of neurotrophic input results in a loss of anterior horn cells. Sarnat et al. (1989) also concluded, based on a histochemical distribution of nucleic acids using fluorochrome acridine orange, that SMA was caused by a disturbance in the genetically coded mechanism which arrests the programmed physiological death of surplus motor neuroblasts at a certain critical time in embryonic life. The postnatal persistence of this cell death, which is normal in the embryonic state, leads to the pathological findings characteristic of SMA.

Fidzianska et al. (1990) performed muscle and sural nerve biopsies and later autopsies on 6-week-old infants who died 8 weeks after birth. They observed numerous spherical or ovoid bodies scattered within muscle cells. These bodies were not stained by PAS, oil red O, acid phosphatase or modified trichrome. Many appeared acidophilic, others basophilic. Electron microscopy revealed numerous immature neurons, a great number of single necrotic fibers and cellular fragments appeared as large round or oval membrane-bound bodies, identifiable as apoptotic bodies. The apoptotic bodies contained structurally intact myofibrils, or degraded myofibrils, or fibrillar or amorphous electron dense material. Kerr et al. (1972) proposed the term 'apoptosis' for cell death that appeared to play a role in animal cell populations. Fidzianska et al. speculated that in a severely growth retarded muscle, the process of muscle cell apoptosis removes the peripheral target of anterior horn cells resulting in secondary motoneuron death. Although further evaluation should be done to explain pathological extension into such regions as the thalamus and Clarke's column, these speculations may shed light on the pathogenesis of SMA.

NEUROPATHOLOGICAL ASPECTS

Changes in motor neurons

The most important change seen in WHD is degeneration of the anterior horn cells which correlates with the course of neurogenic muscular atrophy. WHD is characterized by prominent loss of spinal anterior horn cells. Various stages of degeneration can be seen in remaining motor neurons (Fig. 8a). In the early stage, the nucleus moves from its central position, the Nissl bodies move away from the center and take on a crumbled appearance, and the cytoplasm expands. The nucleus is flattened against the cell wall. There is also a reduction in the size and quantity of the Nissl bodies. The cytoplasm takes on a lightly staining, homogeneous colloidal appearance (central chromatolysis of unknown origin). Small clumps of granules, possibly degenerating organelles, may appear near the center of the cell. Eventually no Nissl bodies remain and the cytoplasm begins to shrink and disappear. Only light spaces outlined by neuropils (empty cell beds) remain (Fig. 8c). Small glial cells gather around the shrunken neurons until they have completely disappeared (neuronophagia). These changes are observable symmetrically at most levels of the spinal cord. This type of cellular

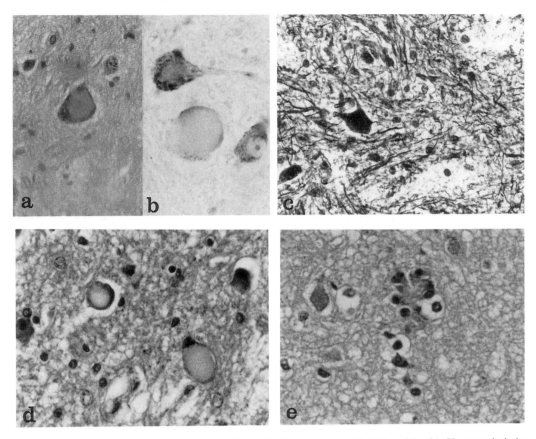

Fig. 8. (a) Chromatolysis in anterior horn cells at the thoracic level. (H & E, ×20). (b) Chromatolysis in Clarke's column. (Nissl, ×40). (c) Empty cell beds in the anterior horn. (Bielschowsky, ×17). (d) Chromatolysis in the thalamus. (H & E, ×20). (e) Neuronophagia in the thalamus. (E & E, ×20).

degeneration is generally observed when axons have been severed, suggesting that the degenerative changes are acute (Werdnig 1891, 1894; Hoffmann 1893, 1897; Conel 1938, 1940).

The same changes are commonly seen in the hypoglossal, ambiguus and facial nuclei. Some authors have also demonstrated changes in the lateral vestibular nuclei as well as the abducens, trochlear, and oculomotor nuclei (Conel 1940). The changes are more severe in lower cranial nerve nuclei.

It is significant that WHD neurons show various stages of degeneration simultaneously, not only within a single anatomical region but also in other areas. It is not uncommon to find normal neurons, chromatolysis and neuronophagia in the same specimen. This fact suggests that the neuronal degeneration in WHD may be due

to primary neuronal dysfunction rather than an extraneuronal factor.

Electronmicroscopic examination of a chromatolytic motoneuron demonstrates a neuronal soma packed with innumerable mitochondria, and a laterally displaced nucleus. Mixed with the mitochondria are small collections of neurofibrillary filaments, clumps of ribosomes without endoplasmic reticulum and small vesicles (Chou and Fakadej 1971). In addition, Fidzianska et al. (1984) reported that there are two different cell populations in WHD, i.e. a relatively immature population and one made up of cells in various stages of degeneration. They speculated that neuronal death might result from failure to make adequate peripheral contact in addition to genetic factor(s).

Sparing of the nucleus of Onuf in the anterior

horn at the S2 level of the spinal cord has been considered a typical finding in WHD. It correlates with clinical symptoms in that sphincter functions are not affected (Iwata and Hirano 1978a; Sung and Mastri 1980). However, Kumagai and Hashizume (1982) and Chou et al. (1982) have reported chromatolysis in the nucleus of Onuf.

The phrenic motoneurons are localized in the anteromedial zone at levels C3, C4 and C5. The preservation of these nuclei was reported by Kuzuhara and Chou (1981), explaining why diaphragmatic movement remains unaffected until the late stages of illness. On the contrary, a few cases of respiratory failure as an early manifestation of SMA have been reported (Mellins et al. 1974; McWilliam et al. 1985; Schapira and Swash 1985). Poets et al. (1990) reported an infant with near-miss sudden infant death syndrome caused by diaphragmatic paralysis, at the age of 7 months, who was later diagnosed as having WHD. Schapira and Swash (1985) suggested the possibility of heterogeneity in SMA I.

Changes in the spinal root

Pathological changes in spinal anterior roots are also important findings in WHD. The anterior roots are atrophic and there is selective loss of large myelinated fibers. Chou and Nonaka (1978) reported axonal degeneration in the proximal portions of anterior spinal roots. However, Ohama and Ikuta (1977) reported a lack of myelin ovoids and an absence of phagocytic activity. Prominent glial bundles in the proximal portions of the anterior roots disappear in a tapering pattern along the periphery (Fig. 9a,c). Electronmicroscopy reveals that each glial bundle is surrounded by basal lamina and composed of aggregates of cylindric processes. There are some desmosome-like junctional complexes between the glial bundles (Fig. 9e, f). The glial bundles are abundant in the cervical and lumbosacral regions. Small glial bundles are also occasionally observed in the posterior roots (Fig. 9b). Chou and Nonaka (1978) proposed that the glial bundles induce retrograde degeneration of the anterior horn cells.

However, this theory does not explain the lesions seen in other structures, such as the thalamus or Clarke's column, which are discussed in more detail below. Ghatak (1978) reported that astrocytic processes with abundant microtubules and prominent junctional devices, resembling those in the subpial region, appeared to have penetrated the ventral roots following axonal degeneration, and that such glial migration was a secondary finding. Iwata and Hirano (1978b) also concluded that the glial bundles were a secondary phenomenon because glial bundles were observed in three cases of paralytic anterior-poliomyelitis long after the occurrence of paralysis. Kimura and Budka (1984) reported that they found glial bundles in various diseases other than WHD and also concluded that this finding is non-specific.

Pathological changes other than those in motor neurons

Although there are no clinical sensory disturbances in WHD, spinal ganglion cells undergo changes nearly identical to those mentioned above, i.e. chromatolysis, residual nodules and a reduction in cell number. Some authors have reported that Clarke's column occasionally shows chromatolysis (Fig. 8b). These degenerative changes are not common, but can often be found if specimens are studied carefully enough.

Chromatolysis is evident in the thalamus as well, particularly in the lateral nucleus (Fig. 8g,e). Conel (1983) reported the distribution of affected nerve cells which is shown in Table 1. This distribution pattern includes the lateral nucleus

TABLE 1
Distribution of affected neurons (Conel 1938).

Ventral column of the spinal cord
Clarke's column
Spinal ganglions
Nucleus supraspinalis
Nucleus ambiguus
Nucleus of the hypoglossal nerve
Formatio reticularis
Lateral vestibular nucleus
Nucleus of the facial nerve
Nucleus of the abducens nerve
Nucleus of the trochlear nerve
Nucleus of the oculomotor nerve
Lateral nucleus of the thalamus

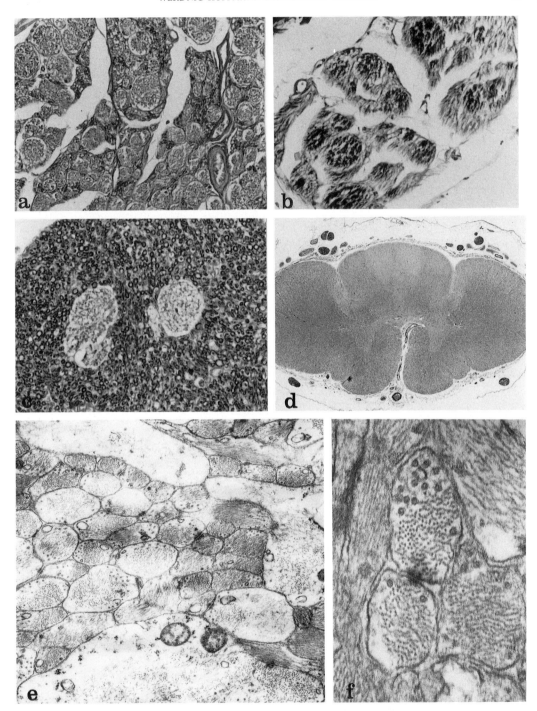

Fig. 9. (a) Cross section of sacral anterior roots reveals numerous glial bundles. (PTHA, ×17). (b) Glial bundles strongly positive with Holzer stain. (×20). (c) Small glial bundles present in posterior roots. (LFB+PAS, ×17). (d) Pallor in the posterior column. (LFB+PAS, ×10). (e), (f) Electronmicroscopy of glial bundles composed of aggregates of cylindric processes of glial fibrils (e) and junctional complexes (f). (e) ×21 000, (f) ×40 000.

TABLE 2
The main neurogenic muscular atrophies caused by defect of anterior horn cells.

	McKusick catalog no.	Mode of inheritance	Age of onset	Prognosis	Mental retardation	DTR	Joint contractures	Other findings
SMA								
Acute WHD (SMA I)	253300	A R	0–3 mos (<9 mos)	Die before 1 yr (≤4 yrs)	(−)	−	Occasionally, and transiently at neonatal period	Confined to bed
Intermediate (SMA II)	253550	A R	6–12 mos	Chronic progressive	(−)	↓/−		Able to maintain sitting posture
WHD c̄ involvement of diaphragm as an initial symptom? (Schapira and Swash 1985)		A R ?	0–4 mos	Poor	?	+	Multiple contractures of proximal joints and fingers	Paradoxical respiration not observed, enough limb movement
Norman's disease		A R	Birth	Progressive	(+) deterioration	+/↑		Involvement of anterior horn cell, brainstem thalamus, cerebellum cortical blindness
SMA c̄ microcephaly and mental subnormality (Spiro et al. 1967)	271110	A R	1 yr		(±)	↓/−		Ambulant (one case) hearing loss tremor
X-linked facio-scapulo-peroneal atrophy (Skre et al. 1978)		X R	Variable	Variable (1 yr to late (teens))				Weakness of shoulder girdle, upper limb, peroneal muscle, cranial nerves unaffected
Progressive distal muscular atrophy (Asano 1960)	253500	A R	≤1		(+)	↑		Impaired sensibility, dysarthria, choreic movements partial optic atrophy
Congenital suprabulbar paresis (Worster-Drought 1956, 1974)	185480	A D	Birth	Variable	+			Dysarthria, difficulty closing the mouth, epilepsy
Fazio-Londe disease	211500	A R	Infancy to childhood	Variable		↑		Diminished diaphragmatic motion, dysphagia, ptosis, facial weakness

(to be continued)

TABLE 2 (continued)

			Early	Slowly progressive	(−)	↑	Ankle	
Juvenile progressive bulbar palsy			Early teens					Daily fluctuation, dysphagia, ptosis, nasal voice, facial weakness
FADS due to defect of anterior horn cells (hydramnion, pulmonary hypoplasia, short umbilical cord, peculiar face, camptodactyly)								
Pena Shokeir I	208150	A R	Fetal onset	Poor			Club foot hip, knee,	
Pena Shokeir II (COFS)	214150	A R	Fetal onset	Poor			Multiple ankyloses	Microcephalus, eye defect
Finnish-type LCCS	253310	A R	Fetal onset	Lethal premature birth			Hip, knee	Fetal hydrops, generalized thinning of tubular bones, esp. ribs, pterygia at neck and elbows, thin spinal cord
LMPS (Chen et al. 1980; Hall et al. 1982)	253290	A R	Embryonal onset?	Poor			Hip, knee shoulder elbow	Multiple pterygia, cleft palate, cardiac hypoplasia
AMC Neurogenic AMC	218100	A R	Birth				Elbows knees	Total absence of muscle spindles (Krugliak et al. 1978)
X-linked AMC c̄ renal and hepatic abnormalities (Nezelof et al. 1979)	301820	X R	Neonatal period	Poor			Clubfoot low implantion of thumb	Jaundice, renal dysfunction
Distal AMC type I	108120	A D	Birth	Relatively good	(−)	+	Variable clubfoot, flexion contractures of fingers	Muscle weakness (−), elevated CK
X-linked distal AMC severe lethal (Hall et al. 1982 Greenberg et al. 1988)	301830	X R	Birth	Die ≤3 mos			Severe contractures	Confined to bed, SMA bone fractures

and Clarke's column, but the changes seen in these regions have not yet been thoroughly investigated. Later Gruner and Bargeton (1952) reported a reduction in the number of neurons, and the appearance of ghost cells and neuronophagia in the posteroventral thalamic nuclei in 9 cases of typical WHD. Recently, Iwata and Hirano (1978c), Shishikura et al. (1983) and Towfighi et al. (1985) have pointed out that the neuronal changes seen in the thalamus and Clarke's column are common findings in WHD. The lateral nuclei of the thalamus receive fibers from the spinothalamic tracts and posterior columns. Clarke's column receives fibers from posterior roots. Marshall and Duchenne (1975) reported myelin pallor in some cases of WHD, especially those with relatively long survival. Shishikura et al. (1986) also observed myelin pallor in a long-surviving case on a respirator (Fig. 9d). However, Maya et al. (1981) found no myelin pallor in such a patient in whom respiratory distress appeared rather later than expected for WHD. Further research should be conducted to clarify whether or not the sensory involvement correlates with progression of the disease process. According to Chou and Nonaka (1978), the total number of posterior root fibers was normal, but large myelinated fibers were decreased in number and small glial bundles were present. Ohama (1982), and Shishikura et al. (1983) also found small glial bundles in the posterior roots. Carpenter et al. (1978) performed sural nerve biopsies on WHD patients under 12 months of age and reported that all showed Wallerian degeneration. Shishikura (1984) reported axonal degeneration in one case. Based on the changes mentioned above, WHD is thought to be a multi-systemic disease involving both motor and sensory systems. Figure 10 illustrates the extent of the main pathological changes in the spinal cord which characterize WHD.

INTER-RELATIONSHIPS BETWEEN PATHO-
PHYSIOLOGICAL CHANGES AND CLINICAL
MANIFESTATIONS

Figure 11 illustrates the complex interrelationships between pathophysiological changes and the clinical manifestations of SMA I and II.

In WHD, degeneration of anterior horn cells leads to generalized neurogenic muscular atrophy and, consequently, to WHDs characteristic clinical features. SMA I shows generalized hypotonia and paralysis of the trunk, lower limbs, except toes and ankles, and upper limbs, though function distal to the elbow is preserved. Axial muscle weakness is so severe that most infants are never able to raise their heads or roll over. Some degree of limitation in external shoulder rotation is common causing the arms to lie in the 'jug handle' position, i.e. internally rotated with hands facing outward. Limitations in hip abduction and knee and elbow extension are also common but not in the distal joints as seen in arthrogryposis.

SMA II, as previously discussed, has a more benign course. Muscle power tends to remain static. An improvement is even possible, presumably due to reinnervation of atrophic muscle by surviving neurons. Some patients, however, experience a definite deterioration in muscle power which may be either gradual or episodic. Skeletal muscle fasciculation is not usually apparent but a fine tremor of the hands and fingers may be observable as a result of reinnervation (Fig. 12). Wheelchair-bound children are at risk for developing flexion contractures of the hips and knees, and scoliosis. Imbalance between peroneal and anterior tibial muscles leads to equinus foot deformities.

Severe weakness of the intercostal muscles with sparing of the diaphragm, reflects relative preservation of phrenic motor neurons in the anteromedial zone of the spinal cord at levels C3–C5 (Kuzuhara and Chou 1981). Breathing becomes almost entirely diaphragmatic, a pattern referred to as paradoxical respiration, with simultaneous abdominal distension and costal recession at inspirium, eventually leading to a bell-shaped chest deformity. Prognosis depends mainly on respiratory function. Extra body type negative pressure ventilators have been reported (Heckmatt et al. 1990; personal experience) to be ineffective in WHD. This may be due to difficulties in fitting the respirator gear exclusively over the thorax; if negative pressure is allowed to reduce abdominal distension, the compensatory function of paradoxical breathing is disturbed. Coughing and crying are weak and ineffectual.

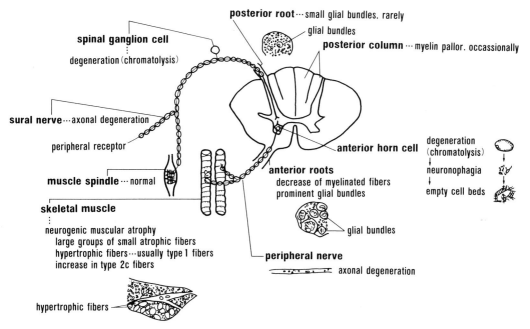

posterior root ··· small glial bundles. rarely

glial bundles

spinal ganglion cell

posterior column ··· myelin pallor. occassionally

degeneration (chromatolysis)

sural nerve ··· axonal degeneration

peripheral receptor

anterior horn cell

degeneration
(chromatolysis)

neuronophagia

empty cell beds

muscle spindle ··· normal

anterior roots
decrease of myelinated fibers
prominent glial bundles

skeletal muscle

neurogenic muscular atrophy
large groups of small atrophic fibers
hypertrophic fibers ··· usually type 1 fibers
increase in type 2c fibers

glial bundles

peripheral nerve
axonal degeneration

hypertrophic fibers

Fig. 10. The extent of the main pathological changes in the spinal cord.

Degeneration of cranial nuclei IX–XI is seen in SMA I and II. Hypoglossal nerve involvement, producing tongue fasciculation and atrophy (more remarkable in later stages of SMA II), is followed by degeneration of the accessory, vagus and glossopharyngeal nerves. The ensuing bulbar palsy results in dysphagia, dysphonia, hyperdiaphoresis and increased mucus secretion, seen after age 2 in SMA II and between 4 and 7 months in SMA I.

Cranial nuclei III–VIII degenerate in some long-surviving cases but no clinical symptoms have been observed to correlate with the pathological findings, possibly because patients do not live long enough for signs and symptoms to manifest. Progressive facial muscle weakness was observed from age 1.5 in an SMA I patient who survived beyond 6 years and facial fasciculation after age 5 in an SMA II patient (personal experience).

Anterior horn cell degeneration also interrupts reflex arcs, causing loss of deep tendon reflexes.

Atrophy of the anterior roots, secondary to horn cell degeneration, causes peripheral nerves to degenerate. The peripheral nerve damage in turn may lead to characteristic hand deformities

such as wrist drop, hyperextension of the MP joints and flexion of the IP joints.

Sensory degeneration may also be seen in SMA I and, possibly, II. Although pathological changes, including spinal ganglion cell chromatolysis, small glial bundles in posterior roots, thalamic chromatolysis, and myelin pallor of the posterior column, have been documented at autopsy and sural nerve degeneration has been demonstrated by biopsy, no corresponding clinical manifestations have been demonstrated. Perhaps because affected infants are too young to report sensory disturbances.

The myocardium is unaffected. Usually, intellect is normal as there are no pathological findings in the cerebral cortex, white matter or cerebellum. The exception is Norman's disease (Norman 1961), a WHD variant, which will be described later.

WHD VARIANTS

Pontocerebellar hypoplasia associated with
infantile motor neuron disease (Norman's disease)

In 1961, Norman reported a 6-month-old infant who had had generalized hypotonia, respiratory

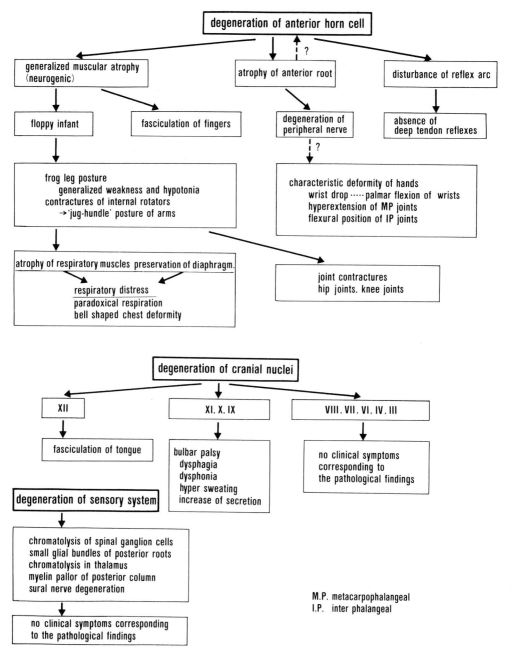

Fig. 11. The complex interrelationships between pathophysiological changes and clinical manifestations of the SMA I and II.

distress and feeding difficulties since birth. Mental deterioration, and adductor spasm of the legs appeared later. Autopsy revealed features of WHD in the spinal cord and brainstem, widespread thalamic degeneration, and symmetrical pallido-Luysial atrophy. Nerve cell losses in the thalamus and pallido-Luysial system were exceptionally severe, but microglial nodules and chromatolysis were features of WHD. The cerebellar folia were poorly developed and there was a loss of Purkinje and granule cells. Myelination was normal except in the thoracic and lumbar regions

SURFACE EMG

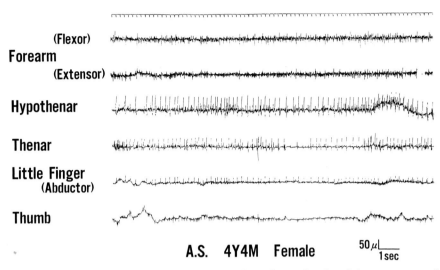

(Flexor)

Forearm

(Extensor)

Hypothenar

Thenar

Little Finger
(Abductor)

Thumb

A.S. 4Y4M Female $50\mu L\underline{}$
                                    1sec

Fig. 12. Surface EMG activity recorded from muscles of the finger, hand and forearm at rest. Note the rhythmic, repetitive firing indicative of fasciculation. Tremor-like movement of the fingers upon extension is characteristic of SMA II. Spiro (1970) described non-rhythmic jerky movements, which he termed 'minipoly-myoclonus', a useful diagnostic sign in SMA II.

where there was a well demarcated pallor in the corticospinal tracts. Norman and Kay (1965) reported another case, a 23-month-old infant, with this unusual variant of WHD. Spasticity rather than hypotonia was the salient feature, and the diagnosis of WHD was not suggested during life. Neuropathologic findings included acute changes in anterior horn cells, brainstem nuclei, and the thalamus, which were histologically typical of WHD. The cerebellum was small and showed mild diffuse atrophy of the granular layer, marked shrinkage of the Purkinje cells and gliosis in both the cortical layer and the white matter. The thalami showed an extensive loss of nerve cells with corresponding gliosis, particularly of the lateral nuclei.

Such a combination, though rare, has been reported by other authors, who considered it to be a separate disease entity. It has been called 'cerebellar hypoplasia in Werdnig-Hoffmann disease' (Weinberg and Kirkpatrick 1975), 'anterior horn cell disease associated with pontocerebellar hypoplasia' (Goutieres et al. 1977), 'amyotrophic cerebellar hypoplasia' (Leon et al. 1984), and 'pontocerebellar hypoplasia associated with in-

fantile motor neuron disease', also termed 'Norman's disease' (Kamoshita et al. 1990).

Steinman et al. (1980) reported 2 infants who clinically resembled WHD, but were found at autopsy to have a more extensive neuronal disorder, involving the thalamus, pons, cerebellum, and spinal cord. The cerebellar atrophy was mild. This disorder was thought to be distinct from WHD and was termed 'infantile neuronal degeneration'. Further evaluation is needed to determine whether this combination is a specific entity or a variant of WHD.

SMA with microcephaly and mental subnormality (271110)

Spiro et al. (1967) reported on a family in which 3, possibly 4, brothers were afflicted with a slowly progressive variant of SMA associated with subnormal mentality. The parents and 3 sisters were normal. The second brother, an 8-year-old, had suffered progressive muscle weakness since age 1. He had always had a waddling gait, never learned to ride a bicycle or run, and had difficulty rising from a recumbent position and climbing stairs. Bilateral hearing loss of unknown type, microcephalus, symmetrical proximal muscle atrophy, normal cranial nerve function except sternocleidomastoid, normal coordination with slight tremor, diminished or absent deep tendon reflexes, normal peripheral sensory function and a normal MCV were demonstrated. Neither myotonia nor fasciculations were present. The EMG and muscle biopsy were consistent with a neuropathy. The other 2 boys had similar signs and symptoms. The oldest brother, a 10-year-old, had more severe retardation and had never been ambulant. The clinical courses of these patients were similar to that of SMA II/KW except for microcephalus, hearing loss and subnormal mentality. The relationship of this disease to KW or SMA II parallels that of WHD and its variants.

Other spinal muscular atrophies

X-linked facio-scapulo-peroneal atrophy. Skre et al. (1978) reported an apparently XR neural muscular atrophy with clinical presentations ranging from mild shoulder girdle and peroneal weakness to paralysis extensive enough to be diagnosable as WHD. The proband noticed nocturnal drooling, forward displacement of the shoulders and protruding scapulae in his late teens. Progressive shoulder girdle, upper limb and peroneal muscle atrophy followed. A cousin had proximal muscle weakness around age 1 and WHD was diagnosed at 4. By age 20 there was total paralysis of the limb girdles and trunk and marked distal weakness. Cranial nerves I through X were unaffected. EMG and muscle biopsy pointed to a neurogenic etiology. The predominantly facio-scapulo-humoral and peroneal distribution of the atrophy, probable mode of inheritance and clinical variability, led the authors to propose an atypical SMA, possibly a WHD variant.

Congenital suprabulbar paresis (185480). Congenital suprabulbar paresis was originally described by Klippel and Weil (1909) in a 50-year-old man with subnormal intellect, speech difficulties, monotonous voice, slow bulbar movements and limited ability to chew. Family history indicated an autosomal inheritance pattern.

In the early 1950s Worster-Drought, studying this syndrome in a number of families described it as weakness or paralysis of the muscular structures innervated via the medulla or bulb. Involvement of the orbicularis oris muscle, tongue and soft palate, leading to dysarthria and difficulty closing the mouth, are present from birth. Patton and Brett (1986), in a follow-up of Worster-Drought's patients, suggested AD inheritance with variable expression and penetrance. Though the presentation does share features with WHD, such as hypotonia and feeding difficulties, there is also mental retardation and dysarthria.

Juvenile progressive bulbar atrophy (JPBA). Albers et al. (1983), reporting on 2 young women with JPBA, described the clinical and electrodiagnostic features which differentiate this disorder from other neuromuscular diseases. The symptoms, first noted in adolescence, included dysphagia, ptosis, speech difficulties and extreme facial weakness. Although JPBA primarily involves bulbar muscles, pontine, spinal and even midbrain motor neurons may be affected. Juvenile and childhood forms are far more rare than the adult form and there is considerable overlap with regard to onset.

In Fazio-Londe, the childhood form, brainstem motorneuron lesions are identical to those of WHD spinal anterior horn cells. Symptoms, which manifest between 2 and 12 years, include facial paralysis, dysarthria and dysphagia. The course can be either acute or chronic. Its predominantly bulbar symptoms distinguish it from WHD which presents with more severe lower limb involvement followed much later by bulbar symptoms.

Progressive distal muscular atrophy (253500). Asano et al. (1960) reported a progressive distal muscular atrophy which manifested before age 1. Unlike WHD, prominent features included choreic movements of the arms and face, partial optic atrophy, mental deficiency and increased deep tendon reflexes.

As mentioned by Dubowitz (1975), diseases which present as a floppy infant syndrome include congenital myotonic dystrophy, Prader-Willi syndrome, myasthenia gravis, congenital myopathies, congenital muscular dystrophies, and should be considered in the differential diagnosis of WHD. Once we diagnose the condition as neurogenic, a number of rare, early-onset, neurogenic muscular atrophies with clinical presentations similar to that of WHD which have been reported, must be considered in the differential diagnosis.

Fetal akinesia deformation sequence (FADS) due to defective anterior horn cells

Pena-Shokeir syndrome (208150). Pena and Shokeir (1974a) reported on 2 sisters who had hip and knee ankyloses, severe camptodactyly, clubfeet, characteristic facies and pulmonary hypoplasia. Facial anomalies included low set ears, hypertelorism, small mandible, and a depressed tip of the nose. Hydramnios, as opposed to the oligohydramnios seen in Potter's syndrome, had developed during both pregnancies. Punnet et al. (1974) reported similar cases. Subsequently, Pena and Shokeir (1976) reported 3 additional cases suggesting AR inheritance rather than a teratogenic etiology. The designation 'Pena-Shokeir I syndrome', has caused some confusion because Pena and Shokeir (1974b) also reported a similar syndrome, i.e. the cerebro-oculo-facial-skeletal (COFS) syndrome (214150). The latter is characterized by microcephaly, hypotonia, failure to thrive, arthrogryposis, eye defects, prominent nose, large ears, overhanging upper lips, micrognathia, widely spaced nipples, kyphoscoliosis, and osteoporosis. This constellation, clearly a different entity, has been called Pena-Shokeir II (Smith 1982).

The heterogeneity of Pena-Shokeir I syndrome, however, has been troublesome. Lazjuk et al. (1978) reported 4 cases with pulmonary hypoplasia, multiple ankyloses and camptodactyly. They analyzed the phenotypical pictures of their own cases and those in the literature and concluded that differences in the severity of pulmonary

hypoplasia and associated malformations suggest that this syndrome is a complex of related entities. Paralysis in rat fetuses, induced by daily transuterine injections of curare from day 18 of gestation until term, produces multiple anomalies; multiple joint contractures, pulmonary hypoplasia, micrognathia, fetal growth retardation, short umbilical cord, and hydramnios. These abnormalities are attributable to curare's paralytic effect (Moessinger 1983) and bear a striking resemblance to Pena-Shokeir I syndrome. The phenotype may actually represent a fetal akinesia deformation sequence (FADS), explaining the heterogeneity in the literature. Hall (1986) also pointed out interfamilial differences in prognosis and pathological findings.

Neuropathological findings include a marked decrease in anterior horn cells (Moerman et al. 1983; Chen et al. 1983), calcified vessels in the second and third cortical layers with spongiosa changes, loss of cerebellar Purkinje cells (Lindhout et al. 1985), septum pellucidum agenesis (Hageman et al. 1987), gross absence of cortical sulci and Rolandic fissures and complete loss of pyramidal cells with a cortical cytoarchitecture corresponding to 10–15 weeks' gestation (Bisceglia et al. 1987), hydroanencephaly (Lindhout et al. 1985; Hageman et al. 1987), and persistent fetal meningeal circulation (Hageman et al. 1987). Reported muscle abnormalities also indicate heterogeneity; neurogenic atrophy and/or fatty replacement (Pena and Shokeir 1974a; Moerman et al. 1983; Hageman et al. 1987), regeneration and type I fiber predominance suggesting myopathy (Mease et al. 1976), and immaturity with persistent myotubes (Bisceglia et al. 1987). Several case reports suggesting AR inheritance point to a genetic origin for akinesia in utero.

Prenatal diagnosis is possible by US examination (MacMillan et al. 1985; Muller and de Jong 1986; Ohlsson et al. 1986) in the second trimester. The diagnosis is based on hydramnios, restricted limb movements, decreased respiration-like movements of a small chest, arthrogryposis, clubfoot, camptodactyly, etc. Muller (1986) has pointed out the similarities between Pena-Shokeir I and trisomy 18, but so far chromosome analyses have been normal (Chen et al. 1983; Moerman et al. 1983; Bisceglia 1987; Honda et al. 1987; Yamada et al. 1987; Itoh et al. 1990), except one case with 46 XY q + (Dimmick et al. 1977).

Finnish-type lethal congenital contracture syndrome (LCCS) (253310). Herva et al. (1985) reported a disorder characterized by multiple congenital contractures, failure of postnatal survival, facial abnormalities, marked micrognathia, hypertelorism, short neck, and low set ears. The 16 cases belonged to 10 sibships, the grandparents of 8 of which were from neighboring communities. Neuropathologic findings of 5 fetuses, prenatally diagnosed by hydrops on US, included pulmonary hypoplasia, malpositioning of hips and knees, occasional pterygia of the neck and elbows, hypoplastic muscle with myotubes, spinal cord thinning, a paucity of anterior horn cells, and generalized thinning of tubular bones, especially the ribs (Herva et al. 1988). The authors speculated that the loss of axons and anterior horn motoneurons, in the brainstem as well as the spinal cord, pointed to a degenerative process rather than dysmorphogenesis. Controversy remains as to whether LCCS is the same as Pena-Shokeir I syndrome or not. Differential diagnosis (Herva et al. 1985, 1988) is based on the presence of massive fetal hydrops from the 15th week, lethality with a mean gestational age of 29 weeks and the absence of cerebral and/or cerebellar malformation or degeneration.

A lethal recessive syndrome with multiple pterygia (LMPS) (253290). A lethal recessive syndrome with multiple pterygia and/or cardiac hypoplasia as well as features in common with Pena-Shokeir I syndrome, including congenital contractures, hydramnios, intrauterine growth retardation, craniofacial abnormalities, prematurity, lung hypoplasia and lethality has been described (Chen et al. 1980, 1983; Hall et al. 1982a). This syndrome may represent a severe form of Pena-Shokeir I (Chen et al. 1983).

Davis and Kalousek (1988) identified 16 cases of FADS among 948 previable fetuses. In the 8 fetuses, ranging from 13 to 20 weeks in gestational age, with peripheral joint contractures, micrognathia and pulmonary hypoplasia, the akinesia was attributed to restriction of fetal movement. Chromosome analyses, done in 6 of

the 8, were normal. Among the other 8 fetuses, 2, aborted at 14 and 15 weeks, were identified as having LMPS which is characterized by joint contractures, particularly of the shoulders, hips, knees, elbows, and ankles, with prominent pterygium formation. Other anomalies such as fetal hydrops, posterior cystic cervical hygroma, cleft palate, and pulmonary and cardiac hypoplasia are also common. The 6 remaining fetuses, described as having FADS and early multiple joint contractures in both upper and lower limbs with pterygia, were only 8–9 weeks in developmental age. All 6 had short umbilical cords and short small intestines. Three also had bilateral cervical pterygium, microcephaly, and delayed or absent eyelid development. The other 3 formed a heterogeneous group. The presence of pterygia across immobilized joints in all of the 8–9-week fetuses led the authors to speculate that joint fixation and pterygium formation had resulted from joint immobility which began during the embryonic period. Fetal immobility, on the other hand, produced joint contractures without pterygia.

Arthrogryposis multiplex congenita (AMC)

Neurogenic AMC. Weissman et al. (1963) reported an arthrogryposis-like disorder characterized by flexion contractures at the elbows and knees but without hip dislocation. Lebenthal (1970), observing 23 kindred cases in an inbred Arab population, speculated that the disorder was myopathic in origin. Six of these 23 cases also had congenital heart disease.

Krugliak et al. (1978) autopsied 3 Bedouin Arab infants who had died shortly after birth and concluded that the etiology of the syndrome was neuropathic, terming it the Neurogenic form of AMC (208100). In all 3 cases contractures of knee and ankle joints were found and in 2 of them there were flexion contractures of the hip and IP joints. However, neither scoliosis nor cranial deformity were present. Case 1, whose parents were first cousins, died of respiratory distress at 6 h of age. There were no indications of bone fracture or congenital heart disease. Other organs were normal except for subcapsular hepatic hematomas. Case 2 had a fractured right humerus. Fresh hemorrhages were found. The authors asserted, on the basis of pathological evidence, that this form of AMC was neuropathic rather than myopathic. Depletion of spinal motor neurons not accompanied by gliosis, large groups of small muscle fibers, and replacement of muscle by fat varying from mild to marked in different areas were found at autopsy. There was no histological evidence of attempted reinnervation. In addition, a complete absence of muscle spindles was found in one infant. This was confirmed by examining several hundred serial paraffin sections taken from 36 samples from both upper and lower limb muscles, covering 11 different levels.

X-linked recessive AMC (301830).   Hall et al. (1982c) described 3 X-linked recessive forms of AMC. Type 1, termed severe lethal, was characterized by severe contractures, femoral fracture at birth, scoliosis, chest deformities, hypotonia, and characteristic facies. A decreased number of cells in the lateral anterior horn was confirmed at autopsy in one case who died due to respiratory failure at 3 months of age. Greenberg et al. (1988) reported 4 similar cases of apparently X-linked recessive (XR) infantile SMA. Three of these infants had fractures, of both femurs and the humerus in 2 cases and 1 femur in another case, and all had contractures, areflexia, and hypotonia at birth. In 2 of the pregnancies decreased fetal movement had been noted. Three of the boys died before age 2. Though 3 infants were not autopsied, so anterior horn cell disease could not be confirmed, muscle biopsy demonstrated neurogenic atrophy in all 4. One had a normal sural nerve biopsy and EMG evidence of a neurologic process, suggestive of a progressive SMA. Analysis of this patient's pedigree strongly suggested XR inheritance. The authors speculated that their cases constituted a distinct clinicopathological entity based on the presence of contractures and fractures at birth. It was concluded that XR infantile SMA can be distinguished from the autosomal form on the basis of family history and the presence of contractures and fractures at birth. Greenberg et al. (1988) has suggested that the existence of this type of SMA may explain the predominance of male

patients among cases of SMA I which has led some authors to propose the possibility of the existence of a phenocopy of acute WHD.

Type II, termed moderately severe, was characterized by severe contractures, bilateral ptosis, cryptorchidism, inguinal hernia and normal intelligence and might have been due to a nonprogressive intrauterine myopathy.

Type III, termed resolving, was characterized by mild contractures at birth which subsequently improved dramatically and may have been due to light connective tissue or misplaced tendons (Hall et al. 1982c).

X-linked AMC with hepatic and renal dysfunction (301820).   Nezelof et al. (1979) described a probably XR AMC in four brothers of North African descent, all of whom died in infancy, with a unique combination of pigmented liver disease, cholestatic jaundice and renal dysfunction. All 4 had AMC, developed jaundice within days or weeks, failed to thrive and died of a hemorrhagic disorder (eldest boy), respiratory failure (second) or infection (third and fourth). Abnormal hepatic and renal functions were documented in the 2 youngest cases. A history of similar disorders in several males in the mother's family pointed to XR inheritance. Autopsy of the 2 youngest cases demonstrated extreme atrophy of peripheral muscles, rarefaction and changes in motor neurons in spinal anterior horn cells, capsulo-synovial fibrosis of the joints, renal tubular cell degeneration with nephrocalcinosis, and granular deposits of pigment within liver cells. The combination of hepatic, renal and neurologic abnormalities led the authors to propose that a primary hepatic defect in conjugation or elimination of a toxic endogenous metabolite might have given rise to the renal and neurologic abnormalities.

Distal AMC (108120; 108130; 108140).   Hall et al. (1982b), reporting on a distal AMC, concluded that there is at least one autosomal dominant (AD) form. The cases of a father and his 2 daughters (Daentl 1974) and an affected father, son and daughter (McCormack 1980) were cited. Both forms of familial AMC were characterized by clubfoot deformities, flexion contractures of the fingers, and normal intelligence. In addition, members of the former family had hip dislocation, inguinal hernia, small mandible, limited ankle, knee, elbow, wrist and shoulder motion, short neck and elevated CK. The latter showed some intrafamilial variability in terms of hand deformities, ulnar deviation of the fingers, and mild muscle weakness but all had small stature, narrow shoulders and delayed carpal ossification. This type of AMC can be distinguished from other forms by its AD inheritance, characteristic hand, finger and foot deformities and fairly good response to physical therapy.

Hageman et al. (1984) reviewed 4 cases of distal AMC including a Dutch boy with congenital myopathy with core-like structures, a girl (age 8) with congenital hypertrophic neuropathy and perinatal-in-origin encephalopathy, a girl (age 11) with encephalopathy of prenatal origin and mild axonal neuropathy, and a girl (age 4) with anterior horn cell degeneration and encephalopathy of both pre- and perinatal origin. The author asserted that AMC, whether distal or not, is not a clinical entity but rather a manifestation of various cerebral, neuromuscular and connective tissue disorders.

OTHER RARE CONDITIONS WHICH SHOULD BE CONSIDERED IN THE DIFFERENTIAL DIAGNOSIS

Infant botulism can resemble acute WHD. Botulinal toxin blocks acetylcholine release producing hypotonia and a symmetrical, descending flaccid paralysis. The spectrum of clinical presentations ranges from very mild symptoms to death generally due to acute respiratory failure, which is virtually indistinguishable from SIDS.

Epidemiologic studies (Spika et al. 1989) have implicated honey and corn syrup consumption as well as chronic constipation as risk factors for infant botulism. For infants less than 2 months of age living in a rural, farming community appeared to be a risk factor. *Clostridium botulinum* spores have been isolated from up to 25% of honey (Sugiyama et al. 1978; Midura et al. 1979; Huhtanen et al. 1981) and in 0.5% of corn syrup samples (Kautter et al. 1982). However, Spika et al. (1989) speculated that chronic constipation, i.e. less than one bowel movement every

3 days, was the most significant risk factor. They suggested that *Clostridium botulinum* may proliferate in the bowel under these conditions, be absorbed, and ultimately lead to symptoms. In addition retrograde movement of the organism and/or its toxin into the ileum may occur.

A careful history can help to distinguish botulism from other neuromuscular disorders. Most affected infants are under 6 months and were previously healthy. Almost all have a recent history of constipation, sleepiness, lethargy, listlessness, decreased appetite, weak crying and reduced spontaneous movement. Cranial nerve involvement and fatigability are initially subtle. Dysphagia, manifesting as drooling, may occur. As bulbar muscles become increasingly affected, ptosis and loss of head control become evident. Physical exam may reveal easy fatigability, with suck and gag reflexes being diminished by repeated stimulation (Arnon 1980). Decreased anal sphincter tone and an impaired pupillary light reflex (shining a bright light into the eye repetitively for 1–3 min produces increasingly sluggish pupillary constriction) are often present (Arnon 1980). Though generalized flaccid paralysis usually takes a few days to develop, some infants have severe weakness within hours of onset.

Two EMG patterns have been demonstrated in infant botulism cases (Clay et al. 1977). The most common is brief, small, abundant motor-unit action potentials. Post-tetanic facilitation patterns have also been recorded. Definitive diagnosis depends upon identification of *Clostridium botulinum* in fecal or autopsy specimens. Therapy is supportive and generally followed by gradual recovery.

An infant with an *inborn error of metabolism in the leucine degradation pathway (210200)* was reported by Eldjarn et al. (1970) with signs of delayed motor development, hypotonia, muscular atrophy, hyperdiaphoresis, tongue fasciculations and absent deep tendon reflexes. Unpleasant smelling urine pointed to a metabolic etiology. Two abnormal metabolites, β-hydroxyisovaleric acid and β-methyl-crotonylglycine, were found in the urine.

Incontinentia pigmenti (IP; Bloch-Sulzberger syndrome) (308300), the hallmark of which is whorled pigmentary skin lesions early in life, can present with subnormal mentality, seizure disorders and malformations. One case of IP, a newborn with generalized weakness, hypotonia, large vesicular-bullous lesions on the legs and decreased fetal movement, associated with anterior horn cell degeneration has been reported (Larsen et al. 1987). Autopsy revealed a significant decrease in spinal anterior horn cells, loss of Nissl substance, and central chromatolysis, as in WHD.

Infantile neuroaxonal dystrophy (Seitelberger disease) (256600). Sporadic cases of infantile neuroaxonal dystrophy, a degenerative encephalopathy which usually becomes symptomatic in adolescence, have also been reported. Affected infants may present with hypotonia, poor sucking and decreased deep tendon reflexes. In contrast to WHD, these cases have hypothalamic and infundibular involvement. The case of Nagashima et al. (1985), had hypothalamic hypothyroidism and diabetes insipidus. Mental retardation, and epilepsy were present in the 2 siblings reported by Crome and Weller (1965). Hunter et al. (1987) termed this clinical picture prenatal or connatal neuroaxonal dystrophy.

Phenylketonuria (PKU) (261600). Some of the many typical findings of PKU include disturbances of gait, stance and sitting posture. Meier et al. (1975) reported a boy who developed tremor of the extremities and inability to remain upright, after age 1, which subsided with dietary therapy. Though this clinical presentation could be confused with SMA II, PKU should be suspected from family history, mental retardation, and reduced pigmentation. Hyperphenylalaninemia is confirmatory.

MANAGEMENT

Medical management of WHD is essentially supportive: adequate nutrition, prevention of respiratory complications, scoliosis, contractures, accidents and obesity, and psychosocial support. Vitamin E, 800–1000 IU daily, is recommended as many infants have low plasma levels (Shapira et al. 1981).

Respiratory therapy is primarily a parental responsibility. Percussion and postural drainage clear secretions, reducing the risks of bronchial infection and atelectasis. Scoliosis reduces vital capacity so its prevention is an integral part of maintaining respiratory function.

Scoliosis should be aggressively managed in its earliest stages, especially in SMA II. Wheelchair-bound patients can generally wear light, plastic orthoses. Bracing is more difficult in mobile patients who may move in ways requiring spinal flexibility, e.g. flexing and extending their legs while sitting on the floor and sliding on the buttocks. Some authors (Evans et al. 1981; Shapiro and Bresnau 1982) recommend surgery for severe scoliosis but its feasibility depends on individual prognosis. A customized wheelchair with adequate back support can prevent kyphosis. Frequent positional changes and discouraging sole use of the dominant hand can also retard the development of scoliosis. Parents should be encouraged to do physical therapy, i.e. stretching exercises, to maintain range of motion and prevent flexion contractures. Infants with extensive paralysis must be turned frequently to prevent lung atelectasis and decubiti, though the latter are rarely seen in SMA even in long-surviving cases (personal experience). As paralysis worsens, metabolic and vascular derangements such as nephrolithiasis, postural hypotension and constipation, become frequent. Clinicians must be alert to the possibility of life-threatening complications like thrombophlebitis, pulmonary embolism and pneumonia.

Genetic counselling has been offered to couples with an affected child, the risk of recurrence in future siblings being one in four, or, if the SMA is XR, 50% for males. Couples must consider how much they want more children, the prognosis of their affected child and coping with a possibly protracted course in future offspring. Though prenatal diagnosis will make this a moot point for many couples, counselling will still be part of a physician's overall management of families touched by SMA.

Finally, physicians must be a constant source of emotional support to families living with the stress of having a severely handicapped child with a terminal illness.

CONCLUSION

The aim of this chapter has been to review the clinical features, pathology and differential diagnosis of WHD. In addition we have touched upon recent advances in molecular genetics which will undoubtedly necessitate extensive revision of this text in the upcoming decade. Although hope of some form of gene therapy now exists, we must not forget that careful management and prevention of complications are still the mainstays of treatment for WHD.

Acknowledgements
The authors are very grateful to Profesor Yukio Fukuyama (our head at Tokyo Women's Medical College), Dr. Paul Fleury (Amsterdam) and Professor Harvey B. Sarnat (Calgary) for their invaluable advice. The authors would like to thank Mrs. Mesako Hirano for typing the manuscript.

REFERENCES

ALBERS, J. W., S. ZIMNOWODZKI, C. M. LOWREY and B. MILLER: Juvenile progressive bulbar palsy: Clinical and electrodiagnostic findings. Arch Neurol. 40 (1983) 351–353.

ARNON, S. S.: Infant botulism. Annu. Rev. Med. 31 (1980) 541–560.

ASANO, N., M. KIZU, T. YAMADA, N. ASANO and C. KIJIMA: A peculiar type of progressive muscular atrophy. J. Hum. Genet. 5 (1960) 139–146.

BISCEGLIA, M., L. ZELANTE, C. BOSMAN, R. CERA and B. DALLAPICCOLA: Pathologic features in two siblings with the Pena-Shokeir I syndrome. Eur. J. Pediatr. 146 (1987) 283–287.

BOUWSMA, G. and N. J. LESCHOT: Unusual pedigree patterns in seven families with spinal muscular atrophy, further evidence for the allelic model hypothesis. Clin. Genet. 30 (1986) 145–149.

BRZUSTOWICZ, L. M., T. LEHNER, L. H. CASTILLA, G. K. PENCHASZADEH, K. C. WILHELMSEN, R. DANIELS, K. E. DAVIES, M. LEPPERT, F. ZITER, D. WOOD', V. DUBOWITZ, K. ZERRES, I. HAUSMANOWA-PETRUSEWICZ, J. OTT, T. L. MUNSAT and T. C. TILLAM: Genetic mapping of chronic childhood-onset spinal muscular atrophy to chromosome 5q11.2–13.3. Nature 344 (1990) 540–541.

BULCKE, J. A.: Commentary: Ultrasound and CT scanning in the diagnosis of neuromuscular disease. In: I. Gamstorp and H. B. Sarnart (Eds.), Progressive Spinal Muscular Atrophies. New York, Raven Press (1984) 153–162.

BYERS, R. K. and B. Q. BANKER: Infantile muscular atrophy. Arch. Neurol. (Chic.) 5 (1961) 140–164.

CARPENTER, S., G. KALPATI, S. ROTHMAN, G. WATTERS and F. ANDERMANN: Pathological involvement of primary sensory neurons in Werdnig-Hoffmann disease. Acta Neuropathol. (Berl.) 42 (1978) 91–97.

CARRIER, N. N. and G. BERHILLIER: Carnitine levels in normal children and in patients with diseased muscle. Muscle Nerve 3 (1980) 326–334.

CHEN, H., C. G. CHANG, R. P. MISRA, H. A. PETERS, N. S. GRIJALVA and J. M. OPITZ: Multiple pterygium syndrome. Am. J. Med. Genet. 7 (1980) 91–102.

CHEN, H., B. BLUMBERG, L. IMMKEN, R. LACHMAN, D. RIGHTMIRE, M. FOWLER, R. BACHMAN and F. A. BEEMER: The Pena-Shokeir syndrome: report of five cases and further delineation of the syndrome. Am. J. Med. Genet. 16 (1983) 213–224.

CHOU, S. M. and A. V. FAKADEJ: Ultrastructure of chromatolytic motoneurons and anterior spinal roots in a case of Werdnig-Hoffmann disease. J. Neuropathol. Exp. Neurol. 30 (1971) 368–379.

CHOU, S. M. and I. NONAKA: Werdnig-Hoffmann disease: Proposal of a pathogenetic mechanism. Acta Neuropathol. 41 (1978) 45–54.

CHOU, S. M., S. KUZUHARA and I. NONAKA: Involvement of the onuf nucleus in Werdnig-Hoffmann disease. Neurology 32 (1982) 880–884.

CLAY, S. A., C. RAMSEYER, L. S. FISHMAN and R. P. SEDGWICK: Acute infantile motor unit disorder. Infantile Botulism? Arch. Neurol. 34 (1977) 246–249.

CONEL, J. L.: Distribution of affected nerve cells in a case of amyotonia congenita. Arch. Neurol. Psychiatry 40 (1938) 337–351.

CONEL, J. L.: Distribution of affected nerve cells in amyotonia congenita (Second case). Arch. Pathol. 30 (1940) 153–164.

CROME, L. and S. D. V. WELLER: Infantile neuroaxonal dystrophy. Arch. Dis. Child. 40 (1965) 502–507.

CUNNINGHAM, M.: Werdnig-Hoffmann disease. The effects of intrauterine onset on lung growth. Arch. Dis. Child. 53 (1978) 921–925.

DAENTL, D. L., B. O. BERG, R. B. LAYZER and C. J. EPSTEIN: A new familial arthrogryposis without weakness. Neurology 24 (1974) 55–60.

DAVIS, J. E. and D. K. KALOUSEK: Fetal akinesia deformation sequence in previable fetuses. Am. J. Med. Genet. 29 (1988) 77–87.

DIMMICK, J. E., K. BERRY, P. M. MACLEOD and D. F. HARDWICK: Syndrome of ankylosis, facial anomalies, and pulmonary hypoplasia: A pathologic analysis of one infant. Birth Defect 13 (1977) 133–137.

DUBOWITZ, V.: Infantile spinal muscular atrophy. Handbook of Clinical Neurology, Vol 22. System Disorders and Atrophies — Part II (1975) 81–101.

DUBOWITZ, V.: Muscle Disorders in Childhood. Philadelphia, Saunders (1978).

ELDJARN, L., E. JELLUM, O. STOKKE, H. PANDE and P. E. WAALER: β-Hydroxyisovaleric aciduria and β-methylcrotonylglycinuria; a new inborn error of metabolism. Lancet Sept 5 (1970) 521–522.

EVANS, G. A., J. C. DRENNAN and B. S. RUSSMAN: Functional classification and orthopaedic management of spinal muscular atrophy. J. Bone Joint Surg. 63-B (1981) 516–522.

FARRINGTON, F. W. and J. W. CORCORAN: Werdnig-Hoffmann disease — Infantile progressive muscular atrophy — report of case. J. Dent. Child. 42 (1975) 49–52.

FIDZIANSKA, A.: Ultrastructural changes in muscle in spinal muscular atrophy — Werdnig-Hoffmann's disease. Acta Neuropathol. (Berl.) 27 (1974) 247–256.

FIDZIANSKA, A.: Morphological differences between the atrophied muscle fiber in amyotrophic lateral sclerosis and Werdnig-Hoffmann disease. Acta Neurol. Pathol. (Berl.) 34 (1976) 321–327.

FIDZIANSKA, A., J. RAFAXOWSKA and A. GLINKA: Ultrastructural study of motoneurons in Werdnig-Hoffmann disease. Clin. Neuropathol. 3 (1984) 260–265.

FIDZIANSKA, A., H. H. GOEBEL and I. WARLO: Acute infantile spinal muscular atrophy. Brain 113 (1990) 433–445.

FRIED, K. and G. MUNDEL: High incidence of spinal muscular atrophy type I (Werdnig-Hoffmann disease) in the Kararite community in Israel. Clin. Genet. 12 (1977) 250–251.

FRIED, K. and A. E. H. EMERY: Spinal muscular atrophy type II. A separate genetic and clinical entity from type I (Werdnig-Hoffmann disease) and type III (Kugelberg-Welander disease). Clin. Genet. 2 (1971) 203–209.

GHATAK, N. R.: Spinal roots in Werdnig-Hoffmann disease. Acta. Neuropathol. 41 (1978) 1–7.

GILLIAM, T. C., L. M. BRZUSTOWICZ, L. H. CASTILLA, T. LEHNER, G. K. PENCHASZADEH, R. J. DANIELS, B. C. BYTH, J. KNOWLES, J. E. HISLOPS, Y. SHAPIRA, V. DUBOWITZ, T. L. MUNSAT, J. OTT and K. E. DAVIES: Genetic homogeneity between acute and chronic forms of spinal muscular atrophy. Nature 345 (1990) 823–825.

GOUTIERES, F., J. AICARDI and E. FARKAS: Anterior horn cell disease associated with pontocerebellar hypoplasia in infants. J. Neurol. Neurosurg. Psychiatry 40 (1977) 370–378.

GREENBERG, F., K. R. FENOLIO, J. F. HEJTMANCIK and D. ARMSTRONG: X-Linked infantile spinal muscular atrophy. Am. J. Dis. Child. 142 (1988) 217–219.

GRUNER, G. E. and E. BARGETON: Lesions thalamiques dans la myatonie du nourrisson. Rev. Neurol. 65 (1952) 46–52.

HAGEMAN, G., F. G. I. JENNEKENS, J. K. VETTE and J. WILLEMSE: The heterogeneity of distal arthrogryposis. Brain Dev. 6 (1984) 273–283.

HAGEMAN, G., J. WILLEMSE, B. A. VAN KETEL, P. G. BARTH and D. LINDHOUT: The heterogeneity of the Pena-Shokeir syndrome. Neuropediatrics 18 (1987) 45–50.

HALL, J. G.: Analysis of Pena-Shokeir phenotype. Am. J. Med. Genet. 25 (1986) 99–117.

HALL, J. G., S. D. REED, K. N. ROSENBAUM, J. GERSHANIK, H. CHEN and K. M. WILSON: Limb pterygium syndromes: a review and report of eleven patients. Am. J. Med. Genet. 12 (1982a) 377–409.

HALL, J. G., S. D. REED and G. GREENE: The distal arthrogryposes; delineation of new entities—review and nosologic discussion. Am. J. Med. Genet. 11 (1982b) 185–239.

HALL, J. G., S. D. REED, C. I. SCOTT, J. G. ROGERS, K. L. JONES and A. CAMARANO: Three distinct types of X-linked arthrogryposis seen in 6 families. Clin. Genet. 21 (1982c) 81–97.

HAUSMANOWA-PETRUSEWICZ, I.: Infantile and juvenile spinal muscular atrophy. In: J. N. Walton, N. Canal and G. Scarlato, (Eds.), Muscle Diseases. Proc. Internat. Congr., Milan 1969. Amsterdam, Excerpta Medica ICS No. 199 (1970) 558–567.

HAUSMANOWA-PETRUSEWICZ, I. and A. FIDZIANSKA-DOLOT: Clinical features of infantile and juvenile spinal muscular atrophy. In: I. Gamstorp and H. B. Sarnat (Eds.), Progressive Spinal Muscular Atrophies: International Review of Child Neurology Series. New York, Raven Press (1984) 31–42.

HAUSMANOWA-PETRUSEWICZ, I. and A. KARWANSKA: Electromyographic findings in different forms of infantile and juvenile proximal spinal muscular atrophy. Muscle Nerve 9 (1986) 37–46.

HAUSMANOWA-PETRUSEWICZ, I., J. PROT and E. SAWICKA: Le probleme des formes infantiles et juveniles de l'atrophie musculaire spinale. Rev. Neurol. 114 (1966) 295–306.

HECKMATT, J. Z. and V. DUBOWITZ: Diagnosis of spinal muscular atrophy with pulse echo ultrasound imaging. In: I. Gamstorp and H. B. Sarnat (Eds.), Progressive Spinal Muscular Atrophies. New York, Raven Press (1984) 141–152.

HECKMATT, J. Z., V. DUBOWITZ and S. LEEMAN: Detection of pathological change in dystrophic muscle with B-scan ultrasound imaging. Lancet 1 (1980) 1389–1390.

HECKMATT, J. Z., L. LOH and V. DUBOWITZ: Night-time nasal ventilation in neuromuscular disease. Lancet 335 (1990) 579–582.

HENDERSON, C. E. and M. FARDEAU: Les facteurs de croissance nerveuse: une hypothèse sur leur role dans la pathogénie des amyotrophies spinales infantiles. Rev. Neurol. (Paris) 144 (1988) 730–736.

HERVA, R., J. LEISTI, P. KIRKINEN and U. SEPPANEN: A lethal autosomal recessive syndrome of multiple congenital contractures. Am. J. Med. Genet. 20 (1985) 431–439.

HERVA, R., N. G. CONRADI, H. KALIMO, J. LEISTI and P. SOURANDER: A syndrome of multiple congenital contractures; neuropathological analysis on five fetal cases. Am. J. Med. Genet. 29 (1988) 67–76.

HOFFMANN, J.: Ueber chronische spinale Muskel-atrophie im Kindesalter auf familiarer Basis. Dtsch. Z. Nervenheik. 3 (1893) 427–470.

HOFFMANN, J.: Weitere Beiträge zur Lehre von der hereditären progressiven spinalen Muskelatrophie im Kindesalter. Dtsch. Z. Nervenheilk. 10 (1897) 292.

HONDA, T., J. ABE, H. YOSHIZAWA, K. KIMURA, N. MOTANI, N. TAKAUCHI, S. TAKEUCHI and M. ISHIDA: Two cases of Pena-Shokeir syndrome. Acta. Neonat. Jpn. (in Japanese) 23 (1987) 662–669.

HORIKAWA, H., M. KONAGAYA, T. TAKAYANAGI and H. OTSUJI: The muscle CT of thigh in chronic Werdnig-Hoffmann disease. Clin. Neurol. (in Japanese) 26 (1986) 490–497.

HUHTANEN, C. N., D. KNOX and H. SHIMANUKI: Incidence and origin of *Clostridium botulinum* spores in honey. J. Food Protect. 44 (1981) 812–813.

HUNTER, A. G. W., C. L. JIMENEZ, B. F. CARPENTER and I. MACDONALD: Neuraxonal dystrophy presenting with neonatal dysmorphic features, early onset of peripheral gangrene, and a rapidly lethal course. Am. J. Med. Genet. 28 (1987) 171–180.

ITOH, K., N. YOKOYAMA, A. ISHIHARA, S. KAWAI, S. TAKADA, M. NISHINO, Y. LEE, H. NEGISHI and H. ITOH: A case of Pena-Shokeir syndrome. Acta Neonatol. Jpn. (in Japanese) 26 (1990) 468–473.

IWATA, M. and A. HIRANO: Sparing of the Onufrowicz nucleus in sacral anterior horn lesions. Ann. Neurol. 4 (1978a) 245–259.

IWATA, M. and A. HIRANO: 'Glial bundles' in the spinal cord late after paralytic anterior poliomyelitis. Ann. Neurol. 4 (1978b) 562–563.

IWATA, M. and A. HIRANO: A neuropathological study of the Werdnig-Hoffmann disease (in Japanese). Neurol. Med. 8 (1978c) 40–53.

KAMOSHITA, S., Y. TAKEI, M. MIYAO, M. YANAGISAWA, S. KOBAYASHI and K. SAITO: Pontocerebellar hypoplasia associated with infantile motor neuron disease (Norman's disease). Pediatr. Pathol. 10 (1990) 133–142.

KAUTTER, D. A., T. LILLY, H. M. SOLOMON and R. K. LYNT: *Clostridium botulinum* spores in infant foods: a survey. J. Food Protect. 45 (1982) 1028–1029.

KELLEY, R. I. and J. T. SLADKY: Dicarboxylic aciduria in an infant with spinal muscular atrophy. Ann. Neurol. 20 (1986) 734–736.

KERR, J. F. R., A. H. WHYLLIE and A. R. CURRIE: Apoptosis: a basic biological phenomenon with wide-ranging implications in tissue kinetics. Br. J. Cancer 26 (1972) 239–257.

KIMURA, T. and H. BUDKA: Glial bundles in spinal nerve roots. An immunocytochemical study stressing their non-specificity in various spinal cord and peripheral nerve diseases. Acta Neuropathol. 65 (1984) 46–52.

KLIPPEL, M. M. and M. PIERRE-WEIL: Syndrome labio-glosso-laryngé pseudo-bulbaire héréditaire et familial. Rev. Neurol. (Paris) 17 (1909) 102–103.

KRUGLIAK, L., N. GADOTH and A. J. BEHAR: Neuropathic form of arthrogryposis multiplex congenita: report of 3 cases with complete necropsy, including the first reported case of agenesis of muscle spindles. J. Neurol. Sci. 37 (1978) 179–185.

KUMAGAI, T. and Y. HASHIZUME: Morphological and morphometric studies on the spinal cord lesion in Werdnig-Hoffmann disease. Brain Dev. 4 (1982) 87–96.

KUZUHARA, S. K. and S. M. CHOU: Preservation of the phrenic motoneurons in Werdnig-Hoffmann disease. Ann. Neurol. 9 (1981) 506–510.

LARSEN, R., S. ASHWAL and N. PECKHAM: Incontinentia pigmenti; association with anterior horn cell degeneration. Neurology 37 (1987) 446–450.

LAZJUK, G. I., E. D. CHERSTVOY, I. W. LURAIE and M. K. NEDZVED: Pulmonary hypoplasia, multiple ankyloses and camptodactyly: one syndrome or some related forms? Helv. Paediatr. Acta 33 (1978) 73–79.

LEBENTHAL, E., S. B. SHOHET, A. ADAM, M. SEELENFREUND, A. FRIED, T. NAJENSON, U. SANDBANK and Y. MATOTH: Arthogryposis multiplex congenita — 23 cases in an Arab kindred. Pediatrics 46 (1970) 891–899.

LEON, G. A. DE, W. D. GROVER and C. A. D'CRUZ: Amyotrophic cerebellar hypoplasia. Acta. Neuropathol. 63 (1984) 282–286.

LINDHOUT, D., G. HAGERMAN, G. A. BEEMER, P. F. IPPEL, L. BRESLAU-SIDERIUS and J. WILLEMSE: The Pena-Shokeir syndrome: report of nine Dutch cases. Am. J. Med. Genet. 21 (1985) 655–668.

MACMILLAN, R. H., G. M. HARBERT, W. D. DAVIS and T. E. KELLY: Prenatal diagnosis of Pena-Shokeir syndrome type. 1. Am. J. Med. Genet. 21 (1985) 279–284.

MARSHALL, A. and L. W. DUCHEN: Sensory system involvement in infantile spinal muscular atrophy. J. Neurol. Sci. 26 (1975) 349–359.

MAYA, K., K. INOUE and A. HIRANO: Pathological findings of a prolonged case of the Werdnig-Hoffmann disease. An autopsied case of 12-year-old boy. Neurol. Med. 14 (1981) 243–225.

MCCORMACK, M. K., P. J. COPPOLA-MCCORMACK and M.-L. LEE: Autosomal-dominant inheritance of distal arthrogryposis. Am. J. Med. Genet. 6 (1980) 163–169.

MCWILLIAM, R. C., D. GARDNER-MEDWIN, D. DOYLE and J. B. P. STEPHENSON: Diaphragmatic paralysis due to spinal muscular atrophy. Arch. Dis. Child. 60 (1985) 145–149.

MEASE, A. D., G. W. YEATMAN, G. PELTELL and G. B. MERENSTEIN: A syndrome of ankylosis, facial anomalies and pulmonary hypoplasia secondary to fetal neuromuscular dysfunction. Birth Defects Orig. Art. Ser. XII (5) (1976) 193–200.

MEIER, C., J. LUTSCHG and F. A. BISCHOFF: Progressive neural muscular atrophy in a case of phenylketonuria. Dev. Med. Child Neurol. 17 (1975) 625–630.

MELKI, J., S. ABDELHAK, P. SHETH, M. F. BACHELOT, P. BURLET, A. MARCADET, J. AICARDI, A. BAROIS, J. P. CARRIERE, M. FARDEAULL, D. FONTAN, G. PONSOT, T. BILLETTE, C. ANGELINI, C. BARBOSA, G. FERRIERE, G. LANZILL, A. OTTOLINI, M. C. BABRON, D. COHEN, A. HANAUER, F. CLERGET-DARPOUX, M. LATHROP, A. MUNNICH and J. FREZEL: Gene for chronic proximal spinal muscular atrophies maps to chromosome 5q. Nature (Lond.) 344 (1990a) 767–768.

MELKI, J., P. SHETH, S. ABDELHAK, P. BURLET, M.-F. BACHELOT, M. G. LATHROP, J. FREZAL and A. MUN-

NIGH: Mapping of acute (type I) spinal muscular atrophy to chromosome 5q12–q14. Lancet 336 (1990b) 271–273.

MELLINS, R. B., A. P. HAYS, A. P. GOLD, W. E. BERDON and J. D. BOWDLER: Respiratory distress as the initial manifestation of Werdnig-Hoffmann disease. Pediatrics 53 (1974) 33–40.

MIDURA, T. F., S. SNOWDEN, R. M. WOOD and S. S. ARNON: Isolation of *Clostridium botulinum* from honey. J. Clin. Microbiol. 9 (1979) 282–283.

MOERMAN, P. H., J. P. FRYNS, P. GODDEERIS and J. M. LAUWERYNS: Multiple ankyloses, facial anomalies, and pulmonary hypoplasia associated with severe antenatal spinal muscular atrophy. J. Pediatr. 103 (1983) 238–241.

MOESSINGER, A. C.: Fetal akinesia deformation sequence: an animal model. Pediatrics 72 (1983) 857–863.

MULLER, L. M. and G. DE JONG: Prenatal ultrasonographic features of the Pena-Shokeir I syndrome and the trisomy 18 syndrome. Am. J. Med. Genet. 25 (1986) 119–129.

NAGASHIMA, K., S. SUZUKI, E. ICHIKAWA, S. UCHIDA, T. HONMA, T. KUROUME, J. HIRATO, A. OGAWA and Y. ISHIDA: Infantile neuroaxonal dystrophy; perinatal onset with symptoms of diencephalic syndrome. Neurology 35 (1985) 735–738.

NEZELOF, C., M. C. DUPART, F. JAUBERT and E. ELIACHAR: A lethal familial syndrome associating arthrogryposis multiplex congenita, renal dysfunction, and a cholestatic and pigmentary liver disease. J. Pediatr. 94 (1979) 258–260.

NORMAN, R. M.: Cerebellar hypoplasia in Werdnig-Hoffmann disease. Arch. Dis. Child. 36 (1961) 91–101.

NORMAN, R. M. and J. M. KAY: Cerebello-thalamo-spinal degeneration in infancy: An unusual variant of Werdnig-Hoffmann disease. Arch. Dis. Child. 40 (1965) 302–308.

OHAMA, E.: Some aspects of anatomy and pathology of nerve roots. Adv. Neurol. Sci. 26 (1982) 737–752.

OHAMA, E. and F. IKUTA: The morphopathogenesis of Werdnig-Hoffmann disease. Presence of axons and Schwann cells of fetal type. Neurol. Med. (in Japanese) 6 (1977) 494–501.

OHLSSON, A., K. W. FONG, T. H. ROSE and D. C. MOORE: Prenatal sonographic diagnosis of Pena-Shokeir syndrome type I, or fetal akinesia deformation sequence. Am. J. Med. Genet. 29 (1988) 59–65.

OKADA, S., I. NONAKA and S. M. CHOU: Muscle fiber type differentiation and satellite cell population in normally growing and neonatally denervated muscles in the rat. Acta Neuropathol. (Berl.) 65 (1984) 90–98.

PASCALET-GUIDON, M.-J., E. BOIS, J. FEINGOLD, J.-F. MATTEI, J. C. COMBES and C. HAMON: Cluster of acute infantile spinal muscular atrophy (Werdnig-Hoffmann disease) in a limited area of Reunion Island. Clin. Genet. 26 (1984) 26.

PASCUAL CASTROVIEJO, I.: Commentary: Clinical aspects of spinal muscular atrophy. In: I. Gamstorp and H. B. Sarnat (Eds.), Progressive Spinal Muscular Atrophies. New York, Raven Press (1984) 43–54.

PATTON, M. A. and E. M. BRETT: A family with congenital suprabulbar paresis (Worster-Drought syndrome). Clin. Genet. 29 (1986) 147–150.

PEARN, J. H., C. O. CARTER and J. WILSON: The genetic identity of acute infantile spinal muscular atrophy. Brain 96 (1973) 463–470.

PENA, S. D. J. and M. H. K. SHOKEIR: Syndrome of camptodactyly, multiple ankyloses, facial anomalies and pulmonary hypoplasia: a lethal condition. J. Pediat. 85 (1974a) 373–375.

PENA, S. D. J. and M. H. K. SHOKEIR: Autosomal recessive cerebro-oculo-facio-skeletal (COFS) syndrome. Clin. Genet. 5 (1974b) 285–293.

PENA, S. D. J. and M. H. K. SHOKEIR: Syndrome of camptodactyly, multiple ankyloses, facial anomalies and pulmonary hypoplasia — further delineation and evidence for autosomal recessive inheritance. Birth Defects Orig. Art. Ser. XII(5) (1976) 201–208.

POETS, C., R. HEYER, H. V. D. HARDT and G. F. WALTER: Akute respiratorische Insuffizienz als klinische Erstmanifestation der spinalen Muskelatrophie. Monatsschr. Kinderheilkd. 138 (1990) 157–159.

PUNNETT, H. H., M. L. KISTENMACHER, M. VALDES-DAPENA and R. T. ELLISON, JR.: Syndrome of ankylosis, facial anomalies and pulmonary hypoplasia. J. Pediatr. 85 (1974) 375–377.

RYNIEWICZ, B. and M. PAWINSKA: Immunological studies in SMA. Eur. J. Pediatr. 128 (1978) 57–62.

SAITO, Y.: Muscle fiber type differentiation and satellite cell population in Werdnig-Hoffmann disease. J. Neurol. Sci. 68 (1985) 75–87.

SARNAT, H. B., P. JACOB and C. JIMENEZ: Atrophie spinal musculaire: l'évanouissement de la fluorescence à l'arn des neurones moteurs en dégénérescence: Une étude à l'acridine-orange. Rev. Neurol. (Paris) 145 (1989) 305–311.

SAWCHAK, J. A., B. BENOFF, J. H. SHER and S. A. SHAFIQ: Werdnig-Hoffmann disease: myosin isoform expression not arrested at prenatal stage of development. J. Neurol. Sci. 95 (1990) 183–192.

SCHAPIRA, D. and M. SWASH: Neonatal spinal muscular atrophy presenting as respiratory distress: A clinical variant. Muscle Nerve 8 (1985) 661–663.

SCHMALBRUCH, H.: The effect of peripheral nerve injury on immature motor and sensory neurons and on muscle fibres; Possible relation to the histogenesis of Werdnig-Hoffmann disease. Rev. Neurol. (Paris) 144 (1988) 721–729.

SHAPIRA, Y., R. AMIT and E. RACHNILEWITZ: Vitamin E deficiency in Werdnig-Hoffmann disease. Ann. Neurol. 10 (1981) 266–268.

SHAPIRO, F. and M. BRESNAN: Current concepts review: Orthopaedic management of childhood neuromuscular disease. J. Bone Joint Surg. 64-A (1982) 785–789.

SHISHIKURA, K.: Peripheral nerves in Werdnig-Hoffmann disease. Neuropathology 5 (1984) 325–336.

SHISHIKURA, K., M. HARA, Y. SASAKI and K. MISUGI: A

neuropathologic study of Werdnig-Hoffmann disease with special reference to the thalamus and posterior roots. Acta. Neuropathol. 60 (1983) 99–106.

SHISHIKURA K., Y. ARAI, K. SUGITA, K. YAZIMA, M. OSAWA, H. SUZUKI, Y. FUKUYAMA, J. SATO and M. HARA: Sensory system involvement in Werdnig-Hoffmann disease type I. Xth International Congress of Neuropathology, Stockholm, 1986.

SKRE, H., S. I. MELLGREN, P. BERGSHOLM and J. E. SLAGSVOLD: Unusual type of neural muscular atrophy with a possible X-chromosomal inheritance pattern. Acta Neurol. Scand. 58 (1978) 249–260.

SMITH, D. W.: Recognizable Patterns of Human Malformations, 3rd ed. Philadelphia, W. B. Saunders (1982) 136.

SNOW, M. H.: A quantitative ultrastructural analysis of satellite cells in denervated fast and slow muscles of the mouse. Anat. Rec. 207 (1983) 593–604.

SPIKA, J. S., N. SHAFFER, N. HARGRETT-BEAN, S. COLLIN, K. L. MACDONALD and P. A. BLAKE: Risk factors for infant botulism in the United States. Am. J. Dis. Child. 143 (1989) 828–832.

SPIRO, A. J.: Minipolymyoclonus. A neglected sign in childhood spinal muscular atrophy. Neurology (Minneap.) 20 (1970) 1124–1134.

SPIRO, A. J., M. H. FOGELSON and A. C. GOLDBERG: Microcephaly and mental subnormality in chronic progressive spinal muscular atrophy of childhood. Dev. Med. Child. Neurol. 9 (1967) 594–601.

STALBERG, E. and P. R. W. FAWCETT: Electrophysiological methods for the study of the motor unit in spinal muscular atrophy. In: I. Gamstorp and H. B. Sarnat (Eds.), Progressive Spinal Muscular Atrophies: International Review of Child Neurology Series. New York, Raven Press (1984) 111–134.

STEINMAN, G. S., L. B. RORKE and M. J. BROWN: Infantile neuronal degeneration masquerading as Werdnig-Hoffmann disease. Ann. Neurol. 8 (1980) 317–324.

SUGIYAMA, H., D. C. MILLS and L. J. C. KUO: Number of Clostridium botulinum spores in honey. J. Food Protect. 41 (1978) 848–850.

SUNG, J. H. and A. R. MASTRI: Spinal autonomic neurons in Werdnig-Hoffmann disease, mannosidosis, and Hurler's syndrome: Distribution of autonomic neurons in the sacral spinal cord. J. Neuropathol. Exp. Neurol. 39 (1980) 441–451.

TOWFIGHI, J., R. S. K. YOUNG and R. M. WARD: Is Werdnig-Hoffmann disease a pure lower motor neuron disorder? Acta Neuropathol. 65 (1985) 270–280.

VISSER, M. DE and B. VERBEETEN JR: Computed tomography of the skeletal musculature in Becker-type muscular dystrophy and benign infantile spinal muscular atrophy. Muscle Nerve 8 (1985) 435–444.

WALSH, F. S., S. E. MOORE and B. D. LAKE: Cell adhesion molecule N-CAM is expressed by denervated myofibres in Werdnig-Hoffmann and Kugelberg-Welander type spinal muscular atrophies. J. Neurol. Neurosurg. Psychiatry 50 (1987) 439–442.

WEINBERG, A. G. and J. B. KIRKPATRICK: Cerebellar hypoplasia in Werdnig-Hoffmann disease. Dev. Med. Child. Neurol. 17 (1975) 511–516.

WEISSMAN, S. L., C. KHERMOSH and A. ADAM: Arthrogryposis in an Arab family. In: E. Goldschmidt (Ed.), Genetics of Migrant and Isolate Populations. Baltimore, Williams and Wilkins (1963) 313.

WERDNIG, G.: Zwei frühinfantile hereditäre Fälle von progressiver Muskelatrophie unter dem Bilde der Dystrophie, aber auch neurotischer Grundlage. Arch. Psychiat. Nervenkr. 22 (1891) 437–481.

WERDNIG, G.: Die frühinfantile progressive spinale Amyotrophie. Arch. Psychiat. Nervenkr. 26 (1894) 706–744.

WORSTER-DROUGHT, C.: Congenital suprabulbar paresis. J. Laryng. Otol. 70 (1956) 453–463.

WORSTER-DROUGHT, C.: Suprabulbar paresis and its differential diagnosis with special reference to acquired suprabulbar paresis. Dev. Med. Child Neurol. 16 (Suppl.) (1974) 1–33.

YAMADA, T., H. NISHIDA, T. ARAI, K. NOSE, M. FUKUDA, K. IGUCHI and S. SAKAMOTO: A case of Pena-Shokeir I syndrome. Acta Neonatol. Jpn. (in Japanese) 23 (1987) 883–887.

Handbook of Clinical Neurology, Vol. 15 (59): Diseases of the Motor System
J.M.B.V. de Jong, editor
© Elsevier Science Publishers B.V., 1991

CHAPTER 6

Wohlfart-Kugelberg-Welander disease

STEPHAN ZIERZ[1] and KLAUS ZERRES[2]

[1]*Neurologische Universitätsklinik and* [2]*Institut für Humangenetik der Universität Bonn, Bonn, FRG*

Wohlfart (1942), Wohlfart et al. (1955) and Kugelberg and Welander (1954, 1956) drew attention to a hereditary disease with proximal muscular atrophy and weakness in young adults clinically resembling progressive proximal muscular dystrophy but with fasciculations. EMG and muscle biopsy proved a neurogenic origin of the disease.

DEFINITION

The Wohlfart-Kugelberg-Welander disease represents an inherited group of chronic childhood spinal muscular atrophy that is characterized by a progressive disease of anterior horn cells with initial involvement of proximal muscle groups. In contrast to acute infantile spinal muscular atrophies (type Werdnig-Hoffmann) and to spinal muscular atrophies of adult onset, the age of onset ranges from about 3 years to the third decade of life. As in other forms of spinal muscular atrophies there is no evidence of pyramidal tract involvement and signs of sensory deficits are absent. The peripheral motor and sensory nerve conduction velocities are normal. It is, however, debated if the Wohlfart-Kugelberg-Welander disease represents a homogeneous nosological entity, especially because there seem to be different modes of inheritance in families with this disease. Different attempts at classification of the chronic childhood proximal spinal muscular atrophies have been proposed.

CLASSIFICATION

Classification of spinal muscular atrophies is still controversial. Most classification systems are clinically orientated with classification parameters such as age of onset, ability to sit or walk, and age of death. Table 1 summarizes the characteristics of common classification systems. Most classifications differentiate at least between infantile, intermediate, juvenile, and adult types. The groups defined in these classifications, however, differ significantly, especially with respect to the intermediate and juvenile types. Cases with the Wohlfart-Kugelberg-Welander disease are comprised under the intermediate, the juvenile, and the early adult types. Thus, the Wohlfart-Kugelberg-Welander disease presenting with a chronic course would cover an age of onset between 6 months and 15–20 years. Even with this very wide definition, however, the distinction of more severely affected and adult cases can sometimes be difficult.

A severe problem for an adequate classification is the determination of the prognosis of the disease. It is a frequently encountered problem that patients with prolonged survival are differently classified at different stages of their

TABLE 1

Different classifications of proximal spinal muscular atrophy. Cases with Wohlfart-Kugelberg-Welander disease are mostly comprised in the intermediate and juvenile types and in the early adult types.

Author	Classification parameters	Infantile SMA	Intermediate SMA	Juvenile SMA	Adult SMA	
Pearn (1980)	Type	I	Chronic childhood	SMA	IV a.r.	a.d.
	Age of onset	95% <3 mo	<3 y	3–8 y	15–50 y	25–50 y
	Life expectancy	mean: 7 mo	18 mo to adulthood	variable	normal	reduced
Emery (1971)	Type	I	II	III	IV a.r./a.d./x-rec.	
	Age of onset	<12 mo	3–18 mo	>2 y	>30 y	
	Life expectancy	<4 y	>4 y	adulthood	about 50 y	
Harding (1984)	Type	I (acute infantile)	II (chronic childhood)	IV (juvenile onset)	III a.r.	V a.d.
	Age of onset	<6 mo	3 mo–15 y	6 mo–15 y	15–60 y	25–65 y
	Life expectancy	7–18 mo	18 mo–40 y	normal?	normal	20 y after onset
Bundey and Brett (1985)	Type	I	II	III	IV a.r.	IV a.d.
	Age of onset	<5 mo	<2 y	>2 y	15–50 y	20–50 y
	Life expectancy	<3 y	adulthood	adulthood	normal	reduced
Hausmanowa-Petrusewicz (1989)	Type	Ia Ib	II	III		
	Age of onset	at birth	2–6 y	2–15 y		
	Life expectancy	2–4 y >10 y	long	long		

Abbreviations: a.d.: autosomal dominant, a.r.: autosomal recessive, X-rec.: X-linked recessive, mo: months, y: years.

lives, changing from the Werdning-Hoffmann type to the intermediate type and finally to the Wohlfart-Kugelberg-Welander type. Thus, age of onset of the disease alone does not sufficiently allow a prediction of the prognosis for single patients. We are aware of patients who presented with severe hypotonia at birth and who reached the 4th decade of life. Despite all these difficulties a clinically oriented classification is necessary for practical reasons.

Because of its common use in the literature Emery's classification is outlined in detail. Emery (1971) defined 4 different types of proximal spinal muscular atrophy (Table 2) including the juvenile and adult autosomal dominant types as well as the X-linked type Kennedy. Kennedy's disease, however, is today regarded as a separate entity different from classical proximal spinal muscular atrophy (Harding 1984; Warner et al. 1990). The considerable clinical variability in proximal spinal atrophy makes the precise classification of a single case with respect to the course of the

disease nearly impossible. This is reflected by the fact that several different clinical classification systems have been proposed. Recent results in the molecular biology of proximal spinal muscular atrophy indicate that most cases with proximal spinal muscular atrophy represent a broad clinical spectrum probably due to different mutations of a single gene.

GENETICS

Proximal spinal muscular atrophy is a genetically heterogeneous condition with different types following different modes of inheritance.

Autosomal recessive inheritance (253400)

The vast majority of cases with affected persons within one sibship follows an autosomal recessive mode of inheritance. There are, however, numerous families with affected persons that show

TABLE 2
Classification of proximal spinal muscular atrophies according to Emery (1971). Cases with Wohlfart-Kugelberg-Welander disease are comprised in group III.

Type		Age		Ability to sit without support
		onset	death	
I	Infantile (severe) autosomal recessive	< 12 months	< 4 years	Never
II	Intermediate autosomal recessive (?)	3–18 months	> 4 years	Usually
III	Juvenile (relatively benign) 1. autosomal recessive 2. autosomal dominant	> 2 years	adulthood	Always
IV	Adult 1. autosomal recessive 2. autosomal dominant 3. X-linked recessive	> 30 years	adulthood	Always

TABLE 3
Genetic studies on chronic childhood spinal muscular atrophies.

Study	Number of patients	Interpretations of the results
Brandt (1951)	52 families	most cases transmitted by a.r. genes with some dominant spontaneous mutations
Winsor et al. (1971)	60 families	a.r.
Bundey and Lovelace (1975)	33 families	Most cases a.r., single cases a.d., spontaneous mutations among the cases of later onset
Emery et al. (1976)	376 sibships	a.r., a few sporadic cases possibly due to new dominant mutations
Pearn et al. (1978a) Pearn (1978a)	124 cases	a.r. in 75% of cases. Spontaneous mutations of phenocopies in late-onset cases
Hausmanowa-Petrusewicz et al. (1985)	354 cases	Age of onset 3–9 months: a.r. Age of onset 10–36 months: a.d., spontaneous mutations or phenocopies, age of onset 37 months–18 years: a.r.
Zerres (1989)	243 cases	Never able to walk: a.r. Able to walk, age of onset 6 months–3 years a.d., spontaneous mutations or phenocopies in a number of cases

Abbreviations: a.r.: autosomal recessive, a.d.: autosomal dominant.

other modes of inheritance. On the basis of segregation analyses, 7 out of 8 studies on the genetics of proximal spinal muscular atrophies showed a significant deviation from the assumption of autosomal recessive inheritance (Table 3). The genetic interpretation of these cases with a chronic course deviating from the assumption of recessive inheritance might be that there is a

certain amount of autosomal dominant spontaneous mutations or phenocopies. There is no
evidence for non-genetically determined phenocopies.

Autosomal dominant inheritance (158600)

Pearn (1978b) defined two different forms of
autosomal dominant proximal spinal muscular
atrophy: (i) a childhood form with an onset of
disease between birth and the age of 8 years and
a benign course; and (ii) an adult form with age
of onset not before 20 years and with a more
rapid progress of symptoms. Both forms are very
rare. The first form accounts for about 2–4% of
all cases with childhood onset, whereas the
second form represents about 30% of cases with
adult onset. Most likely, the penetrance of both
forms is complete. Our own experience with 6
families with this type of proximal spinal muscular atrophy shows that in 3 families the affected
persons show variable courses of the disease that
does not allow a clear distinction between the 2
forms (Rietschel et al. 1990).

X-linked recessive inheritance

Besides Kennedy's type of bulbospinal muscular
atrophy (313200) there is no conclusive evidence
for the existence of an X-linked proximal spinal
muscular atrophy. In several families that have
previously been diagnosed as spinal muscular
atrophy the detection of deletions in the
dystrophin gene led to the final diagnosis of
muscular dystrophy (Clarke et al. 1989; Lunt et
al. 1989). Greenberg et al. (1988) described a
family with probably X-linked proximal spinal
muscular atrophy of very early onset that cannot
be classified so far.

Unusual pedigrees

There are pedigrees with two or more affected
persons most often in different branches of the
pedigree, which cannot be explained by 'simple'
Mendelian modes of inheritance without additional postulations (Zerres and Grimm 1983;
Bouwsma and Leschot 1986; Zerres et al. 1987).
Becker (1964) postulated the 'allelic model' as a

possible explanation: Affected persons of these
families should carry two alleles a^+ and a' (a^+:
'pathogenic' allele, a': 'modifying' allele) at the
same disease locus. Healthy parents usually carry
a normal allele a besides a mutant allele a^+ or
a'. The existence of different alleles a', a" etc.
might explain the different manifestations within
these families. In these sibships the recurrence
risk for further affected siblings is 25%, which is
in accordance with 'normal' autosomal recessive
inheritance. In contrast to the 'typical' autosomal
recessive mode of inheritance with a low gene
frequency of usually one mutated gene Becker's
model postulates a rare frequency of the allele
a^+ and a high frequency of the a' allele. This
model might explain why in a family with the
a^+ allele affected persons can be found in
different parts of the pedigree. It is possible that
one spouse independently carries a' as well, if
this gene frequency is high in the population.
Based on 6 of our own and on 43 pedigrees
reported in the literature the gene frequency of
the hypothetical allele a' is about 10% (Zerres
1989). The risk for children of other family
members is usually low (most often below 5%)
but still higher than in 'classical' autosomal
recessive inheritance.

Genetic counselling

Unless the genetic identification of a single case
is clearly possible, genetic counselling may be
very difficult. In cases with a positive family
history of one affected parent autosomal dominant inheritance can usually be assumed. Cases
with very early onset (acute fatal spinal muscular
atrophy, Werdnig-Hoffmann type) as well as
adult cases without family history are considered
to be recessively inherited. Baraister (1990) summarized the empirical risk figures for genetic
counselling as follows: If the age of onset is
before 3 years and if the proband has a chronic
form of proximal spinal muscular atrophy (type
II or type III according to Emery's classification)
the risk for another affected sib is 1 in 5. If the
onset is later than 3 years the risk for a sib is 1
in 10. A similar risk exists for sibs of patients
with an age of onset later than 5 years but in
this group there may be new dominant mutations

with a risk to offspring of about 1 in 10. In patients with an age of onset between 10 and 36 months (Hausmanowa-Petrusewicz et al. 1985) and in patients with onset of the disease before 3 years of life who were able to walk (Zerres 1989), the risk for siblings might be much lower.

Future identification of the mutation on the DNA basis might make genetic counselling more accurate. As long as the genetic heterogeneity is questionable and unless information on the DNA level of at least 2 affected siblings or of parental consanguinity is available, prenatal diagnosis on the DNA level should be made available with caution only, especially in those families with milder forms of proximal spinal muscular atrophy.

Molecular biology

Recently the gene for acute, intermediate, and juvenile proximal spinal muscular atrophy has been mapped to chromosome 5q (Brzustowicz et al. 1990; Melki et al. 1990a,b; Gilliam et al. 1990). The combined lod scores were 11.41 (Gilliam et al. 1990) and 11.08 (theta 0.03) (Melki et al. 1990b). Gilliam et al. (1990) mentioned 2 families who did not show evidence for linkage with markers on chromosome 5q. These data indicate that a genetic homogeneity exists between the familial cases of acute and chronic forms of proximal spinal muscular atrophy. With the exception of one study on the acute form of proximal spinal muscular atrophy (Melki et al. 1990b) all other studies investigated families with more than one affected sibling or with parental consanguity. Thus, the results are consistent with an autosomal recessive mode of inheritance but so far the linkage data do not exclude genetic heterogeneity (Zerres et al. 1990). Therefore, further linkage studies are necessary.

CLINICAL FEATURES

The clinical features and severity of the Wohlfart-Kugelberg-Welander disease are very variable. Symptoms may be first noted between 3 years and the third decade. Before the onset of the disease all patients are able to walk. Usually, proximal muscle weakness begins gradually, but sudden exacerbations, especially associated with intercurrent infections or vaccination are also possible. Sometimes the first symptoms are observed after confinement to bed because of other diseases. Periods of immobilization seem to have an adverse effect on the course of the disease. In the beginning typically the muscles around the hips are affected causing difficulties in walking and running. Later on the patients show difficulties in climbing stairs and a tendency to fall. Gait becomes waddling, arising from the ground might be performed using Gowers' maneuver. The distribution of affected muscles is symmetrical and accompanied by clear atrophy. About a fourth of the patients exhibit a hypertrophy of some muscle groups, especially of the calves (Namba et al. 1970; Bouwsma and van Wijngaarden 1980) (Fig. 1). So far, it has not been established whether this is a real hypertrophy of the calf muscles to compensate for the loss of function of other weakened muscles, whether it is a pseudohypertrophy due to an increase of fatty connective tissue as it is seen in Duchenne muscular dystrophy or whether the atrophic thigh muscle causes the calves to appear hypertrophic. The disappearance of the calf hypertrophy after prolonged immobilization seems to support the hypothesis of a compensatory hypertrophy (Bouwsma and van Wijngaarden 1980). The tonus of affected muscle groups is decreased. In the later course of the disease muscle groups of the shoulder and proximal upper limbs are affected causing difficulties in raising the arms as in combing one's hair, and winged scapulae may be noted. Movement of hand and fingers are usually spared even in advanced stages. In about a third of the patients tremor of the fingers can be observed.

Fasciculations are found in about 30–50% of the patients, predominantly in the tongue, shoulder girdle and upper arm muscles and are more noticeable in adolescence and early adult life than later on (Namba et al. 1970). Fasciculations seem to occur more often in males than in females, probably because they are easier to detect due to less subcutaneous fatty tissue in males. Fasciculations of the eyelids may be an additional clue to the clinical diagnosis of spinal muscular atrophy, but so far they have only

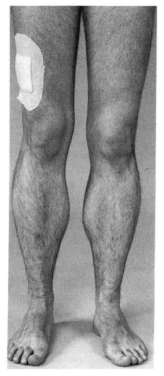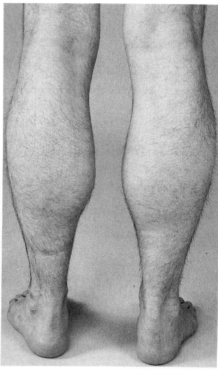

Fig. 1. Wohlfart-Kugelberg-Welander disease with proximal atrophy of the legs and hypertrophy of the calves.

been documented in patients with the severe Werdnig-Hoffmann form and with the 'intermediate' form (Skouteli and Dubowitz 1984). *Painful muscle cramps* are also frequently encountered complaints (Zerres 1989).

Reflexes

Tendon reflexes are diminished or absent in some or all muscles, particularly in the lower limb. Rarely, some tendon reflexes are hyperactive although other reflexes are diminished or absent. In contrast to distal spinal muscular atrophy and to peripheral neuropathy the ankle jerk is preserved until late, corresponding to the usually good preservation of the calf muscle. Babinski's reflex may occur in 3–9% of the patients (Namba et al. 1970; Gardner-Medwin et al. 1967). Therefore, the presence of extensor plantar reflexes alone does not necessarily contradict the diagnosis of spinal muscular atrophy.

Contractures

As the disease progresses more muscles become weak or paralytic and contractures of the hip, knee, elbows, and ankle may develop. Foot deformations, however, can also be observed at an early stage of the disease. Compensatory lordotic posture of the spine may result in severe scoliosis and thoracic malformations (Eng et al. 1984).

Cranial nerve involvement

About one third of patients with chronic proximal spinal muscular atrophy also showed weakness of muscles innervated by cranial nerves. The cranial nerve manifestations include mild dysphagia, which is, however, not severe enough to interfere with adequate food intake, weakness and fasciculations of the tongue, dysarthria and nasal voice, weakness of the facial and masseter muscles, weakness and wasting of the sternoclei-domastoid, the neck, and the trapezius muscles, ptosis and bilateral ophthalmoplegia (Gardner-Medwin et al. 1967; Aberfeld and Namba 1969; Namba et al. 1970; Wallar et Reece, 1978; Gruber et al. 1983; Barois et al. 1989; Zerres 1989).

Sensory involvement

The absence of sensory disturbance is considered to be an indispensable diagnostic criterion for spinal muscular atrophy. Rare reports on minimal sensory disturbances in distal parts of the limbs in chronic progressive spinal muscular atrophy are rather tenuous (Winder and Auer 1989; Tsukagoshi et al. 1965).

Cardiac involvement

Rarely, cardiac involvement has been reported in patients with Wohlfart-Kugelberg-Welander disease. There are no specific pathological features but usually a congestive cardiomyopathy can be found (Okazaki et al. 1976; Gilbert 1987). It is, however, difficult to understand how the motor neuron disease could be pathogenetically linked to a cardiomyopathy. Some of the cases with cardiomyopathy probably represent the X-linked form of scapuloperoneal myopathy associated with cardiomyopathy described by Rotthauwe et al. (1972) and Thomas et al. (1972).

Other complications

Complications reported in patients with Wohlfart-Kugelberg-Welander disease include diabetes mellitus, dwarfism, and convulsions (Namba et al. 1970). These complications, however, are very rare and it is possible that they are merely coincidental. Patients with Wohlfart-Kugelberg-Welander disease do not show intellectual deficits. Rare cases with spinal muscular atrophy and mental retardation are mentioned in the section 'Differential diagnosis' of this chapter. Gynaeco-mastia has been frequently reported in patients with spinal muscular atrophy, particularly in Kennedy's bulbospinal muscular atrophy and in the X-linked adult form of spinal muscular atrophy (Warner et al. 1990; Hausmanowa-

Petrusewicz et al. 1983) indicating an imbalance of the estrogen-androgen system (see below).

Course of the disease

Usually the disease begins gradually but the course of the disease is quite variable. Long static periods may occur in the progression of weakness (Pearn et al. 1978b; Gardner-Medwin et al. 1967; Russman et al. 1983; Namba et al. 1970). The progression of the disease may be slow enough to allow the patient to walk, although with difficulties, 10 to even 30 years after the onset of the disease. The majority of patients, however, use a wheelchair in their 30s. Usually, patients with a later onset of symptoms are seen to have a longer period of deterioration (Barois et al. 1989). The duration of the disease may be longer than 30 years and the patient's life span might also be normal (Namba et al. 1970). The long-term survival mostly depends on respiratory function.

Sex influence

There have been conflicting reports about sex influence on frequencies of the disease in males and females, the age of onset, and the severity of symptoms (Pearn et al. 1978b; Hausmanowa-Petrusewicz et al. 1979; Namba et al. 1970; Zerres 1989; Hausmanowa-Petrusewicz et al. 1984).

LABORATORY STUDIES

With a few exceptions, biochemical investigations of serum of patients with proximal spinal muscular atrophies have revealed no specific abnormalities. Increased serum activities of creatinphosphokinase (CK) were found in about half of the patients with chronic proximal spinal muscular atrophy (Namba et al. 1979). Usually the values are twice as high as normal but may also be 5–10 times of the upper normal limit. The very high CK levels commonly seen in Duchenne muscular dystrophy are rarely observed. Serum activities of other muscle-specific enzymes (e.g. aldolase and lactate dehydrogenase) may also be elevated, as well as the activities of glutamic pyruvic transaminase and glutamic oxa-lacetic transaminase (Namba et al. 1970). There are neither systematic studies on the correlation of elevated CK levels with age, duration or severity of the disease, and morphological changes in muscle, nor on isoenzyme patterns of CK. It has, however, been suggested that the level of CK elevation remains steady or even increases with the progression of the disease rather than decreases as it does in Duchenne muscular dystrophy (Brooke 1986). This might be related to the greater mobility of patients with the chronic form of this disease. The elevated levels of serum CK imply an abnormal permeability of the sarcolemma and are likely to reflect the non-specific 'myopathic' changes often described in muscle biopsies of patients with spinal muscular atrophies (see below). In a study on 100 patients with proximal spinal muscular atrophy elevated serum CK was found in all of the 23 males, who presented with hypertrophy of the calves. The CK activities in these patients were up to 1830 U/l with a mean value of 643 U/l (Bouwsma and van Wijngaarden 1980). Serum creatine and urinary excretion of creatine was frequently increased, and urinary excretion of creatinine was decreased in patients with chronic proximal spinal muscular atrophy indicating the degree of muscle wasting (Namba et al. 1970). Cerebrospinal fluid was usually normal (Gardner-Medwin et al. 1967; Aberfeld and Namba 1969).

Serum estrone (E_1) levels were significantly increased in males with Kugelberg-Welander disease but also in patients with amyotrophic lateral sclerosis, and the X-linked bulbospinal muscular atrophy type Kennedy (Usuki et al. 1989). This might explain why in some patients with Wohl-fart-Kugelberg-Welander disease gynecomastia has been observed (Namba et al. 1970). Usuki et al. (1989) speculated that the elevated serum E_1 concentrations resulted from increased peripheral androgen-to-estrogen conversion. The pathophys-iological mechanism, however, remains obscure.

ELECTROPHYSIOLOGICAL STUDIES

As in other diseases of the lower motor neuron, the electromyogram (EMG) shows changes of denervation and reinnervation. Fasciculations, fibrillations, positive sharp waves, and, particu-larly in patients with longer duration of the disease process, increased motor unit duration and territory with giant polyphasic potentials may all be found. The characteristic EMG criteria have been summarized in detail by Buchthal and Kamieniecka (1982). In older patients with ad-vanced atrophy 'myopathic' changes with short duration and low amplitude polyphasic action potentials can be observed as well (Gath et al. 1969; Mastaglia and Walton 1971). This might indicate the existence of motor units composed of only a few functioning fibers or of low fiber density (Kugelberg 1975). It was, however, also suggested that the separation of late components of motor unit potentials as distinct short and low potentials was the reason for this 'pseudo-myopathic' EMG changes, but this assumption does not sufficiently explain all histopathological and EMG changes (Buchthal and Kamieniecka 1982; Hausmanowa-Petrusewicz and Karwanska 1986).

Hausmanowa-Petrusewicz and Karwanska (1986) studied the electromyographic changes in 223 patients with different forms of infantile and juvenile proximal spinal muscular atrophy. The patients were classified into 3 groups: group A comprised acute infantile forms of the Werdnig-Hoffmann type; group B represented chronic intermediate forms; and group C comprised the Kugelberg-Welander forms. Fibrillations, positive sharp waves, and fasciculations were found in all groups, but fasciculations were rare in group A. Spontaneous rhythmically firing motor units were only found in group A and B but not in group C, whereas pseudomyotonic discharges were only observed in patients of group B and C with a longer duration of the disease. The parameters of individual motor unit potentials also differed in the different groups. Group A showed bimodal distributions for amplitude and duration with both high amplitude potentials of long duration and short low amplitude potentials. In patients of groups B and C with duration of the disease longer than 4 years the values of these parameters were shifted toward higher amplitudes and longer durations. In long duration cases of group C, however, there were also short low amplitude potentials and 'linked' potentials. It was suggested that in the early stage of proximal spinal muscular atrophy these EMG findings do not only have diagnostic but also prognostic implications.

Motor nerve conduction velocities are normal but in patients with severe neuronal loss can be slightly reduced, suggesting a selective loss of fast-conducting large-diameter nerve fibers (Moose and Dubowitz 1976). *Sensory nerve conduction velocities* are normal in proximal spinal muscular atrophy. *Vasomotor reflexes* seemed also to be normal (Gardner-Medwin et al. 1967).

MUSCLE BIOPSY

The muscle biopsy shows the typical changes of chronic denervation with a mixture of large and small groups of atrophic fibers and of variable groups of normal and markedly hypertrophied fibers. Target fibers, usually found in other denervating diseases occur only in the minority of cases. The characteristic myopathological features of denervation in general are summarized by Banker (1986) and Jennekens (1982). Although there are no changes specific to Wohlfart-Kugelberg-Welander disease in contrast to other anterior horn cell disorders, the myopathological features particularly found in patients with Wohlfart-Kugelberg-Welander spinal muscular atrophy have been extensively studied by Mastaglia and Walton (1971) and are outlined as follows.

Typically, small and large groups of uniformly atrophic fibers are found between fascicles with normal fibers and some markedly hypertrophic fibers. Randomly scattered small angulated fibers may be observed in earlier stages of denervation. In long-standing cases the degree of atrophy may be so extreme that only the dark pyknotic nuclei of the fibers are seen. Both atrophic and non-atrophic fibers are arranged in groups that are usually of uniform fiber type based on the myosin ATPase reaction. Atrophic fibers are usually of type II but sometimes also entirely of type I (Hausmanowa-Petrusewicz et al. 1968, 1980; Mastaglia and Walton 1971; Heene 1970). However, heterogeneous fiber types in atrophic fascicles are also observed. In groups of the more grossly atrophied fibers, which are especially

found in very chronic cases, it is virtually impossible to type a proportion of the fibers. Non-atrophic fibers are also uniformly either type I or type II fibers (Dubowitz 1966; Mastaglia and Walton 1971). Identification of fiber type by combined criteria of myosin ATPase, phosphorylase, and succinate-dehydrogenase reactions revealed 'hybrid' fibers with histochemical characteristics of both type I and type II fibers (Mastaglia and Walton 1971).

It has been suggested that the grouping of fibers of uniformly histochemical type may indicate a preferential susceptibility of either type I or type II motor neurons to the underlying degenerative disorder. However, it is now generally accepted that the grouping of fibers of single histochemical type is due to extensive collateral reinnervation of previously denervated muscle fibers by sprouts from surviving motor nerve fibers. The reinnervated muscle fibers are then transformed to histochemical conformity with the new 'parent' unit. Thus, the normal histochemical mosaic pattern of non-atrophic fibers will be replaced by non-atrophic fibers of uniform histochemical type. When the newly formed grouped fibers are again denervated as the disease progresses the atrophic fibers will also be of uniform fiber type. Heterogeneous fiber types in atrophic fascicles may be due to incomplete collateral sprouting from intact motor units.

Non-atrophic fibers in biopsies of patients with Wohlfart-Kugelberg-Welander disease also frequently show markedly hypertrophic fibers, excessive variation in fiber size of surviving fibers, excessive numbers of internal nuclei, fiber splitting, ringbinden and lobulated fibers, degenerative changes with necrosis, regenerative fibers with basophilic sarcoplasma and large vesicular nuclei with prominent nucleoli, and proliferation of interstitial connective and fatty tissue (Mastaglia and Walton 1971; Guerard et al. 1985; Namba et al. 1971). Since these changes are the characteristic features of primary degenerative myopathies, their occurrence in spinal muscular atrophy has been interpreted as 'pseudomyopathic' changes or as non-specific concomitant myopathy. In general, myopathic changes tend to be more prominent in older patients but there was no definite correlation with the clinical duration of the disease or with the particular muscle biopsied. Myopathic changes, however, were especially found in cases with high serum levels of CK activity. Conversely, the changes were less marked in patients with low serum CK levels (Mastaglia and Walton 1971). Myopathic changes have also been found in cases with adult forms of spinal muscular atrophy and in other chronically denervating diseases such as in patients with poliomyelitis in the remote past, and they have also been observed in experimentally denervated extraocular muscles (Mastaglia and Walton 1971; Brenni et al. 1981; Drachman et al. 1967, 1969). However, they are usually not observed in rapidly progressive cases with spinal muscular atrophy of the type Werdnig-Hoffmann. 'Myopathic' changes are therefore the rule rather than the exception in these conditions and when found in combination with groups of extremely atrophic fibers in a muscle biopsy are diagnostic of a chronic anterior horn cell disorder. Some authors even strongly object to the use of the misleading terms 'myopathic' or 'pseudomyopathic' for these histological features in chronic spinal muscular atrophies (Brenni et al. 1981). Rather than being non-specific or an indication for an associated separate myopathic process, the 'myopathic' changes may be considered to be secondary to the neurogenic disorder. The degenerative changes might be the result of excessive work load on fibers that had undergone compensatory hypertrophy. Other hypotheses for the occurrence of hypertrophic and degenerating fibers postulate an imbalance of trophic nerve factors released by the degenerating nerves or a chronic hypoxia causing the degeneration because the muscle fiber area per number of capillaries was significantly increased in hypertrophic fibers (Brenni et al. 1981).

PATHOLOGY AND PATHOGENESIS

So far, the underlying biochemical defect and the gene product defective in spinal muscular atrophies have not been identified. As in other motor neuron diseases the underlying pathological mechanism in spinal muscular atrophies is a selective loss of anterior horn cells of the spinal cord. The general pathological features of degeneration of anterior horn cells in motor neuron disease are outlined in detail by Banker (1986). There is only a limited number of post-mortem studies on the spinal cord in Wohlfart-Kugelberg-Welander disease (Gardner-Medwin et al. 1967; Kohn 1968; Aberfeld and Namba 1969; Namba et al. 1970; Welander 1955). The patients showed degeneration and almost complete loss of anterior horn cells, the degenerating cells being small and pyknotic. Occasionally degenerated ganglion cells could also be seen in the lateral and posterior horns. The anterior horns showed gliosis. Within the spinal cord no areas of demyelination were found. The anterior roots, however, showed a distinct demyelination in comparison to the normal posterior roots (Kohn 1968). Anterior horn cell loss and degeneration of the sensory neurons in the lumbar dorsal root ganglia with Wallerian degeneration of the fasciculus gracilis were found in a patient with the putative clinical diagnosis of Wohlfart-Kugelberg-Welander disease and with marked decrease of vibration sense in the legs (Winder and Auer 1989). In this case, however, predominant proximal muscle involvement has not been documented and the diagnosis is questionable. Aberfeld and Namba (1969) reported a patient with Wohlfart-Kugelberg-Welander disease and progressive ophthalmoplegia. Pathological examination revealed degeneration of the anterior horn cells and of motor nuclei of the cranial nerves including the oculomotor nucleus. Murakami (1990) studied spinal anterior horn cells from C6 of spinal cord from 6 patients with motor neuron disease including one 27-year-old patient with spinal muscular atrophy using microdensitophotometrical techniques. A significant decrease of cellular RNA content was found in both large and small anterior horn cells suggesting that an abnormality of RNA synthesis precedes the light microscopic changes seen in nerve cells.

Hausmanowa-Petrusewicz et al. (1980) postulated that the Wohlfart-Kugelberg-Welander disease is related to a defect of innervation during the course of myogenesis in fetal life. This was mainly based on ultrastructural and biochemical data. It was shown that: (i) the morphological features of the atrophic fibers suggested that the smallness of these fibers reflects lack of development rather than atrophy of previously normally innervated fibers; and (ii) the relative amounts of actin and of myosin light chain 3 were significantly decreased in muscle from Wohlfart-Kugelberg-Welander patients similar to the protein composition found in normal fetal muscle. The authors hypothesized that the presence of immature fibers impairs the normal development of muscle fibers and prevents an increase in number of mature fibers. It was speculated that the defect of innervation in Wohlfart-Kugelberg-Welander disease occurs at a later stage of the fetal development than in Werdnig-Hoffmann disease, which might explain the clinical and morphological differences between the 2 diseases.

Walsh et al. (1987) studied the expression of the neural cell adhesion molecule (N-CAM) in muscle biopsies from patients with Werdnig-Hoffmann disease and with Wohlfart-Kugelberg-Welander disease. N-CAM is believed to be involved in controlling cell-cell interactions in a variety of tissues. In skeletal muscle N-CAM is normally expressed by myoblasts and developing muscle, and in cell culture by myotubes. It is, however, not expressed in normal innervated adult muscle fibers. In Werdnig-Hoffmann disease all muscle fibers (atrophic and non-atrophic) were positive for N-CAM, whereas in Wohlfart-Kugelberg-Welander disease only the atrophic fibers were positive but not the non-atrophic and hypertrophic fibers. These findings indicate an unstable innervation of all muscle fibers in Werdnig-Hoffmann disease but a more stable innervation in Wohlfart-Kugelberg-Welander disease. So far, it is not known whether this difference in N-CAM expression reflects different pathogenetic mechanisms in the 2 diseases.

Sillevis Smitt and de Jong (1989) reviewed 38 animal models for amyotrophic lateral sclerosis and spinal muscular atrophies. Most of these models reproduce certain structural or physiological aspects of the human diseases, but none of the models provides an exact copy of a specific human motor neuron disease. The animal models comprise experimentally induced diseases as well as hereditary models.

DIFFERENTIAL DIAGNOSIS

Myopathies

Because of the clinical features of the Wohlfart-Kugelberg-Welander disease with involvement of predominantly proximal muscle groups of the upper and lower limbs the clinically and pathologically heterogeneous group of 'limb-girdle muscular dystrophies' represents an important clinical differential diagnosis. These myopathies include the scapulo-humeral muscular dystrophy (Erb), childhood muscular dystrophy with autosomal recessive inheritance, the dominant late onset form of proximal muscular dystrophy, Becker muscular dystrophy, Emery-Dreifuss muscular dystrophy, manifesting carriers of the gene for Duchenne or Becker dystrophy, the quadriceps myopathy, several forms of congenital myopathies, metabolic myopathies, and various non-genetic myopathies including polymyositis, sarcoidosis, and hyperthyroidism (Walton and Gardner-Medwin 1988).

The EMG is still the most simple and valuable diagnostic tool for differentiating the spinal muscular atrophies from myopathies. Serum CK levels might often not be very helpful because some myopathies usually do not show extremely increased CK levels and because CK can also be elevated in Wohlfart-Kugelberg-Welander disease. The precise diagnosis usually requires the muscle biopsy.

Other spinal muscular atrophies

Besides the diseases mentioned in Table 1 numerous other forms of spinal muscular atrophies can be differentiated from Wohlfart-Kugelberg-Welander disease. This is based on clinical and pathological criteria, age of onset, and the mode of inheritance. Several of these forms are summarized in other chapters of this volume and some spinal muscular atrophies mainly of the juvenile and adult age are outlined as follows.

Bulbospinal muscular atrophy type Kennedy (X-linked recessive). Although characterized by involvement of predominantly proximal muscle groups, this disease is classified separately from the proximal muscular atrophies mentioned above. Usually, age of onset is between 20 and 40 years. Muscle cramps may precede the onset of weakness for years. Proximal weakness develops first in the lower limbs and then spreads to the shoulder girdle, face and bulbar muscles. Dysphagia and dysarthria develop years later and muscle weakness finally spreads to also involve distal muscle groups. Frequently associated symptoms include gynecomastia and testicular atrophy (Kennedy et al. 1968; Warner et al. 1990). Gene linkage studies indicated a proximity of the gene for Kennedy disease and the gene encoding for the androgene receptor. Five out of 23 families with Kennedy disease showed defects of specific exons of the androgen receptor gene (Fischbeck et al. 1990).

Other spinal muscular atrophies with predominant bulbar palsy. The autosomal recessive bulbopontine paralysis with deafness, *Vialetto-van Laere syndrome*, usually presents between 20 and 40 years with facial weakness, dysphagia and dysarthria associated with sensorineural deafness. The weakness later spreads to affect the limbs. Survival is usually not beyond the fourth decade. The autosomal recessive *Fazio-Londe* disease does not show deafness and generally begins in early childhood (see chapter 9 of this volume). The *hereditary progressive bulbar palsy of adulthood* usually presents in late life with bulbar involvement and mild limb weakness (Schiffer et al. 1986). Matsunaga et al. (1973) described 7 patients with proximal or distal spinal muscular atrophy, ophthalmoplegia, and bulbar involvement. The mode of inheritance and the nosological entity of this *oculopharyngeal spinal muscular atrophy*, however, is debated (Zerres 1989). It remains open if the case described by Aberfeld and Namba (1969) with proximal spinal muscular atrophy and ophthalmoplegia represents a variant of Wohlfart-Kugelberg-Welander disease with cranial nerve involvement as the authors suggested, or if this case has to be classified separately (Baraister 1990). Dobkin and Verity (1976) described an autosomal dominant progressive bulbar and distal spinal muscular atrophy of juvenile onset and late progression of the distal muscle weakness with '*ragged-red fibers*' in

the muscle biopsy. Although 'ragged-red fibers' are typically found in mitochondrial encephalomyopathies indicating an accumulation of morphologically abnormal mitochondria, this myopathological feature is not specific to mitochondrial myopathies.

Spinal muscular atrophies with predominant scapular involvement. The *scapuloperoneal spinal muscular atrophy* with autosomal dominant, autosomal recessive, and X-linked recessive forms is outlined in chapter 4 of this volume. The autosomal *facioscapulohumeral spinal muscular atrophy* resembles facioscapulohumeral muscular dystrophy and comprises rapidly (Cao et al. 1976) and slowly (Fenichel et al. 1967) progressive forms of spinal muscular atrophies with an age of onset usually in the second decade of life. It is debated if the rapidly progressive scapulohumeral spinal muscular atrophy described by Jansen et al. (1986) with an age of onset in the fourth to sixth decade of life represents a separate nosological entity. The eponyms Vulpian-Bernhardt for the scapulohumeral muscular atrophy especially found in the older German literature are no longer used.

Other forms of proximal spinal muscular atrophy. Rare cases with chronic neurogenic atrophy of predominant quadriceps involvement (*neurogenic quadriceps amyotrophy*) have been considered to be 'formes frustes' of Wohlfart-Kugelberg-Welander disease (Furukawa et al. 1977). Other hereditary forms of proximal spinal muscular atrophy include non-progressive spinal muscular atrophy with non-progressive *mental retardation* (Staal et al. 1975) and progressive spinal muscular atrophy with *microcephaly and mental retardation* (Spiro et al. 1967).

Distal spinal muscular atrophies. Chronic distal spinal muscular atrophies represent a clinically and genetically heterogeneous group of disorders (Meadows and Marsden 1969; Harding and Thomas 1980; Harding 1984). They account for about 10% of all cases with spinal muscular atrophies (Pearn and Hudgson 1979). The age of onset in the autosomal dominant and autosomal recessive forms ranges from early childhood to adolescence. Distal atrophy and weakness affects predominantly lower limbs and only in about one-quarter of patients also upper limbs. The condition is relatively benign. Adult cases may be indistinguishable from early forms of scapuloperoneal spinal muscular atrophies. Distal spinal muscular atrophy has to be separated from the neuronal forms of peroneal muscular atrophy (HMSN types I and II, neuronal type of Charcot-Marie-Tooth disease) and is often also called spinal form of Charcot-Marie-Tooth disease or spinal muscular atrophy of the Charcot-Marie-Tooth type. Although the neuronal forms often also show clinically normal sensory function, the spinal forms can be differentiated from the neuronal forms by the relative preservation of the tendon reflexes and by normal sensory nerve action potentials. The eponym Duchenne-Aran often found in the German literature for the adult forms of distal spinal muscular atrophies is not used in the English literature.

Other forms of distal spinal muscular atrophy are associated with optic atrophy and neural deafness (Iwashita et al. 1979; Chalmers and Mitchell 1987), with ataxia, retinitis pigmentosa and diabetes mellitus (Furukawa et al. 1968), with vocal cord paralysis (Young and Harper 1980), or with hereditary stimulus-sensitive myoclonus (Jankovic and Rivera 1979). The sporadic *juvenile distal and segmental muscular atrophy* of upper extremities (Hirayama disease, monomelic amyotrophy, chronic asymmetric spinal muscular atrophy; see chapter 8 of this volume) is characterized by unique distribution of the muscular atrophy in the hand and forearm. Usually only one arm is affected, but the contralateral arm and the legs might also be slightly involved (Gourie-Devi et al. 1984; Oryema and Spiegel 1990).

Motor neuron disease with GM$_2$ gangliosidosis

Besides other central nervous system syndromes including Tay-Sachs disease, deficiency of lysosomal hexosaminidase A (GM$_2$ gangliosidosis) may also produce symptoms of motor neuron disease. Johnson et al. (1982) reported a 24-year-old patient with progressive proximal limb weakness without other features of central nervous system involvement that resembled Wohlfart-

Kugelberg-Welander disease. Two other family members had Tay-Sachs disease. Biochemically there was a severe deficiency of hexosaminidase A and the rectal ganglion cells were filled with classic membranous cytoplasmatic bodies of Tay-Sachs disease.

REFERENCES

ABERFELD, D. C. and T. NAMBA: Progressive ophthalmoplegia in Kugelberg-Welander disease. Arch. Neurol. 20 (1969) 253–256.

BANKER, B.: The pathology of the motor neuron disorders. In: A. G. Engel and B. Q. Banker (Eds.), Myology. Basic and Clinical. New York, McGraw-Hill (1986) 2031–2067.

BARAISTER, M.: The Genetics of Neuromuscular Disorders. Oxford, Oxford University Press (1990).

BAROIS, A., B. ESTOURNET, G. DUVAL-BEAUPERE, J. BATAILLE, and D. LECLAIR-RICHARD: Amyotrophie spinale infantile. Rev. Neurol. (Paris) 145 (1989) 299–304.

BECKER, P. E.: Atrophia musculorum spinalis pseudomyopathica. Z. Menschl. Vererb. u. Konstitutionslehre 37 (1964) 193–220.

BOUWSMA, G. and N. J. LESCHOT: Unusual pedigree patterns in seven families with spinal muscular atrophy: further evidence for the allelic model hypothesis. Clin. Genet. 30 (1986) 45–149.

BOUWSMA, G. and G. K. VAN WIJNGAARDEN: Spinal muscular atrophy and hypertrophy of the calves. J. Neurol. Sci. 44 (1980) 275–279.

BRANDT, S.: Werdnig-Hoffmann's Infantile Progressive Muscular Atrophy. Copenhagen, Ejnar Munksgaard (1951).

BRENNI, G., F. JERUSALEM and H. SCHILLER: Myopathologie chronischer Denervationsprozesse. Nervenarzt 52 (1981) 692–702.

BROOKE, M. H.: A Clinician's View of Neuromuscular Diseases, 2nd. ed. Baltimore, Williams and Wilkins (1986).

BRZUSTOWICZ, I. M., T. LEHNERT, L. H. CASTILLA, G. K. PENCHSZADEH, K. C. WILHELMSEN, R. DANIELS, K. E. DAVIES, M. LEPPERT, F. TITER, D. WOOD, V. DUBOWITZ, K. ZERRES, I. HAUSMANOWA-PERTSEWICZ, J. OTT, T. L. MUNSAT and T. C. GILLIAM: Genetic mapping of chronic childhood-onset spinal muscular atrophy to chromosome 5q11.2-13.3. Nature 344 (1990) 540–541.

BUCHTHAL, F. and Z. KAMIENIECKA: The diagnostic yield of quantified electromyography and quantified muscle biopsy in neuromuscular disorders. Muscle Nerve 5 (1982) 265–280.

BUNDEY, S. and E. M. BRETT: Genetics and Neurology. Edinburgh, Churchill Livingstone (1985).

BUNDEY, S. and R. E. LOVELACE: A clinical and genetic study of chronic proximal spinal muscular atrophy. Brain 98 (1975) 455–472.

CAO, A, C. CIANCHETTI, L. CALISTI and W. TANGHERONI: A

family of juvenile proximal spinal muscular atrophy with dominant inheritance. J. Med. Genet. 13 (1976) 131–135.

CHALMERS, W. R. and D. MITCHELL: Optico-acoustic atrophy in distal spinal muscular atrophy. J. Neurol. Neurosurg. Psychiatry 50 (1987) 238–239.

CLARKE, A., K. E. DAVIES, D. GARDNER-MEDWIN, J. BURN and J. HUDGSON: Xp21 DNA probe in diagnosis of muscular dystrophy and spinal muscular atrophy. Lancet 1 (1989) 443.

DOBKIN, B. H. and M. A. VERITY: Familial progressive bulbar and spinal muscular atrophy: Juvenile onset and late morbidity with ragged-red fibers. Neurology 26 (1976) 754–763.

DRACHMAN, D. A., S. R. MURPHY, M. NIGAM and J. R. HILLS: 'Myopathic' changes in chronically denervated muscle. Arch. Neurol. 16 (1967) 14–24.

DRACHMAN, D. A., N. WITZEL, M. WASSERMAN and H. NAITO: Experimental denervation of ocular muscles. Arch. Neurol. 21 (1969) 170–183.

DUBOWITZ, V.: Enzyme histochemistry of skeletal muscle. part 3 (neurogenic muscular atrophies). J. Neurol. Neurosurg. Psychiatry 29 (1966) 23–28.

EMERY, A. E. H: The nosology of spinal muscular atrophies. J. Med. Genet. 8 (1971) 481–495.

EMERY, A. E. H.: Duchenne Muscular Dystrophy. Oxford, Oxford University Press (1978).

EMERY, A. E. H., A. M. DAVIE, S. HOLLOWAY and R. SKINNER: International collaborative study of the spinal muscular atrophies. Part II. Analysis of genetic data. J. Neurol. Sci. 30 (1976) 375–384.

ENG, G. D., H. BINDER and B. KOCH: Spinal muscular atrophy: experience in diagnosis and rehabilitation management of 60 patients. Arch. Phys. Med. Rehabil. 65 (1984) 549–553.

FENICHEL, G. M., E. S. EMERY and P. HUNT: Neurogenic atrophy simulating facioscapulohumeral dystrophy. Arch. Neurol. 17 (1967) 257–260.

FISCHBECK, K. H., D. SOUDERS, E. M. WILSON and D. B. LUBAHN: The androgene receptor gene in X-linked spinal muscular atrophy. Neurology 40 (Suppl. 1) (1990) 163–164.

FURUKAWA, T., A. TAKAGI, K. NAKAO, H. SUGITA, H. TSUKAGOSHI and T. TSUBAKI: Hereditary muscular atrophy with ataxia, retinitis pigmentosa, and diabetes mellitus. Neurology 18 (1968) 942–947.

FURUKAWA, T., N. AKAGAMI and S. MARUYAMA: Chronic neurogenic quadriceps amyotrophy. Ann. Neurol. 2 (1977) 528–530.

GARDNER-MEDWIN, D., P. HUDGSON and J. N. WALTON: Benign spinal muscular atrophy arising in childhood and adolescence. J. Neurol. Sci. 5 (1967) 121–158.

GATH, I., O. SJAASTAD and A. C. LOKEN: Myopathic electromyographic changes correlated with histopathology in Wohlfart-Kugelberg-Welander disease. Neurology 19 (1969) 344–352.

GILBERT, E. F.: The effects of metabolic diseases on the cardiovascular system. Am. J. Cardiovasc. Pathol. 1 (1987) 189–213.

GILLIAM, T. C., L. BRZUSTOWICZ, L. H. CASTILLA, T. LEHNER, G. K. PENCHASZADEH, R. J. DANIELS, B. C. BYTH, J. KNOWLS, J. E. HISLOP, Y. SHAPIRA, V. DUBOWITZ, T. L. MUNSAT, J. OTT and K. E. DAVIES: Genetic homogeneity between acute and chronic forms of spinal muscular atrophy. Nature 345 (1990) 823–825.

GOURIE-DEVI, M., T. G. SURESH and S. K. SHANKAR: Monomelic amyotrophy. Arch. Neurol. 41 (1984) 388–394.

GREENBERG, F., K. R. FENOLIO, J. P. HEJTMANCIK, D. ARMSTRONG, J. K. WILLIS, E. SHAPIRA, H. W. HUNTINGTON and R. L. HAUN: X-linked infantile spinal muscular atrophy. Am. J. Dis. Child. 142 (1988) 217–219.

GRUBER, H., J. ZEITLHOFER, J. PRAGER and P. PILS: Complex oculomotor dysfunctions in Kugelberg-Welander disease. Neuroophthalmology 3 (1983) 125–128.

GUERARD, M. J., C. A. SEWRY and V. DUBOWITZ: Lobulated fibers in neuromuscular diseases. J. Neurol. Sci. 69 (1985) 345–356.

HARDING, A. E.: Inherited noeronal atrophy and degeneration predominantly of lower motor neurons. In: P. J. Dyck, P. K. Thomas, E. H. Lambert and R. Bunge (Eds.), Peripheral Neuropathy. Philadelphia, W. B. Saunders (1984) 1537–1556.

HARDING, A. E. and P. K. THOMAS: Hereditary distal spinal muscular atrophy. J. Neurol. Sci. 45 (1980) 337–348.

HAUSMANOWA-PETRUSEWICZ, I.: A research strategy for the resolution of childhood spinal muscular atrophy. In: L. Merlini, C. Granata and V. Dubowitz (Eds.), Current Concepts in Childhood Spinal Muscular Atrophy. Wien, New York, Springer (1989) 21–32.

HAUSMANOWA-PETRUSEWICZ, I., W. ASKANAS, B. BADURSKA, B. EMERYK, A. FIDZIANSKA, W. GARBALINSKA, L. HETNARSKA, H. JEDRZEJOWSKA, Z. KAMIENIECKA, I. NIEBROJ, J. PROT and E. SAWICKA: Infantile and juvenile spinal muscular atrophy. J. Neurol. Sci. 6 (1968) 269–287.

HAUSMANOWA-PETRUSEWICZ, I., J. ZAREMBA and J. BORKOWSKA: Chronic form of childhood spinal muscular atrophy. J. Neurol. Sci. 43 (1979) 313–327.

HAUSMANOWA-PETRUSEWICZ, I., A. FIDZIANSKA, I. NIEBROJ-DOBOSZ and M. H. STRUGALSKA: Is Kugelberg-Welander spinal muscular atrophy a fetal defect? Muscle Nerve 3 (1980) 389–402.

HAUSMANOWA-PETRUSEWICZ, I., J. BORKOWSKA and Z. JANCZEWSKI: X-linked adult form of spinal muscular atrophy. J. Neurol. 229 (1983) 175–188.

HAUSMANOWA-PETRUSEWICZ, I., J. ZAREMBA, J. BORKOWSKA and W. SZIRKOWIEC: Chronic proximal spinal muscular atrophy of childhood and adolescence: sex influence. J. Med. Genet. 21 (1984) 447–450.

HAUSMANOWA-PETRUSEWICZ, I., J. ZAREMBA and J. BORKOWSKA: Chronic proximal spinal muscular atrophy of childhood and adolescence: problems of classification and genetic counselling. J. Med. Genet. 22 (1985) 350–353.

HAUSMANOWA-PETRUSEWICZ, I. and A. KARWANSKA: Electromyographic findings in different forms of infantile and juvenile proximal spinal muscular atrophy. Muscle Nerve 9 (1986) 37–46.

HEENE, R.: Histologisch-histochemische Untersuchungen zur Typologie der Muskelveränderungen bei Atrophia musculorum spinalis pseudomyopathica (Wohlfahrt-Kugelberg-Welander). Z. Neurol. 198 (1970) 291–304.

IWASHITA, H., N. INOUE, S. ARAKI and Y. KUROIWA: Optic atrophy, neural deafness and distal neurogenic amyotrophy. Report of a family with two affected siblings. Arch. Neurol. 22 (1979) 35–64.

JANKOVIC, J. and V. M. RIVERA: Herditary myoclonus and progressive distal spinal muscular atrophy. Ann. Neurol. (1979) 227–231.

JANSEN, P. H. P., E. M. G. JOOSTEN, H. H. J. JASPER and H. M. VINGERHOETS: A rapidly progressive autosomal dominant scapulohumeral form of spinal muscular atrophy. Ann. Neurol. 20 (1986) 538–540.

JENNEKENS, F. G. I.: Neurogenic disorders of muscle. In: F. L. Mastaglia and J. N. Walton (Eds.), Skeletal Muscle Pathology. Edinburgh, Churchill-Livingstone (1982) 204–324.

JOHNSON, W. G., J. WIGGER, H. R. KARP, L. M. GLAUBIGER and L. P. ROWLAND: Juvenile spinal muscular atrophy: a new hexosaminidase deficiency phenotype. Ann. Neurol. 11 (1982) 11–16.

KENNEDY, W. R., M. ALTER and J. H. SUNG: Progressive proximal spinal and bulbar muscular atrophy of late onset: a sex-linked recessive trait. Neurology 18 (1968) 671–680.

KOHN, R.: Postmortem findings in a case of Wohlfart-Kugelberg-Welander disease. Confin. Neurol. (Basel) 30 (1968) 253–260.

KUGELBERG, E.: Chronic proximal (pseudomyopathic) spinal muscular atrophy: Kugelberg-Welander syndrome. In: P. J. Vinken and G. W. Bruyn (Eds.), Handbook of Clinical Neurology, Vol 22: System Disorders and atrophies. Amsterdam, North-Holland Publ. (1975) 67–80.

KUGELBERG, E. and L. WELANDER: Familial neurogenic (spinal?) muscular atrophy simulating ordinary proximal dystrophy. Acta Psychiatr. Scand. 29 (1954) 42–43.

KUGELBERG, E. and L. WELANDER: Heredofamilial juvenile muscular atrophy simulating muscular dystrophy. Acta Neurol. Psychiatr. 75 (1956) 500–509.

LUNT, P. W., W. J. K. CUMMING, H. KINGSTON, A. P. READ, R. C. MOUNTFORD, M. MAHON and R. HARRIS: DNA probes in differential diagnosis of Becker muscular dystrophy and spinal muscular atrophy. Lancet 1 (1989) 46–47.

MASTAGLIA, F. L. and J. N. WALTON: Histological and histochemical changes in skeletal muscle from cases of chronic juvenile and early adult spinal muscular atrophy (the Kugelberg-Welander syndrome). J. Neurol. Sci. 12 (1971) 15–44.

MATSUNAGA, M., T. INOKUCHI, A. OHNISHI and Y. KUROIWA: Oculopharyngeal involvement in familial neurogenic muscular atrophy. J. Neurol. Neurosurg. Psychiatry 36 (1973) 104–111.

MEADOWS, J. C. and C. D. MARSDEN: A distal form of chronic spinal muscular atrophy. Neurology 19 (1969) 53–58.

MELKI, J., S. ABDELHAK, P. SHETH, M. F. BACHELOT, P. BURLET, A. MARCADET, J. AICARDI, A. BAROIS, J. P. CARRIERE, M. FARDEAU, D. FONTAN, G. PONSOT, T. BILLETE, C. ANGELINI, D. BARBOSA, G. FERRIERE, G. LANZI, A. OTTOLINI, M. C. BABRON, D. COHEN, A. HANAUER, F. CERGET-DARPOUX, M. LATHROP, A. MUNNICH and J. FREZAL: Gene for chronic proximal spinal muscular atrophies maps to chromosome 5q. Nature 344 (1990a) 767–768.

MELKI, J., P. SHETH, S. ABDELHAK, P. BURLET, M. F. BACHELOT, M. G. LATHROP, J. FREZAL and A. MUNNICH: Mapping of acute (type I) spinal muscular atrophy to chromosome 5q12-q14. Lancet 2 (1990b) 271–273.

MOOSA, A. and V. DUBOWITZ: Motor nerve conduction velocity in spinal muscular atrophy of childhood. Arch. Dis. Child. 51 (1976) 974–977.

MURAKAMI, T.: Motor neuron disease: quantitative morphological and microdensitophotometric studies of neurons of anterior horn and ventral root of cervical spinal cord with special reference to the pathogenesis. J. Neurol. Sci. 99 (1990) 101–115.

NAMBA, T., D. C. ABERFELD and D. GROB: Chronic proximal spinal muscular atrophy. J. Neurol. Sci. 11 (1970) 401–423.

OKAZAKI, K., S. SAKATA and T. SAITO: Anesthetic management of patients with Kugelberg Welander disease accompanied by cardiac abnormalities. Jpn. J. Anesth. 25 (1976) 398–401.

ORYEMA, J., P. ASHBY and S. SPIEGEL: Monomelic atrophy. Cam. J. Neurol. Sci. 17 (1990) 124–130.

PEARN, J.: Segregation analysis of chronic childhood spinal muscular atrophy. J. Med. Genet. 15 (1978a) 418–423.

PEARN, J.: Autosomal dominant spinal muscular atrophy. J. Neurol. Sci. 38 (1978b) 263–275.

PEARN, J.: Classification of spinal muscular atrophies. Lancet 1 (1980) 919–922.

PEARN, J. and P. HUDGSON: Distal spinal muscular atrophy: a clinical and genetic study of 8 kindreds. J. Neurol. Sci. 43 (1979) 183–191.

PEARN, J., S. BUNDEY, C. O. CARTER, J. WILSON, D. GARDNER-MEDWIN and J. N. WALTON: A genetic study of subacute and chronic spinal muscular atrophy in childhood. J. Neurol. Sci. 37 (1978a) 227–248.

PEARN, J. H., D. GARDNER-MEDWIN and J. WILSON: A clinical study of chronic childhood spinal muscular atrophy: a review of 141 cases. J. Neurol. Sci. 38 (1978b) 23–37.

RIETSCHEL, M., S. RUDNICK-SCHÖNEBORN and K. ZERRES: Clinical variability in autosomal dominant proximal SMA. J. Neurol. Sci. 98 (Suppl.) (1990) 54–55.

ROTTHAUWE, H. W., W. MORTIER and H. BEYER: Neuer Typ einer recessiv X-chromosomal vererbten Muskeldystrophie: Scapulo-humero-distale Muskeldystrophie mit frühzeitigen Kontrakturen und Herzrhythmusstörungen. Humangenetik 16 (1972) 181–200.

RUSSMAN, B. S., R. MELCHREIT and J. C. DRENNAN: Spinal

muscular atrophy: the natural course of disease. Muscle Nerve 6 (1983) 179–181.

SCHIFFER, D., F. BRIGNOLIO, A. CHIO, M. T. DIORDANA and A. MIGHELI: Clinical-anatomical study of a family with bulbospinal muscular atrophy in adults. J. Neurol. Sci. 73 (1986) 11–22.

SILLEVIS SMITT, P. A. E. and J. M. B. V. DE JONG: Animal models of amyotrophic lateral sclerosis and the spinal muscular atrophies. J. Neurol. Sci. 91 (1989) 231–258.

SKOUTELI, H. and V. DUBOWITZ: Fasciculation of the eyelids: an additional clue to clinical diagnosis in spinal muscular atrophy. Neuropediatrics 15 (1984) 145–146.

SPIRO, A. J., M. H. FOGELSON and A. C. GOLDBERG: Microcephaly and mental sub normality in chronic progressive spinal muscular atrophy of childhood. Dev. Med. Child Neurol. 9 (1967) 594–601.

STAAL, A., L. N. WENT and F. M. BUSCH: An unusual form of spinal muscular atrophy with mental retardation occurring in an inbred population. J. Neurol. Sci. 25 (1975) 57–64.

THOMAS, P. K., D. B. CALNE and C. F. ELLIOT: X-linked scapuloperoneal syndrome. J. Neurol. Neurosurg. Psychiatry 35 (1972) 208–215.

TSUGKAGOSHI H., T. NAKANISHI, K. KONDO, T. TSUBAKI: Hereditary proximal neurogenic muscular atrophy in adult. Arch. Neurol. 12 (1965) 597–603.

USUKI, F., O. NAKAZATO, M. OSAME and A. IGATA: Hyperestrogenemia in neuromuscular diseases. J. Neurol. Sci. 89 (1989) 189–197.

WALLAR, P. H. and J. M. REECE: Ocular findings in a patient with Kugelberg-Welander syndrome: a case report. J. Pediatr. Ophthalmol. Strabismus 15 (1978) 15–18.

WALSH, F. S., S. E. MOORE and B. D. LAKE: Cell adhesion molecule N-CAM is expressed by denervated myofibres in Werdnig-Hoffmann and in Kugelberg-Welander type spinal muscular atrophies. J. Neurol. Neurosurg. Psychiatry 50 (1987) 439–442.

WALTON, J. N. and D. GARDNER-MEDWIN: The muscular dystrophies. In: J. Walton (Ed.), Disorders of Voluntary Muscle. Edinburgh, Churchill-Livingstone (1988) 519–568.

WARNER, C. L., S. SERVIDEI, D. J. LANGE, E. MILLER, R. E. LOVELACE and L. P. ROWLAND: X-linked spinal muscular atrophy (Kennedy's syndrome): A kindred with hypobetalipoproteinemia. Arch. Neurol. 47 (1990) 1117–1120.

WELANDEER, L.: Aussprache zu H. Becker: Zur Klinik der Myopathien. Dtsch. Z. Nervenheilkunde 173 (1955) 480–481.

WINDER, T. R. and R. N. AUER: Sensory neuron degeneration in familial Kugelberg-Welander disease. Can. J. Neurol. Sci. 16 (1989) 67–70.

WINSOR, E. J., E. G. MURPHY, M. W. THOMPSON and T. E. REED: genetics of childhood spinal muscular atrophy. J. Med. Genet. 8 (1971) 143–148.

WOHLFART, G: Zwei Fälle von Dystrophia musculorum progressiva mit fibrillären Zuckungen und atypischem Muskelbefund. Dtsch. Z. Nervenheilkunde 153 (1942) 189–204.

WOHLFART, G., J. FEX and S. ELIASSON: Hereditary spinal muscular atrophy: a clinical entity simulating progressive muscular dystrophy. Acta Psychiatr. Neurol. Scan. 30 (1955) 395–406.

YOUNG, I. D. and P. S. HARPER: Hereditary distal spinal musculare atrophy with vocal cord paralysis. J. Neurol. Neurosurg. Psychiatry 43 (1980) 413–418.

ZERRES, K.: Klassifikation und Genetik spinaler Muskelatrophien. Stuttgart, New York, Georg Thieme Verlag (1989).

ZERRES, K. and T. GRIMM: Genetic counselling in families with spinal muscular atrophy type Kugelberg-Welander. Hum. Genet. 65 (1983) 74–75.

ZERRES, K., M. STEPHAN, U. KEHREN and T. GRIMM: Becker's allelic model to explain unusual pedigrees with spinal muscular atrophy. Clin. Genet. 31 (1987) 276–277.

ZERRES, K., S. RUDNIK-SCHÖNEBORN and M. RIETSCHEL: Heterogeneity in proximal spinal muscular atrophy. Lancet 2 (1990) 749–750.

Handbook of Clinical Neurology, Vol. 15 (59): Diseases of the Motor System
J.M.B.V. de Jong, editor
© Elsevier Science Publishers B.V., 1991

Spinal muscular atrophy of infantile and juvenile onset, due to metabolic derangement

JACOB TROOST

Department of Neurology, Medisch Spectrum Twente, Enschede, The Netherlands

Spinal muscular atrophies (SMAs) are the second most frequent neuromuscular diseases in childhood after Duchenne's muscular dystrophy.

SMA due to a specific metabolic derangement, however, is a very rarely occurring disorder, with the possible exception of a syndrome resembling juvenile SMA in hexosaminidase A deficiency, an autosomal recessive disorder common in Ashkenazi Jews.

Much genetic and clinical heterogeneity exists among the SMAs. They all share, however, progressive muscular weakness and atrophy of striated muscles, due to anterior horn cell degeneration (Gamstorp and Sarnat 1984; Dubowitz 1975; Hausmanowa-Petrusewicz 1978).

Muscular weakness occurs in all types of SMA. The distribution of the weakness (proximal, distal, bulbar) determines the type of SMA. Other clinical features occurring in SMA are atrophy of the involved muscles, hypotonia, hyporeflexia and fasciculations. Symptoms due to a lesion of the central motor system do not belong to the syndrome nor lesion of the sensory system. Contractures and skeletal deformities are common.

A classification based on age of onset, distribution and progression is given in Table 1.

The diagnostic differentiation from clinically similar hypotonic children with muscular weakness is possible with the aid of neurophysiological and neuropathological investigations. Neurophysiological investigations show evidence of denervation in the form of fibrillations, fasciculations and positive sharp waves. Sensory and motor nerve conduction velocities are normal. Pathologically the large motor neurons in the spinal cord are degenerated. The lost neurons are replaced by glia.

Other ancillary investigations are seldom helpful, although the serum creatine kinase level is sometimes elevated.

PROXIMAL SPINAL MUSCULAR ATROPHIES

Type I: acute infantile SMA or Werdnig-Hoffmann's disease (253300)

History. The history of progressive spinal muscular atrophy starts with the descriptions of Werdnig (Werdnig 1891; Werdnig 1894) and Hoffmann (Hoffmann 1893; Hoffmann 1897; Hoffmann 1900) of 6 families, with 10 affected children, who, however, presented their first symptoms at the age between 5 and 9 months. Their patients belong therefore to type II SMA of the late infantile or early childhood group.

Genetics. The disorder is transmitted by an autosomal recessive gene in almost all cases.

Clinical symptoms. The onset is early, usually in utero, with diminished fetal movements, or

TABLE 1
Spinal muscular atrophies.

Proximal
    Type I: acute infantile (Werdnig-Hoffmann disease)
    Type II: late infantile
                juvenile (Kugelberg-Welander disease)
                chronic childhood SMA
    Type III: adult
Bulbospinal
Distal
Scapuloperoneal
Other: e.g. Ryukyuan
SMAs due to metabolic derangement, e.g. hexosaminidase A deficiency, Tay Sachs disease, or Sandhoff disease.

within the first 2–3 months. The infant shows generalised hypotonia and there is a generalised paralysis of the limbs and trunk. The intercostal muscles are severely affected and breathing is almost entirely diaphragmatic. This gives the chest a bell-shaped appearance together with a distension of the abdomen. Fasciculation of the muscles and contractures of any severity are seldom seen. The infants suffer from recurrent respiratory infections and rarely survive the first year of life.

Neurophysiology. A normal nerve conduction is found. Needle electromyography shows some fibrillations and fasciculations.

Pathology. Loss of motor neurons in all levels of the spinal cord and also of the lower cranial nerves (V–XII).

Muscle biopsy. Large groups of rounded atrophic type 1 and 2 fibers are seen.

Type II: late infantile and juvenile onset SMA (253400)

This constitutes the largest group of childhood SMAs. Recent genetic studies suggest, that this group of childhood SMAs as well as the chronic type of childhood SMAs includes cases of arrested Werdnig-Hoffmann disease.

Genetics. The disorder is inherited as a simple autosomal recessive gene. A major advance has been the genetic mapping of chronic SMA to chromosome 5q (Brzustowicz et al. 1990; Melki et al. 1990). Juvenile- and late-onset forms map to the same region.

Interesting is the fact, that the enzymne hexosaminidase-B also maps to the same region. This may be coincidental; however, hexosaminidase-A and -B deficiency (chronic GM_2 gangliosidosis) sometimes presents with progressive neurogenic atrophy, which is clinically indistinguishable from chronic SMA. Two percent of the cases is inherited in an autosomal dominant form.

Clinical findings. The age of onset varies from 1 month to 17 years. The majority of cases occurs by 5 years of age. The disease runs a more benign course than the type I SMA. Sudden deterioration due to intercurrent infections is not uncommon. The median age of death exceeds 12 years in the recessive cases. In the dominant cases the course of the disease is even more benign and the muscle weakness is usually generalised and mild. The muscular weakness in the recessive cases concerns especially the shoulder and pelvic girdle and is accompanied by severe muscular atrophy. Fasciculations are more common than in type I. Skeletal abnormalities are also more often found.

Neurophysiology. Electromyography shows signs of denervation and reinnervation. Nerve condition is normal.

Pathology. Apart from fiber-type grouping and pseudo-myopathic features in muscle biopsy, the neuropathological changes are similar to those described in type I.

Type III: adult onset SMA

The onset of symptoms is usually between the second and fifth decades, with proximal symmetrical muscle weakness, especially in the lower extremities.

Genetically the syndrome is caused by 3 distinct genes: autosomal recessive, dominant and X-linked.

The course of the disease is relatively benign.

BULBOSPINAL MUSCULAR ATROPHY (313200)

This disease is in all cases inherited in an X-linked recessive way. The gene is linked to the $DXYS_1$ marker on the proximal part of the long arm of the X chromosome (Fishbeck et al. 1986). The onset of muscular weakness is usually between the ages of 20 and 40 years.

Involvement of the bulbar muscles predisposes to recurrent aspiration pneumonia. Fasciculations, tremors of the hands and gynaecomastia are common. Diabetes mellitus and infertility occur in patients and their normal relatives in some families. The disorder is slowly progressive.

DISTAL SPINAL MUSCULAR ATROPHY

Probably 4 genetically different subtypes exist: Two autosomal dominant and 2 autosomal recessive types. The disorder starts usually at birth or soon afterwards. Particularly weakness, distal atrophy and hypotonia, especially in the lower limbs are noted. The disease must be differentiated from the hereditary motor and sensory neuropathies types I and II.

Sensation in distal SMA is not affected contrary to HMSN types I and II. The motor conduction velocity of the peripheral nerves is normal in distal SMA and severely affected in HMSN I and II.

SCAPULOPERONEAL SPINAL MUSCULAR ATROPHY

There are several genetic types of scapuloperoneal SMA. Difficulty in walking is usually the presenting sign. Weakness and atrophy of the legs are always present. In the X-linked recessive inherited type of the disorder cardiac conduction defects have been described. The other, rarely occurring SMAs are not summarized here.

SPINAL MUSCULAR ATROPHY IN METABOLIC DERANGEMENT

Hexosaminidase A deficiency (272800)

Hexosaminidase A (Hex A) deficiency leads to the neuronal accumulation of GM_2 gangliosides

and has been shown to cause Tay-Sachs disease in juveniles and adults.

Hexosaminidase is a lysosomal enzyme involved in the degradation of several natural substances including GM_2 ganglioside, asialo GM_2 ganglioside GA_2, globoside and oligosaccharides (Bach and Suzuki 1975).

The lysosomal enzyme hexosaminidase (N-acetyl-β-hexosaminidase (Hex)) comprises 3 isozymes (Hex A, B and S) of 2 subunits (α and β). Hex A, the only Hex isozyme capable of degrading ganglioside, is composed of alpha and beta chains, whereas Hex B and Hex S are composed of beta and alpha chains respectively.

The alpha and beta chain are coded respectively on chromosomes 15 and 5. Thus a mutation of the alpha locus gives rise to deficiency of Hex A and a mutation at the beta locus to a deficiency of Hex A and Hex B. Hex A deficiency occurs in Tay-Sachs disease and has been reported in adult GM_2 gangliosidosis (Argov and Navon 1984). Hex A and Hex B deficiency occurs in Sandhoff disease.

Tay-Sachs disease, or infantile GM_2 gangliosidosis, is caused by a severe deficiency of Hex A. The juvenile and adult types of GM_2 gangliosidosis are caused by a decreased Hex A activity, with some residual function.

Tay-Sachs disease, or infantile GM_2 gangliosidosis, is the classical ganglioside storage disease. The first symptoms of the disease are between the third and sixth month of life. The presenting symptoms are: marked irritability and excessive startle responses. By the age of 6 months the infants become hypotonic, followed by a regression of the development. This is followed by progressive blindness, deafness, seizures, spasticity and macrocephaly. In fundo oculi a cherry-red spot is seen in almost all cases. Most children die before the age of 3 years. There is a marked predilection for Ashkenazi Jews.

The diagnosis is made by measuring hexosaminidase activity in leucocytes, fibroblasts or other tissues.

In the last 2 decades, however, it became clear that Hex A deficiency not only causes Tay-Sachs disease, but a whole group of progressive neurologic diseases. Among those are progressive cerebellar ataxia, resembling the Ramsay-Hunt syndrome, Friedreich ataxia, olivopontocerebellar ataxia or typical spinocerebellar ataxia. Patients are described with upper and lower motor neuron disease, resembling amyotrophic lateral sclerosis, whereas others have the symptomatology of spinal muscular atrophies. Resting tremor, dystonia, spastic pareses and psychoses have also been described (Johnson 1981).

In 1982 Johnson et al. described a 24-year-old patient with progressive difficulty in walking. In retrospect he always had a peculiar gait and always had difficulty in walking. His legs became weak at the age of 15. Then he noted progressive difficulty rising from low chairs, climbing stairs or jumping. For many years he suffered cramps of the feet, calves or lower posterior thighs. His family had no neurologic disorders. The parents were of Ashkenazi background and unrelated.

On examination the patient was tall and thin, with high pedal arches and no scoliosis. He was intelligent. Cranial nerve functions were normal except that the tongue seemed to be twitching. his neck and arms were strong. In walking, he fixed each leg in a position of genu recurvatum. He could walk on his heels but not on his toes. He could not rise from a standard chair without using his hands. Fasciculations were active and widespread in the deltoid, quadriceps and gastrocnemius muscles. Diffuse weakness of all the muscles of the legs was noted. His extremities were thin, but there was no focal wasting.

The biceps, triceps jerks were normally active. Left knee and ankle jerks were hypoactive, while the right knee jerk showed only trace activity and the right ankle jerk was absent. There were no Babinski signes.

Routine blood chemistry and CSF examination were not revealing.

Evoked responses were normal.

Electromyography showed fibrillations and fasciculations with decreased number of motor unit potentials.

Nerve conduction velocities were normal.

Muscle biopsy showed group atrophy and target fibers; atrophic fibers included both fiber types equally.

The hexosaminidase values were indistinguishable from those of patients with infantile Tay-Sachs disease. The parents showed values in the range of Tay-Sachs carriers.

This patient had juvenile spinal muscular atrophy as judged by clinical, electrophysiological and muscle biopsy criteria.

Inheritance was compatible with an autosomal recessive mode. The diagnosis was also compatible with Kugelberg-Welander syndrome.

The patient had a severe deficiency of Hex A, while his parents had activities in the heterozygote range. A rectal biopsy showed ganglion cells, filled with membranous cytoplasmic bodies indistinguishable from those seen in neurons from classic infantile Tay-Sachs disease.

Thus anterior horn cells disease giving rise to a spinal muscular atrophy may be a feature of a GM_2 gangliosidosis.

It is not known why some cases have an isolated SMA while other cases have anterior horn cell disease as a component of a more generalized neurologic disorder, for instance an amyothrophic lateral sclerosis.

Jellinger et al. (1982) described postmortem material of a 67-year-old woman with proximal muscle weakness in the legs, slurred speech and mental subnormality. The symptoms began at age 19. A brother suffered from a similar chronic neuromuscular disease.

Abundant lipid accumulation in CNS neurons and severe cerebellar cortical atrophy were found.

Skeletal muscle showed a terminal stage of denervation atrophy with severe lipomatosis. Complex lamellar cytoplasmic inclusions were seen in cells of the brain. In addition there were various lipopigment bodies, fingerprint profiles, rare polyglucosan bodies, rod-like structures and filamentous sheaves.

Accumulation of GM_2 and GA_2 in the cerebral cortex was demonstrated by thin-layer chromatography. Determination of hexosaminidase was not possible.

This observation can also be classified as a motor neuron disease phenotype of GM_2 gangliosidosis.

Argov and Navon (1984) described 8 patients in which there was evidence of lower-motor neuron disease in all but one patient. Furthermore periods of psychosis occurred in all patients. The motor symptoms in these patients appeared at the age of 9–30 years, mostly in the late teens. The borderlines between the juvenile and adult disorders are not clear.

Other neurological symptoms in these patients were spinocerebellar disorder, pyramidal signs with lower motor neuron disorder, an ALS phenotype, dentatorubral tremors and dementia.

Dale et al. (1983) described a familial Hex A deficiency with Kugelberg-Welander phenotype and mental change. The disorder in this family also started in their teens.

Mitsumoto et al. (1985) added 3 more patients with motor neuron disease. One of them had features similar to the case of Johnson et al. (1982). The CT scan of this patient showed, however, cerebral atrophy, suggesting that this patient had multisystem degeneration.

Parnes et al. (1985) described 3 patients from 2 families with an unusual phenotypical variant. The clinical picture was dominated by spinal motor neuron involvement, mimicking juvenile SMA. Atypical features included prominent muscle cramps, postural and action tremor, recurrent psychosis, incoordination, corticospinal and corticobulbar involvement and dysarthria.

According to the authors the presence of these atypical features should raise the suspicion of the presence of Hex A deficiency and GM_2 gangliosidosis.

A case of SMA of late onset with cerebellar atrophy and Hex A deficiency was described by Karni et al. (1988).

Specola et al. (1990) reported on 5 other patients. Two patients were examples of the juvenile form of Hex A deficiency. In this group the clinical phenotype was homogeneous. Language retardation, dysarthria and gait instability appeared around age 3, followed by ataxia and pyramidal tract signs. The motor signs were compatible with anterior horn cell disease. Intellectual deterioration was present at the end stage. Some patients had seizures and a few dystonia or choreo-athetotic movements. No ethnic predilection existed, contrary to the adult form, which affects predominantly Ashkenazi Jews.

The age of onset in the adult form is variable, 35% of the reported patients had their first clinical manifestations of their disorder before age 10.

Therefore the term adult form according to the authors is misleading and should be replaced by 'chronic' type.

In the chronic type, speech difficulties are commonly the first sign followed by gait difficulties, pyramidal tract signs, ataxia, lower motor neuron involvement and psychiatric disorders. Intellectual impairment is frequent. Other neurological symptoms are dystonia, tremor, supranuclear ophthalmoplegia.

Five of the 7 patients of Specola had a different enzymatic pattern, compatible with the B1 variant of GM_2 gangliosidosis. In this variant there is a severe deficiency of the Hex A activity toward the artificial substrate 4-methyl-umbelliferyl-β-N-acetylglucosaminide-6-sulfate (4MUGlcNAc-6-S), whereas the deficiency toward the conventional synthetic substrate 4-methyl-umbelliferyl-β-N-acetylglucosaminide (4MUGlcNAc) is only partial.

In conclusion one can state that the occurrence of lower motor neuron disease in patients with a Hex A deficiency, whether or not in combination with other neurologic dysfunction is not as rare as has been thought before. Spinal muscular atrophies with atypical features should therefore necessitate the measurement of hexosaminidase A.

Hexosaminidase A and B deficiency (268800)

Sandhoff disease. As has been described above, a mutation at the beta locus on chromosome 5 leads to a deficiency of Hex A and Hex B.

A deficiency of Hex A and B typically presents as infantile Sandhoff disease, which is phenotypically similar to Tay-Sachs disease except that there is also visceral involvement. The onset is between 3 and 6 months. These patients suffer from hyperacusis, motor weakness, spasticity and progressive mental and motor decline. A cherry-red spot and blindness are frequent findings. Accumulation of the intra-cytoplasmic inclusions occur in liver, kidneys, lymph nodes, spleen and lung as well as in the central nervous system. Macrocephaly is present.

Infrequent atypical presentations have been reported, including juvenile or adult onset cerebellar ataxia and/or motor neuron disease. The latter variety is extremely rare.

In 1986 Cashman et al. described a patient with partial deficiency of hexosaminidase due to a N-acetyl-β-hexosaminidase β locus defect with a juvenile motor neuron disease. The patient was a woman of 26 years of mixed German, Norwegian and Irish ancestry with no family history of neurological disease. Her early development was normal. From about age 7 her speech became progressively dysarthric. From age 10 she developed leg weakness with intermittent muscle twitches.

Neurological evaluation at age 13 revealed pseudobulbar dysarthria. Hip flexors and ankle dorsiflexors were mildly weak. Quadriceps fasciculations were recorded. Deep tendon reflexes were brisk. Equivocal extensor responses were present. A tremor of the outstretched hands was observed. Sensation was normal. Her gait had a steppage quality.

By age 15 her palate did not move on phonation. She could not climb stairs and her leg reflexes were depressed. An EMG revealed positive waves, fibrillations, fasciculations and a decreased interference pattern. Nerve conduction velocities were normal.

Gastrocnemius muscle biopsy demonstrated severe neurogenic atrophy. A sural nerve biopsy was normal.

By age 21, dysphagia and tongue fasciculations were noted. Over the next year she developed loss of pain and temperature sensitivity on the lower half of the calf. Sensory nerve conduction velocities were normal. The peroneal motor nerve conduction velocities were slowed (34 m/s).

Routine blood and CSF examinations were normal.

A rectal section biopsy at age 24 showed cells in the submocosal plexus swollen with granules that appeared light blue in semithin epon sections stained with toluidine blue. Ganglion cells demonstrated membrane bound concentric lamellar inclusions suggestive of cytoplasmic bodies observed in the gangliosidoses.

Total hexosaminidase activity was 10–15% of control values. The Hex B activity in serum, leucocytes and fibroblasts was nearly absent with partial Hex A activity in serum and leucocytes and low normal Hex A activity in fibroblasts. The authors found that fibroblast Hex A in their patient was less active to the natural substrate GM_2 as compared to 4MU-βGlcNAc supporting the hypothesis that defective ganglioside catabolism is the biochemical determinant of progressive motor neuron disease in association with Hex deficiencies.

In 1988 Michael Rubin (Rubin et al. 1988) described 2 sisters with progressive muscle cramps, as well as wasting and weakness of the legs with onset after age 20. They also showed intention tremor of the upper extremities and dysarthria starting during the first decade. The older patient also had fasciculations; the younger hyperreflexia.

A severe deficiency of both Hex A and Hex B was found. The phenotypic expression of this disease is similar to motor neuron disease due to alpha locus mutations, suggesting that the Hex A deficiency, even though only a partial one, may be the important pathogenic factor according to the authors.

This is remarkable as the locus for the alpha unit is located on chromosome 15 and the locus for the beta unit on chromosome 5, whereas the gene locus for the inherited chronic and juvenile SMAs without Hex deficiency is located on chromosome 5.

Thomas et al. (1989) described a 32-year-old male with an onset of upper limb postural tremor in adolescence followed by muscle cramps. Progressive proximal amyotrophy and weakness in the limbs developed late in the third decade.

Enzyme studies revealed hexosaminidase A and B deficiency.

Lysosomal sialidase deficiency (256550)

A deficiency of sialidase, which constitutes the biochemical basis of genetic metabolic disorders known as mucolipidosis I, the cherry-red spot myoclonus syndrome and the Goldberg syndrome with elevated amounts of gangliosides GM_3, GD_3 and probably GM_4 and LM_1 (Svennerholm 1977) does not cause motor neuron diseases.

The mucolipidoses and especially mannosidosis and fucosidosis show signs of lower motor neuron disease, which are, however, caused by a lesion of the peripheral nerves and not of anterior horn cell disease.

Sea-blue histiocytosis with progressive anterior horn cell degeneration

Sea-blue histiocytosis has been associated with several childhood neurological syndromes including myoclonic seizures, progressive dementia, macular degeneration, vertical ophthalmoplegia, posterior column disease, Niemann-Pick disease and neuronal ceroid-lipofuscinosis. Ashwal et al. (1984) described a 15-year-old girl with a history of progressive generalized weakness and sea-blue histiocytes. The patient showed a slowly progressive central nervous system disorder and at age 13 developed rapidly progressive generalized weakness with marked fasciculations and atrophy. Laboratory investigation revealed the presence of sea-blue histiocytes in the bone marrow without evidence of a disorder of sphingolipid metabolism or neuronal ceroid lipofuscinosis.

Muscle and sural nerve biopsy revealed findings typical of anterior horn cell degeneration.

Only one previous patient has been reported to have sea-blue histiocytosis in association with peripheral nerve system disease. This was a 17-year-old boy with easy bruisibility, epistaxis, splenomegaly and sea-blue histiocytes on bone marrow examination who developed hypaesthesia of the left leg. Motor nerve conduction velocities of peroneal and posterior tibial nerves were decreased and a sural nerve biopsy demonstrated mild axonal degeneration (Blankenship 1973). The authors believe, that the presence of sea-blue histiocytes does reflect an underlying lipid, glycolipid, phospholipid or ceroid metabolism disturbance.

Juvenile neurogenic muscle atrophy with multiple lysosomal enzyme deficiencies

Goto et al. (1983) described an 18-year-old boy with childhood onset of mental retardation, neurogenic muscle atrophy with hyperreflexia, Marfan like features, multiple epiphyseal dysplasia, increased urine excretion of dermatan sulfate and decreased lysosomal enzyme activities in β-galactosidase, β-glucuronidase and N-acetyl-β-D-glucosaminidase.

According to the clinical and biochemical results, the case may be considered to be one of the mucopolysaccharidoses.

Spinal muscular atrophy with β-hydroxyisovaleric aciduria and β-methylcrotonylglycinuria (210200)

Eldjarn et al. (1970) described in 1970 in a letter to the Lancet a 4.5-month-old girl whose mother and father were first cousins. The infant showed signs of retarded motor development, muscular hypotonia and muscular atrophy. Already in the second week after birth there were feeding difficulties. The clinical picture resembled that of an infantile spinal muscular atrophy.

In addition, however, her urine had an unpleasant smell like that of a cat's urine. In the urine large amounts of β-hydroxy-isovaleric acid and β-methyl-crotonylglycine were found. In the urine of the mother, father and 2 elder brothers significant amounts of β-methyl-crotonylglycine were found.

The accumulation of both substances is probably due to a failure in the leucine degradation. Because of this probability the child was placed on a diet with the minimum requirements of the essential amino-acids. After introduction of this diet the excretion of β-isovaleric acid dropped rapidly, whereas only a slight reduction in the excretion of β-methyl-crotonylglycine was seen. The clinical features didn't change. The child died at the age of 8 months.

Neurogenic muscular atrophy and hyperlipidemia

Oishi et al. (1977) described 2 families with chronic neurogenic proximal muscular atrophy with hyperlipidemia.

The onset of the disorder was between 17 and 46 years. The cause was of a chronic progressive type, with generalized fasciculations, hyporeflexia and also atrophy of the muscles in the hands and tongue. In all cases there existed a hyperlipidemia. Lipid analysis of the muscles showed an elevated amount of phospholipids and a reduction of sphingomyelins and unsaturated fatty acids. Similar cases were described by Quarfordt et al. (1970) and Dahl et al. (1975).

REFERENCES

ARGOV, Z. and R. NAVON: Clinical and genetic variations in the syndrome of adult GM$_2$ gangliosidosis resulting from hexosaminidase A deficiency. Ann. Neurol. 16 (1984) 14–20.

ASHWAL, S., T. V. TRASHER, D. R. RICE and D. A. WENGER: A new form of sea-blue histiocytes with acid phosphatemia: a syndrome resembling Gaucher disease: the Lewis variant. Ann. Neurol. 16 (1984) 184–192.

BACH, G. and K. SUZUKI: Heterogeneity of human hepatic N-acetyl-β-D-hexosaminidase A activity toward natural glycosphingolipid substrates. J. Biol. Chem. 250 (1975) 1328–1332.

BLANKENSHIP, R. M., B. R. GREENBURG and R. N. LUCAS: Familial sea-blue histiocytes with acid phosphatemia: a syndrome resembling Gaucher disease: the Lewis variant. J. Am. Med. Assoc. 225 (1973) 54–56.

BRZUSTOWICZ, L. M., T. LEHNER, L. H. CASTILLA, G. K. PENCHASZADEH, K. C. WILHELMSEN, R. DANIELS, K. E. DAVIES, M. LEPPERT, F. ZITER, D. WOOD, V. DUBOWITZ, K. ZERRES, I. HAUSMANOWA-PETRUSEWICZ, J. OTT, T. L. MUNSAT and T. C. GILLIAM: Genetic mapping of chronic childhood-onset spinal muscular atrophy to chromosome 5q 11.2–13.3. Nature 344 (1990) 540–541.

CASHMAN, N. R., J. P. ANTEL, L. W. HANOCH, G. DAWSON, A. L. HORWITZ, W. G. JOHNSON, P. R. HUTTENLOCHER and R. L. WOLLMANN: N-acetyl-β-hexosaminidase β locus defect and juvenile motor neuron disease: A case study. Ann. Neurol. 19 (1986) 568–572.

DAHL, D. S. and H. A. PETERS: Lipid disturbances with spinal muscular atrophy. Arch. Neurol. 32 (1975) 195–203.

DALE, A. J. D., A. G. ENGEL and N. L. RUDD: Familial hexosaminidase A deficiency with Kugelberg-Welander phenotype and mental change (abstract). Ann. Neurol. 14 (1983) 109.

DUBOWITZ, V.: Infantile spinal muscular atrophy. In: P. J. Vinken and G. W. Bruyn (Eds.), Handbook of Clinical Neurology, Vol. 22, System Disorders and Atrophies, Part II. Amsterdam, North Holland Publ. Co. (1975) 81–101.

ELDJARN, L., E. JELLUM, O. STOHHE, H. PANCKE and P. E. WAALER: β hydroxyisovaleric aciduria and β methylcrotonylglycinuria: a new inborn error of metabolism. The Lancet (1970) 521–522.

FISHBECK, K. H., V. IONASESCU and A. W. RITTER: Localization of the gene for X-linked spinal muscular atrophy. Neurology 36 (1986) 1595–1598.

GAMSTORP, I. and H. B. SARNAT: Progressive Spinal Muscular Atrophies. New York, Raven Press (1984).

GOTO, I., H. NAKAI, T. TABIRA, N. SHINNO, Y. TANAKA, H. SHIBASAKI and H. KUROIWA: Juvenile neurogenic muscle atrophy with lysosomal enzyme deficiencies: new disease or variant of mucopolysaccharidosis? J. Neurol. 229 (1983) 45–54.

HAUSMANOWA-PETRUSEWICZ, I.: Spinal muscular atrophy: Infantile and juvenile type. Published for the National Library of Medicine and the National Science Foundation. Washington DC (1978).

HOFFMANN, J.: Uber chronische spinale Muskelatrophie im Kindesalter, auf familiarer Basis. Dtsch. Z. Nervenheilk. 3 (1893) 427–470.

HOFFMANN, J.: Weiterer Beitrag zur Lehre von der hereditaren progressiven spinalen Muskelatrophie im Kindesalter. Dtsch. Z. Nervenheilk. 10 (1897) 292–320.

HOFFMANN, J.: Dritter Beitrag zur Lehre von der hereditären progressiven spinalen Muskelatrophie im Kindesalter. Dtsch. Z. Nervenheilk. 18 (1900) 217–224.

JELLINGER, K., A. P. ANZIL, D. SEEMANN and H. BERNHEIMER: Adult GM$_2$ gangliosidosis masquerading a slowly progressive muscular atrophy: motor neuron disease phenotype. Clin. Neuropathol. 1 (1982) 31–44.

JOHNSON, W. G.: The clinical spectrum of hexosaminidase deficiency diseases. Neurology 31 (1981) 1453–1456.

JOHNSON, W. G., H. J. WIGGER, H. R. KARP, L. M. GLAUBIGER and L. P. ROWLAND: Juvenile spinal muscular atrophy: a new hexosaminidase deficiency phenotype. Ann. Neurol. 11 (1982) 11–16.

KARNI, A., R. NAVON and M. SADEH: Hexosaminidase A deficiency manifesting as spinal muscular atrophy of late onset. Ann. Neurol. 24 (1988) 451–453.

MELKI, J., S. ABDELHAK, P. SHETH, M. F. BACHELOT, P. BURLET, A. MARCADET, J. AICARDI, A. BAROIS, J. P. CARRIERE, M. FARDEAU, D. FONTAN, G. PONSOT, T. BILLETTE, C. ANGELINI, C. BARBOSA, G. FERRIERE, G. LANZIL, A. OTTOLINI, M. C. BABRON, D. COHEN, A. HANAUER, F. CLERGET-DARPOUX, M. LATHROP, A. MUNNICK and J. FREZAL: Gene for chronic proximal spinal muscular atrophies maps to chromosome 5q. Nature 344 (1990) 767–768.

MITSUMOTO, H., R. J. SLIMAN, I. A. SCHAFER, C. S. STERNICK, B. KAUFMAN, A. WILBOURN and S. J. HORWITZ: Motor neuron disease and adult hexosaminidase A deficiency in two families: Evidence for multisystem degeneration. Ann. Neurol. 17 (1985) 378–385.

OISHI, M., S. NAKAJIMA, T. IWAGAKI, M. TOYODA, S. EBIHARA and F. GOTOH: 2 Sippen von chronischer neurogener proximaler Muskelatrophie mit Hyperlipämie. Nervenarzt 48 (1977) 386–390.

PARNES, Sh., G. KARPATI, S. CARPENTER, N. M. K. NG YING KING, L. S. WOLFE and L. SURANYI: Hexosaminidase A deficiency presenting as atypical juvenile-onset spinal muscular atrophy. Arch. Neurol. 42 (1985) 1176–1180.

QUARFORDT, S. H., D. C. DE VIVO, W. K. ENGEL, R. I. LEVY and D. S. FREDERICKSON: Familial adult-onset proximal spinal muscular atrophy. Arch. Neurol. 22 (1970) 541–549.

RUBIN, M., G. KARPATI, L. S. WOLFE, S. CARPENTER, M. H. KLAVINS and D. J. MAHURAN: Adult onset motor neuropathy in the juvenile type of hexosamindase

A and B deficiency. J. Neurol. Sci. 87 (1988) 103–119.

SPECOLA, N., M. T. VANIER, F. GOUTIERES, J. MIKOL and J. AICARDI: The juvenile and chronic form of GM$_2$ gangliosidosis: clinical and enzymatic heterogeneity. Neurology 40 (1990) 145–150.

SVENNERHOLM, S.: In JUPAC-JUB commission on biochemical nomenclature (CBN): The nomenclature of lipids. Recommendation 1976. Eur. J. Biochem. 79 (1977) 11–21.

THOMAS, P. K., E. YOUNG and R. H. M. KING: Sandhoff disease mimicking adult-onset bulbospinal neuropathy. J. Neurol. Neurosurg., Psychiatry 52 (1989) 1103–1106.

WERDNIG, G.: Zwei frühinfantile hereditären Fälle von progressiver Muskelatrophie unter dem Bilde der Dystrophie, aber auf neurotischer Grundlage. Arch. Psychiatr. Nervenkr. 22 (1891) 437–480.

WERDNIG, G.: Die frühinfantile progressive spinale Amyotrophie. Arch. Psychiatr. Nervenkr. 26 (1894) 706–744.

Handbook of Clinical Neurology, Vol. 15 (59): Diseases of the Motor System
J.M.B.V. de Jong, editor
© Elsevier Science Publishers B.V., 1991

Non-progressive juvenile spinal muscular atrophy of the distal upper limb (Hirayama's disease)

KEIZO HIRAYAMA

Department of Neurology, School of Medicine, Chiba University, Chiba, Japan

This disease occurs in young persons, predominantly in males ranging from 15 to 25 years of age. The main clinical features include predominantly unilateral weakness of the fingers and hand, and atrophy of the hand and forearm, the so-called oblique amyotrophy. The clinical course is non-progressive.

Twelve patients with this disease were first reported by Hirayama et al. (1959), who distinguished this disorder clinically from previously known degenerative and progressive motor neuron diseases or any known category of diseases causing muscular atrophy. The clinical features have been further clarified by a report on 20 patients (Hirayama et al. 1963) and one on 38 patients (Hirayama 1972). More than 150 clinical cases have since been reported in Japan, including a series of 71 patients by Sobue et al. (1978).

Other countries providing clinical reports include Denmark (Pilgaard 1968), Holland (Compernolle 1973), Singapore (Loong et al. 1975), USA (Adornato et al. 1978), India (Singh et al. 1980; some of the patients of Gourie-Devi et al. 1984), Malaysia (Tan 1985), France (Leys and Petit 1987; Chaine et al. 1988; Gaio et al. 1989; Biondi et al. 1989), and Sri Lanka (some of the patients of Peiris et al. 1989).

Because of its benign course pathological study of this disease was not possible until 1982, a quarter of a century after its first clinical description, when we had the first opportunity to perform an autopsy in a patient with this disease who died from lung cancer (Hirayama et al. 1987). This showed a specific and asymmetrical lesion localized in the anterior horns of the cervical enlargement, most severe at C7 and C8.

CLINICAL FEATURES

Age of onset and sex

The disease develops in young persons between 15 and 25 years, mostly in males. Almost all cases are sporadic.

Familial incidence

Seems to be decidedly low. One family with a father and son, and two families with brothers have been reported (Hirayama 1972).

Insidious onset and cold paresis

The onset is insidious, with weakness of fingers and atrophy of hand and forearm muscles without precipitant infection or trauma. Many patients report that the weakness of the fingers easily worsens in a somewhat cold environment, and improves in a warm one. We have termed this phenomenon cold paresis. The patients often first notice their disease during the winter. The cold paresis continues after the arrest of the atrophic process.

Oblique amyotrophy and unilateral predominance

Atrophy of the muscles is confined to the hand and forearm, not accompanied by fasciculations at rest. In the hand, the thenar, hypothenar and interosseous muscles are involved to various degrees. In the forearm, the distal and ulnar part is involved but the brachioradialis muscle is spared. The border zone of the atrophy runs obliquely over both palmar and dorsal surfaces of the forearm from the middle portion of the radial border to the elbow (Fig. 1). The distribution of atrophy clearly shows that the lesion is in the C7 through T1 spinal segments.

Weakness of the fingers and wrist develops in both the extensor and flexor muscles. Usually the extensors of the fingers and the flexors of the wrist are more markedly involved. Relative sparing of the extensor carpi radialis leads to stronger radial than ulnar extension of the wrist. Pronation of the forearm is usually preserved.

There may be very mild atrophy of the triceps muscle, which is innervated by the C6 through C8 segments. The brachioradialis, biceps and shoulder girdle muscles innervated by the C5, C6 and/or higher segments are spared. There is no atrophy of the face, neck, trunk or leg muscles.

The muscular weakness and atrophy occur predominantly on one side and are entirely unilateral in more than half of patients. The atrophy of the less-affected arm is not only less marked but also more confined to certain hand muscles. Symmetrical involvement is very rare.

As will be stated later, the final degree of weakness and atrophy of the involved muscles differs from patient to patient once the disease process has been arrested.

Fine tremulous movements and fascicular twitchings on contraction

Although there is no involuntary movement in the resting fingers, moderate extension of the fingers will produce fine, fast, irregular and non-synchronous tremulous movements in each finger on the affected side. Simultaneously with these movements there may be irregular, recurrent, short twitchings in the dorsolateral portion of the forearm, similar to the fasciculation at rest seen in neurogenic atrophy. We call these fascicular twitchings on contraction. These tremor-like involuntary movements of the fingers and the fascicular twitchings on contraction of the extensor muscles of the forearm seem to be 2 different expressions of the same phenomenon of abnormal activity of the diseased muscles.

Slight and inconstant autonomic disturbances

Autonomic disturbances may be the cause of the cold paresis described above, which is seen in more than 4/5 of the patients. Other phenomena such as easily becoming cold, cyanosis, livedo reticularis and chilblains of the involved fingers in cold environment are also suggestive of vasomotor disturbance. In rare cases there is palmar hyperhidrosis. Horner's syndrome is very rarely seen, and anisocoria is noted by exception.

Negative symptoms and signs

Careful examination does not reveal sensory impairment in the involved hand and fingers in most cases. Occasional slight hypesthesia in a localized area on the dorsum of the hand, of which patients never complain spontaneously, can be found in rare cases.

Stretch reflexes of the arm muscles are all within normal range.

Fig. 1. Distribution of muscular atrophy. (A) Moderately affected patient at 20 years of age (2 years after onset). The left hand and forearm show oblique amyotrophy. The dorsal interosseous muscles of the right hand are also mildly atrophic. (B) Severely affected patient (autopsied). Atrophy is limited to the left hand and forearm, sparing the brachioradialis muscle. The right intrinsic hand muscles also show mild atrophy. (Reproduced from Hirayama et al. (1987) by courtesy of the Editors of *J. Neurol., Neurosurg. Psychiatry.*)

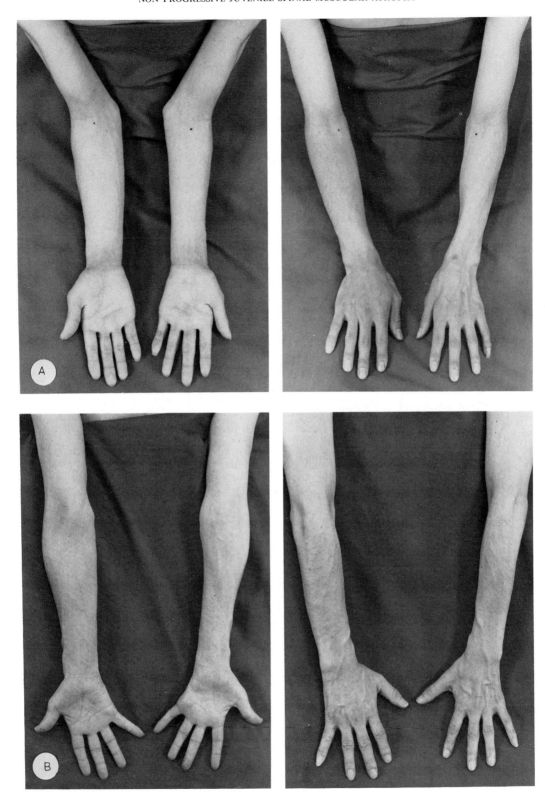

Except for a rare increase in the patellar and Achilles tendon reflexes, the legs exhibit no pyramidal tract signs or central motor deficit. There is no disturbance of micturition.

Non-progressive course

Insidious onset and slow initial development are characteristic of this disease. Its progression is arrested as a rule within 1–3 years and in rare cases within several years. The upper arm may show slight disuse atrophy in patients with severe weakness of the fingers.

LABORATORY DATA

Electrophysiological examinations

Electromyography (EMG) shows a typical neurogenic pattern in the atrophied muscles, indicating anterior horn involvement. The frequency of the neurogenic pattern varies in each muscle (Nagaoka et al. 1980) (Fig. 2). The muscles of the contralateral unaffected side in patients with unilateral involvement may also show neurogenic changes in the homonymous muscles, indicative of subclinical involvement of the lower motor

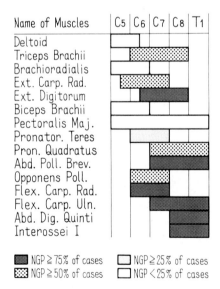

Fig. 2. Frequency of the neurogenic pattern (NGP) in EMG at each segmented level (26 patients). (Modified from Nagaoka et al. (1980) by courtesy of the Editors of *Brain and Nerve*.)

neuron in the non-atrophic arm. In about 10% of the patients, however, the muscles of the non-affected side show no abnormality. In less than 25% of the patients, the brachioradialis, biceps and deltoid muscles may show abnormal discharge. In more than half the cases, the triceps brachii shows a neurogenic pattern in the absence of atrophy. EMG of the legs shows a normal patterns. *Motor nerve conduction velocities* of the ulnar and median nerves are normal. These findings indicate segmental spinal cord involvement of C6 through C8 and T1.

Muscle biopsy

The atrophic flexor carpi ulnaris muscle shows typical neurogenic changes with small angular fiber grouping mixed with large type grouping indicative of reinnervation.

Cerebrospinal fluid (CSF)

CSF cells are normal. CSF protein may show a slight increase (40–60 mg/dl; normal range below 40) in half the patients. The standard Queckenstedt test may not show any abnormalities, but a more elaborate pressure curve analysis, in which the CSF pressure is measured serially for 60 s, before and after releasing bilateral jugular vein compression for 30 s, often shows a moderate but slight delay in the rise and fall of the pressure when the neck is flexed, while there is only a slight delay when the neck is in the neutral or extended position (Yamazaki and Hirayama 1990) (Fig. 3). The increase in CSF protein is noted in patients with the abnormal pressure curve in neck flexion. The abnormality in the Queckenstedt test is important and may well explain the pathophysiology of the disease process together with the characteristic cervical myelographic abnormalities which will be shown below. These abnormalities improve over the years after onset and finally disappear in patients whose clinical course exceeds 20 years (Tokumaru and Hirayama 1989; Yamazaki and Hirayama 1990).

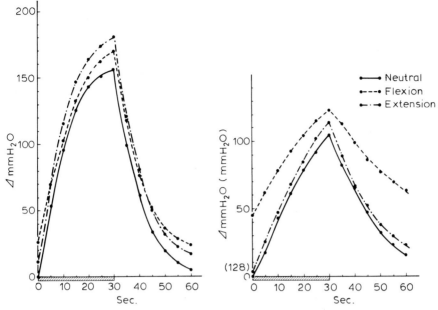

Fig. 3. Pressure curve analysis in the Queckenstedt test. (A) Normal control (average of 18 persons). See text concerning the method. (B) Hirayama's disease in early stage. A moderate delay in rise and fall of the pressure is noted with the neck flexed, but slight delay in the neutral or extended position (average of 20 persons). (Reproduced from Yamazaki and Hirayama (1990) by courtesy of the Editors of *Clinical Neurology*.)

Neuroradiological studies

X-ray studies of the cervical spine show no abnormalities, and specifically no spondylotic changes, dislocation, or vertebral canal stenosis. Flexion and extension of the neck are not restricted.

Myelographic studies have recently been reported which may elucidate the pathophysiology of the myelopathy of this disease (Yada et al. 1982; Mukai et al. 1985; Kikuchi et al. 1987). Our study (Tokumaru and Hirayama 1989) shows that the posterior wall of the lower cervical dural canal moves forward from the posterior surface of the vertebral canal when the neck is flexed (Fig. 4B). Thus, the dural canal is flattened anteroposteriorly and the spinal cord is compressed against the vertebral body (see also Fig. 6B). The forward displacement of the dural canal is most marked at the C6 vertebral body but disappears when the neck is in the neutral position (Fig. 4A). Extension of the neck would not displace the dural canal. These findings

correspond to the results of the pressure curve analysis of the Queckenstedt test mentioned above.

CT with intrathecal contrast medium (metrizamide) shows anteroposterior flattening of the cord at the level of the C5–C7 vertebral bodies, most marked at C6 level (Fig. 5). Flexion of the neck results in marked flattening, often with asymmetry, even when this is not apparent in the neutral position (Fig. 6). The flattening is more marked on the side with predominant muscular atrophy. Less marked flattening without muscular atrophy indicates subclinical anterior horn lesions, which cause a neurogenic EMG pattern.

Magnetic resonance imaging (MRI) shows the lower cervical epidural space between the posterior surface of the vertebral canal and the posterior wall of the dural canal, resulting from the forward displacement of the latter in flexion of the neck, as a high intensity area in the T1 and T2 weighted spin-echo images (Mukai et al. 1987) (Fig. 7). This finding is considered to represent congestion of the epidural venous plexus, the mechanism of which is yet unknown.

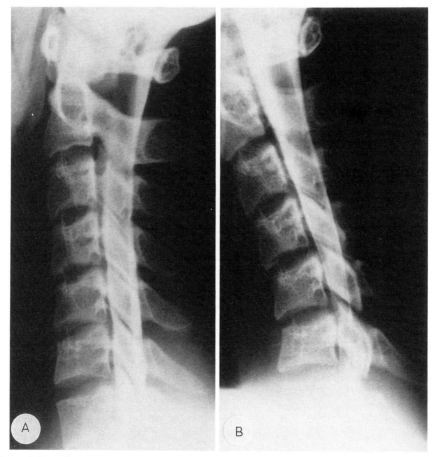

Fig. 4. Lateral view of myelogram. (A) The cervical dural canal is of normal position and diameter when the neck is in the neutral position. (B) Forward displacement of the dural canal with anteroposterior flattening and compression of the spinal cord at the level of the C5, C6 and C7 vertebral bodies in neck flexion.

Autonomic nervous system function tests

Several autonomic function tests were performed to clarify the cause of the cold paresis. Unpublished data on our 20 patients showed decreased skin temperature in the distal portion of the affected arm in thermography. Plethysmography showed a decreased amplitude in the same area, indicative of increased vasomotor function. Thermal sweating tests showed hyperhidrosis in the same area, indicative of sudomotor hyperfunction, in about 40% of the patients. Similar findings were reported by Kihara et al. (1988).

In about 10 patients there was no supersensitive dilatation of the pupils by epinephrine, which indicates normal postsynaptic sympathetic ocular function. In about 20% of the patients, decreased

dilatation of the pupil by cocaine was noted, suggestive of central sympathetic ocular nervous dysfunction.

These findings suggest that damage of the descending spinal sympathetic pathway is a possible cause of the cold paresis of the fingers.

CLINICAL DIAGNOSIS

The clinical features and laboratory studies in the diagnosis of this disease can be summarized as follows. (1) Incidence: It occurs almost exclusively in males of 15–25 years and is usually non-familial. (2) Main features: Insidious onset of weakness of the fingers and muscular atrophy of the hand, unilateral in many cases or asymmetric when bilaterally affected, often associated

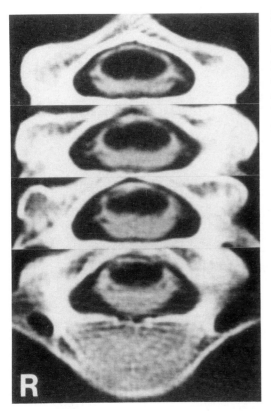

Fig. 5. CT with intrathecal contrast medium in the neutral position of the neck. Anteroposterior flattening of the spinal cord at the level of the C5 and C6 vertebral bodies, most marked at C6 level. More marked atrophy on the right (R) side with predominant muscular atrophy. From the top: C4, C5, C6 and C7 vertebral levels, respectively.

with cold paresis. Characteristic oblique amyotrophy in the forearm and hand. (3) Associated features: Fine, tremulous, irregular and asynchronous involuntary movements of the fingers on moderate extension, associated with fascicular twitchings on contraction in the extensor muscles of the forearm. (4) Negative features: In general, absence of sensory disturbance, reflex changes in the arms, pyramidal tract signs in the legs (very little, if any), ocular sympathetic dysfunction or urinary disturbance. (5) Clinical course: Non-progressive course and arrest within a few years after onset. (6) Laboratory data: Neurogenic changes in EMG and muscle biopsy in the atrophic muscles. Often subclinical EMG changes in the homonymous muscles of the non-atrophic side. Normal motor nerve conduction velocity.

Finally, localized and asymmetrical atrophy of the spinal cord at the lower cervical levels (most severe at the C6 vertebral body), forward displacement of the posterior wall of the lower cervical dural canal in neck flexion and absence of abnormalities of the vertebral canal verified by myelography, CT with intrathecal contrast medium or MRI.

Several conditions which may cause localized amyotrophy of the distal arm should be differentiated from this disease.

Syringomyelia

Oblique amyotrophy may occur in the early stage of syringomyelia involving the lower cervical cord with little or no sensory disturbance. MRI or CT with intrathecal contrast medium easily displays the syrinx.

Amyotrophic lateral sclerosis (ALS)

Similar distal amyotrophy of the hand and forearm may occur in the early stage, but oblique amyotrophy never occurs in the presence of brachioradialis atrophy in ALS. The muscular atrophy extends to the upper arm diffusely, together with typical fasciculations at rest. The EMG shows more diffuse, early neurogenic changes, even in the subclinical stage. The age of onset of ALS is much later and early onset of this disease is exceptional.

Cervical spondylosis associated with myelopathy

Distribution of the muscular atrophy is usually proximal, though it may rarely be distal and show oblique amyotrophy. The presence of sensory disturbance, older age onset and radiological studies usually allow easy differentiation.

Anterior tephromalacia (Marie and Foix 1912)

This condition may be similar in that the atrophy is localized to the hand muscles and the onset is insidious without sensory disturbance. However, the atrophy does not extend to the forearm and onset is generally after the age of 45.

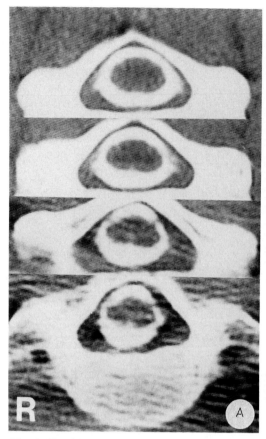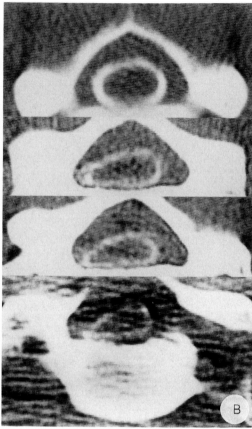

Fig. 6. CT with intrathecal contrast medium in flexion of the neck. Although the flattening of the cervical cord is not obvious in the neutral position (A), flexion of the neck results in forward displacement of the dural canal accompanied by marked and asymmetrical flattening of the spinal cord at the level of the C5–C7 vertebral bodies (B). From the top: C4, C5, C6 and C7 vertebral levels, respectively.

Acute anterior poliomyelitis

There is no history of polio, antecedent fever or signs of infection in our cases.

Spinal cord tumor

If this is localized in the lower cervical cord, distribution of the muscle atrophy may be similar. It is usually accompanied by sensory disturbance in the arm and pyramidal tract signs in the leg and CSF and radiological studies will easily differentiate the tumor.

Traumatic myelopathy

This has a different symptomatology, whether severe (Foo et al. 1982), mild (Hughes et al.

1964) or delayed (Lhermitte et al. 1942; André 1946). The possible relationship of the onset of Hirayama's disease to trauma will be discussed in the section on pathogenetic mechanisms.

Cervical rib and carpal tunnel syndrome

Muscular atrophy is usually confined to the small hand muscles, not extending to the forearm. Radiological or electrophysiological studies can easily differentiate them.

NEUROPATHOLOGICAL FINDINGS

The first autopsy study of the disease was reported by us (Hirayama et al. 1987). In this patient, weakness began in the left hand at the

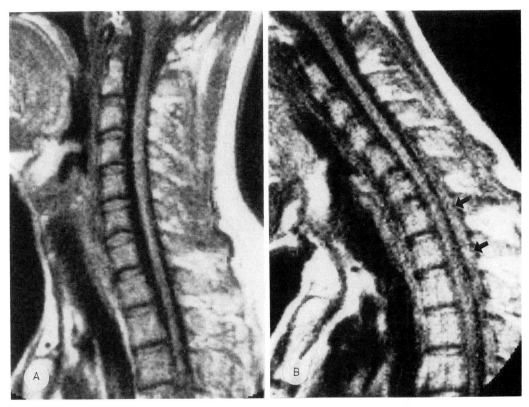

Fig. 7. Lateral view of MRI in flexion of the neck. (A) Neutral position of the neck. (B) Flexion of the neck results in widening of the posterior epidural space at the lower cervical level, shown as a high intensity area (arrows) in spin-echo image (TR 400 ms, TE 40 ms).

age of 15. The progress of the disease was arrested about 1 year after onset. At 34 the patient noticed mild fatigue and atrophy in the right hand, the onset of which had been insidious. The patient showed the typical symptoms and signs at 35, when he was admitted to our hospital (Fig. 1B). At 38 the patient died of lung cancer.

Macroscopically, the spinal cord showed evident anteroposterior flattening at C7 and C8. The anterior roots at C5 through T1, especially C7 through T1, were thin on both sides.

Microscopically, the anterior horns of the lower cervical cord were reduced to less than half the normal anteroposterior size on both sides, most severely at C7 and C8, with moderate asymmetry. The lesions in the anterior horns were also distributed bilaterally, and less markedly, up to C5 and down to T1 levels (Fig. 8). The center of

the lesions showed mild necrosis without cavity formation. At its periphery, there was a decreased number of large and small nerve cells. The surviving cells showed various degenerative changes such as lipofuscin accumulation, chromatolysis and shrinkage (Fig. 9). We noted mild astrogliosis without macrophage infiltration. The corresponding anterior roots, both intramedullary and extramedullary, were thin and sparse, more marked on the left where the muscular atrophy was more prominent. Mild thickening of the arachnoid was noted over the anterior surface of the cord, especially over the root exit zone. A mild decrease of lower cervical sympathetic ganglion cells was also noted.

The white matter, posterior horns and posterior roots were normal, as were the intramedullary and extramedullary vessels. No degenerative change of the spine was observed.

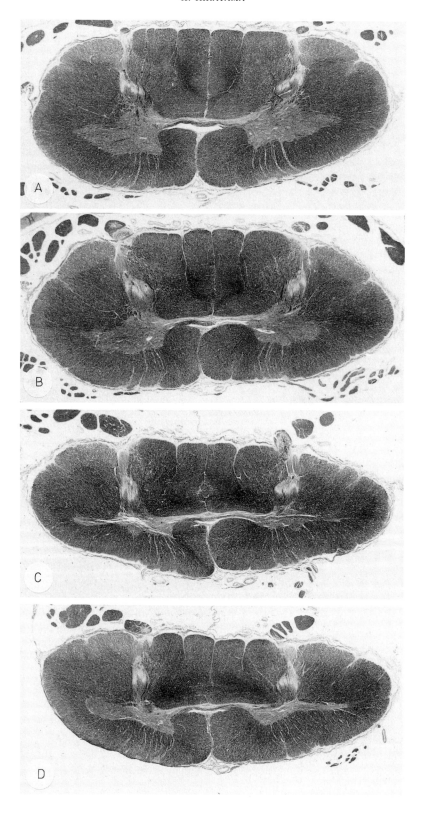

The second autopsy study was recently reported by Araki et al. (1989) in Japan. The onset was at age 24 and the patient died at 76. The symptoms and signs in early life were characteristic of the disease and limited to the right side. The neuropathological findings were compatible with those of the first patient except for a higher extension of the lesion up to C2, which was interpreted as secondary to concomitant cervical spondylosis.

POSSIBLE PATHOGENETIC MECHANISM AND PROPOSED THERAPY

The forward displacement of the posterior wall of the dural canal at the lower cervical level on neck flexion is presumed to be a main possible pathogenetic mechanism of this disease. This displacement is clearly shown by current neuroradiological techniques, but its etiology and time of occurrence are unknown. Because of the displacement, the spinal cord is compressed anteroposteriorly at the C7 and C8 segmental levels in neck flexion, as is also shown by CT with intrathecal contrast medium (Fig. 6) and the pressure pattern analysis of the Queckenstedt test (Fig. 3). This compression may cause microcirculatory disturbances in the territory of the anterior spinal artery or in the anterior portion of the spinal cord. The chronic circulatory disturbance resulting from repeated flexion or sustained flexed posture of the neck may produce necrosis of the anterior horns, which are most vulnerable to ischemia (Lapresle 1969).

The cold paresis of the fingers in most of the patients may be due to sympathetic nervous dysfunction, perhaps secondary to lesions of the descending spinal sympathetic pathway in the lateral funiculus (Kerr et al. 1964; Nathan et al. 1987), which surrounds the cervical anterior horn. The lesions may be the cause of the cold paresis. Selective involvement of the sympathetic pathways may spare the ocular sympathetic fibers, so that Horner's syndrome is very rarely noted in this disease.

Very few of our patients appeared to develop the disease after trauma. Trauma has been postulated in the etiology (Tsukagoshi et al. 1971), but clinical features similar to those of this disease have not been reported in cases of spinal cord injury. If there is a preexisting subclinical displacement of the posterior portion of the spinal dural canal, minor additional trauma may easily cause the myelopathy.

At present we have no explanation for the forward displacement of the dural canal. The displacement and the resultant compression of the spinal cord may be self-limiting, decreasing with the passage of time and finally disappearing after more than 20 years (Tokumaru and Hiray-

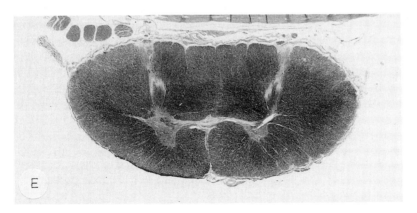

Fig. 8. Neuropathology of the cervical enlargement in the patient shown in Fig. 1B. (A) C5; (B) C6; (C) C7; (D) C8; (E) T1. The anterior horns at C7 and C8 are markedly shrunken anteroposteriorly on both sides. Lesser changes extend up to C5 and down to T1. Note the asymmetry of the lesion: more marked on the right (patient's left), on the side of predominant muscular atrophy. Klüver-Barrera stain. (Reproduced from Hirayama et al. (1987) by courtesy of the Editors of *J. Neurol., Neurosurg. Psychiatry.*)

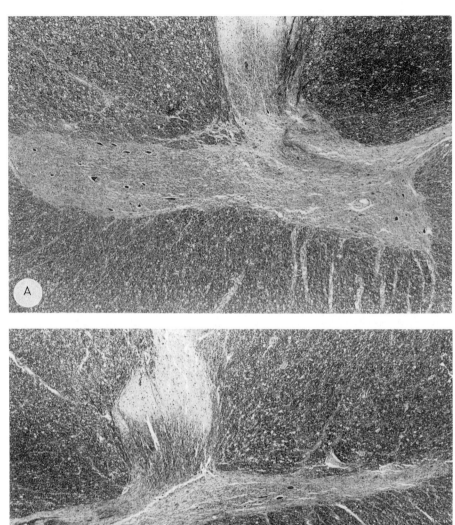

Fig. 9. Neuropathology of the anterior horns of C8. (A) right, (B) left. Shrinkage of gray matter with decreased numbers of large and small neurons, more marked in the central part where mild necrosis without cavity formation is observed. Klüver-Barrera stain. (Reproduced from Hirayama et al. (1987) by courtesy of the Editors of *J. Neurol., Neurosurg. Psychiatry*.)

ama 1989; Yamazaki and Hirayama 1990). This may explain the clinical course of initial progression and subsequent arrest.

Therefore, treatment should be aimed at early prevention and correction of this forward displacement of the dural canal, including devices that prevent excessive neck flexion, such as a neck collar.

REFERENCES

ADORNATO, B. T., W. K. ENGEL, J. KUCERA and T. E. BERTORINI: Benign focal amyotrophy. In: the 30th Annual Meeting of the American Academy of Neurology, Los Angeles, 27–30 April 1978. Neurology 28 (1978) 399 (Abstr).

ANDRÉ, M.: Sur une nécrose oedémateuse de la moëlle, plusieurs jours après un traumatisme fermé apparemment sans gravité. J. Belge Neurol. Psychiatrie 46 (1946) 439–449.

ARAKI, K., Y. UEDA, C. MICHINAKA, M. TAKAMASU, T. TAKINO and H. KONISHI: An autopsy case of juvenile muscular atrophy of unilateral upper extremity (Hirayama's disease). J. Jpn. Soc. Intern. Med. 78 (1989) 674–675.

BIONDI, A., D. DORMONT, I. WEITZNER JR., P. BOUCHE, P. CHAINE and J. BORIES: MR imaging of the cervical cord in juvenile amyotrophy of distal upper extremity. Am. J. Neuroradiol. 10 (1989) 263–268.

CHAINE, P., P. BOUCHE, J. M. LÉGER, D. DORMONT and H. P. CATHALA: Atrophie musculaire progressive localisée à la main — forme monomélique de maladie du motoneurone? Rev. Neurol. 144 (1988) 759–763.

COMPERNOLLE, T.: A case of juvenile muscular atrophy confined to one upper limb. Eur. Neurol. 10 (1973) 237–242.

FOO, D., A. BIGNAMI and A. B. ROSSIER: Posttraumatic anterior spinal cord syndrome — pathological studies of two patients. Surg. Neurol. 17 (1982) 370–375.

GAIO, J. M., B. LECHEVALIER, M. HOMMEL, F. VIADER, F. CHAPON and J. PERRET: Amyotrophie spinale chronique des membres supérieurs de l'adulte jeune (syndrome de O'Sullivan et McLeod) — étude en IRM de la moelle cervicale. Rev. Neurol. 145 (1989) 163–168.

GOURIE-DEVI, M., T. G. SURESH and S. K. SHANKAR: Monomelic amyotrophy. Arch. Neurol. 41 (1984) 388–394.

HIRAYAMA, K.: Juvenile non-progressive muscular atrophy localised in the hand and forearm — observation in 38 cases. Clin. Neurol. 12 (1972) 313–324.

HIRAYAMA, K., M. TOMONAGA, K. KITANO, T. YAMADA, S. KOJIMA and K. ARAI: Focal cervical poliopathy causing juvenile muscular atrophy of distal upper extremity: a pathological study. J. Neurol., Neurosurg. Psychiatry 50 (1987) 285–290.

HIRAYAMA, K., Y. TOYOKURA and T. TSUBAKI: Juvenile muscular atrophy of unilateral upper extremity — a new clinical entity. Psychiatr. Neurol. Jpn. 61 (1959) 2190–2197.

HIRAYAMA, K., T. TSUBAKI, Y. TOYOKURA and S. OKINAKA: Juvenile muscular atrophy of unilateral upper extremity. Neurology 13 (1963) 373–380.

HUGHES, J. T. and B. BROWNELL: Cervical spondylosis complicated by anterior spinal artery thrombosis. Neurology 14 (1964) 1073–1077.

KERR, F. W. L. and S. ALEXANDER: Descending autonomic pathways in the spinal cord. Arch. Neurol. 10 (1964) 249–261.

KIHARA, M., H. WATANABE, T. TOMITA, Y. SUZUKI and M. TAKAMIYA: A case of juvenile muscle atrophy of unilateral extremity, with special reference to local sweat rate and skin temperature. Autonomic Nerv. Syst. 25 (1988) 61–66.

KIKUCHI, S., K. TASHIRO, M. KITAGAWA, Y. IWASAKI and H. ABE: A mechanism of juvenile muscular atrophy localized in the hand and forearm (Hirayama's disease) — flexion myelopathy with tight dural canal in flexion. Clin. Neurol. 27 (1987) 412–419.

LAPRESLE, J.: Sur quelques aspects neuropathologiques des troubles de la circulation dans la moelle épinière. Bull. Schweiz. Akad. Med. Wiss. 24 (1969) 512–529.

LEYS, D. and H. PETIT: Amyotrophie juvénile distale chronique unilatérale localisée à un membre supérieur (type Hirayama) — un cas européen. Rev. Neurol. 143 (1987) 611–613.

LHERMITTE, J. and B. DE ROBER: La myelomalacie tardive par effort. Rev. Neurol. 74 (1942) 175–176.

LOONG, S. C., M. H. L. YAP and I. P. NEI: An unusual form of motor neuron disease. In: 4th Asian and Oceanic Congress of Neurology, 16–21 November, 1975. Bangkok, Sompong Press (1975) 35 (Abstr).

MARIE, P. and CH. FOIX: L'atrophie isolée non progressive des petits muscles de la main — téphromalacie antérieure. Nouv. Iconographie Salpêtrière 25 (1912) 353–363, 427–453.

MUKAI, E., T. MATSUO, T. MUTO, A. TAKAHASHI and I. SOBUE: Magnetic resonance imaging of juvenile-type distal and segmental muscular atrophy of upper extremities. Clin. Neurol. 27 (1987) 99–107.

MUKAI, E., I. SOBUE, T. MUTO, A. TAKAHASHI and S. GOTO: Abnormal radiological findings on juvenile-type distal and segmental muscular atrophy of upper extremities. Clin. Neurol. 25 (1985) 620–626.

NAGAOKA, M., K. HIRAYAMA, T. CHIDA, M. YOKOCHI and H. NARABAYASHI: Electromyographic analysis on juvenile muscular atrophy of unilateral upper extremity. Brain Nerve 32 (1980) 821–828.

NATHAN, P. W. and M. C. SMITH: The location of descending fibres to sympathetic preganglionic vasomotor and sudomotor neurons in man. J. Neurol., Neurosurg. Psychiatry 50 (1987) 1253–1262.

PEIRIS, J. B., K. N. SENEVIRATNE, H. R. WICKREMASINGHE, S. B. GUNATILAKE and R. GAMAGE: Non familial juvenile distal spinal muscular atrophy of upper extremity. J. Neurol., Neurosurg. Psychiatry 52 (1989) 314–319.

PILGAARD, S.: Unilateral juvenile muscular atrophy of upper limbs. Acta Orthop. Scand. 39 (1968) 327–331.

SINGH, N., K. K. SACHDEV and A. K. SUSHEELA: Juvenile muscular atrophy localised to arms. Arch. Neurol. 37 (1980) 297–299.

SOBUE, I., N. SAITO, M. IIDA and K. ANDO: Juvenile type of distal and segmental muscular atrophy of upper extremities. Ann. Neurol. 3 (1978) 429–432.

TAN, C. T.: Juvenile muscular atrophy of distal upper extremities. J. Neurol., Neurosurg. Psychiatry 48 (1985) 285–286.

TOKUMARU, Y. and K. HIRAYAMA: Anterior shift of posterior lower cervical dura mater in patients with juvenile muscular atrophy of unilateral upper extremity. Clin. Neurol. 29 (1989) 1237–1243.

TSUKAGOSHI, H., T. MANNEN and Y. TOYOKURA: Causative mechanism of juvenile muscular atrophy of unilateral upper extremity (Hirayama). In: the 36th Kanto regional meeting of the Japanese Society of Neurology, Tokyo, 27 February, 1971. Clin. Neurol. 11 (1971) 771 (abstr).

YADA, K., S. TACHIBANA and K. OKADA: Spinal cord lesion due to relative imbalance of cervical spine and cervical cord. In: 1981 Annual Report of Prevention and Treatment for the Congenital Anomalies of the Spine and Spinal Cord, The Minist. Health Welfare Jpn. (1982) 48–55.

YAMAZAKI, M. and K. HIRAYAMA: Queckenstedt test — pressure pattern analysis in spinal subarachnoideal block. Clin. Neurol. 30 (1990) 247–253.

Handbook of Clinical Neurology, Vol. 15 (59): Diseases of the Motor System
J.M.B.V. de Jong, editor
© Elsevier Science Publishers B.V., 1991

CHAPTER 9

Progressive bulbar paralysis of childhood

MANUEL R. GOMEZ

Section of Pediatric Neurology, Department of Neurology, Mayo Clinic and Mayo Medical School, Rochester, MN, USA

Progressive bulbar paralysis of childhood (PBPC) (211500) is an autosomal recessive heritable disease of the brain stem motoneurons manifested by progressive paralysis of muscles innervated by the cranial nerves. It is lethal in 7–25 months. Because isolated degeneration of the motoneurons of the brain stem in children is extremely rare and because it may be part of other disorders, the idea that PBPC is a clinical entity is not as yet universally accepted. The question is whether it is part of the spinal muscular atrophies (SMA), namely infantile progressive spinal muscular atrophy (SMA 1 or Werdnig-Hoffmann disease), SMA 2, and juvenile progressive spinal muscular atrophy (SMA 3 or Wohlfart-Kugelberg-Welander disease) (Namba et al. 1970); or the same entity as familial chronic progressive bulbopontine paralysis with deafness (Van Laere 1968)? Other disorders having common clinico-pathologic features with PBPC are juvenile motor neuron degeneration with or without corticospinal tract involvement and the Madras variety of progressive motor neuron disease (Spillane 1972). The controversy is apt to continue until the exact nature of the gene product of PBPC and related disorders is understood.

HISTORICAL REVIEW

Duchenne (de Boulogne) recognized progressive muscular atrophy in 1849 (Poore 1883) and de-scribed progressive bulbar paralysis with the name glosso-labio-laryngeal palsy in 1860. Duchenne's ideas of progressive disease of the motoneurons of the spinal cord and brain stem were supported by the pathologic findings first described by Charcot and Joffroy (1869). Bernhardt (1889) pointed out the hereditary nature of progressive bulbar paralysis of adults. Berger (1876) reported a 12-year-old child with progressive paralysis of muscles innervated by V, VI, X, and XII and bilateral pyramidal tract signs. Fazio (1892) de-scribed a 22-year-old woman and her 4.5-year-old son, both of whom had progressive bulbar paralysis. The boy had paralysis of the lower facial muscles and paresis of the superior facial muscles and tongue, dysarthria, dysphonia and difficulty breathing. He had no dysphagia. Deep reflexes and sensation were normal. His mother had similar symptoms since the age of 18 years.

Londe (1893, 1894) reported 2 brothers born to second cousin parents and first cousin paternal grandparents. The oldest boy had developed weakness of the orbicularis oculi and oris and of the tongue at the age of 6 years. His face was expressionless and he had weakness of the levators palpebrae, fasciculations about the chin, lower eyelids, zygomatic muscles and labial commis-sures, atrophy and fasciculations of the tongue, difficulty swallowing liquids and paresis of the vocal cord abductors. Deep reflexes and sensation were normal. His younger brother developed left facial paralysis starting in the upper facial muscles and hypoesthesia of the left cornea, palate and pharynx. The course of the disease was progressive for both patients but their fate is unknown.

Thomson (1891) reported a child with onset of bulbar signs at the age of 2 years and involvement of V, VI, and VII and Hoffmann (1891) another child with onset at 11 years and involvement of VII, X, XI and XII, respectively. There is not enough information to be absolutely certain that these 2 patients had PBPC rather than the more common pontine glioma, or other brain stem lesion. Equally questionable are the cases reported by Remak (1892), Filatow (1894), Tromner (1905), Marinesco (1915), Paulian (1922), and Van Bogaert (1925, 1951).

The patient reported by Van Bogaert in 1925 deserves special mention. This was a 7.5-year-old boy who gradually developed weakness of muscles innervated by VII so that the first symptom was difficulty pronouncing labial and lingual consonants. This was followed by atrophy of lips and cheeks and loss of facial expression.

Later, as his speech became unintelligible, he had dysphagia, atrophy and fasciculations of facial muscles, weakness of masseters and pterygoids, paresis of orbicularis oculi and palpebral ptosis, palatal paralysis and fasciculations and atrophy of the right side of the tongue. There were no pyramidal signs. The boy died at the age of 12.5 years or 5 years after the onset of symptoms and no autopsy was done. Van Bogaert reported in 1951 that at the time the case was published the patient had 2 older brothers, 9 and 10 years of age. The oldest brother was examined at the age of 12 years and followed clinically by Van Bogaert for 15 years. During this time he developed muscular atrophy and tendon hyperreflexia and later cerebellar and dorsal column signs, and spinal amyotrophy. The deep tendon reflexes disappeared and there were bilateral Babinski signs. His tongue became atrophic and there was a symmetric paralysis of the soft palate. Thus, although the younger brother appeared to have PBPC, the older had a more diffuse disease with involvement of the anterior horn cells, and the lateral and dorsal columns of the spinal cord.

TABLE 1

Verified cases of progressive bulbar paralysis of childhood.

Author	Year	Sex	Affected sibling	Age at onset (mos)	Symptoms at onset	Sequence of CN involvement	Duration (mos)	Age at death (yrs)
Gomez et al.	1962	F	Brother	33	Dyspnea, stridor, ptosis	X, III, VII, V, XII, VI	17	4.2
Alexander et al.	1976	M	None (1st child)	25	Dyspnea, stridor	X, V, VII, XII, III, VI	17	3.6
Della Giustina et al.	1979	F	None (4th child)	66	Dyspnea, stridor	X, VII, XII, III, IV, V	25	7.6
Benjamins	1980	M	Sister	29	Stridor, ptosis	X, III, XII, VII	7	3

TABLE 2

Unverified probable cases of progressive bulbar paralysis of childhood.

Author	Year	Sex	Affected sibling	Age at onset (yrs)	Symptoms at onset	Sequence of CN involvement	Duration (mos)
Londe	1893	M	Brother	5	Bil. facial palsy	VII, X, XII, V	?
Londe	1894	M	Brother	8	Bil. facial palsy	VII, X, XII, V	?

TABLE 3
Unverified cases of progressive bulbar paralysis of childhood.

Author	Year	Sex	Family history	Age at onset (yrs)	Symptoms at onset	Cranial nerves involved	Course
Berger	1876	M	–	12	Motor V, dysphagia, dysarthria	V, VII, X, XII	Unknown
Hoffman	1891	M	Neg.	11	Nasal voice and regurgitation	V, X, XI, XII	Died in 12 mos
Thomson	1891	F	–	2	–	V, VI, VII	Unknown
Fazio	1892	M	Mother PBP	4	–	VII, X, XII	Unknown
Remak	1892	F	Oldest of 3 sibs	12	Left facial palsy	VII, X, XII	Died in 9 mos
Filatow	1894	–	–	11	Weakness of lips	XII, X, VII, V	Unknown
Trömner	1905	M	–	10	Dysphagia, dysmetria, dysarthria	V, VII, X, XII	Unknown
Marinesco	1915	M	Sister	12	Dysarthria, dysphagia	V, VII, X, XI, XII	Alive 8 yrs later
Marinesco	1915	F	Brother	8	Chewing difficulty	III, V, VII, X, XI, XII	Alive 8 yrs later
Paulian Van Bogaert	1922 1925	M	Brother	7	Dysarthria, facial weakness	VI, VII, IX, X, XI, XII	Died in 12 mos
Van Bogaert	1951	M	Brother				
Beauvais	1988	F	Parent, consanguin.	10	Dysphonia, strabismus, dysphagia	VI, IX, X	Stable 5 yrs later

 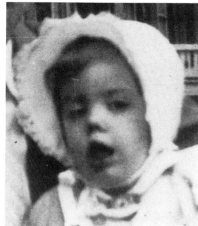

Fig. 1. Facial appearance of patient at age 6 months (left) when she was free of symptoms of PBPC, and at age 4 (right) 2 months before her death. (Gomez et al. 1962.)

Since 1962, 3 children have been reported who progressively developed bulbar paralysis and postmortem examination demonstrated isolated degeneration of the cranial nerve motoneurons (Gomez et al. 1962; Alexander et al. 1976; Della Gustina et al. 1979). In addition, in the sibship of the first verified case, a child was born who developed the same symptoms as his late sister and died with what can be accepted as typical PBPC (Benjamins, 1980). The 4 patients are listed in Table 1 as verified PBPC cases.

The 2 brothers reported by Londe are accepted only as probable cases of PBCP and are listed in Table 2. Although the older brother appears to be another case of PBPC, the younger brother had hypoesthesia of the cornea, palate and pharynx apparently due to involvement of the sensory neurons or their brain stem nuclei unlike proven cases of PBPC.

The mother and son reported by Fazio and 11 additional cases described in the medical literature since 1876 can only be classified as possible because other pathology was not ruled out by pathologic examination or by neuroimaging methods. There are so many unanswered questions about these 13 patients that it is preferable not to include them here; they have been listed in Table 3 and a bibliographic reference is given. Other reported cases of PBPC are completely excluded from this chapter because the information given by the authors is insufficient to place them in any of the 3 aforemen-

tioned groups. Some of them, like the patients reported by Brown (1894), Vialetto (1936), Arnould et al. (1968) and Trillet et al. (1970), are cases of Van Laere's disease (1968).

The eponym Fazio-Londe disease is no longer necessary and historically incorrect. For those who prefer to use eponyms, 'Londe's disease', as proposed by Della Gustina et al. (1979) should be more acceptable. Since Fazio (1892) described progressive bulbar paralysis in a woman and her son, his name could be appropriately applied to a variety of this disease with autosomal dominant inheritance.

NATURAL HISTORY OF PROGRESSIVE BULBAR PARALYSIS OF CHILDHOOD

A description of the clinical and pathological features of the only 4 known patients with unquestionable PBPC listed in Table 1 follows:

Case 1

Gomez et al. (1962) reported a white girl who had begun to drool after her first birthday and at age 33 months had a generalized convulsion. She had right palpebral ptosis; intramuscular neostigmine and intravenous edrophonium chloride tests were negative. Findings at age 34 months were: right palpebral ptosis, bilateral facial weakness, inspiratory stridor, hoarse voice, croupy cough, generalized hypotonia, and wide-based gait. Laryngoscopy disclosed paretic vocal cords. At age 38 months left palpebral ptosis appeared and was unable to close her eyes tightly; she drooled and was dysphagic. The EMG demonstrated no abnormalities in motor units innervated from spinal segments C5–T1 and L2–S2. Biopsy of gastrocnemius muscle showed no pathologic changes. At age 39 months she required nasogastric tube feeding, had bilateral ptosis, right abducens paralysis, weakness of the jaw muscles, bilateral facial paralysis, absent gag reflex, dysphonia, atrophy and fasciculations of the tongue, generalized hypotonia and weakness, symmetric tendon hyperreflexia, and flexor plantar reflexes. At 3.5 years she underwent a tracheostomy following an episode of non-convulsive apnea and cyanosis. She grew progressively weaker and developed bilateral unsustained ankle clonus. Figure 1 shows photographs taken at age 6 months and on her fourth birthday. At age 4 years 2 months she became febrile and lethargic and died.

Postmortem findings included aspiration pneumonia, malnutrition, and a dysplastic kidney. Microscopic examination of the central nervous system demonstrated depletion of nerve cells in cranial nerve nuclei III, IV, VI, VII, X, and XII (Fig. 2). Many neurons were small and pyknotic, and in the nucleus of XII there was ballooning of cells without gliosis. There were depletion and degeneration of neurons in the dentate nuclei. Degenerating nerve cells in various stages of shrinkage, nuclear swelling, and pyknosis were present throughout the ventral horns in the cervical and upper thoracic spinal cord (Fig. 3). In the dorsal spinocerebellar tract and part of the restiform body there were slight swelling of medullated fibers and small collections of neutral fat thought to have been due to acute terminal changes from hypoxia. The pyramidal tracts were intact. In the extraocular muscles there was marked variation in diameter of muscle fibers and fiber type grouping. Atrophy was less severe in cervical, intercostal, and diaphragmatic muscles but in these there was also variation in muscle fiber diameter.

Case 2

Alexander et al. (1976) reported a boy born after a full-term gestation to a 14-year-old primipara with no history of neuromuscular disease, he had developed normally although did not walk until age 17 months. At age 24 months he had progressive exertional dyspnea and inspiratory stridor unassociated with fever. Laryngoscopy showed decreased vocal cord movement. Tracheostomy relieved the dyspnea and he remained dependent on the tracheostomy. The neurologic examination was unremarkable. Seven weeks later he developed facial weakness over a 1-week period and the examination showed weakness and atrophy of the masseter, temporalis, facial, sternocleidomastoid and tongue muscles. An edrophonium chloride test was negative. In the subsequent 2 weeks he had tremulous arms, ataxic gait, and dysphagia. Further examination disclosed left palpebral ptosis, right lateral rectus palsy, bilateral mild intention tremor of the arms, and titubating broad-based gait. Laryngoscopy disclosed complete vocal cord paralysis. By age 13 months he was having recurrent aspiration and had developed pneumonia. The neurologic examination was unchanged. The EMG of upper and lower limbs and the motor nerve conduction velocity of median and common peroneal nerves were normal. There was no decremental response to repetitive stimulation of the median nerve with recording in the thenar muscle. During the final 2 months of his life he was fed by gastrostomy and suffered repeated respiratory infections and pulmonary insufficiency. He died at the age of 42 months.

Pathologic examination disclosed aspiration of gastric contents and early focal pneumonitis. Gross examination of the CNS and trunk musculature was unremarkable. Microscopic examination of the XII nucleus and motor nuclei of VII and V nerves showed 'virtual absence of neurons and variable amounts of glial proliferation'. The motor neurons of VI were reduced in number and those present showed swelling, dissolution of Nissl substance, eccentric nuclei with vacuolar change and shrunken neurons with pyknotic nuclei. The nuclei of III and IV appeared normal. The nucleus ambiguus was not clearly identified,

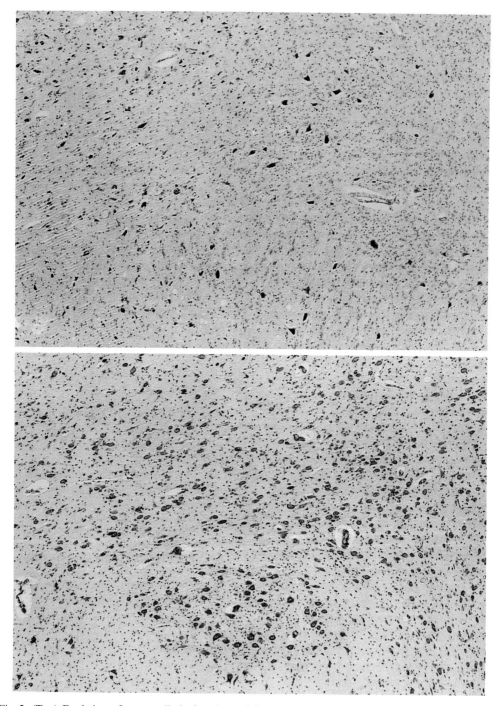

Fig. 2. (Top) Depletion of nerve cells in fourth cranial nerve nucleus. (Bottom) Nucleus of same nerve of normal child of same age. Nissl stain. (Gomez et al. 1962.)

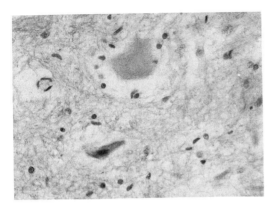

Fig. 3. 'Balloon' cell in ventral horn of cervical spinal cord. Hematoxylin-eosin stain. (Gomez et al. 1962.)

probably due to degeneration. The remainder of the brain stem structures were unremarkable. There were scattered areas of loss of Purkinje cells in the cerebellum, some of which were undergoing degeneration. The dentate nucleus and cerebellar peduncles were unremarkable. There were no hypoxic changes in the hippocampus or any abnormality of the pyramidal tract or cerebral cortex including the motor area. Microscopic examination of the upper cervical spinal cord showed loss and degeneration of anterior horn cells but in the thoracic and lumbosacral regions there was an adequate number of anterior horn cells which were morphologically normal. The intrinsic laryngeal muscles showed microscopic changes compatible with denervation and similar changes were found in the cervical strap muscles.

Case 3

Della Gustina et al. (1979) reported a 5.5-year-old girl, the fourth child of healthy parents, who developed acute respiratory distress and inspiratory stridor. She had been in good health and did not have previous hoarseness, cough or symptoms of respiratory infection. Laryngoscopy showed paralysis of the vocal cord abductor. She also had left facial and palatal weakness, and tongue atrophy. The respiratory difficulty improved slowly and after 3 months the endotracheal tube was removed. Over the next 2 years the patient developed bilateral facial paralysis,

dysarthria, and dysphagia. She then developed inspiratory stridor and had complete paralysis of the palate, and vocal cords, facial diplegia, atrophy and fasciculation of the tongue and variable paralyses of III and IV with ptosis and strabismus. There was no paralysis or atrophy of the limb or trunk muscles but there were fasciculations of the shoulder muscles. An electromyogram showed partial denervation of the upper and lower limb muscles. Motor nerve conduction velocity was normal. Two months later at the age of 7.5 years, the patient died of respiratory arrest.

Pathological examination showed no gross abnormalities of the brain, brain stem, cerebellum or spinal cord. Microscopic examination showed neuronal depletion in the caudal part of III and nearly complete depletion in IV. The Edinger-Westphal, Darkschewitsch and Cajal nuclei red nucleus and substantia nigra were normal. The severely affected motor nuclei of V and VII only had a few neurons left and these displayed central chromatolysis and eccentric nuclei. There was moderate neuronal loss in the nucleus of VI and ambiguus and complete neuronal loss in the XII nerve nucleus. The dorsal nucleus of X and the gracilis, cuneate and inferior olive nuclei were normal. There was moderate loss of anterior horn cells associated with glial reaction.

Case 4

Benjamins (1980) reported the brother of Case 1. He was born 5 years after his mother gave birth to a set of twin girls who were healthy, and 6 weeks prematurely. Birth weight was 2386 g. There were no neonatal difficulties. Early motor development was normal: he sat alone at 7 months and walked alone at 10 months. At 29 months he developed inspiratory stridor. On examination he had mild bilateral ptosis, fasciculations of the tongue, hyperactive deep tendon reflexes and normal plantar responses. Laryngoscopy demonstrated paralysis of the right and paresis of the left vocal cord. He underwent a tracheostomy. Needle examination of the orbicularis oculi and oris demonstrated increased number of positive sharp waves; there were no fibrillations. By the age of 32 months he had developed dysphagia, choking episodes, bilateral ptosis, right greater than left, and facial diplegia. He had no gag reflex and had deep tendon hyperreflexia. At age 33 months fluoroscopy demonstrated bilaterally diminished diaphragmatic movements. As the respiratory problems increased, his muscle strength decreased, he underwent generalized muscle wasting and died at the age of 36 months, 7 months after onset of symptoms. No postmortem examination was done. Figure 4 shows the pedigree of Cases 1 and 4.

The onset of symptoms in the proven cases has always been before or at the age of 5 years [33 months (Gomez et al. 1962), 29 months (Alexander et al. 1976), 5 years (Della Gustina et al. 1979), and 25 months (Benjamins 1980)]. The probable cases, those 2 brothers reported by Londe (1983, 1984) had their clinical commencement at age 5 years and 8 years. In all 6 cases laryngeal stridor was either the presenting or one of the first clinical features. When first examined, 2 of these 6 patients had ptosis, 3 had dyspnea and 2 had facial diplegia. All 3 patients (Cases 1, 2 and 3) pathologically examined had died between 17 and 25 months after onset of symptoms. The first symptoms of Case 4, the brother of Case 1, were ptosis and stridor and he only lived 7 months afterwards while his sister lived 17 months after onset of symptoms.

All 4 unquestionable cases of PBPC had gradual onset of symptoms and signs of motor neuron loss in all cranial nerves III–XII, starting with X alone or X and III and followed by VII, XII, V, VI and IV. Possibly, motor V involvement occurred earlier than recognized and the same

can be said of cranial nerves IV, VI, VII and XII. It should be noted that it is easier to find palpebral ptosis and laryngeal stridor due to vocal cord paresis than it is to detect early weakness of masseters, temporal, superior oblique, lateral rectus or facial muscles or even tongue fasciculations in a patient in respiratory distress or still too young to cooperate. Although cranial nerve paralysis is eventually bilateral, levator palpebrae, vocal cord, facial muscles, masseter and temporal muscle paresis have been often unilateral for months before becoming bilateral. Dysphagia and aspiration of secretions or food was severe enough in all patients to require a permanent nasogastric tube, suctioning, and tracheostomy 6 months to 1 year after onset of symptoms.

The EMG showed signs of denervation of the facial muscles in 2 patients. The patient reported by Della Gustina et al. (1979) had EMG signs of denervation in muscles of upper and lower extremities. This was not present in the patients reported by Gomez et al. (1962) and Alexander et al. (1976).

Death occurred at the age of 3, 3.6, 4.2 and 7.6 years from respiratory insufficiency often complicated by pneumonia.

NOSOLOGY

Is PBPC an entity of its own different from SMA 1? The pathology in the 3 patients with autopsy was monotonously similar to each other. The changes in the motor nuclei of the brain stem are also similar to those seen in the spinal anterior horn cells of patients with any of the 3 varieties of spinal muscular atrophy. The autosomal recessive inheritance in Cases 1 and 4 and in the probable cases reported by Londe (1893, 1894) place this disease nosographically close to other motoneuron degenerations of the young. But this is not to say that this disease is identical in its pathology to SMA 1 (Werdnig Hoffmann's disease). As shown in Table 4, the motoneurons primarily affected in PBPC are in the mesencephalic-ponto-bulbar nuclei while in SMA 1 are those of the anterior horn cells and only late in the course of the disease are the lower cranial nerve nuclei affected. The form of onset in PBPC

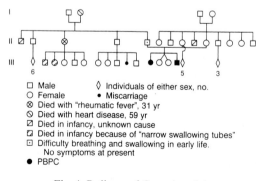

Fig. 4. Pedigree of Cases 1 and 4.

TABLE 4
Montoneurons affected in PBPC, SMA 1 and Van Laere's

	III	N	Vm	VI	VII	VIII	Xa	XI	XII	AHC
PBPC	+	+	+	+	+		+		+	±
SMA	±		+	±	+		+	+	+	+
Van Laere	±		+	±	+	+	+	±	+	+

is abrupt, always between the patient's second and sixth birthdays and the disease is fatal within 2 years. From the clinical point of view there are no other important differences: they are both of autosomal recessive inheritance. Pathologically, the differences are in the distribution of the lesions as shown in Table 4. In addition, the thalamus, Clarke's column and dorsal root ganglion cells may undergo degeneration in SMA 1 (Banker and Engel 1986).

No such changes were present in the 3 cases of PBPC although slight swelling of the medullated fibers of dorsal spinocerebellar tract was found in Case 1. This was associated with neural fat collection and there were also similar focal changes in a small part of a restiform body and cerebellar vermis which were attributed to acute terminal hypoxia and being artificially ventilated terminally. The clinical evidence of anterior horn cell involvement comes late and is meagre and the microscopic changes are less severe in PBPC than in SMA 1. In brief, the 2 diseases are different clinicopathological entities although their similarities are striking.

Are PBPC and progressive bulbopontine paralysis with deafness (211530) the same disease? The latter, also known as Van Laere's syndrome (1966) was reported earlier by Brown (1894) and Vialetto (1936). Its clinical features are distinct: the first symptom to appear is deafness. Secondly, facial paralysis appears which may be more severe in the upper face (Alberca et al. 1980). Thirdly, the disease is chronic despite an early onset of bulbar signs (Alberca et al. 1980). As the disease progresses, there is atrophic paralysis greater in the proximal than the distal muscles of all extremities and the muscles innervated by the lower cranial nerves (Robain et al. 1981). The muscle stretch reflexes may be initially exaggerated and plantars may be extensor. There is involvement of the vestibular in addition to

the cochlear nerve nuclei and the sensory nucleus of V may also be affected. Once signs of X nuclei involvement appear, the progression may be rapid towards death. Although there is involvement of cranial nerve nuclei of III, V, and VI both clinically and pathologically, it is less severe than in PBPC and SMA 1. On the other hand, the corticospinal tract is definitely involved (Alberca et al. 1980). There have been patients who, in addition, had optic nerve atrophy (Khaldi et al. 1987), retinitis pigmentosa (Alberca et al. 1980), or cerebellar ataxia (Alberca et al. 1980). It seems clear that, proposed by several authors (Della Gustina et al. 1979; Alberca et al. 1980), this is a nosological entity different from PBPC. It is also of hereditary nature and apparently autosomal recessive with female predominance.

Another disorder to be considered is juvenile type of slowly progressive bulbar palsy. Markand and Daly (1971) described a 35-year-old woman who had onset of dysphagia at 13 and, within 3 years, developed nasal speech and weakness of the orbicularis oculi. At the age of 25 years she had weakness of the muscles of the neck and of mastication. Masseters and temporalis muscles were wasted, the soft palate was immobile, and her voice was weak and hoarse. There was also wasting of the upper limb muscles, greater proximally. In brief, the patient had late onset of a slowly progressive disease of bulbopontine motor nuclei. Markand and Daly (1971) compared their patient to patients with the juvenile type of progressive spinal muscular atrophy described by Wohlfart (1942) and by Kugelberg and Welander (1956).

Finally, a unique form of disease of the motoneurons has been reported by Yokochi et al. (1989). This was a boy who developed progressive weakness of muscles innervated by the lower cranial nerves and spasticity of the lower extremities since the age of 3 years and died of bronchopneumonia at the age of 5 years. Pathologic examination demonstrated degeneration of upper and lower motoneurons and spinocerebellar and olivocerebellar tracts. These authors found Lewy body-like intraneuronal inclusions with trilaminar membranous profile on electromicroscopic examination in the anterior horn cells and neurons of the dorsal nucleus of Clarke, facial nerve nucleus, inferior olive and substantia nigra. It can be speculated that this isolated and extensive form of neuronal degeneration is possibly of viral etiology and different from PBPC.

REFERENCES

ALBERCA, R. and C. MONTERO: Paralisis bulbar cronica infantil esporadica, amiotrofia espinal, sordera degenerativa y retinitis pigmentaria. Arch. Neurobiol. 36 (1973) 3–14.

ALBERCA, R., C. MONTERO, A., IBAÑEZ, D. I. SEGURA and G. MIRANDA-NIEVES: Progressive bulbar paralysis associated with neural deafness. A nosological entity. Arch. Neurol. 37 (1980) 214–216.

ALEXANDER, M. P., E. S. EMERY and F. C. KOERNER: Progressive bulbar paresis in childhood. Arch. Neurol. 33 (1976) 66–68.

ARNOULD, G., P. TRIDON, M. LAXENAIRE, L. PICARD, M. WEBER and B. BRICHET: Paralysie bulbo-pontine chronique progressive avec surdite: à propos d'une observation de syndrome de Fazio-Londe. Rev. Oto-neuro-ophtal. 40 (1968) 158–161.

BANKER, B. Q.: The Pathology of Motor Neuron Disorders, Chapter 71, In: A. G. Engel, and B. Q. Banker, (Eds.) Myology. New York, McGraw-Hill Book Company, 1986.

BEAUVAIS, P., A. ROUBERGUE, T. B. DE VILLEMENT, and J. M. RICHARDET: Paralysie bulbopontine progressive de l'enfant. Arch. Fr. Pédiatr. 45 (1988) 653–655.

BENJAMINS, D.: Progressive bulbar palsy of childhood in siblings. Ann. Neurol. 8 (1980) 203.

BERGER: Schlesische Gesellschaft für vaterländische Cultur. Berl. Klin. Wschr. 13 (1876) 234.

BERNHARDT, M.: Über eine hereditäre Form der progressiven spinalen mit Bulbarparalyse complicierten Muskelatrophie. Virchows Arch. Pathol. Anat. 115 (1889) 197–216.

BROWN, C. H.: Infantile amyotrophic lateral sclerosis of the family type. J. Nerv. Ment. Dis. 21 (1894) 707–716.

CHARCOT, J. M. and A. JOFFROY: Deux cas d'atrophie musculaire progressive: avec lésions de la substance grise. Arch. Physiol. Norm. Pathol. 2 (1869) 354–367; 629–649; 744–760.

DELLA GUSTINA, F., G. FERRIERE, N. EVROIRD and G. LYON: Progressive bulbar paralysis in childhood (Londe syndrome). A clinico-pathological report. Acta Paediatr. Belg. 32 (1979) 129–133.

DUCHENNE, G. B. A.: Paralysie musculaire progressive de la langue, du voile de palais et des lèvres: affection non encore décrite comme espèce morbide distincte. Arch. Gen. Med. 106 (1860) 283–296; 431–445.

FAZIO, F.: Ereditarieta della paralisi bulbare progressiva. Rif. Med. 4 (1892) 327.

FILATOR, N.: Ein Fall von Bulbärparalyse bei einem 11 jährigen Knaben. Neurol. Cbl. 13 (1894) 717–718.

GOMEZ, M. R., V. CLERMONT and J. BERNSTEIN: Progressive bulbar paralysis in childhood (Fazio-Londe disease): report of a case with pathologic evidence of nuclear atrophy. Arch. Neurol. (Chic.) 6 (1962) 317–323.

HOFFMANN, J.: Ein Fall von chronischer progressiver Bulbärparalyse im kindlichen Alter. Dtsch. Z. Nervenheilk. 1 (1891) 169–172.

KHALDI, F., B. BENNACEUR, N. KANOUN, D. DEY, F. R. COUERFALA and S. SAMMOND: Un nouveau cas de paralysie bulbo-pontine progressive avec surdite et atrophique optique. Ann. Pediatr. 34 (1987) 731–733.

KUGELBERG, E. and L. WELANDER: Heredofamilial juvenile muscular atrophy. Arch. Neurol. (Chic.) 75 (1956) 500–509.

LONDE, P.: Paralysie bulbaire: progressive, infantile et familiale. Rev. Med. (Paris) 13 (1893) 1020–1030.

LONDE, P.: Paralysie bulbaire: progressive, infantile et familiale. Rev. Med. (Paris) 14 (1894) 212–254.

MARINESCO, G.: Sur deux cas de paralysie bulbaire progressive, infantile et familiale. C. R. Soc. Biol. (Paris) 78 (1915) 481–483.

MARKAND, O. N. and D. D. DALY: Juvenile type of slowly progressive bulbar palsy: report of a case. Neurology (Minneap.) 21 (1971) 753–758.

NAMBA, T., D. C. ABERFELD and D. GROB: Chronic proximal spinal muscular atrophy. J. Neurol. Sci. 11 (1970) 401–423.

PAULIAN, E. D.: Contributions cliniques à l'étude de la paralysie bulbaire infantile familiale. Rev. Neurol. 38 (1922) 275–278.

POORE, G. V.: Selections from the Clinical Works of Dr. Duchenne (de Boulogne). London, The New Sydenham Society (1883).

REMAK, E.: Zur Pathologie der Bulbarparalyse. Arch. Psychiatr. Nervenkr. 23 (1892) 919–960.

ROBAIN, O., G. PONSOT, R. HULIN and J. ARTHUIS: Les paralysies bulbaires progressives juvéniles. Arch. Franc. Pediatr. 38 (1981) 19–24.

SPILLANE, J. D.: The geography of neurology. Br. Med. J. 2 (1972) 506–512.

THOMSON, J.: Exhibition of patients: chronic cerebral lesions. Edinburgh Med. J. 37/1 (1891) 262–263.

TRILLET, M., P. F. GIRARD, B. SCHOTT, P. RAMEL and R. WOEHRLE: La paralysie bulbo-pontine chronique progressive avec surdit: à propos d'une observation clinique. Lyon Med. 223 (1970) 145–153.

TRÖMNER: Infantiler progressiver Bulbarparalyse. Neurol. Cbl. 24 (1905) 72.

VAN BOGAERT, L.: La sclerose laterale amyotrophique et la paralysie bulbaire progressive chez l'enfant. Rev. Neurol. 41 (1925) 180–192.

VAN BOGAERT, L.: Sur ces formes d'hérédoataxie de l'enfant et de l'adolescent qui comportent une atteinte grave des noyaux moteurs spinobulbomésencephaliques. Rev. Neurol. 84 (1951) 121–130.

VAN LAERE, J.: Over een nieuw geval van chronische bulbopontiene paralysis met doofheid. Verh. Vlaam. Akad. Geneesk. Belg. 30 (1968) 288–308.

VIALETTO, E.: Contributo alla forma ereditaria della paralisi bulbare progressiva. Riv. Sper. Freniat. 60 (1936) 1–24.

WOHLFART, G.: Zwei Fälle von Dystrophia musculorum progressiva mit fibrillären Zückungen und atypischem Muskelbefund. Ein Beitrag zur Frage des Vorkommens von Übergangsformen zwischen progressiver Muskelatrophie und neuraler progressiver Muskelatrophie. Dtsch. Z. Nervenheilk. 153 (1942) 189–204.

YOKOCHI, K., M. ODA, J. SATOH and Y. MORIMATSU: An autopsy case of atypical infantile motor neuron disease with hyaline intraneuronal inclusions. Arch. Neurol. 46(1) (1989) 103–107.

Handbook of Clinical Neurology, Vol. 15 (59): Diseases of the Motor System
J.M.B.V. de Jong, editor
© Elsevier Science Publishers B.V., 1991

CHAPTER 10

Progressive dysautonomias

OTTO APPENZELLER

Lovelace Medical Foundation, Biomedical Research Division, Albuquerque, NM, USA

CLASSIFICATION

Proper diagnosis is the first step in accurate management of any disease, and autonomic failure (AF) is no exception. Most forms of autonomic failure, however, are insidious in onset and non-specific in symptomatology. Autonomic dysfunction involves failure of autonomic reflex function, which preserves homeostasis of the organism and its response to changes in environment and stress. Usually when diseases comprising a major diagnostic category (e.g., diabetes) are diagnosed, a diagnosis of secondary autonomic failure is made in association with that disease. Thus, in patients with diabetic autonomic failure, laboratory tests and associated symptoms and signs point to structural abnormalities in autonomic reflexes and in their pathways, caused by or at least commonly associated with diabetes. There are many other patients, however, whose pathology or biochemical deficits are less well defined but whose autonomic symptoms are similar to those of diabetics.

Signs of autonomic failure may be discovered with special tests during life. After death, pathologic findings at postmortem may suggest a preferential degeneration of parts of the autonomic nervous system. These disorders are classified, for want of a better term, as primary autonomic diseases. Autonomic failure commonly

occurs in diabetes and alcoholism as well as in a large number of chronic, acute, and subacute neuropathies. They are easily distinguishable from primary autonomic failure, an unexplained and often selective damage to the autonomic structures, because autonomic failure in this setting is classified by association with other diseases. In primary autonomic failure, however, selective damage to autonomic structures and degeneration are usually found without underlying disease. Without any other neurologic disease or signs of nervous system dysfunction, autonomic failure, a rare disorder formerly called idiopathic orthostatic hypotension (Bradbury and Eggleston 1925), is classified as 'pure'. These authors suggested that because postural hypotension was the chief and most disabling symptom in autonomic failure, it was the only feature of this disease. They ignored the less apparent neurologic disturbances associated with autonomic failure: disturbances of the bladder, sexual dysfunction, and heat intolerance due to abnormalities in sweating. For this rare disease, the term 'pure autonomic failure' is at present preferable, even though the pathogenesis and, for that matter, the postmortem pathologic features are not well known.

Shy and Drager (1960) reported for the first time the syndrome that was later named after them. The syndrome, which they studied in 2 patients, consisted of the following symptoms: orthostatic hypotension, urinary and rectal incontinence, loss of sweating, iris atrophy, external ocular palsies, rigidity of limbs, distal tremor, loss of arm movements associated with walking, impotence, loss of rectal sphincter tone, atonic bladder, fasciculations, and wasting of distal muscles. They also found evidence of neuropathic lesions in the electromyogram, suggesting involvement of the anterior horn cells, and a neuropathic pattern on muscle biopsy. They observed that the onset of the disease is usually after the fifth decade of life. Johnson et al. (1966) were the first to note that a degeneration of intermediolateral cell columns accompanies postural hypotension, and from their pathologic material, they suggested that olivopontocerebellar atrophy should be linked with the Shy-Drager syndrome.

Autonomic failure frequently accompanies otherwise clinically typical Parkinson's disease (Appenzeller and Goss 1971; Vanderhaegan et al. 1970). The classic pathologic feature of Parkinson's disease, the Lewy body, is a hyaline eosinophilic cytoplasmic neuronal inclusion. This inclusion is also found in patients with pure autonomic failures in some neurons, particularly in those of the hypothalamus. Thus, the diagnosis of Parkinsonism may sometimes be premature if symptoms of autonomic failure have not yet made their appearance. Early on, the symptoms of Parkinson's disease may be indistinguishable from those of the disorder previously called Shy-Drager syndrome and now called multiple system atrophy. Pure autonomic failure is also difficult to distinguish in its early stages from Parkinson's disease. It may be erroneously diagnosed in patients who actually suffer from multiple system atrophy only recognizable after some period of disability. In patients with olivopontocerebellar atrophy (OPCA), on the other hand, significant postural hypotension may develop early or late in the disease. Nevertheless, a careful clinical examination of a number of such patients (Miyazaki 1978) shows a high incidence of postural hypotension and of urinary sphincter disturbances. In such patients, pyramidal and extrapyramidal symptoms and signs were common, whereas in those with pure olivopontocerebellar atrophy without major autonomic deficits, such symptoms were infrequent.

Evidence is emerging that all patients with primary autonomic failure (with the exception of pure autonomic failure) have loss of intermediolateral cell columns, which are the final common pathways for the sympathetic nervous system. The pathogenetic mechanism for this atrophy, however, is unclear. Different insults to these neurons (found in Parkinson's disease and MSA) may account for the onset of atrophy in both diseases. That is because cell loss may be associated both with neuron destruction in the intermediolateral cell columns and with the diversity of clinical manifestations in patients with primary autonomic failure. In pure autonomic failure, there is an additional loss of ganglionic neurons, which remain relatively intact in multiple system atrophy. This more distal involvement (in the hierarchy of the autonomic nervous system) in pure autonomic failure is in accord with the evidence from examination of plasma norepinephrine (NE) levels, which are low in PAF, whereas the levels in MSA are normal (Bleddyn-Davies and Sever 1988).

Patients with primary autonomic failure are uncommon. It is vital, therefore, if treatments and results of special investigations are to be compared, that their classification be precise and fairly uniform. Physiologic and pharmacologic tests and careful clinical examination, in addition to non-invasive tests of autonomic function, are necessary during life, and careful criteria for the postmortem diagnosis are also important in assessing the results of therapeutic trials in retrospect. Thus, a classification of primary autonomic failure that is now proposed is useful for further study (Table 1).

TABLE 1
Classification of primary autonomic failure.

1. Pure autonomic failure (PAF), referred to in the past as idiopathic orthostatic hypotension (Bradbury and Eggleston 1925).
2. Autonomic failure (AF) with multiple system atrophy (MSA), including striatonigral degeneration (SND), and olivopontocerebellar atrophy (OPCA); AF with MSA, also known as Shy-Drager syndrome (1960) (146500).
3. Autonomic failure (AF) with Parkinson's disease (PD) (Fichefet et al. 1965).
(Adapted from Bannister 1988)

INCIDENCE OF MULTIPLE SYSTEM ATROPHY WITH
AUTONOMIC FAILURE (AF)

A review of the literature on MSA suggests that
most of it has been written by those interested in
autonomic nervous system function. The MSA
literature alone, however, fails to accurately
represent the number of existing MSA patients,
a large number of whom do in fact exist. They
have been found in Parkinson's disease clinics
throughout the United States. Clinical and pathological evidence clearly indicates that these patients suffered from MSA with AF. A small
series of pathologic examinations of Parkinson's
patients showed that patients unrecognized as
MSA sufferers during life actually did suffer from
MSA with AF. This finding suggests that MSA
is far commoner in Parkinson's patients than
was previously believed. A review of 12 patients
from a larger series who died 1 week to 4 years
after stereotactic operation for Parkinson's disease (before the introduction of levo-dopa)
showed in all degeneration of the substantia
nigra. Lewy bodies were found in only 5 of these
patients; 2 of the remaining 7 cases had multiple
system atrophy (Tygstrup and Norholm 1963).
An examination of 89 brains from patients with
clinical Parkinsonism showed in 7.9% the pathologic findings of multiple system atrophy (Takei
and Mira 1973). Clinically, on the other hand,
5–6% of all Parkinsonian patients showed features of OPCA as well (Duvoisin 1984). In a
large series of 600 brains examined between 1957
and 1987 from patients with Parkinsonism, 5.1%
had pathological features of MSA (Quinn 1989).
A recently established Parkinsonism brain bank
showed that 11% of 83 brains purportedly
received from Parkinsonian patients had MSA,
and another 7% did not have idiopathic Parkinson's disease (Quinn 1989). If these figures are
representative of the true incidence of MSA in
Parkinsonian populations, then a prevalence of
16.4 per 100 000 has been estimated for the
disease that is twice that of Huntington's chorea
in the United Kingdom (Quinn 1989).

Confusion continues on how the term 'Shy-
Drager syndrome' should be used or if it should
be used at all. The 2 patients described by Shy
and Drager (1960) had the combination of

autonomic failure with neurologic signs, and in
postmortem examination evidence of widespread
neuronal loss was found. These patients were
clinically clearly distinguishable from those with
idiopathic Parkinson's disease. Therefore, patients with Parkinsonism and autonomic failure
cannot be classed as suffering from the Shy-
Drager syndrome, even though they may have
Lewy body disease or MSA. It is also not certain
whether autonomic failure has to precede neurologic features for patients to be classed as
suffering from the Shy-Drager syndrome. One
patient, who was first described as suffering from
primary orthostatic hypotension, developed other
neurologic signs 3.5 years later and at autopsy
had the classic pathologic findings of the Shy-
Drager syndrome (Oppenheimer 1982). Whether
the Shy-Drager syndrome term should only apply
to patients with the full clinical symptomatology
or to the pathologic findings of multiple system
atrophy is yet to be determined. Some (Quinn
1989) have gone so far as to suggest that 'if the
term is to be retained at all, then it should refer
to the clinical picture of Parkinsonism with
autonomic failure plus features incompatible with
idiopathic Parkinson's disease, namely, lack of
response to levo-dopa or the presence of pyramidal or cerebellar signs'.

A new term, 'progressive autonomic failure'
(PAF), used to denote patients with idiopathic
orthostatic hypotension, has already been replaced and changed to 'pure autonomic failure'
(Bannister 1988). One could well question how
pure and how severe autonomic failure must be
before it qualifies to be included under the term
of PAF. After all, it is well recognized that
autonomic dysfunction can be found on special
tests in the majority of patients with Parkinson's
disease at some stage of the illness (Appenzeller
and Goss 1971), and also, to a milder degree, in
some normal subjects, particularly after 1 or 2
days of bed rest (Appenzeller 1988).

The definition of autonomic failure is not as
yet at hand, though it is clear that certain tests,
particularly of cardiovascular autonomic control,
have been well validated. For these tests it is
possible to state the limits of normalcy in various
age groups of controls. What is not clear,
however, is the number and combination of

abnormal test results that must be found before a diagnosis of autonomic failure can be made. Abnormal results in 2 or more tests of autonomic function, as a requirement for the diagnosis of autonomic failure (McLeod and Tuck 1987), have been proposed, and these include the response of blood pressure to change in posture and isometric muscle contraction, heart rate responses to standing, and variation in heart rate with respiration. The Valsalva ratio, plasma NE measurement, and sweat tests are also listed. The latter 2 have not been well characterized in normal subjects and particularly not studied across age groups. Such tests, moreover, are incapable of separating those patients with autonomic failure based on MSA from those with autonomic failure based on Lewy body disease, though some help may be obtained from measuring circulating NE concentrations.

No unanimity on defining orthostatic hypotension is found either. Some require accompanying symptoms of cerebral hypoperfusion with definite numerical falls in systolic and diastolic blood pressure (Schatz et al. 1963; Thomas and Schirger 1970). Others simply require a definite fall in systolic blood pressure only (Johnson et al. 1965). Yet others consider the presence of orthostatic symptoms with confirmatory sphygmomanometric measurement, a reduction in blood pressure of 30 mmHg or more, and mean arterial pressure drops of more than 20 mmHg (Cohen et al. 1987). And some (McLeod and Tuck 1987) require a symptomatic drop of 20 mmHg in systolic blood pressure only before further investigations are undertaken. Yet, in the management of patients with orthostatic hypotension, blood pressure alone is often considered without attention to symptoms, which, from experience with such patients, clearly indicates that major drops in blood pressures, both systolic and diastolic, can at times be tolerated without accompanying clinical evidence of cerebral hypoperfusion. Moreover, the well-recognized inconsistency in recording orthostatic hypotension is not included in any of the definitions.

While the diagnosis of autonomic failure on clinical grounds is often uncertain because of lack of definitions, the diagnosis of MSA is similarly unclear. Some criteria require 'the pres-ence of autonomic failure when striatonigral or olivopontocerebellar involvement is confirmed on neurologic examination' (see Quinn 1989). It is obvious that these criteria can include in the same diagnostic category Lewy body Parkinson's disease with autonomic failure and MSA with autonomic failure. In order to encompass this category the term 'Parkinson-plus syndrome' has been proposed to cover patients with atypical Parkinsonism having clinical features incompatible with the idiopathic variety of the disease. But an essential component to this definition in some views is the lack of response to levo-dopa therapy, because in Lewy-body Parkinson's disease without autonomic failure, a response to this therapy is invariably seen (Gibb 1988). A poor, incomplete, or totally absent response is found in those MSA patients with clinical features of Parkinson's disease, with some exceptions (Kosaka et al. 1984).

Clinically, in patients with MSA, there is a striking preservation of the intellect, quite different from those with true Parkinsonism, where dementia is not infrequent, particularly in advanced cases. Some patients with coexisting Alzheimer's disease and MSA have been reported, but this seems to be purely coincidental (Trotter 1973). The presence of dementia seems to rule out, for practical purposes, uncomplicated MSA. Nevertheless, some cases of Parkinsonism with dementia and some with autonomic dysfunction with inadequate responses to levo-dopa have been reported that were shown at autopsy to suffer from diffuse (including cortical neurons) Lewy-body disease (Burkhard et al. 1988). It is clear that Lewy bodies can be found in association with some minor cell loss in the brains of many subjects who had no clinical evidence of Parkinson's disease during life. However, it should be emphasized that idiopathic Parkinson's disease is characterized pathologically by Lewy bodies in the substantia nigra and substantial cell loss in these structures. Without additionally examining the rest of the striatum, a number of patients may be classified as having idiopathic Parkinson's disease, when in fact the striatum also showed cell loss, and in that case MSA is the correct pathologic diagnosis (Quinn 1989).

A family reported by Lewis (1964) is cited in

support of the view that MSA can be inherited. There were 4 cases in this family with clinical disease lasting from 6 months to 34 years. Clinically there was orthostatic hypotension, extrapyramidal, pyramidal, and cerebellar signs, but also amyotrophy, fasciculations, and foot drop. None of these patients came to autopsy, and the unusual clinical picture suggests that this was not a family with MSA (Quinn 1989). Nevertheless, some patients whose MSA was proven at postmortem examination had other family members affected by Parkinsonism. But this association of cases is thought to be attributable to chance alone. The current view is that neither MSA nor SND is inherited and that most cases represent sporadic disease.

Mild autonomic failure and extrapyramidal signs (clinically characterized by spasticity, ophthalmoplegia, peripheral neuropathy, and cerebellar signs), constitute a dominantly inherited disorder known as Joseph's disease (109150) (Rosenberg 1984). While this disease is clearly inherited, neither the extrapyramidal nor the autonomic symptoms are overwhelming, and the pathology is characterized by symmetric degeneration of the thoracic cord and cerebellum, variable involvement of the basis pontis, ocularmotor nuclei, and the nigra and striatum, but always sparing the inferior olives. Clearly, the pathology and clinical manifestations are different, even though some autonomic impairment is found in some patients.

The inheritance of OPCA is complex, and adult-onset patients have a dominantly inherited disease in about half the cases. The remainder of cases of clinical adult-onset OPCA are sporadic. Clinical features and pathology are not distinct enough between the 2 varieties of OPCA to allow clear recognition of sporadic or hereditary cases. A review of 117 pathologically confirmed cases of OPCA (Berciano 1982) showed 39% of 54 familial cases to have Parkinsonian features and 55% of 63 sporadic cases to have also had Parkinsonian signs. Pyramidal signs were present in 50% of familial cases and in 46% of sporadic cases; cerebellar signs in 95% of familial and 87% of sporadic cases; sphincter disturbance in 39% of familial and 48% of sporadic cases. Pathologically, in addition to the usual involve-

ment of the olives, pons, and cerebellum, the striatum was involved in 22% of familial cases and 38% of sporadic cases, and the substantia nigra in 46% and 48% of cases, respectively. No clear distinction could therefore be made on pathologic grounds between sporadic and familial cases, and clinically neither Parkinsonian features nor sphincter disturbances distinguished the 2 groups. Nevertheless, there is an earlier onset of disease in hereditary cases than in sporadic patients, 39 years versus 49 years (Harding 1981). The mean disease duration is longer in familial cases (14.9 years) than in sporadic cases (6.3 years) (Berciano 1982). These patients clearly do not have multiple system atrophy, for the selection for inclusion into these reports was only cerebellar ataxia, and those with a course of less than 2 years were also not included. Nevertheless, it is to be noted that no effort was made to test autonomic function in these patients with OPCA, though it is likely that some abnormalities would have been found had they been sought. In the sporadic cases, ophthalmoplegia was frequent when compared with its incidence in familial cases in which patients often had pigmentary retinal degeneration and optic atrophy. Any patients with OPCA and clinically significant autonomic failure qualifying for inclusion into the MSA syndrome would only be acceptable if they were sporadic (Quinn 1989).

Abnormal eye movements have been reported in a number of MSA patients. The most common abnormality is impairment of convergence, hypometric voluntary saccades, or saccadic pursuits, and occasionally a supranuclear gaze palsy for upward gaze (Lepore 1984). Thus, a few patients with MSA may fulfill the criteria for the Steele-Richardson-Olszewski syndrome (Golbe et al. 1988). But a predominant downward gaze palsy should exclude the diagnosis of MSA.

The latest proposal for diagnostic criteria for inclusion in multiple system atrophy is given in Table 2. There are a number of patients with clinical idiopathic Parkinson's disease, but the diagnosis becomes less certain if there is, besides a poor response to levo-dopa therapy, also clinical evidence of pyramidal or cerebellar signs in addition to major symptoms of autonomic failure. Such patients may at autopsy prove to have MSA.

TABLE 2
Multiple system atrophy: proposed diagnostic criteria.

SND type (predominantly parkinsonism)		OPCA type (predominantly cerebellar)
Sporadic adult-onset non/poorly levodopa responsive parkinsonism*	Possible	–
Above, plus severe symptomatic autonomic failure† autonomic failure +/or cerebellar signs +/or pyramidal signs	Probable	Sporadic adult-onset cerebellar ± pyramidal syndrome* with severe symptomatic autonomic failure +/or parkinsonism
Confirmed at p/m	Definite	Confirmed at p/m

*Without DSM III dementia, predominant downgaze PSNP or other identifiable cause.
†Postural syncope or presyncope and/or marked urinary incontinence or retention not due to other causes.
Sporadic: one other case of typical clinical IPD among 1st or 2nd degree relatives allowable.
Adult onset: onset age 30 years or above.
(Reproduced from Quinn 1989 by courtesy of the Author and Editors of *J. Neurol. Neurosurg. Psychiatry*.)

Difficulty with comparing clinical series arise from the dating of onset of disease. It is not agreed on whether impotence and sphincter disturbances, which may be non-specific and seen in a number of other disorders, should be taken as clinical onset or the beginning of postural hypotension, together with symptoms of motor dysfunction. In idiopathic Parkinson's disease, the disorder is known to antedate the clinical onset of symptoms by as much as 20 years because of the redundancy margin of the central neuronal population. It is likely that in MSA, disease onset may also antedate by many years the beginning of clinical deficits.

Several clinical distinguishing features between idiopathic Parkinsonism and MSA have been recognized. Contractures of the hand, well-described in Parkinsonism-dementia cases and in patients with postencephalitic Parkinsonism, are more common in cases of idiopathic Parkinsonism. They rarely occur in MSA (Quinn et al. 1988). In MSA, however, a fixed and disproportionate antecollis, often interfering with communication, vision, and feeding, is frequent (Caplan 1984). Vocal cord paralysis manifested by inspiratory stridor, particularly at night, is frequently encountered in MSA and only rarely in idiopathic Parkinson's disease (Williams et al. 1979). Speech, often affected in Parkinsonism, is more severely impaired in MSA. Though muscle atrophy and fasciculations have been described in MSA (Shy and Drager 1960), these are not commonly found. Pain, however, often in the most severely affected limbs, is a frequent symptom in idiopathic Parkinsonism and usually improves with levo-dopa treatment in such patients, whereas those suffering from MSA have no relief with this therapy (Caplan 1984).

A review of autonomic symptoms in 34 patients diagnosed as suffering from idiopathic Parkinson's disease showed that 85% had nocturia, 74% had daytime frequency, and 71% had some incontinence. Twenty-five percent complained of absent sweating of hands and feet in hot and humid environments, and 15% had faintness before the beginning of dopaminergic therapy, whereas 37.5% had faintness after its commencement. Twenty-three percent had no or only occasional erections. Whether these symptoms in Parkinsonism suggest that the majority of these patients suffered from MSA is by no means clear and requires further study of the rates of these symptoms in age-matched normal subjects (Quinn 1989).

LABORATORY INVESTIGATION

Autonomic failure can be demonstrated by a number of tests, but these do not identify the pathogenesis of the disorder. These tests have been reviewed (Appenzeller 1986). More recently, the significance of resting supine plasma-norepinephrine levels and the responsiveness of NE to standing have been employed to differentiate between patients with MSA and those with pure autonomic failure. At rest, patients with MSA have normal or slightly elevated plasma NE levels, whereas those with PAF have low or undetectable levels (Polinsky 1988). In both groups, unlike in normal subjects, standing does not increase plasma NE levels. Nevertheless, the clinical usefulness in individual patients for diagnosis between an inexorably progressive nervous system disease (MSA) and autonomic failure with a better prognosis (PAF) remains to be determined (Polinsky 1988).

Other tests also are of limited use in these conditions. Computed tomography (CT) does not yield specific findings, but in those patients with MSA who have cerebellar symptoms and signs, a non-specific cerebellar or brain stem atrophy may be demonstrable. In those who present with Parkinsonian features, the CT scan is often normal. Of interest is that striatal imaging in all cases of MSA is usually unhelpful because it is within normal limits (Quinn 1989). Magnetic resonance imaging (MRI) suggests that striatal pathology can be demonstrated in some patients with MSA, but there are no detectable abnormal signals in idiopathic Parkinson's disease. Those patients with Parkinson's disease that is rapidly progressive and poorly responsive to treatment showed different signals, particularly from the putamen, and this was attributed to deposition of iron in that structure. However, whether iron content in the putamen is indeed present in such patients and results in the different images on MRI is not certain (Borit et al. 1975). Nevertheless, the abnormal signals on MRI scanning do seem to correspond to putaminal pathology characteristically found on histo-

logic examination in striatonigral degeneration, resulting clinically in autonomic failure and MSA, so that this scanning technique might be helpful in differentiating these patients from those with idiopathic Parkinsonism (Drayer et al. 1986; Rutledge et al. 1987).

Few patients with MSA have been submitted to positron emission tomography (PET) scans. ^{18}F-fluorodopa scans have, as expected, shown low concentration of activity in the striatum, similar to the findings in idiopathic Parkinson's disease (Brooks and Frackowiak 1989). On the other hand, with more modern scanners, it is possible to show a lower caudate uptake in MSA patients when compared to the relatively preserved caudate uptake in idiopathic Parkinson's disease (Rutledge et al. 1987). PET imaging with ^{11}C-nomifensine (a measure of dopamine reuptake site function) shows diminished striatal binding in MSA when compared to controls, suggesting that a parallel loss of dopamine reuptake sites accompanies the diminished striatal dopamine storage capacity. This finding has been interpreted to reflect loss of nigro-striatal nerve terminals (Brooks and Frackowiak 1989). The post-synaptic dopamine D_2 receptors in MSA have not as yet been examined, but because of the known poor responsiveness of these patients to levo-dopa and the pathologically recognizable striatal degeneration, one would expect low densities of striatal D_2 receptor sites in these brains (Brooks and Frackowiak 1989).

ELECTROPHYSIOLOGIC INVESTIGATIONS

These in general are unhelpful. Peripheral nerve conduction velocities may rarely show a subclinical polyneuropathy (Cohen et al. 1987) and occasionally electromyography suggests neurogenic atrophy (Shy and Drager 1960). More modern studies, such as thermal thresholds, visual and somatosensory evoked responses, magnetic central motor conduction times or electrical motor conduction times have not been adequately examined. Controversy continues about brain stem auditory evoked responses, one group having demonstrated abnormal latencies and amplitude ratios of waves V/I in 11 of 13 patients (Prasher and Bannister 1986). Others, however, have not been able to confirm these abnormalities (Quinn 1989). However, while these conventional electrophysiologic studies are unhelpful in the diagnosis, the recording of individual motor units from the striated component of the urethral sphincter were consistently abnormal, showing polyphasic and long duration potentials in all of 14 patients studied who suffered from MSA (Kirby et al. 1986). A further study of such subjects showed that this investigation has a specificity (0.92) but less sensitivity (0.62) in distinguishing between probable MSA and probable idiopathic Parkinson's disease (Quinn 1989).

There is no agreed-upon schema for the diagnosis of multiple system atrophy, unlike the schema for multiple sclerosis. No single clinical feature or investigation at present can make a definite diagnosis of MSA, and only the complete clinical picture, together with the results of a number of investigations, can help in ascertaining the diagnosis before pathologic examination.

TREATMENT

The many drugs offered on good theoretical grounds for the treatment of MSA have, unfortunately, not been shown to be useful for most patients. In some, levo-dopa may transiently give some improvement, and in those who do not appear to benefit from this treatment, withdrawal may sometimes be associated with worsening of extrapyramidal signs. Also, some patients develop intractable nausea and vomiting and more frequent symptomatic postural hypotension when given this drug (Lees 1988). Nevertheless, all patients should be given a trial, and the occasional one may show useful benefits. Manipulations such as head-up tilt of the bed at night, elastic thigh-length support stockings, and atrial pacing may help maintain reasonable flow of blood to the brain. Fluorocortisone and/or indomethacin have been tried and may be helpful for some time. Caffeine, together with ergotamine (given intramuscularly in the morning), has been shown to be helpful in a proportion of patients. One should pay particular attention, however, to postprandial hypotension, which must be taken into account for timing of drug administration.

The oral administration of L-threo-3,4-dihy-droxyphenylserine (L-threo-DOPS), a precursor of NE, has been found beneficial in one patient with MSA and severe postural hypotension (Kachi et al. 1988). In this patient, muscle sympathetic activity was measured by neuronography. Muscle action potentials were few in the supine position, and with severe orthostatic hypotension only a slight increase in frequency in these potentials was found during the head-up position to 40°. After the oral administration of 200 mg of L-threo-DOPS in the 40° head-up position, a considerable increase in muscle sympathetic discharge rates was found 30 min after the administration of the drug, and orthostatic hypotension was improved. Three hours after the administration of L-threo-DOPS, the discharge rate decreased and orthostatic hypotension reappeared. At that time, the plasma concentration of NE was as its highest levels. Orthostatic hypotension improved concomitantly with an increase in muscle sympathetic discharge rates after the administration of L-threo-DOPS but at a time when plasma catecholamine concentrations were not at their highest level. When the highest levels of catecholamines were measured in the circulation, orthostatic hypotension reappeared. It seems, therefore, that the drug in this patient activated sympathetic outflow, evidenced by an increase in muscle sympathetic discharges proximal to the sympathetic ganglia.

Beneficial effects of selegiline hydrochloride given alone or in combination with carbidopa–levodopa have been noted. Serious hypertensive reactions to this combination have, however, been recorded.

Many bladder infections due to incomplete emptying and retention with overflow should be managed not only with antibiotics but also with intermittent self-catheterization, if possible. Occasionally an in-dwelling suprapubic or urethral catheter may be necessary. Constipation can often be managed with high-fiber diet, laxatives, suppositories, or enemas. Respiratory stridor, often at night, is a bad prognostic sign, and it is debatable whether tracheostomy should be carried out to prolong a miserable existence. In the same category are cricopharyngeal myotomy and gastrostomy. The mainstays of management remain social work, physiotherapy, occupational therapy, and speech therapy, more often useful in improving swallowing than in communication.

PATHOLOGY OF PRIMARY AUTONOMIC FAILURE

Shy and Drager's (1960) first report included neuropathologic examination of a 56-year-old man who had died after 6.5 years of orthostatic hypotension and disturbances in micturition. They found in many sites in the central nervous system cell loss and glyosis of varying extent. Sites included the caudate nuclei, the substantia nigra, the Purkinje cells in the cerebellum, the inferior olives, the dorsal vagal nuclei, and the lateral horns of the thoracic spinal cord (Shy and Drager 1960). Five years later Fichefet et al. (1965) published a second report about a patient examined after death at age 72. The patient had suffered for about 2 years with an illness that began with orthostatic hypotension and that later had features of Parkinsonism. At autopsy pathologically typical changes of Parkinson's disease were found, along with a marked loss of the pigmented cells from the substantia nigra and locus ceruleus and numerous Lewy cytoplasmic inclusions in the remaining pigmented cells. Some cell loss in the anterior and lateral horns of spinal cord was also present. The next major advance in neuropathological examination of such cases was made in 1966 (Johnson et al.). Two patients examined after death showed a marked cell loss in the lateral horns of the thoracic spinal cord and very little abnormality in any other part of the autonomic nervous system. One patient had clinically 'pure' autonomic failure (i.e. autonomic failure without system involvement) of 4 years duration before death. Autonomic failure in that patient was manifested by orthostatic hypotension, sexual impotence, and loss of sweating. The other had, in addition to these symptoms, other symptomatic system involvement. In both patients careful examination at 12 thoracic levels with age-matched controls showed an almost 90% loss of preganglionic sympathetic neurons in the intermediolateral columns of the spinal thoracic cord. There were some Lewy inclusions in the substantia nigra and locus ceruleus, though no histologically detectable loss of pigmented cells in these

structures was found. The second patient died at age 54 after 4 years of illness characterized by autonomic failure and severe motor and cerebellar disturbances. Pathologically, OPCA was found with additional loss of cells in the putamen, in the vestibular nuclei and pigmented nuclei, and in the intermediolateral columns. No Lewy bodies were found, and the cell loss in the lateral horns was somewhat less striking, only of about 75%. The conclusion reached by Johnson et al. (1966) was that intermediolateral cell column loss had caused autonomic failure, which had been an important and disabling disorder during life. Similar cases, reported by Schwarz (1967) and Nick et al. (1967), were characterized by cell loss in the pigmented nuclei and putamen, the dorsal vagal nuclei, the olives, the cerebellar cortex, and the intermediolateral cell columns. Two cases had loss of nuclei in the pontine region. Lewy bodies were not present. A possible relationship of these cases to those reported under the heading of striatonigral degeneration (SND) by Adams et al. (1961, 1964) was noted. In addition, cases in which lesions of OPCA were combined with those of SND were also found (Graham and Oppenheimer 1969), and in these patients, as well as in a patient with SND, autonomic failure was present. Graham and Oppenheimer (1969) proposed that cases of SND and OPCA were pathologically and clinically a manifestation of the same disorder. Bannister and Oppenheimer (1972) proposed the term 'multiple system atrophy' (MSA) for these diseases, which were attributed to a familial or sporadic degeneration that had its clinical onset in middle life and pathologically affected the following structures: the pigmented nuclei (locus ceruleus and substantia nigra); the striatum; pontine nuclei; the inferior olives; the dorsal, vagal, and vestibular nuclei; and the cerebellar Purkinje cells. When the intermediolateral cell columns were affected pathologically, there was invariably also autonomic failure during life. Graham and Oppenheimer (1969) pointed out and Vanderhaeghen et al. (1970) confirmed that there were 2 types of MSA from the pathological point of view. One belonged to a group of patients whose symptoms were similar to those of Parkinson's disease and the second belonged to patients with clinical autonomic failure who on pathologic examination showed lesions of MSA. These 2 groups were also evidently separate clinically because in those cases where Lewy bodies were found (with affinities to Parkinson's disease), the age of onset was older (usually about 65, with death at age 70), whereas cases with MSA had an age of onset at about 49, with death at about 55. In those with older onset, autonomic failure may have been isolated, and in patients who had additional symptoms, these were usually Parkinsonian-like. Those with earlier onset had many additional neurologic symptoms and signs, which also included Parkinsonian features. In both groups, if these were present, a loss of pigmented cells in the substantia nigra was found, but the cell loss in the intermediolateral cell columns was equal in both groups.

Full histologic studies in cases of progressive autonomic failure in about 67 patients showed these to be clearly divisible into a group containing Lewy bodies in the pigmented nuclei and elsewhere. The patients had Parkinsonism clinically as well as pathologic evidence for this disease. In those who had no clinically recognizable Parkinsonian features but Lewy bodies at autopsy, the substantia nigra did not show cell loss. The intermediolateral cell columns were affected in the majority of these patients, and a large number also had loss of dorsal vagal nuclei. None had features of SND or OPCA pathologically. A large second group of about 46 patients were classed as having multiple system atrophy. Thirty-five of these had clinically Parkinsonian features, and all had some damage to the substantia nigra. Similar changes were found in the putamen. In addition, in more than half of the patients there was evidence of olivopontocerebellar atrophy, and in the majority of these there was pathologic evidence of striatonigral degeneration. Lewy body inclusions were seen in only 2 cases, but the association of lesions in the olives and cerebellum and striatum suggest that these were patients who had multiple system atrophy rather than Parkinson's disease. Where counting techniques with appropriate controls were employed all these cases were found to show substantial losses of cells in the intermediolateral cell columns, which is a hallmark of those cases

where autonomic failure was prominent during life (see Oppenheimer 1988).

The rest of the central nervous system has not been intensively studied pathologically in patients with autonomic failure during life. The vestibular nuclei anterior horn cells and motor cells and the cortical spinal tract have been examined. These structures are not important in causing autonomic dysfunction, yet several abnormalities have been found. Loss of pigmented cells in the locus ceruleus has been reported in material from patients who also had Lewy bodies, but the symptoms of this type of patient are similar to those of patients with Parkinson's disease, so that the cell loss in the pigmented nuclei is in keeping with the cliical picture during life. But in autonomic failure and MSA the locus ceruleus was found to be affected in 28 cases and the vestibular nuclei in 15. In this group the anterior horn and pyramidal tracts showed some degenerative changes in 19 cases. What must be stressed, however, is that lesions in these sites have been reported in patients with MSA who clinically did not show evidence of autonomic failure. Thus, the specificity of cell loss and degenerative changes in these structures is, in retrospect, doubtful.

Peripheral nerve lesions are found in some but not all cases of MSA and in many with MSA without autonomic dysfunction. The peripheral autonomic ganglia have not been carefully examined in at least half of the reported cases, but in 8 a mild loss of cells in the sympathetic ganglia was found. The presence of hyaline Lewy cytoplasmic inclusions in the sympathetic chain ganglia was originally described by Bethlem and Den Hartog Jager (1960) in patients with Parkinson's disease and was subsequently found in the sympathetic ganglia of such patients and other elderly subjects (Appenzeller 1966). But in patients with autonomic failure associated with Lewy bodies, only 4 showed these inclusions in the sympathetic ganglia without extensive damage. Extensive damage in sympathetic ganglia was reported only by Rajput and Rodzilsky (1976). In this report of 6 patients with autonomic failure and Parkinson's disease the intermediolateral cell columns were depleted as expected and the sympathetic ganglia showed a severe loss of cells, degenerating cells, and many Lewy inclusion bodies. No reports of abnormalities in parasympathetic ganglia, visceral nerve plexuses, or other autonomic nerves have yet appeared. There are few convincing reports of degenerative changes in somatic nerves in patients with Parkinson's disease or in patients with MSA without clinical autonomic failure. It is clear that pathologic examination has so far failed to unravel the nature of the disease, and it is therefore still classed as a primary degenerative disorder of the nervous system.

No familial incidence has been recognized, though one family with Parkinsonism, orthostatic hypotension in combination with ataxia (4 cases), has been described by Lewis (1964). An HLA predisposition has been reported (Bannister et al. 1983), but unfortunately not confirmed in a larger study (Nee et al. 1989). No provoking factors have been recognized, such as infections or metabolic disturbances, and no epidemiologic clues of environmental toxins have been found.

The neuropathologic examination is of course far removed from the functional disturbance, but it has clearly shown that in the majority of patients autonomic failure can be attributed to loss of preganglionic autonomic neurons in the intermediolateral cell columns. However, there are 2 recorded cases in which autonomic failure had occurred without cell loss in the intermediolateral columns (Hughes et al. 1970; Evans et al. 1972). Other structures important in autonomic function may have been affected in these patients; yet the clinical manifestations of autonomic deficits may have been exactly the same as those in patients with the more usual cell loss in the intermediolateral columns. Moreover, it is difficult to conclude from normal histologic appearance or normal cell numbers in the intermediolateral columns that their function was normal, since appearances may not always give a clue to the integrity of function of the cells.

The human cortical spinal tract has been repeatedly evaluated for fiber size and fiber population. A large variation in the density and type of fibers was found but largely discounted. The variation was attributed to different study methods used rather than to an actual variation. A comparison between the morphometric results on myelinated fibers in the cortical spinal tract at the seventh thoracic spinal segment was carried out in patients with MSA and amyotrophic lateral sclerosis and with patients who died from non-neurologic disease. A depletion of small-sized myelinated fibers was found in MSA, whereas large-sized myelinated fibers were largely preserved. In amyotrophic lateral sclerosis the large myelinated fibers were predominantly lost, however. This differential vulnerability of cortical spinal tract fibers may be related to autonomic spinal pathways, since these are largely unaffected in amyotrophic lateral sclerosis, and profound autonomic dysfunction is the hallmark of MSA (Sobue et al. 1987).

Successful clinical studies require that patients be correctly identified and placed in appropriate groups. In studies comparing Parkinson's disease with multiple system atrophy, patients should be divided into those with autonomic failure and features of Parkinson's disease and those with multiple system atrophy with few Parkinsonian features and autonomic failure. Also, these patients tend not to respond to anti-Parkinsonian drugs containing levo-dopa, whereas those with mainly Parkinsonian features and autonomic failure may well do so. To study pharmacologic responses to these drugs a study of cardiovascular responses to infusion of dopamine or NE was conducted. Five normal subjects, 5 patients with Parkinson's disease and 5 patients with Shy-Drager syndrome (MSA), were observed after receiving injections of the 2 drugs. The responses of patients with Parkinson's disease were in general the opposite of those with Shy-Drager syndrome (Wilcox and Aminoff 1976). No group of subjects with pure autonomic failure (PAF) (previously called idiopathic orthostatic hypotension) has been well studied pathologically. The few pathologic studies on such patients have not been extensive, since life expectancy of these patients does not seem to be shortened by autonomic failure (Polinsky et al. 1989). It will be some time, therefore, before a definitive statement on the pathology of pure autonomic failure will be made.

Peripheral autonomic ganglia

Difficulties with the pathologic analysis of peripheral ganglia in autonomic failure have not as yet been overcome. The available evidence suggests a primary defect in preganglionic intermediolateral neurons or in the preganglionic-postganglionic interaction to account for the autonomic failure observed in MSA and in Parkinson's disease with autonomic failure. Pure autonomic failure has been less well studied, however.

Changes in the sympathetic ganglia obtained at postmortem from patients with autonomic failure are not clear-cut. In one case of autonomic failure with Parkinson's disease reported by Rajput and Rodzilsky (1976), a marked neuronal loss and degeneration in the sympathetic ganglia, together with numerous Lewy bodies, was found. The cell loss from the intermediolateral columns in this case was minimal.

The levels of choline acetyltransferase (CAT) in the superior cervical ganglia of 4 cases with autonomic failure was normal (Petito and Black 1978). CAT is the transmitter-synthesizing enzyme of preganglionic nerve endings, suggesting that these were unaffected in these cases. Also, the number of neurons in the intermediolateral cell column was only slightly reduced in 2 of the 4 cases, but no formal counts were made. The superior cervical ganglia were also normal in tyrosine hydroxylase (TOH) content, but dopamine beta hydroxylase (DBH) was very low or not detectable. Both enzymes are involved in the synthesis of NE by the adrenergic neurons. One interpretation of these findings was that there was a compensatory intraganglionic sprouting of nerve fibers from terminals that had survived the cell loss from the intermediolateral cell columns. Histologically, evidence of chromatolysis of ganglion cells in 3 cases and an increase in satellite cells in all, including perivascular mononuclear infiltrations and Lewy-like hyaline interstitial bodies in 1 case were found. Ultrastructurally enlarged swollen axons filled with proliferated filaments and focal accumulation of mitochondria typically found in dystrophic axons (Appenzeller et al. 1988) were also described.

Neuropathologic examination is often difficult to correlate with clinical deficits. This is particularly so in diseases of the autonomic nervous system, since the clinical deficits may result from biochemical, metabolic, or regulatory influences that may not have structural correlates. More-

over, the postmortem examination of pathologic specimens is often from patients with advanced disease, which is usually complicated by superimposed failure of other organs, making interpretation of autonomic deficits even more difficult. Anatomically, the autonomic nervous system is also devoid of clear differentiation between pre- and postganglionic, and upper and lower motor neurons. There is a convergence and also a divergence of preganglionic neurons onto neurons of the ganglia, and functionally the latter receive inputs that act at a subthreshold level before bringing the neuron to threshold firing. Postganglionic peripheral effectors are smooth muscle or cardiac muscle and glands and blood vessels. These neuroeffector contacts are distant from the ganglia. In addition, there is typically electrotonic coupling in the effector organs (distinct from the somatic system). Therefore, slowly progressive changes induced by a variety of injurious agents are usually not clinically evident, unless the organism is stressed and until the underlying pathology is severe.

Despite these difficulties some material is available, including control sympathetic ganglia that have been carefully examined (Matthews 1988). In subjects who suffered from MSA, neurons were normal in number in the sympathetic ganglia, but differed from controls in the reduced incidence of distinct Nissl granules in neuronal cell bodies. No abnormalities in the blood vessels supplying the ganglia or in adventitial cells were found, but in MSA patients some perivenular lymphocytic infiltration was present. But such perivenular infiltration with lymphocytes is often seen in patients with diabetic autonomic neuropathy and autonomic failure as well (Fig. 1). On the other hand, in the sympathetic ganglia from 2 subjects with pure autonomic failure during life, who had Lewy bodies on pathologic examination, a reduction in the densities of nucleated neuronal profiles in the neuropil was well documented. The overall impression was of a severe loss of neurons with evidence of neuronophagia in all examined ganglia. The Lewy bodies were found in the somata of neurons and sometimes in enlarged neuronal processes (Fig. 2). The surviving neurons, however, showed high counts of visible Nissl granules and no increase in

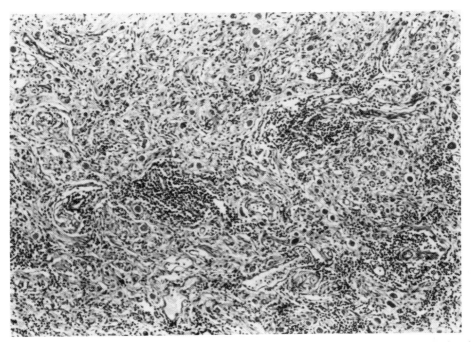

Fig. 1. Paravertebral sympathetic chain from patient with longstanding diabetic neuropathy and sudden death. There is widespread, mainly perivascular, lymphocytic infiltration suggesting the possibility of an immunologic reaction. (Reproduced from Appenzeller 1986.)

lipofuscin bodies in the MSA subjects when compared to controls. In the ganglia of patients with pure autonomic failure, therefore, the prevailing appearance is of a severe loss of neurons and the presence of Lewy bodies. On formal counting of neuronal nuclei of serial sections of the entire superior cervical ganglion, a 75% loss of neurons was found, and this was far above that expected for age of patients in normal superior cervical ganglia (Matthews 1988). In general, ultrastructural examination confirmed light microscopy with few additional findings. In one subject, the axodendritic synapses in the superior cervical ganglion showed depletion of synaptic vesicles and coated vesicles, which suggested extensive release of neurotransmitters. This subject suffered from MSA. In another similar patient, presynaptic profiles were swollen and depleted of synaptic vesicles also. In yet another subject with MSA an appreciable loss of axons in the cervical sympathetic trunk was discernible and collagen-filled channels with reduplicated basal lamina were seen, suggestive of degeneration.

Neurochemical analysis of peripheral autonomic ganglia

Histochemical and immunohistochemical analysis of the thoracic sympathetic ganglia from one subject with pure autonomic failure showed normal acetylcholinesterase activity. Immunofluorescence microscopy of ganglia from one subject with MSA and one with PAF disclosed persistent varicose nerve fibers immunoreactive for SP and others for CGRP. Encephalin was demonstrable in the celiac ganglion. Less, however, was found in the paravertebral sympathetic chain. Slight immunoreactivity for tyrosine hydroxylase in most neurons was found and some neurons contained encephalin immunoreactivity. The celiac ganglion of one patient suffering from PAF showed NPY in some neurons and in dystrophic neurites, including some small cells which were possibly small, intensely fluorescent cells (SIF-cells). On the other hand, catecholamine fluorescence (usually colocalized with NPY) was hardly found, and this was attributed to perhaps low initial levels and postmortem diffusion (Matthews 1988).

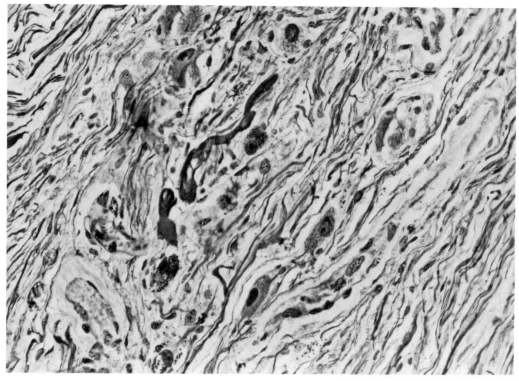

Fig. 2. Enlarged neuronal processes in sympathetic chain. These enlargements may be artifactual and can be found in many paravertebral sympathetic chains of elderly subjects. From a patient with malignancy-associated hypercalcemia (Bodian protargol 580 ×).

CENTRAL CHEMICAL PATHOLOGY OF AUTONOMIC FAILURE

The spectacular success of unraveling the central neurochemistry of Parkinson's disease has not been duplicated in investigations of the neurochemistry of autonomic failure. The neurochemistry of Parkinson's disease was determined by postmortem studies of dopamine and other neurotransmitter content of the basal ganglia which has led to subsequent beneficial therapeutic manipulations. Postmortem study of autonomic failure is limited, however, because of the relative rarity of patients with the condition. To date, only a small number of neurochemical examinations of brain tissue obtained at autopsy and of spinal fluid obtained from patients with the condition have been conducted.

In 4 autopsied brains neurochemical measurements were made. In addition to autonomic failure, the patients had pyramidal, extrapyrami-dal and cerebellar signs; clinically, they fitted best into the MSA group. Low dopamine (DA) in the substantia nigra, corpus striatum, hypothalamus, septal nuclei, and locus ceruleus were found (Spokes et al. 1979), and similar low values in the same areas were reported by Kwak (1985) from 7 such cases. The low DA content of the basal ganglia, which resulted from striatonigral degeneration, was consistent with the clinical Parkinsonian features of these patients. The loss of DA from the hypothalamus, nucleus accumbens, and the septal nuclei also is seen in Parkinson's disease and suggests involvement of mesolimbic and intrinsic hypothalamic DA systems. The dropout of dopaminergic neurons in Parkinson's disease and striatonigral degeneration is widespread and also reflected in the low CSF levels of homovanillic acid, a metabolite of dopamine (Williams 1981). While it is well recognized that depletion of catecholamines (CA) is associated with severe motor deficits, it is less clear how such a loss of neurotransmitters affects autonomic function, particularly as a loss of DA from the hypothalamus and the nigrostriatalmesolimbic systems, commonly found in Parkinson's disease, is not associated with autonomic deficits. It is not known, however, whether DA fibers in the medulla are affected in Parkinson's disease or in multiple system atrophy with striatonigral degeneration.

Reduced NE levels have been reported in MSA. Most markedly depleted were the septal nuclei, locus ceruleus, and hypothalamus, areas rich in this amine (Spokes et al. 1979; Kwak 1985). However, the losses of NE from the basal ganglia were not different from those found in patients with Parkinson's disease (Spokes and Bannister 1981). In other regions the loss of NE was greater than the losses found in Parkinson's disease. Significantly low levels of NE in the CSF from patients with Parkinsonism and those with MSA, more marked in the latter group, were also reported (Williams 1981). While in general biochemical changes in the central nervous system accompany histologically recognizable cell loss, it is surprising that in MSA depletion of NE from the locus ceruleus can be found chemically in the absence of a marked dropout of cells, as judged on histologic examination, suggesting the low NE levels in these patients are not merely secondary to cell death (Spokes et al. 1979). Tyrosine hydroxylase (TOH), an enzyme important in the production of NE, has been found deficient in the locus ceruleus of 3 patients who had autonomic failure. Two of those probably were Parkinson's-associated MSA and one pure MSA (Black and Petito 1976). The medullary reticular formation is involved in MSA because both NA and TOH activity are reduced in the septal nuclei and in the hypothalamus, so that the brain stem nuclei and intermediolateral cell columns of the spinal cord are more likely to be deficient in noradrenergic innervation as well. However, actual chemical determination of NE and TOH activity in these structures has not as yet been reported.

The serotonin (5-HT) content of various brain structures in autonomic failure has not been examined, though a 50% reduction in striatal 5-HT has been found in patients with Parkinson's disease without accompanying autonomic failure (Bernheimer et al. 1961). In one case of striatonigral degeneration without autonomic impairment, a normal 5-HT content of the striatum has been recorded (Sharpe et al. 1973). The CSF 5-hydroxyindoleacetic acid, thought to be an index of central 5-HT turnover, has been found to be normal in Parkinson's disease, but a tendency to lower levels in patients with MSA (though still within the normal range) has been found (Williams 1981). No definitive evidence of involvement of serotonergic fiber systems in MSA is yet at hand.

Acetylcholine

The integrity of central cholinergic neurons can be gauged from analysis of choline acetyltransferase (CAT) activity. There is no correlation with histologic changes in the basal ganglia and alteration in CAT. Patients with MSA have changes in these areas that parallel those found in patients with Parkinson's disease (Spokes et al. 1979; Kwak 1985). These findings suggest that a reduction in CAT activity does not imply cell death but reflects functional changes in cholinergic neurons. Because similar losses of CAT occur in the basal ganglia in Huntington's chorea, Parkinson's disease, and MSA, it is very unlikely that they are connected with the prominent autonomic failure in MSA.

In the dentate, red, olivary, and pontine nuclei, low CAT levels were found in MSA. Those with the most marked depletion had features of olivopontocerebellar atrophy (OPCA) often found in patients with MSA and autonomic failure. These nuclei are closely interconnected by projections and are linked with the hypothalamus, where CAT loss was consistently found in MSA subjects. CAT activity was also decreased in the septum and hippocampus in 3 patients (Spokes et al. 1979). This was not only observed in such patients but can frequently be found in patients with Huntington's chorea or Alzheimer's disease, where autonomic dysfunction is not usually present. The levels of CAT in CSF were found to be low in MSA (Polinsky et al. 1989). There was, however, no correlation between low CAT and low monoamine CSF levels in these patients. In PAF, by contrast, CAT levels in CSF were normal, suggesting that autonomic failure in these patients was the result of peripheral rather than central pathology.

Amino acids

Glutamic acid decarboxylase (GAD) is a marker for gamma amino butyric acid (GABA)-containing neurons. GAD is markedly affected, however, by agonal states which reduce the activity of this enzyme in the brain. GAD is markedly reduced in MSA (Spokes et al. 1979; Kwak 1985), but because of the complicating influences of slow death, the significance of this loss in the genesis of symptoms of MSA is not clear. The measurement of GABA concentration, on the other hand, may give a better indication of the functional integrity of GABAergic neurons. Normal levels of GABA were reported by Kwak (1985) in the basal ganglia, but a great reduction of GABA in the dentate nucleus was found. A reduction in glutamate and aspartate in the cerebellum correlated with cell loss in the inferior olivary nucleus. This nucleus contains neurons whose ascending climbing fibers contain these amino acids. A loss of glutamate-containing granule cells in the cerebellar cortex was also found in keeping with the low amino acids assayed.

Blood pressure regulation circuits are affected because of cerebellar input on circulatory reflex centers. A loss of pathways from the cerebellum associated with depletion of GAD, GABA, aspartate, and glutamate may, hypothetically, contribute to the postural hypotension of MSA patients.

Substance-P

The only peptide of numerous peptide neurotransmitters to have been measured in MSA is Substance-P (SP). Markedly reduced levels were found in the basal ganglia, particularly the substantia nigra. The corpus striatum was less affected (Kwak 1985). CSF-SP was also reduced to 50% of normal in MSA patients (Williams 1981). However, the pattern of loss is similar in Huntington's chorea, and it is thought to reflect a loss of SP-containing striatopallidonigral fibers.

The connection between autonomic failure and SP-containing fibers is therefore tenuous.

From the foregoing it is clear that much exploration of biochemical abnormalities in autonomic failure remains to be done if a rational substitution therapy on the line of that available for Parkinson's disease is to be found. While preliminary studies reviewed here have shown changes in noradrenergic, cholinergic, and dopaminergic systems and in amino acid neurotransmitters as well as in peptides, the relationship of these changes to the clinical manifestations and pathologic findings in MSA remains controversial.

Neuropeptides in autonomic pathways of the spinal cord

The peptide concentrations in postmortem spinal cord (Anand et al. 1988) show a regional distribution in the dorsal horn, dorsal columns, lateral horn, dorso-lateral columns, and ventral horns. Significant decreases in SP in all 5 sites in spinal cords of patients with MSA were found. Substance K was low only in the lateral horn, dorsal columns, and dorsal horns. Calcitonin-gene-related-peptide (CGRP) was significantly decreased in the dorsal horn, dorsal column, and dorsolateral column, whereas somatostatin and galanin were significantly low only in the dorsal horns. Vasoactive intestinal polypeptide (VIP) was below detection levels in all samples. The reduction in SP is particularly significant since immunohistochemical staining of this peptide is entirely confined to fibers, mostly projections to preganglionic cell bodies. No SP immuno-staining in the cell bodies of the thoracic autonomic efferents has been recognized. The suggestion has therefore been made that depletion of SP provides a specific marker for fiber loss or dysfunction, particularly in the dorsal and lateral horns, in patients with autonomic failure (see Anand 1988). In cerebrospinal fluid, a reduction of 50% of SP has also been reported in MSA patients (Williams 1981). Important though these quantitative data are, it must be understood that they are from patients without sensory symptoms or clinical evidence of peripheral neuropathy. Since these peptides are accepted as important in sensory-afferent function, the lack of such symptoms in the face of the marked depletion needs to be further investigated. Since the flare response in the skin induced by histamine injection is normal in patients with MSA (Anand et al. 1988), the proposal has been made that the depletion of SP and CGRP is exclusively confined to visceral, rather than somatic afferents. The significance of changes in levels of neuropeptides, both in nervous tissue and CSF, in patients with autonomic failure, is at this early stage not yet assessable.

PERIPHERAL CHEMICAL PATHOLOGY OF AUTONOMIC FAILURE

The release of neurotransmitter substances bridges the gap between neurons and postsynaptic tissues and occurs in response to neuronal action potentials. Studies of peripheral neurotransmitter function have yielded most information because of access to neurotransmitters and to their removal mechanisms. On the other hand, central nervous system metabolic assessment of neurotransmitters requires indirect strategies which have not as yet been well developed.

The cell bodies of peripheral noradrenergic neurons are located primarily in sympathetic ganglia. Synthetized NE is stored in vesicles and released into the synaptic cleft with stimulation. The release of NE is followed by activation of pre- and post-synaptic receptors. The pre-synaptic receptors also function as a negative feedback control and inhibit further NE release. Since the most devastating symptom in autonomic failure is postural hypotension, the post-synaptic receptor stimulation on vascular neuroeffector junction, which results in vasoconstriction, is of great interest. The main mechanism for inhibiting actions of NE is neuronal re-uptake, but the density of innervation of blood vessels and the geometry of the synaptic cleft are important in determining the kinetics of neuronal re-uptake. A small amount of NE, however, escapes into the circulation and makes it likely that plasma NE levels correlate with neuronal activity. The plasma levels are a result of the balance between NE entry into the circulation after release by sympathetic nerve endings and eventual removal

from the plasma of this neurotransmitter. In autonomic failure differences in pressor responses are the characteristic manifestation of post-synaptic stimulation and depend to some extent on the site of the lesion.

Many approaches to the studies of NE metabolism, such as the measurement of urinary metabolites, as well as circulating NE, are used, but each is complementary in assessment of autonomic function and the site of lesions in autonomic failure. The measurement of NE plasma levels is fraught with difficulties because the site of the sampling, the time of day, the position of the subjects, and other variables may influence measurements. Standardized blood sampling methods have not been generally used and studies are difficult to compare.

Plasma NE levels in patients with pure autonomic failure (PAF) and in those with MSA have been measured. MSA patients have normal or slightly elevated plasma NE levels. There are, however, some exceptions found by Bannister et al. (1977) and Sasaki et al. (1983). There is no unanimity of opinion about whether peripheral nerve involvement may occur in MSA and thus account for the reduced plasma NE as found in other neuropathic conditions, for example, diabetes mellitus, in the few patients showing reduced levels. Most investigators have found reduced NE levels in PAF, except for the study by Esler et al. (1980), which also demonstrated a decrease in clearance of NE in his patients. This reduced clearance may be responsible for the relatively normal or elevated plasma levels as found in this study (Esler et al. 1980).

Stressing the noradrenergic mechanism by increasing plasma NE through standing, a standard stimulus, allows comparison between different studies, since each subject is his own control. No significant increases in plasma NE concentration after 5 or 10 min standing in patients with PAF or MSA were reported by Ziegler et al. (1977), and this was confirmed by Bannister et al. (1979). The interpretation of this finding is that in both types of autonomic failure the lesion is in the baroreflex arc which interferes with sympathetic nervous system responses to postural change. In PAF the low supine NE levels are consistent with a peripheral localization of the lesion, whereas the supine levels in MSA, which are normal or slightly elevated, are compatible with a failure to activate mostly intact postganglionic sympathetic neurons (Polinsky 1988). Unfortunately, the usefulness of measurements of plasma NE levels in the diagnosis of MSA or PAF has not been demonstrated. Nevertheless, in large groups of patients with these 2 conditions it has been found that if basal values are normal a diagnosis of MSA is likely, whereas low basal plasma NE levels are more in keeping with the diagnosis of pure autonomic failure (Polinsky 1988). The considerable overlap in measured values, however, necessitates the use of a number of other clinical and neuropharmacologic tests to help distinguish the 2 conditions.

The intravenous administration of trace amounts of radiolabeled NE at constant rate is a test to assess clearance and secretion. After steady-state conditions have been achieved, kinetic processes that determine plasma concentration of NE become equilibrated so that the rate of infusion of radiolabeled NE is equal to the sum of the rates by which NE leaves the circulation. Thus the endogenous secretion rate into the plasma and clearance of NE can be estimated. Using such rather complicated techniques, it has been shown that a reduction in clearance of NE is only found in patients with PAF. There is a normal clearance of NE in MSA patients, which supports the intactness of postganglionic sympathetic neurons in the latter condition (Esler et al. 1980; Polinsky et al. 1985).

Because isoproterenol is cleared from the plasma through extraneuronal mechanisms only, its disappearance after steady-state infusion is much slower when compared to NE. A defect in neuronal uptake is observed with isoproterenol steady-state infusion in PAF when comparison is made with disappearance rates of NE in this condition. Thus, very low plasma levels and reduced clearances of NE in PAF is thought to be a reflection of more severe disease of longer duration (Polinsky 1988). There is, however, no correlation between the plasma levels and clearance of NE, and this may be a reflection of diminished NE release. Defective neuronal uptake and delayed removal of NE could explain the occasionally observed prolongation of pressor effects of exogenously administered NE in some patients with PAF (Bannister et al. 1979).

Excretion of metabolites

Released NE is taken up by sympathetic neurons and is deaminated by monoamine oxidase. Further metabolic conversion results in the production of vanillymandelic acid (VMA) and 3-methoxy, 4-hydroxyphenylglycol (MHPG). Any released NE not taken up by sympathetic nerve endings is converted to normetanephrine. Most of this substance is deaminated, but a small amount is excreted in the urine as conjugated NE, and this reflects the activity of sympathetic neurons. The function of sympathetic neurons can thus be gauged from the urinary excretion and compartmentalization of metabolites of NE (Kopin et al. 1983a). In PAF, all NE metabolites are low in the urine in keeping with a decrease in the number of functioning noradrenergic neurons. In MSA, on the other hand, normetanephrine is reduced out of proportion to VMA or MHPG. This, together with the slight reduction in NE synthesis and the relative decrease in normetanephrine excretion, is a sign of a failure to activate the intact peripheral sympathetic nervous system.

Responses to pressor drugs

Chronic postganglionic denervation increases the response to NE, but there is a reduction in the pressor effects of indirectly acting sympathomimetic drugs. By contrast, after decentralization there is only a modest increase in pressor response when compared to peripheral denervation, and this increase is not accompanied by a decrease in peripheral NE stores. In decentralization the response to tyramine, an indirect sympathomimetic, remains normal. Denervation interrupts central pathways to end-organs as well, so that the observed responses in lesions of the central nervous system may be the result of local effects and may also be influenced by changes in reflex modulation. In addition, responses to drugs can result from abnormalities in absorption, elimination, and impairment, with action of the drug on target organs. Impaired reflex mechanisms and local changes at the cellular level of drug action influence the responses. In each case, many of these mechanisms may be influential in shaping results of pressor administration. In patients with autonomic dysfunction, therefore, the characteristics of pressor responses vary according to the site of the lesions. Thus, postganglionic sympathetic dysfunction may induce exaggerated responses to NE; decreased peripheral stores may also be found, and impaired neuronal uptake or modulation of pressor responses by baroreflexes can also be involved. On the other hand, preganglionic central lesions may induce increased pressor responses to NE because of impairment of baroreflex arcs.

In orthostatic hypotension, blood pressure responses to pharmacologic agents have been examined in an attempt to localize lesions responsible for orthostatic hypotension. Forearm blood-flow measurement during intraarterial infusion of NE and tyramine in patients with pure autonomic failure showed a decrease in flow response to NE, but not tyramine. Whereas in patients with MSA normal forearm blood flow reduction in response to both drugs occurred (Kontos et al. 1976). Histologically, Kontos et al. (1975) showed an absence of catecholamine fluorescence of perivascular plexuses in vessel walls of skeletal muscles in PAF, but such a loss has also been reported in those with MSA (Bannister et al. 1981a). However, absent catecholamine fluorescence must be interpreted with caution since histofluorescent studies may give variable results even in the same laboratories. Two large series of patients with autonomic failure have been investigated, and amongst those 8 with MSA and 2 with pure autonomic failure (Bannister et al. 1979) showed greater than normal blood pressure responses to NE. The degree of responsiveness also correlated with baroreflex impairment. These studies were confirmed and amplified. Blood pressure responses and plasma catecholamine levels before and during infusion of NE, angiotensin, and tyramine were examined in 8 patients with PAF and 9 with MSA. In both conditions, an increased pressure response was observed with NE and angiotensin, a non-adrenergic pressor agent. This reponse is in keeping with the impaired baro-

reflexes of these patients, causing defective modulation of blood pressure. When the dose response curve to NE was examined, however, only patients with PAF had a shift to the left, evidence of true adrenergic receptor supersensitivity. Thus, these studies support the different level of impairment in these patients, PAF causing denervation and the central nervous system lesions in MSA leading to decentralization.

Peripheral neuronal uptake and stores of NE can be indirectly measured by the increments in plasma NE after the intravenous administration of tyramine (Polinsky et al. 1981a). In PAF a reduction in the increment in plasma NE compared to MSA or control subjects is found. This normal responsiveness to tyramine in MSA further attests the functional integrity of postganglionic NE-containing neurons. However, other investigators of MSA (Bannister et al. 1979) did not find significant increases in plasma NE in response to tyramine infusion.

Other drugs and hormones have also been used to assess cardiovascular responses in autonomic failure and some of these may have therapeutic benefit. An exaggerated pressor response to somatostatin (Hoeldtke et al. 1986) and its analog has been found, and enhanced pressor responses to antidiuretic hormone have been demonstrated in PAF (Mohring et al. 1980). Analogs of antidiuretic hormone with long action suitable for intranasal administration have been developed, thus allowing possible therapeutic manipulation of such patients.

Methoxamine administration is associated with an increased pressor response (Parks et al. 1961) and increased chronotropic and vasodepressor responses to isoproterenol have been found in MSA (Mathias et al. 1977; Bannister et al. 1981b) and in PAF (Robertson et al. 1984). Platelet adrenergic receptors can be measured by the use of dihydroergocryptine, and an increased affinity and number of alpha receptors in MSA have been reported (Bannister et al. 1980). This has been confirmed and similar findings have been reported in PAF (Kafka et al. 1984).

In general, pharmacologic studies have lent support to the clinical and biochemical distinction between MSA and PAF, which suggest dysfunction at a postganglionic noradrenergic level in

PAF and are consistent with a central involvement of the sympathetic nervous system in MSA. While the pathologic examination of MSA patients is in keeping with these clinical and pharmacologic findings, the pathology of PAF has not been as yet fully elucidated and the postganglionic lesion has not been demonstrated histologically.

Epinephrine

Stress causes release of epinephrine from the adrenal medulla, which is a specialized postganglionic neuron. Epinephrine is involved in circulatory shock and vascular collapse, but it is more intimately connected to metabolic homeostasis, such as an increase in hepatic glucose output, activation of lipolysis and inhibition of insulin secretion. Hypoglycemia induced by insulin is a potent stimulus for epinephrine secretion. An absent response of the adrenal in insulin-induced hypoglycemia was reported in 2 patients with postural hypotension (Luft and Von Euler 1953). A study of patients with MSA and normal controls showed that in normal subjects glucose recovery curves after the nadir of hypoglycemia induced by insulin is biphasic, showing an initial rapid rise, followed by a slower rate of return towards normal values (Polinsky et al. 1980). In patients with adrenergic failure, the initial rapid phase of glucose recovery is not present. The implication is that the rapid phase in glucose recovery is the result of epinephrine secretion. In normal subjects there is also a rapid rise in plasma epinephrine and a lesser increase in NE in response to insulin infusion. The peak levels of epinephrine are achieved within minutes after the nadir of insulin-induced hypoglycemia. In patients with either MSA or PAF the catecholamine response to hypoglycemia is deficient. This deficiency may be due to a lesion at any point in the pathways from receptors for glucose in the hypothalamus or brain stem to the adrenal medulla. Such impaired responses to insulin-induced hypoglycemia have also been found in patients following sympathectomy, splanchnicectomy, or in adrenalectomized patients (Polinsky 1988) and may be seen in patients with diabetic autonomic neuropathy. These results show that,

while the response to hypoglycemic stress is impaired in those with autonomic failure under discussion here, this impairment is not of localizing value in determining the site of the lesion. Normally epinephrine is very effective in quickly restoring glucose to euglycemic levels after induced hypoglycemia. But this mechanism does not seem essential in MSA or PAF, because patients suffering from these disorders do not have hypoglycemic episodes. MSA and PAF are therefore quite distinct from acute pandysautonomia, in which hypoglycemia is not infrequent.

Acute autonomic failure is generally accompanied by deficits of autonomic function. In chronic and insidious onset of autonomic disease clinical autonomic dysfunction is more frequent when special stress is placed on the diseased system. Thus, in acute pandysautonomia, occurring in younger individuals more susceptible to physical stress, and its demands on glucose metabolism, hypoglycemia is frequently seen, whereas in MSA and PAF occurring in older individuals less prone to physical stress, even though the setting for hypoglycemia is present, it is clinically unimportant.

METABOLISM OF CENTRAL NERVOUS SYSTEM
NEUROTRANSMITTERS

The metabolism of brain neurotransmitters is important in autonomic function, but this is currently indirectly approached by measurement of neurotransmitters and their metabolite levels in the cerebrospinal fluid and tentatively also by the administration of labeled ligands or precursors prior to positron emission tomography.

Norepinephrine

The innervation with NE-containing nerve fibers of the hypothalamus, an important structure in autonomic control, comes from neurons whose cell bodies are located in the locus ceruleus. (These neurons also supply perivascular noradrenergic nerve fibers to a large part of the cerebral vasculature.) From this nucleus there are also noradrenergic projections to the cerebellum, hippocampus, and cerebral cortex. Additional pathways with diffuse origin in the brain stem and projecting to the hypothalamus, limbic system, and other brain stem nuclei are also found. Descending NE-containing pathways are found in the bulbo-spinal tracts, which innervate preganglionic sympathetic neurons and anterior horn cells. 3-Methoxy-4-hydroxyphenylglycol (MHPG) is the main brain metabolite of NE (Kopin et al. 1983b). MHPG levels in the CSF need to be correlated with levels found in free plasma because MHPG crosses the blood-brain barrier, before MHPG can be used as an index of central NE metabolism. A formula for correcting total CSF MHPG for the contribution from plasma has been developed (Kopin et al. 1983b).

The total CSF MHPG levels in patients with MSA and PAF is decreased (Polinsky et al. 1984). Since only PAF patients have reduced levels of plasma MHPG, the corrected CSF MHPG level is lower than normal only in MSA patients, showing that in these cases the central nervous system NE metabolism is diminished. The low MHPG levels in the CSF of patients with PAF results from the very small contribution from peripheral NE metabolism, but the central component of MHPG in these patients is normal. These metabolic studies support the pathologic findings in the brain of patients with MSA that show extensive involvement of noradrenergic structures. Thus, the decreased MHPG production in MSA patients is caused by degeneration of noradrenergic structures in the central nervous system.

Serotonin and dopamine

In patients with MSA the non-autonomic neurologic deficits could be related to dopamine since extrapyrmidal features are often found in such patients. Because serotoninergic neurons are found in the brain stem and the pathology shows extensive involvement of the brain stem in MSA, it is unlikely that this neurotransmitter is also abnormal in PAF patients.

Homovanillic acid (HVA) is the primary metabolite of dopamine in humans, whereas 5-hydroxyindoleacetic acid (5-HIAA) is the primary metabolite of serotonin. Because both these metabolites are removed from the CSF by the same mechanism, these neurotransmitters can be

discussed together. Low CSF HVA levels in MSA have been reported (Moskowitz and Wurtman 1975). This has been confirmed with different methods (Williams et al. 1979). Additionally in MSA, CSF HVA, but not 5-HIAA, is lowered by the dopamine agonist bromocryptine, which suggests that dopaminergic receptors remain functional in this disorder. In a larger series of patients, a 50% reduction of both HVA and 5-HIAA, compared with reductions of these substances in controls was found (Polinsky 1988), whereas patients with PAF have normal CSF HVA and 5-HIAA. The elevation of both metabolites by the administration of probenecid in MSA suggests that the low levels result from a decrease in turnover rate rather than from an abnormality in transport of the metabolites out of CSF compartments. The significant correlation between HVA and 5-HIAA levels in the CSF of patients with MSA suggests that both dopaminergic and serotoninergic structures are equally affected in these patients.

Hormones and peptides

The widespread influence of the autonomic nervous system is in part mediated through a number of hormones and peptide systems that are involved in numerous actions, including metabolic and stress responses and digestion.

All hormonal and peptide responses usually studied in patients with autonomic failure have been examined after insulin-induced hypoglycemia, with the exception of melatonin. While this might not be frequently observed in practice, insulin administration nevertheless provides a standardized way of measuring hormone responses to hypoglycemic stress. The restoration of normal blood sugars after insulin administration does not depend entirely on the secretion of epinephrine (see above). The other protective mechanisms can easily be demonstrated in patients with adrenergic failure where prolonged and dangerously low blood sugar levels are not usually observed. In patients with autonomic failure, therefore, the hypoglycemic counterregulatory hormone responses can be observed in the absence of catecholamine secretion to the same stimulus and can therefore give an insight into the relationship between the autonomic nervous system and the release of hormones and peptides.

Glucagon

Glucagon synthesized by pancreatic alpha-cells is released in response to various stressors and during hypoglycemia. Its hyperglycemic effects follow the release of catecholamines. In patients with MSA and PAF where catecholamine responses are absent, the baseline plasma glucagon, the maximum levels achieved in response to hypoglycemia, and those found during the nadir of hypoglycemia were normal (Polinsky et al. 1981b). Other authors have found absent responses in 4 patients with MSA (Sasaki et al. 1983). These different reports are difficult to reconcile at present. In general, it is thought that patients with MSA and PAF, in spite of an absence of catecholamine response to hypoglycemia, have normal glucagon release.

Growth hormone

Stress that causes release of catecholamine also stimulates the release of growth hormone. In 4 patients with MSA, insulin hypoglycemia stress did not lead to growth hormone release. These patients were deficient in catecholamine, glucagon, prolactin, and cortisol responses to the induced hypoglycemia. On the other hand, other patients with MSA had normal cortisol and growth hormone responses to insulin hypoglycemia (Polinsky et al. 1981b), which suggests a varying extent of central nervous system lesions in MSA. In PAF patients there was also a normal growth hormone response during hypoglycemia. Like any lesions beyond the hypothalamus and in the peripheral sympathetic nervous system, dysfunction at peripheral levels disrupts only catecholamine response to hypoglycemia but leaves other central mediated hypothalamopituitary hormonal responses intact.

Cortisol

The adrenal cortex releases cortisol during stress. This response, however, is superimposed upon circadian rhythms in the level of this hormone.

In patients with autonomic failure, the basal cortisol levels are higher than in controls. The higher levels have been attributed to the stress of postural hypotension before this symptom is adequately controlled (Polinsky et al. 1981b). Insulin hypoglycemia causes a normal increase in plasma cortisol in patients with MSA (Wilcox et al. 1975) and in those with orthostatic hypotension and adrenergic insufficiency (Polinsky et al. 1981b). These studies are in contrast to those of Sasaki et al. (1983), but the basal levels of cortisol in their patients were not elevated; therefore, it seems that increases in plasma catecholamine levels in response to stress are unnecessary for stress-induced cortisol release.

Gut peptides

Many peptides are secreted by cells throughout the gastrointestinal tract, but only 2 have hitherto been considered worthy of study in the context of autonomic failure. The release of pancreatic polypeptide is likely to be under vagal-cholinergic control (Polinsky 1988). Measurement of pancreatic polypeptide levels in the plasma gives a measure of vagal (parasympathetic) function. Simultaneous measurements of catecholamines during insulin-induced hypoglycemia can therefore give additionally an approximation of sympathetic and parasympathetic nervous responses and adrenal medullary responses. During hypoglycemia a marked increase in plasma levels of pancreatic polypeptides occurs in normal subjects. This response is absent in patients with PAF and in the majority, though not all, of patients with MSA (Polinsky et al. 1982). No correlation, however, between catecholamine and pancreatic polypeptide responses during hypoglycemia have been found. This supports the finding of parasympathetic involvement in MSA, which may respond to cholinomimetic treatment (Khurana et al. 1980). Thus the impaired responses of both pancreatic polypeptide and catecholamine to insulin hypoglycemia are evidence of widespread sympathetic and parasympathetic involvement in PAF and MSA.

Gastrin

This gut peptide increases antral motility and acid secretion. Vagus (parasympathetic) stimulation also increases gastrin release, but basal gastrin levels are increased after vagotomy, presumably because of the resulting distension of the stomach and increased pH. Infusion of epinephrine causes an increase in gastrin release through beta-adrenergic stimulation. A correlation between plasma epinephrine and gastrin increases after hypoglycemia has also been found. Abnormal basal and stimulated gastrin levels in patients with autonomic failure have been reported (Polinsky et al. 1988). In PAF there are high basal gastrin levels in keeping with peripheral vagal dysfunction. Hypoglycemia in PAF patients causes a significantly greater than normal increase in gastrin, but MSA patients show a blunted response. In PAF the enhanced responsiveness to hypoglycemia has been attributed to adrenergic supersensitivity and in MSA the reduced response to a decrease in catecholamines with normal receptor sensitivity in this disease. Whether the abnormal gastrin responses in autonomic failure have clinical correlates in these conditions has not been determined, but they may play a role in postprandial fullness and in postprandial hypotension often found in such patients.

CENTRAL NERVOUS SYSTEM RESPONSES IN AUTONOMIC FAILURE

Insulin-induced hypoglycemia, in addition to evoking peripheral peptide and hormonal responses, also releases pituitary peptides, including beta-endorphin and ACTH. The effect of ACTH is through the secondary release of cortisol, whereas beta-endorphin causes an increase in blood sugar through its stimulating effect on the secretion of glucagon and inhibition of somatostatin. Pharmacologic studies have shown that arecoline, a cholinomimetic, stimulates the release of beta-endorphin and that atropine, a cholinergic inhibitor, blocks the hypoglycemia-induced release of this peptide. Therefore, a central cholinergic pathway is involved in this response.

Insulin-induced hypoglycemia does not produce the normal increase in beta-endorphin and ACTH in patients with MSA, but these responses remain normal in PAF. In the latter condition, the absent catecholamine response to hypoglycemia and the normal beta-endorphin and ACTH responses suggest that peripheral catecholamine release is not required for a normal ACTH beta-endorphin response. Thus, these studies confirm the almost pure involvement of peripheral autonomic structures in PAF. The central abnormalities in MSA, which involve the hypothalamus, also are reflected in the absent peptide release to hypoglycemia in these patients. This is supporting evidence that cholinergic central pathways are affected in this disease.

Melatonin

A circadian rhythm for release of melatonin is induced by the suprachiasmatic nucleus, and environmental light is the Zeitgeber for inhibiting the secretion of melatonin by the pineal gland. In this response retinohypothalamic projections are involved. The efferent pineal innervation takes origin from the medial forebrain bundle and midbrain reticular formation. From there fibers pass to the intermediolateral cell columns and on to the superior cervical ganglion, which provides the postganglionic innervation of the pineal through cervical sympathetic nerve fibers (see Appenzeller 1982). Stimulation of beta-adrenergic receptors in the pineal increases synthesis and release of melatonin. Melatonin is metabolized to 6-hydroxymelatonin, after conjugation, it is excreted in the urine. The measurement of urinary 6-hydroxymelatonin conjugates is therefore an index of sympathetic nervous system activity that influence the pineal as well as an index of the gland's function.

The circadian rhythmicity of 6-hydroxymelatonin excretion in normal subjects and patients with MSA and PAF was studied (Tetsuo et al. 1981). The highest level occurs normally between 2400 and 0600 hours, whereas the lowest level is measurable between 1200 and 1800 hours. The normal diurnal pattern is found in patients with PAF, though the total urinary excretion of 6-hydroxymelatonin was significantly reduced. In patients with MSA, the diurnal pattern was markedly abnormal. Most excreted high levels of urinary 6-hydroxymelatonin during the day, although their overall level of excretion (for at least half the MSA patients) was low. Thus, in

PAF the preserved circadian rhythmicity of excretion reflects an intact central innervation, but the low nighttime levels an impairment of pineal postganglionic sympathetic innervation; whereas in MSA, the disrupted rhythmicity is more in keeping with the central lesions in these patients. A summary of the pharmacologic and biochemical distinction between PAF and MSA is shown in Table 3.

POSTPRANDIAL HEMODYNAMICS

The ingestion of food induces autonomic and hormonal changes that have hemodynamic consequences. A number of gut peptides with cardiovascular effects are also released and may modulate autonomic nervous system activity. Vasomotor function after the ingestion of glucose was first reported in 1970 (Appenzeller and Goss). After the ingestion of food, there is a considerable increase in intestinal blood flow (Quamar and Read 1986). Systemic blood pressure is usually maintained, but baroreflexes, as evidenced by the overshoot after Valsalva's maneuver, become impaired even in normal subjects after the ingestion of glucose. Vasoactive substances are released in response to food ingestion. Other responses (in normal subjects) include increases in heart rate, cardiac output, and stroke volume. Together, these constitute a significant readjustment in the circulation to maintain blood flow to the brain. Plasma NE and renin activity increase (Mathias et al. 1986, 1987), but there is an overall fall in peripheral vascular resistance, presumably attributable to the impairment of neurogenic vasomotor tone. The overall effect, though, of these multiple influences on the circulation is the maintenance of blood pressure in normal subjects.

In patients with impaired autonomic function, the ingestion of glucose can totally abolish baroreceptor function and lead to severe postural hypotension (Appenzeller and Goss 1970). Similar observations were made in hypertensive patients given ganglion blocking agents. Symptomatic postprandial hypotension preventing maintenance of blood pressure was reported by Seyer-Hansen (1977) in a patient with Parkinsonism with autonomic failure. The patient complained of dizziness and visual obscuration after

TABLE 3
Biochemical and pharmacologic distinction between PAF and MSA.

	PAF	MSA
Biochemical		
Supine plasma NE	Low	Normal
Urinary NE metabolites		
Total	Decreased	Normal
NM/Total	Normal	Decreased
Responses to hypoglycaemia		
β-endorphin/ACTH	Normal	Decreased
Gastrin	Increased	Decreased
CNS neurotransmitter metabolism		
NE	Normal	Decreased
DA	Normal	Decreased
5-HT	Normal	Decreased
Pharmacologic		
Pressor responses		
NE	Markedly increased	Increased
Angiotensin	Increased	Increased
NE release by tyramine	Low	Normal
NE receptor sensitivity	Increased	Normal
NE clearance	Decreased	Normal

NE, norepinephrine; DA, dopamine; NM, normetanephrine.
(Reproduced from Polinsky 1988 by courtesy of the Author and Publisher.)

almost every meal. Postural hypotension could be provoked in the patient by oral glucose administration. A number of patients with autonomic dysfunction were reported showing profound falls in systolic and diastolic blood pressures after food ingestion (Robertson et al. 1981). In these patients the blood pressures remained low, even after they lay down for up to 3 h. This fall in blood pressure after food ingestion may occur within 10–15 min and reaches its nadir about 60 min after food intake (Bannister et al. 1984; Mathias et al. 1986). The fall in blood pressure is independent of changes in forearm and skin blood flow, clearly showing that hemodynamic adjustments are not present in spite of the severe fall in blood pressure in patients with autonomic failure. In such patients, the assumption of the upright posture, like in those studied in a parachute harness and then tilted upright (Appenzeller and Goss 1970), may lead to symptomatic impairment of cerebral perfusion.

Measurement of plasma epinephrine and NE after food ingestion in patients with autonomic failure showed no changes in the levels of these catecholamines, confirming the impaired modulation of blood pressure by the sympathetic nervous system after food ingestion in these patients. Normally, there is also a small but definitive increase in NE in response to food (Mathias et al. 1988). But, plasma renin activity increased to the same extent in both normal subjects and patients with autonomic failure. Since the fall in blood pressure in the patients is very much greater than changes in controls in response to food the release in renin is inappropriate in magnitude when compared with the fall in blood pressure. Renin release depends upon beta-receptor stimulation and because of the marked fall in blood pressure and the anticipated impairment of beta-receptor activity the increase in renin levels is inappropriate, considering the magnitude of blood pressure change. Hematocrit, plasma electrolytes, and osmolality remain unaffected and thus do not contribute to the observed postprandial hypotension in those with autonomic failure.

The levels of gastrin, motilin, vasoactive intestinal polypeptide, somatostatin, and cholecysto-

kinin are similar in controls and patients (Mathias et al. 1986). However, after food ingestion, patients with autonomic failure have higher levels of enteroglucagon, pancreatic polypeptide, and neurotensin. Only neurotensin has vasodilatory effects, which may be important in causing symptomatic post-prandial falls in blood pressure in patients with autonomic failure. While many gut peptides have potent vasodilatory effects, their unchanged levels (compared to those of controls) in the circulation after food ingestion does not exclude regional (e.g., splanchnic circulatory effects) influences that may affect blood pressure.

Insulin can also lower blood pressure substantially in patients with autonomic failure and even without a change in blood glucose. The bolus administration of intravenous insulin causes hypotension without change in forearm muscle or cutaneous blood flows (Mathias et al. 1988). Studies of animals whose carotid arteries were clamped for observation of their subsequent hypertensive response showed that administering insulin and glucose to maintain euglycaemia is sufficient to abolish this response. The results of sections carried out at various levels of the nervous system in such animals are most consistent with a block of neurovascular transmission induced by isulin, which inhibits the normal hypertensive response to carotid clamping (Burks and Appenzeller 1971). In other studies carried out in humans, cardiac output, muscle, and cutaneous blood flow remained unchanged after food ingestion, suggesting that splanchnic vasodilation is the cause of hypotension. Insulin may bring about hypotension after ingestion of food because of its reactive release in response to the ingested glucose, and studies on diabetics given insulin are consistent with this interpretation (Page and Watkins 1976).

Food components have been studied by comparing the blood pressure responses to isocaloric, isovolumic, and mostly comparable isotonic solutions of different types of food and after ingestion of mixed food (protein, fat, and carbohydrate). Lipids have a lesser and less sustained effect, and protein alone has no definitive influence (Bannister et al. 1987). The hypotensive changes after glucose ingestion are not produced when the sugar is substituted with an isocaloric, isoosmotic, and isovolumic solution of xylose (Mathias et al. 1988).

The rate of gastric emptying has been measured in autonomic failure (Mathias et al. 1986). Patients with autonomic failure typically experience accelerated gastric emptying, together with postprandial hypotension. Some, however, have normal emptying, and occasionally, as was the case with 2 patients with autonomic failure due to amyloidosis, gastric emptying is markedly delayed, while postprandial hypotension is severe (Mathias et al. 1988). Whether accelerated gastric emptying in patients with autonomic failure contributes to the genesis of postprandial hypotension in such patients remains to be elucidated.

In the management of symptomatic postprandial hypotension, a number of drugs have been used and one, indomethacin (50 mg by mouth), attenuated the hypotensive response, suggesting that prostaglandins might be involved with their vasodilatory action in contributing to this effect of food. No long-term studies with this drug have been reported to date. Because caffeine activates the renin-angiotensin system and stimulates sympathetic nervous activity, it helps to maintain blood pressure in autonomic failure in some patients and has been found effective in preventing postprandial hypotension. This beneficial effect is independent of renin angiotensin stimulation or sympathetic nervous system activation and has been attributed to caffeine's ability to block vasodilatory adenosine receptors. Thus, 2 cups of coffee, which contain approximately 500 mg of caffeine, have been proposed as an effective treatment in the long-term management of such patients (Mathias et al. 1988).

The somatostatin analog (SMS201–995) also has been shown to be effective in the long-term management of symptomatic postprandial hypotension. Its effectiveness is attributed to the drug's capacity to block the release of a number of vasodilatory peptides important in maintenance of blood pressure in patients with chronic autonomic failure (Hoeldtke et al. 1986).

The role of systemic hypotension in the genesis of transient ischemic attacks has been controversial and until recently largely discounted. Nevertheless, a systematic search for postural hypotension in patients with transient ischemic attacks suggests that this might be of pathogenetic significance in those with marked carotid occlusive vascular disease, particularly if the disease occurs in association with diabetes mellitus or in patients with hypertension who are given hypotensive medication. Focal cerebral hypoperfusion resulting from a combination of occlusive vascular disease and orthostatic hypotension is a treatable cause of transient ischemic attacks. By preventing recurrent hypotension in such patients permanent focal neurologic deficits may be avoided (Dobkin 1989).

Autonomic dysfunction, to varying extent, has been demonstrated in elderly patients also. A large number develop postural hypotension if the study group is confined to patients over the age of 80 years (Lipschitz et al. 1983). In these patients, a fall in systolic blood pressure was found to be particularly common after food ingestion if the patient maintained a sitting position. This hypotensive response to meals, together with the frequent increased susceptibility to the action of drugs that impair autonomic function, may make elderly people particularly susceptible to postural hypotension after meals and its consequent cerebral ischemia (Appenzeller and Goss 1970). Thus, post-prandial hypotension is not confined to MSA but may occur in the elderly or in those with other diseases affecting the autonomic nervous system.

SLEEP AND RESPIRATORY ABNORMALITIES IN AUTONOMIC FAILURE

During rapid-eye-movement sleep (REM sleep) there is a burst of cholinergic hyperactivity and aminergic inhibition. The serotoninergic neurons in the raphe nuclei of the brainstem are intimately concerned with the production of non-REM sleep. These neurons initiate and maintain the non-REM state by inhibiting cholinergic ascending reticular activating structures. They also are the triggering neurons for REM sleep.

Two systems that act independently are responsible for breathing: a metabolic influence or automatic system and a voluntary or behavioral system. During wakefulness, both these systems are active, but during sleep respiration depends on the rhythmicity of the automatic respiratory control, which is located in the medulla oblongata. In addition, the reticular arousal system exerts a tonic influence on brainstem respiratory neurons (McNicholas et al. 1983). The mechanisms responsible for respiration during sleep are the loss of reticular inhibition and the absence of wakefulness stimulus on automatic respiration. Close relations in the pontomedullary region are found between the respiratory, the central autonomic, and the hypnogenic neurons in the lower brainstem. These close relations are further influenced by projections from the caudal portion of the nucleus tractus solitarius to the hypothalamic nuclei and fibers from the hypothalamus to the nucleus of the tractus solitarius (Brodal 1981). These intimate relationships hint at a fine balance between the neurons of the autonomic nervous system controlling respiration and those involved in sleep-wakefulness stages.

Patients with MSA develop progressive respiratory disturbances and changes in wakefulness and sleep patterns, adding to the progressive disability of the advanced stage of the disease, so that progressive autonomic and somatic neurologic dysfunction is complicated by respiratory failure. Ventilatory disturbances have been recorded during wakefulness and sleep and those respiratory disturbances present during wakefulness become much aggravated during sleep, a result of the physiologic vulnerability of brain stem respiratory neurons during sleep. Clinically, respiratory abnormalities are manifested as restlessness, daytime somnolence, early morning headache, fatigue during the day, and disturbed sleep at night. Symptoms are complicated by intellectual failure. On examination, irregular rate and rhythm of respiration may be found in those with respiratory muscle-motor-neuron disorders, and restlessness is invariably present. However, restlessness is not a symptom of those with dysfunction of the medullary or respiratory control system, since the sensation of restlessness depends on respiratory excursion of the chest and diaphragm. Many patients with advanced MSA die from hypoventilation during sleep, resulting eventually in fatal sleep apnea. Cor pulmonale and congestive heart failure may develop, and these are the result of the recurring

and prolonged hypoxemia, respiratory acidosis, and hypercapnia, particularly during sleep, leading to pulmonary hypertension.

A wide variety of respiratory abnormalities have been recognized in MSA. These include upper airway obstructive and mixed apneas, both in non-REM and REM sleep, associated with significant oxygen desaturation. Abnormal amplitude, rate, and rhythm of respiration, both with and without significant oxygen desaturation, usually becomes worse during sleep. Transient occlusion of the upper airway and loss of integration of diaphragmatic and intercostal muscle activities cause failure of ventilation. In some patients there are prolonged periods of central apnea, with only minimal oxygen desaturation, occur during wakefulness. Cheyne-Stokes pattern and an accentuation of this respiratory abnormality during sleep is often found. The postural fall in blood pressure may be accompanied by periodic breathing and inspiratory gasps. Apneustic-like breathing has also been found. Transient and sudden respiratory arrest also occurs. During stages 1 and 2 of non-REM sleep and during REM sleep, recurring episodes of apnea have been found in patients with MSA (Chokroverty 1986).

Various types of apnea have been found in patients with MSA: central apnea, where respiratory muscle activity does not occur and air is not moved through the nose or mouth; obstructive apnea, characterized by absence of air exchange, but a persistence of diaphragmatic and intercostal muscle activities; and, lastly, mixed apneas, characterized by beginning central apnea that is followed by obstructive apnea before a resumption of normal respiration. Inspiratory gasps, which resembled apneustic breathing, have been noted clinically (Bannister and Oppenheimer 1972). In another patient with MSA cluster breathing with periods of apnea were recorded during wakefulness and longer such apneic periods were recorded during sleep (Lockwood 1976). Pathologic examination of this patient revealed characteristics of MSA, but gliosis in the pontine tegmentum and medulla involving the reticular formation was prominent. In this patient the periodic apnea without impaired ventilatory responses to the inhalation of CO_2

suggested that the respiratory neurons functioned independently from the medullary chemoreceptors, which control ventilation. In another patient with alveolar hypoventilation, arrhythmic breathing, and episodes of apnea, including apneustic breathing, autopsy showed the typical changes of MSA and widespread degeneration in the brainstem and spinal cord.

Laryngeal stridor due to laryngeal abductor paralysis and excessive snoring during sleep have been described in many cases with MSA (Bannister et al. 1981b; Israel and Marino 1977; Guilleminault et al. 1977; Williams et al. 1979). The clinical manifestation of the stridor in these patients is a peculiar noisiness, particularly at night, which has been likened to a 'donkey braying'. This abnormality was found in 8 of 12 cases with MSA (Williams et al. 1979). In 3 cases with laryngeal stridor that required tracheostomy, the pathology showed atrophy of the posterior cricoarytenoid muscles, which was consistent with denervation atrophy. But cell counts of the nucleus ambiguus in these patients were normal, implying that perhaps a biochemical defect is responsible for the histologic appearance of the abductor muscles of the vocal cords (Bannister et al. 1981b). In many patients, hypersomnolence during the day is associated with apneic episodes, often of an obstructive nature during sleep, and some may require resuscitation at night because of respiratory arrest (Brisken et al. 1978).

The hypoxic ventilatory responses may be blunted and the hypercapnic ventilatory response is often also impaired, suggesting a metabolic defect in the control system (McNicholas 1983). In these patients, very irregular breathing patterns during sleep were found, as well as many apneic episodes without oxygen desaturation. A larger study of 9 patients with MSA showed a number of respiratory dysrhythmias, including central apnea, Cheyne-Stokes, or Cheyne-Stokes-like breathing, upper airway obstruction, and mixed apneas, often accompanied by desaturation, and usually found during non-REM sleep, stages 1 and 2, and REM sleep (Chokroverty 1988). In these patients, there was also evidence of cardiac autonomic denervation, because heart-rate variation during apneic-hyperpneic cycles did not occur. These findings, together with other studies, suggested an impaired metabolic respiratory control system.

Extensive pathologic and clinical studies of patients with MSA suggest that the pathogenetic mechanisms of respiratory abnormalities are based on the following. There may be involvement of medullary respiratory neurons, with consequences in pattern- and rhythm-generating mechanisms, without changes in central chemoreceptors. The arousal system may be impaired through direct involvement of the ascending reticular activating system. Non-respiratory motor neurons may be involved in the brainstem, as well as respiratory motor neurons, and motor weakness may contribute to upper airway obstructive apnea. Impulse traffic along the phrenic and intercostal nerves may be markedly reduced because of involvement of anterior horn cells in the cervical and thoracic spinal cord, and this involvement may cause a central type of apnea. Moreover, involvement of higher centers in the forebrain and midbrain may interfere with input to the medullary respiratory neurons and cause dysrhythmic and apneustic breathing. This may also be the manifestation of involvement of hypothalamic projections to respiratory neurons and to the nucleus tractus solitarius and nucleus ambiguus. A peripheral vagal afferent involvement may reduce input to the central respiratory neurons and cause respiratory rhythm abnormalities and sympathetic denervation of the nasal mucosa can increase nasal resistance to air flow and promote obstructive apnea. Lastly, the neurochemical changes in MSA may also interfere with normal respiration (Chokroverty 1988).

Experimentally, norepinephrine, serotonin, and dopamine have an important role in the control of breathing (Dempsey et al. 1986). Experimental evidence also shows that dopamine can affect respiration. This action can be on central structures when the neurotransmitter stimulates breathing or peripheral structures through the carotid body chemoreceptor, where dopamine is inhibitory on respiration. Imbalance between these 2 effects on respiration may contribute to respiratory dysrhythmia.

The management of respiratory abnormalities in MSA is symptomatic and may be inadvisable in the later stages of the disease. Beyond general measures, such as avoidance of sedatives and alcohol or hypnotic drugs, which further depress respiratory neuron activity, mechanical appliances (and including tracheostomy) to promote airway patency have been advocated with varying success.

PARANEOPLASTIC SYNDROMES (PANEOS) AND AUTONOMIC FAILURE

A number of PANEO syndromes defined as remote effects of cancer have been associated with clinical autonomic failure. These disorders are not attributable to direct effects, such as pressure on normal tissues or tissue destruction by the tumor or the tumor's metastasis. A number of potential mechanisms for these disorders include the release by the tumor of peptides with endocrine properties, for example, ACTH; release of cytokines or growth factors; and immunologic cross-reactivity between the tumor and normal cell components. In such conditions, the neurologic PANEOS may result from sharing of antigens between tumors and parts of the nervous system. There are numerous examples of PANEOS, but the one most pertinent to autonomic failure is the Lambert-Eaton myasthenic syndrome (LEMS). This syndrome is characterized by proximal muscle weakness, a reduction of or absent tendon reflexes, and dysfunction of the autonomic nervous system, characterized by dry mouth, sexual impotence, and constipation. It is associated with small-cell lung carcinoma in about 60% of patients (O'Neill et al. 1988). Applying surface electrodes to such patients and recording from their affected muscles shows an abnormally small compound muscle-action potential (CMAP) in response to a single supramaximal nerve stimulus. On the other hand, during high-frequency nerve stimulation or immediately following a brief period of maximal voluntary contraction of a weak muscle, a large increase occurs in the CMAP and also in muscle strength, and this posttetanic potentiation is a characteristic electrophysiologic feature of the disease (Lambert et al. 1956).

In LEMS there is a reduced quantal release of acetylcholine from the motor nerve terminals. By analogy another neuromuscular transmission disorder, myasthenia gravis, has by now a well-understood autoimmunologic cause. The evidence for a similar autoimmunologic mechanism in LEMS is only now appearing and is based on the following: Injection of LEMS IgG passively transfers into recipient mice the physiologic and morphologic abnormalities observed in the human disease. Most notably there is a disorganization and paucity in the arrangement of active-zone particles that are believed to represent the voltage-gated calcium channels (VGCCs). Also, calcium influx through VGCCs into small-cell lung carcinoma cells (often associated with LEMS) is reduced by LEMS IgG. This suggests that in SCLS-associated LEMS the antibody response may be triggered by VGCCs that are expressed on the cells and hence by inference that immunologic cross-reactivity causes the clinical neurologic abnormality. The proposal that similar VGCCs on neuronally derived cells may trigger the disorder in patients with LEMS not associated with tumor has also been made. Thus, LEMS now provides yet another example of complicated relations between nervous and immune systems and tumor processes (see Vincent et al. 1989).

The evidence that LEMS with associated cancer is accompanied by detectable antibodies to VGCC epitopes suggests that such a mechanism of putative cross-reactivity with VGCCs on cells of the autonomic nervous system could explain the autonomic dysfunction in LEMS. The fact that most patients with progressive autonomic failure have no associated small-cell lung carcinoma does not negate this possibility. Autoantibodies to VGCCs may be important in the genesis of some progressive autonomic failures just as they are thought to arise in patients with LEMS not occurring in association with SCLCs, though the etiology of such antibodies remains enigmatic. One possibility is that in such patients occult tumors successfully dealt with by the subject's immune response have led to a persistence of the progressive autonomic failure syndrome. The availability of tests for IgG binding on VGCCs in a variety of neuronal cells may be explored in a further molecular analysis of progressive autonomic failure. Clinically, support for this interpretation comes from the occurrence of pandysautonomic syndromes attributed to autoimmune disease (Appenzeller and Kornfeld 1973) and of similar syndromes in association with small-cell carcinoma (Ivy 1961).

WAARDENBURG SYNDROME (193500)

The Waardenburg syndrome is a congenital ectodermal germ layer defect. It is inherited in an autosomal dominant form, with variable phenotypic expressivity (Schweitzer and Clack 1984). The syndrome is characterized by dystopia canthorum, broad nasal root, hypoplasia of the medial eyebrows, heterochromia iridis, white forelock, and congenital deafness. Many additional features have been reported, which include meningocele, atresia of the esophagus, and Hirschsprung's disease. In addition, a case with anal atresia, esophageal atresia, and tracheal esophageal fistula has also been described (Nutman et al. 1986). The disease is found in many groups. An inuit (Eskimo) boy with this disease had in addition vascular abnormalities, including left pulmonary artery stenosis, ocular ptosis, and unilateral duplication of the renal collecting system. On histologic examination hypoganglionosis, hyperganglionosis, and ectopic ganglia in the lamina propria (neuronal colonic dysplasia) were found in the rectum. The hypopigmented skin was devoid of melanocytes, whereas the hyperpigmented skin patches contained melanin throughout the basal layer, but the melanocytes were not evenly distributed (Caplan and DeChaderevian 1988). Another patient with the syndrome had bilateral sectoral iris heterochromia and fundus bicolor. Complete iris heterochromia also is found in some patients (Nork et al. 1986). The cervical sympathetic system is linked to the pigmentary abnormalities and inner ear anomalies and this, together with aganglionosis, suggests a disorder of the neural crest (Schweitzer and Clack 1984).

In 3 South African families of Afrikaner descent, the syndrome has been tracked back for 12 generations. The earliest known family member with this syndrome was born in 1842.

Elsewhere, a patient with features of Waardenburg syndrome had in addition features not previously associated with the disorder. These included marked mental and motor retardation and neurologic disorders, including dystonia, muscular stiffness, and peripheral neuropathy. The sural nerve biopsy in this patient showed chronic injuries evidenced by the formation of lesions resembling onion bulbs (Kawabata et al. 1987).

This rare inherited disorder shows many features of autonomic dysfunction involving the eye, ear, and enteric nervous system. However, no functional autonomic tests have been performed on such patients, and the disease is still in the early descriptive state. Little pathogenetic knowledge exists, but the disorder is thought to result from a failure of migration of neural crest cells.

DOPAMINE-β-HYDROXYLASE DEFICIENCY (223360)

This deficiency is a distinct autonomic disorder. It is possible to recognize this disease in infants presenting with delayed eye opening, hypoglycemia, hypothermia or hypotension, or both. A simple and definite assay of plasma NE and dopamine can clinch the diagnosis at this early stage and help management with appropriate therapy now available.

The description of a 42-year-old man with a lifelong history of orthostatic intolerance (Biaggioni et al. 1990) is illustrative of the history of this only recently recognized disorder. This patient had a negative family history, bilateral ptosis, and frequent 'fainting spells', which during childhood were worsened by exercise and treated as epileptic. Though sexual maturation was normal, sexual function was characterized by retrograde ejaculation, and difficulty in maintaining erection. Mental and physical development were otherwise entirely normal. Orthostatic symptoms became more incapacitating after heavy meals or alcohol. Hot environments causing vasodilatation contributed to the worsening of the symptoms, as did micturition, also associated with vasodilatation. Orthostatic hypotension documented at age 33 was treated unsuccessfully with fludrocortisone, pseudoephedrine, indomethacin, and tranylcypromine. The monoamine oxidase inhibitor produced paranoid thinking and tonic clonic seizures with postictal symptoms, which occurred while assuming the upright posture. Phenytoin had no effect on seizures. On neurologic examination, the only signs were postural hypotension and blepharoptosis. Pupils remained 2 mm in diameter, even in dim light, but were normally reactive, and a slight reduction in tendon reflexes was observed. The rest of the examination was entirely normal, including electroencephalography. Special autonomic tests showed normal sinus arrhythmia, indicating intact parasympathetic innervation of the heart, an absent overshoot after Valsalva's maneuver, suggesting impaired reflex vasoconstriction. Beta blockade did not affect resting heart rate, suggesting diminished sympathetic drive, but atropine caused a marked increase in heart rate, supporting the intactness of parasympathetic innervation to the heart. Tyramine had no effect on blood pressure, heart rate, or plasma NE levels, but increased plasma dopamine by 37%. CSF and plasma dopamine levels at rest were also markedly elevated. In addition, phenylephrine, 0.5% instillation into the conjuctival sacs, produced mydriasis, indicating noradrenergic hypersensitivity of the iris.

These features are characteristic of a selective failure of noradrenergic function, with preservation of sympathetic cholinergic and parasympathetic function. Biochemically, the most important feature is the virtual absence of plasma and CSF NE and epinephrine and their metabolites, and a marked increase in plasma and CSF levels of dopa and dopamine and their metabolites. Adrenoceptors are markedly hypersensitive because of the absence of NE, and since the administration of tyramine increased plasma dopamine and not NE, the clinical syndrome and biochemical abnormalities are consistent with the proposition that instead of NE, dopamine is present in presynaptic sympathetic terminals. The virtual absence of 3-4-dihydroxylphenylglycol supports this interpretation. In this patient the administration of d,l-DOPS, a drug endogenously converted into NE, increased his blood pressure and plasma NE levels, thus further providing evidence for an isolated defect in dopamine β-hydroxylase. Of interest is also the

possibility that the increased dopamine release from sympathetic terminals contributes to orthostatic hypotension, since a reduction in dopamine production by metyrosine was significantly related to an increase in blood presure (Biaggioni et al. 1987).

The syndrome of congenital dopamine β-hydroxylase deficiency is an experiment of nature that allows important deductions about the physiologic role of various catecholamines. In this syndrome the depletion of NE results in severe orthostatic hypotension. The impaired co-innervation by noradrenergic fibers of motor fibers may be responsible for the mild ptosis of eyelids and weakness of the facial musculature, including muscle hypotonia and decreased reflexes. The occurrence of spontaneous hypoglycemia with bursts of hyperinsulinism probably results from lack of epinephrine due to the failure of the adrenal medulla.

The excessive production of dopamine can account for the reduced REM sleep periods and increase in slow-wave sleep. This also may underlie the hypoprolactinemia and the sodium wasting through the kidney in spite of low arterial pressures. The relative deficit of NE and epinephrine, when compared to the dopamine excess in the production of these clinical abnormalities is further supported by the evidence from substitution of the missing catecholamines by the administration of an alternate substrate for catecholamine synthesis that bypasses the dopamine β-hydroxylase deficiency and the success of d,1-threodihydroxyphenylserine (d,1-DOPS) that can be decarboxylated to form NE in the treatment and maintenance of blood pressure of these rare patients (Man in't Veld et al. 1988).

REFERENCES

ADAMS, R. D., L. VAN BOGAERT and H. VAN DER EECKEN: Dégénérescences nigro-striées et cérébello-nigro-striées. Psychiatr. Neurol. 142 (1961) 219.

ADAMS, R. D., L. VAN BOGAERT and H. VAN DER EECKEN: Striato-nigral degeneration. J. Neuropathol. Exp. Neurol. 23 (1964) 584–608.

ANAND, P.: Neuropeptides in the spinal cord in multiple system atrophy. In: R. Bannister (Ed.),

Autonomic Failure. Oxford, Oxford University Press (1988) 511–520.

ANAND, P., R. BANNISTER, G. P. MCGREGOR ET AL.: Marked depletion of dorsal spinal cord Substance P and calcitonin gene-related peptide with intact skin flare responses in multiple system atrophy. J. Neurol. Neurosurg. Psychiatry 51 (1988) 192–196.

APPENZELLER, O.: The human sympathetic chain in health and disease and experimental autonomic neuropathy, an immunologically induced disorder of reflex vasomotor function. Thesis, University of Sydney, Sydney, Australia (1966).

APPENZELLER, O.: Clinical Autonomic Failure: Practical Concepts. Amsterdam, Elsevier (1982).

APPENZELLER, O.: The Autonomic Nervous System. Amsterdam, Elsevier Biomedical Press (1986).

APPENZELLER, O.: Neurology of endurance training. In: O. Appenzeller (Ed.), Sports Medicine: Fitness, Training, Injuries, 3rd ed. Baltimore, Urban and Schwarzenberg (1988) 35–71.

APPENZELLER, O. and J. E. GOSS: Glucose and baroreceptor function in cerebrovascular disease and in other disorders with baroreceptor reflex block. Arch. Neurol. 23 (1970) 137–146.

APPENZELLER, O. and J. E. GOSS: Autonomic deficits in Parkinson's syndrome. Arch. Neurol. 24 (1971) 50–57.

APPENZELLER, O. and M. KORNFELD: Acute pandysautonomia. Arch. Neurol. (Chic.) 29 (1973) 334–339.

APPENZELLER, O. and E. J. MCANDREWS: The influence of the central nervous system on the triple response of Lewis. J. Nerv. Ment. Dis. 143 (1966) 190–194.

APPENZELLER, O., M. KORNFELD and M. APPENZELLER: Autonomic dystrophy in copper-deficient mice. Ann. Neurol. 24 (1988) 301.

BANNISTER, R.: Introduction and classification. In: R. Bannister (Ed.), Autonomic Failure, 2nd ed. Oxford, Oxford University Press (1988) 1–20.

BANNISTER, R. and D. R. OPPENHEIMER: Degenerative diseases of the nervous system associated with autonomic failure. Brain 95 (1972) 457–474.

BANNISTER, R., P. SEVER and M. GROSS: Cardiovascular reflexes and biochemical responses in progressive autonomic failure. Brain 100 (1977) 327–344.

BANNISTER, R., R. CROWE, R. EAMES ET AL.: Defective cardiovascular reflexes and supersensitivity to sympathomimetic drugs in autonomic failure. Brain 102 (1979) 163–176.

BANNISTER, R., R. CROWE, E. HOLLY ET AL.: Different alpha receptor properties and baroreflex loss in alpha-adrenergic denervation supersensitivity in man. J. Physiol. (Lond.) 308 (1980) 44–45P.

BANNISTER, R., A. W. BOYLSTON, A. W. DAVIES ET AL.: Beta-receptor numbers and thermodynamics in denervation supersensitivity. J. Physiol. (Lond.) 319 (1981a) 369–377.

BANNISTER, R., W. GIBSON, L. MICHAELS ET AL.: Laryngeal abductor paralysis in multiple system atrophy. Brain 104 (1981b) 351–368.

BANNISTER, R., R. CROWE, R. EAMES ET AL.: Defective catecholamine fluorescence in the sympathetic perivas-

cular nerve plexuses in autonomic failure. Neurology (Minneap.) 31 (1981) 1501–1506.

BANNISTER, R., J. MOWBRAY and A. SIDGWICK: Genetic control of progressive autonomic failure: evidence for an association with an HLA antigen. Lancet 1 (1983) 1017.

BANNISTER, R., N. J. CHRISTENSEN, D. F. DA COSTA ET AL.: Mechanisms of post-prandial hypotension in autonomic failure. J. Physiol. (Lond.) 349 (1984) 67P.

BANNISTER, R., D. F. DA COSTA, S. FORSTER ET AL.: Cardiovascular effects of lipid and protein meals in autonomic failure. J. Physiol. (Lond.) 377 (1987) 62P.

BERCIANO, J.: Olivopontocerebellar atrophy. A review of 117 cases. J. Neurol. Sci. 53 (1982) 253–272.

BERNHEIMER, H., W. BIRKMAYER and O. HORNYKIEWICZ: Verteilung des 5-hydroxytryptamins (Serotonin) im Gehirn des Menschen und sein Verhalten bei Patienten mit Parkinson-Syndrom. Klin. Wochenschr. 39 (1961) 1056–1059.

BETHLEM, J. and W. A. DEN HARTOG JAGER: The incidence and characteristics of Lewy bodies in idiopathic paralysis agitans (Parkinson's disease). J. Neurosurg. Psychiatry 23 (1960) 74–80.

BIAGGIONI, I., A. S. HOLLISTER and D. ROBERTSON: Dopamine in dopamine β-hydroxylase deficiency. N. Engl. J. Med. 317 (1987) 1415–1416.

BIAGGIONI, I., D. S. GOLDSTEIN, T. ATKINSON ET AL.: Dopamine β-hydroxylase deficiency in humans. Neurology (Minneap.) 40 (1990) 370–373.

BLACK, I. and C. PETITO: Catecholamine enzymes in the degenerative neurological disease idiopathic orthostatic hyptension. Science NY 192 (1976) 910–912.

BLEDDYN-DAVIES, I. and P. S. SEVER: Adrenoceptor function. In: R. Bannister (Ed.), Autonomic Failure, 2nd ed. Oxford, Oxford University Press (1988) 348–366.

BORIT, A., L. J. RUBINSTEIN and H. URICH: The striatonigral degenerations. Putaminal pigments and nosology. Brain 98 (1975) 101–112.

BRADBURY, S. and C. EGGLESTON: Postural hypotension: autopsy on a case. Am. Heart J. 1 (1925) 73–86.

BRISKIN, J. G., K. L. LEHRMAN and C. GUILLEMINAULT: Shy-Drager syndrome and sleep apnea. In: C. Guilleminault and W. C. Dement (Eds.), Sleep Apnea Syndromes. New York, Alan R. Liss (1978) 316–322.

BRODAL, A.: Neurologic Anatomy, 3rd ed. Oxford, Oxford University Press (1981).

BROOKS, D. J. and R. S. J. FRACKOWIAK: PET and movement disorders. J. Neurol. Neurosurg. Psychiatry Special Suppl. (1989) 68–77.

BURKHARD, C. R., C. M. FILLEY, B. K. KLEINSCHMIDT-DE-MASTERS ET AL.: Diffuse Lewy body disease and progressive dementia. Neurology 38 (1988) 1520–1528.

BURKS, T. F. and O. APPENZELLER: Insulin attenuation of pressor responses to carotid artery occlusion. Proc. West Pharmacol. Soc. 14 (1971) 76–77.

CAPLAN, L. R.: Clinical features of sporadic (Dejerine-

Thomas) olivoponto-cerebellar atrophy. Adv. Neurol. 41 (1984) 217–224.

CHOKROVERTY, S.: Sleep and breathing in neurological disorders. In: N. H. Edelman and T. V. Santiago (Eds.), Breathing Disorders of Sleep. New York, Churchill Livingstone (1986) 225–264.

CHOKROVERTY, S.: Sleep apnoea and respiratory disturbances in multiple system atrophy with autonomic failure. In: R. Bannister (Ed.), Autonomic Failure, 2nd ed. Oxford, Oxford University Press (1988) 432–450.

COHEN, J., P. LOW, R. FEALEY ET AL.: Somatic and autonomic function in progressive autonomic failure and multiple system atrophy. Ann. Neurol. 22 (1987) 692–699.

DEMPSEY, J. A., E. B. OLSON JR. and J. B. SKATRUD: Hormones and neurochemicals in the regulation of breathing. In: A. F. Fishman, N. S. Cherniak and J. G. Widdicombe (Eds.), Handbook of Physiology, Section 3: The Respiratory System, Vol. II, Part I. Bethesda, Maryland, Amer. Physiol. Soc. (1986) 181–221.

DOBKIN, B. H.: Orthostatic hypotension as a risk factor for symptomatic occlusive cerebrovascular disease. Neurology (Minneap.) 39 (1989) 30–40.

DRAYER, B. P., W. OLANOW, P. BURGER ET AL.: Parkinson plus syndrome: diagnosis using high field MR imaging of brain iron. Radiology 159 (1986) 493–498.

DUVOISIN, R. C.: An apology and an introduction to the olivopontocerebellar atrophies. Adv. Neurol. 41 (1984) 5–12.

ESLER, M., G. JACKMAN, D. KELLEHER ET AL.: Norepinephrine kinetics in patients with idiopathic autonomic insufficiency. Circ. Res. 46 (Suppl. 1) (1980) 47–48.

EVANS, D., P. LEWIS, O. MALHOTRA ET AL.: Idiopathic orthostatic hypotension. Report of an autopsied case with histochemical and ultrastructural studies of the neuronal inclusions. J. Neurol. Sci. 17 (1972) 69–218.

FICHEFET, J. P., J. E. STERNON, L. FRANKEN ET AL.: Etude anatomo-clinique d'un cas d'hypotension orthostatique 'idiopathique'. Considerations pathogénique. Acta cardiol. 20 (1965) 332–348.

GIBB, W. R. G.: The Lewy body and autonomic failure. In: R. Bannister (Ed.), Autonomic Failure, 2nd ed. Oxford, Oxford University Press (1988) 484–497.

GOLBE, L. I., P. H. DAVIS, B. S. SCHOENBERG ET AL.: Prevalence and natural history of progressive supranuclear palsy. Neurology 38 (1988) 1031–1034.

GRAHAM, J. G. and D. R. OPPENHEIMER: Orthostatic hypotension and nicotine sensitivity in a case of multiple system atrophy. J. Neurol. Neurosurg. Psychiatry 32 (1969) 28–34.

GUARD, O., M. SINDOU and H. CARRIER: Syndrome de Shy-Drager; une nouvelle observation anatomoclinique. Lyon Med. 231 (1974) 1075–1084.

GUILLEMINAULT, C., A. TILKIAN, K. LEHRMAN ET AL.: Sleep apnoea syndrome: states of sleep and autonomic dysfunction. J. Neurol. Neurosurg. Psychiatry 40 (1977) 718–725.

HARDING, A. E.: Idiopathic late onset cerebellar ataxia. A clinical and genetic study of 36 cases. J. Neurol. Sci 51 (1981) 259–271.

HOELDTKE, R. D., T. M. O'DORISIO and G. BODEN: Treatment of autonomic neuropathy with a somatostatin analogue SMS-201-995. Lancet 2 (1986) 602–605.

HUGHES, R. C., N. E. F. CARTLIDGE and P. MILLAC: Primary neurogenic orthostatic hypotension. J. Neurol. Neurosurg. Psychiatry 33 (1970) 363–371.

ISRAEL, R. H. and J. M. MARINO: Upper airway obstruction in the Shy-Drager syndrome. Ann. Neurol. 2 (1977) 83.

IVY, H. K.: Renal sodium loss and bronchogenic carcinoma: associated autonomic neuropathy. Arch. Intern. Med. 108 (1961) 47–55.

JOHNSON, R. H., A. C. SMITH, M. M. E. SPALDING ET AL.: Effect of posture on blood pressure in elderly patients. Lancet 1 (1965) 731–733.

JOHNSON, R. H., G. DE J. LEE, D. R. OPPENHEIMER ET AL.: Autonomic failure with orthostatic hypotension due to intermediolateral column degeneration. Q. J. Med. 138 (1966) 276–292.

KACHI, T., S. IWASE, T. MANO ET AL.: Effect of L-threo-3,4-dihydroxy-phenylserine on muscle sympathetic nerve activities in Shy-Drager syndrome. Neurology (Mineapp.) 38 (1988) 1092–1094.

KAFKA, M. S., R. J. POLINSKY, A. WILLIAMS ET AL.: Alpha-adrenergic receptors in orthostatic hypotension syndromes. Neurology (Minneap.) 34 (1984) 1121–1125.

KAPLAN, P. and J. P. DE-CHADEREVIAN: Piebaldism-Waardenburg syndrome: histopathologic evidence for a neural crest syndrome. Am. J. Med. Genet. 31 (1988) 679–688.

KAWABATA, E., N. OHBA, A. NAKAMURA ET AL.: Waardenburg syndrome: a variant with neurological involvement. Ophthalmic Paediatr. Genet. 8 (1987) 165–170.

KHURANA, R. K., E. NELSON, B. AZZARELLI ET AL.: Shy-Drager syndrome: diagnosis and treatment of cholinergic dysfunction. Neurology (Minneap.) 30 (1980) 805–809.

KIRBY, R., C. FOWLER, J. GOSLING ET AL.: Urethro-vesical dysfunction in progressive autonomic failure with multiple system atrophy. J. Neurol. Neurosurg. Psychiatry 49 (1986) 554–562.

KONTOS, H. A., D. W. RICHARDSON and J. E. NORVELL: Norepinephrine depletion in idiopathic orthostatic hypotension. Ann. Intern. Med. 82 (1975) 336–341.

KONTOS, H. A., D. W. RICHARDSON and J. E. NORVELL: Mechanism of circulatory dysfunction in orthostatic hypotension. Trans. Am. Clin. Climatol. Assoc. 87 (1976) 26–33).

KOPIN, I. J., R. J. POLINSKY, J. A. OLIVER ET AL.: Urinary catecholamine metabolites distinguish different types of sympathetic neuronal dysfunction in patients with orthostatic hypotension. J. Clin. Endocrinol. Metab. 57 (1983a) 632–635.

KOPIN, I. J., E. K. GORDON, D. C. JIMERSON ET AL.: Relation

between plasma and cerebrospinal fluid levels of 3-methoxy-4-hydroxyphenyl-glycol. Science 219 (1983b) 73–75.

KOSAKA, K., M. YOSHIMURA, K. IKEDA ET AL.: Diffuse type of Lewy body disease: progressive dementia with abundant cortical Lewy bodies and senile changes of varying degree — a new disease! Clin. Neuropathol. 3 (1984) 185–192.

KWAK, S.: Biochemical analysis of transmitters in the brain of multiple system atrophy. No Shinkei 37 (1985) 691–694.

LAMBERT, E. H., L. M. EATON and E. D. ROOKE: Defect of neuromuscular conduction associated with malignant neoplasms. Am. J. Physiol. 187 (1956) 612–613.

LEES, A. J.: The treatment of multiple system atrophy: striatonigral degeneration and olivopontocerebellar degeneration. In: R. Bannister (Ed.), Autonomic Failure, 2nd ed. Oxford, Oxford University Press (1988) 596–604.

LEPORE, R. E.: Disorders of ocular motility in the olivopontocerebellar atrophies. Adv. Neurol. 41 (1984) 97–103.

LEWIS, P.: Familial orthostatic hypotension. Brain 87 (1964) 719–728.

LIPSCHITZ, L. A., R. H. NYQUIST ET AL.: Postprandial reduction in blood pressure in the elderly. N. Engl. J. Med. 309 (1983).

LOCKWOOD, A. H.: Shy-Drager syndrome with abnormal respirations and antidiuretic hormone release. Arch. Neurol. 33 (1976) 292–295.

LUFT, F. and V. VON EULER: Two cases of postural hypotension showing a deficiency in release of norepinephrine and epinephrine. J. Clin. Invest. 32 (1953) 1065–1069.

MAN IN 'T VELD, A., F. BOOMSA, J. LENDERS ET AL.: Patients with congenital dopamine-β-hydroxylase deficiency. A lesson in catecholamine physiology. Am. J. Hypertens. 1 (1988) 231–238.

MATHIAS, C. J., W. B. MATTHEWS and J. M. K. SPALDING: Postural changes in plasma renin activity and responses to vasoactive drugs in a case of Shy-Drager syndrome. J. Neurol. Neurosurg. Psychiatry 40 (1977) 138–143.

MATHIAS, C. J., D. F. DA COSTA, P. FOSBRAEY ET AL.: Postcibal hypotension in autonomic failure. In: N. J. Christensen, O. Henricksen and N. A. Lassen (Eds.), The Sympatho-Adrenal System, Alfred Benzon Symposium No. 23, Munksgaard, Copenhagen (1986) 402–413.

MATHIAS, C., D. DA COSTA and R. BANNISTER: Postcibal hypotension in autonomic disorders. In: R. Bannister (Ed.), Autonomic Failure. Oxford, Oxford University Press (1988) 367–380.

MATHIAS, C. J., D. F. DA COSTA, P. FOSBRAEY ET AL.: Hypotensive and sedative effects of insulin in autonomic failure. Br. Med. J. 295 (1987) 161–163.

MATTHEWS, M. R.: Assessing the peripheral ganglia in autonomic failure. In: R. Bannister (Ed.), Autonomic Failure, 2nd ed. Oxford, Oxford University Press (1988) 521–543.

MCLEOD, J. G. and R. R. TUCK: Disorders of the autonomic nervous system: Part 2, investigation and treatment. Ann. Neurol. 21 (1987) 519–529.

MCNICHOLAS, W. T., R. RUTHERFORD, R. GROSSMAN ET AL.: Abnormal respiratory pattern generation during sleep in patients with autonomic failure. Am. Rev. Respir. Dis. 128 (1983) 429–433.

MIYAZAKI, M.: Shy-Drager syndrome: a nosological entity? The problem of orthostatic hypotension. In: I. Sobue (Ed.), Spinocerebellar Degenerations. Baltimore, University Park Press (1980) 35–44.

MOHRING, J., K. GLANZER, J. A. MACIEL ET AL.: Greatly enhanced pressor response to antidiuretic hormone in patients with impaired cardiovascular reflexes due to idiopathic orthostatic hypotension. J. Cardiovasc. Pharmacol. 2 (1980) 367–376.

MOSKOWITZ, M. A. and R. J. WURTMAN: Catecholamines and neurologic diseases. N. Engl. J. Med. 293 (1975) 274–280, 332–338.

NEE, L. E., R. T. BROWN and R. J. POLINSKY: HLA in autonomic failure. Arch. Neurol. (Chic.) 46 (1989) 758–759.

NICK, J., F. CONTAMIN, R. ESCOUROLLE ET AL.: Hypotension orthostatique idiopathique avec syndrome neurologique complexe a predominance extra-pyramidale. Rev. Neurol. (Paris) 116 (1967) 213–227.

NORK, T. M., Z. M. SHIHAB, R. S. YOUNG ET AL.: Pigment distribution in Waardenburg's syndrome: a new hypothesis. Graefes. Arch. Clin. Exp. Ophthalmol. 224 (1986) 487–492.

NUTMAN, J., R. STEINHERZ, Y. SIVAN ET AL.: Possible Waardenburg syndrome with gastrointestinal anomalies. J. Med. Genet. 23 (1986) 175–178.

O'NEILL, J. H., N. M. F. MURRAY and J. NEWSOM-DAVIS: The Lambert-Eaton myasthenic syndrome: a review of 50 cases. Brain 111 (1988) 577–596.

OPPENHEIMER, D.: Neuropathology of progressive autonomic failure. In: R. Bannister (Ed.), Autonomic Failure. Oxford, Oxford University Press (1982) 267–283.

OPPENHEIMER, D.: Neuropathology and neurochemistry of autonomic failure. A neuropathology of autonomic failure. In: R. Bannister (Ed.), Autonomic Failure, 2nd ed. Oxford, Oxford University Press (1988) 451–463.

PAGE, M. N. and P. J. WATKINS: Provocation of postural hypotension by insulin in diabetic autonomic neuropathy. Diabetes 25 (1976) 90–95.

PARKS, V. J., A. G. SANDISON, S. L. SKINNER ET AL.: Sympathomimetic drugs in orthostatic hypotension. Lancet 1 (1961) 1133–1136.

PETITO, C. K. and I. B. BLACK: Ultrastructure and biochemistry of sympathetic ganglia in idiopathic orthostatic hypotension. Ann. Neurol. 4 (1978) 6–17.

POLINSKY, R. J.: Neurotransmitter and neuropeptide function in autonomic failure. In: R. Bannister (Ed.), Autonomic Failure: A Textbook of Clinical Disorders of the Autonomic Nervous System, 2nd ed. Oxford, Oxford University Press (1988) 321–347.

POLINSKY, R. J., I. J. KOPIN, M. H. EBERT ET AL.: The adrenal medullary response to hypoglycemia in patients with orthostatic hypotension. J. Clin. Endocrinol. Metab. 51 (1980) 1401–1406.

POLINSKY, R. J., I. J. KOPIN, M. H. EBERT ET AL.: Hormonal responses to hypoglycemia in orthostatic hypotension patients with adrenergic insufficiency. Life Sci. 29 (1981a) 417–425.

POLINKSY, R. J., I. J. KOPIN, M. H. EBERT ET AL.: Pharmacologic distinction of different orthostatic hypotension syndromes. Neurology (Minneap.) 31 (1981b) 1–7.

POLINSKY, R. J., I. L. TAYLOR, P. CHEW ET AL.: Pancreatic polypeptide responses to hypoglycemia in chronic autonomic failure. J. Clin. Endocrinol. Metab. 54 (1982) 48–52.

POLINSKY, R. J., D. C. JIMERSON and I. J. KOPIN: Chronic autonomic failure: CSF and plasma 3-methoxy-4-hydroxyphenylglycol. Neurology (Minneap.) 34 (1984) 979–983.

POLINSKY, R. J., R. T. BROWN, R. S. BURNS ET AL.: Cerebrospinal fluid monoamine metabolites in patients with progressive autonomic failure. J. Neurol. 232 (Suppl.) (1985) 71.

POLINSKY, R. J., I. L. TAYLOR, V. WEISE ET AL.: Gastrin responses in patients with adrenergic insufficiency. J. Neurol. Neurosurg. Psychiatry (1988) 67–71.

POLINSKY, R. J., D. V. HOLMES, R. T. BROWN ET AL.: CSF acetylcholinesterase levels are reduced in multiple system atrophy with autonomic failure. Neurology 39 (1989) 40–44.

PRASHER, D. and R. BANNISTER: Brain stem auditory evoked potentials in patients with multiple system atrophy with progressive autonomic failure (Shy-Drager syndrome). J. Neurol. Neurosurg. Psychiatry 49 (1986) 278–289.

QAMAR, M. I. and A. E. READ: The effect of feeding and sham feeding on the superior mesenteric blood flow in man. J. Physiol. (Lond.) 1377 (1986) 59P.

QUINN, N.: Multiple system atrophy — the nature of the beast. J. Neurol. Neurosurg. Psychiatry Special Suppl. (1989) 78–89.

QUINN, N. P., H. RING, M. HONAVAR ET AL.: Contractures of the extremities in Parkinsonian subjects. A report of three cases with a possible association with bromocriptine treatment. Clin. Neuropharmacol. 11 (1988) 268–277.

RAJPUT, A. H. and B. RODZILSKY: Dysautonomia in Parkinsonism: a clinico-pathological study. J. Neurol. Neurosurg. Psychiatry 39 (1976) 1092–1100.

ROBERTSON, D., D. WADE and R. M. ROBERTSON: Post-prandial alterations in cardiovascular hemodynamics in autonomic dysfunctional states. Am. J. Cardiol. 48 (1981) 1048–1052.

ROBERTSON, D., A. S. HOLLISTER, E. L. CAREY ET AL.: Increased vascular beta 2-adrenoceptor responsiveness in autonomic dysfunction. J. Am. Coll. Cardiol. 3 (1984) 850–856.

ROSENBERG, R. N.: Joseph disease: an autosomal dominant motor system degeneration. Adv. Neurol. 41 (1984) 179–193.

RUTLEDGE, J. N., S. K. KILAL, A. J. SILVER ET AL.: Study of movement disorders and brain iron by M. R. Am. J. Neuroradiol. 8 (1987) 397–411.

SASAKI, K., A. MATSUHASHI, S. MURABAYASHI ET AL.: Hormonal response to insulin induced hypoglycemia in patients with Shy-Drager syndrome. Metabolism 32 (1983) 977–981.

SCHATZ, I. J., S. PODOLSKY and B. FRAME: Idiopathic orthostatic hypotension. J. Am. Med. Assoc. 186 (1963) 537–540.

SCHWARZ, G. A.: The orthostatic hypotension syndrome of Shy-Drager. Arch. Neurol. (Chic.) 16 (1967) 123–139.

SCHWEITZER, V. G. and T. D. CLACK: Waardenburg's syndrome: a case report with CT scanning and cochleovestibular evaluation. Int. J. Pediatr. Otorhinolaryngol. 7 (1984) 311–322.

SEYER-HANSEN, K.: Post-prandial hypotension. Br. Med. J. 2 (1977) 1262.

SHARPE, J., N. REWCASTLE, K. LLOYD ET AL.: Striatonigral degeneration. Response to levodopa therapy with pathological and neurochemical correlation. J. Neurol. Sci. 19 (1973) 275–286.

SHY, G. M. and G. A. DRAGER: A neurologic syndrome associated with orthostatic hypotension. Arch. Neurol. (Chic.) 2 (1960) 511–527.

SOBUE, G. Y., T. HASHIZUME, T. MITSUMA ET AL.: Size-dependent myelinated fiber loss in the corticospinal tract in Shy-Drager syndrome and amyotrophic lateral sclerosis. Neurology (Minneap.) 37 (1987) 529–532.

SPOKES, E. and R. BANNISTER: Catecholamines and dopamine receptor binding in Parkinsonism. In: R. Capildeo (Ed.), Research Progress in Parkinson's Disease. London, Pitman (1981) 195–204.

SPOKES, E., R. BANNISTER and D. R. OPPENHEIMER: Multiple system atrophy with autonomic failure. Clinical, histological and neurochemical observations on four cases. J. Neurol. Sci. 43 (1979) 59–82.

TAKEI, Y. and MIRA, S. S.: Striatonigral degeneration: A form of multiple system atrophy with clinical parkinsonism. In: A. M. Zimmerman (Ed.), Progress in Neuropathology, Vol. 2, New York, Grune and Stratton (1973) 217–251.

TETSUO, M., R. J. POLINSKY, S. P. MARKEY ET AL.: Urinary 6-hydroxy-melatonin excretion in patients with orthostatic hypotension. J. Clin. Endocrinol. Metabol. 53 (1981) 607–610.

THOMAS, J. E. and A. SCHIRGER: Idiopathic orthostatic hypotension. A study of its natural history in 57 neurologically affected patients. Arch. Neurol. (Chic.) 22 (1970) 289–293.

TROTTER, J. L.: Striato-nigral degeneration. Alzheimer's disease and inflammatory changes. Neurology 23 (1973) 1211–1216.

TYGSTRUP, I. and T. NORHOLM: Neuropathological findings in 12 patients operated for Parkinsonism. Acta. Neurol. Scand. 39 (Suppl. 4) 1963, 188–195.

VANDERHAEGEN, J. J., O. PERIER and J. E. STERNON: Pathological findings in idiopathic orthostatic hypotension. Arch. Neurol. 22 (1970) 207–214.

VINCENT, A., B. LANG and J. NEWSOM-DAVIS: Autoimmunity to the voltage-gated calcium channel underlies the Lambert-Eaton myasthenic syndrome, a paraneoplastic disorder. Trends. Neuro. Sci. 12 (1989) 496–502.

WILCOX, C. S. and M. J. AMINOFF: Blood pressure responses to noradrenalin and dopamine infusions in Parkinson's disease and the Shy-Drager syndrome. Br. J. Clin. Pharmacol. 3 (1976) 207–214.

WILCOX, C. S., J. J. AMINOFF, G. B. KEENAN ET AL.: Circulating levels of corticotrophin and cortisol after infusion of L-DOPA, dopamine and noradrenaline in man. Clin. Endocrinol. 4 (1975) 191–198.

WILLIAMS, A.: CSF biochemical studies on some extrapyramidal diseases. In: F. C. Rose and R. Capildeo (Eds.), Research Progress in Parkinson's disease. London, Pitman (1981) 170–180.

WILLIAMS, A., D. HANSON and D. B. CALNE: Vocal cord paralysis in the Shy-Drager syndrome. J. Neurol. Neurosurg. Psychiatry 42 (1979) 151–153.

ZIEGLER, M. G., C. R. LAKE and I. J. KOPIN: The sympathetic nervous system defect in primary orthostatic hypotension. N. Engl. J. Med. 296 (1977) 293–297.

Handbook of Clinical Neurology, Vol. 15 (59): Diseases of the Motor System
J.M.B.V. de Jong, editor
© Elsevier Science Publishers B.V., 1991

CHAPTER 11

Amyotrophic lateral sclerosis

H. RICHARD TYLER and JEREMY SHEFNER

Department of Neurology, Harvard Medical School, Brigham and Women's Hospital, Boston, MA, USA

The terms amyotrophic lateral sclerosis and motor neuron disease have been used interchangeably to describe a disease characterized by progressive motor dysfunction leading to death over a number of years. The main attraction of the term motor neuron disease is that it incorporates a number of different clinical syndromes which are now thought to be variants of the same pathophysiological process. Thus, motor neuron disease (MND) now encompasses the clinical syndromes of primary muscular atrophy, progressive bulbar palsy, primary lateral sclerosis, and amyotrophic lateral sclerosis. However, the above named syndromes may be quite distinct in their clinical courses, as well as having somewhat different epidemiologies. Another objection to the term motor neuron disease is, that detailed pathologic study often reveals abnormalities that are not limited to motor systems.

Although MND has proved to be a useful umbrella term, we prefer to distinguish it from specific syndromes such as amyotrophic lateral sclerosis (ALS). ALS is reserved for a syndrome that includes signs of both upper and lower motor neuron abnormalities, eventually including both somatic and bulbar musculatures. Primary muscular atrophy implies primarily lower motor neuron signs and often minimal to absent bulbar signs. Progressive bulbar palsy describes a syndrome of primarily lower motor neuron signs limited to the cranial nerves, and progressive pseudo-bulbar palsy refers to an upper motor neuron syndrome limited to the bulbar musculature. Abnormalities of control of laughing and crying are not uncommon. As all of these syndromes progress they can involve axial musculature to some degree.

In the United States, most of the above syndromes have been called ALS or Lou Gehrig's disease. Until specific tests are developed to distinguish between these syndromes, a lumping approach may be a practical way to conceptualize this disease. However, it is hoped that the experienced neurologist will be aware of the significant differences among specific syndromes, especially in view of the varying prognostic implications. These differing clinical courses are of great importance both in the care of individual patients as well as in the design and assessment of therapeutic trials.

HISTORY

When Bell (1824, 1830) formulated his ideas on the nervous system he separated motor and sensory function and believed they were handled by separate nerve systems. He reported a number of cases in which there were purely motor or purely sensory abnormalities. One of his cases (Case 47) was 'Mrs. G.', a lady of 'upwards of

50'. She initially presented with left leg weakness, and subsequently developed dysphagia and dysarthria. The left arm, right leg and right arm subsequently weakened and the patient eventually became quadriplegic. She had increasing trouble with her secretions. There was 'twitching of the muscles'. Memory and sensation were normal.

Upon her death, her physician Thomas Ingle described his findings at autopsy: 'the anterior half of the cord ... in a semifluid state approaching nearly to the consistency of cream whilst the posterior portion possessed its usual firmness'.

Other isolated cases of slowly progressive motor dysfunction were reported. Cruveilhier had lectured about such cases although he was not the first to write about them; for that reason it was called 'Cruveilhier's atrophy' (Hammond 1876). Both Duchenne and Aran were aware of Cruveilhier's lectures (Hammond 1876). Duchenne (1847) subsequently reported one case of pure motor degeneration. Aran (1850) subsequently published a report reviewing the literature on these cases. He called them progressive muscular atrophy.

Duchenne (1855) clearly described cases in some detail in his book 'L'Electrisation Localisée'. The early history is reviewed by Charcot and Marie (1885), Charcot (1962) and Astier (1856). Norris (1975) and Bonduelle (1990) have reviewed these early reports and felt that Aran really made most of the contributions, but Duchenne got credit due to his prominence. Cruveilhier (1853) described atrophy of the anterior rootlets and the motor nerves to the extremities in one of Duchenne's cases that was described in Aran's (1850) publication. This was the expected lesion based on the Bell-Magendie doctrine. He suspected there was a lesion in the anterior gray matter but never demonstrated it. Luys (1860) demonstrated atrophy and loss of the anterior horn cells. Charcot and Marie (1885) and Hayem (1869) found widespread loss of motor neurons throughout the spinal cord. Thus these early descriptions refer to a syndrome of progressive motor dysfunction primarily due to anterior horn cell degeneration.

When Charcot (1886) (Fig. 1) reviewed these cases he considered them 'protopathic' and distinguished them from other cases in which the gray

LE PROFESSEUR J.-M. CHARCOT

Fig. 1. Prof. J. M. Charcot. Pen-and-ink sketch by F. Desmoulin, Collection at the Académie nationale de Médecine Paris. Previously published in W. Haymaker, *Founders of Neurology*, 1953 courtesy of Charles C. Thomas, Publisher, Springfield, IL.

matter lesion was secondary to changes in the white matter, nerve or roots. These cases he called 'deuteropathic'. He described 'amyotrophic lateral sclerosis' as one recognizable form of the deuteropathic spinal muscle atrophies (1962). In recognition of his classic and complete descriptions of this disorder it is often called Charcot's disease (Charcot and Joffroy 1869; Charcot and Marie 1885; Dejerine and André-Thomas 1909).

Charcot noted that patients with ALS had a worse prognosis than those with progressive muscle atrophy. The usual case lived only 1–3 years as compared to much longer survival seen with the Duchenne-Aran disorder. He also noted that bulbar involvement was unusual in progressive muscle atrophy but common in ALS and noted that involvement of the legs was much quicker to follow arm involvement in ALS. Other differences included muscle rigidity and pain secondary to contractures which were frequently seen in patients with ALS but not in patients with progressive muscle atrophy. Muscle tenderness was also frequently noted in ALS patients.

Charcot also observed that while patients with ALS often presented with relatively diffuse paresis and atrophy, in patients with progressive muscle atrophy there was usually a very focal muscle wasting. The wasted muscle or muscle group stood out in comparison to its neighbors which would often be normal; often wasting occurred proximally with distal muscles spared. 'The ... feature is in some sort characteristic: it is writes M. Duchenne (de Boulogne), "the facies of the disease' ". (Charcot 1862). In contrast, the paresis in ALS 'extended a little everywhere in a uniform manner, so to speak, from the extremity of the member to its root. We no longer observe, here, the individual atrophy of muscles, which we noticed in connection with the common muscle atrophy: on the contrary, we see a kind of general emaciation of atrophy, en mass' (Charcot 1962).

It is probable that a number of cases of primary muscle atrophy were in fact syringomyelia or chronic peripheral neuropathy as the concepts of these entities was to develop some 20–30 years later. There is no doubt, however, that some of the cases were truly progressive muscular atrophy as conceived by Duchenne and Aran.

Progressive bulbar palsy was described by Duchenne (1860), Duchenne and Joffroy (1870), Charcot and others. Charcot considered it a distinct entity as did Duchenne; however, Charcot recognized that some cases developed significant changes in arms and legs and considered these examples of amyotrophic lateral sclerosis (Hammond 1876). Others considered these changes secondary and maintained the separation from ALS (Hammond 1876). Leyden (1870) reported on white matter tract lesions in cases of primary muscle atrophy and progressive bulbar palsy relating these to ALS. Dejerine (1883) made the point that autopsy findings in these 3 illnesses were often indistinguishable, and diagnosed cases of progressive bulbar palsy (labio-glosso-laryngeal paralysis) as ALS.

Charcot (1865) and subsequently Erb (1875,

1902) reported cases of what appeared to be pure spastic quadraparesis, with pathology seemingly limited to white matter tracts. This syndrome was named primary lateral sclerosis. There remain serious concerns as to the nature of the illnesses that have been described under this term, whether they relate to the other syndromes, or even if there is a discrete clinical pathological entity that warrants a unique classification.

Dejerine's idea that all cases of progressive muscle atrophy, progressive bulbar palsy, primary lateral sclerosis and ALS should be considered a unitary entity has significant support from neuropathological studies (Chou 1978; Brownell et al. 1970) as well as from clinical observations. It is this viewpoint that provides the basis for the use of the term 'motor neuron disease' to describe all of the above entities. Support for Duchenne's formulation also comes from the series of patients reported by Mackay (1963) who followed 70 cases of which 23 were originally diagnosed as progressive muscle atrophy, 11 primary lateral sclerosis and 36 had a syndrome of spasticity and amyotrophy (ALS). All patients were followed until their death, by which time 61 patients had findings consistent with ALS, 8 patients still had only lower motor neuron signs and thus were diagnosed as having progressive muscle atrophy, and only 1 had a pure upper motor neuron syndrome consistent with primary lateral sclerosis. Mackay felt that the progression of signs noted in his patients suggested that the different clinical syndromes noted in his patients represented varying expressions of the same disease process. Nonetheless, the different clinical courses followed by patients with different initial syndromes provide a strong reason for continuing to consider the different clinical subtypes at least somewhat independently.

Primary lateral sclerosis

Six years after Charcot and Joffroy (1869) described the syndrome of ALS, Erb (1875) noted the existence of a syndrome of a slowly progressive pyramidal tract syndrome without concurrent muscle atrophy, referring to this condition as primary lateral sclerosis. He subsequently discussed 10 cases of primary lateral sclerosis for which pathology was available (Erb 1903); in these cases, the primary abnormality was severe sclerosis of the lateral columns of the spinal cord, with sparing of the ventral gray matter. However, for many years, the only descriptions of patients with isolated progressive spastic disorders lacked any pathological verification (Stark and Moersch 1945; Wechsler and Brody 1946). Most patients in these series had a progressive spastic paraparesis involving the lower extremities, with the arms involved or affected to a much lesser degree. Anal and urinary sphincter disturbances were uncommon but did occur. Bulbar dysfunction was not initially described.

Despite the fact that patients with the clinical syndrome described above appeared to follow a much more protracted course than that usually seen in ALS, it has been generally felt that primary lateral sclerosis represented a form fruste of ALS, in which the signs of lower motor neuron disease had not yet developed (Mackay 1963; Goldblatt 1969). As evidence for this contention, Mackay followed 11 patients who intially displayed only upper motor neuron abnormalities; over a period of years, 10 patients eventually developed signs of muscle atrophy.

Recently, however, both clinical and pathological cases have been reported that suggest that primary lateral sclerosis may in fact be a distinct disease entity. Fisher (1977) described 6 cases of non-inherited pure spastic paralysis, 1 of which had pathological verification. The patient in whom an autopsy was performed experienced a 3-year progressive course of spastic quadraparesis, complicated by a pontine infarct. Pseudobulbar palsy was observed in this patient, but only after his stroke; there was no sensory disturbance, and intelligence was preserved. At autopsy, both lateral corticospinal tracts were pale, but anterior horn cells were reported to be of normal frequency and morphology. Autopsies were lacking in Fisher's other 5 cases, but 3 of them were reported to have had normal EMGs, providing some evidence for the integrity of anterior horn cells. Another pathologically studied case of primary lateral sclerosis was reported by Beal and Richardson (1981). Similarly to the case of Fisher, the patient progressed over 3–5 years to a state of spastic quadraparesis

and pseudo-bulbar palsy; in this patient, however, spastic dysarthria was the presenting symptom. Autopsy showed loss of myelinated fibers in corticospinal tracts, with normal numbers of anterior horn cells.

Other clinical studies also provide support for the existence of primary lateral sclerosis as a disease separate from ALS. Ungar-Sargon et al. (1980) evaluated a series of 672 patients with spastic paraparesis, and found 44 for which no anatomic lesion or underlying disease could be found to account for the spasticity. Twenty-two of these patients had EMGs, which were reported to be normal in all cases. Russo (1982) reported 4 patients with a syndrome of isolated spastic paraparesis, all of whom had had symptoms for at least 5 years. EMG and nerve conduction studies were within normal limits for each patient. Except for the fact that sphincter disturbances were reported in all cases, these patients resembled others previously discussed.

In summary, there is clinical evidence for a syndrome of isolated spastic paraparesis without lower motor neuron dysfunction, sensory disturbances, cerebellar or mental status abnormalities. This syndrome is distinct from ALS in that muscle atrophy may be absent even after many years of disease, in addition to the fact that the course is considerably more indolent. Pathologic evidence of this syndrome, while not profuse, none the less exists.

ETIOLOGY

Viral

Speculation regarding a possible viral etiology for ALS has centered primarily around the possible relationship between poliovirus infection and subsequent motor system degeneration. As early as 1875, Charcot described progressive muscular atrophy in a patient with a remote history of poliomyelitis. In recent years, there has been a profusion of reports describing a relatively benign, progressive lower motor neuron disorder occurring approximately 40 years after affliction with poliomyelitis (Campbell et al. 1969; Mulder et al. 1972; Palmucci et al. 1980; Palahas et al. 1985; Dalakas et al. 1986a, b). This syndrome has been called Post Polio Muscular Atrophy; its lack of upper motor neuron involvement, benign course, and clear relationship to prior severe polio infection clearly separate it from Motor Neuron Disease as discussed elsewhere in this chapter.

More interesting is whether there is a relationship between classical ALS and either chronic polio infection or remote polio. The concept of chronic polio infection causing a slowly progressive motor neuron syndrome has received support from rare case reports. Alajouanine (1934) described a patient with a 15-month history of progressive lower motor disease whose central nervous system tissue was subsequently injected intrathecally into a rabbit. Within 6 months the rabbit developed a quadriparesis and subsequently died. More recently, type 2 poliovirus was islated from a woman with an 18-month history of progressive weakness (Lepage et al. 1975). However, clear evidence of upper motor neuron disease was lacking in both of the above cases. Viral-like inclusions have been noted in the cell bodies of anterior horn cells of several patients with ALS (Oshiro et al. 1976; Sun et al. 1975), but virus from these patients was not isolated.

Most recent reports on the relationship of polio to ALS have been concerned with whether remote polio infection could be a causative factor (Brown and Weiner 1984). Epidemiological studies have suggested a relationship, with a history of prior polio found in 2–5% of ALS patients; this is about 10 times the expected incidence (Zilkha 1962; Poskanzer et al. 1969; Norris et al. 1975). However, in a case controlled study of ALS in Rochester, Minnesota, no cases of prior polio were noted in 35 patients. In a detailed virological and histopathological study of a patient with classical ALS who had had polio 30 years previously, no evidence for viral inclusion bodies or ongoing polio infection was noted (Roos et al. 1980).

Studies attempting to demonstrate poliovirus antibodies in ALS patients have been generally unsuccessful. Neither neutralising (Lehrich et al. 1974; Jokelainen et al. 1977; Kurent et al. 1979) nor complement fixing (Cremer et al. 1973;

Lehrich et al. 1974) antibodies to polio have been found to be increased in the serum of ALS patients. CSF studies have been similarly unrevealing (Jokelainen et al. 1977; Kurent et al. 1979). More recently, nucleic acid hybridization studies using complementary DNA probes specific for polio virus have also been unsuccessful in demonstrating differences between ALS patients and control populations (Miller et al. 1980; Kohne et al. 1981; Brahic et al. 1985).

More suggestive of a possible relationship between poliovirus and ALS are reports of circulating (Bartfield et al. 1989) and tissue bound (Oldstone et al. 1976; Palo et al. 1978; Pertschuk et al. 1979) immune complexes found in ALS patients. Enteroviral related antigens have been isolated in some of these immune complexes (Bartfield et al. 1989; Petschuk et al. 1979) but not in others (Oldstone et al. 1976). The antigen isolated in serum immune complexes was found to react to a wide range of enterovirus infected cells, with the response to coxsackie virus infected cells being stronger than the response to poliovirus infected cells (Bartfield et al. 1989). Thus, the circulating immune complexes noted in the above reports may reflect an immune response not specifically directed at poliovirus.

Cell mediated immune responses have also been studied. Bartfield et al. (1982) found an increase in cell mediated immunity to poliovirus in patients with ALS but not in controls. Similarly, lymphocytes from ALS patients have been found to produce lymphokines (Bartfield et al. 1989) and a migration inhibition factor (Kott et al. 1979) in response to challenge from oral polio vaccine or enterovirus infected cells. These results suggest a prior sensitization to polio, perhaps resulting in a subsequent autoimmune response.

In summary, there is little direct evidence to implicate poliovirus as a direct cause of ALS. The cell mediated immune responses described above do suggest that prior viral sensitization may be related to a subsequent immune response seen in ALS patients. However, the response is not specific to poliovirus, with other enteroviruses causing similar effects (Bartfield et al. 1989).

The search for other viral causes of ALS has been unrevealing (Weiner et al. 1980) There have been several reports of viral transmission of ALS to experimental animals. Zil'ber et al. (1963) reported that CNS extracts from patients with ALS produce an ALS like syndrome in monkeys. However, no virus was isolated, and the study was not successfully reproduced in other laboratories (Gibbs and Gajdusek 1972; Brody 1965). Muller and Schaltenbrand (1979) isolated a virus from a patient with clinical ALS and were able to transmit a motor neuron syndrome in hamster. However, the patient from whom the virus was isolated had a persistent CSF pleiocytosis, and thus probably did not have classical ALS. Similarly, another patient with an ALS like syndrome but persistent CSF pleiocytosis was found to harbor a tick borne flavivirus (Muller and Hilgenstock 1975).

Surveys investigating antibody response to a large number of DNA and RNA viruses have yielded negative results (Harter 1982). Electron microscope studies occasionally reveal viral like structures in neuronal cytoplasm (Norris et al. 1975; Pena 1977), as well as muscle (Oshiro et al. 1976). However, these abnormalities have not proved consistent and do not provide evidence for a disease process affecting the bulk of ALS patients.

Immunological abnormalities

Antel et al. (1982) and Drachman and Kuncl (1989) have well summarized the evidence for an autoimmune etiology in ALS. Two lines of evidence will be mentioned here. First, humoral substances that react with neurons have been found in the sera of ALS patients. In some studies, cytotoxic effects on neurons have been reported from direct application to neurons (Wolfgram and Myers 1973; Horwich et al. 1974; Wolfgram 1976; Roisen et al. 1982), while in another study, inhibition of neuronal sprouting was caused by sera from ALS patients but not by sera from controls (Gurney et al.1983, 1984). Brown et al. (1987) demonstrated the presence of antineural antibodies in the sera of ALS patients; their presence was also noted in sera of some controls though with significantly less frequency in a recent study.

A population of motor neurons in spinal cords stained positively for IgG in 13/15 spinal cords and pyramidal cells in 6/11 motor cortices. This was not seen in controls. Reactive macrophages and microglia were seen in degenerating pyramidal tracts and ventral horns (Engelhardt and Appel 1990).

Another suggestion of an immunologic relation to ALS is the presence of paraproteinemia in a significant percentage of ALS patients. Some of these patients presented primarily with a motor neuropathy, but classical ALS in the presence of paraproteinemia has been described (Latov 1982; Krieger and Melmed 1982; Shy et al. 1986). In one series, abnormal paraproteins were noted in approximately 5% of patients with motor neuron disease (Shy et al. 1986). The percentage of ALS patients found to have abnormal paraproteins varies with the method of determination. A recent study found 3% of 120 patients with ALS to have paraproteins when cellulose acetate gels were used (as in the study by Shy et al. 1986). However, when determinations were made by immunofixation electrophoresis, 9% of patients had paraproteins (Younger et al. 1990). Paraproteinemia was more common in patients that had increased CSF protein levels.

Perhaps more interesting is the association recently noted between antiganglioside antibodies and motor neuron disease. Gangliosides are known to be present in high concentrations on neuronal membranes. Antibodies to gangliosides GM_1, GD_{1a}, and GD_{1b} have been detected in sera of ALS patients, either in the presence or absence of paraproteinemia (Nardelli et al. 1987; Latov et al. 1988; Shy et al. 1989; Pestronk et al. 1989). The percentage of ALS patients in whom ganglioside antibodies have been detected has ranged from 10.5% to more than 75% in different series (Shy et al. 1987; Pestronk et al. 1989).

The specific antibody detected may vary according to the clinical syndrome. Pestronk et al. (1989) found that when patients exhibited prominent upper motor neuron signs, antibodies to GD_{1a} were likely to be present. In contrast, patients with primarily lower motor neuron disease were more likely to have anti GM_1 antibodies. Pestronk et al. (1990) further divided patients with lower motor neuron syndromes.

Patients with electrophysiological evidence of multifocal motor neuropathy with conduction block and patients with a predominantly distal motor neuropathy without conduction block both tended to have high titers of anti GM_1 antibodies, while those with a predominantly proximal lower motor neuron syndrome were much less likely to have antibodies. Patients with mixed upper and lower motor neuron disease have been found to have antibodies to both GM_1 and GD_{1a}. This specificity suggests that these antibodies may have a pathogenic role in ALS, although this is by no means proven.

If antiglycolipid antibodies are important in the pathogenesis of ALS, they are not likely to be specific for this disease. In a study comparing anti GM_1 antibody levels in patients with ALS as well a number of other neurologic diseases, the percentage of patients with positive titers was found to be 23% for ALS patients, 19% in patients with neuropathy, and 7% in age matched controls with both neurologic and non-neurologic diseases (Nobile-Orazio et al. 1990a, b). In another laboratory, 19% of patients with ALS had antiganglioside antibodies, as compared to 10% of patients with other neurologic diseases (Salazar-Grueso et al. 1990).

In summary, the relationship of antiganglioside antibodies to ALS is not clear. They are present in patients with ALS to a degree that varies considerably from laboratory to laboratory. In addition, they are present in nearly equal frequency in patients with other neurological diseases, although the frequency of occurrence seems highest in patients with neuropathy. A small subgroup of patients with a lower motor neuron syndrome causing either multifocal motor neuropathy with conduction block or an isolated motor neuropathy without conduction block may have extremely high titers of antibodies (Salazar Grueso et al. 1990; Santoro et al. 1990; Shy et al. 1990; Nobile-Orazio et al. 1990a, b; many others); such patients may improve with therapeutic efforts directed towards immunosuppression. Whether such patients should be included under the diagnosis of ALS is, however, open to debate. A recent experimental autoimmune model of motor neuron disease has been reported (Engelhardt et al. 1989).

Genetic abnormalities

Although most cases of ALS are sporadic, a family history can be obtained in approximately 5–10% of patients (Emery and Holloway 1982; Figlewicz et al. 1989). There have been reported two sets of monozygotic twins in which both members acquired ALS (Williams et al. 1989); occurrence in pairs of dizygotic twins has also been reported (Estrin 1977). In many families, the disease follows an autosomal dominant pattern with incomplete penetrance. Mulder et al. (1986) analysed the characteristics of 329 patients with ALS from 72 families. Average age of onset was 48.3 years, with disease duration of 2.4 years. The male/female ratio was 1.19. These values are quite similar to those seen in sporadic disease: however, other series of familial cases suggest a slightly more aggressive course and younger age of onset (Figlewicz et al. 1989). Clinical presentation does not appear to differ between the familial and sporadic cases.

The fact that familial ALS does not clinically differ from sporadic disease offers an important opportunity to gain insight into its pathophysiology. Large families with multiple affected members allow the potential use of genetic probes and linkage analyses; so far, such studies have been unrevealing (Siddique et al. 1989), but the potential exists for discovering the gene responsible for familial ALS and determining its product. More recent studies have raised interest in chromosome 21 (Siddique et al. 1990).

Indirect evidence for a genetic component to sporadic ALS derives from studies of HLA incidence. Several studies have noted an increased incidence of HLA-A3 in ALS patients (Antel et al. 1976; Kott et al. 1979), with a suggestion that patients with the A3 antigen may have a more rapid course. Other studies, however, have not shown this pattern (Terasaki and Mickey 1975; Pederson et al. 1977). A more benign form of ALS has been associated with the HLA-B12 antigen (Antel et al. 1976). An argument against a strong relationship of specific histocompatibility antigens with ALS is that the presence of specific antigens varies systematically among different ethnic groups, while ALS appears to be remarkably constant across such groups.

Many diseases with clear genetic inheritance patterns have motor system degeneration as a significant clinical component. Adult GM_2 gangliosidosis secondary to hexosaminidase deficiency is an autosomal recessive disorder that can include both upper and lower motor neuron degeneration (Jellinger et al. 1976; Mitsumoto et al. 1985). Patients with this disorder are distinguished from classical ALS by the frequent association with degeneration of non-motor systems, early age of onset and longer disease duration. Although clinically hexosaminidase deficiency (Johnson 1982) can be confused with ALS, no cases were found in a survey of 350 ALS patients (Gudesblatt et al. 1988), so it is unlikely that this disease is a significant cause of ALS.

Toxins

Interest in metal intoxication as a potential cause for ALS has been present since Wilson (1907) described patients with lead poisoning and both upper and lower motor neuron abnormalities. Environmental studies have often demonstrated an increased incidence of lead exposure in ALS patients (Currier and Haerer 1968; Felmus et al. 1976). More recently, CSF and serum lead levels have been repeatedly measured in ALS patients, with some studies showing significant increases in ALS patients (Conradi et al. 1976, 1978, 1980) and others showing no difference from controls (House et al. 1978; Stober et al. 1983). In one study, lead content in anterior horn cells was found to be significantly elevated in ALS patients (Petkau et al. 1974).

Although there is no conclusive evidence that lead toxicity is an etiologic factor for the majority of ALS patients, there seems no doubt that it can cause a syndrome quite similar to ALS. Patients with a predominantly lower motor neuron degeneration secondary to lead have been reported by multiple authors (Wilson 1940; Boothby et al. 1974), and cases of ALS following lead exposure by many years have been reported (Wilson 1940; Livesly and Sisson 1968; Campbell et al. 1970; Simpson et al. 1974).

Mercury intoxication is also known to produce an ALS like syndrome (Brown 1954; Kantarjian 1961; Barber 1978). Most of these cases have been in farmers exposed to organic mercury compounds. Mercury levels have not been systematically measured in ALS patients.

Intoxication with a wide variety of trace metals has been considered in ALS. Aluminum has attracted significant interest as a neurotoxin both in motor system disease and Alzheimer's disease. Significant elevations in tissue levels have been noted in patients with ALS from the pacific islands (Yoshimasu et al. 1980, 1983), but are not known to exist in ALS patients from other geographical locations. Other trace metals that have been considered without strong supporting evidence include manganese, magnesium, and selenium (Mitchell 1987).

More recently, detailed studies of epidemiology as well as soil analyses in the Western Pacific regions in which ALS and Parkinsonism-dementia (PD) cluster have demonstrated high concentrations of aluminum and manganese coupled with low concentrations of calcium and magnesium leading to the proposal that mineral deficiency states lead to metal intoxication (Yase 1989). These findings were confirmed and extended by Gadjusek and colleagues (Gadjusek and Salazar 1982; Garruto and Yase 1986), who proposed that chronic calcium and magnesium deficiency led to secondary hyperparathyroidism and increased absorption of toxic metals. In support of this concept, increased amounts of manganese and aluminum have been found in spinal cords of ALS patients from Japan and Guam (Yase 1972; Perl et al. 1982). The decline of ALS and PD on Guam as nutrition improves is also consistent with the above hypothesis (Garruto and Yase 1986).

There is another potential toxin that may account for the increased incidence of ALS in Guam, Japan and New Guinea. The seed of the false sago palm (*Cycas circinalis*) was used for food and traditional medicine preparations in all 3 of the above areas, with its use declining after World War 2 (Spencer et al. 1989). This seed has been found to contain a number of neurotoxic compounds, particularly α-amino-β-methylaminopropionic acid (L-BMAA). Extracts of the seed have been observed to produce a combined upper and lower motor neuron degeneration when fed to monkeys. In an important study, synthetic L-BMAA was given to 13 monkeys, who developed abnormalities in pyramidal function, parkinsonian features, and behavioral changes (Steele and Guzman, 1987; Spencer et al. 1989). These animal findings, coupled with extensive epidemiological studies, made cycad intoxication an attractive possibility to account for the ALS/PD complex of the western pacific. However, other studies make this hypothesis less tenable. The diet followed by Guamanians in the regions with high incidences of ALS/Parkinsonism/Dementia has recently been shown to have such small levels of BMAA that the likelihood of toxicity seems extremely remote (Duncan et al. 1990). In addition, the course of the response to BMAA in monkeys and other animals is subacute, with progression occurring only while the toxin is being given, and some animals showing signs of recovery after the toxin is removed (Norris 1975; Hooper 1978; Gajdusek 1990).

DNA and RNA hypothesis

The hypothesis is that a defect of DNA repair mechanism in motor neurons is the primary abnormality in ALS (Mann and Yates 1981; Bradley and Krasin 1982). Mann and Yates (1981) have suggested that the initial change in motor neurons affected in ALS is hyperchromatic condensation of the nucleus associated with shrinkage of the nucleus. There is a loss of basophilia and decrease in cytoplasmic RNA content. They raise the question of a pathogen such as a slow virus. Davidson and Hartman (1981) and Davidson et al. (1981) noted a 30–40% reduction in RNA content of motor neurons. Bradley suggested that a deficiency of normal DNA repair mechanism resulted in accumulation of damaged DNA, which results in abnormal transcription of RNA. A new selective assay revealing a DNA metabolism abnormality was recently reported by Tanz et al. (1990) and will have to be further evaluated.

Endocrine

Patten and Engel (1982) have suggested a role for the parathyroid hormone possibly combined with an environmental factor such as calcium or aluminum. They note many patients with primary or secondary hyperparathyroidism can have weakness, wasting, fasciculations and pyramidal signs mimicking motor neuron disease. About 17% of their patients with ALS had calcium abnormalities. Patients we have studied with ALS have not had any parathyroid hormone or calcium abnormality suggestive of significant hyperparathyroidism. We believe that this relationship is not significant.

Trophic deficiency theory

Appel (1981) has suggested ALS can be caused by a lack of a disorder specific neurotrophic hormone. This hormone would arise in the target of the nerve cell and be released by that cell. The support for this idea comes from the fact that protein or protein extract from muscle can promote neuron survival and encourage neuritic outgrowth in neuronal cultures and can increase choline acetyltransferase (CAT) activity. One would be required to postulate as part of this hypothesis a reason for the cessation of production of this factor. Giller et al. (1977) had demonstrated a soluble muscle factor that enhances CAT in motor neuron cultures.

Dendritic abnormality or damage leading to nerve cell death

One of the earliest changes seen in neurons is a depletion of neurofilaments leading to dendritic atrophy and vulnerability of dendrites to break (Karpati et al. 1988). There are also demonstrated defects in both anterograde and retrograde axoplasmic flow in nerves from patients with ALS (Breuer et al. 1987, 1989). The hypothesis suggests that those types of deficits will lead to death of nerve cells.

Altered receptors

Sar and Stumpf (1977) and Weiner (1980) noted that there were androgen receptors on the surface of motor neurons. The extraocular cranial nerve nuclei (3, 4, 6) and the sphincteric (Onuf's) nuclei in the sacral cord did not share these receptors.

The fact that they are not affected in ALS raised the question of whether androgen receptors may play a major role in establishing nerve cell vulnerability to this disease.

Pancreatic abnormalities in ALS

A search for a metabolic dysfunction in extra neural organs which might play a role in the etiology of ALS has interested many investigators. Wechsler et al. (1944) noted that 16 of 68 cases had significant gastrointestinal problems or dietary deficiency. Ask-Upmark and Meurling (1955); Ask-Upmark (1969) observed that 5 of 20 cases had a history of gastric resection. Only 1–2 cases might have been expected suggesting that some form of deficiency might have played a role in the higher than expected number of patients who developed the disease.

Defects in liver function tests and pyruvate metabolism were noted in 15/36 patients by Cumings (1962). Poser and Bunch (1966) found that changes in some amino acids in American but not Guamanian cases.

Studies in carbohydrate metabolism showed a higher than expected incidence of diabetes and/ or elevated blood sugars in patients (Steinke and Tyler 1944). Nine of 11 patients had a glucose abnormality that was mediated through the pancreas and could not be accounted for by defective muscle utilization. Others (Ionasescu and Luca 1964) have confirmed that muscular carbohydrate metabolism was not altered in ALS — and basically confirmed Steinke and Tyler's results. Quick (1969) also demonstrated pancreatic gland dysfunction in this disease but treatment with pancreatic extracts did not affect the course of the illness.

Trauma

There is a small percentage of cases that give a history of the illness starting with a specific trauma. Anecdotal references to prize fighters, football players etc. who have been subject to repeated trauma and then develop ALS. In at least one retrospective study (Gallagher and Sanders 1987) a history of trauma was obtained in 58% of patients. This was 3 times the level reported in multiple sclerosis controls. It has not been possible to relate this trauma history to the basic disease process.

Excitatory amino acid hypothesis

Recent evidence has suggested that the excitatory neurotransmitters, aspartate and glutamate, may exert toxic effects on neurons, and that neuronal degeneration associated with processes as diverse as stroke and Huntington's Disease may occur as a result of excitotoxins. A number of studies have suggested that glutamate metabolism may be abnormal in ALS. Plaitakis and Caroscio (1985; 1987) measured fasting plasma glutamate levels in 22 patients with ALS and compared the results with control subjects. They found an approximately 100% increase in glutamate in the ALS patients; this was by far the largest discrepancy for any amino acid. Oral glutamate loading produced more of an elevation of plasma levels in patients than controls. In contrast, glutamate levels were found to be significantly reduced in brains of ALS patients (Plaitakis et al. 1988); CSF glutamate levels have been variably reported as increased (Rothstein et al. 1990) or unchanged (Perry et al. 1990). Based primarily on these data, Plaitakis (1990) has proposed that glutamate metabolism is abnormal in ALS, and that specific neuronal degeneration occurs as a result of excitatory amino acid neurotoxicity.

In this formulation, an explanation must be provided to account for the fact that ALS is a disease that affects only a limited number of brain areas. Plaitakis proposed that a local process whose presence was limited to motor systems might make these areas selectively sensitive to toxic effects. Glycine, an inhibitory amino acid neurotransmitter primarily found in the spinal cord and brainstem, was suggested as a possible cofactor. Glycine potentiates glutaminergic transmission at the N-MDA receptor, thought to be the site of glutamate's neurotoxic effects (Johnson and Ascher 1987; Choi 1988). Perhaps the presence of glycine in certain brain areas might make those areas selectively vulnerable.

Although the above hypothesis is intriguing, many questions remain. CSF and plasma levels of glutamate have not been universally found to be elevated in ALS patients, as the above hypothesis suggests should be the case (Patten et al. 1978; Perry et al. 1990). In addition, if the presence of glycine produces selective vulnerability in certain brain areas, then degeneration should be seen in many locations such as the optic nerve nuclei and other brainstem areas (Young 1990). Further study is obviously necessary before the usefulness of this hypothesis can be conclusively assessed.

EPIDEMIOLOGY

Studies on the epidemiology of ALS may be difficult to interpret because of uncertainties in criteria used in making diagnoses. When ALS is strictly defined as a disease involving both upper and lower motor neuron dysfunction, studies often show a slightly lower incidence of disease than those using a broader definition of motor neuron disease (MND). Overall, the incidence rate for motor neuron disease is approximately 1.5/100 000 population among whites, with the rate perhaps dropping to 1.2/100 000 if a more strict definition of ALS is used (Kurtzke 1982).

Geographical distribution and incidence by age

The incidence of MND is remarkably similar worldwide. Rates range between 0.4 and 1.9/100 00 population (Kurtzke 1982); the lower figures reported for Mexico and Europe probably represent incomplete case ascertainment (Juergens and Kurland 1980). For ALS, the reported incidence rates vary between 0.7 and 1.5/100 000 (Kurtzke 1982, Table 6). A slightly higher incidence of 2.6/100 000 was noted in Sweden and related to 'heavy manual labor' (Gunnarsson and Palm 1985). There are, however, geographic foci where disease incidence is 50–100 times higher than the world average. On Guam, the Kii peninsula of Japan, and in Irian Jaya, incidence rates as high as 147/100 000 have been reported (Arnold et al. 1953; Mulder and Kurland 1954; Gadjusek and Salazar 1982). Interestingly, the areas of increased incidence are quite distinctly demarcated, lying nearby to other areas where

incidence rates among the same ethnic groups are normal (Yuasa et al. 1976).

Except for the clusters of high concentration of ALS mentioned above, the disease appears to be distributed fairly evenly throughout the population. Incidence is roughly equal in northern and southern hemispheres. In population studies within the United States and Denmark, little evidence for foci of high incidence was found (Kurland and Mulder 1955). Two exceptions to this pattern have been noted in a town in northeastern Wisconsin (Taylor and Davis 1989; Sienko et al. 1990) and a single county in Sweden (Gunnarsson et al. 1990), where increased incidences have been reported. There appears to be no predilection for ALS either in urban or rural environments (Kondo and Tsubaki 1981). In addition, race does not appear to be a significant factor. Although incidence rates are higher for whites than blacks in the United States (Kurland et al. 1973a, b), the similarity in the sex ratio and in the incidence of disease as a function of age makes it likely that the difference in reported incidence has more to do with unequal access to health care than to the distribution of illness. In Hawaii, no differences were noted in disease prevalence between Caucasian and Japanese urban dwellers (Matsumoto et al. 1972).

ALS is a disease primarily of late middle age. Average age of onset varies in different series from 52 to 66 years, with the average age at death being approximately 62 (Kurland et al. 1969). In most studies of age specific rates, incidence is very low until about age 40, and then rises sharply to a maximum at approximately age 65 (Kurland et al. 1973; Jokelainen 1977; Rosati et al. 1977; Juergens and Kurland 1980; Kurtzke 1982). After 65 years, most studies show a decline in incidence rates; an exception to this is a study performed in Rochester Minnesota where incidence continued to rise with advancing age (Juergens and Kurland 1980).

Sex ratio. In all studies, the incidence of ALS in males is greater than in females. Differences in sex ratios vary from a high of 2.9 reported in Ireland (Kurland et al. 1969), to 1.2 in France (Bonduelle et al. 1970). This difference is maintained through all ages and ethnic groups. In the clusters of high disease concentration in the Western Pacific islands, sex ratios approximate those seen in other parts of the world. The only groups for which an increased incidence in males is not seen is familial ALS. In a study of 90 families, a male/female ratio of 1.03 was found, reflecting nearly equal incidence in male and female populations (Figlewicz et al. 1989).

Disease prevalence and duration. Prevalence rates for ALS vary from series to series, with an average of about 4–6/100 000 (Kurland et al. 1973a, b; Jokelainen 1977; Juergens and Kurland 1980; Kurtzke 1982); such a prevalence would suggest that 12 000–16 000 patients in the United States are affected with ALS (Kurland et al. 1969). If an average incidence rate of 1.5/100 000 is assumed, disease duration should be approximately 3 years. This is in general agreement with studies of illness duration, which report mean durations from 12 to 92 months (Gudmundsson 1968; Mulder and Howard 1976; Rosati et al. 1977; Juergens and Kurland 1980; Christensen et al. 1990). It is felt that, on average, survival is prolonged in younger patients (Kondo, 1975; Caroscio et al. 1984). Kondo, for example, found that disease duration declined from 45 months for patients with onset in their 4th and 5th decades, to 20–25 months for those with onset in their 70's. In addition, a significant percentage of patients live significantly longer than might be expected: 8–10% of patients live longer than 10 years after diagnosis. The course of the disease is remarkably variable; Appel et al. (1987) and Jablecki et al. (1989), using a quantitative scale to measure disease progression, found a 20–60-fold difference in rate of progression between the fastest and slowest progressing patient. These scales were designed to quantify short-term progression and survival and may be of use in specific situations such as therapeutic trials.

Related syndromes. In a study of more than 400 patients with motor neuron disease in the United States and Great Britain, Norris (1975) found that 9% fulfilled the clinical criteria for progressive muscular atrophy. These patients differed from classical ALS patients in several ways. The male/female ratio was 5.6/1 significantly more weighted towards males than in ALS. In addition, the average age on onset was younger (48 years) and duration of illness longer (7.5 years) than what has been reported in most series of ALS patients. The more benign course of progressive muscular atrophy was also confirmed in other series (Zilkha 1962; Kondo 1984). Such differences raise the question of whether progressive muscular atrophy is merely a subset of ALS or is a separate disease in its own right. (See also Chapter 2 by F. H. Norris.)

Patients presenting with the syndrome of progressive bulbar palsy generally have a worse prognosis than those with classical ALS (Bonduelle 1975; Kondo 1984). Patients presenting with progressive bulbar palsy tend to be older than other ALS patients, however (Daube 1985; Jablecki et al. 1989). If this age difference is taken into account, the prognosis in these patients may be the same as other ALS patients. There appears to be a somewhat different sexual predilection, with different series suggesting a female predominance (Rose 1977) or a nearly equal male/female incidence ratio (Kondo 1975).

Familial ALS (105400; 205100). It has long been recognized that a higher than expected number of ALS patients have a family history of the disorder. The percentage of patients with familial ALS has not been clearly determined, but may be as high as 10% (Norris et al. 1989). Whether such patients differ from patients with sporadic ALS is in some dispute. Emery and Holloway (1982) summarized the clinical features of 967 patients with sporadic ALS, and 231 patients with familial disease. They found a significantly earlier age of onset and shorter disease duration in the familial group. In contrast, in a study of 90 families with familial ALS, Figlewicz et al. (1989) found that the only significant difference between sporadic and familial cases was that there was a nearly equal male/female ratio among the familial patients.

Guamian amyotrophic lateral sclerosis

Espinosa et al. (1962) has described 3 forms of amyotrophic lateral sclerosis; (i) sporadic; (ii) hereditary; and (iii) 'litico', which is the form seen in Guam.

The high incidence 'amyotrophic lateral sclerosis' in Guam was first noted by Arnold et al. (1953), Koerner et al. (1952) and Mulder and Kurland (1954).

The presumption was that a full investigation of the disorder in Guam might provide clues to the etiology of ALS.

The sex distribution and site of onset of weakness appear to be the same (Mulder and Espinosa 1969). ALS appeared to affect primarily native Guamians. The Guamanian cases have a younger average age onset (about 10 years). A positive family history was present in 57% of the initial Guam study group as compared to 6% of the Rochester, Minnesota control group. The Guamanian cases frequently had dementia and/or Parkinsonism which are unusual but not unknown in the sporadic American cases (Elizan et al. 1966a, b). In most American series there are some that could be considered examples of primary muscle atrophy. These have not been seen in the Guamanian patients all having hyperactive reflexes including Babinski signs and palmomental reflexes. About 10% of Guamanian patients have spasticity before muscle wasting becomes apparent. This is unusual in sporadic cases (Elizan et al. 1966a, b).

Three of 62 original Guamanian cases were alive in 1967. They all had bulbar signs but it appeared that the disease may have arrested. Elizan et al. (1960a, b) who have studied patients with ALS in 1966 in Guam, noted that there were a few of non-Guamanian descent, 1 Japanese, 1 Filipino, 1 Caucasian and 1 Korean. We have seen 1 patient (Caucasian) who was affected with the sporadic American illness who resided in Guam as a member of the military.

PATHOLOGY

In the early 19th century Bell and Magendie independently conducted investigations which separated motor from sensory function. When primary muscle atrophy was described by Aran (1850) and Duchenne (1855) as affecting purely motor function it was quickly recognized that the motor rootlets were atrophied and smaller than normal (Cruveilhier 1853). The quick recognition of this pathology was a direct result of

Bell's earlier work demonstrating that the anterior roots dealt with motor function and the posterior roots with sensory function.

With the recognition that anterior horn cells (Luys 1860) are involved in this disorder and the fact that it affected motor cranial nerve function, the clinical disorders we call primary muscle atrophy, primary bulbar paralysis and ALS were quickly characterized pathologically.

Excellent general reviews of pathology include Bertrand and Van Bogaert (1925) and Hirano et al. (1969).

Grossly there may be atrophy of a precentral gyrus of the cerebrum, thin hypoglossal nerves and small anterior roots emanating out of the spinal cord (Fig. 2). There is gross shrinkage and sclerosis of the anterolateral tracts which lead to a loss of volume of these spinal areas in contrast to the normal size of the posterior columns of the cord. On cut transverse sections these tracts appear whiter than normal. There is also atrophy and shrinkage of somatic musculature.

Cell changes

Histologically there is extensive nerve cell loss with some astrocytic gliosis in the anterior horns

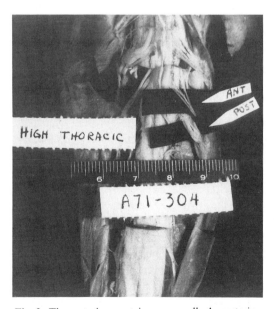

Fig. 2. The anterior root is very small. A posterior root is demonstrated at the same level for comparison.

and in some of the cranial nerve nuclei. In some cases the total number of nerve cells in a spinal section is normal but there is a decrease in large neurons and an increase in small neurons. In rare cases no loss is noted even though the patient had florid clinical symptoms. This suggests that loss of function can precede death or disappearance of nerve cells.

Involvement is usually most pronounced in the cervical enlargement and the 10th, 11th and 12th cranial nerve nuclei. V and VII are only moderately affected. There is no relation to blood supply of the cord.

Both upper and lower motor neurons are affected. The Betz cells of the precentral gyrus are affected. There is shrinkage of the nucleus and cytoplasm often with accumulation of lipofuscin. There are 'ghost cells' which degenerate without eliciting a neurologic reaction. There are swollen and ballooned proximal portions of motor neuron axons seen in the gray matter; they have been called spherites by Carpenter (1968) (Delisle and Carpenter 1984). These are full of neurofibrillary material.

Mann and Yates (1974) describe changes in the nuclei of anterior horn cells with clumping of DNA into heterochromatic granules. There are usually no viral inclusion bodies.

Cell change almost always spares the oculomotor, trochlear and abducens nuclei and the Onufrowicz nucleus of the spinal cord (Mannen et al. 1975, 1977; Hirano and Iwata 1978). There may be striking variations in the neuronal changes in a given nucleus; affected cells can lie adjacent to normal cells. There is not a uniform change as seen in the lipoidoses or in aluminum intoxication for example. (The range of change can extend from complete cell loss to normal appearing cells.)

The most common neuronal change is 'simple atrophy'. This is characterized by shrunken, dark, basophilic cytoplasm and pyknosis of the nucleus and corkscrew deformity of the dendrites (Hirano and Iwata 1978). Frequently there are nerve cells with lipofuscine granules. Bunina (1962) described the presence of small eosinophilic cytoplasmic inclusions, which have been observed in sporadic, familial and Guamanian cases

(Hirano et al. 1969). They are present in about 50% of cases. These inclusions do not appear to contain viral material. Kato et al. (1989) record a case with round eosinophilic hyaline inclusion bodies with halos in some of the anterior horn cells. These stained for a ubiquitin antibody (FD2). In this case ubiquitination was associated with the pathologic change in anterior horn cells.

Murayama et al. (1990) using antiubiquitin antibodies described 2 types of filaments in anterior horn cells which they felt might be precursors to some of the inclusions seen. These might be very early changes in the process.

Other changes include central chromatolysis, neurophagia and vacuolation which all occur rarely in ALS. Isolated cases in which unusual argyrophilic inclusions, Lewy bodies, Lafora bodies and viral particles have all been reported (Hirano and Iwata 1978).

Recently studies using MAbs(6AZ) (a specific neural surface antibody which reacts with specific subsets of neurons) demonstrated that patients with ALS have heterotypically localized neurons in white matter tracts subserving motor function. This suggests to the author, an abnormal developmental substrate to this disease (Kozlowski et al. 1989). The same technique has confirmed that there are changes of the neuronal surface in cells of the spinocerebellar system in ALS patients (Williams et al. 1990).

There is other biochemical evidence that the disease process may be more widespread than classic neuropathology suggests (Dawson and Stefansson 1984; Rapport et al. 1985).

White matter changes

The pattern of white matter degeneration in ALS is different from pure degeneration of the corticospinal tract as seen in hemispherectomy cases but bears a resemblance to degeneration seen with pontine lesions and spinal cord injury in which all descending tracts are affected (Ikuta et al. 1978). The diffuse pallor and significant white matter change which is so obvious in routine myelin stains highlights the difficulty in the use of the term 'motor neuron disease' which is not based on pathology but a clinical overview.

One can trace demyelination and axonal loss in the crossed and direct corticospinal tracts. At the level of the internal capsule the posterior portion is primarily affected. It is most severe and less diffuse in the medullary pyramids (Fig. 3). Demyelination is usually denser in the cervical cord than the medulla. The spinocerebellar tracts may also be involved, usually less severely than the pyramidal tract and often segmentally. The whole myelin of the cord is pale except in the posterior columns. The generalized loss of staining for myelin is not often emphasized (Fig. 4).

Because the lesion is more obvious distally in the corticospinal tract, one has to assume either that fibers are damaged more distally, or that more fibers are added to the pyramidal tract as it ascends. The ventral roots show loss of axons, especially the larger ones.

Changes in blood vessels and lymphocyte infiltration

Occasionally one can see lymphoid cells around blood vessels, especially with recent lesions. This is more commonly found in the anterior horn or subtantia gelatinosa. There is often perivascular tissue disintegration in white matter. Troost et al. (1989) noticed such infiltrates in 79% of cases (38/48) when studied with monoclonal antibodies to macrophage and T and B cells. This is not seen with the usual Wallerian degeneration and suggests some other mechanism may be involved.

Fibrosis and hyaline degeneration of vessels and occasional amyloid bodies are not infre-

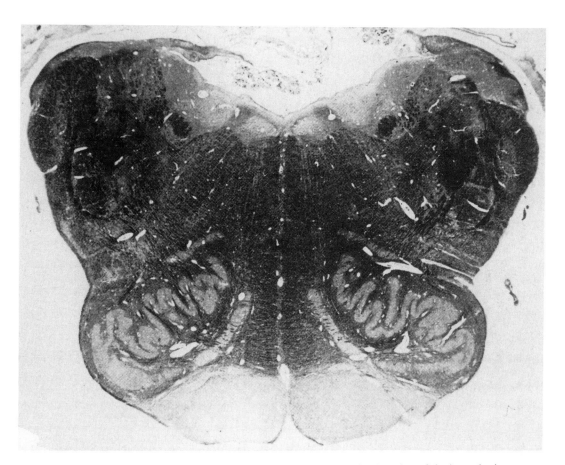

Fig. 3. Pallor of the pyramids is notable bilaterally in this myelin stained section of the lower brain stem.

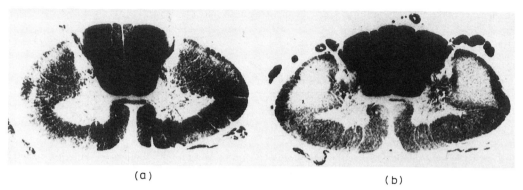

(a)   (b)

Fig. 4. There is pallor of the pyramidal tracts especially noted in the cervical area (b). This is seen in the dorsolateral area and in the ventral tract. There is loss of myelin staining everywhere but in the posterior column. The changes are less marked in the thoracic cord (a). (Reproduced from W. Blackwood et al., in *Greenfield's Neuropathology*, 1963 by courtesy of The Williams and Wilkins Co., Baltimore, MD.)

quently seen especially in the posterior column (Van Bogaert and Bertrand 1926).

Changes in muscle

The gross appearance of skeletal muscle suggests wasting and fibrosis. Histologically one notes atrophic fibers and well preserved fibers in each field. Muscle biopsy has been useful in confirming denervation in muscle or in establishing its presence before clinical signs are present. It is also useful in ruling out other forms of muscle disease, especially primary muscle disease such as polymyositis (Dubowitz and Brooke 1985). If taken too early from an affected muscle if can prove to be of no help. There are some instances in which changes can be interpreted as myopathic and provide misleading information. The changes are similar to those seen in other denervating diseases. The typical denervated muscle fiber becomes small and angulated in cross section and may have concave borders. Early in the illness denervated fibers are scattered, but groups of atrophic fibers appear as disease progresses. The number of atrophic fibers increases within a group until an entire fascicle may be involved (Engel and Brooke 1969).

The usual biopsy shows signs of denervation and reinnervation of both type 2 and type 1 muscle fibers. There is some hypertrophy of type 2 but not type 1 fibers. These observations suggest that cells innervating type 1 muscle units may be affected first.

Brooke and Engel (1966) have compared these changes in ALS to other chronic peripheral neuropathies. In ALS there is less fibrosis, necrosis and phagocytosis of fibers.

Guamian ALS

In Guamian ALS there is Alzheimer's neurofibrillary degeneration in nerve cells. This is especially true of the cerebral cortex, hypothalamus, basal ganglia including amygdaloid, substantia innominata, substantia nigra, locus coeruleus, reticular formation, dentate nucleus and reticular nucleus of the spinal cord. Granulovacuolar changes in the pyramidal cells of the Ammon's horn and occasional eosinophilic rod-like inclusions were also noted (Hirano et al. 1969).

The neuropathology of the Kii Peninsula cases is similar to the Guam cases in that there was the usual expected pathology and distribution of typical ALS but there was also Alzheimer's neurofibrillary degeneration. In addition the Kii cases had an increased number of eosinophilic inclusion bodies in the pigmented cells of the substantia nigra and the globus pallidus (Shiraki 1965).

Some of the familial cases have milder involvement of upper motor neurons or involvement of other tracts including posterior columns, Clarke's columns and spinocerebellar tracts. Cytoplasmic hyaline inclusion (Bunina) bodies are more common (Hirano et al. 1969).

ELECTROPHYSIOLOGY OF ALS

Electromyography

The electromyogram (EMG) has proved to be an important tool both in establishing the diagnosis of ALS as well as providing insight into its pathophysiology. The abnormalities seen in ALS are similar to that seen in other forms of neurogenic disease. Evidence of ongoing denervation is derived from the presence of abnormal spontaneous activity such as fibrillation potentials. As a consequence of denervation, surviving motor axons reinnervate muscle fibers, producing characteristic abnormalities in motor unit morphology as recorded by conventional concentric EMG electrodes as well as by newer techniques such as single fiber- and macro-EMG. While the abnormalities seen in ALS resemble those present in other diseases of the anterior horn cell and motor axon, the distribution of abnormal findings as well as the pace of disease progression are often quite useful in distinguishing ALS from other diseases.

Spontaneous activity

Fibrillation potentials. These potentials, as well as positive sharp waves, reflect activity of individual muscle fibers that have lost their synaptic contact with motor neurons. The frequency with which fibrillations are observed in ALS varies according to the amount of weakness or atrophy of the muscle being studied, the duration of the disease, as well as location in the body. Thus, Lambert (1969) found fibrillations in only 24% of muscles of ALS patients that were judged to be of clinically normal strength, while they were noted in 54% of muscles with 3/4 strength, and 97% of muscles with less than 1/2 normal strength. Similarly, Erminio et al. (1959) noted fibrillations in 31% of mildly affected muscles, increasing to 73% as weakness increased.

Fibrillations are seen more commonly as disease duration increases; this is to be expected from the above discussion as muscle weakness also increases with disease progression. Farago and Hausmanowa-Petrusewicz (1977) noted that 19% of patients with ALS for less than 1 year

lacked fibrillation potentials, decreasing to 9% in patients with ALS for more than 1 year. Within individual patients, fibrillations become more evident with time; in a longitudinal study, 3 of 4 patients in whom fibrillation potentials were absent initially were found to have them on retest 10–17 months later (Kuncl et al. 1988). Thus it appears that fibrillations become more frequent as the process of regeneration is no longer able to keep up with the loss of motor fibers.

Despite the clear variability of clinical presentation of ALS, fibrillation potentials are more likely to be seen in certain muscle groups than others. In the limbs, fibrillations are present more often in distal than proximal muscles (Kuncl et al. 1988). Facial and tongue muscles are less likely than limb muscles to show fibrillations. Interestingly, fibrillations are noted in thoracic paraspinal muscles in 84% of patients. Diagnostically, this is a useful finding as spondylosis of the thoracic spine is quite uncommon as compared to cervical and lumbar spondylosis, and abnormalities noted in the thoracic region are thus less likely to reflect root compression.

Fasciculations. One of the most characteristic abnormalities seen in patients with ALS is the presence of fasciculations. Clinically, they are seen in the majority of patients; electrophysiologically, they are noted almost without exception, even in sites where involuntary movements are not appreciated (Hjorth et al. 1973). Early electrophysiological studies described very large, polyphasic fasciculations as being characteristic of ALS (Denny-Brown and Pennybacker 1938); however, Trojaborch and Buchthal (1965) were unable to distinguish between the morphology of fasciculations observed in ALS patients and those seen in muscle disease or radiculopathy. The only consistent difference was that fasciculations in ALS patients tended to fire less frequently than those seen in other diseases.

Fasciculations are likely to originate at variable levels from the anterior horn cells to the neuromuscular junction. Spinal anesthesia has been reported to reduce the frequency of fasciculation by 55–60% (Swank and Price 1943), suggesting a contributory but not crucial role for the anterior horn cell. In studies where the axillary or femoral nerves were sectioned, fasciculations initially increased in frequency over several days before becoming abolished over several weeks (Forster et al. 1946), again providing evidence for a peripheral location of at least some fasciculations. In contrast, when Norris (1965) placed EMG electrodes in different muscles innervated by the same nerve root level, he noted that fasciculations were frequently synchronized; that is, they would occur nearly simultaneously in the separate muscles. Such a finding can be easily accounted for only by a generator central to the anterior horn cell.

More recently, collision techniques have been employed to study the sites of origination of individual fasciculations. Underlying this technique is the finding that, at the site of origin of a fasciculation, an action potential is generated in the motor neuron that travels in both antidromic and orthodromic directions. In collision studies, the occurrence of a fasciculation triggers a supramaximal stimulus to the nerve innervating the muscle being recorded. If the fasciculation originates at a distal site, the antidromically propagated action potential collides with the potential produced by the supramaximal stimulus, and a second muscle potential is not generated. However, if the fasciculation originates centrally to the site of nerve stimulation, the stimulus should produce a muscle action potential resembling the fasciculation just recorded (Fig. 5). If a variable delay is added to the stimulus onset, the location of the site producing the fasciculation can be more precisely estimated. Using this type of technique, Wettstein (1979) found that of 25 fasciculation potentials recorded in hand or forearm muscles, 2 originated at a site distal to the wrist, while 8 were produced by sites that were proximal to the mid-forearm. For 15 motor units, fasciculations seemed to come from multiple sites both in distal and proximal locations. Using a similar technique, Conradi et al. (1982) recorded fasciculations in severely atrophic foot muscles from which only single motor units were present. In this situation, all fasciculations were found to generate quite distally, probably at the region of the neuromuscular junction. Thus, it appears that fasciculations are generated

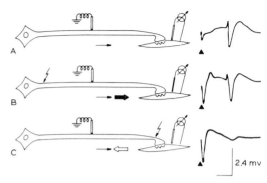

Fig. 5. Illustration of the collision technique for demonstrating the location of the site of fasciculation, for a fasciculation that is generated at variable locations along the axon. A. Stimulation at the location shown produces a muscle action potential as shown. B. The stimulus is triggered by the occurrence of a fasciculation; in this case, the stimulus evokes a response, indicating that the fasciculation is generated proximal to the stimulus. C. The fasciculation triggered stimulus does not generate a response because of collision, suggesting a distal location for the generator. (Reproduced with permission from A. Wettstein, *Annals of Neurology* 5 (1979) 295–300.)

at multiple sites along the anterior horn cell, and that a single motor fiber may have more than one site responsible for producing a spontaneous discharge.

Characteristics of motor units

The changes seen in motor unit morphology in patients with ALS are qualitatively similar to those that occur as a consequence of any form of neurogenic atrophy. As anterior horn cells are lost, viable motor axons establish synaptic contact with muscle fibers that have been denervated. The electrical correlate of this process is an increase in amplitude of the motor unit action potential. On average, motor unit amplitude is increased by about a factor of 4 over normal (Lambert 1969), with units in severely involved muscles having amplitudes of approximately 10 times normal and amplitudes from units in only slightly affected muscles increased by a factor of 3 (Erminio et al. 1959). Although the increase in amplitude is often quite striking, it is not as extreme as that seen in more chronic causes of neurogenic atrophy such as peroneal muscular atrophy (Erminio et al. 1959; Buchthal and Behse 1977).

Several theories have been proposed to account for the increased compound motor action potential amplitude of motor units in ALS. Buchthal (Buchthal and Clemmeson 1941; Erminio et al. 1959) felt that the motor unit amplitude was too large to be generated by an individual motor unit, and proposed that synchronous firing of multiple units was occurring. Denny-Brown (1949) felt that the abnormally large motor action potential could reflect activity by large motor units not activated in normal muscle. The view that is currently most accepted is that the extensive collateral sprouting at distal portions of surviving axons noted by Wohlfart (1957, 1960) leads to single motor units large enough to produce the large action potentials seen in ALS patients. This position has been supported recently by studies showing decreasing numbers of motor units in diseased muscle (McComas et al. 1971a, b).

Motor unit duration increases along with the increase in amplitude. In patients with very slowly progressive disease, motor unit potentials of increased amplitude may have only mildly increased duration; this reflects mature remodeling of the motor unit so that synchronization of firing of muscle fibers is nearly normal (Bonduelle 1975). However, as the disease continues and viable motor axons must innervate an increasing number of fibers, synchronization is lost and motor unit duration becomes more prolonged. The loss of synchrony also is reflected in increased polyphasia of motor units (Lambert 1969). In very wasted muscles, potentials may actually become shorter in duration (Lambert 1969). Overall, the average increase in motor unit duration is about 35% (Lambert 1969; Farago and Hausmanowa-Petrusewicz 1977).

Motor unit stability. With disease progression, motor neurons degenerate and new connections are made by surviving axon terminals; these new synaptic contacts are measurably less stable than normal. This has been documented in two ways. When compound motor action potentials are evoked by repetitive nerve stimulation at rates of 1 to 10 Hz, an abnormal decrement in response

amplitude is seen in ALS patients (Denys and Norris 1979; Carlton and Brown 1979; Bernstein and Antel 1981); similar to but less extreme than the decrement seen in patients with myasthenia gravis. This decrement most likely reflects unstable conduction either in terminal axon twigs or at the neuromuscular junction. That it occurs primarily in motor units that are actively reinnervating is indicated by the fact that decrement is almost extreme in severely atrophic muscles and in patients with rapidly progressive disease. Decrement is seen more often in actively fasciculating muscles (Denys and Norris 1979), suggesting that, in those muscles, fasciculations also represent instability in the distal motor unit.

Single fiber EMG provides another method of evaluating stability of conduction in the terminal axon or in the neuromuscular junction. In a jitter study, two single muscle fibers from the same motor unit are recorded simultaneously while the motor unit is being voluntarily activated. The variability in latency between the 2 fiber action potentials is called jitter; an increase in jitter represents variability in conduction in one or both of the distal neurons contacting the fibers. Occasionally, one of the muscle fibers may not fire at all in response to a motor neuron action potential; this is called blocking and is always abnormal. In ALS patients, jitter has consistently been found to be increased, and blocking has been noted as well (Stålberg et al. 1975; Swash and Schwartz 1982, Stålberg 1982). Jitter is abnormally high in more than 30% of muscles that are clinically unaffected, and is almost always increased in weak or atrophic muscles. Early in the course of the disease, jitter increases over time, but may actually decrease in endstage disease (Swash and Schwartz 1982). This decrease may represent the maturation of previously formed synaptic connections combined with the failure of the few surviving motor units to make new synapses.

Motor unit territory. The increased amplitude and duration of individual motor units is a result of changes in the structure of the motor unit. Erminio et al. (1959) used a multilead EMG electrode to more fully evaluate these changes. Leads were separated by 2.5 mm, and the electrode was advanced into a motor unit so that the center leads recorded the largest voltage when the motor unit fired; by this method, the center leads were felt to be in the center of the motor unit. Examination of the fall off in amplitude in leads as a function of distance from the motor unit center led to an estimate of the spatial distribution of the motor unit along the axis of the electrode. Multiple passes through the motor unit at different angles allowed the entire spatial extent to be mapped. Motor units in ALS patients were found to have a territory 80–140% larger than those measured in normal subjects, with territories in severely affected muscles larger than those in less affected muscles (Fig. 6). In the biceps brachii for example, motor unit territories were 12.2 mm^2.

More recent studies evaluating the spatial distribution of motor units have employed macro EMG techniques (Stålberg 1980, 1982a, b, 1983), whereby a single fiber EMG electrode is situated in the center of a macroelectrode with a very large recording surface. The single muscle fiber

response recorded by the single fiber electrode is used to trigger an averager into which the macroelectrode response is fed; by averaging many responses, a compound motor action potential is derived from the motor unit to which the single fiber recorded by the single fiber electrode belongs. Since the recording electrode is so large, this compound action potential should be an accurate estimate of the response of all muscle fibers within a motor unit. Using this technique, Tackmann and Vogel (1988) found abnormally large macro EMG potentials in 46 of 51 muscles studied from patients with ALS.

In addition to being increased in area, motor units in ALS patients are also more densely packed; that is, neighboring muscle fibers are much more likely to be innervated by the same motor axon than in normal muscle. The most direct electrophysiological measurement of this phenomenon is fiber density, which is recorded with a single fiber EMG electrode. Fiber density is defined as the average number of single muscle fibers from which responses can be recorded after a single fiber electrode is situated close to a muscle fiber. Due to properties of the electrode as well as the amplifier filter setting, a single fiber electrode will pick up potentials from muscle fibers within an approximately 300 μm area. In normal subjects, fiber density in most muscles is less than 1.5; that is, in a single electrode penetration, the single fiber electrode will record responses from either 1 or 2 fibers. In ALS, fiber density is increased dramatically, usually from 2 to 4 times normal (Stålberg et al. 1975; Stålberg 1982a, b; Swash and Schwartz 1982; Tackmann and Vogel 1988). Similarly to the changes described for single fiber jitter, fiber density tends to increase early in the course of the disease, but may decrease later. The late decrease may represent failure of diseased motor units to maintain previously formed connections.

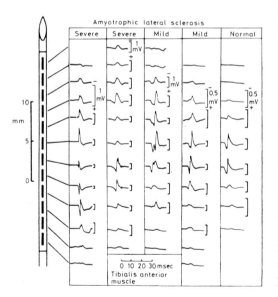

Fig. 6. Motor unit territories in a normal subject and ALS patients. The right column shows a multi-electrode recording from a normal tibialis anterior motor unit. The spatial distribution of the motor unit can be seen to be significantly smaller than those recorded in muscles from mildly and severely affected ALS patients, as shown in the other columns. (Reproduced with permission from F. Erminio et al., *Neurology 9* (1959) 657–671.)

Number of motor units. The increased motor unit territory, fiber density, and compound motor action potential amplitude are all reflections of reinnervation by viable motor axons after the death of other anterior horn cells. Thus, the number of functioning motor units in a given muscle should decrease with disease progression.

A method for estimating the number of motor units was initially developed for a distal foot muscle (McComas et al. 1971a, b; Hansen and Ballentyne 1978), and later extended to larger proximal muscles (Brown et al. 1988; Strong et al. 1988). In normal subjects, estimates of number of motor units ranged from about 200 in the extensor digitorum brevis (McComas et al. 1971a) to more than 900 in the biceps brachii (Brown et al. 1988). In ALS patients, the number of motor units falls dramatically, to an average of 14 in the extensor digitorum brevis; however, the force of an evoked muscle twitch remains normal until the number of remaining motor units declines below 10% of normal (McComas et al. 1971b). In a similar study performed on hand muscles, number of motor units was followed over time (Brown and Jatoul 1974). Clinical weakness and muscle wasting was only apparent after the total motor unit count fell below 20% of normal; in proximal arm muscles, normal strength was occasionally seen with motor unit counts falling to 6% of normal (Strong et al. 1988). Thus, reinnervation can effectively maintain muscle strength despite the loss of the majority of motor fibers.

Nerve conduction studies

Motor nerve conduction. In studies of motor conduction, the minimum latency of a compound motor action potential evoked by supramaximal stimulation to the nerve innervating that muscle is used as a measure of conduction time. This reflects conduction only in the fastest of motor fibers. Using this measure, motor conduction velocity in ALS patients has consistently been shown to be normal or near normal until muscle atrophy becomes extreme. Thus in 322 patients with ALS, conduction velocity in the ulnar nerve was reduced from the normal average by only 8% (Lambert 1962, 1969). Motor conduction velocity was also normal in 40 patients studied by Hausmanowa-Petrusewicz and Kopec (1970), as well as in the majority of patients studied by Eretekin (1967) and Denys (1980a). As amplitude of the evoked compound motor action potential decreases with increasing muscle atrophy, there is a trend toward reduction in conduction velocity

(Lambert 1969; Denys 1980a). The decrease in conduction velocity may be due in part to a selective loss of large diameter motor fibers; however methods of estimating minimum motor conduction velocity also indicate that smaller diameter, slower conducting motor fibers are also lost late in the disease course (Hausmanowa-Petrusewicz and Kopec 1973).

Even when motor conduction velocity is normal, distal motor latency is often prolonged in ALS patients (Hausmanowa-Petrusewicz and Kopec 1970; Lambert 1969; Denys 1980a). Distal latencies that are prolonged out of proportion to proximal conduction velocity is a finding frequently associated with dying back neuropathies where the terminal axon is more affected than the cell body of proximal axon. Although ALS primarily causes a motor neuronopathy, recent morphologic studies do suggest a component of dying back axonopathy (Bradley et al. 1983).

Studies investigating proximal motor conduction velocity also tend to demonstrate abnormalities in ALS patients even when motor conduction velocities are normal. F-wave latencies, which provide a measure of conduction along the entire length of the motor axon from periphery to spinal cord, have consistently demonstrated moderate slowing (Albizzati et al. 1976; Argyropolous et al. 1978; Petajan 1985). Latencies are increased to approximately the same level as is seen in other diseases causing neurogenic atrophy (Petajan 1985). It is probable that the increased sensitivity of F-waves as compared to motor conduction velocity in detecting abnormalities in ALS patients simply represents the fact that a longer segment of nerve is being evaluated.

Methods capable of assessing motor conduction in central pathways have recently been developed. Either motor cortex or spinal cord can be stimulated directly using electrical stimuli, and compound motor action potentials recorded from extremity muscles (Hugon et al. 1987; Ugawa et al. 1988). By subtracting the latency of the response evoked by spinal cord stimulation from the response latency to cortical stimulation, a measure of motor conduction velocity along central nervous system pathways can be calculated. In a study of 13 ALS patients with and without clinical evidence of pyramidal tract dysfunction, slowing of central motor conduction was observed in at least one muscle in 12 patients (Ugawa et al. 1988). Abnormalities were more often seen when recording from lower extremity muscles; in addition, slowing was more pronounced when recording from the side of the body associated with an extensor plantar response. This technique holds considerable promise for more clearly defining the physiology underlying pyramidal tract lesions in ALS as well as other motor system diseases.

Intense, brief magnetic fields have also been used to stimulate motor cortex. This type of stimulus has great advantages over electrical stimulation of the brain as it involves significantly less patient discomfort. Cortical stimulation studies with magnetic fields have been performed on ALS patients, with results that are consistent with those reported for electrical stimulation (Mills et al. 1987; Barker et al. 1987).

Sensory conduction. Studies of peripheral sensory conduction in ALS patients have consistently demonstrated either no abnormality or extremely mild dysfunction in the face of profound motor neuron loss. In an early study, Gilliat et al. (1961) recorded compound sensory action potentials from the superficial peroneal nerve in 5 patients who had profound distal atrophy. Potentials were obtained in all cases; a moderate reduction in sensory conduction velocity was felt to be secondary to reduced temperature. Other studies (Fincham and Van Allen 1964; Ertekin 1967) also demonstrated normal sensory conduction in median and ulnar nerves of ALS patients. Studies of the purely sensory sural nerve have also been normal (Lambert 1962, 1969). However, all of these studies measured conduction of only the largest diameter, fastest conducting sensory fibers in a nerve. By averaging the responses to many stimulus presentations and using very high amplification, reduced conduction velocity in a subpopulation of smaller diameter myelinated fibers has been noted in approximately 50% of ALS patients so studied (Shefner, Tyler and Krarup, unpublished data). This finding is consistent with many morphologic studies demonstrating abnormalities in peripheral sensory fibers in

ALS patients (Kawamura et al. 1981; Bradley et al. 1983; Ben Hamida et al. 1987).

Somatosensory evoked potentials (SSEPs) have recently been used to study central sensory pathways in ALS patients, with inconclusive results. Some studies find no abnormalities (Chiappa, 1983; Dioszeghy et al. 1987); however, other studies have reported abnormalities in SSEPs in 30–59% of ALS patients (Cosi et al. 1984; Radtke et al. 1986; Matheson et al. 1986; Subramaian and Yiannikas 1990). In these studies, peripheral sensory conduction was within normal limits; abnormalities were most often limited to potentials reflecting conduction in central pathways. In one study, conduction in cortical or subcortical pathways was felt to be most affected (Bosch et al. 1985).

Suggested electrophysiologic criteria for the diagnosis of ALS

Denys (1989) has proposed a set of general criteria needed to support the diagnosis of ALS. Although these similar criteria will not be fulfilled in every case of ALS, they are useful guidelines to be aware of. They suggest that nerve conduction studies should be performed in one upper extremity and one lower extremity, in nerves innervating muscles that are clinically weak. Motor nerve conduction studies should be within normal limits, except for reduction in compound motor action potential amplitude in atrophied muscles. Sensory nerve conduction studies should be within normal limits unless a coincident neuropathy is present.

EMG should be performed in at least 3 extremities, with the face and tongue counting as one extremity. Evidence of ongoing denervation should be noted in all 3 extremities, consisting of positive sharp waves and fibrillations. In addition, there should be diffuse motor unit abnormalities consistent with neurogenic disease.

CLINICAL

The initial diagnosis of amyotrophic lateral sclerosis is often made when minimal symptoms are present. Motor weakness in the absence of sensory loss which develops primarily in one extremity is usually the first symptom and brings the patient to the physician.

In the upper extremity there is often a stereotyped pattern of muscle involvement. Patients develop weakness of dorsiflexion of wrist and fingers, with weakness of the abductor pollicis brevis and the interossii. Flexors of the fingers and wrist remain relatively normal. If there is minimal weakness in comparable muscles on the opposite side one can be quite confident of the diagnosis even early in the illness.

Thenar and first dorsal interosseous wasting is noted early (Fig. 7). This produces a flat hand with the thumb in the plane of the fingers — the so called 'main du singe', or flat primate appearing hand. As the disease progresses there is clawing of the fingers as the flexors are stronger than the extensors and the metacarpalphalangeal joints become hyperextended — the 'main en griffe' (Fig. 8). More proximally the biceps, deltoid and infraspinati muscles are usually affected before the triceps. Weakness is usually asymmetric.

About 40–60% of patients will first develop signs in arm musculature. About 25–30% will develop bulbar symptoms initially, and 20% will develop symptoms initially in the legs.

Neck extensors frequently weaken, causing the head to drop forward. If this appears early or is severe it has been called the Vulpian-Bernhardt syndrome (Hammond 1876). This can be so severe that the head has to be supported by bracing. The upper part of the trapezius usually remains intact even late in the illness. In the lower extremity, weakness initially involves dorsiflexors, inverters and everters of the foot. Plantar flexion is often relatively spared. The patient will note a tendency for the foot to slap when walking. Eventually foot drop occurs. The thigh muscles and hip flexors are usually involved later. Though the weakness is often very asymmetrical some changes in the corresponding muscles of the contralateral limb is common.

Asymptomatic loss of chest expansion or loss of intercostal function in the face of extremity weakness may provide another early clue to the diagnosis. Conversely, occasional patients may present with primary respiratory failure without bulbar or extremity weakness. Respiratory abnor-

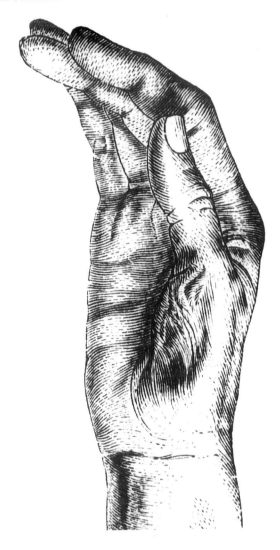

Fig. 7. From Duchenne, *De l'Electrisation localisée*, 3rd ed. Paris, J. D. Baillière et Fils (1872).

malities may be easily seen by observing the expansion of the mid and lower chest wall. Following a patient's course with a tape measure to document chest expansion is an economical and objective way of following respiratory function. Among other respiratory function tests, the maximal expiratory flow volume may help identify the patient with more severe expiratory muscle weakness. These cases may be predicted to sharply drop their vital capacity and flow with changes in resistance (Kreitzer et al. 1978),

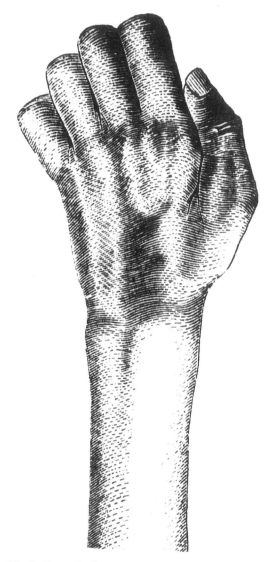

Fig. 8. From Duchenne, *De l'Electrisation localisée*, 3rd ed. Paris, J. D. Baillière (1872).

failure of progression for long periods of time can make one question the accuracy of diagnosis. Conversely, the onset of symptoms can be so sudden as to raise the question of a neuritis or infective etiology, especially if it is followed by a long clinical plateau. Uncommonly, patients may actually improve slightly over weeks to a month. It is possible that such improvement may reflect a period of disease inactivity, during which reinnervation of denervated muscle fibers can produce an increase of strength.

Fasciculation

Fasciculations may be floridly present with sheets of rippling muscles present throughout somatic musculature, or may be so rare as to escape observation by a good clinical observer. Once present they tend to be persistent — they do not vary in frequency or morphology from hour to hour, but can vary over months. Early in the illness, fasciculations tend to be small and infrequent. With disease progression, fasciculations usually get larger, reflecting increasing size of reinnervated motor units. Patients are often aware of fasciculations, but usually do not find them painful or distressing.

Fasciculations usually are single but may occur in pairs or triplets. Repetitive tetanic bursts, commonly seen in root lesions, are not found in ALS (Tyler, unpublished data). Specific units will discharge at a rhythm which is usually irregular. There can be 'fast' and 'frequent' or 'very infrequent' discharging units. The frequency distribution of the discharge is not normal.

Rare patients can present with gross fasciculations and twitching. Clinically weakness is usually noted by the time fasciculations are obvious. Rowland (1980) notes no documented case of fasciculation without weakness has been verified as having ALS. However, one of us has recently seen a patient initially presenting with gross fasciculations but an otherwise normal motor exam. The fasciculations were noted by the patient's internist on a routine examination. When seen 1 month later, weakness was now present, and within 3 months weakness was more severe. Until this example presented Rowland's observation was concordant with our experience.

and thus be more susceptible to respiratory failure secondary to mild upper airway infections.

Andres et al. (1987), on the basis of quantified muscle testing has suggested that there is a linear rate of decline of muscle strength in most patients. We have been less impressed with this but have not followed patients quantitatively. It is clear that patients can experience long clinical plateaus. We have a number of patients who demonstrated no progression for up to 3 years, with further decline subsequently noted. The

It remains an excellent generalization and it is important for the clinician who sees such patients to recognize that weakness is usually present in a fasciculating muscle. The diagnostic implications of clinical fasciculations have been reviewed by Layzer (1982).

Cramps

Some degree of muscle cramping is common. It often occurs early in the course of the illness and often in muscles at the early stages of involvement. Many patients have cramps in the legs but cramping in the abdomen, chest areas and hands also occurs. One possible etiology for cramps is that they reflect an overuse phenomenon at a time when the muscle group in question is still strong enough to be used in a normal fashion. As weakness increases, the patient may make adjustments so that the weakened muscle group is used differently. There is no clear relationship between the presence of cramping and other lower or upper motor neuron abnormalities. It is therefore felt to be a non-specific finding.

Cranial nerve abnormalities

Usually the first sign of bulbar involvement is a problem with articulation. This is usually traceable to a difficulty with tongue movement. The first sounds with which a patient has difficulty have lingual or dental components. Progression to complete anarthria is common. Dysphagia is often noted as dysarthria becomes significant, and progresses to become a limiting component in maintaining nutrition. Characteristically the tongue shows wasting with fasciculations (Fig. 9). One can recognize fibrillary twitching in the tongue because epithelium directly overlies the muscle. This is the only muscle in which fibrillations can be regularly recognized at the bedside by visual means.

Early bifacial weakness is common. This is noted by weakness around the orbicularis oculi and orbicularis oris muscles. The lips are characteristically everted with drooping of the lower lip. Mouth closure is often incomplete, often due both to weakness of facial muscles as well as masseter weakness. This gives the face a very characteristic

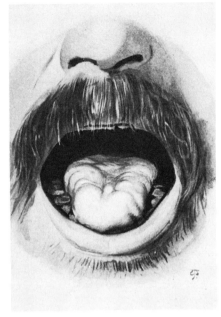

Fig. 9. From B. Bramwell, *Atlas of Clinical Medicine.* Edinburgh, T. and A. Constable at University Press (1892).

appearance (Fig. 10). There is wasting of the mentalis muscle with loss of the rounded protuberance in the anterior chin. It is easy to see fasciculations here. The palate fails to elevate. Because of the difficulty with swallowing secretions accumulate and drooling is common. Head control may be impaired because of sternocleidomastoid weakness and neck extensor weakness.

Extraocular motility and pupillary responses are unaffected. In approximately 50% of patients, a mild reduction of Bell's phenomenon is noted. Rarely, patients will have nystagmus.

Tone, pyramidal dysfunction and reflex abnormalities

Some evidence for upper motor neuron involvement is the rule in ALS. Commonly, however, such abnormalities are not as functionally significant as concurrent lower motor neuron abnormalities. Deep tendon reflexes are often brisk in the presence of profound atrophy; less often, tone may be markedly increased. Severe spasticity can be seen and in a few patients be the primary disability with very labored gait. In the usual

Fig. 10. From B. Bramwell, *Atlas of Clinical Medicine*. Edinburgh, T. and A. Constable at University Press (1892).

patient it is less frequently functionally disabling as compared to lower motor neuron weakness. Baclofen can provide a significant improvement in function in some patients where spasticity is interfering with limb performance.

A number of observers have commented that a Babinski sign is less frequently seen in the patient with ALS than in other pyramidal tract diseases, even though other signs of upper motor neuron involvement may be present. This can be due to concurrent weakness in the extensor hallucis muscles, but this explanation is not valid in most patients and the explanation underlying this observation is unknown. When one cannot obtain pathological reflexes, slowing of repetitive movements like fingertapping may help to distinguish ALS from SMA.

Upper motor neuron bulbar signs are often quite obvious, and may be among the earliest abnormalities noted. An increase in the jaw jerk,

positive corneomandibular reflexes, enhanced gag reflex, suck and snout reflexes may all be seen before significant facial weakness is appreciated. Usually, however, these signs are seen along with significant tongue wasting. Forced yawning is frequent. The phenomenon of pseudo-bulbar affect is also quite commonly seen. In some patients there is a loss of control of crying or laughing and these emotional functions seem overactive and inappropriate to the stimuli; less commonly, apparently spontaneous outbursts of emotionality are observed.

Fatigability (myasthenia)

Some patients exhibit a marked fatiguing of motor functions. It usually relates to specific symptoms, i.e. increasing dysarthria after talking or decreased ability to walk for more than a fixed distance without tiring. Fatigue can be documented electrophysiologically with repetitive stimulation as discussed in the section on neurophysiology. On occasion treatment with prostigmine or pyridostigmine can give partial symptomatic relief.

Occasionally, patients may present in such a way that the clinical distinction between ALS and myasthenia gravis is quite difficult. Such patients often have bulbar symptoms, no signs of atrophy or fasciculation, and abnormal decrement with repetitive nerve stimulation studies. Initially they may respond well to anticholinergics and be given a diagnosis of myasthenia gravis. However, the response to the mestinon is typically transient and further clinical deterioration subsequently makes the appropriate diagnosis apparent. Such cases have been noted by Wilson (1940) and we have seen a few examples: in one case, the diagnosis only became obvious after a year of treatment for myasthenia gravis.

Mental changes

A number of authors have reported cases of dementia associated with ALS (Van Bogaert 1925; Michaux et al. 1955). Pathological study in these cases has revealed lesions more widespread than those classically seen in ALS. A number of cases with ALS and pathological and clinical evidence of Alzheimer's disease have also been reported. In general, clinical evidence of dementia is unusual and probably affects 2 or 3% of patients.

The association of Pick's disease with subsequent development of ALS has been reported a number of times and was reviewed by Bonduelle (1975). In some of these cases changes in the substantia nigra resembling Parkinsonian pathology were noted. Bonduelle et al. (1959) had previously reported 2 cases in which ALS was associated with dementia and an extrapyramidal syndrome. In these cases the microscopic pathology was primarily frontal, with significant loss of cortical neurons and a spongiform change in cortex.

Neary et al. (1990) have reported 4 cases in which similar frontal distribution and spongiform changes occur. This was also present in a large number of cases reported from Japan (Mitsuyama 1984; Morita et al. 1987). In Neary's case SPECT scanning showed reduced tracer uptake in the frontal area. The 'frontal' dementia with spongiform changes seem to be the most frequent type of dementia seen in these patients but coexistence with Alzheimer's disease and more widespread disease has also been noted.

Many patients will show disorders of communication in that there will be frequent misspelling of common words and their sentence structure will be more telegraphic with loss of the descriptive or connecting words. This is usually seen in those patients that are anarthric and communicate by writing. This may be simply an effort saving device but this does not explain the misspelling or duplication of letters frequently noted. We have seen this phenomenon in about 20% of patients.

Psychological status

There is no evidence to support the notion that any specific personality type is more likely to get the illness (Gould 1980). The average patient when informed he has this illness often reacts in a stereotypical fashion. He is usually accepting of the diagnosis, but because of its seriousness will frequently seek a confirmatory opinion with an expert in the disease. This reaction is not dissimilar to that seen in other illnesses in which a patient has to accept the fact that he is dying.

With the realization there is no treatment, the patient often becomes depressed. He often temporarily withdraws from medical interaction. This is difficult for family and friends to deal with. After a variable period of time he usually accepts the inevitable and comes to peace with himself.

Gould (1980) stresses how these patients redefine reality to maintain some hope.

Some of the psychological characteristics were reviewed by Houp et al. (1977) and Gould (1980).

Some generalized cognitive impairment has been noted in some series when patients have been formally tested. Dementia is usually not a problem (Iwawaki et al. 1990).

Sensory symptoms

Bedside techniques disclose sensory abnormalities in ALS patients very rarely. Normal sensation is one of the major clinical aids to diagnosis. Although objective signs of impairment are uncommon, a number of patients will note paresthesia of the extremities, especially the feet. These are usually described as 'pins and needles' or tingling. Some elderly patients may have minimal disturbance of vibration sense in the legs, perhaps correlating with the pathologic changes occasionally seen in the posterior columns. This rarely is progressive and if severe should make one seek alternative reasons: i.e., cervical spondylosis, diabetic neuropathy etc.

One can rarely see sensory levels or demonstrate disassociated sensory loss that suggests an intramedullary pathology. This is most often seen in younger patients as well as very rapidly progressing cases. Such sensory losses are usually transient and are not noted over long periods of time. They can be found by quite experienced examiners, however, and can be quite confusing diagnostically.

Mulder et al. (1983) have tested patients with a quantitative sensory examination methodology. Eighteen per cent of his patients had some sensory disturbance when tested in this formal way. It most commonly was vibration sense and touch that were abnormal.

Autonomic dysfunction

Most patients will develop some autonomic dysfunction in the extremities including some blueish discoloration of dependent feet or edema. These findings suggest loss of sympathetic innervation.

Abnormalities of other autonomic processes are unusual. Bladder function is usually normal; complaints of urinary incontinence should prompt a search for other diseases. Bowel function is similarly unaffected, although significant constipation may occur when abdominal musculature becomes so weak that it is impossible to increase intraabdominal pressure. Even in this situation, however, the anal sphincter is usually uninvolved. Sexual function is similarly uninvolved.

Laboratory investigation

The cerebrospinal fluid (CSF) cell count and total protein are usually within normal limits in ALS. However, in a minority of cases the protein may be elevated. CSF protein over 75 mg/dl was noted in 5% of 129 ALS patients combined from 3 series (Merritt and Fremont-Smith 1938; Swank and Putnam 1943; Guiloff et al. 1980). More recently, increased CSF protein over 50 mg/dl was noted in 25% of 120 patients, and levels over 75 mg/dl were seen in 4% (Younger et al. 1990). In this series, patients with the highest CSF protein levels were the most likely to have paraproteinemia; 2 of the patients with the highest CSF protein levels were also found to have lymphoma. Oligoclonal bands are also noted occasionally in the CSF of ALS patients. In the series reported by Younger et al. (1990), 12% of their patients had oligoclonal bands.

In contrast to CSF protein which may occasionally be elevated in ALS, CSF pleiocytosis is very unusual in ALS. Merritt and Fremont-Smith (1938) found CSF white cell count to be less than $5/mm^3$ in 29 of 33 patients, with the remaining 4 patients having less than 12 cells. Similarly, in another series, 82/83 patients had cell counts less than 10 white cells/mm^3 (Laterre 1975). Thus a significant increase in CSF white

cell count should prompt a search for an alternative disease process.

The value of CSF examination in ALS is not that it provides confirmatory information that aids in the diagnosis of ALS, but rather that it helps to eliminate other potential causes of extremity weakness. Although now extremely rare, the general paresis of neurosyphilis is characterized by a CSF pleiocytosis of 25–75 cells/mm^3, as well as total protein usually greater than 50 mg/dl (Merritt et al. 1946). Acute poliomyelitis also produces a significant inflammatory reaction; in the acute phase, less than 2% of patients have CSF cell counts less than 10 cells/mm^3 (Merritt and Fremont-Smith 1938). Lyme disease, which can produce a variety of neurological syndromes including a polyradiculopathy with significant lower motor neuron weakness, is less easy to distinguish from ALS on the basis of CSF profile. While CSF from patients with Lyme disease related encephalomyelitis usually shows both an increased white cell count and total protein (Kohler et al. 1988), patients with predominantly neuropathic symptoms usually have normal CSF cell counts and protein levels (Logigian et al. 1990).

TREATMENT

Symptomatic respiratory symptoms

The common causes of death are respiratory insufficiency and infection secondary to incompetent airway. Whether to use assisted ventilation or not in patients is a decision that must be made by the physician and patient in each individual case (Sivak and Streib 1980). Early discussion with the patient and family is desirable. Decisions to support respiration often have to be made precipitously and reasoned discussion at a time of crisis may not be possible. It is important to note that a decision to use ventilation does not imply that the requirement for assistance will be permanent. Frequently, respiratory failure is due to relatively minor infections and pulmonary function returns to its previous baseline with appropriate therapy. In one series of 16 patients treated with assisted respiration 9 were able to be weaned from the ventilator (Goulon and Goulon-Goeau 1989).

The question of early use of a tracheostomy for airway control must also be made on an individual basis. Frequently, patients present with recurrent pulmonary infections due to aspiration at a time when their pulmonary function is still adequate. In these patients, tracheostomy combined with placement of a gastrostomy or jejunostomy tube can significantly prolong life and improve its quality. However, even minor abdominal procedures are associated with risk, especially since diaphragmatic splinting after abdominal surgery reduces already compromised pulmonary function. The newer technique of placing a gastrostomy tube endoscopically probably reduces this risk. When one does a gastronomy in a patient with severely compromised diaphragms or very low vital capacity gastronomy may have an increased risk. It is more difficult to breathe or expel air without a diaphragm and excess air in the stomach can lead to further compromise of pulmonary function. In these cases a nasogastric tube is the only alternative.

Rehabilitation

In the early stages of the illness there is little need of rehabilitative intervention. Patients should maintain their normal activity. Moderate exercise programs are useful; however, intensive training programs are rarely productive.

If there is specific muscle weakness, appropriate bracing may be helpful; i.e. a dorsiflexion orthosis for the foot if there is a foot drop, or splinting of the wrist and fingers for weakness of extension. Lightweight or plastic braces are desirable. If the patient develops finger weakness, occupational therapy is often useful in designing alternative strategies for accomplishing activities of daily life.

Physical therapy is very important in maintaining the normal range of motion in weakened limbs. Frozen joints secondary to disuse are a significant cause of pain in ALS; this is a particularly unfortunate complication as it is completely preventable. In addition, ease of transfer is significantly enhanced if joint con-

tractures are not present. Special attention should be paid to the shoulders and the ankles.

Problems in communication are a major concern for the patient with bulbar signs. Initially a speech therapist can be helpful mainly by having the patient slow his speech to increase intelligibility or suggest oral devices which can help airway control. As speech is lost most patients will switch to written communication. Communication systems employing eye movements have been developed for patients with bulbar dysfunction and quadriplegia (Figs. 11 and 12). The reader is referred to excellent reviews by Aronson (1980), Elliott and Hudson (1980) and Denys (1980b).

It is surprising how few of these patients develop bed sores. This was specifically noted by Toyokura (1978). They are often severely wasted, have hypertonia, are bedfast — yet unlike comparable patients with multiple sclerosis for example, this is not a common complication. This may relate to the presence of normal sensation and the relatively intact mental status of the patients. There have been dermal abnormalities described in ALS (Fullmer et al. 1960; Beache et al. 1986) and they may also play a role in the resistance to bed sores.

Non-specific drug therapy

Fatiguability can be helped on occasion with neostigmine and pyridostigmine. Baclofen may help spastic symptoms and hypertonia. Rarely it will help with dysarthria or dysphagia. X-ray treatment and section of the chordae tympani have also been suggested for excessive salivation. We have not found this useful. Salivation and drooling is helped with amitriptyline, methantheline, propantheline or methylphenidate. These medications may make secretions more viscous, however, and patient acceptance is not universal. Meperidine has been suggested for dysphagia (Norris 1980).

Some have tried injections into the vocal cords and cricopharyngeal myotomy to prevent aspiration secondary to dysphagia. Our experience with these methods has been very limited and unrewarding. Cramps may respond at times to Dilantin or Tegretol.

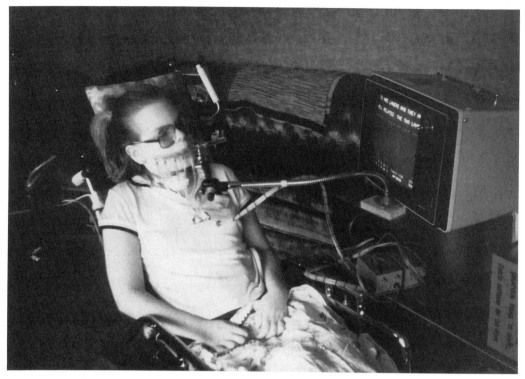

Fig. 11. Patient controls the computer by eye movement.

Specific drug therapy

Table 1 shows a partial listing of therapeutic trials that have been carried out in ALS. It is clear that no specific therapy has been established as being effective either in improving symptoms of patients with ALS or in delaying the rate of clinical progression. Multiple experimental designs have been employed with varying degrees of sophistication.

In designing a clinical trial for a disease such as ALS, many factors have to be considered. One important requirement is that inclusion criteria for entrance into a study must be carefully specified. Such criteria should be sufficiently rigorous so that the appropriate diagnosis of all patients included should not be in question. It should be clear whether all patients have to have abnormalities in both upper and lower motor neuron function, or if specific clinical syndromes such as progressive muscular atrophy or primary lateral sclerosis are included. In addition, criteria

for the severity of symptoms at study onset must be specified.

Critical in the design of a study is the measures that will be used to judge efficacy. As discussed previously, progression of ALS is extremely variable both in rate and quality. Baseline measurements should include quantitative estimates of muscle strength from upper and lower extremity muscles as well as facial musculature. Functional measurements such as speed of walking, ability and speed of stair climbing, as well as ability to perform fine movements in the distal upper extremities may all be useful. In addition, pulmonary function tests are important quantitative measures that should be part of any assessment. Even with the use of multiple measures such as those mentioned above, assessment of progression can be difficult. Comparison of videotapes of patient activities and speech recordings may provide information over and above that provided by more quantitative assessments. Formal rating scales have been devised that

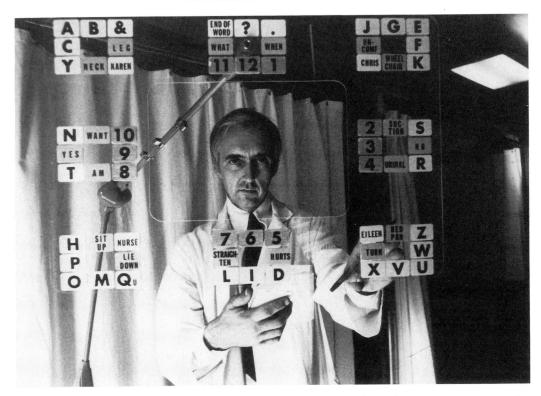

Fig. 12. By looking at the appropriate part of the plastic letter screen the patient's eye movements can be interpreted by the observer and can indicate either a request or letter of the alphabet.

include formal muscle testing, quantitated estimates of tone and reflex abnormalities (Appel et al. 1987). Although the use of such scales is important in the serial assessment of patient progression, they do not replace the overall clinical impression that may be obtained from simple observation.

It is imperative that any study judging the efficacy of therapy in ALS be double blinded and placebo controlled. Placebo effects are powerful in many experimental situations, but are particularly important in studies of a slowly progressive, invariably fatal disease. Thus, patients must be randomly assigned to either treatment or control group. It is equally important that both the experimenter and the patient be blinded as to the patient assignment, as placebo pressures affect the investigator as well as the patient. Designs which employ crossover from placebo to treatment group and vice versa are particularly useful, as they allow patients to be assured of receiving any experimental treat-

ment despite the chance of being initially assigned to a control group.

Another consideration in the design of a therapeutic trial is length of time for each treatment. Given the extreme variability of ALS, a treatment must be given for a relatively long time. However, competing pressures include the natural desire of a patient to have the opportunity to switch from a potential placebo to an active treatment. Given the average expected life span of 3 years from diagnosis, treatment periods on the order of years are inappropriate. A 6-month treatment period might be an appropriate compromise.

Schoenberg (1982), as well as others have reviewed the requirements for design of appropriate treatment protocols in ALS. The reader is referred to these sources for a more detailed discussion.

Patients who are in experimental drug trials will often take 6–12 weeks before they can identify progression. About 70% of cases will go

TABLE 1
Recent therapeutic trials in ALS.

Interferon (Rissanen et al. 1980)
Idoxuridine (Liversedge et al. 1970)
Amantadine (Engel et al. 1969)
Transfer factor (Jones et al. 1978; Olarte et al. 1979)
Isoprinosine (Fareed and Tyler 1971; Percy et al. 1971)
Tilerone (Olson et al. 1978)
Plasmapheresis (Silani et al. 1980)
Guanidine (Norris et al. 1971; Norris 1973)
Penicillamine (Currier and Haerer 1968)
Modified snake neurotoxin (Rivera et al. 1979; Tyler 1979)
Levamisol (Olarte 1982)
Lipoic acid (Engel et al. 1969)
Vitamin E (Denker and Sheinman 1941)
Thyrotrophin releasing hormone (TRH); fresh human plasma (Engel et al. 1969)
Prednisone or hydrocortisone (Pieper and Fields 1959; Quick and Greer 1967; Engel et al. 1969)
Pancreatin, Vitamin B-12, Bile salt and Liver extract capsules (Engel et al. 1969)
Penicillamine (Tyler unpublished)
Vitamin E and pancreatic extract (Dorman et al. 1969)
Magnesium pemoline (Engel et al. 1969; Tyler unpublished)
B-12 (Pieper and Fields 1959)
Steroids (Pieper and Fields 1959)
Baclofen (Norris et al. 1979)
Branched chain amino acids (Plaitakis 1990)

6 months before they can be certain they are getting worse. Untreated patients will often perceive a daily progress. This is knowing they have an untreated progressive illness. As soon as they are given something which 'might help' they often stabilize and 25% will claim some improvement (even on placebo) over the next few weeks. In some cases the improvement with placebo can be striking. Small variations in muscle strength from day to day is not unusual.

DIFFERENTIAL DIAGNOSIS

In most cases an experienced neurologist can make a definitive diagnosis from the history and neurological examination. The presence of a characteristic distribution of weakness, fasciculations and increased deep tendon reflexes without bladder or bowel difficulties does not present a diagnostic dilemma. However, early or atypical cases require distinction from a number of different disease entities. Some specific diseases that are often considered to need clinical distinction from ALS are discussed below.

Lead poisoning

The typical case of lead poisoning rarely poses problems in differential diagnosis; however, occasionally it can mimic Primary Muscle Atrophy. The question of whether lead intoxication can produce upper motor neuron abnormalities remains open; however, if so, it must occur exceedingly rarely. We have not observed such cases but they have been the subject of case reports (Boothby et al. 1974). Given the fact that lead intoxication is a treatable illness it seems prudent to explore histories for lead exposure and to measure lead levels in all patients.

Mercury poisoning

We have never seen a patient with mercury intoxication present with a syndrome that would be confused with ALS. The typical clinical picture is one of tremor, dementia, paresthesia and ataxia. The cases that have been described only have pathologic (rather than clinical) changes in nerve cells. Rare cases have been described by Brown (1954), Kantarjian (1961) and Shirakawa et al. (1976) which superficially resemble the lower motor neuron form of ALS.

Manganese intoxication

Manganese toxicity usually causes psychiatric disorders and extrapyramidal syndrome. There is one case reported of a manganese miner with ALS (Voss 1939) which probably was incidental.

Viral disorders and other infections

Rarely in ALS, weakness of a limited number of muscles can progress so rapidly (i.e. involvement of tongue over 3–5 days) that poliomyelitis or some other viral illness is considered. This can be ruled out by spinal fluid examination or appropriate blood tests. Usually time is the most effective test; when such rapid progression occurs in the elderly (in whom paralytic poliomyelitis is extremely rare) it is usually a presenting symptom of Progressive Bulbar Palsy.

Alaoui-Faris et al. (1990) have reported 5 cases of possible syphilitic origin with spinal fluid hypercellularly and increased protein that closely mimicked ALS and responded to treatment. This should be considered in all cases with abnormal CSF.

Diabetic amyotrophy

Diabetic amyotrophy, or proximal diabetic polyradiculopathy, can sometimes resemble predominantly lower motor neuron forms of ALS. The entity is not clearly described and different authors include different syndromes — but all include some focal wasting and usually a Babinski reflex. The pattern of weakness is helpful as most cases of diabetic amyotrophy show a relation to an upper lumbar root or portion of the lumbosacral plexus. The presence of pyramidal findings or fasciculations are strongly in favor of ALS. Many diabetic patients also have findings of a distal diabetic neuropathy. Garland (1955), who originally described the syndrome, included patients with progressive course, fasciculations and death which would be difficult to separate from ALS in a diabetic.

Others have tended to emphasize the pain, lower extremity predominance and areflexia that are characteristic of diabetic amyotrophy. Many of the cases we have seen would fall under the definition of a diabetic plexopathy with an isolated pyramidal tract findings. The latter could have a number of explanations including cervical spondylosis.

Degenerative disease of the spine

When signs of upper and lower motor neuron involvement are only noted in the extremities, it is crucial to rule out a combination of cervical and lumbosacral spine disease producing a combined myelopathy and lower motor neuron syndrome. The incidence of degenerative joint disease of the spine increases with age, and is not necessarily associated with pain. Cervical disc herniation commonly causes a myelopathy affect-ing arms and legs; if multiple radiculopathies are superimposed, the resulting clinical syndrome may be indistinguishable from ALS.

Appropriate imaging studies of both the cervical and lumbosacral area are therefore a necessary part of the evaluation of a patient with ALS without bulbar signs. Currently, MRI is the least invasive and most useful method. Even with good imaging studies, however, the question may not be adequately resolved. Many patients with ALS have abnormal spine MRI scans; it is often quite difficult to decide if the abnormalities noted on MRI are sufficient to cause the observed syndrome. It is for this reason that some patients with ALS ultimately are diagnosed only after unsuccessful spinal decompression. One should always be cautious of the diagnosis of ALS in the absence of signs above the neck.

Muscle disease

Polymyositis and other myopathies of middle and late life often progress with a time course similar to ALS. Usually the clinical presentation is distinct, with a proximal symmetric weakness noted in the presence of normal reflexes. Sedimentation rate and CPK abnormalities often help to make the diagnosis. Occasionally, however, blood tests may be normal and clinical presentation may be atypical. As was noted previously CPK may be significantly elevated in a small number of patients with ALS. The explanation for this is not clear. In these cases electromyography can usually distinguish clearly between these disease entities.

Multiple sclerosis

Distinguishing ALS from multiple sclerosis which presents as a primary bilateral spastic quadriparesis or spastic paraparesis may occasionally be difficult. Usually, distinction can be made on the basis of the relapsing/remitting course often seen in multiple sclerosis, as well as the failure to note lower motor neuron abnormalities. Absence of confirmatory MRI findings, spinal fluid abnormalities and normal evoked potentials should always raise one's suspicions in these cases. Electromyography of muscles looking for evidence of lower motor neuron disease usually helps one suspect the correct diagnosis.

Parathyroid and ALS

Muscle weakness is a primary symptom in patients with hyperparathyroidism. This usually shows up earliest in the proximal muscles of the lower extremities. This can approach 80% (Patten and Engel 1982). In our experience this is usually part of a generalized weakness and some weakness at the shoulder and the neck flexors also exists. Hyperreflexia and atrophy were noted by Vicale (1949); he also noted the absence of clonus, and absence of unequivocal corticospinal tract signs, bulbar weakness and fasciculations. Patten and Engel (1982), however, noted Babinski signs in 3 patients and aphonia and hoarseness in 4. The EMG, however, did not resemble the usual ALS case and no giant units were noted. We have not encountered many patients presenting as hyperparathyroidism who have been seriously considered as having ALS or consequently misdiagnosed ALS that was found to be hyperparathyroidism. It may be harder to separate from SMA.

Thyroid disease

The older literature describes cases of thyroid disease with weakness and fasciculations. These cases were not seen with any frequency in recent years. A recent case of Serradell et al. (1990), however, had pyramidal signs and fasciculations and had a diagnosis of ALS. Treatment of hyperthyroidism reversed the situation. Because this is a treatable condition it should be considered in all cases.

Radiation myelopathy and plexopathy

After radiation treatment of the spine a segmental lower motor neuron syndrome compatible with involvement of the segment of spinal cord that was irradiated can rarely develop. This usually occurs after a time interval of 1–3 years (Greenfield and Stark 1948; Maier et al. 1969; Somasoundarum et al. 1975; Sadowsky et al. 1976; Kristensen et al. 1977; Rowland 1980). This is purely a motor syndrome of lower motor neuron type with wasting and slowly progressive weakness often over many years. In its initial stages it may be very difficult to separate from early motor neuron disease — but usually its slow course and the absence of any spread beyond the geography of the segment of cord that received X-rays indicate its true nature. It is not clear whether this is a motor radiculopathy or an anterior horn cell syndrome.

A slowly progressive plexopathy can be seen, usually occurring months to years after radiation therapy. As compared to neoplastic infiltration, this disorder is relatively painless, and its progressive nature and multiple root distribution may allow for confusion with early ALS. Its strict unilaterality, local tissue changes and confinement to the area irradiated all help making this diagnosis. EMG may aid in the diagnosis, as bizarre high frequency complex repetitive discharges are characteristic of radiation plexopathy.

ALS and cancer

There has been no clear evidence to support a direct relation between cancer and ALS syndromes. In every large series there may be some patients with neoplasm but it does not appear to be more frequent than expected in the general population.

Norris (1965) reported an increased incidence of neoplasms in a series of patients with ALS. Brain et al. (1965), Norris and Engel (1965) and Norris et al. (1969) also noted some cases. In those cases where both illnesses have appeared together we have not been impressed that treatment of the tumor has any effect on the course of ALS. Evans et al. (1990) reported a case with kidney carcinoma with some amelioration of the symptoms with treatment of the cancer. This is obviously worth noting to see if any future reports of similar cases occur. We have one patient in whom removal of his kidney tumor did not affect the disease process.

ALS and extrapyramidal syndrome

A few cases of Parkinson's disease developing with ALS have been described (Bonduelle et al.

1959; Legrand et al. 1959). There have been a number of isolated case reports of other unique patients with findings suggestive of an extrapyramidal disorder (Patrikios 1951; Greenfield and Matthews 1954; Hufschmidt 1960) and most of these patients have been thought to have had post encephalitic Parkinsonism with superimposed ALS. Many patients with ALS who become demented seem to have extrapyramidal features (Bonduelle 1975). Bonduelle also notes that some families have siblings or parents who have Parkinsonism with one member developing ALS. This has been confirmed in other reports (Van Bogaert and Radermecker 1954); we have reported one such family (Tyler 1975). In our case the pathology was typical of ALS.

Some of the patients with dementia, extrapyramidal features and amyotrophy may represent a form of Creutzfeldt-Jakob disease. Bonduelle (1975) notes his cases had a different type of spongiform cortical change and lacked other features of Creutzfeldt-Jakob pathology.

Creutzfeldt-Jakob disease, dementia and ALS

The original description of Jakob and Creutzfeldt allows for amyotrophy and fasciculations but always in the context of much wider clinical dementing syndromes (Tyler 1982). Meyer (1929) described an 'ALS variant'. Siedler and Malamud (1963) noted some evidence of anterior horn cell disease in 50% of patients. There have been 3 distinct pathologies reported in demented patients.

1. Spongiform cortical changes with loss of ganglion cells and atrophy but no senile plaques.
2. Widespread neuronal disorder with intracytoplasmic Bunina-like bodies and extensive gliosis.
3. Changes typical of Alzheimer's disease.

Agents which can be transmitted have been sought but not found.

In some cases, not associated with spongiform encephalopathy, there may be selective temporal lobe atrophy.

Human retrovirus and ALS

HTLV-1 (human retrovirus) associated myelopathy can mimic ALS (Evans et al. 1989; Vernant et al. 1989). This is especially true when there is an associated polymyositis. The myelopathies are usually more chronic and are associated with bladder and/or bowel symptoms. The lower extremity symptoms often precede upper extremity symptoms. Bulbar signs are usually not present, an unusual occurrence in far advanced spastic ALS. Low grade pleocytosis in cerebrospinal fluid is often present and oligoclonal bands on electrophoretic studies are common. Some sensory symptoms are often present. Electromyographic studies may be confusing as they can show findings 'consistent with anterior horn cell disease' because of the associated radiculopathy. Lab tests include HTLV-1 antibody testing and HTLV-1 PCR testing for HTLV-1 DNA in whole blood cells. Tropical spastic paraparesis is probably the same illness.

At this time, only one case has been reported of ALS and HIV infection (Hoffman et al. 1985). In this case, a 26-year-old man presented with a 1-year history of both upper and lower motor neuron abnormalities. At the time of presentation, the patient was noted to have antibodies to HIV. He had enlarged cervical lymph nodes but no other signs of the acquired immunodeficiency syndrome.

Hoffman et al. (1985) also investigated serum from 14 patients from Guam with ALS, and found no serologic evidence of HIV infection. Similarly, Dalakas and Pezeshkpour (1988) report that antibodies to both HIV and HTLV-1 were absent in 17 ALS patients, as well as 25 patients with postpolio progressive muscular atrophy. At the present time, therefore, the propensity for HIV infection to produce an ALS-like syndrome must be considered quite weak.

Fasciculation syndromes

There are some benign syndromes with fasciculations which have to be differentiated from ALS. The muscle-pain/fasciculation syndrome usually has symptoms primarily in the lower extremities; pain is often an aching, bruising sort. Fasciculations may be generalized but are primarily noted in the legs. Patients can complain of a feeling of weakness but objective muscle testing is usually normal (Hudson et al. 1978).

Isaacs (1961) described a condition with muscle stiffness, fasciculation and sweating responsive to dilantin. EMG shows continuous activity of normal appearing motor units.

REFERENCES

ALAJOUANINE, T.: La poliomyélite antérieure subaigüe progressive. Rev. Neurol. (Paris) 11 (1934) 225–265.

ALAOUI-FARIS, M. E., A. MEDEJEL, K. A. ZEMMOURI, M. YAHY-AOUI and T. CHKILI: Le syndrome de sclérose latérale amyotrophic d'origin syphilitique. Etude de 50 cas. Rev. Neurol. (Paris) 146 (1990) 41–44.

ALBIZZATI, M. G., S. BASSI, D. PASSERINI and V. CRESPI: F-wave velocity in motor neuron disease. Acta Neurol. Scand. 54 (1976) 269–277.

ANDRES, P. L., L. M. THIBODEAU, L. J. FUESON and T. L. MUNSAT: Quantitative assessment of neuromuscular deficit in Amyotrophic Lateral Sclerosis. Neurol. Clin. 5 (1987) 125–142.

ANTEL, J. P., B. G. W. ARNASON, T. C. FULLER and J. R. LEHRICH: Histocompatibility typing in Amyotrophic Lateral Sclerosis. Arch. Neurol. 33 (1976) 423–425.

ANTEL, J. P., A. B. C. NORONHA, J. J. F. OGAR and B. G. W. ARNASON: Immunology of Amyotrophic Lateral Sclerosis. In: L. P. Rowland (Ed.), Human Motor Neuron Diseases, New York, Raven Press (1982) 395–402.

APPEL, S. H.: A unifying hypothesis for the cause of Amyotrophic Lateral Sclerosis, Parkinsonism and Alzheimer's Disease. Ann. Neurol. 10 (1981) 499–505.

APPEL, V., S. S. STEWART, G. SMITH and S. H. APPEL: A rating scale for Amyotrophic Lateral Sclerosis: description and preliminary experience. Ann. Neurol. 22 (1987) 328–333.

ARAN, F. A.: Recherches sur une maladie non encore décrite du système musculaire (Atrophie musculaire progressive): Arch. Gén. Méd. 24 (1850) 172–214.

ARGYROPOULOS, C. J., C. P. PANAYIOTOPOULOS and S. SCARPALEZOS: F- and M-wave conduction velocity in Amyotrophic Lateral Sclerosis. Muscle Nerve 1 (1978) 479–485.

ARNOLD, A., D. C. EDGRAN and V. S. PALLADINO: Amyotrophic Lateral Sclerosis: fifty cases observed in Guam. J. Nerv. Ment. Dis. 117 (1953) 135.

ARONSON, A. E.: Definition and scope of the communication disorder. In: D. W. Mulder (Ed.), The Diagnosis and Treatment of Amyotrophic Lateral Sclerosis. Boston, Houghton Mifflin (1980) 217–224.

ASK-UPMARK, E.: Precipitating factors in the pathogenesis of Amyotrophic Lateral Sclerosis. Acta Med. Scand. 170 (1969) 717–723.

ASK-UPMARK, E. and S. MEURLING: On the presence of a deficiency factor in the pathogenesis of Amyotrophic Lateral Sclerosis. Acta Med. Scand. 152 (1955) 217–222.

ASTIER, H. A.: Etude sur l'atrophie musculaire progressive, (Thèse pour le doctorat en médecine), Paris, Regnoux (1856).

BARBER, T. E.: Inorganic mercury intoxication reminiscent of Amyotrophic Lateral Sclerosis. J. Occup. Med. 20 (1978) 667–669.

BARKER, A. T., I. L. FREESTON, R. JALINOUS and J. A. JARRATT: Magnetic stimulation of the human brain and peripheral nervous system: An introduction and the results of an initial clinical evaluation. Neurosurgery 20 (1987) 100–109.

BARTFELD, H., C. DHAM, H. DONNENFELD, L. JASHNANI, R. CARP, R. KASCSAK, J. VILCEK, M. RAPPORT and S. WALLENSTEIN: Immunological profile of Amyotrophic Lateral Sclerosis patients and their cell mediated immune responses to viral and CNS antigens. Clin. Exp. Immunol. 48 (1982) 137–147.

BARTFELD, H., C. DHAM, H. DONNENFELD, R. A. OLLAR, M. T. DE MASI and R. KASCSAK: Enteroviral-related antigen in circulating immune complexes of Amyotrophic Lateral Sclerosis patients. Intervirology 30 (1989) 202–212.

BEAL, M. F. and E. P. RICHARDSON: Primary lateral sclerosis, a case report. Arch. Neurol. 38 (1981) 630–633.

BEACH, R. L., E. T. REYES, B. W. FESTOFF, R. YAMAGIHARA, D. C. GADJUSEK and J. S. RAO: Collogenase activity in skin fibroblasts in patients with amyotrophic lateral sclerosis. J. Neurol. Sci. 72 (1986) 49–60.

BELL, C.: An Exposition of the Natural System of the Nerves of the Human Body. London, A. R. Spottiswoode (1824).

BELL, C.: The Nervous System of the Human Body. London. Longman, Rees, Orme, Brown and Green (1830).

BEN HAMIDA, M., F. LETAIEF, F. HENTATI and C. BEN HAMIDA: Morphometric study of sensory nerve in classical (or Charcot disease) and juvenile Amyotrophic Lateral Sclerosis. J. Neurol. Sci. 78 (1987) 313–329.

BERSTEIN, L. P. and J. P. ANTEL: Motor Neuron Disease: Decremental responses to repetitive nerve stimulation. Neurology 31 (1981) 204–207.

BERTRAND, I. and L. VAN BOGAERT: Rapport sur la Sclérose Latérale Amyotrophique anatomie pathologique. Rev. Neurol 52 (1925) 779–806.

BONDUELLE, M.: Amyotrophic Lateral Sclerosis. In: P. J. Vinken and G. W. Bruyn (Eds), Handbook of Clinical Neurology. Amsterdam, North Holland Publishing Co., (1975) 281–338.

BONDUELLE, M.: Aran-Duchenne? La Querelle de l'Atrophie Musculaire Progressive. Rev. Neurol. (Paris) 146 (1990) 97–106.

BONDUELLE, M., P. BOUYGUES, J. DELAHOUSSE and

C. FAVERET: Evolution simultanée chez un même malade d'une maladie de Parkinson et d'une sclérose latérale amyotrophique. Rev. Neurol. (Paris) 101 (1959) 63–66.

BONDUELLE, M., P. BOUYGUES, G. LORMEAU et al.: Etude clinique et évolutive de cent vingt-cinq cas de sclérose latérale amyotrophique. Presse Méd. 78 (1970) 827–832.

BOOTHBY, J. A., P. V. DEJESUS and L. P. ROWLAND: Reversible forms of motor neuron disease: lead 'neuritis.' Arch. Neurol. 31 (1974) 18–23.

BOSCH, E. P., T. YAMADA and J. KIMURA: Somatosensory evoked potentials in motor neuron disease. Muscle Nerve 8 (1985) 556–562.

BRADLEY, W. G. and F. KRASIN: DNA hypothesis of Amyotrophic Lateral Sclerosis. In: L. P. Rowland (Ed.), Human Motor Neuron Disease. New York, Raven Press (1982) 493–502.

BRADLEY, W. G., P. GOOD, C. G. RASOOL and L. S. ADELMAN: Morphometric and biochemical studies of peripheral nerves in amyotrophic lateral sclerosis. Ann. Neurol. 14 (1983) 267–277.

BRAHIC, M., R. A. SMITH, C. J. GIBBS JR., R. M. GARRUTO, W. W. TOURTELLOTTE and E. CASH: Detection of picornavirus sequences in nervous tissue of Amyotropic Lateral Sclerosis and control patients. Ann. Neurol. 18 (1985) 337–343.

BRAIN, LORD, P. B. CROFT and M. WILKINSON: Motor Neuron disease as a manifestation of neoplasm. Brain 88 (1965) 479–500.

BREUER, A. C., M. S. LYNN, M. B. ATKINSON, S. M. CHOU, A. J. WILBOURN, K. E. MARKS, J. E. CULVER and E. J. FLEEGLER: Fast axonal transport in Amyotrophic Lateral Sclerosis: an intraaxonal organelle traffice analysis. Neurology 37 (1987) 738–748.

BREUER, A. C., H. MITSUMOTO and M. B. ATKINSON: Alleviation of axonal transport in motor neuron disease may be caused by endogenous agents. International ALS–MND Update, April (1989) 19.

BRODY, J. A.: Chronic sequelae of tick-borne encephalitis and Vilyuish encephalitis. In: D. C. Gajdusek, C. J. Gibbs and M. Alpers (Eds), Slow, Latent and Temperate Virus Infections NINB Monograph No. 2. Washington, D.C., U.S. Government Printing Office (1965) 111–113.

BROOKE, M. H. and W. K. ENGEL: The histological diagnosis of neuromuscular disease. Arch. Phys. Med. 47 (1966) 99–121.

BROWN, I. A.: Chronic Mercurialism, a case of the clinical syndrome of Amyotrophic Lateral Sclerosis. Arch. Neurol. Psychiatry 72 (1954) 674–681.

BROWN, R. and H. WEINER: The relation between polio virus and Amyotrophic Lateral Sclerosis. In: F. Clifford Rose (Ed.), Recent Progress in Motor Neuron Diseases. London, Pittman Press (1984) 349–359.

BROWN, R. H., D. JOHNSON, M. OGONOWSKI and H. L. WEINER: Antineural antibodies in the serum of patients with Amyotrophic Lateral Sclerosis. Neurology 37 (1987) 152–155.

BROWN, W. F. and N. JAATOUL: Amyotrophic Lateral Sclerosis. Electrophysiologic study (number of motor units and rate of decay of motor units). Arch. Neurol. (Chic.) 30 (1974) 242–248.

BROWN, W. F., M. J. STRONG and R. SNOW: Methods for estimating numbers of motor units in biceps-brachialis muscles and losses of motor units with aging. Muscle Nerve 11 (1988) 423–432.

BROWNWELL, B., R. OPPENHEIMER and J. T. HUGHES: The central nervous system in motor neuron disease. J. Neurol. Neurosurg. Psychiatry 33 (1970) 338–357.

BUCHTHAL, F. and F. BEHSE: Peroneal Muscular Atrophy and related disorders. Clinical manifestations as related to biopsy findings, nerve conduction and electromyography. Brain 100 (1977) 41–66.

BUCHTHAL, F. and S. CLEMMENSEN: On differentiation of muscle atrophy by electromyography. Acta Psychiatr. Scand. 16 (1941) 143–181.

BUNINA, T. L.: Intracellular inclusions in familial Amyotrophic Lateral Sclerosis. Zh. Neuropathol. Pskyhiat. 62 (1962) 1293–1299. (cited by Hirano 1967).

CAMPBELL, A. M. G., E. R. WILLIAMS and J. PEARCE: Late motor neuron degeneration following poliomyelitis. Neurology 19 (1969) 1101–1106.

CAMPBELL, A. M. G., E. R. WILLIAM and D. BARLTROP: Motor Neuron Disease and exposure to lead. J. Neurol. Neurosurg. Psychiatry 33 (1970) 877–885.

CARLTON, S. A. and W. F. BROWN: Changes in motor unit populations in motor neuron disease. J. Neurol. Neurosurg. Psychiatry 42 (1979) 42–51.

CAROSCIO, J. T., W. F. CALHOUN and M. D. YAHR: Prognostic factors in motor neuron disease: A prospective study of survival. In: F. C. Rose (Ed.), Research Progress in Motor Neuron Diseases. Bath, Pitman (1984) 34–43.

CARPENTER, S.: Proximal axonal enlargement in motor neuron disease. Neurology 18 (1968) 841–851.

CHARCOT, J. M.: Lectures in the Diseases of the Nervous System, 2nd series. NY, Hafner Publishing Co., (1862).

CHARCOT, J. M.: Sclérose des cordons latéraux de la moelle épinière chez une femme hystérique atteinte de contracture permanente des quatres membres. L'Union Méd. 25 (1865) 451–467, 461–472.

CHARCOT, J. M.: Observation communiqué en 1875. à la Société de Biologie par M. Raymond. Paralysie essentielle de l'Enfance: Atrophie musculaire consécutive. Gaz. Méd. (Paris) (1875) 225–226.

CHARCOT, J. M.: Oeuvres completes: leçons sur les maladies du système nerveux. Delahaye and Lecrosnier (Paris) (1886) 212–297.

CHARCOT, J. M. and A. JOFFROY: Deux cas d'atrophie musculaire progressive avec lésions de la substance grise et des faisceaux antérolateraux de la moelle épinière. Arch. Physiol. 2 (1869) 354–367, 629–760.

CHARCOT, J. M. and P. MARIE: Deux nouveaux cas de sclérose latérale amyotrophique suivis d'autopsie. Arch. Neurol (Paris) 10 (1885) 168–186.

CHIAPPA, K. H.: Evoked Potentials in Clinical Medicine. New York, Raven Press (1983).

CHOI, D. W.: Glutamate neurotoxicity and diseases of the nervous system. Neuron 1 (1988) 623–634.

CHOU, S. M.: Pathognomy of intraneuronal inclusions in ALS. In: T. Tsubuki and Y. Toyokura (Eds), Amyotrophic Lateral Sclerosis, Baltimore, University Park Press (1978) 135–174.

CHRISTENSEN, P. B., E. HOJER-PEDERSEN and N. B. JENSEN: Survival of patients with amyotrophic lateral sclerosis in 2 Danish counties. Neurology 40 (1990) 600–604.

CONRADI, S., L. RONNEVI and O. VESTERBERG: Abnormal tissue distribution of lead in Amyotrophic Lateral Sclerosis. J. Neurol. Sci. 29 (1976) 259–265.

CONRADI, S., L. RONNEVI and O. VESTERBERG: Increased plasma levels of lead in patients with Amyotrophic Lateral Sclerosis compared with control subjects as determined by flameless atomic absorption spectrophotometry. J. Neurol. Neurosurg. Psychiatry 41 (1978) 389–393.

CONRADI, S., L. O. RONNEVI, G. NISE and O. VESTERBERG: Abnormal distribution of lead in Amyotrophic Lateral Sclerosis: Reestimation of lead in the cerebrospinal fluid. J. Neurol. Sci. 48 (1980) 413–418.

CONRADI, S., L. GRIMBY and L. LUNDEMO: Pathophysiology of fasciculations in ALS as studied by electromyography of single motor units. Muscle Nerve 5 (1982) 202–208.

COSI, V., M. POLONI, L. MAZZINI and R. CALLIECO: Somatosensory evoked potentials in Amyotrophic Lateral Sclerosis. J. Neurol. Neurosurg. Psychiatry 47 (1984) 857–861.

CREMER, N. E., L. S. OSHIRO and F. H. NORRIS, JR. and E. H. LENNETTE: Culture of tissues from patients with Amyotrophic Lateral Sclerosis. Arch. Neurol. 29 (1973) 331–333.

CRUVEILHIER J.: Sur la paralysie musculaire progressive atrophique. Arch. Gen. Med. 91 (1853) 561–603.

CUMINGS, J. N.: Biochemical aspects. Proc. R. Soc. Med. 55 (1962) 1023–1024.

CURRIER, R. D. and A. F. HAERER: Amyotrophic Lateral Sclerosis and Metalic Toxins. Arch. Environ. Health 17 (1968) 712–719.

DALAKAS, M. C. and G. H. PEZESHKPOUR: Neuromuscular diseases associated with human immunodeficiency virus infection. Ann. Neurol. 23 (1988) 538–548.

DALAKAS, M. C., J. L. SEVER, M. FLETCHER et al.: Neuromuscular symptoms in patients with old poliomyelitis: clinical, virological and immunological studies. In: L. S. Halstead and D. O. Weichers (Eds), Late Effects of Poliomyelitis. Miami, Symposium Foundation (1985) 73–89.

DALAKAS, M. C., G. ELDER, G. CUNNINGHAM and J. L. SEVER: Morphologic changes in the muscle of patients with post poliomyelitis muscular atropy (PPMA) analysis of 38 biopsies. Neurology 36, Suppl 1, (1986a) 137.

DALAKAS, M. C., G. ELDER, M. HALLET, J. RAVITS, M. BAKER, N. PAPADOPOULOS, P. ALBRECHT and J. SEVER: A long term follow-up study of patients with post polio-

myelitis neuromuscular symptoms. N. Engl. J. Med. 314 (1986b) 959–963.

DAUBE, J. R.: Electrophysiologic studies in the diagnosis and prognosis of motor neuron disease. In: M. J. Aminoff (Ed.), Neurologic Clinics, Symposium of Electrodiagnosis. Philadelphia, W. B. Saunders 3 (1985) 477–493.

DAVIDSON, T. J. and H. A. HARTMAN: Base composition of RNA obtained from motor neurons in Amyotrophic Lateral Sclerosis. J. Neuropathol. Exp. Neurol. 40 (1981) 193–198.

DAVIDSON, T. J., H. A. HARTMAN and P. C. JOHNSON: RNA content and volume in motor neurons in Amyotrophic Lateral Sclerosis. J. Neuropathol. Exp. Neurol. 40 (1981) 32–36; 187–192.

DAWSON, G. and K. STEFANSSON: Gangliosides of human spinal cord: aberrant composition of cords from patients with Amyotrophic Lateral Sclerosis. J. Neurosci. Res. 12 (1984) 213–220.

DEJERINE, J.: Etude anatomique et clinique sur la paralysie labio-glosso-laryngée. Arch. Physiol. Norm. Pathol. 2 (1883) 180–227.

DEJERINE, J. and ANDRÉ-THOMAS: Maladie de la moelle épinière. In: Nouveau traité de médecine et de thérapeutique, Paris. Bailliere et fils (1909) 517–518, 527.

DELISLE, M. B. and S. CARPENTER: Neurofibrillary axonal swellings and amyotrophic lateral sclerosis. J. Neurol. Sci. 63 (1984) 241–250.

DENKER, P. G. and L. SCHEINMAN: Treatment of amyotrophic lateral sclerosis with vitamin E. J. Am. Med. Ass. 116 (1941) 1893–1805.

DENNY-BROWN, D.: Interpretation of the electromyogram. Arch. Neurol. Psychiat. 61 (1949) 99–128.

DENNY-BROWN, D. and J. F. PENNYBACKER: Fibrillation and fasciculation in voluntary muscle. Brain 61 (1938) 311–334.

DENYS, E. H.: Motor nerve conduction velocities. In: D. W. Mulder (Ed.), The Diagnosis and Treatment of Amyotrophic Lateral Sclerosis. Boston, Houghton Mifflin (1980a) 105–118.

DENYS, E. H.: Living without speech. In: D. W. Mulder (Ed.), The Diagnosis and Treatment of Amyotrophic Lateral Sclerosis. Boston, Houghton Mifflin (1980b) 235–240.

DENYS, E. H.: The role of EMG in ALS treatment trials. Committee Report. International ALS–MND Update Aug (1989) 11–12.

DENYS, E. H. and F. H. NORRIS: Amyotrophic Lateral Sclerosis. Impairment of neuromuscular transmission. Arch. Neurol. 36 (1979) 202–205.

DIÓSZEGHY, P., A. ÉGERHÁZI and F. MECHLER: Somatosensory evoked potentials in Amyotrophic Lateral Sclerosis. Electroenceph. clin. Neurophysiol. 27 (1987) 163–167.

DORMAN, J. D., W. K. ENGEL and D. M. FREID: Therapeutic trial in amyotrophic lateral sclerosis. J. Am. Med. Assoc. 209 (1969) 257–258.

DRACHMAN, D. B. and R. W. KUNCL: Amyotrophic Lateral Sclerosis: An unconventional autoimmune disease? Ann. Neurol. 26 (1989) 269–274.

DUBOWITZ, V. and M. H. BROOKE: Muscle Biopsy a Practical Approach, 2nd ed. London, Balliere Tindall (1985) 223–227.

DUCHENNE, G.: Recherche faites à l'aide du galvanisme sur l'état de contractilité et de la sensibilité électromusculaire dans les paralysies des membres supérieurs. Compt. Rend. Acad. Sci. 29 (1847) 667–670.

DUCHENNE, G.: De l'Electrisation Localisée. Paris, Baillière, 1855.

DUCHENNE, G.: Paralysie musculaire progressive de la langue, du voile du palais et des lèvres. Arch. Gén. Méd. 16 (1860) 283–296, 431–445.

DUCHENNE, G. and A. JOFFROY: De l'atrophie aigue et chronique des cellules nerveuses de la moelle et du bulbe rachidien. Propos d'une observation de paralysie labio-glosso-laryngée. Arch. Physiol. 3 (1870) 499.

DUNCAN, M. W., J. C. STEELE, I. J. KOPIN and S. P. MARKEY: 2-Amino-3-(methylamino)-propanoic acid (BMAA) in cycad flour: An unlikely cause of amyotrophic lateral sclerosis and parkinsonism-dementia of Guam. Neurology 40 (1990) 767–772.

ELIZAN T. S., A. H. HIRANO, B. M. ABRAMS, R. L. HEED, C. VAN NUIS and L. T. KURLAND: Amyotrophic Lateral Sclerosis and Parkinsons-Dementia Complex of Guam. Arch. Neurol. (Chic.) 14 (1966a) 356–368.

ELIZAN, T. S., K. M. CHEN, K. U. MATHAL, D. DUNN and L. T. KURLAND: Amyotrophic Lateral Sclerosis and Parkinson Dementia Complex. Arch. Neurol. (Chic.) 14 (1966b) 347–355.

ELLIOTT, J. L. and A. J. HUDSON: Treatment of communication disorders. In: D. W. Mulder (Ed.), The Diagnosis and Treatment of Amyotrophic Lateral Sclerosis. Boston, Houghton Mifflin (1980) 225–234.

EMERY, A. E. H. and S. HOLLOWAY: Familial motor neuron diseases. In: L. P. Rowland (Ed.), Human Motor Neuron Disease. New York, Raven Press (1982) 139–148.

ENGEL, W. K. and M. H. BROOKE: Muscle biopsy in ALS and other motor neuron diseases. In: F. H. Norris, Jr. and L. T. Kurland (Eds), Motor Neuron Diseases. New York, Grune and Stratton (1969) 154–159.

ENGEL, W. K., L. A. HOGENHUIS, W. J. COLLIS, D. S. SCHLACH, M. H. BARLOW, G. N. GOLD and J. D. DORMAN: Metabolic Studies and Therapeutic Trials in Amyotrophic Lateral Sclerosis. In: F. H. Norris, Jr. and L. T. Kurland (Eds), Motor Neuron Diseases. New York, Grune and Stratton (1969) 199–223.

ENGLEHARDT, J. I. and S. H. APPEL: IgG reactivity in the spinal cord and motor cortex in Amyotrophic lateral sclerosis. Arch. Neurol. 47 (1990) 1210–1218.

ENGELHARDT, J. I., S. H. APPEL and J. M. KILLIAN: Experimental autoimmune motor neuron disease. Ann. Neurol. 26 (1989) 368–376.

ERB, W. H.: Über einen wenig bekannten spinalen Symptomen-Complex. Berl. Klin. Wochenschr. 12 (1875) 357–359.

ERB, W. H.: Über die spastische Spinalparalyse und ihre Existenzberechtigung. Neurol. Zbl. 22 (1902) 606.

ERB, W. H.: Concerning spastic and syphilitic spinal paralysis. W. Lond. Med. J. (1903).

ERETEKIN, C.: Sensory and motor conduction in motor neuron disease. Acta. Neurol. Scand. 43 (1967) 499–512.

ERMINIO, F., F. BUCHTHAL and P. ROSENFALCK: Motor unit territory and muscle fiber concentration in paresis due to peripheral nerve injury and anterior horn involvement. Neurology 9 (1959) 657–671.

ESTRIN, W. J.: Amyotrophic Lateral Sclerosis in dizygotic twins. Neurology 26 (1977) 692–694.

EVANS, B. K., I. GORE and L. E. HARRELL, T. ARNOLD and S. JOH: HLTV-1 Associated Myelopathy and Polymyositis in a U.S. Native. Neurology 39 (1989) 1572–1575.

EVANS, B. K., C. FAGAN, T. ARNOLD, E. J. DROPCHOS and S. J. OH: Paraneoplastic motor neuron diseases and renal cell carcinoma: Improvement after neoplasty. Neurology 40 (1990) 960–962.

FARAGO, A. and I. HAUSMANOWA-PETRUSEWICZ: The analysis of EMG findings in Amyotrophic Lateral Sclerosis. Electroenceph. clin. Neurophysiol. 17 (1977) 157–166.

FAREED, G. and H. R. TYLER: The use of isoprinosine in patients with amyotrophic lateral sclerosis. Neurology 21 (1971) 937–940.

FELMUS, M. T., B. M. PATTEN and L. SWANKE: Antecedent events in Amyotrophic Lateral Sclerosis. Neurology 26 (1976) 167–172.

FIGLEWICZ, D. A., D. MCKENNA-YASEK, R. HORVITZ, G. A. ROULEAU and R. H. BROWN: Epidemiological analysis of 90 families with hereditary Amyotrophic Lateral Sclerosis. International ALS–MND Update (1989) 9–10.

FINCHAM, R. W. and M. W. VAN ALLEN: Sensory nerve conduction in Amyotrophic Lateral Sclerosis. Neurology 14 (1964) 31–33.

FISHER, C. M.: Pure spastic paralysis of corticospinal origin. Can. J. Neurol. Sci. 4 (1977) 251–258.

FORSTER, F. M., W. J. BORKOWSKI and B. J. ALPERS: Effects of denervation on fasciculations in human muscle. Arch. Neurol. Psychiat. (Chic.) 56 (1946) 276–283.

FULLMER, H. M., S. and D. SEIDLER, R. S. KROOTH and L. T. KURLAND: A cutaneous disorder of connective tissue in amyotrophic laterial sclerosis. Neurology 10 (1960) 717–724.

GAJDUSEK, D. C.: Cycad toxicity not the cause of high incidence of amyotrophic lateral sclerosis/parkinsonism-dementia on Guam, Kii peninsula of Japan or in West New Guinea. In: A. J. Hudson (Ed.), Amyotrophic Lateral Sclerosis: concepts in pathogenesis and etiology. Toronto, University of Toronto Press (1990) 317–325.

GAJDUSEK, D. C. and A. M. SALAZAR: Amyotrophic Lateral Sclerosis and Parkinsonian syndromes in high incidence among the Auyu and Jakai people of West New Guinea. Neurology 32 (1982) 107–126.

GALLAGHER, J. P. and M. SANDERS: Trauma and amyotrophic lateral sclerosis. A report of 78 patients. Acta Neurol. Scand. 75 (1987) 145–150.

GARLAND, H.: Diabetic Amyotrophy. Br. Med. J. 2 (1955) 1287.

GARRUTO, R. M. and Y. YASE: Neurodegenerative disorders of the western Pacific: The search for mechanism of pathogenesis. TINS (1986) 368–374.

GIBBS, C. J. and D. C. GAJDUSEK: Amyotrophic Lateral Sclerosis, Parkinson's Disease and the Amyotrophic Lateral Sclerosis — Parkinsonism — dementia complex in Guam. J. Clin. Pathol. 25 (Suppl.) 6 (1972) 132–140.

GILLER, E. L., J. H. NEALE, P. N. BULLOCK, B. K. SCHRIER and P. G. NELSON: Choline acetyltransferase activity of spinal cord cultures increased by co-culture with muscle and by muscle conditioned medium. J. Cell Biol. 74 (1977) 16–29.

GILLIATT, R. W., H. V. GOODMAN and R. G. WILLISON: The recording of lateral popliteal nerve action potentials in man. J. Neurol. Neurosurg. Psychiatry 24 (1961) 305–318.

GOLDBLATT, D.: Motor Neuron Disease: Historical introduction. In: F. H. Norris, Jr. and L. T. Kurland (Eds), Motor Neuron Diseases. New York, Grune and Stratton (1969) 3–11.

GOULD, B. S.: Psychiatric aspects. In: D. W. Mulder (Ed.), The Diagnosis and Treatment of Amyotrophic Lateral Sclerosis. Boston, Houghton Mifflin (1980) 157–168.

GOULON, M. and C. GOULON-GOËAU: Ventilatory assistance in Amyotrophic Lateral Sclerosis. Rev. Neurol. (Paris) 145 (1989) 293–298.

GREENFIELD, J. G. and W. B. MATTHEWS: Post encephalitic Parkinsonism with amyotrophy. J. Neurol. Neurosurg. Psychiatry 17 (1954) 50–56.

GREENFIELD, M. M. and F. M. STARK: Post-irradiation neuropathy. Am. J. Roentgenol. 60 (1948) 617–622.

GUDESBLATT, M., M. D. LUDMAN, J. A. COHEN, R. J. DESNICK, S. CHESTER, G. A. GRABOWSKI and J. T. CAROSCIO: Hexoseaminidase A activity and Amyotrophic Lateral Sclerosis. Muscle Nerve 11 (1988) 227–230.

GUDMUNDSSON, K. R.: The prevalence of some neurologic diseases in Iceland. Acta Neurol. Scand. 44 (1968) 57–69.

GUILOFF, R. J., B. MCGREGOR, W. BLACKWOOD and E. PARRE: Motor neuron disease with elevated cerebrospinal fluid protein. J. Neurol. Neurosurg. Psychiatry 43 (1980) 390–396.

GUNNARSSON, L. G. and R. PALM: Motor neuron disease and heavy manual labor: An epidemiological survey of Varnland County, Sweden. Neuroepidemiology 3 (1985) 195–205.

GUNNARSSON, L. G., G. LINDBERG, B. SODERFELT and D. AXELSON: The mortality of motor neuron disease in Sweden. Arch. Neurol. 47 (1990) 42–46.

GURNEY, M. E., A. C. BELTON, N. CASHMAN and J. P. ANTEL: Sera from Amyotrophic Lateral Sclerosis patient block terminal sprouting at the neuromuscular function. Neurology 33 (Suppl.) (1983) 155–156.

GURNEY, M. E., A. C. BELTON, N. CASHMAN and J. P. ANTEL: Inhibition of terminal axonal sprouting by serum

from patients with Amyotrophic Lateral Sclerosis. N. Engl. J. Med. 311 (1984) 933–939.

HAMMOND, W. A.: A Treatise on the Diseases of the Nervous System. New York, Appleton Co. (1876).

HANSEN, S. and J. P. BALLANTYNE: A quantitative electrophysiological study of motor neurone disease. J. Neurol. Neurosurg. Psychiatry 41 (1978) 773–783.

HARTER, D. H.: Viruses other than poliovirus in human Amyotrophic Lateral Sclerosis. In: L. P. Rowland (Ed.), Human Motor Neuron Disease. New York, Raven Press (1982) 339–342.

HAUSMANOWA-PETRUSEWICZ, I. and J. KOPEC: Motor nerve conduction velocity in anterior horn lesions. Electromyography 10 (1970) 227–237.

HAUSMANOWA-PETRUSEWICZ, I. and J. KOPEC: Motor nerve conduction velocity in anterior horn lesions. In: J. E. Desmedt (Ed.), New Developments in Electromyography and Clinical Neurophysiology. Basel, Karger (1973) 298–305.

HAYEM: Note sur un cas d'atrophie musculaire progressive avec lésions de la moelle. Arch. Physiol. 2 (1869) 263, 391.

HIRANO, A. and M. IWATA: Pathology of motor neurons with specific reference to Amyotrophic Lateral Sclerosis and related diseases. In: T. Tsubaki and Y. Toyokura (Eds), Amyotrophic Lateral Sclerosis. Baltimore, University Park Press (1978) 107–133.

HIRANO, A., L. T. KURLAND and G. P. SAYRE: Familial Amyotrophic Lateral Sclerosis. Arch. Neurol. 16 (1967) 232–243.

HIRANO, A., N. MALAMUD, L. KURLAND and H. M. ZIMMERMAN: A review of the pathologic findings in Amyotrophic Lateral Sclerosis. In: F. H. Norris, Jr. and L. T. Kurland (Eds), Motor Neuron Diseases. New York, Grune and Stratton (1969) 51–60.

HJORTH, R. J., J. C. WALSH and R. G. WILLISON: The distribution of frequency of spontaneous fasciculations in motor neuron disease. J. Neurol. Sci. 18 (1973) 469–474.

HOFFMAN, P. M., B. W. FESTOFF, L. C. GIRON, L. S. HOLLENBACK, R. M. GARRUTO and F. W. RUSCETTI: Isolation of LAV/HTLV 3 from a patient with Amyotrophic Lateral Sclerosis. N. Engl. J. Med. 313 (1985) 324–325.

HOOPER, P. T.: Cycad poisoning in Australia-etiology and pathology. In: R. F. Keeler, K. R. Van Kampen and L. F. James (Eds), Effects of Poisonous Plants on Livestock. New York, Academic Press (1978) 337–347.

HORWICH, M. S., W. K. ENGEL and P. B. CHAUVIN: Amyotrophic lateral sclerosis sera applied to cultured motor neurons. Arch. Neurol. (Chic.) 30 (1974) 332–333.

HOUP, J. L., B. S. GOULD and F. H. NORRIS: Psychological characteristics of patients with Amyotrophic Lateral Sclerosis. Psychosom. Med. 39 (1977) 299–303.

HOUSE, A. O., R. J. ABBOTT, D. L. DAVIDSON, I. T. FERGUSON

and J. A. LENNON: Response to penicillamine of lead concentrations in CSF and blood of patients with motor neuron disease. Br. Med. J. 2 (1978) 1684.

HUDSON, A. J., W. F. BROWN and J. GILBERT: The muscle pain fasciculation syndrome. Neurology 28 (1978) 1105–1109.

HUFSCHMIDT, H. J., G. SCHALTENBRAND and H. SOLCHER: Über Muskelatrophien im Zusammenhang mit postencephalischem Parkinsonismus. Dt. Z. Nervenheilk. 181 (1960) 335–344.

HUGON, J., M. LUBEAU, F. TABARAUD, F. CHAZOT, J. M. VALLAT and M. DUMAS: Central motor conduction in motor neuron disease. Ann. Neurol. 22 (1987) 544–546.

IKUTA, F., T. MAKIFUCHI and T. ICHIKAWA: Comparative studies of tract degeneration in Amyotrophic Lateral Sclerosis and other disorders. In: T. Tsubaki and Y. Toyokura (Eds), Amyotrophic Lateral Sclerosis. Baltimore, University Park Press (1978) 177–200.

IONASESCU, V. and N. LUCAS: Studies in carbohydrate metabolism in Amyotrophic Lateral Sclerosis and heredity proximal spinal muscle atrophy. Acta Neurol. Scand. 40 (1964) 47–57.

ISAACS, H.: A syndrome of continuous muscle fiber activity. J. Neurol. Neurosurg. Psychiatry 24 (1961) 319–325.

IWASAKI, Y., M. KINOSHITA, K. IKEDA, K. TAKAMIYA and T. SHIOJIMA: Cognitive impairment in amyotrophic lateral sclerosis and its relation to motor disabilities. Acta Neurol. Scand 81 (1990) 141–143.

JABLECKI, C. K., C. BERRY and J. LEACH: Survival prediction in Amyotrophic Lateral Sclerosis. Muscle Nerve 12 (1989) 883–841.

JELLINGER, K., A. P. ANZIL, D. SEEMANN and H. BERNHEIMER: Adult GM_2 gangliosidoses masquerading as slowly progressive muscle atrophy: motor neuron disease phenotype. Clin. Neuropathol. 1 (1982) 31–44.

JOHNSON, R. T.: Virological studies on amyotrophic lateral sclerosis: an overview. In: J. M. Andrews, R. T. Johnson and M. A. Brazier (Eds), Amyotrophic Lateral Sclerosis. New York, Academic Press (1976) 173–178.

JOHNSON, W. G.: Hexoseaminadase deficiency: a cause of recessively inherited motor neuron disease, In: L. P. Rowland (Ed.), Human Motor Neuron Diseases. New York, Raven Press (1982) 159–164.

JOHNSON, J. W. and P. ASCHER: Glycine potentiates the N-MDA receptor in cultured mouse brain neurons. Nature 325 (1987) 529–531.

JOKELAINEN, M.: Amyotrophic Lateral Sclerosis in Finland. I: An epidemiologic study. Acta Neurol. Scand. 56 (1977) 185–192.

JOKELAINEN, M., A. TILIKAINEN and K. LAPINLEIMU: Polio antibodies and HLA antigens in Amyotrophic Lateral Sclerosis. Tissue Antigens 10 (1977) 259–266.

JONAS, S., M. WICHTER and L. SPITLER: Amyotrophic Lateral Sclerosis: Failure of Transfer Factor Therapy. Ann. Neurol. 6 (1979) 84.

JUERGENS, S. M. and L. T. KURLAND: Epidemiology. In: D. M. Mulder (Ed.), The Diagnosis and Treatment of Amyotrophic Lateral Sclerosis. Boston, MA Houghton Mifflin (1980) 35–46.

KANTARJIAN, A. D.: A Syndrome clinically resembling Amyotrophic Lateral Sclerosis following chronic mercurialism. Neurology 11 (1961) 639–644.

KARPATI, G., S. CARPENTER and H. DURHAM: A hypothesis for the pathogenesis of Amyotrophic Lateral Sclerosis. Rev. Neurol. (Paris) 144 (1988) 672–675.

KATO, T., T. KATAGIRI, A. HIRANO, T. KAWANAMI and H. SASAKI: Lewy body-like hyaline inclusions in sporadic motor neuron disease are ubiquitinated. Acta Neuropathol. 77 (1989) 391–396.

KAWAMURA, Y., P. J. DYCK, J. SHIMONO, H. OKAZAKI, J. TATEISHI and H. DOI: Morphometric comparison of the vulnerability of peripheral motor and sensory neurons in Amyotrophic Lateral Sclerosis. J. Neuropathol. Exp. Neurol. 40 (1981) 667–675.

KILNESS, A. W. and F. H. HICHBERG: Amyotrophic Lateral Sclerosis in a high selenium environment. J. Am. Med. Ass. (1977) 2843–2844.

KOERNER, D. R.: Amyotrophic lateral sclerosis in Guam. Ann. Intern. Med. 37 (1952) 1204.

KOHLER, J., U. KERN, J. KASPER, B. RHESE-KUPPER and U. THODEN: Chronic central nervous system involvement in Lyme borreliosis. Neurology 38 (1988) 863–867.

KOHNE, D. E., C. J. GIBBS, L. WHITE, S. M. TRACY, W. MEINKE and R. A. SMITH: Virus detection by nucleic acid hybridization: Examination of normal and ALS tissues for the presence of poliovirus. J. Gen. Virol. 56 (1981) 223–233.

KONDO, K: Clinical variability of motor neuron disease. Neurol. Med. 2 (1975) 11–16.

KONDO, K: Epidemiology of motor neuron disease: ageing and exhaustion hypotheses revisited. In: F. ŁC. Rose (Ed.), Research Progress in Motor Neuron Diseases. Bath, Pittman (1984) 20–23.

KONDO, K. and T. TSUBAKI: Case-control studies of motor neuron disease, association with mechanical injuries. Arch. Neurol. (Chic.) 38 (1981) 220–226.

KOTT, E., E. LIVNI, R. ZAMIR and A. KURITSKY: Cell mediated immunity to polio and HLA antigens in Amyotrophic Lateral Sclerosis. Neurology 29 (1979) 1040–1044.

KOZLOWSKI, M. A., C. L. WILLIAMS, D. R. HINTON and C. A. MILLER: Heterotopic neurons in ALS spinal cord. Neurology 39 (1989) 644–648.

KREIGER, C. and C. MELMED: Amyotrophic Lateral Sclerosis and paraproteinemia. Neurology 32 (1982) 896–898.

KREITZER, S. M., N. A. SAUNDERS, H. R. TYLER and R. H. INGRAM, JR: Respiratory muscle function in Amyotrophic Lateral Sclerosis. Am. Rev. Res. Dis. 117 (1978) 437–447.

KRISTENSEN, O., B. MELGAARD and A. V. SCHIODT: Radiation myelopathy of the lumbosacral spinal cord. Acta Neurol. Scand. 56 (1977) 217–222.

KUNCL, R. W., D. R. CORNBLATH and J. W. GRIFFIN: Assessment of thoracic paraspinal muscles in the diagnosis of Amyotrophic Lateral Sclerosis. Muscle Nerve 11 (1988) 484–492.

KURENT, J. E.,   B. R. BROOKS,   D. L. MALDEN,   J. L. SEVER and W. K. ENGEL: CSF viral antibodies: evaluation in Amyotrophic Lateral Sclerosis and late onset post poliomyelitis progressive muscular atrophy. Arch. Neurol. (Chic.) 36 (1979) 269–273.

KURLAND, L. T. and D. W. MULDER: Epidemiologic investigation of Amyotrophic Lateral Sclerosis. Neurology 5 (1955) 182–196, 249–268.

KURLAND, L. T., N. W. CHOI and G. P. SAYRE: Implications of incidence and geographic patterns in the classification of Amyotrophic Lateral Sclerosis. In: F. H. Norris, Jr. and L. T. Kurland (Eds), Motor Neuron Diseases. New York, Grune and Stratton (1969) 28–50.

KURLAND, L. T., J. F. KURTZKE and I. D. GOLDBERG: In: Epidemiology of neurologic disorders. Vital and Health Statistics Monographs. American Public Health Association. Harvard University Press (1973a) 108.

KURLAND, L. T.,   J. F. KURTZKE,   I. D. GOLDBERG   and N. W. CHOI: Amyotrophic Lateral Sclerosis and other motor neuron diseases. In: L. T. Kurland, J. F. Kurtzke and I. D. Goldberg (Eds), Epidemiology of Neurologic and Sense Organ Disorders. Cambridge, MA, Harvard University Press (1973b) 108–127, 350–354.

KURTZKE, J. F.: Epidemiology of Amyotrophic Lateral Sclerosis. In: L. P. Rowland (Ed.), Human Motor Neuron Diseases. New York, Raven Press (1982) 281–302.

KUSHNER, M. J.,   M. PARRISH,   A. BURKE,   M. BEHRENS, A. P. HAYS, B. FRAME and L. P. ROWLAND: Nystagmus in motor neuron disease: Clinicopathologic study of two cases. Ann. Neurol. 16 (1984) 71–77.

LAMBERT, E. H.: Diagnostic value of electrical stimulation of motor nerves. Electroenceph. Clin. Neurophysiol. Suppl. 22 (1962) 9–16.

LAMBERT, E. H.: Electromyography in Amyotrophic Lateral Sclerosis. In: F. H. Norris, Jr. and L. T. Kurland (Eds), Motor Neuron Diseases. New York, Grune and Stratton (1969) 135–153.

LATERRE, E. C.: Cerebrospinal fluid. In: P. J. Vinken and G. W. Bruyn (Eds), Handbook of Clinical Neurology, Vol. 19. Amsterdam, North Holland Publ. Co. 1975, 125–138.

LATOV, N.: Plasma cell dyscrasia and Motor Neuron Disease. In: L. P. Rowland (Ed.), Human Motor Neuron Disease. New York, Raven Press (1982) 273–279.

LATOV, N., A. P. HAYS, P. D. DONOFRIO et al.: Monoclonal IgM with unique specificity to gangliosides GM_1 and GD_{1b} and to lacto-N-tetraose associated with human motor neuron disease. Neurology 38 (1988) 763–768.

LAYZER, R. B.: Diagnostic implications of clinical fasciculation and changes. In: L. P. Rowland (Ed.), Human Motor Neuron Disease. New York, Raven Press (1982) 23–30.

LEGRAND, R.,   M. LINQUETTE,   J. DELAHOUSSE   and

A. GÉRARD: A propos d'un nouveau cas d'association d'une maladie de Parkinson et d'une sclérose latérale amyotrophique. Rev. Neurol. (Paris) 101 (1959) 191–193.

LEHRICH, J. R., J. OGER and B. G. W. ARNASON: Neutralizing antibodies to poliovirus and mumps virus in Amyotrophic Lateral Sclerosis. J. Neurol. Sci. 23 (1974) 537–540.

LEPAGE, M., M. FROISSART and P. GALIBERT: Poliomyélite antérieure subaigüe. Rev. Neurol. (Paris) 131 (1975) 721–724.

LEYDEN, E.: Ueber progressive Bulbar-paralysie. Arch. Psychiatr. ii (1870) 648, 657, iii 338.

LIVERSEDGE, L. A.,   W. R. SWINBURNE   and   G. M. QUILL: Idoxuridine and motor neuron disease. Br. Med. J. (1970) 745–750.

LIVESLEY, B. and C. E. SISSON: Chronic lead intoxication mimicking motor neuron disease. Br. Med. J. 4 (1968) 387–388.

LOGIGIAN, E. L., R. F. KAPLAN and A. C. STEERE: Chronic neurologic manifestations of Lyme Disease. N. Engl. J. Med. 323 (1990) 1438–1444.

LOIZOU, L. A., M. SMALL and G. A. DALON: Cricopharyngeal myotomy in motor neuron disease. J. Neurol. Neurosurg. Psychiatry 43 (1980) 42–45.

LUYS, J.: Atrophie musculaire progressive. Gaz. Méd. Fr. 3/4 (1860) 505.

MACKAY, R. P.: Course and prognosis in Amyotrophic Lateral Sclerosis. Arch. Neurol. 8 (1963) 117–127.

MAIER, J. G., R. H. PERRY, W. SAYLOR and M. H. SULAK: Radiation myelitis of the dorsolumbar spinal cord. Radiology 93 (1969) 153–160.

MANN, D. M. A. and P. O. YATES: Motor Neuron Disease: The nature of the pathogenic mechanism. J. Neurol. Neurosurg. Psychiatry 37 (1974) 1036–1046.

MANNEN, T.,   M. IWATA,   Y. TOYOKURA and K. NAGASHIMA: Pathology of anterior horn of the sacral cord in cases of Amyotrophic Lateral Sclerosis and its clinical significance. Neurol. Med. 3 (1975) 169–175.

MANNEN, T., M. IWATA, Y. TOYOKURA and K. NAGASHIMA: Preservation of a certain motoneuron group of the sacral cord in amyotrophic lateral sclerosis: Its clinical significance. J. Neurol. Neurosurg. Psychiatry 40 (1977) 464–469.

MANNEN, T., M. IWATA, Y. TOYOKURA and K. NAGASHIMA: The Onuf's nucleus and the external anal sphincter muscles in amyotrophic lateral sclerosis and the Shy-Drager Syndrome. Arch. Neuropathol. 58 (1982) 255–260.

MATHESON, J. K., H. J. HARRINGTON and M. HALLETT: Abnormalities of multimodality evoked potentials in Amyotrophic Lateral Sclerosis. Arch. Neurol. (Chic.) 43 (1986) 338–340.

MATSUMOTO, N.,   R. M. WORTH,   L. T. KURLAND   and H. OKAZAKI: Epidemiologic study of Amyotrophic Lateral Sclerosis in Hawaii, identification of high incidence among Filipino men. Neurology 22 (1972) 934–940.

MCCOMAS, A. J.,   P. FAWCETT,   M. J. CAMPBELL   and

R.E.P.SICA: Electrophysiological estimation of the number of motor units within a human muscle. J. Neurol. Neurosurg. Psychiatry 34 (1971a) 121–131.

MCCOMAS, A.J., R.E.P.SICA, M.J.CAMPBELL and A.R.M. UPTON: Functional compensation in partially denervated muscles. J. Neurol. Neurosurg. Psychiatry 34 (1971b) 453–460.

MERRITT, H.H. and F.FREMONT-SMITH: The Cerebrospinal Fluid. Philadelphia, W. B. Saunders (1938).

MERRITT, H.H., R.D.ADAMS and H.C.SOLOMON: Neurosyphilis. New York, Oxford University Press (1946).

MEYER, A.: Über eine der amyotrophischen Lateralsklerose nahestehende Erkrankung mit psychischen Störungen. Z. Ges. Neurol. Psychiat. 121 (1929) 107–138.

MICHAUX, L., M.SAMSON, J.M.HARP and J.GRONER: Évolution démentielle de cas de sclérose latérale amyotrophique accompagnés d'aphasie. Rev. Neurol. 92 (1955) 357–367.

MILLER, J.R., R.V.GUNTAKA and J.C.MYERS: Amyotrophic Lateral Sclerosis search for poliovirus by nucleic acid hybridization. Neurology 30 (1980) 884–886.

MILLS, K.R., N.M.F.MURRAY and C.W.HESS: Magnetic and electrical transcranial brain stimulation: Physiological mechanisms and clinical applications. Neurosurgery 20 (1987) 164–168.

MITCHELL, J.D.: Heavy metals and trace elements in amyotrophic lateral sclerosis. Neurol. Clin. 5 (1987) 43–60.

MITCHELL, J.D.: Trace element, free radical and Amyotrophic Lateral Sclerosis. International ALS-MND Update April (1989) 14–15.

MITCHELL, J.D., I.A.HARRIS, B.W.EAST and B.PENTLAND: Trace elements in cerebrospinal fluid in motor neuron disease. Br. Med. J. 288 (1984) 1791–1792.

MITSUMOTO, H., R.J.SLIMAN, I.A.SCHAFER, C.S.STERNICK, B.KAUFMAN, A.WILHOURN and S.J.HORWITZ: Motor neuron disease and adult hexoseaminidase A deficiency in two families: evidence for multisystem degeneration. Ann. Neurol. 17 (1985) 378–385.

MITSUYAMA, K.: Presenile dementia with motor neuron disease in Japan. Clinco-pathological review of 26 cases. J. Neurol. Psychiatry 47 (1984) 953–959.

MORITA, K., H.KAIYA, T.IKEDA and M.NAMBA: Presenile dementia combined with amyotrophic lateral sclerosis: A review of 34 Japanese cases. Gerontol. Geriatr. 6 (1987) 263–277.

MULDER, D.W. and R.E.ESPINOSA: Amyotrophic Lateral Sclerosis: Comparison of the Clinical Syndrome in Guam and the United States. In: F.H. Norris Jr. and L.T. Kurland (Eds), Motor Neuron Diseases. New York, Grune and Stratton (1969) 12–19.

MULDER, D.W. and R.M.HOWARD: Patient resistance and Prognosis in Amyotrophic Lateral Sclerosis. Mayo Clin. Proc. 51 (1976) 537–541.

MULDER, D.W., R.A.ROSENBAUM and D.D.LAYTON, JR: Late Progression of poliomyelitis of forme fruste Amyotrophic Lateral Sclerosis. Mayo clin. Proc. 47 (1972) 756–761.

MULDER, D.W., W.BUSHEK, E.SPRING, J.KARNES and P.J.DYKE: Evaluation of detection of thresholds of cutaneous sensation. Neurology 33 (1983) 1625–1627.

MULDER, D.W., L.T.KURLAND, K.P.OFFORD and M.BEARD: Familial adult motor neuron diseases: Amyotrophic Lateral Sclerosis. Neurology 36 (1986) 511–517.

MULLER, W.K. and F.HILGENSTOCK: An uncommon case of Amyotrophic Lateral Sclerosis with isolation of a virus from the CSF. J. Neurol. 211. (1975) 11–23.

MULLER, W.K. and G.SCHALTENBRAND: Attempts to reproduce Amyotrophic Lateral Sclerosis in Laboratory animals by inoculation of Schu virus isolated from a patient with apparent Amyotrophic Lateral Sclerosis. J. Neurol. 22 (1979) 1–19.

MURAYAMA, S., H.MORI, Y.IHARA, T.W.BOULDIN, K.SUZAKI and M.TOMONAGA: Immunocytochemical and ultrastructural studies of lower motor neurons in Amyotrophic Lateral Sclerosis. Ann. Neurol. 27 (1990) 137–148.

NARDELLI, E., A.T.STECK, M.SCHLUEP, K.FELGENHAUER and F.S.JERUSALEM: Neuropathy and monoclonal IgM M-protein with antibody activity against gangliosides (abstract). J. Neuroimmunol. 16 (1987) 131.

NEARY, D., J.S.SNOWDEN, D.M.A.MANN, B.NATHAN, P.J.GOULDING and N.MACDERMOTT: Frontal lobe dementia and motor neuron diseases. J. Neurol. Neurosurg. Psychiatry 53 (1990) 23–32.

NOBILE-ORAZIO, E., G.LEGNAME, R.DAVERIO, M.CARPO, A.GUILIANI, S.SONNINO and G.SCARLATO: Motor neuron disease in a patient with monoclonal IgMk directed against GM_1, GD_{1b}, and high-molecular weight neural specific glycoproteins. Ann. Neurol. 28 (1990a) 190–194.

NOBILE-ORAZIO, E., M.CARPO, G.LEGNAME, N.MEUCCI, S.SONNINO and G.SCARLATO: Anti-IgM antibodies in motor neuron disease and neuropathy. Neurology 40 (1990b) 1747–1750.

NORRIS JR, F.H.: Synchronous fasciculation in motor neuron disease. Arch. Neurol. 13 (1965) 495–500.

NORRIS JR, F.H.: Guanidine in ALS. N. Engl. J. Med. 288 (1973) 690–691.

NORRIS JR, F.H.: Adult spinal motor neuron disease. In: P.J. Vinken and G.W. Bruyn (Eds), Clinical Neurology. Amsterdam, North Holland Publishing Co. (1975) 1–56.

NORRIS JR, F.H. and W.K.ENGEL: Carcinomatous Amyotrophic Lateral Sclerosis. In: Brain, the Lord, and F.H. Norris Jr. (Eds), The Remote Effects of Cancer on the Nervous System. New York, Grune and Stratton (1965) 24–41.

NORRIS JR, F.H., W.H.MCMENEMEY and R.O.BARNARD: Anterior Horn Pathology in Carcinomatous Neuromyopathy compard with other forms of Motor

Neuron Disease. In: F.H. Norris Jr. and L.T. Kurland (Eds), Motor Neuron Diseases. New York, Grune and Stratton (1969) 100–111.

NORRIS, F. H., P. R. CALANCHINI, R. J. FALLAT et al.: The administration of guanidine in Amyotrophic Lateral Sclerosis. Neurology 24 (1971) 721–728.

NORRIS, F. H., M. J. AGUILAR, R. P. COLTON, M. B. A. OLDSTONE and N. E. CREMER: Tubular particles in a case of recurrent lymphocytic meningitis followed by Amyotrophic Lateral Sclerosis. J. Neuropathol. Exp. Neurol. 34 (1975) 133–147.

NORRIS JR, F. H., B. SACHNUS and M. CAREY: Trial of Baclofen in Amyotrophic Lateral Sclerosis. Arch. Neurol. 36 (1979) 715–716.

NORRIS, JR, F. H., E. H. DENYS, I. IKS and E. MUKAI: Populations study of Amyotrophic Lateral Sclerosis — 4 California clusters. International ALS-MND Update (1989) 12–13.

OLARTE, M. R.: Therapeutic Trials in Amyotrophic Lateral Sclerosis. In: L. P. Rowland (Ed.), Human Motor Neuron Disease. New York, Raven Press (1982) 555–558.

OLARTE, M. R., J. C. GERSTEN, J. ZABRISKIE and L. P. ROWLAND: Transfer Factor is Ineffective in Amyotrophic Lateral Sclerosis. Ann. Neurol. 5 (1979) 385–388.

OLDSTONE, M. B. A., C. B. WILSON, L. H. PERRIN and F. H. NORRIS: Evidence for immune-complex formation in patients with Amyotrophic Lateral Sclerosis. Lancet 2 (1976) 169–172.

OLSON, W. H., J. A. SIMONS and G. W. HALAAS: Therapeutic Trial of Tilerone in ALS: lack of benefit in a double blind, placebo controlled study. Neurology 28 (1978) 1293–1298.

OSHIRO, L. S., N. E. CREMER, F. H. NORRIS, JR. and E. H. LENNETTE: Viruslike particles in muscle from a patient with Amyotrophic Lateral Sclerosis. Neurology 26 (1976) 57–60.

PALAHAS, M. C., J. L. SEVER, D. L. MADDEN et al.: Neuromuscular symptoms on patients with old poliomyelitis: Clinical virological and immunological studies. In: L. S. Halstead and D. O. Weichers (Eds), Late Effects of Poliomyelitis. Miami, Symposium Foundation (1985) 73–89.

PALO, J., A. RISSANEN, E. JOKINEN, J. LAHDEVIRTA and O. SALA: Kidney and skin biopsy in Amyotrophic Lateral Sclerosis. Lancet 1 (1978) 1270.

PALMUCCI, L., A. BERTOLOTTO, C. DORIGUZZI, T. MONGINI and D. SCHIFFER: Motor neuron disease following poliomyelitis. Eur. Neurol. 19 (1980) 414–418.

PATRIKIOS, J.: Sclerose Laterale Amyotrophique avec mouvements involontaires des doigts et du poignet gauches de caractere extrapyramidal. Rev. Neurol. 85 (1951) 60–62.

PATTEN, B. M.: ALS associated with aluminium deposition in CNA. International ALS-MND Update April (1989) 17.

PATTEN, B. M. and W. K. ENGEL: Phosphate and parathyroid disorder associated with the syndrome of Amyotrophic Lateral Sclerosis. In: L. P. Rowland (Ed.), Human Motor Neuron Diseases. New York, Raven Press (1982) 181-200.

PATTEN, B. M., T. HARATI, L. ACOSTA et al.: Free amino acid levels in amyotrophic lateral sclerosis. Ann. Neurol. 3 (1978) 305–309.

PEDERSEN, L., P. PLATZ, C. JERSILD and M. THOMSEN: HLA in patients with Amyotrophic Lateral Sclerosis. J. Neurol. Sci. 31 (1977) 313–318.

PENA, C. E.: Virus-like particles in Amyotrophic Lateral Sclerosis: Electron microscopical study of a case. Ann. Neurol. 1 (1977) 290–297.

PERCY, A. K., L. E. DAVIS, D. M. JOHNSTON and D. B. DRACHMAN: Failure of Isoprinosine in Amyotrophic Lateral Sclerosis. N. Engl. J. Med. 285 (1971) 689.

PERL, D. P., D. C. GAJDUSEK, R. M. GARRUTO, R. T. YANAGIHARA and C. J. GIBBS: Intraneuronal alumium accumulation in Amyotrophic Lateral Sclerosis and Parkinsonism-dementia of Guam. Science 217 (1982) 1053–1055.

PERRY, T. L., C. KREIGER, S. HANSEN and A. EISEN: Amyotrophic Lateral Sclerosis: Amino acid levels and plasma cerebrospinal fluid. Ann. Neurol. 28 (1990) 12–17.

PERTSCHUK, L. P., A. W. COOK, J. K. GUPTA, J. D. BROOME, D. J. BRIGATI, J. C. VALENTIN, E. A. RAINFORD, D. S. KIM and F. NIDSGORSKI: Jejuenal immunopathology in Amyotrophic Lateral Sclerosis and multiple sclerosis. Lancet 1 (1977) 1119–1123.

PERTSCHUK, L. P., D. S. KIM, I. PRASAD, A. W. COOK, J. K. GUPTA and J. D. BROOME: Jejunal mucosa in motor neuron disease and other chronic neurologic disorders. In: P. O. Behan and F. C. Rose (Eds), Progress in Neurologic Research. Pitman Medical (1979) 44–61.

PESTRONK, A., R. N. ADAMS, D. CORNBLATH, R. W. KUNCL and D. DRACHMAN: Patterns of serum Igm antibodies to GM_1 and GD_{1a} ganglioside in Amyotrophic Lateral Sclerosis. Ann. Neurol. 25 (1989) 98–102.

PESTRONK, A., V. CHAUDHRY, E. L. FELDMAN, J. W. GRIFFIN, D. R. CORNBLATH, E. H. DENYS, M. GLASBERG, R. W. KUNCL, R. K. OLNEY and W. C. YEE: Lower motor neuron syndromes defined by patterns of weakness, nerve conduction abnormalities, and high titers on antiglycolipid antibodies. Ann. Neurol. 27 (1990) 316–326.

PETAJAN, J. H.: F-waves in neurogenic atrophy. Muscle Nerve 8 (1985) 690–696.

PETKAU, A., A. SAWATZKY, C. R. HILLIER et al.: Lead content of neuromuscular tissue in Amyotrophic Lateral Sclerosis: Case report and other considerations. Br. J. Ind. Med. 31 (1974) 275–287.

PIEPER, S. J. and W. S. FIELDS: Failure of Amyotrophic Lateral Sclerosis to respond to intrathecal steroids and vitamin B_{12} therapy. Neurology 9 (1959) 522–526.

PLAITAKIS, A.: Glutamate dysfunction and selective motor neuron degeneration in Amyotrophic Lateral Sclerosis: a hypothesis. Ann. Neurol. 18 (1990) 3–8.

PLAITAKIS, A. and J. J. CAROSCIO: Abnormal glutamate metabolism in ALS. Ann. Neurol. 18 (1985) 16T.

PLAITAKIS, A. and J. J. CAROSCIO: Abnormal glutamate

metabolism in Amyotrophic Lateral Sclerosis. Ann. Neurol. 22 (1987) 575–579.

PLAITAKIS., E. CONSTANTAKAKIS and J. SMITH: The neuroexcttoxic amino acids, glutamate and aspartate are altered in the spinal cord and brain in Amyotrophic Lateral Sclerosis. Ann. Neurol. 24 (1988) 446–449.

POSER, C. M. and L. D. BUNCH: Serum amino acid studies in Amyotrophic Lateral Sclerosis. Arch. Neurol. 14 (1966) 305–312.

POSER, C. M., M. JOHNSON and L. D. BUNCH: Serum amino acid studies in Amyotrophic Lateral Sclerosis. Arch. Neurol. 12 (1965) 604–609.

POSKANZER, D. C., H. M. CANTOR and G. S. KAPLAN: The frequency of preceeding poliomyelitis in Amyotrophic Lateral Sclerosis. In: F. H. Norris Jr. and L. Kurland (Eds), Motor Neuron Diseases. New York, Grune and Stratton (1969) 286–290.

QUICK, D. T.: Pancreatic dysfunction in Amyotrophic Lateral Sclerosis. In: F. H. Norris, Jr. and L. T. Kurland (Eds), Motor Neuron Diseases. New York, Grune and Stratton (1969) 189–198.

QUICK, D. T. and M. GREER: Pancreatic dysfunction in patients with Amyotrophic Lateral Sclerosis. Neurology 17 (1967) 112–116.

RADTKE, R. A., A. ERWIN and C. W. ERWIN: Abnormal sensory evoked potentials in Amyotrophic Lateral Sclerosis. Neurology 36 (1986) 796–801.

RAPPORT, M. M., H. DONNENFELD, W. BRUNNER, B. HUNGUND and H. BARTFELD: Ganglioside patterns in Amyotrophic Lateral Sclerosis brain regions. Ann. Neurol. 18 (1985) 60–67.

RISSANEN, A., J. PALO, G. MYLLYLA and K. CANTELL: Interferon therapy for ALS. Ann. Neurol. 7 (1980) 392.

ROISEN, F. J., H. BARTFELD, H. DONNENFELD and J. BAXTER: Neuron specific in vitro cytotoxicity of sera from patients with Amyotrophic Lateral Sclerosis. Muscle Nerve 5 (1982) 48–53.

ROOS, R. P., M. V. VIOLA, R. WOOLMAN, M. H. HATCH and J. P. ANTEL: Amyotrophic Lateral Sclerosis with antecedent poliomyelitis. Arch. Neurol. 37 (1980) 312–313.

ROSATI, G., L. PINNA, E. GRANIERI, I. AIELLO, R. TOLA, V. AGNETTI, A. PIRISI and P. BASTIANI: Studies on epidemiological, clinical and etiological aspects of ALS disease in Sardinia, Southern Italy. Acta Neurol. Scand. 55 (1977) 231–234.

ROSE, F. C.: Clinical aspects of motor neuron diesease. In: F. C. Rose (Ed.), Motor Neuron Disease. New York, Grune and Stratton (1977) 1–13.

ROTHSTEIN, J. D., G. TSAI, R. W. KUNCL, L. LAWSON, D. R. CORNBLATH, D. B. DRACHMAN, A. PESTRONK, B. STAUCH and J. D. COYLE: Abnormal excitatory amino acid metabolism in Amyotrophic Lateral Sclerosis. Ann. Neurol. 28 (1990) 18–25.

ROWLAND, L. P.: Motor neuron diseases: The clinical syndrome. In: D. Mulder (Ed.), The Diagnosis and Treatment of Amyotrophic Lateral Sclerosis. Boston, Houghton-Mifflin (1980) 7–27.

RUSSO, L. S.: Clinical and electrophysiological studies in primary lateral sclerosis. Arch. Neurol. 39 (1982) 662–664.

SADOWSKY, C. H., E. SACHS and J. OCHOA: Post-radiation motor neuron syndrome. Arch. Neurol. 33 (1976) 786–787.

SALAZAR-GRUESO, E. F., L. M. E. GRIMALDI, R. P. ROOS, R. VARIAKOIS, B. JUBELT and N. R. CASHMAN: Isoelectric focusing studies of serum and cerebrospinal fluid in patients with antecedent poliomyelitis. Ann. Neurol. 26 (1989) 709–713.

SALAZAR-GRUESO, E. F., M. J. ROUTBORT, J. MARTIN, G. DAWSON and R. P. ROOS: Polyclonal IgM anti-GM_1 ganglioside antibody in patients with motor neuron disease and variants. Ann. Neurol. 27 (1990) 558–563.

SANTORO, M., F. P. THOMAS, M. E. FINK, D. J. LANGE, A. UNCINI, N. H. WADIA, N. LATOV and A. P. HAYS: IgM deposits at Nodes of Ranvier in a patient with Amyotrophic Lateral Sclerosis, anti-GM_1 antibodies, and multifocal motor conduction block. Ann. Neurol. 28 (1990) 373–377.

SAR, M. and W. E. STUMPF: Androgen concentration in motor neurons of cranial nerves and spinal cord. Science 197 (1977) 77–79.

SCHOENBERG, B.: Controlled therapeutic trials in motor neuron disease: methodologic considerations. In: L. P. Rowland (Ed.), Motor Neuron Diseases. New York, Raven Press (1982) 547–554.

SERRADELL, A. P., J. R. GONZALEZ, J. M. TORRES, J. L. TRULL, J. O. BIELSA and A. U. ELOLA: Syndrome de sclérose latérale amyotrophique et hyperthyroïdie. Rev. Neurol. (Paris) 146 (1990) 219–220.

SHIRAKAWA, K., T. YUASA, K. HIROTA, T. TSUBAKI and M. HOSHI: A case of methylmercury poisoning with onset of a clinical sydrome resembling Amyotrophic Lateral Sclerosis. Neurol. Med. (Tokyo) 4 (1976) 58–62.

SHIRAKI, H.: The neuropathology of Amyotrophic Lateral Sclerosis (ALS) in the Kii peninsula and other areas of Japan. In: F. H. Norris Jr. and L. Kurland (Eds), Motor Neuron Diseases. New York, Grune and Stratton (1965) 80–84.

SHY, M. E., L. P. ROWLAND, T. SMITH, W. TROJABORG, N. LATOV, W. SHERMAN, M. A. PESCE, R. LOVELACE and E. F. OSSERMAN: Motor neuron disease and plasma cell dycrasia. Neurology 36 (1986) 1429–1436.

SHY, M. E., V. A. EVANS, F. D. LUBLIN, R. L. KNOBLER, T. HEIMAN-PATTERSON, A. J. TALMOUSH and G. PARRY: Anti-GM_1 antibodies in motor neuron disease patients without plasma cell dyscrasia. Ann. Neurol. 22 (1987) 167.

SHY, M. E., V. A. EVANS, F. D. LUBLIN, R. L. KNOBLER, T. HEIMAN-PATTERSON, A. J. TAIMOUSH, G. PARRY, P. SCHECK and T. G. DERYK: Antibodies to GM_1 and GD_{1b} in patients with motor neuron disease without plasma cell dyscrasia. Ann. Neurol. 25 (1989) 511–513.

SHY, M. E., T. HEIMAN-PATTERSON, G. J. PARRY, A. TAHMOUSH, V. A. EVANS and P. K. SCHICK: Lower motor neuron disease in a patient with autoantibodies against Gal (B1-3)GalNAAc in gangliosides GM$_1$ and GD$_{1b}$: improvement following immunotherapy. Neurology 40 (1990) 842–844.

SIDDIQUE, T., M. Q. PERICA-VANCE, B. R. BROOKS, R. P. ROOS, G. NICHOLSON, F. MOORE et al.: Identification of the gene defect in familial Amyotrophic Lateral Sclerosis as a strategy to understand the mechanism of motor neuron degeneration. International ALS–MND Update April (1989) 8–9.

SIDDIQUE, T., M. A. PERICA-VANCE, R. P. ROOS, H. P. WATKINS, W. Y. HUNG, B. R. BROOK, F. NORE, R. TANDEN, G. NICHOLSON, D. WILLIAM, J. BEBORT, M. ZEIDMAN, J. ANTELA and T. MUNSAT: Chromosome 21 markers in familial Amyotrophic Lateral Sclerosis. Neurology (Suppl.) 40 (1990) 310.

SIEDLER, H. and H. MALAMUD: Creutzfeldt-Jakob's Disease. J. Neuropathol. Exp. Neurol. 22 (1963) 381–402.

SIENKO, D. G., J. P. DAVIS, J. A. TAYLOR and B. R. BROOKS: Amyotrophic lateral sclerosis. A case control study following detection of a cluster in a small Wisconsin community. Arch. Neurol. 47 (1990) 38–41.

SILANI, V., G. SCARLATO, G. VALLI and M. MARCONI: Plasma exchange ineffective in Amyotrophic Lateral Sclerosis. Arch. Neurol. 37 (1980) 511–513.

SIMPSON, J. A., D. A. SEATON and J. F. ADAMS: Response to treatment with chelating agents of anemia, chronic encephalopathy and myelopathy due to lead poisoning. J. Neurol. Neurosurg. Psychiatry 27 (1974) 536–543.

SIVAK, E. D. and E. W. STREIB: Management of hypoventilation in motor neuron disease presenting with respiratory insufficiency. Ann. Neurol. 7 (1980) 188–191.

SOMASOUNDARUM, M., E. S. CHO and J. B. POSNER: Anterior horn cell degeneration as a 'remote effect' of lymphoma. Trans. Am. Neurol. Assoc. 100 (1975) 144–148.

SPENCER, P. S., V. PALMER and G. KISBY: Western Pacific Amyotrophic Lateral Sclerosis and exposure to untreated Cycad seed. International ALS — MND Update (1989) 30–31.

STALBERG, E.: Electrophysiological studies of reinnervation in ALS. Adv. Neurol. 36 (1982a) 47–89.

STALBERG, E.: Electrophysiological studies of reinnervation in ALS. In: L. P. Rowland (Ed.), Human Motor Neuron Disease. New York, Raven Press (1982b) 47–59.

STALBERG, E., M. S. SCHWARTZ and J. V. TRONTELJ: Single fiber electromyography in various processes affecting the anterior horn cell. J. Neurol. Sci. 24 (1975) 403–415.

STARK, F. M. and F. P. MOERSCH: Primary lateral sclerosis. J. Nerv. Ment. Dis. 102 (1945) 332–337.

STEELE, J. C. and T. GUZMAN: Observations about Amyotrophic Lateral Sclerosis and the Parkinsonism dementia complex of brain with regard to epidemiology and etiology. Can. J. Neurol. Sci. 14 (1987) 358–362.

STEINKE, J. and H. R. TYLER: The association of Amyotrophic Lateral Sclerosis and carbohydrate intolerance. Metabolism 13 (1964) 1376–1381.

STOBER, T., W. STELLE and K. KUNZE: Lead concentrations in blood, plasma erythrocytes and cerebrospinal fluid in Amyotrophic Lateral Sclerosis. J. Neurol. Sci. 61 (1983) 21–26.

STRONG, M. J., W. F. BROWN, A. J. HUDSON and R. SNOW: Motor unit estimates in the biceps-brachialis in Amyotrophic Lateral Sclerosis. Muscle Nerve 11 (1988) 415–422.

SUBRAMANIAM, J. S. and C. YIANNIKAS: Multimodality evoked potentials in motor neuron disease. Arch. Neurol. 47 (1990) 989–994.

SUN, C. N., C. AREOZ, G. LUCAS, P. N. MORGAN and H. J. WHITE: Amyotrophic Lateral Sclerosis: Inclusion bodies in a case of the classical sporadic form. Ann. Clin. Lab. Sci. 5 (1975) 38–44.

SWANK, R. L. and J. C. PRICE: Fascicular muscle twitchings in Amyotrophic Lateral Sclerosis. Their origin. Arch. Neurol. Psychiat. 49 (1943) 22–26.

SWANK, R. L. and T. J. PUTNAM: Amyotrophic Lateral Sclerosis. Arch. Neurol. Psychiatr. 49 (1943) 151–177.

SWASH, M. and M. S. SCHWARTZ: A longitudinal study of changes in motor units in motor neuron disease. J. Neurol. Sci. 56 (1982) 185–197.

TACKMANN, W. and P. VOGEL: Fibre density, amplitudes of macro-EMG motor unit potentials and conventional EMG recordings from the anterior tibial muscle in patients with Amyotrophic Lateral Sclerosis. J. Neurol. 235 (1988) 149–154.

TANZ, W. S., G. J. TSONGALIS, K. SUMNER and W. C. LAULET: A new selective assay reveals a major abnormality in DNA metabolism in amyotrophic lateral sclerosis. Neurology 40 (1990) 315.

TAYLOR, J. A. and J. P. DAVIS: Evidence for clustering of Amyotrophic Lateral Sclerosis in Wisconsin. J. Clin. Epidemiol. 42 (1989) 569–575.

TERASAKI, P. I. and M. R. MICKEY: HLA haplotype of 32 diseases. Transplant Rev. 22 (1975) 105–191.

TOYOKURA, Y: Negative Features in ALS. In: T. Tsubaki and Y. Toyokura (Eds), Amyotrophic Lateral Sclerosis. Baltimore, University Park Press (1978) 53–60.

TROJABORG, G. W. and F. BUCHTHAL: Malignant and benign fasciculations. Acta Neurol. Scan. Suppl. 13, 41 (1965) 251–254.

TROOST, D., J. J. VAN DEN OORD, J. M. B. V. DE JONG and D. F. SWAB: Lymphocytic infiltration in the spinal cord of patients with Amyotrophic Lateral Sclerosis. Clin. Neuropathol. 8 (1989) 289–294.

TYLER, H. R.: Nerve cell disorders in a family with variable clinical manifestations including presenile dementia, Parkinsonism and motor neuron disease. Excerpta Medica. International Congress Series 296 (1975) 65.

TYLER, H. R.: Double-blind study of modified neurotoxin in motor neuron disease. Neurology 29 (1979) 77–81.

TYLER, H. R.: Nonfamilial amyotrophy with dementia or multisystem degeneration and other neurologic disorders. In: L. P. Rowland (Ed.), Human Motor Neuron Disease. New York, Raven Press (1982) 173–180.

UNGAR-SARGON, J.Y., R.E. LOVELACE and J.C.M. BRUST: Spastic paraplegia-paraparesis. J. Neurol. Sci. 46 (1980) 1–12.

UGAWA, Y., T. SHIMPO and T. MANNEN: Central motor conduction in cerebrovascular disease and motor neuron disease. Acta. Neurol. Scand. 78 (1988) 297–306.

VAN BOGAERT, L.: Les troubles mentaux dans la sclérose latérale amyotrophique. Encéphale 20 (1925) 27–47.

VAN BOGAERT, L. and I. BERTRAND: Pathologic changes of senile type in Charcot's disease. Arch. Neurol. Psychiatr. (Chic.) 16 (1926) 263–284.

VAN BOGAERT, L. and M.A. RADERMECKER: Scléroses Latérales Amyotrophiques typique et paralysies agitantes héréditaires, dans un même famille, avec une forme de passage possible entre les deux affections. Mschr. Psychiat. Neurol. 127 (1954) 185–203.

VERNANT, J.C., G. BUISSON, R. BELLANCE, M.A. FRANCOIS, O. MADKA and O. ZAVARO: Pseudo Amyotrophic Lateral Sclerosis, Peripheral Neuropathy and Chronic Polyradiculoneuritis in patients with HTLV-1 associated paraplegia. In: G.C. Roman, J.C. Vernant and M. Osame (Eds), HTLV-1 and the Nervous System. New York, Alan R. Liss publisher (1989) 361–365.

VICALE, C.T.: The diagnostic features of muscular syndrome resulting from hyperparathyroidism, osteomalacia owing to renal tubular acidosis and perhaps related disorders of calcium metabolism. Trans. Am. Neurol. Assoc. 74 (1949) 143–147.

VOSS, H.: Progressive Bulbärparalyse und Amyotrophische Lateralsklerose nach chronischer Manganvergiftung. Arch. Gewerbepathol. Gewerhe Hygiene 9 (1939) 464–476.

WECHSLER, I.S. and S. BRODY: The problem of primary lateral sclerosis. J. Am. Med. Assoc. 130 (1946) 1195–1198.

WECHSLER, I.S., M.R. SAPIRSTEIN and S. STEIN: Primary and symptomatic Amyotrophic Lateral Sclerosis. Am. J. Med. Sci. 208 (1944) 70–81.

WILLIAMS, C., M.A. KOZLOWSKI, D.R. HINTON and C.A. MILLER: Degeneration of spinocerebellar neurons in amyotrophic lateral sclerosis. Ann. Neurol. 27 (1990) 215–225.

WEINER, L.P.: A possible role of androgen receptors in Amyotrophic Lateral Sclerosis. A hypothesis. Arch. Neurol. 37 (1980) 129–131.

WEINER, L.P., S.A. STOHLMEN and R.L. DAVIS: Attempts to demonstrate virus in Amyotrophic Lateral Sclerosis. Neurology 30 (1980) 1319–1322.

WETTSTEIN, A.: The origin of fasciculations in motor neuron disease. Ann. Neurol. 5 (1979) 295–300.

WILLIAMS, D.B., R.S. PAMPHLETT and R.J.A. TRENT: Monozygotic twins and familial Amyotrophic Lateral Sclerosis. International ALS–MND Update (1989) 11–12.

WILSON, S.A.K.: The amyotrophy of chronic lead poisoning: Amyotrophic Lateral Sclerosis of toxic origin. Rev. Neurol. Psychiatr. 5 (1907) 441–445.

WILSON, S.A.K.: Neurology London. Edward Arnold (1940) 730.

WOHLFART, G.: Collateral regeneration from residual motor nerve fibers in Amyotrophic Lateral Sclerosis. Neurology 7 (1957) 124–134.

WOLFGRAM, F.: Blind studies in the effect of amyotrophic lateral sclerosis sera in motor neuron in vitro. In: J.M. Andrew, R.T. Johnson and M.A.B. Brazier (Eds), Amyotrophic Lateral Sclerosis Research Trends. New York, Academic Press (1976) 145–451.

WOLFGRAM, F. and L. MYERS: Amyotrophic Lateral Sclerosis: effect of serum in anterior horn cells in tissue culture. Science 179 (1973) 579–580.

YASE, Y.: The pathogenesis of Amyotrophic Lateral Sclerosis. Lancet 2 (1972) 292–296.

YASE, Y.: Role of metal/mineral metabolism in the ALS process, epidemiological and environmental background. International ALS–MND Update April (1989) 15–16.

YOSHIMASU, F., M. YASUI, Y. YASE, S. IWATA, D.C. GAJDUSEK, C.J. GIBBS and K.M. CHEN: Studies on Amyotrophic Lateral Sclerosis by neutron activation analysis 2: Comparative study on analytical results on Guam PD, Japanese ALS and Alzheimer diseases cases. Folia Psychiatr. Neurol. (Jpn.) 34 (1980) 75–82.

YOSHIMASU, F., M. YASUI, Y. YASE, Y. UEBAYASHI, S. TANAKA, S. IWATA, K. SASJIMG, D.C. GAJDASEK and K.M. CHEN: Studies on Amyotrophic Lateral Sclerosis by neutron activation analysis 3: Systematic analysis of metals on Guamian ALS and PD cases. Folia Psychiatr. Neurol. (Jpn.) 36 (1983) 173–180.

YOUNG, A.B.: What the excitement about excitatory amino acids in Amyotrophic Lateral Sclerosis. Ann. Neurol. 28 (1990) 9–11.

YOUNGER, D.S., L.P. ROWLAND, N. LATOV, W. SHERMAN, M. PESCE, D.J. LANGE, W. TROJABORG, J.R. MILLER, R.E. LOVELACE, A.P. HAYES and T.S. KIM: Motor neuron disease and amyotrophic lateral sclerosis. Relation of high CSF protein content to paraproteinemia and clinical syndromes. Neurology 40 (1990) 595–599.

YUASA, R., H. HIYAMA and T. HASHIMOTO: Motor neuron disease in the Osaka prefecture, Japan. Clin. Neurol. (Tokyo) 16 (1976) 207–212.

ZIL'BER, L.A., Z.L. BAGDAKOVA, A.N. GARDASJAN, N.V. KONOVALOV, T.L. BUNINA and E.M. RARABADZE: Study of the etiology of Amyotrophic Lateral Sclerosis. Bull. WHO 29 (1963) 449–456.

ZILKHA, K.J.: Discussion of Motor Neuron Disease. Proc. R. Soc. Med. 55 (1962) 1028–1029.

Handbook of Clinical Neurology, Vol. 15 (59): Diseases of the Motor System
J.M.B.V. de Jong, editor
© Elsevier Science Publishers B.V., 1991

CHAPTER 12

Progressive bulbar palsy in adults

G. W. BRUYN

Department of Neurology, Academic Hospital, State University Leiden, The Netherlands

The constellation of symptoms and signs first defined by Duchenne de Boulogne (1860) as 'paralysie musculaire progressive de la langue, du voile du palais et des lèvres', is not a nosological entity. It would therefore scarcely merit a separate chapter in a book on nervous system disorders, were it not for its characteristic clinical features and its grim connotation.

Also known by such awkward termini technici as 'labiofacioglossopharyngolaryngeal paralysis', or as 'labiopalatoglossal' or 'glossolabiopharyngeal palsy', the syndrome rarely occurs in its own right. In the great majority of instances it forms part of what in modern terms is denoted by the as nondescript as sweeping term 'motor neuron disease' which, however, is incorrect, because not only the motor neurons are affected. Wherever there are motor neurons in the central nervous system, they may fall ill, as parts of a population, or as entire populations, whether cortical, basal ganglionic, mesencephalic, bulbo-pontine or spinal, and whether they are somato-motor or visceral-motor. In a bygone clinical era, distinction was therefore made between Erb's spastic spinal paralysis (primary lateral sclerosis), Aran's progressive spinal muscular atrophy, Duchenne's progressive bulbar palsy (PMB), Charcot's amyotrophic lateral sclerosis (ALS being a combination of the previous two or three), Charcot-Marie-Tooth-Hoffman's peroneal (or spinal) muscular atrophy, Mills' unilateral or hemiplegic ALS-type, later joined by Shy-Drager syndrome and other dysautonomias, all members of a large spectrum of disorders produced by progressive death of motor neurons.

Notwithstanding the clinical empirism, that progressive bulbar palsy *qua talis* may manifest as more or less an integral part of an inordinate number of nosologically different diseases, ranging from toxic, to infectious, to metabolic, to what for want of known causation is phrased as 'degenerative-abiotrophic', the primary significance of the syndrome derives from its constituency within the realm of Charcot's disease. Indeed, many years ago, Professor Forbes H. Norris wrote to the present author, that among the 700+ cases of ALS seen by him, not more than 2 or 3 might have been classified as progressive bulbar palsy. The editor of the present volume (JMBV de Jong) observed one case among 400. Clearly, this testifies to the significance of the strictness of diagnostic criteria with respect to the various clinical types; it may, however, also be attributable (even if hypothetically so) to a temporal difference in disease course: if a case presenting as progressive bulbar palsy dies before symptoms of cervical or lumbar motoneuron degeneration have a chance to become manifest, though at autopsy their degeneration is established, it is a matter of clinical habit or diagnostic preference to classify that case as PBP or ALS.

Charcot's amyotrophic lateral sclerosis (ALS) is a disease in which progressive death of both 'upper' (or first) and 'lower' (or second) motor neuron populations is combined in a variable temporal sequence. As a consequence, the disease may manifest initially with the clinical features of depletion of the lumbar ventral horn motoneurons (which used to be called the 'pseudopolyneuritic type'), or with those of cervical ventral motorneuron death (known as Aran's type), or alternatively with fall-out of the corticospinal pyramidal cells and tract (known as the pyramidal syndrome of primary lateral sclerosis), or alternatively, with masking of signs of upper motor neuron death by subsequent lower motor neuron death or vice versa, or alternatively initially manifest with bulbar palsy. The rich variety of symptoms occurring in this way, however, will not remain beyond the diagnostic competence of the wary physician. Diagnostic correctness on initial presentation leaves room for improvement in the so called pseudopolyneuritis and Mills types, where follow-up proves diagnosis to err in about two-thirds of the cases (O'Reilly et al. 1982).

Numerous reviews that have been devoted to it in the past were synoptically discussed by Colmant (1975). For a fascinating historical review of the *auctores intellectuales* of the disorder (which opposes the interpretation of Mackay 1963) as well as a still modern clinical-pathological review on ALS, the reader is referred to Norris (1975).

SYMPTOMATOLOGY

The patient with progressive bulbar palsy develops difficulties in speech, chewing, swallowing and phonation ('dysarthria-dysphagia-dysmasia-dysphonia'). Usually, but not invariably, disturbed speech is the presenting symptom, often noticed by family or friends before the patient does. The speech acquires a thick and indistinct ('slurred') quality, usually with nasal overtones, and gradually becomes a tiring effort so that the patient tends to avoid the execution of long and complex sentences. Frank dysarthria is often present. The palsy, resulting from cell death in the XIIth, Xth, IXth and VIIth motor nuclei, affects the muscles of the tongue, palate, lips, cheeks, and late if at all, the pharynx. As a result, the generation of articulative vowels and consonant, becomes progressively impaired: the *labials* (b, p, f, m, v, w, o, u), the *dentolabials* and *linguals* (d, l, n, t, r, s), the *dentals* (sh, j, zh), the *palatals* (g, k, the French ng), and *gutturals* (German/Dutch k, g, ch) (Critchley and Kubik 1925). Poor closure of the glottis during speech results in escape of air, necessitating abrupt changes in breathing pattern. Weakness and atonia of 'Riolan's bouquet' (m. styloglossus, stylopharyngeus and stylohyoideus) cause backward/downward displacement of the tongue, affecting also respiration.

More or less synchronously, the voice loses its rhythm, pitch, modulation, melody, and volume leading to slowing of speech and a monotonous voice (aprosodia). Purity of tone is usually lost (rendering the voice a rough or hoarse quality) (Chevrie-Müller et al. 1968, 1970). Whistling, blowing-up the cheeks, sucking, kissing, etc. becomes impossible. In the course of time, due to the inherent relentless progression of the disorder, the patient will be merely able to produce unintelligible noises ('èh, òh, àh') and becomes anarthric and ultimately aphonic. The tongue displays atrophy and fasciculations and lies immobile on the floor of the mouth. The palatal arches are immobile, the pharyngeal reflex (gag-reflex), though surprisingly long elicitable, with protracted disease course is eventually abolished. Of all the affected muscles, those of the tongue show the most marked impairment (De Paul et al. 1988) (Fig. 1).

Not only articulation disappears; the facial muscles due to progressive denervation predominantly of the lower facial half, but ultimately also of the upper facial half, are no longer able to produce the movements of mimic rendering a mask-like immobility to the lower half of the face. The face acquires an expressionless appearance, and in cases with a protracted course, eye-closure becomes impossible.

Soon, the impairment in articulation and phonation is associated with disturbed deglutition. The patient can no longer protrude his tongue

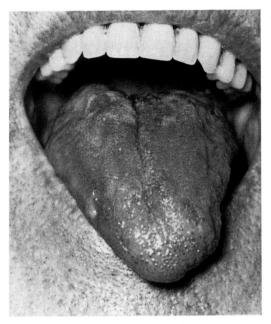

Fig. 1. Male 58 years. Atrophy of tongue, predominating on right side.

or move solid bits of food about within the oral cavity, or push it back to the pharynx, and deglutition of fluids due to absent closure of the choanae by the soft palate (atonic palsy of tensor and levator veli palatini and eventually reduced action of the superior pharyngeal constrictor), which results in regurgitation through the nose. Swallowing of solid morsels of food is difficult, as the food often remains stuck in the piriform recesses and epiglottic vallecula (Bosma and Brodie 1969; Robbins 1987). This causes the patient to choke and to try promotion of the passage of the food by anteflexing the head. The elevation of the hyoid at onset of the swallowing reflex is deficient because of the atonia and weakness of Riolan's nosegay muscles, as fully reviewed by Colmant (1975). Pharyngeal retention ('stasis') and repeated piecemeal swallowing, pharyngeal dilatation, and aspiration of the bolus is seen on roentgen-contrast-studies. The risk of aspiration-pneumonia is always kept in mind by the physician in care of such a patient. Cachexia due to insufficient uptake of nourishment (inanition) develops in most patients. Due to the defective closure of the lips and the impaired

swallowing mechanism, constant salivation adds to the patient's irritation and misery.

The masticatory muscles become weak and atrophic due to neuronal depletion in the trigeminal motor nucleus. The patient is unable to bite or chew nutrients of hard consistency. In the course of the disease the mouth tends to fall open and the patient has to support the mandibula, a sign also often observable in myasthenia gravis in the elderly. The atrophic buccinator and temporal muscles render the face a sunken, haggard appearance; deficient or absent contraction on command can be easily palpated. The masseter-reflex may remain brisk for a long period however, particularly when there is also upper motor neuron degeneration.

Due to ultimate and additional involvement of the diaphragm (Neau et al. 1988) the aphonic patient develops dyspnoea. The sternocleidomastoid and trapezius muscles (ncl. XI) start to waste and no longer assist as auxiliary respiratory factors (for the same reason, patients can no longer keep their head upright or turn it sideways; their heads hang in anteflexion). The bulbar palsy may also initially present with respiratory insufficiency (Mussini et al. 1984; Meyrignac et al. 1985). Vocal cord abductor paralysis may cause life-threatening asphyxia. A rare familial disorder of chronic spinal muscular atrophy associated with bilateral paralysis of the vocal cords due to neuronal death in the nucleus ambiguus (Young-Harper syndrome) has been reported by Serratrice et al. (1984). Ordinary ailments, such as catching a common cold or having a bronchitis, being a mere bother to an otherwise healthy person, constitute a serious and near-catastrophic danger to the patient whose abilities to cough, to swallow and to clear the throat are substantially jeopardised.

Not only the bulbopontine somatomotor, but also the visceromotor neurons degenerate, producing reduced oesophageal and gastric (Boudin et al. 1971) mobility with dilatation. Tachycardiac crises due to involvement of the ncl. IX (ambiguus) and ncl. dorsalis X were pointed out by Marburg (1936), circulation disturbances are common (Gadermann and Mertens 1953). Sphincter dysfunction is proverbially rare, occurring ultimately in ± 10% of the patients that live long enough. In this context it should be pointed out that one of the decisive neuropathological differences between ALS and the Shy-Drager syndrome is the indemnity of nucleus X of Onufrowicz at the S_{2-3} level (Onuf's nucleus) innervating the bulbo-ischiocavernosus-sphincter-, and pelvic floor musculature (Schrøder and Reske-Nielsen 1984; Konno et al. 1986).

In cases with a protracted course, the oculomotor nuclei (ncl. III) may also become involved (see Colmant 1975, pp. 140–141).

Sensory impairment is not part of the clinical picture, which explains the rarity of decubitus ulcers. Another remarkable feature, known to every neurologist, is that the cutaneous abdominal reflexes remain elicitable to the end.

Within 2–3 years the condition of the patient is no longer compatible with life. Deprived of speech, quasi-unable to eat or drink, and dyspnoeic, the patient whose consciousness and cognitive, emotional and voluntary capacities remain fully normal, has to live through an unimaginable mental ordeal, before death because of respiratory failure, or aspiration-pneumonia, or cardiac failure mercifully ends the misery.

Auxiliary techniques are partly non-contributory, such as CSF examination and EEG, and partly redundant because the results reveal what one should expect (denervation signs on EMG and localised atrophy on CT or NMR).

EPIDEMIOLOGY

Exact determination of the incidence and prevalence of 'pure', i.e. 'non ALS-contaminated' progressive bulbar palsy is illusory. Moreover, the syndrome may manifest as part of the symptomatology of a group of heterogeneous disorders. In by far the greater majority of instances, progressive bulbar palsy is either the initial, middle-phase, or penultimate syndrome of manifestation of ALS.

The available epidemiological data on ALS confirm that ALS ('motor neuron disease') occurs in every race, at every longitude or latitude and that the annual incidence ranges between 0.9 and 2.7 per 100 000 population, with a mean at around 1.3 and prevalence rates between 2 and 7 per 100 000 population. Populations with unusu-

ally high annual incidences and prevalences, such as Guam, Kii peninsula, southwest New Guinea, Filipinos in Hawaii (Matsumoto et al. 1972), and central Finland (Murros and Fogerholm 1983) and east of Oslo (Lundar 1978) as well as with unusually low rates, such as in Mexico City (Olivares et al. 1972), Czechoslovakia (Sercl and Kovarik 1967; Kovarik 1974), Iceland (Gudmundsson 1968), Israel (Kahana et al. 1976), Poland (Cendrowski et al. 1970), Canada (Hudson 1986), and some parts of Italy (Rosati et al. 1977; Bracco et al. 1979; Leone et al. 1983; Salemi et al. 1989) have been object of intensive study. Also, the death rate in MND which apparently is rising over the recent decades (Jokelainen 1976; Kurtzke 1982; Buckley et al. 1983; Gunnarson and Palm 1984; Hudson 1986) has been well analysed.

Discussion of the wealth of reports on the epidemiology, having been reviewed by Bobowick and Brody (1975) and Rowland (1982), falls outside the scope of this chapter and section, which remains restricted to progressive bulbar paralysis.

With all mental reserves one should court *vis à vis* epidemiological data on this syndrome, because of differences in application of diagnostic criteria by various neurologists in various parts of the world, the following may be offered. The ratio of PBP to ALS, because of divergent diagnostic preferences ranges widely from 1% to 2% (Swank and Putnam 1943; Mortara et al. 1984), via 7–15% (Tans 1950; Mackay 1963; Olivares et al. 1972; Zack et al. 1977; Bracco et al. 1979; Kondo and Tsubaki 1981; David et al. 1981; Caroscio et al. 1987; Granieri et al. 1988) to 19–43% (Boman and Meurman 1967; Nishigaki 1970; Müller-Jensen and Bernhardt 1973; Rosati et al. 1977; Rosen 1978; Kristensen and Melgaard 1977; Mortara et al. 1981; Forsgren et al. 1983; Murros and Fogerholm 1983; Gunnarson and Palm 1984; Woo et al. 1986; Caroscio et al. 1987; Lopez-Vega et al. 1988). The highest ratios were reported by Chazot et al. (1989) to occur in the departments surrounding Limoges: 45% and by de Domenico et al. (1988) in the province of Messina: 56%; of all clinical types, the bulbar type manifested in the oldest age-group. Proportionally, bulbar paralysis manifests in an average of 16% of the cases, which would — given an annual ALS incidence of 1.3×10^{-5} — yield an annual PBP incidence of 0.20×10^{-5}.

If one examines the male-female ratio of PBP, keeping in mind this ratio in ALS (MND) *grosso modo* being about 1.45:1, the following figures strongly suggest a value-reversal in PBP. Boman and Meurman (1967) reported a male-female ratio of 1.3:1 versus a ratio of 2:1 in all their 120 cases; Sercl and Kovaric (1967) 1:1.6; Vejjajiva (1967) 1:1; Dazzi et al. (1969) 1:1.5; Rosati et al. (1977) 1:3; Kristensen and Melgaard (1977) 1.2:1 among a total material ratio of 2:1; David et al. (1981) 1.4:1 against a 2.3:1 ratio among the total material; Forsgren et al. (1983) 1:1.5; Caroscio et al. (1987) 1:1.3; Granieri et al. (1988) 1:1.5, and Salemi et al. (1989) 1:2. In contrast, Tans (1950) found a male to female ratio of 3:1, Girke and Kovarik (1970) 2:1, Rai and Jolly (1971) 7:1, Nishigaki (1970) 3:1; in these last 3 sources the high disproportion suggests a selection factor. Accordingly, PBP appears quite more frequent in female patients.

Also, PBP more frequently has a higher age at onset in females (Kristensen and Melgaard 1977; David et al. 1981; Forsgren et al. 1983) and, onset of ALS in females appears to be at a higher age (Zack et al. 1977; Rosen 1978; Juergens et al. 1980; Mortara et al. 1981; Murros and Fogerholm 1983). Finally, PBP (because of its tendency to manifest at a higher age?) has a poorer prognosis because of a faster course (average 18–24 months), the survival rates of it being 3–5 times smaller than for the cervical and, in particular, the lumbar clinical types of ALS (Rosati et al. 1977; Rosen 1978; Kristensen and Melgaard 1977; Juergens et al. 1980; Mortara et al. 1981; David et al. 1981; Forsgren et al. 1983; Murros and Fogerholm 1983; Gunnarson and Palm 1984; Carroscio et al. 1987; Granieri et al. 1988). It goes without saying that a higher age at onset implies a shorter life expectancy though that cannot fully explain the difference. In this respect the diagrams provided by Boman and Meurman (1967) clearly show that the shortest survival times in PBP are independent from age at onset, whereas the cervical and spinal types of ALS unequivocally demonstrate the linear relationship between shorter duration of disease (life expectancy), and higher age at onset. In the ALS/MND group a younger onset carries a better prognosis, the progressive spinal muscular atrophy subpopulation having the 'best' prognosis. Most reports containing pertinent survival data apparently indicate that the ALS population to contain 2 subpopulations, one with a fast downhill, and one with a protracted course. These, of course, represent the bulbar-type and the progressive spinal muscular atrophy type ALS subpopulations (Fig. 2); whereas the bulbar group has a maximum survival of 4–5 years, the other ALS-types may well run a course of 10–20 years.

NEUROPATHOLOGY

The histological changes encountered in PBP are essentially those found in ALS. One might expect them to be restricted to the medulla oblongata and lower pons, but such a simple approach fails to catch reality. If cell degeneration in the bulbar area has reached the critical threshold below which symptoms manifest, the degeneration of Betz-cells and the pyramidal tract may well have been going on and remained occult because the

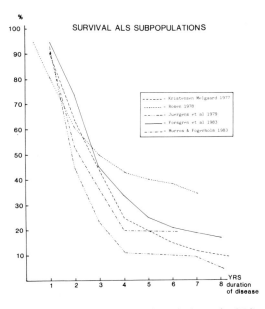

Fig. 2. Composite diagram of survival rate in ALS. The steep down-hill grades represent the bulbar type subpopulation.

critical line between latency and overt manifestation has not been crossed. Quite often, but
particularly in familial cases, the neuropathologist
observes demyelination in the tracts of Flechsig
and Gowers and in the funiculus gracilis without
finding the mention of sensory impairment in the
clinical history. The same, *a pari passu*, obtains
for lumbar motor neuron loss in the lumbar
spinal cord (Helfand 1933; Lawyer and Netsky
1953; Fotopulos et al. 1958).

Gross findings on autopsy include greyish
shrinking of the hypoglossal nerve, and possibly
of ventral roots of the cervical and lumbar
segments. Light microscopy reveals (Hassler 1953;
Colmant 1958) depletion as well as signs of
degeneration (paleness, shrinking, marginal accumulation of Nissl substance, eccentric displacement of the nucleus, exceptionally chromatolysis
and swelling) of the motorneurons of the ncl.
XII, ncl. ambiguus, motor ncl. X, ncl. VII
(Figs. 3 and 4), and of course of the ventral
horn motorneurons (Fig. 5). As emphasised by

Greenfield (1963) the size and shape of the
ventral horns remain unchanged, and the motor
neuron loss predominates in the dorsolateral cell
groups. Clarke's column cells have been said to
remain immune, but may also be involved (Kato
et al. 1987). There may be moderate gliosis and
slight perivascular infiltration. The degeneration
in the caudal cranial nerves is primarily axonal,
followed by segmental degeneration; retrograde
axonal transport speed is diminished (Breur et
al. 1987). Examination of the hypoglossal nerve
reveals reduction of the large myelinated fibers
(Atsumi and Miyatake 1987). Although Guillain
et al. (1925) observed degeneration of Roller's
nucleus, Greenfield (1963) denied involvement of
this structure.

In some instances, proximal axonal enlargements due to (phosphorylated) neurofilament
accumulation may be observed (Carpenter 1968;
Chou et al. 1970; Kurachi et al. 1979; Hirano et
al. 1984; Mizusawa et al. 1989). In a personal
case of a woman aged 67 with a classic pro-

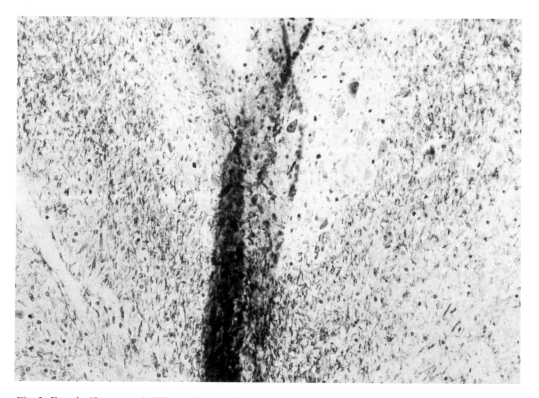

Fig. 3. Female 69 years. ncl. XII quasi-total cell-depletion on the right side, ghost cells on the left. Klüver-
Barrera 40 ×.

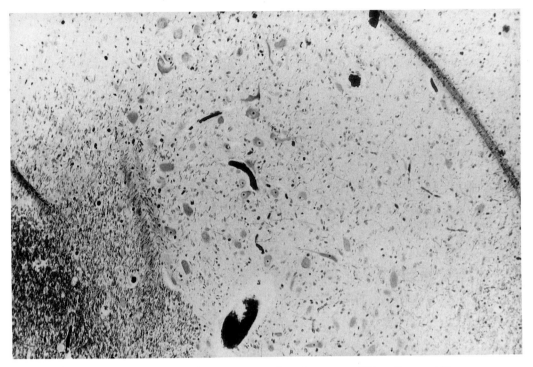

Fig. 4. Same case. Ncl. dorsalis X. Degeneration of neurones. Klüver-Barrera 150 ×.

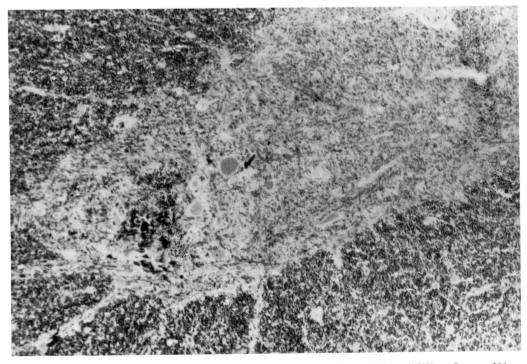

Fig. 5. Same case. A few remaining and degenerating ventral horn cells at C III level. Klüver-Barrera 200 ×.

gressive bulbar palsy, who died within 10 months, marked cell-loss was seen in the inferior olivary nucleus and locus coeruleus with depigmentation. Increase of lipofuscin was not conspicuous, except in the extremely shrunken neuronal elements.

The disorder is not restricted to involvement of the motor neurons, thus testifying to the fact that the term 'motor neuron disease' is a misnomer. Bertrand and van Bogaert (1925) pointed out the demyelination of the rubrospinal tract. Not exceptionally, demyelination is seen of the spinocerebellar tracts, of the gracile funiculus and of the medial longitudinal fasciculus (Hirano et al. 1967, Castaigne et al. 1971). Not exclusively in hereditary or familial instances of ALS, but also in sporadic cases, this widespread CNS involvement is associated with the presence of hyaline eosinophilic, Lewy-body-like cytoplasmic inclusions (Kato et al. 1988; Mizusawa et al. 1989).

Close observation in PBP will reveal craniocaudally increasing demyelination of the crossed and uncrossed pyramidal tracts, due to loss of the large-calibre fibres, with the qualification that the pyramidal degeneration does not reach as high (internal capsule) as it may in 'common' ALS (Hirayama et al. 1962; Shibasaki 1968; Nishigaka 1970). Examination of paralysed muscle reveals the typical neuronal type of atrophy with islands of small shrunken fibres and 'type-grouping'.

In a number of familial cases, Bunina's eosinophilic cytoplasmic inclusion bodies have been noted (Bunina 1962; Hirano et al. 1967) as well as ballooning of motor neurons.

DIAGNOSIS AND DIFFERENTIATION

The thesis that PBP may occur (by decomposition from ALS) as a segregated clinical form in its own right (i.e. without admixture of pyramidal signs or symptoms of cervico-lumbar ventral horn cell depletion) is to be regarded with skeptic reserve. The famous case of Guillain et al. (1925) had muscle wasting of shoulder girdle, arms and hands. The life-expectancy of patients with PBP, being the shortest of that of the upper (Erb) and lower (Aran/Duchenne) components of Charcot's amyotrophic lateral sclerosis, may preclude the

other cerebral and spinal components, of which the underlying pathological process develops slower, from becoming clinically manifest (Hemmer 1951; Norris and Kurland 1969). Even so, if one finds, upon EMG examination of a patient with 'pure' PBP, unequivocal signs of denervation in a distal muscle (e.g. interosseus or extensor digitorum brevis), it becomes a matter of nosologic rectitude or adherence to rigid principles to diagnose such a case as ALS. This obtains even more when a sign such as a Babinski reflex (indicative of pyramidal tract dysfunction) is elicited in a case of 'pure' PBP.

Instances have been reported of hereditary PBP and hereditary ALS. Within the first-named category the infantile form (Fazio-Londe disease) (211500) seems fairly well established; indisputable evidence is scarce for the adult form, of which quite a few instances were described before EMG became a widely used tool in neurological diagnosis. The only instances of hereditary PBP that may come into consideration are those reported by Lovell (1932), Dittel (1940) and Robertson (1953). The report by Sack (1932) on 'Betjugen' (as hereditary bulbar palsy is called by the inhabitants) in Siberic yakuts, seems to describe the autosomal dominant type. In the family reported by Cooper (1933) both PBP and PSMA occurred, and in the Robertson report 2 of the 3 cases had spinal muscular atrophy in addition. Takikawa (1953) reported a sex-linked recessive type. The wealth of reports on familial/ hereditary ALS (Hudson, 1981) and its particular features that may lead one to suspect multi-system degeneration (Tanaka et al. 1984) and of progressive spinal muscular atrophy warrants a separate chapter in this volume and will be not reviewed here, even if these hereditary cases may include instances with (predominant) bulbar manifestation. Nor will a review be made of bulbar palsy in progressive spinal muscular atrophies, even if the latter show a more clear-cut (albeit with a variable scala of modes) genetic predisposition. Progressive bulbar paralysis in the Vulpian-Bernhardt type, the X-linked bulbospinal types, the facioscapulohumeral and scapuloperoneal forms are discussed in the pertinent chapters within this volume. The same obtains for PBP in Strümpell-Lorrain disease, and the dysautonomias including Shy-Drager's and Riley-Day syndromes.

The cases of PBP associated with *deafness* (211530) as reviewed by Gomez (1975) and Colmant (1975) should, in the present writer's interpretation, be classified as Fazio-Londe disease.

Following Colmant's 1975 review of PBP associated with palatal *myoclonus* (due to lesion in either the olivary-cerebellar fibres, the dentato-rubral tract, or the central tegmental tract) or myoclonus of trunk and limbs, no new instances of this combination have been reported to the present writer's awareness.

Contrariwise, the association between ALS and *dementia* is much more conspicuous, particularly in hereditary ALS, and is not exceptionally combined with *Parkinsonism*. This is not surprising in view of the ALS–Parkinson-Dementia complex of the Australasian theater (Alter and Schaumann 1976a; Burnstein and Ananth 1980; Schmitt et al. 1984). The association with Pick's disease certainly is not exceptional, one of the latest reports being that by Constantinidis (1987). Familial non-specific dementia was reported by Ojeda et al. (1984). The ALS-P-D association is often familial (Hudson 1981). However, there is no remarkably frequent PBP-P-D association. A warning against posing a diagnosis of 'ALS-dementia' or 'PBP-dementia' without neuropathological verification is to be found in the reports from the Leiden University clinic (Staal and Went 1968; Staal and Bots 1969; Bots and Staal 1973) in which histopathological work-up revealed familial cases of 'ALS-dementia' to have suffered from Hallervorden-Spatz disease.

Without unequivocal post-mortem neuropathological changes in the cerebellum, the association of PBP and *cerebellar symptoms*, as reviewed by Colmant (1975) will remain dubious. Particularly in familial ALS cases, degeneration has been often observed of the middle root zone of the posterior funiculi, spinocerebellar tracts, Clarke's columns, cuneate and dentate nucleus (Engel et al. 1959; Horton et al. 1976; Makifuchi and Ikata 1977; Hirano et al. 1984; Kato et al. 1987).

The presence of Progressive Bulbar Palsy has been signalled, as a symptomatic feature, in a wide variety of other diseases: malignant disease, gangliosidosis, Waldenström's macroglobulinemia, and paraproteinemias (Smith and Alvarez 1987, reply by Shy 1987), Sjögren's syndrome (Yamashita et al. 1988), basilar impression, chorea-acanthocytosis, botulism, diphtheria, tick paralysis (neuroborreliosis; Pearn 1977; Tibballs and Cooper 1986), trichinosis, (post-)poliomyelitis, myasthenia gravis, and heavy metal intoxication. In these rare instances, associated clinical features usually keep the alert physician on the straight and narrow path to correct diagnosis.

THERAPY

This is palliative. A full survey of measures to be taken to make the patient's remaining days lighter and to support the patient where possible, was provided by Norris et al. (1985). These are of value not only in the group of patients whose life expectancy is short; there is a subgroup of 15–20% whose duration of illness may extend over a decade or more; it is the physician in charge who should be aware of the multiple aids, measures, and medications that indeed constitute a help for the patient.

REFERENCES

ALTER, M. and B. SCHAUMANN: A family with ALS and Parkinson. J. Neurol. 212 (1976a) 281–284.

ALTER, M. and B. SCHAUMANN: Hereditary ALS: a report of two families. Eur. Neurol. (Basel) 14 (1976b) 247–255.

ATSUMI, T. and T. MIYATAKE: Morphometry of the degenerative process in the hypoglossal nerve in ALS. Acta Neuropathol. 73 (1987) 25–31.

BERTRAND, I. and L. VAN BOGAERT: Rapport sur la SLA. Rev. Neurol. (Paris) I (1925) 779–906.

BOBOWICK, A. R. and J. A. BRODY: Epidemiology of neurodegenerative system disorders. In: P. J. Vinken and G. W. Bruyn (Eds), Handbook of Clinical Neurology, Vol. 21, Ch. 2. Amsterdam, N.Y., North-Holland Publ. Co. (1975) 3–52.

BOMAN, K. and T. MEURMAN: Prognosis of amyotrophic lateral sclerosis. Acta Neurol. Scand. 43 (1967) 489–498.

BOSMA, J. F. and D. R. BRODIE: Disabilities of the pharynx in ALS as demonstrated by cineradiography. Radiology 92 (1969) 97–103.

BOTS, G. TH. A. M. and A. STAAL: Amyotrophic lateral sclerosis-dementia complex, neuroaxonal dystrophy, and Hallervorden-Spatz disease. Neurology 23 (1973) 35–39.

BOUDIN, G., B. PÉPIN and J. C. VERNANT: Cas familial de paralysie bulbopontine chronique progressive avec surdité. Rev. Neurol. (Paris) 124 (1971) − 90–92.

BRACCO, L., P. ANTUONO and L. AMADUCCI: Study of epidemiological and etiological factors of amyotrophic lateral sclerosis in the province of Florence, Italy. Acta Neurol. Scand. 60 (1979) 112–124.

BREUR, A. C., M. P. LYNN and B. ATKINSON: Fast axonal transport in ALS. Neurology 37 (1987) 738–48.

BUCKLEY, J., C. WARLOW, P. SMITH, D. HILTON-JONES, S. IRVINE and J. R. TEW: Motor neuron disease in England and Wales, 1959–1979. J. Neurol. Neurosurg. Psychiatry 46 (1983) 197–205.

BUNINA, T. L.: On intracellular inclusion in familial ALS. Korsakov's J. Nevropat. Psikhiat. 60 (1962) 1293.

BURNSTEIN, M. H. and J. ANANTH: ALS, dementia and psychosis. Psychiatry J. Univ. Ottawa 5 (1980) 166–167.

CAROSCIO, J. T., M. N. MULVIHILL, R. STERLING and B. ABRAMS: Amyotrophic lateral sclerosis. Its natural history. Neurol. Clin. 5 (1987) 1–8.

CARPENTER, S.: Proximal axonal enlargement in motoneuron disease. Neurology 18 (1968) 84.

CASTAIGNE, P., P. J. CAMBIER and R. ESCOUROLLE: Sclerose Laterale Amyotrophique et les lesions degeneratives des cordons posterieurs. J. Neurol. Sci. 13 (1971) 125–35.

CENDROWSKI, W., M. WENDER and M. OWSIANOWSKI: Analyse épidémiologique de la sclérose latérale amyotrophique sur le territoire de la grande-Pologne. Acta Neurol. Scand. 46 (1970) 609–617.

CHAZOT, F., F. TABARAUD, J. M. BOULESTEIX, J. HUGON, J. M. VALLAT and M. DUMAS: Epidémiologie de la sclérose latérale amyotrophique en limousin. Rev. Neurol. (Paris) 145 (1989) 408–410.

CHEVRIE-MÜLLER, C., N. DORDAIN and F. GRÉMY: Etude de la voix et de la parole aux cours de syndromes bulbaires. J. Franç. OtoRhinoLaryng. 17 (1968) 225–232.

CHEVRIE-MÜLLER, C., N. DORDAIN and F. GRÉMY: Etude phoniatrique clinique et instrumentale des dysarthries. Rev. Neurol (Paris) 122 (1970) 123–138.

CHOU, S. M., J. D. MARTIN and J. A. GUTROCHT: Axonal balloons in subacute motoneuron disease. J. Neuropathol. Exp. Neurol. 29 (1970) 141–142.

COLMANT, H. J.: Die amyotrophische laterale sklerose. In: O. Lubarsch, F. Henke and R. Rossle (Eds), Handb. d. spez. Pathol. Anat. u. Histol., Vol. 13, Part 2B. Berlin, Springer (1958) 2624–2692.

COLMANT, H. J.: Progressive bulbar palsy in adults. In: P. J. Vinken and G. W. Bruyn (Eds), Handbook of Clinical Neurology, Vol. 22, Ch. 6. Amsterdam, N. Y., North-Holland Publ. Co. (1975) 111–156.

CONSTANTINIDIS, J.: Syndrome familiale: association de maladie de Pick et de sclérose latérale amyotrophique. Encéphale 13 (1987) 215–293.

COOPER, H. J.: Progressive bulbar paralysis with familial occurrence. Arch. Neurol. (Chic.) 30 (1933) 696–699.

CRITCHLEY, M. and C. S. KUBIK: The mechanism of speech and deglutition in progressive bulbar paralysis. Brain 48 (1925) 492–534.

DAVID, P., M. LOMONACO and G. PALIERI: Clinical features of ALS. Ital. J. Neurol. Sci. 2 (1981) 113–117.

DAZZI, P., F. S. FINIZIO and A. MERCURIALE: Rilievi clinico-statistici su 150 casi di sclerosi laterale amiotrofici. G. Psichiat. Neuropat. 97 (1969) 711–728.

DE DOMENICO, P., C. E. MALARA, L. MARABELLO ET AL.: Amyotrophic lateral sclerosis: An epidemiological study in the province of Messina, Italy, 1976–1985. Neuroepidemiology 7 (1988) 152–158.

DE PAUL, R., J. H. ABBS and M. CALIGIARI: Hypoglossal, trigeminal and facial motoneuron involvement in ALS. Neurology 38 (1988) 281–283.

DITTEL, R.: Beitrag zur Frage der Erblichkeit der amyotrophischen Lateralsklerose. Nervenarzt 13 (1940) 121–123.

DUCHENNE DE BOULOGNE, G.: Paralysie musculaire progressive de la langue, du voile du palais et des lèvres. Arch. Gen. Med. 16 (1860) 283–296, 431–445.

ENGEL, K., L. T. KURLAND and I. KLATZO: An inherited disease similar to amyotrophic lateral sclerosis with a pattern of posterior column involvement. An intermediate form? Brain 82, part II (1959) 203–220.

FORSGREN, L., B. G. L. ALMAY, G. HOLMGREN and S. WALL: Epidemiology of motor neuron disease in northern Sweden. Acta Neurol. Scand. 68 (1983) 20–29.

FOTOPULOS, D., J. BLUMENTHAL and H. SCHÜTZ: Kasuistischer Beitrag zur Frage der progressiven Bulbärparalyse als Teilsyndrom der ALS. Psychiatr. Neurol. Med. Psychol. (Leipzig) 10 (1958) 260–264.

GADERMANN, E. and H. G. MERTENS: Über die Kreislaufregulation bei der Bulbärparalyse. Verh. Dtsch. Ges. Kreislaufforsch. 19 (1953) 229–233.

GIRKE, W. and J. KOVARIK: Frühformen der myatrophischen Lateralsklerose mit besonderer Berücksichtigung des bulbärparalytischen Initialsyndroms. Arch. Psychiat. Nervenkrkh. 213 (1970) 105–120.

GOMEZ, M. R.: Progressive bulbar paralysis of childhood. Fazio-Londe disease. In: P. J. Vinken and G. W. Bruyn (Eds), Handbook of Clinical Neurology, Vol. 22, Ch. 5. Amsterdam, North-Holland Publ. Co. (1975) 103–156.

GRANIERI, E., M. CARRERAS, R. TOLA, E. PAOLINO, G. TRALLI, R. ELEPRA and G. SERRA: Motor neuron disease in the province of Ferrara, Italy, in 1964–1982. Neurology 38 (1988) 1604–168.

GREENFIELD'S NEUROPATHOLOGY: In: W. Blackwood et al. (Eds.). London, Edw. Arnold Publ. (1963).

GUDMUNDSSON, K. R.: The prevalence of some neurological disease in Iceland. Acta Neurol. Scand. 44 (1968) 57–69.

GUILLAIN, G., TH. ALAJOUANINE and I. BERTRAND: Sur un cas de paralysie bulbaire chronique avec des lésions nucleaires pures. Rev. Neurol. (Paris) 1 (1925) 577–585.

GUNNARSSON, L. G. and R. PALM: Motor neuron disease

and heavy metal manual labor: an epidemiologic survey of Warmland County, Sweden. Neuroepidemiology 3 (1984) 195–206.

HASSLER, R.: Die progressie Bulbärparalyse. In: J. V. Bergman (Ed.), Handbuch fur innere Medizin, Vol. V/3. Berlin, Springer Verlag (1953) 552.

HELFAND, M.: Progressive bulbar paralysis. Its pathology and relation to ALS. J. Nerv. Ment. Dis. 78 (1933) 362–380.

HEMMER, R.: Krankheitsdauer und Prognose verschiedener Formen der amyotrophischen Lateralsklerose und spinalen Muskelatrophie nach katamnestischen Untersuchungen. Nervenarzt 22 (1951) 427–430.

HIRANO, A., L. T. KURLAND and G. P. SAYRE: Familial ALS. A subgroup characterized by posterior and spinocerebellar tract involvement and hyaline inclusions. Arch. Neurol. (Chic.) 16 (1967) 232–242.

HIRANO, A., I. NAKANO and L. T. KURLAND: Fine structure study of neurofibrillary changes in a family with ALS. J. Neuropathol. Exp. Neurol. 43 (1984) 471–480.

HIRAYAMA, K., T. TSUBAKI and Y. TOYOKURA: The representation of the pyramidal tract in the internal capsule and basis pedunent. Neurology 12 (1962) 337–342.

HORTON, W. A., R. ELDRIDGE and J. A. BRODY: Familial MND. Evidence for at least 3 types. Neurology 26 (1976) 460–465.

HUDSON, A. J.: ALS and its association with dementia, parkinsonism and other neurological disorders. Brain (1981) 217–247.

HUDSON, A. J.: The incidence of ALS in South West Ontaria, Canada. Neurology 36 (1986) 1524–1528.

JOKELAINEN, M.: The epidemiology of amyotrophic lateral sclerosis in Finland. J. Neurol. Sci. 29 (1976) 55–63.

JUERGENS, S. M., L. T. KURLAND, H. OZAKI and D. W. MULDER: ALS in Rochester, Minnesota, 1925–1977. Neurology 30 (1980) 463–470.

KAHANA, E., M. ALTER and S. FELDMAN: Amyotrophic lateral sclerosis. A population study. J. Neurol. 212 (1976) 205–213.

KATO, T., A. HIRANO and L. T. KURLAND: Asymmetric involvement of the spinal cord, involving both large and small anterior horn cells in a case of familial ALS. Clin. Neuropathol. 6 (1987) 67–70.

KATO, T., T. KATAGIRI, A. HIRANO, H. SASAKI and S. ARAI: Sporadic lower motor neuron disease with Lewy body-like inclusions: a new subgroup? Acta Neuropathol. (Berl.) 76 (1988) 208–211.

KONDO, K. and T. TSUBAKI: Case-control studies of motor neuron disease. Association with mechanical injuries. Arch. Neurol. (Chic.) 38 (1981) 220–226.

KONNO, H., T. YAMAMOTO, Y. IWASAKI and H. IIZUKA: Shy-Drager syndrome and amyotrophic lateral sclerosis. Cytoarchitectonic and morphometric studies of sacral autonomic neurons. J. Neurol. Sci. 73 (1986) 193–204.

KOVARIK, J.: Unsere Erfahrungen auf dem Forschungsgebiet der myatrophischen Lateralsklerosen. Sborn. Ver. Praci Lek. Univ. Karlovy Hradci Kralove (1974) 233–246.

KRISTENSEN, O. and B. MELGAARD: Motor neuron disease. Prognosis and epidemiology. Acta Neurol. Scand. 56 (1977) 299–308.

KURACHI, M., T. KOIZUMI, R. MATSUBARA, K. ISAKI and H. OIWAKE: Amyotrophic lateral sclerosis with temporal lobe atrophy. Folia Psychiatr. Neurol. Jpn. 33(2) (1979) 205–215.

KURTZKE, J. F.: Epidemiology of amyotrophic lateral sclerosis. In: L. P. Rowland (Ed.), Human Motor Neuron Diseases. New York, Raven Press (1982) 281–302.

LAWYER, T. and M. G. NETSKY: Amyotrophic lateral sclerosis. A clinico anatomic study of fifty-three cases. Arch. Neurol. Psychiatr. (Chic.) 69 (1953) 171–192.

LEONE, M., A. CHIO and P. MORTON: Motor neuron disease in the province of Turin, Italy, 1971–1931. Acta Neurol. Scand. 68 (1983) 316–327.

LOPEZ-VEGA, J. M., J. CALLEJA, O. COMBARROS, J. M. POLO and J. BERCIANO: Motor neuron disease in Cantabria. Acta Neurol. Scand. 77 (1) (1988) 1–5.

LOVELL, H. W.: Familial progressive bulbar paralysis. Acta Neurol. Psychiatr. 28 (1932) 394–398.

LUNDAR, T.: Amyotrofisk lateralsklerose. 35 tilfelle fra Sentralsykehuset i Akershus. Tidsskr. Nor. Laegeforen 98 (1978) 678–681.

MACKAY, R. P.: Course and prognosis in amyotrophic lateral sclerosis. Arch. Neurol. (Chic.) 8 (1963) 117–127.

MAKIFUCHI, T. and F. IKATA: Degeneration of the accessory cuneate nucleus in familial ALS with middle root involvement. Brain Nerve (Tokyo) 29 (1977) 1332–1334.

MARBURG, O.: Die chronisch-progressiven nuclearen Amyotrophien. In: O. Bumke and O. Foerster (Eds.), Handbuch der Neurologie. Vol. 16. Berlin, Springer (1936) 524–663.

MATSUMOTO, N., R. M. WORTH, L. T. KURLAND and H. OKAZAKI: Epidemiologic study of amyotrophic lateral sclerosis in Hawaii. Identification of high incidence among Filipino men. Neurology 22 (1972) 934–940.

MEYRIGNAC, C., J. POIRIER and J. D. DEGOS: ALS presenting with respiratory insufficiency. Eur. Neurol. (Basel) 24 (1985) 115–120.

MIZUSAWA, H., S. MATSUMOTO, S. H. YEN, A. HIRANO, R. R. ROJAS-CORONA and H. DONNENFELD: Focal accumulation of phosphorylated neurofilaments within anterior horn cell in familial amyotrophic lateral sclerosis. Acta Neuropathol. (Berl.) 79 (1989) 37–43.

MORTARA, P., D. BARDELLI, M. LEONE and D. SCHIFFER: Prognosis and clinical varieties of ALS disease. Ital. J. Neurol. Sci. 2 (1981) 237–241.

MORTARA, P., A. CHIÒ, M. G. ROSSO, M. LEONE and D. SCHIFFER: Motor neuron disease in the province of Turin, Italy, 1966–1980. Survival analysis in an unselected population. J. Neurol. Sci. 66 (1984) 165–173.

MÜLLER-JENSEN, A. and W. BERNHARDT: Unsere Erfahrungen bei der myatrophischen Lateralsklerose. Nervenarzt 44 (1973) 143–149.

MURROS, K. and R. FOGERHOLM: Amyotrophic lateral sclerosis in middle — Finland: an epidemiological study. Acta Neurol. Scand. 67 (1983) 41–47.

MUSSINI, J. M., O. TALEB and C. BEAUVILLAIN DE MONTREUIL: Manifestations otorhinolaryngologique isolées revelant une sclérose latérale amyotrophique. Revue ONO 56 (1984) 329–323.

NEAU, J. P., R. ROBERT, S. ANTON: Les paralysies diaphragmatiques revelées par une insuffisance respiratoire. A propros d'un cas de SLA. Rev. Méd. Interne 9 (1988) 260–262.

NISHIGAKI, S.: Zur Klinik und Pathologie verschiedener Formen der myatrophischen Lateralsklerose. Arch. Psychiatr. Nervenkrkh. 213 (1970) 121–138.

NORRIS, F. H.: Adult spinal motor neuron disease. Progressive muscular atrophy (Aran's disease) in relation to amyotrophic lateral sclerosis. In: P. J. Vinken and G. W. Bruyn, (Eds.) Handbook of Clinical Neurology Vol. 22: Amsterdam, North-Holland Publ. Co., (1975) 1–56.

NORRIS, F. H. and L. T. KURLAND: Motor Neuron Disease. New York, Grune and Stratton (1969).

NORRIS, F. H. JR, M. J. AGUILAR, R. P. COLTON, M. B. OLDSTONE and N. E. CREMER: Tubular particles in a case of recurrent lymphocytic meningitis followed by amyotrophic lateral sclerosis. J. Neuropathol. Exp. Neurol. 34(2) (1975) 133–147.

NORRIS, F. H., R. A. SMITH and E. H. DENYS: Motor neurone disease: towards better care. Br. Med. J. 291 (1985) 259–262.

OJEDA, V. J., K. M. R. GRAINGER and T. J. DAY: Familial motor neurone disease associated with non-specific organic dementia. A clinico-pathological study of a family. Med. J. Aust. 141 (1984) 430–433.

OLIVARES, L., E. SAN ESTÉBAN and M. ALTER: Mexican 'Resistance' to amyotrophic lateral sclerosis. Arch. Neurol. (Chic.) 27 (1972) 397–402.

O'REILLY, D. F., BRAZIS, F. W. and F. A. RUBINO: The misdiagnosis of unilateral presentations of ALS. Muscle Nerve 5 (1982) 724–726.

PEARN, J.: Neuromuscular paralysis caused by tick envenomation. J. Neurol. Sci. 34 (1977) 37–42.

PEDERSEN, L., P. PLATZ, C. JERSILD and M. THOMSEN: HLA (SD and LD) in patients with amyotrophic lateral sclerosis (ALS). J. Neurol. Sci. 31 (1977) 313–318.

RAI, B. and S. S. JOLLY: Motor neuron disease. A clinical study. J. Indian. Med. Assoc. 57 (1971) 315–318.

ROBBINS, J.: Swallowing in ALS and motoneuron disease. Neurol. Clin. 5 (1987) 213–229.

ROBERTSON, E. E.: Progressive bulbar paralysis showing heredofamilial incidence and intellectual impairment. Arch. Neurol. Psychiatr. (Chic.) 69 (1953) 197–207.

ROSATI, G., L. PINNA, E. GRANIERI, I. AIELLO, R. TOLA, V. AGNETTI, A. PIRISI and P. DE BASTIANI: Studies on epidemiological, clinical and etiological aspects of ALS disease in Sardinia, Southern Italy. Acta Neurol. Scand. 55 (1977) 231–244.

ROSEN, A. D.: Amyotrophic lateral sclerosis. Clinical features and prognosis. Arch. Neurol. 35 (1978) 638–642.

ROWLAND, L. P. (Ed.): Advances in Neurology, Vol. 36. Human Motor Neuron Diseases. N.Y., Raven Press (1982).

SACK, M.: 'Betjugen' — eine eigenartige hereditäre Form der Bulbärparalyse bei Jakuten. Sovet. Nevropat. 1 (1932) 814–817.

SALEMI, G., B. FERRO and A. ARCARA: ALS in Palermo, Italy. Ital. J. Neurol. Sci. 10 (1989) 505–509.

SCHMITT, H. P., W. EMSER and C. HEIMER: Familial occurrence of ALS, Parkinson and Dementia. Ann. Neurol. 16 (1984) 642–648.

SCHRØDER, H. D. and E. RESKE-NIELSEN: Preservation of the nucleus X-pelvic floor motor system in amyotrophic lateral sclerosis. Clin. Neuropathol. 3(5) (1984) 210–216.

SERCL, M. and J. KOVARIK: The clinical forms and course of ALS. Sborn Ved. Praci. lek. Hradci Kralove 10 (1967) 411–416.

SERRATRICE, G., J. F. PELLISSIER, J. E. GASTAUD and C. DESNUELLES: Amyotrophie Spinale Chronique avec paralysie des cordes vocales: syndrome de Young et Harper. Rev. Neurol. (Paris) 140 (1984) 657–658.

SHIBASAKI, H.: Fibre-analytical study on pyramidal tracts in CVD and motor neuron disease. Folia Psychiatr. Neurol. Jpn. 22 (1968) 205–226.

SHY, M. E.: Reply to Smith/Alvarez. Neurology 37 (1987) 1688–1689.

SMITH, D. B. and R. ALVAREZ: Progressive bulbar palsy and gammopathy. Neurology 37 (1987) 1688.

STAAL, A. and L. N. WENT: Juvenile amyotrophic lateral sclerosis-dementia complex in a Dutch family. Neurology 18 (1968) 800–806.

STAAL, A. and G. TH. A. M. BOTS: A case of hereditary juvenile amyotrophic lateral sclerosis complicated with dementia. Clinical report and autopsy. Psychiatr. Neurol. Neurochir. 72 (1969) 129–135.

SWANK, R. L. and T. J. PUTNAM: ALS and related conditions. A clinical analysis. Arch. Neurol. Psychiatr. (Chic.) 49 (1943) 151–177.

TAKANA, J., H. NAKAMURA, Y. TABUCHI and K. TAKAHASHI: Familial ALS, features of multisystem degeneration. Acta Neuropathol. (Berl.) 64 (1984) 22–29.

TAKIKAWA, K.: A pedigree with progressive bulbar palsy and sex-linked recessive inheritance. Jpn. J. Genet. 28 (1953) 116–119.

TANS, J. M. J.: Amyotrophische lateraalsclerose. Een klinisch-anatomische studie. Thesis. Amsterdam, Scheltema and Holkema (1950).

TIBBALLS, J. and S. J. COOPER: Paralysis with Ixodes cornuatus envenomation. Med. J. Aust. 145(1) (1986) 37–8.

VEJJAJIVA, A., J. B. FOSTER and H. MILLER: Motor neuron disease. A clinical study. J. Neurol. Sci. 4 (1967) 299–314.

WOO, E., S. NIGHTINGALE, D. J. DICK, T. J. WALLS, J. F. FRENCH and D. BATES: A study of histocompatibility antigens in patients with motor neuron disease in the northern region of England. J. Neurol. Neurosurg. Psychiatry 49 (1986) 435–437.

YAMASHITA, K., S. KOBAYASHI, S. YAMAGUCHI, K. OKADA, S. ARIMOTO, S. FUJIHARA, K. SHIMODE, K. IMAOKA,

T. TSUNEMATSU: A case of bulbar palsy associated with Sjogren syndrome. Nippon Naika Gakkai Zasshi (IPZ) 77(8) (1988) 1280–1281.

ZACK, M. M., L. P. LEVITT and B. SCHOENBERG: Motor neuron disease in Leheigh county, Pennsylvania: an epidemiologic study. J. Chron. Dis. 30 (1977) 813–818.

Handbook of Clinical Neurology, Vol. 15 (59): Diseases of the Motor System
J.M.B.V. de Jong, editor
© Elsevier Science Publishers B.V., 1991

CHAPTER 13

Dementia and parkinsonism in amyotrophic lateral sclerosis

ARTHUR J. HUDSON

University of Western Ontario, Ontario and University Hospital, London, Ontario, Canada

The association of dementia with some cases of amyotrophic lateral sclerosis (ALS) has been recognized for almost as long as ALS has been known (Westphal 1886; Westphal 1925). This association has had many different explanations and a popular misconception, still, is that ALS is a disease with preserved intellect and 'ALS-dementia', if not a coincidence, is either a separate entity or a variant of some other disorder (Westphal 1925; Meyer 1929; Allen et al. 1971; Mitsuyama and Takamiya 1979). It has been widely considered that ALS-dementia is actually an amyotrophic form of Creutzfeldt-Jakob disease (CJ; Davison 1932; Sherratt 1974) but ALS in this or any form is not transmissible to non-human primates (Salazar et al. 1983). Moreover, such typical features of CJ as myoclonus and choreoathetosis, are not present in ALS-dementia and the course of dementia in ALS is usually much longer than in CJ. Dementia accompanying ALS can sometimes be accompanied by psychosis and in this aspect it can resemble CJ.

There are a number of clues in support of ALS-dementia and/or parkinsonism being closely related to classical ALS rather than some other entity. Possibly the strongest evidence is the recognized relationship between ALS, parkinsonism and dementia (ALS-PD) in Guam and other western Pacific areas where these diseases occur in high incidence and can even be found in the same household. The endemic occurrence of the all 3 disorders within the same localities in the western Pacific is to some extent reflected in familial ALS with dementia and/or parkinsonism in some family members. Dementia and parkinsonism may appear either independently or in combination with ALS in a single family, providing some evidence that they may be merely different aspects of the same disease. About 3.5% of cases with sporadic ALS and up to 7% of familial ALS cases show dementia; about 1.5% of all ALS patients have parkinsonian features (Hudson et al. 1986).

DEFINITION OF TERMS

Because of some differences in the use of terms an explanation of their use is helpful. Amyotrophic lateral sclerosis (ALS) as employed here is equivalent to 'motor neuron disease' in the English usage and includes the syndromes of progressive bulbar palsy (PBP), progressive muscular atrophy (PMA) and 'ALS' in the older sense of PMA plus upper motor neuron signs. Classical ALS refers to the sporadic (non-familial) form of the disease as distinct from the clinically identical (although to some extent pathologically different) familial and western Pacific conditions. Some earlier authors used the term 'extrapyramidal' to describe symptoms of rigidity, regular tremor, festinating gait, etc. that are characteristic of parkinsonism.

AMYOTROPHIC LATERAL SCLEROSIS AS A 'SYSTEM' DISEASE

Although ALS, clinically, may appear to be an exclusively upper and lower motor neuron disease in which the extraocular muscles, bladder and bowel and sensory system are not affected, nevertheless, neuronal systems other than the primary motor system are sometimes impaired. Mild eye movement disorder is recognized in ALS (Leveille et al. 1982; McGlone and Hudson 1983) but ophthalmoplegia is very rare except when patients have been allowed to survive on a respirator well beyond the time when respiratory muscle paralysis would have caused their demise (Hayashi et al. 1987). When it occurs there may be extensive neuronal degenerative changes and gliosis of the oculomotor, trochlear and abducens nuclei (Harvey et al. 1979). Sensory symptoms also have been viewed as rare in ALS but paresthesia and reduced vibration and tactile sense in the lower extremities may be present in 20% of classical and familial cases (Mulder et al. 1983, 1986; Radtke et al. 1986).

Saper et al. (1987) and others have proposed that ALS belongs to a group of conditions termed 'system degenerations' wherein a specific population of interconnected and functionally related neurons are affected by a disease process (von Matt 1964; Horoupian et al. 1984; Saper et al. 1987). The connectivity of neurons is of importance in such disorders. For example, in Alzheimer's disease, that many regard as this type of disorder, there is loss of the nucleus basalis of Meynert and other cell groups in the basal forebrain (collectively, the magnocellular basal nuclei) that are interconnected and are 'the major source of cholinergic innervation of the cerebral cortex in mammals, including humans' (Saper et al. 1985). The possibility that the system degenerations are spreading degenerative processes from one neuron to the next across normal synaptic connections suggests a number of possible causes, e.g. the transfer of toxins or infective agents. In ALS substances could be taken up at the axon terminal of a lower motor neuron and retrogradely transported to the cell body and then delivered across synaptic spaces to internuncial and upper motor neurons via their axon terminals, as has been demonstrated experimentally in the case of some toxic proteins (Schwab and Thoenen 1976; Schwab et al. 1979). Neurotropic viruses are also known to spread within the nervous system in this manner (Ugolini et al. 1989).

Central nervous system growth factors have, as yet, been poorly identified but their existence is demonstrated by the degeneration of neurons which follows their deafferentation, e.g. neuronal degeneration in the lateral geniculate body with sectioning of the optic nerve. Loss of trophic influence of one neuron upon another could provide another explanation for the loss of neurons that occurs in the system degenerations. An argument in opposition to this view in ALS is that structures are sometimes involved (e.g. thalamus and subthalamic nucleus) that do not appear to have direct connection with either the corticospinal and spinal motor neurons that are predominantly affected (Brownell et al. 1970). It is also puzzling why Onufrowicz (Onuf's) nucleus consisting of lower motor neurons that innervate the pelvic floor musculature, including the sphincters of the bowel and bladder, is spared. However, the pelvic sphincter neurons may be the only lower motor neurons, apart from the extraocular muscle motor nuclei, that do not receive direct afferents from the cerebral cortex and, in such case, they would not be transsynaptically affected by the corticospinal tracts in an anterograde system degeneration (Hudson and Kiernan 1988). Another possibility is that Onuf's nucleus may be spared only because it is relatively resistant to whatever factor is responsible for the demise of other lower motor neurons.

As stated earlier, there may be pathological changes in *classical* sporadic ALS in other parts of the nervous system in addition to the lower motor neurons or corticospinal tracts (Castaigne et al. 1972). They occur infrequently and variably and are usually mild in comparison with those in ALS-dementia and/or parkinsonism but, sometimes, changes are marked and in either case there is no clinical indication of their presence.

Some authors view them as rare or atypical but Smith (1960) observed degenerating myelin fibers throughout the cerebral cortex, thalamus, basal ganglia, substantia nigra, and other regions in 7 cases of classical ALS. Brownell et al. (1970) found degenerative changes and gliosis in the regions such as cerebral cortex, thalamus, corpus striatum, globus pallidus, subthalamic nucleus and substantia nigra in 36 cases of 'typical' ALS. *Familial* ALS cases may also show such changes in the absence of any clinical evidence other than typical upper and lower motor neuron signs. These include degenerative changes in the dorsal columns, Clarke's column, spinocerebellar tract, cerebral cortex, globus pallidus, substantia nigra, subthalamic nucleus (corpus Luysii), inferior olivary complex, posterior horns in the spinal cord and spinothalamic pathway (Hasaerts-van Geertruyden 1955; Magee 1960; Kubo et al. 1967; Moya et al. 1969; Metcalfe and Hirano 1971; Hudson 1981).

ALS in the *western Pacific*, especially Guam, has been thoroughly studied (Hirano et al. 1966; Gajdusek 1982). The extensive involvement of cortical and subcortical structures with no clinical indication of this is similar to what has been said of the classical and familial diseases. Although clinically very similar to the classical disease Guamanian ALS differs in the Alzheimer's neurofibrillary changes that accompany neuronal degeneration and occur extensively throughout the nervous system.

Thus, ALS in general is a disease that affects much more than the primary motor system and degenerative changes can be found in the cerebral cortex and subcortical nuclei. Moreover, parkinsonism and dementia can accompany each of the different types of ALS. As proposed by a number of workers, these features suggest that ALS is a system degeneration although the various types of ALS show some pathological differences.

SPORADIC AMYOTROPHIC LATERAL SCLEROSIS AND DEMENTIA/PARKINSONISM

Dementia in sporadic ALS

Clinical features. The age of onset (early sixth decade) and male to female ratio of approxi-

mately 2 to 1 for the entire group of ALS-dementia and/or parkinsonism cases are not different than in patients with ALS only. Also, life expectancy and cause of death that are very largely determined by the effects of complete and widespread skeletal muscle loss in ALS are similar.

ALS that is accompanied by dementia shows the typical clinical appearance of the classical disease with the superimposition of impaired intellect. The patient may present with either ALS or dementia but these usually appear within a year or two of each other (Hudson 1981). If onset is with ALS the patient complains of either focal limb weakness or dysarthria. With limb weakness the physician observes, sooner or later, the classical diagnostic triad of weakness, wasting and fasciculation in one or more extremities but sometimes there is, initially, only mild spasticity. The most notable cases of spasticity are those whose illness begins with dysarthria and for many months the extremities may be spared as in cases of bulbar paralysis without dementia. The dementia, whether beginning before or after the onset of ALS, generally resembles Alzheimer's or Pick's disease and usually presents with loss of recent memory. This is followed by confusion and disorientation in time and place with, in some cases, psychotic features consisting of hallucinations, severe depression, paranoia, behavioral disorder, etc. (van Bogaert 1925; Zeigler 1930; Wechsler and Davison 1932; de Morsier 1967). An unusual case of ALS-dementia with a severe psychological disorder consistent with Kluver-Bucy syndrome was described with autopsy findings by Dickson et al. (1986). The syndrome was present for 7 years to the time of death during which time the patient developed features of ALS (see Pathology).

Clinical investigation with brain scanning techniques such as magnetic resonance imaging and single proton emission computed tomography show brain atrophy and decreased cerebral metabolism in ALS-dementia (Sawada et al. 1988).

Pathology. The characteristic pathological changes in classical ALS are also found in ALS-dementia, viz. degeneration of lower motor neurons in the segments corresponding to wasted skeletal muscles in the brain stem and spinal cord, loss (or shrinkage) of large pyramidal neurons in the primary motor cortex and degeneration of the corticospinal tracts. Gliosis is present in degenerated areas. Corticospinal tract degeneration is most marked below the medulla. The posterior columns and Clarke's column are usually spared although degeneration of the spinocerebellar tracts is seen in some cases (Hudson 1981). The oculomotor, trochlear and abducens nuclei as in classical ALS are essentially unaffected. Onuf's nucleus should be spared in ALS-dementia if the disorder is a true variant of the classical disease but this feature is seldom reported.

As mentioned earlier, the extensive cortical and subcortical changes that may be present in minor degree in classical ALS without dementia are generally present in more striking degree in ALS-dementia. Degenerative changes with loss of neurons and gliosis in the frontotemporal cortex, especially in layers 2 and 3, are often described. Sometimes a spongy appearance of the cortex is also present in these layers (Delay et al. 1959; Myrianthopoulos and Smith 1962; Mitsuyama and Takamatsu 1971; Wilkstrom et al. 1982; Horoupian et al. 1984). Gliosis, but not spongiform changes, have been reported in the 3rd, 4th and 6th cortical layers (Teichmann 1935). The spongy appearance has been ascribed to the reduction of the dendritic arbor of pyramidal neurons that extend into the upper layers of the cortex (Horoupian et al. 1984). The subcortical structures that may show degenerative changes, notably neuronal loss and/or gliosis, include the caudate nucleus, putamen, globus pallidus, thalamus, subthalamic nucleus, hypothalamus, substantia nigra, red nucleus, inferior olivary complex, dentate nucleus and posterior horn of the spinal cord. Neuritic plaques and neurofibrillary tangles are seldom seen and Pick's bodies and granulovacuolar degeneration are absent (de Morsier 1967).

Neurofilamentous swellings (spheroids consisting of bundles of neurofilaments) in the initial segment of axons and in dendrites of affected neurons are usually present in classical ALS and ALS-dementia (Sasaki et al. 1988; Hirano et al. 1989). Bunina bodies, typical of the classical disease, may also be found in ALS-dementia in the same neurons (Kuroda et al. 1988). Horoupian et al. (1984) investigated 3 patients with dementia that antedated ALS by several years. Despite a clinical appearance of senile dementia of the Alzheimer's type, especially with the long history, there was no loss of cells in the substantia innominata and, in the one case where these were determined in several regions of brain, choline acetyltransferase levels and somatostatin-like immunoreactivity were normal. Spongiform changes were present in the upper layers of the frontotemporal cortex. Of interest was degeneration of the substantia nigra in all 3 cases but only one patient showed parkinsonism. Lewy bodies were absent or rare.

Autopsy in the case of Klüver-Bucy syndrome described by Dickson et al. (see above) showed the typical features of ALS but there were also highly unusual features such as scattered neuritic plaques in the cortex, some with amyloid cores, most numerous in the frontal region. There was also marked neuronal cell loss and gliosis in the amygdala, subiculum and entorhinal cortex that were compatible with Klüver-Bucy syndrome.

Parkinsonism in sporadic ALS

Clinical features. ALS-parkinsonism is less frequent than ALS-dementia and, therefore, quite rare. The 2 diseases appear at approximately the same time but, in the order of their appearance, parkinsonism usually precedes the onset of ALS (Legrand et al. 1959; Brait et al. 1973). The features of ALS and parkinsonism are typical of these diseases as they occur independently (Cordier 1951; Legrand et al. 1959; Bonduelle et al. 1959; Brait et al 1973). The parkinsonism consists of tremor at rest, cogwheel rigidity, micrographia, slowing (shuffling) of gait and other characteristic signs. In the 3 cases of ALS-parkinsonism described by Brait et al. parkinsonian symptoms were impoved by levodopa. None of their cases had histories of encephalitis lethargica or of its late sequelae (oculogyric crises, tics, emotional disturbances etc.).

Movement disorder accompanying ALS that resembles but is not typically parkinsonian or an entirely different movement disorder has also been reported. Patrikios (1951) observed a patient with ALS and intention tremor and Gray et al. (1981, 1985) described 2 unrelated cases of 'pallido-luyso-nigral' atrophy and ALS in which the patients had dystonia and rigidity in one case and torticollis with choreic and ballistic movements in the other (see below). Onset occurred at ages 29 and 32 (duration 6 and 2 years, respectively) and there was a possible familial history in both. The dystonic features apparently were first to appear in both cases.

Pathology. There is a dearth of pathological information on ALS-parkinsonism. One autopsy only, described by Greenfield and Matthews (1954), was found. This case showed marked neuronal loss and gliosis in the globus pallidus, substantia nigra and anterior horns of the spinal cord, with neurofibrillary tangles in the latter. They viewed the findings as post-encephalitic although there was no history of this disorder.

In the two patients with 'pallido-luyso-nigral' atrophy in ALS described by Gray et al. (1981, 1985; above) there was, in addition to the loss of lower motor neurons and fibers in the cortico-spinal pathway, almost complete loss of neurons and marked gliosis in the subthalamic nucleus. There were similarly marked changes in the globus pallidus, notably in the external part, and demyelination of the ansa lenticularis. The substantia nigra showed mild to moderate neuronal loss, gliosis and spongiosis, especially in the medial and internal parts. There were no other degenerative changes of significance, such as Lewy bodies, neurofibrillary changes, neuritic plaques or Pick's bodies, in brain or spinal cord.

Parkinsonism-dementia in sporadic ALS

Clinical features. There are a number of clinical and autopsy reports of cases with a combination of ALS, parkinsonism and dementia (Wechsler and Davison 1932; Caidas et al. 1966; Boudouresques et al. 1967; Bonduelle et al. 1968; Kaiya and Mehraein 1974; Hudson 1981). All 3 disorders usually appear within a few years of each other in most published accounts. Parkinson's and ALS features are typical of these disorders as they occur independently or in combination and as described above for ALS-parkinsonism. Dementia consists of memory loss and other signs of intellectual decline with, sometimes, psychotic features such as delusions and behavioral disturbances as described in cases with ALS-dementia above.

Pathology. When ALS, dementia and parkinsonism occur in combination the cerebral cortical changes are similar to those found in ALS-dementia, notably atrophy of the frontal and to a lesser extent the temporal lobes with loss of neurons and gliosis in layers 1, 2 and 3 (see references above). Status spongiosis in the upper layers of the cortex is observed in some cases. Cases 2 and 3 of Wechsler and Davison (1932) were an exception inasmuch as the neurons of layers 3, 5 and 6 of the frontal and temporal lobes were reduced in number.

Degenerative changes are present in varying degree and extent in subcortical nuclei, notably caudate nucleus, globus pallidus, subthalamic nucleus, substantia nigra, locus coeruleus, accessory olivary nuclei, thalamus, dentate nucleus and posterior horn cells of the spinal cord. Degenerative changes in the substantia nigra are generally marked. However, in the Wechsler and Davison cases 2 and 3 the substantia nigra was reported as intact although the globus pallidus showed degenerative changes. Senile plaques, Pick's bodies and, with a single exception (Bonduelle et al. 1968; case 2), neurofibrillary tangles and granulovacuolar degeneration have not been reported. Lewy bodies are either absent or not reported suggesting that they are not a significant feature.

FAMILIAL AMYOTROPHIC LATERAL SCLEROSIS (105400) AND DEMENTIA/PARKINSONISM

Familial ALS without dementia or parkinsonism is identical, clinically, to classical ALS but there are some pathological differences. Familial ALS as discussed here refers to families in which the disease occurs in 2 or more generations in a single family and, therefore, would be looked upon as dominantly transmitted. In this form it affects up to 7% of cases of ALS (Mulder et al. 1986). The disease is also seen from time to time in 2 or more members of only one generation but it is difficult to determine whether a parent who died from some other disorder might not have had ALS had they survived long enough to develop it. Thus, it is not appropriate to speak of a recessively inherited form of familial ALS at this time. It is quite possible that familial ALS is not simply an inherited disease. Perhaps only a propensity to the disease is inherited and a toxic factor or transmissible agent is also involved.

The age of onset in familial ALS is generally younger than the classical sporadic disease and in a comparison of ALS in families who have dementia and/or parkinsonism with those who do not, the age of onset is similar (late fifth decade). The male to female ratio is close to unity, about 1.2 to 1 and the duration of the disease is also similar.

As in cases of classical (non-familial) ALS there are pathological findings in the nervous system in typical familial ALS in addition to the lower motor neuron and corticospinal tract degeneration that are characteristic of ALS generally. These are not evident clinically. In about 70% of autopsies (approximately 80% of families), there are degenerative changes in the dorsal column (middle root zone of fasciculus cuneatus and/or fasciculus gracilis), Clarke's column and spinocerebellar tract (Hudson 1981; Hirano et al. 1989). Neuronal degeneration and gliosis without clinical signs are found in some cases of familial ALS in the frontotemporal cortex and subcortical structures, notably globus pallidus, subthalamic nucleus, substantia nigra, cerebellum, inferior olives and posterior horns of the spinal cord (Kubo et al. 1967; Moya et al. 1969; Hudson 1981). In familial ALS-dementia and/or parkinsonism these changes are usually found in greater degree corresponding to the clinical findings of dementia, rigidity, etc. An unexpected difference from familial ALS is the absence of dorsal column and spinocerebellar tract features in the autopsied cases of familial ALS-dementia (see Pathology).

Dementia and/or parkinsonism in familial ALS (105550)

Clinical features. In families with ALS, dementia and 'extrapyramidal disorder' (parkinsonism)

may occur independently or in combination with ALS. Usually ALS and dementia are found in combination and the clinical features are much the same as described earlier for the sporadic form of ALS-dementia. Both the ALS and dementia appear at approximately the same time in an affected individual, supporting the notion that they are probably related (Robertson 1953; Campanella and Bigi 1959; Dazzi and Finizio 1969; Finlayson et al. 1973; Pinsky et al. 1975; Hestness and Mellgren 1980). Dementia in some cases consists of a loss of memory and disorientation but can be accompanied by psychotic features. When parkinsonism accompanies ALS it is usually in the company of dementia and, in such cases, all 3 disorders appear at about the same time (Sercl and Kovarik 1963; Yvonneau et al. 1971; Alter and Schaumann 1976a, b). However, parkinsonism may be the sole accompaniment of ALS or may occur independently. Alter and Schaumann (1976a) described a family in which a single member had ALS and 3 members had ALS-parkinsonism. In the family of van Bogaert and Radermecker (1954) with ALS and parkinsonism, 2 members had ALS and 3 had only parkinsonism. One member of this family had parkinsonism and also long standing symmetrical atrophy of the upper extremities. There were no other motor neuron signs and, therefore, it is doubtful that this patient had ALS-parkinsonism.

Deymeer et al. (1989) describes a case of dementia with severe primary degeneration of the thalamus beginning 30 months prior to death with the appearance of classical ALS 10 months before death (see below). A family history was suspected since the mother and 2 maternal uncles had a similar dementing illness.

Pathology. Familial ALS-dementia reported by Robertson (1953), Finlayson et al. (1973) and Pinsky et al. (1975) showed no dorsal column or spinocerebellar tract degeneration. Frontotemporal cerebral cortical atrophy with loss of neurons, gliosis and status spongiosis in layers 2 and 3 was found in all cases. Finlayson et al. also observed pathological changes in the amygdala, locus coeruleus and periaqueductal grey. In a single autopsy case of familial ALS-dementia with features of parkinsonism, Yvonneau et al. (1971) found diffuse degeneration of the cerebral cortex, caudate nucleus and putamen. There was also dorsal column and spinocerebellar tract degeneration as usually observed in familial ALS.

Pathological examination in the case of 'thalamic dementia' described by Deymeer et al. (1989; above) showed profound neuronal loss and reactive gliosis in the mediodorsal nucleus of the thalamus bilaterally with similar but less intense changes in the adjacent ventral anterior and centromedian nuclei. Other thalamic nuclei appeared normal. There was mild to moderate gliosis and microvacuolation of the neuropil of the superficial layers of the frontal and to a lesser extent the temporal cortex but these changes were considered insufficient to account for the patient's dementia. The basal nuclei of Meynert were normal. The medulla and spinal cord showed loss of lower motor neurons and the corticospinal tracts in the medulla and spinal cord showed marked degenerative changes. The authors in this study attempted to transmit the disorder to primates without success and therefore concluded that the patient had a unique, possibly familial multisystem degeneration and not Creutzfeldt-Jakob disease.

WESTERN PACIFIC AMYOTROPHIC LATERAL SCLEROSIS

The high incidence foci of ALS in the western Pacific have attracted attention to the association of ALS with parkinsonism and dementia. These foci are located mainly on the island of Guam, the Kii peninsula of Japan and in the Auyu and Jakai villages of West Irian. Dementia and parkinsonism have shown a marked tendency to occur together but they may also occur in individuals who have ALS. Clinically, ALS in these regions is identical to classical ALS and the parkinsonism and dementia resemble the disorders that accompany the classical disease.

Pathologically, there are a number of similarities between western Pacific and classical ALS such as severe loss of anterior horn cells, degeneration of the corticospinal tracts and sparing of the posterior columns and spinocerebellar tracts (Hirano et al. 1966). Moreover, degenerative

changes are found in the cerebral cortex and the same subcortical nuclei in both conditions but, in contrast to the classical disorder, there is prominent neurofibrillary degeneration that accompanies the neuronal loss and gliosis in western Pacific ALS. The neurofibrillary changes have been found throughout the brain and spinal cord, especially in the frontotemporal cortex, hippocampus, substantia nigra, locus coeruleus and dentate nucleus. Neurofibrillary tangles in neurons are also seen in patients who have parkinsonism and dementia and even clinically normal members of the indigenous population. Granulovacuolar degeneration that is not present in classical or familial ALS has been observed in the pyramidal cells of Ammon's horn (Hirano et al. 1966).

It is noteworthy that post-encephalitic ALS-parkinsonism that followed the epidemics of encephalitis lethargica in the early part of this century also showed extensive involvement of subcortical nuclei with neurofibrillary changes and granulovacuolar degeneration (McMenemey et al. 1967; Hudson 1981).

DISCUSSION AND CONCLUSIONS

The clinical and pathological findings in classical and familial ALS are similar apart from some findings in the familial disorder, such as dorsal column degeneration, that are either infrequently observed or less prominent in the classical disease. The most profound degenerative changes in both conditions are found in the lower motor neurons and corticospinal pathways but there may be pathological alterations of usually mild degree in the cerebral cortex and subcortical nuclei in the absence of any clinical evidence that these structures are affected.

When ALS is accompanied by dementia and parkinsonism alterations in the same cortical and subcortical structures are generally more striking. These changes include degeneration and loss of neurons and gliosis in the frontotemporal cerebral cortex, thalamus, caudate nucleus, globus pallidus, subthalamic nucleus, substantia nigra and locus coeruleus. In the frontotemporal cortex there is often a spongy appearance in the 2nd and 3rd laminae that has been ascribed to the loss of the dendritic arbors of pyramidal cells in the deeper layers. There are, very occasionally, degenerative changes in other subcortical nuclei such as the inferior olivary complex and posterior horn cells of the spinal cord but if other ordinarily unaffected groups of neurons are involved the diagnosis should be questioned. Western Pacific ALS is clinically similar to classical and familial ALS and they have a number of pathological changes in common, notably the cortical regions and subcortical nuclei that are affected, but the prominent neurofibrillary degeneration and granulovacuolar changes in the pryamidal cells of Ammon's horn are found almost exclusively in western Pacific ALS. Only the post-encephalitic form of ALS seems to have these features in common with western Pacific disease.

It is concluded that the different forms of ALS have, both clinically and pathologically, much in common. However, each also has some distinctive pathological findings with the greatest differences being seen in the western Pacific form. The association of dementia and parkinsonism with ALS is probably an extension of the disease process from the predominant involvement of the lower and upper motor neurons. Each of the different types of ALS appears to be a system degeneration wherein specific groups of neurons are involved either through their connectivity and/or a common vulnerability to, as yet, unidentified lethal factors. Because the pathological changes differ in the different types of ALS, especially in the western Pacific ALS as compared to the other types, it is probable that the causal factors (or agents) are also different.

REFERENCES

ALLEN, I. V., E. DERMOTT, J. H. CONNOLLY and L. J. HURWITZ: A study of a patient with the amyotrophic form of Creutzfeldt-Jakob disease. Brain 94 (1971) 715–724.

ALTER, M. and B. SCHAUMANN: A family with amyotrophic lateral sclerosis and parkinsonism. J. Neurol. 212 (1976a) 281–284.

ALTER, M. and B. SCHAUMANN: Hereditary amyotrophic lateral sclerosis. A report of two families. Eur. Neurol. 14 (1976b) 250–265.

BONDUELLE, M., P. BOUYGUES, J. DELAHOUSSE and

C. FAVERET: Evolution simultanée chéz un même malade de Parkinson a d'une sclérose latérale amyotrophique. Discussion. Rev. Neurol. 101 (1959) 63–66.

BONDUELLE, M., P. BOUYGUES, R. ESCOUROLLE and G. LORMEAU: Evolution simultanée d'une sclérose latérale amyotrophique, d'un syndrome Parkinsonnien et d'une démence progressive. A propos de deux observations anatomo-cliniques. Essai d'interprétation. J. Neurol. Sci. 6 (1968) 315–332.

BOUDOURESQUES, J., M. TOGA, J. ROGER, R. KHALIL, R. A. VIGOUROUX, W. PELLET and J. J. HASSOUN: Etat démentiel, sclérose latérale amyotrophique, syndrome extrapyramidal. Etude anatomique. Discussion nosologique. Rev. Neurol. 116 (1967) 693–704.

BRAIT, K., S. FAHN and G. A. SCHWARZ: Sporadic and familial parkinsonism and motor neuron disease. Neurology, (Minneap.) 23 (1973) 990–1002.

BROWNELL, B., D. R. OPPENHEIMER and J. T. HUGHES: The central nervous system in motor neuron disease. J. Neurol., Neurosurg. Psychiatry 33 (1970) 338–357.

CAIDAS, M., V. MARCUTU and O. VUIA: Sclérose latérale amyotrophique associée à la démence et au Parkinsonisme. Acta Neurol. Psychiatr. Belg. 66 (1966) 719–731.

CAMPANELLA, G. and A. BIGI: Su di un caso di sclerosi laterale amiotrofica a carattere familiare. G. Psichiatr. Neuropatol., Ferrara 87 (1959) 804–811.

CORDIER, J.: Syndrome parkinsonien avec des amyotrophies rappelant la sclérose latérale amyotrophique et d'origine post traumatique. Acta Psychiatr. Neurol. Belg. 51 (1951) 194–205.

CASTAIGNE, P., F. LHERMITTE, J. CAMBIER, R. ESCOUROLLE, P. LE BIGOT: Etude neuropathologique de sclérose latérale amyotrophique. Discussion nosologique. Rev. Neurol. 127 (1972) 401–414.

DAVISON, C.: Spastic pseudosclerosis (cortico-pallido-spinal degeneration). Brain 55 (1932) 247–264.

DAZZI, P. and F. S. FINIZIO: Sulla sclerosi laterale amiotrofica familiare contributo clinico. G. Psichiatr. Neuropatol. 97 (1969) 299–337.

DELAY, J., S. BRION, R. ESCOUROLLE and R. MARTY: Sclérose latérale amyotrophique et démence (à propos de deux cas anatomo-cliniques). Rev. Neurol. 100 (1959) 191–204.

DE MORSIER, G.: Un cas de maladie de Pick avec sclérose latérale amyotrophique terminale. Contribution à la sémiologie temporale. Rev. Neurol. 116 (1967) 373–382.

DEYMEER, F., T. W. SMITH, U. DEGIROLAMI and D. A. DRACHMAN: Thalamic dementia and motor neuron disease. Neurology 39 (1989) 58–61.

DICKSON, D. W., D. S. HOROUPIAN, L. J. THAL, P. DAVIES, S. WALKLEY and R. D. TERRY: Klüver-Bucy syndrome and amyotrophic lateral sclerosis: A case report with biochemistry, morphometrics, and Golgi study. Neurology 36 (1986) 1323–1329.

FINLAYSON, M. H., A. GUBERMAN and J. B. MARTIN: Cerebral lesions in familial amyotrophic lateral sclerosis and dementia. Acta Neuropathol. 26 (1973) 237–246.

GAJDUSEK, D. C.: Foci of motor neuron disease in high incidence in isolated populations of East Asia and the Western Pacific. In: L. P. Rowland (Ed.), Human Motor Neuron Diseases. New York, Raven Press (1982) 363–393.

GRAY, F., C. DE BAECQUE, M. SERDARU and R. ESCOUROLLE: Pallido-Luyso-nigral atrophy and amyotrophic lateral sclerosis. Acta Neuropathol. (Berlin) Suppl VII (1981) 348–351.

GRAY, F., J. F. EIZENBAUM, R. GHERARDI, J. D. DEGOS and J. POIRIER: Luyso-pallido-nigral atrophy and amyotrophic lateral sclerosis. Acta Neuropathol. (Berlin) 66 (1985) 78–82.

GREENFIELD, J. G. and W. B. MATTHEWS: Post-encephalitic Parkinsonism with amyotrophy. J. Neurol., Neurosurg. Psychiatry 17 (1954) 50–56.

HARVEY, D. G., R. M. TORACK and H. E. ROSENBAUM: Amyotrophic lateral sclerosis with ophthalmoplegia. A clinicopathologic study. Arch. Neurol. 36 (1979) 615–617.

HASAERTS-VAN GEERTRUYDEN: Sur la sclérose latérale amyotrophique héréditaire. J. Génet. Hum. 4 (1955) 152–163.

HAYASHI, H., S. KATO, T. KAWADA and T. TSUBAKI: Amyotrophic lateral sclerosis: oculomotor function in patients in respirators. Neurology 37 (1987) 1431–1432.

HESTNES, A. and S. I. MELLGREN: Familial amyotrophic lateral sclerosis. Report of a family with predominant upper limb pareses and late onset. Acta Neurol. Scand. 61 (1980) 192–199.

HIRANO, A., N. MALAMUD, T. S. ELIZAN and L. T. KURLAND: Amyotrophic lateral sclerosis and Parkinsonism-dementia complex on Guam. Further pathologic studies. Arch. Neurol. (Chic.) 15 (1966) 35–51.

HIRANO, A., M. HIRANO and H. M. DEMBITZER: Pathological variations and extent of the disease process in ALS. In: A. J. Hudson (Ed.), Amyotrophic Lateral Sclerosis: Current Clinical and Pathophysiological Evidences for Differences in Etiology. Toronto, University of Toronto Press (1990) 166–192.

HOROUPIAN, D. S., L. THAL, R. KATZMAN, R. D. TERRY, P. DAVIES, A. HIRANO, R. DE TERESA, P. A. FULD, C. PETITO, J. BLASS and J. M. ELLIS: Dementia and motor neuron disease: Morphometric, biochemical, and Golgi studies. Ann. Neurol. 16 (1984) 305–313.

HUDSON, A. J.: Amyotrophic lateral sclerosis and its association with dementia, parkinsonism and other neurological disorders; a review. Brain 104 (1981) 217–247.

HUDSON, A. J. and J. A. KIERNAN: Preservation of certain voluntary muscles in motoneurone disease. Lancet i (1988) 652–653.

HUDSON, A. J., A. DAVENPORT and W. J. HADER: The incidence of amyotrophic lateral sclerosis in southwestern Ontario, Canada. Neurology 36 (1986) 1524–1528.

KAIYA, H. and P. MEHRAEIN: Zur Klinik und pathologischen Anatomie des Muskelatrophie-Parkinsonismus-Demenz-Syndroms. Arch. Psychiatr. Nervenkr. 219 (1974) 13–27.

KUBO, H., F. IKUTA and T. TSUBAKI: An autopsy case with a history of familial amyotrophic lateral sclerosis and posterior column involvement. Clin. Neurol. 7 (1967) 45–50.

KURODA, S., Y. HAYASHI and R. NANBA: Bunina bodies in the motor nuclei of the brain stem with motor neuron disease and dementia. Clin. Neurol. 28 (1988) 292–295.

LEGRAND, R., M. LINQUETTE, J. DELAHOUSSE and A. GÉRARD: A propos d'un nouveau cas d'association d'une maladie de Parkinson et d'une sclérose latérale amyotrophique. Rev. Neurol. 101 (1959) 191–193.

LEVEILLE, A., J. KIERNAN, J. A. GOODWIN, J. ANTEL: Eye movements in amyotrophic lateral sclerosis. Arch. Neurol. 39 (1982) 684–686.

MAGEE, K. R.: Familial progressive bulbar-spinal muscular atrophy. Neurology (Minneap.) 10 (1960) 295–305.

MCGLONE, J. and A. J. HUDSON: An eye movement disorder in ALS. Neurology 33 (1983) 254–255.

MCMENEMEY, W. H., R. O. BARNARD and E. H. JELLINEK: Spinal amyotrophy. A late sequel of epidemic encephalitis (von Economo). Rev. Roum. Neurol. 4 (1967) 251–259.

METCALFE, C. W. and A. HIRANO: Amyotrophic lateral sclerosis. Clinicopathological studies of a family. Arch. Neurol. (Chic.) 24 (1971) 518–523.

MEYER, A.: Über eine der amyotrophischen Lateralsklerose nahestehende Erkrankung mit psychischen Störungen. Zugleich ein Beitrag zur Frage der spastischen Pseudosklerose (A. Jakob). Z. Gesamte Neurol. Psychiatr. 121 (1929) 107–138.

MITSUYAMA, Y. and I. TAKAMATSU: An autopsy case of presenile dementia with motor neuron disease. Brain Nerve 23 (1971) 409–416.

MITSUYAMA, Y. and S. TAKAMIYA: Presenile dementia with motor neuron disease in Japan. A new entity? Arch. Neurol. (Chic.) 36 (1979) 592–593.

MOYA, G., G. MIRANDA-NIEVES and M. PEREZ SOTELO: Un cas familial d'amyotrophie spinale progressive montrant une atteinte histologique, cliniquement muette du pallidum, du locus niger, du noyau de Luys et du faiseau de Goll. Acta Neurol. Belg. 69 (1969) 1002–1012.

MULDER, D. W., W. BUSHEK, E. SPRING, J. KARNES and P. J. DYCK: Motor neuron disease (ALS): Evaluation of detection thresholds of cutaneous sensation. Neurology 33 (1983) 1625–1627.

MULDER, D. W., L. T. KURLAND, K. P. OFFORD and C. M. BEARD: Familial adult motor neuron disease: Amyotrophic lateral sclerosis. Neurology 36 (1986) 511–517.

MYRIANTHOPOULOS, N. C. and J. K. SMITH: Amyotrophic lateral sclerosis with progressive dementia and pathologic findings of the Creutzfeldt-Jakob syndrome. Neurology (Minneap.) 12 (1962) 603–610.

PATRIKIOS, M. J.: Sclérose latérale amyotrophique avec

mouvement involontaire des doigts et du poignet gauches de caractère extrapyramidal. Rev. Neurol. 85 (1951) 60–62.

PINSKY, L., M. H. FINLAYSON, I. LIBMAN and B. H. SCOTT: Familial amyotrophic lateral sclerosis with dementia: a second Canadian family. Clin. Genet. 7 (1975) 186–191.

RADTKE, R. A., A. ERWIN and C. W. ERWIN: Abnormal sensory evoked potentials in amyotrophic lateral sclerosis. Neurology 36 (1986) 796–801.

ROBERTSON, E. E.: Progressive bulbar paralysis showing heredofamilial incidence and intellectual impairment. Arch. Neurol. Psychiatry (Chic.) 69 (1953) 197–207.

SALAZAR, A. M., C. L. MASTERS, D. C. GAJDUSEK and C. J. GIBBS: Syndromes of amyotrophic lateral sclerosis and dementia: Relation to transmissible Creutzfeldt-Jakob disease. Ann. Neurol. 14 (1983) 17–26.

SAPER, C. B., D. C. GERMAN and C. L. WHITE III: Neuronal pathology in the nucleus basalis and associated cell groups in senile dementia of the Alzheimer's type: Possible role in cell loss. Neurology 35 (1985) 1089–1095.

SAPER, C. B., B. H. WAINER and D. C. GERMAN: Axonal and transneuronal transport in the transmission of neurological disease: potential role in system degenerations, including Alzheimer's disease. Neuroscience 23 (1987) 389–398.

SASAKI, S., H. KAMEI, K. YAMANE and S. MARUYAMA: Swelling of neuronal processes in motor neuron disease. Neurology 38 (1988) 1114–1118.

SAWADA, H., F. UDAKA, Y. KISHI, N. SERIU, T. MEZAKI, M. KAMEYAMA, M. HONDA and M. TOMONOBU: Single photon emission computed tomography in motor neuron disease with dementia. Neuroradiology 30 (1988) 577–578.

SCHWAB, M. E. and H. THOENEN: Electron microscopic evidence for a transsynaptic migration of tetanus toxin in spinal cord motoneurons: an autoradiographic and morphometric study. Brain Res. 105 (1976) 213–227.

SCHWAB, M. E., K. SUDA and H. THOENEN: Selective retrograde transsynaptic transfer of a protein, tetanus toxin subsequent to its retrograde axonal transport. J. Cell Biol. 82 (1979) 798–810.

SERCL, M. and J. KOVARIK: On the familial incidence of

amyotrophic lateral sclerosis. Acta Neurol. Scand. 39 (1963) 169–176.

SHERRATT, R. M.: Motor neurone disease and dementia: probably Creutzfeldt-Jakob disease. Proc. R. Soc. Med. 67 (1974) 1063–1064.

SMITH, M. C.: Nerve fibre degeneration in the brain in amyotrophic lateral sclerosis. J. Neurol., Neurosurg. Psychiatry 23 (1960) 269–282.

TEICHMANN, E.: Über einen der amyotrophischen Lateralsclerose nahestehenden Krankheitsprozes mit psychischen Symptomen. Z. Gesamte Neurol. Psychiatr. 154 (1935) 32–44.

UGOLINI, G., H. G. J. M. KUYPERS and P. L. STRICK: Transneuronal transfer of herpes virus from peripheral nerves to cortex and brainstem. Science 243 (1989) 89–91.

VAN BOGAERT, L.: Les troubles mentaux dans la sclérose latérale amyotrophique. L'Encephale 20 (1925) 27–47.

VAN BOGAERT, L. and M. A. RADERMECKER: Scléroses latérales amyotrophiques typiques et paralysies agitantes héréditaires, dans une même famille, avec une forme de passage possible entre les deux affections. Monatsschr. Psychiatr. Neurol. 127 (1954) 185–203.

VON MATT, K.: Progressive Bulbärparalyse und dementielles Syndrom. Psychiatr. Neurol., Basel 148 (1964) 354–364.

WECHSLER, I. S. and C. DAVISON: Amyotrophic lateral sclerosis with mental symptoms. A clinicopathologic study. Arch. Neurol. Psychiatry (Chic.) 27 (1932) 857–880.

WESTPHAL, A.: Schizophrene Krankheitsprozesse und amyotrophische Lateralsklerose. Arch. Psychiatr. Nervenk. 74 (1925) 310–325.

WESTPHAL, C.: Fall von amyotrophischer Lateralsklerose mit Bulbärparalyse. Arch. Psychiatr. Nervenkr. 17 (1886) 279–283.

WIKSTRÖM, J., A. PAETAU, J. PALO, R. SULKAVA and M. HALTIA: Classic amyotrophic lateral sclerosis with dementia. Arch. Neurol. 39 (1982) 681–683.

YVONNEAU, M., C. VITAL, C. BELLY and M. COQUET: Syndrome familial de sclérose latérale amyotrophique avec démence. L'Encephale 60 (1971) 449–462.

ZIEGLER, L. H.: Psychotic and emotional phenomena associated with amyotrophic lateral sclerosis. Arch. Neurol. 24 (1930) 930–936.

Handbook of Clinical Neurology, Vol. 15 (59): Diseases of the Motor System
J.M.B.V. de Jong, editor
© Elsevier Science Publishers B.V., 1991

CHAPTER 14

Familial amyotrophic lateral sclerosis

D. B. WILLIAMS

Department of Neurology, John Hunter Hospital, Newcastle, N.S.W., Australia

Since Bonduelle (1975) reviewed familial amyotrophic lateral sclerosis (ALS) in his chapter in Vol. 22 of the original series of this handbook, the neuropathological features of the disease (Horton et al. 1976) and the subject in its entirety have both been extensively reviewed (Emery and Holloway 1982), many new families have been described (Alter and Schaumann 1976; Estrin 1977; Husquinet and Franck 1980; Hawkes et al. 1984; Ojeda et al. 1984; Li et al. 1988; Chio et al. 1987; Selby 1987; Williams et al. 1988; Veltema et al. 1990), the clinical features of the disease have been carefully studied (Giminez-Roldan and Esteban 1977; Mulder et al. 1986; Li et al. 1988), and new methods have been applied to investigate the underlying etiology (Siddique et al. 1989; Williams 1989; Siddique et al. 1990). Despite these efforts, the etiology of the disease, and its relationship to the sporadic form of ALS are, perhaps, even more mysterious than was the case in 1975. Additional information has undoubtedly raised more questions than have yet been answered.

Important problems which remain unsolved include whether familial ALS is homogeneous, or consists of several phenotypically similar but etiologically distinct diseases (Chio et al. 1987); whether any specific clinical features distinguish familial from sporadic ALS; what the clinical and pathological limits of familial ALS are; and

in what ways the etiologies of sporadic and familial ALS may be related to one another (Mulder et al. 1986).

ETIOLOGY

Most reports (Kurland and Mulder 1955; Bonduelle 1975; Emery and Holloway 1982; Mulder et al. 1986; Chio et al. 1987; Li et al. 1988) interpret the observed patterns of familial aggregation as being consistent with autosomal dominant inheritance of a disease gene (105400). It is, however, important to note important reservations concerning that interpretation. In some families one must also postulate diminished penetrance (Kurland and Mulder 1955; Gardner and Feldmahn 1966; Horton et al. 1976; Chio et al. 1987; Williams et al. 1988), and similar aggregations could occur as the result of common exposure to an environmental toxin, or vertical transmission of an infectious agent (Emery and Holloway 1982; Li et al. 1988).

Several recent reports concern families with relatively few affected family members, and it is not clear that all these aggregations represent instances of autosomal dominant inheritance (Chio et al. 1987; Li et al. 1988; Williams, 1989). Frequently, despite careful genealogical research, only 2 or 3 affected individuals can be identified in large pedigrees. In these cases, the interpretation of the familial aggregation may depend on whether the affected individuals are siblings, cousins, or members of different generations within a family. It seems likely that in some families affected first-degree relatives bring a wider, otherwise unrecognized familial aggregation to attention (Selby 1987), while in other families chance, common environmental exposure, or vertical transmission may play a yet-to-be-determined rôle.

The evidence suggesting autosomal recessive inheritance of amyotrophic lateral sclerosis is weak (Dumon et al. 1971; Horton et al. 1976; Estrin 1977), except in juvenile-onset cases, which are often distinguished by prolonged survival, and may be found in a particular ethnic group, or restricted geographical region (Ben Hamida et al. 1990).

Many commentators assume a rôle for environmental modifying factors in determining the characteristics of familial ALS, but the relative contribution of genetic and environmental factors is unknown. Theoretically, careful twin studies could help to determine the relative contribution of genetic and environmental factors. In a genetically determined disease, monozygotic twin pairs should be concordant for the disease twice as frequently as dizygotic twin pairs. Unfortunately, there is no clear interpretation of the limited twin data which is available in ALS.

There are 3 reports of ALS in dizygotic twins but in none of these were other affected family members recorded. In the first report (Dumon et al. 1971) twins were concordant for the disease, but the parents were related, suggesting that the disease in that case may be an autosomal recessive trait. In the second report (Estrin 1977), dizygotic twins developed typical ALS within 2 years of one another. Estrin proposed that ALS was caused by an intra-uterine infective or toxic insult which manifested itself many years later as progressive ALS. However, Jokelainen et al. (1978) reported a set of monozygous female twins who were discordant for ALS 4 years after the onset of the disease in the first twin, which he interpreted as evidence against Estrin's hypothesis of an intra-uterine insult, and more generally as evidence against hereditary etiology in ALS.

In the context of definite familial ALS, 3 pairs of twins have been reported, 2 pairs presumed to have been monozygotic (Mulder et al. 1986; Williams 1989), and the other proven to be so using a range of genetic markers (Williams 1989). The members of the first twin pair were concordant for ALS, the disease duration was less than 3 years in both, and the second member of the pair developed the disease less than 2 years after the first (Mulder et al. 1986). The members of the second twin pair were also concordant for ALS, but disease onset was 11 years later in the second member of the twin pair. The disease lasted less than 3 years in both (Williams 1989). In the only twin pair proven to be monozygotic using genetic markers, the second member of the pair was alive and apparently unaffected 7 years after the onset of the disease in his brother, who survived less than 2 years (Williams 1989). Significant variation in the age at onset in monozygotic twins suggests, but does not prove the existence of important environmental factors in determining the onset, if not the occurrence of familial ALS (Edwards 1969).

Improving knowledge in many areas of basic science will eventually permit tests of specific hypotheses concerning gene-environment interactions. A potentially promising lead is the recent report that patients with ALS are both significantly more likely than controls to have a diminished capacity for sulfation and sulfoxidation (Steventon et al. 1988), and to have an increased capacity for S-methylation (Waring et al. 1989). As sulfation and sulfoxidation (Weinshilboum 1989, 1990) and S-methylation (Keith et al. 1983) are functions of enzymes which are at least partly under genetic control, some instances of familial aggregation of ALS could be due to idiosyncratic metabolism of an agent which is readily detoxified by the majority of the population. Strong evidence suggests that a toxic, intermediary metabolite (oxidized N-methyl-4-phenyl-1,2,3,6-tetrahydropyridine (MPTP)) causes a parkinsonian syndrome following the use of synthetic meperidine analogues (Burns et al. 1983), and patients with Parkinson's disease have significantly diminished capacities for sulfation, sulfoxidation, and S-methylation (Steventon et al. 1989; Waring et al. 1989). However, as enzyme activity is frequently regulated by substrate availability the observed variations in enzyme activity may be a result of ALS rather than its cause. Some distinguishing biochemical features of familial ALS have been reported (Poser et al. 1965), and still others may be postulated from reports of abnormalities in sporadic ALS (Plaitakis et al. 1984; Plaitakis and Caroscio 1987), but the significance of these findings in determining familial aggregation remain to be determined.

In some reports of familial ALS, individuals apparently failed to develop the disease despite the logical necessity of their carrying the disease gene if it is dominantly inherited (Kurland and Mulder 1955; Gardner and Feldmahn 1966; Horton et al. 1976; Mulder et al. 1986; Veltema et al. 1990). In these cases the authors postulated that the individual died 'prematurely' and that they would have developed the disease had they lived longer. This explanation is plausible because of the late onset and short duration of familial ALS, and is supported by additional observations (see below). However, it is unclear to what extent the explanation may be more widely applied to familial aggregations with small numbers of affected individuals, or even some cases of apparently sporadic ALS (Williams et al. 1988; Williams, 1989).

There have been few opportunities to examine or closely follow members who must carry any postulated disease gene, yet were not known to be affected at the time of family ascertainment. However, the available evidence appears to support the hypothesis that some individuals fail to develop familial ALS only because they die prematurely. In 2 cases the postulated gene carrier was found to be affected by ALS (O-II(3) in Williams et al. 1988), or to develop it later (III-3 of Family B in Engel et al. 1959; Dr. L.T. Kurland, pers. commun.). One additional case may have been in a prodromal stage of the illness (O-II(1) in Williams et al. 1988), 1 had clinical features consistent with ALS which were ascribed to 2 separate, unrelated disease processes by the attending physician, and 2 more individuals were unaffected when they were examined at ages 38 and 94, respectively (D.B. Williams, unpublished data).

In each of these families the initially unrecognized affected individuals, and those who later developed ALS, were significantly older than the affected individuals identified when the family was first ascertained. Although it is infrequently recognized, individuals may develop familial ALS in the 8th, 9th or even the 10th decade of life, and in some families failure to include elderly affected individuals could distort the perceived clinical features of the disease in that familial aggregation. If our larger understanding of familial ALS is similarly confounded by failure to include elderly affected family members, and those afflicted by chronic diseases, we do not yet know the magnitude if this effect. If the effect is large it would have important implications for both estimates of the proportion of ALS cases which are actually familial (see below), and all hypotheses concerning disease etiology.

FEATURES DISTINGUISHING FAMILIAL AND
SPORADIC ALS

Although individual cases of familial and sporadic ALS are indistinguishable in the absence of a family history, several clinical features have been reported to distinguish familial ALS (collectively) from sporadic ALS. These are younger age at onset, a tendency for the disease to develop first in the lower limbs, and a sex ratio close to 1:1 (Bonduelle 1975; Rosen 1978; Emery and Holloway 1982; Mulder et al. 1986). In the largest clinical series of familial ALS cases to date (Mulder et al. 1986), each of these clinical features helped to distinguish known familial ALS from sporadic disease. Mulder reported the characteristics of 100 probands from 73 families identified at the Mayo Clinic, and the familial ALS cases were, on average, younger than the sporadic cases. However, the magnitude of that difference depended on the comparison group chosen. The difference was large when comparing familial cases with sporadic cases identified in surveillance of the Olmsted County population, but was much smaller when comparing them with sporadic cases referred to the Mayo Clinic from other regions. When Li et al. (1988) compared the clinical characteristics of familial and sporadic ALS cases referred to London teaching hospitals, they also found only a modest difference between the average age at onset in the familial and the sporadic cases, although the familial cases were, on average, younger.

The authors of the Mayo Clinic study (Mulder et al. 1986) identified and discussed the probable selective recognition of familial disease in their series. Cases had been identified from clinical summaries, and there were several ways in which familial cases may have gone unrecognized. Unfortunately, in the absence of any population-based study of familial ALS, the magnitude and direction of bias arising from this selective recognition cannot be determined with certainty. In general, however, referral bias tends to favour the recognition of younger patients with a more favorable prognosis (Juergens and Kurland 1979).

In the Mayo Clinic series the majority of familial ALS cases developed the disease first in the lower limbs. In clinical series of ALS in general, older age tends to be associated with bulbar onset (Rosen 1978; Gubbay et al. 1985), but Mulder found no relationship between site and age of onset in his familial cases.

By contrast, in the study by Li and colleagues, the site of onset did not differ significantly between familial and sporadic cases. Although a preponderant site of onset did not distinguish familial from sporadic ALS in this study, the finding is difficult to interpret. Almost 25% of their patients were members of one large family, and although clinical features vary within a family, there may be sufficient consistency to help distinguish one family from another (Bonduelle 1975). Long (Espinosa et al. 1962; Horton et al. 1976), or short (Hawkes et al. 1984) disease duration, or a tendency to commence in one anatomical area, such as the bulbar region (Wolfenden et al. 1973), or the shoulder girdle and upper extremities (Veltema et al. 1990), appear to be distinguishing family characteristics in some pedigrees.

The sex ratio of affected individuals in both Mulder's and Li's series did not differ significantly from 1:1, but these results must be interpreted cautiously. The true sex-specific risk of developing familial ALS remains unknown because the sex ratio in the 'at risk' cohort was not reported. In Li's series, the sex ratio among affected individ-

uals in familial ALS did not differ significantly from that in sporadic ALS, although the number of individuals with familial ALS was relatively small.

In another comparison of familial and sporadic ALS (Williams 1989) the author divided 105 familial ALS cases into 2 groups depending on whether the familial aggregation was strong, with multiple affected generations and a ratio of affected individuals approaching 50%, or weak, and therefore less confidently attributable to autosomal dominant inheritance of a disease gene. In that analysis, lower average age of disease onset, a tendency to develop the disease first in the lower limbs, and an equal sex ratio were all associated with strong familial aggregation suggestive of autosomal dominant inheritance. In other forms of familial aggregation, the average age of onset was higher, the lower limbs were not the preponderant site of onset, and there were more affected males than affected females, all characteristics which made these familial cases appear more similar to the comparison cases of sporadic ALS. These results may be interpreted to suggest either that some familial cases are merely chance aggregations of sporadic ALS, or that there is a clinical continuum of familial disease which blends into that of sporadic ALS. If the latter interpretation is adopted, the continuity in clinical phenotype suggests the possibility of common etiological factors in familial and sporadic ALS.

Several clinical features have been found not to be reliable discriminators of familial and sporadic ALS. Because the posterior columns are clearly affected in 'multisystem' familial ALS, clinical sensory disturbance was considered as a potential marker of familial disease. However, in Mulder's large series of Mayo Clinic probands, 20% of familial cases had sensory signs or paresthesias compared with 25% of sporadic referral cases, and the difference was not significant. In Li's series, 14.8% of documented familial cases had either sensory symptoms or signs compared with only 5.2% of sporadic cases, but this difference was also not statistically significant. To date there has been no systematic study of sensory disturbance in familial and sporadic ALS using quantitative methods.

RELATIONSHIP OF FAMILIAL ALS TO OTHER NEURO-DEGENERATIVE DISEASES

Several authors have postulated that there is a relationship between ALS and dementia or parkinsonism, partly because these disorders are associated in the same restricted geographical regions, the same families, and even the same patients in the Mariana Islands of the Western Pacific (Kurland and Mulder 1955; Hirano et al. 1961a, b). There is no clear evidence for a similar association outside the Western Pacific region. An extensive review by Hudson (1981) examined the evidence for an association of these disorders in familial ALS. As presented by Hudson, and confirmed by additional evidence (Ojeda et al. 1984; Schmitt et al. 1984), there are families in which an apparently dominantly inherited phenotype has features of 2 or more of these otherwise distinct syndromes (i.e. ALS, parkinsonism or dementia). These may all appear in each affected individual, or the clinical variation may lead to classification of different family members into different diagnostic categories. The controversy concerns whether these families should be regarded as etiologically distinct, perhaps the result of a unique genetic defect, or as one extreme of the wide clinical spectrum of familial ALS (Hudson 1981). Although Hudson's review of the literature revealed that up to 15% of familial ALS cases had co-existing dementia, there was no statistically significant excess of dementia among familial as compared with sporadic ALS cases at the Mayo Clinic (Mulder et al. 1986). In the light of current evidence, it seems likely that parkinsonism and dementia do occur more frequently than expected in familial ALS, but that this occurs in specific families, and is not a more general phenomenon. Molecular biological laboratory techniques will be required to determine if the observed phenotypic variation correlates with any underlying genetic or etiologic heterogeneity.

DISEASE INCIDENCE

Accepted estimates of the incidence of familial aggregation vary between 5% and 10% (Tandan and Bradley 1985). However, most sources of error in these estimates are likely to cause underestimation of familial disease. The late onset and short duration of the disease mean that the interval between known affected individuals may be several decades, making it more likely that the coincidence of affected individuals will pass unremarked. In addition, it is highly likely that occurrence of the disease in the elderly is underestimated (Mulder 1982; Buckley et al. 1983; Bradley et al. 1987). Over the last 2 decades the age-specific risk of developing and dying of ALS among the elderly increased in both the US (Lilienfeld et al. 1989) and elsewhere (Durrleman and Alperovitch 1989). Although this can be interpreted as being due to increased exposure to a putative environmental etiological agent (Lilienfeld et al. 1989), the changes may also be due to better disease diagnosis among the elderly (Durrleman and Alperovitch 1989; Williams 1989). If one assumes that the latter explanation is true, some apparently sporadic cases diagnosed today may actually be familial cases in which elderly relatives had been misdiagnosed or gone undiagnosed in previous decades. In addition, as most living relatives of ALS patients have not exhausted their risk of developing the disease, it is likely that the familial incidence of the disease is underestimated (Li et al. 1989; Williams et al. 1988). The magnitude of the underestimation is dependent on both the true age-specific risk of developing familial ALS, and the extent to which ALS is under-diagnosed among the elderly. Neither is known, but the former is the subject of current research.

NEUROPATHOLOGY

It has often been considered that pathological involvement of the posterior columns and spinocerebellar tracts are the most important features distinguishing sporadic from familial ALS. The original, now-classic report of this observation in familial ALS (Engel et al. 1959) appeared consistent with the widely held belief that familial and sporadic ALS are etiologically distinct. However, it was soon reported (Hirano et al. 1967; Bonduelle 1975) that this distinctive pattern of pathological involvement occurred in only a minority of familial ALS cases. In the remainder, the pathological features are indistinguishable from those found in sporadic disease. This observation has been confirmed (Horton et al. 1976) and is no longer disputed. In addition, there have been well-documented cases of apparently sporadic ALS in which the posterior columns and spinocerebellar tracts were clearly involved (Hassin 1933; Davison and Wechsler 1936; Lawyer and Netsky 1953). Therefore, no pattern of neuropathological features in ALS can be regarded as an absolute marker of familial disease. The sensitivity and specificity of any given features or combination of features in correctly predicting that a given case is familial remains unknown.

Even in individuals with pathological involvement of the posterior columns and spinocerebellar tracts, considerable variation may occur, which has hampered efforts at classification (Horton et al. 1976). There may appear to be no loss of cells in Clarke's column (Takahashi et al. 1972; Tanaka et al. 1984), or the loss may be moderate (Espinosa et al. 1962), or marked (family B in Engel et al. 1959; Kurent et al. 1975). In some cases, there may be only minimal posterior column involvement (Kurland and Mulder's 'S' family, 1955; Takahashi et al. 1972), while in others spongy degeneration and myelin loss in the posterior columns is marked (Tanaka et al. 1984). Posterior column degeneration classically occurs in the middle-root zones, which, developmentally, are the first areas to be myelinated (Engel et al. 1959; Metcalf and Hirano 1971). However, the degeneration may be more extensive, particularly involving the cuneate fasciculi (Gardner and Feldmahn, 1966; Tanaka et al. 1984). Although various combinations of anatomical abnormalities have been reported, there are no consistent patterns, and no definite association of specific patterns with distinct clinical manifestations.

Unusual pathological features, of unknown significance, have been described in individual patients and families with familial ALS. The features include depigmentation of the substantia nigra (Alter and Schaumann 1976), the presence of amorphous hyaline material in the cerebellum between the Purkinje cell and granular cell layers (Hirano et al. 1967), and Purkinje cell loss affecting the cerebellar vermis (Kurent et al.

1975). In one report of affected sisters there was neuronal degeneration in the oculomotor nuclei and fiber loss from the medial longitudinal fasciculus, as well as neuronal loss from Onufrowicz's nucleus (Tanaka et al. 1984).

An aspect of familial ALS neuropathology which has received little attention is the extent to which specific features vary among affected individuals in the same family. Although pathological features in members of the same family may be very similar (Kurland and Mulder 1955; Engel et al. 1959), some variation undoubtedly occurs. In 1 of 2 families reported by Alter and Schaumann (1976), the variation resulted from pathological involvement or sparing of the corticospinal tracts, while in the other family, the variation was due to involvement or sparing of structures not normally affected in ALS. There are several other reported families in which corticospinal tract degeneration was not identified in all affected family members (Wolfenden et al. 1973; Hawkes et al. 1984; Williams 1989). In some of these familial cases there was also an imperfect correlation between the pathological findings and the observed clinical signs (Espinosa et al. 1962; Bonduelle 1975; Veltema et al. 1990). Such discrepancies may occur when there is a delay between the last clinical examination and death, or when neuropathological examination fails to include sensitive methods for detecting the myelin breakdown products which indicate tract degeneration.

Of potentially greater importance are the reports suggesting that neuropathological variation within families may be as great as that observed between families. These reports are important because of the implications that that observation has for both etiological investigation, and disease classification. Gardner and Feldmahn (1966) reported the features observed in 3 cases from 1 family. All affected family members had neuropathological involvement of structures not normally affected in 'classical' ALS, but the structures involved, and their degree of involvement, varied from case to case. In the first case, in addition to the loss of neurons from the brainstem motor nuclei and anterior horns, only the posterior columns and ventral spinocerebellar tracts were affected. In the second case the lateral corticospinal tracts and dorsal spinocerebellar tracts were additionally affected, and in the third case the tract involvement was detectable much further rostrally than in the other 2 cases, neuronal loss was prominent in Clarke's columns, and was also noticeable in the gracile and cuneate nuclei. In Farmer and Allen's report (1969), the major pathological alteration in one family member was neuronal loss from the anterior horns, with some slight involvement of Clarke's columns. An affected first cousin had, in addition to anterior horn cell loss, marked cell loss in Clarke's columns and mild fiber loss in the posterior columns. In contrast to most other reports of the 'multisystem' form of familial ALS pathology, the pyramidal and spinocerebellar tracts were reportedly normal. Because this family defied classification, Horton (1976) excluded it from consideration in his proposed grouping of familial ALS.

In the first individual of the family reported by Tanaka et al. (1984), there was prominent cell loss in the anterior horns, but not in Clarke's columns. There was some myelin pallor in the anterior and corticospinal tracts, contrasting with marked myelin degeneration and loss in the middle-root zones of the posterior columns. As noted above, the patient's sister had clinical features consistent with ALS and, in addition, 2 unusual features: loss of saccadic eye movement with limitation of horizontal and upward gaze; and disturbed bladder function. As expected, pathological involvement was more extensive than in the first case. Neuronal loss was prominent in the anterior horns and Clarke's columns, but also involved Onufrowicz's nucleus, which is almost always preserved in ALS (Mannen et al. 1977; Konno et al. 1986). In addition, there was marked degeneration of the corticospinal tracts, and myelin loss in the middle-root zones of the posterior columns, combined with myelin loss in the spinocerebellar and spinothalamic tracts. The eye movement abnormalities may have been related to neuronal shrinkage and loss in the oculomotor nuclei, and degeneration of the medial longitudinal fasciculus. Hyaline inclusions resembling Lewy bodies were present in the remaining midbrain oculomotor neurons. Other unusual pathological features for ALS included small infarcted areas in the midbrain tegmentum, associated with gliosis in the red nuclei and substantia nigra. In the cerebellar hemispheres there was a 'paucity' of Purkinje cells, associated with moderate proliferation of the Bergmann glia.

In another family (Williams, 1989), many members of a large pedigree had suffered ALS, and 2 had undergone post-mortem neuropathological examination. One elderly female had degeneration of spinal anterior horn cells and corticospinal tracts in a pattern indistinguishable from the majority of sporadic ALS cases. Her second cousin presented at a relatively young age with progressive wasting and weakness, but in addition complained of progressive sensory symptoms and fecal incontinence, which was initially thought to preclude the diagnosis of ALS. However, at necropsy, the pathological features were those of the 'multisystem' form of ALS, with marked loss of cells in Clarke's columns, and marked fiber loss in the dorsal spinocerebellar tracts and posterior columns (particularly the middle root zones), in addition to the changes of 'classical' ALS noted in the affected cousin. Fecal incontinence is rare in both sporadic and familial ALS, and unfortunately the available sections did not permit clinico-pathological correlation at the appropriate sacral level.

In 1989 this author reviewed all available information concerning the 'B' family (Dr. L.T. Kurland, family members, and other sources). This was the second of the 2 families with 'multisystem' familial ALS pathology first described by Engel and co-authors in 1959. The affected family members described in the original report had similar patterns of neuropathological involvement. An additional family member (B-III(3)) was asymptomatic and unaffected at the age of 67 years when the report was published, but because she had an affected sibling and an affected offspring, she must have carried the disease gene if the disease is dominantly inherited. This woman eventually died with ALS (she had co-existing carcinoma) almost 15 years later, at age 82. At autopsy, despite the knowledge of her relationship to other individuals with the 'multisystem' form of neural degeneration, the attending neuropathologist could find no evidence of pathological involvement outside the anterior horns and corticospinal tracts.

The preceding discussion suggests that at least some aggregations of familial ALS are 'multi-system' pathological disorders, but ones in which some family members may show less extensive pathological involvement than others. However, what appear to be qualitative differences in neuropathological involvement may only be the result of more extreme quantitative variation than that which can be readily detected with standard neuropathological methods. This interpretation is supported by a recent report which demonstrated that the apparent variation in neuropathological involvement in affected individuals from a single family is diminished if one employs more sensitive methods, such as the Haggqvist stain (Veltema et al. 1990). Unfortunately, few investigators have yet attempted to quantify neuronal loss within the central nervous system in ALS. Averback and Crocker's quantitative analysis (1982) did, however, reveal frequent, previously unreported neuronal loss in Clarke's nucleus. In a separate quantitative study, Kozlowski et al. (1989) identified heterotopic neurons with the characteristics of anterior horn cells in the spinal cords of individuals dying of ALS. If confirmed, this report implies that etiological determinants in sporadic ALS must have their effect at a very early stage of spinal cord development. Unfortunately, familial ALS has not yet been the subject of similar neuropathological studies.

The neuropathological spectrum in familial ALS may be regarded as the pathological equivalent of the great inter- and intra-familial variation in clinical characteristics recognized by Kurland and Mulder in their initial report (1955). Where there has been marked intrafamilial pathological variation, the tendency appears to have been for older family members to have less extensive neuropathological involvement, an observation consistent with those made in other late-onset, presumably dominantly inherited neurodegenerative disorders, such as Huntington's disease (Myers et al. 1985). An important consequence of this great inter- and intra-familial clinicopathological variation is that the familial pattern may

not be confidently inferred from examination of a small number of affected individuals, nor can etiological inferences be confidently inferred from either the clinical or the pathological phenotype.

MOLECULAR GENETIC STUDIES

Molecular biological techniques hold the promise of determining the precise defect in genetically determined diseases. This may arise in one conceptual direction when an investigator correctly postulates either the pathophysiological mechanism or the metabolic or structural defect and tests the appropriate 'candidate gene'. Or alternatively, it may arise in the opposite direction when an investigator first identifies the precise location of the genetic defect, and then infers the probable pathophysiological mechanism from the DNA-encoded protein structure. Botstein et al. (1980) proposed using restriction fragment length polymorphisms of DNA to implement the latter method less than 11 years ago, but the technique has already proven successful in determining the underlying defects in some genetically determined diseases. Evidence of linkage can be established most readily when there is a distinct and unique phenotype, and there are sufficient segregation events to examine the statistical association between the linkage marker and the postulated disease gene. In familial ALS the linkage analysis method is compromised by the necessity of inferring essential genetic information about deceased affected individuals, or making assumptions about the unexpired risk of disease development in unaffected but 'at-risk' individuals.

When different families are grouped together for linkage analysis, an implicit assumption is that all have the same disease. From the evidence presented earlier (see above), it seems that in familial ALS such an assumption is difficult to defend, and yet there are no obvious ways of grouping families by phenotypic characteristics so as to increase the likelihood of genetic homogeneity.

Even in large families afflicted with familial ALS, affected individuals seldom suffer the disease contemporaneously, and their parents are usually dead. These limitations markedly diminish the segregation data which are available for analysis. 'At-risk' family members can be assigned an age-dependent estimate of the probability that they carry the disease gene, and in some cases lymphocyte samples from surviving relatives can help to 'reconstruct' an antecedent's genotype. However, the uncertainties inherent in both of these procedures usurp much of the statistical power of the linkage analysis. The case of manic depressive psychosis illustrates the problems which arise when phenotype assignment is uncertain or linkage depends on the correct assignment of risk to 'at-risk' individuals. Following meticulous scientific work, LOD scores strongly suggested that this familial disease in the Amish community was linked to genetic markers on chromosome 11 (Egeland et al. 1987). However, the subsequent development of disease in individuals assumed to have a low probability of carrying the gene markedly lowered the LOD scores, and removed the accumulated evidence for linkage.

Despite the recognized difficulties, two investigators have reported exclusion mapping in familial ALS, in one case using a single large family (Williams, 1989), and in another combining information from a number of families (Siddique et al. 1989). Neither of these studies found evidence for linkage, but a more recent, preliminary report, raises the possibility of linkage in some cases of familial ALS with markers on chromosome 21 (Siddique et al. 1990).

Strategies adopted to circumvent the difficulties associated with linkage analysis in familial ALS include collecting lymphocytes from 'at-risk' individuals and waiting till one or more develop the disease, combining increasing numbers of families, and testing candidate genes on the basis of advances in the understanding of the way in which neurons function, and therefore may be deranged, in familial ALS.

REFERENCES

ALTER, M. and B. SCHAUMANN: Hereditary amyotrophic lateral sclerosis. Eur. Neurol. 14 (1976) 250–256.

AVERBACK, P. and P. CROCKER: Regular involvement of Clarke's nucleus in sporadic amyotrophic lateral sclerosis. Arch. Neurol. 39 (1982) 155–156.

BEN HAMIDA, M., F. C. HENTATI and BEN HAMIDA: Hereditary motor system diseases (chronic juvenile amyotrophic lateral sclerosis). Conditions combining a bilateral pyramidal syndrome with limb and bulbar amyotrophy. Brain 113 (1990) 347–363.

BOLSTEIN, D., R. L. WHITE, M. SKOLNICK and R. W. DAVIS: Construction of a genetic linkage map in man using restriction fragment length polymorphisms. Am. J. Hum. Genet. 32 (1980) 314–331.

BONDUELLE, M.: Amyotrophic lateral sclerosis. In: P. J. Vinken and G. W. Bruyn (Eds.), Handbook of Clinical Neurology. Vol. 22, Amsterdam, North-Holland Publ. Co. (1975) 281–347.

BRADLEY, W. G.: Recent views on amyotrophic lateral sclerosis with emphasis on electrophysiological studies. Muscle Nerve 10 (1987) 490–502.

BUCKLEY, J., C. WARLOW, P. SMITH, D. HILTON-JONES, S. IRVINE and T. R. TEW: Motor neuron disease in England and Wales 1959–1979. J. Neurol., Neurosurg. Psychiatry 46 (1983) 197–205.

BURNS, R. S., C. C. CHIUEH, S. P. MARKEY, M. H. EBERT, D. M. JACOBOWITZ and I. J. KOPIN: A primate model of parkinsonism: selective destruction of dopaminergic neurons in the pars compacta of the substantia nigra by N-methyl-4-phenyl-1,2,3,6-tetrahydropyridine. Proc. Natl. Acad. Sci. (USA) 80 (1983) 4546–4550.

CHIO, A., F. BRIGNOLIO, P. MEINERI and D. SCHIFFER: Phenotypic and genotypic heterogeneity of dominantly inherited amyotrophic lateral sclerosis. Acta Neurol. Scand. 75 (1987) 277–282.

DAVISON, C. and I. S. WECHSLER: Amyotrophic lateral sclerosis with involvement of posterior column and sensory disturbances. Arch. Neurol. Psychiatry 35 (1936) 229–239.

DUMON, J., J. MACKEN and T. H. DE BARSY: Concordance for amyotrophic lateral sclerosis in a pair of dizygous twins of consanguineous parents. J. Med. Genet. 8 (1971) 113–115.

DURRLEMAN, S. and A. ALPEROVITCH: Increasing trend of ALS in France and elsewhere: are the changes real? Neurology 39 (1989) 768–773.

EDWARDS, J. H.: Familial predisposition in man. Br. Med. Bull. 25 (1969) 58–64.

EGELAND, J. A., D. S. GERHARD, D. L. PAULS, J. N. SUSSEX, K. K. KIDD, C. R. ALLEN, A. M. HOSTETTER and D. E. HOUSMAN: Bipolar affective disorders linked to DNA markers on chromosome 11. Nature 325 (1987) 783.

EMERY, E. H. and S. HOLLOWAY: Familial motor neuron diseases. In: L. P. Rowland (Ed.), Human Motor Neuron Diseases. New York, Raven Press (1982) 139–147.

ENGEL, W. K., L. T. KURLAND and I. KLATZO: An inherited disease similar to amyotrophic lateral sclerosis with a pattern of posterior column involvement. An intermediate form? Brain 82 (1959) 203–220.

ESPINOSA, R. E., M. M. OKIHIRO, D. W. MULDER and G. P. SAYRE: Hereditary amyotrophic lateral sclerosis. A clinical and pathological report with comments on classification. Neurology 12 (1962) 1–7.

ESTRIN, W. J.: Amyotrophic lateral sclerosis in dizygotic twins. Neurology 27 (1977) 692–694.

FARMER, T. W. and J. N. ALLEN: Hereditary proximal amyotrophic lateral sclerosis. Trans. Am. Neurol. Assoc. 94 (1969) 140–144.

GARDNER, J. H. and A. FELDMAHN: Hereditary adult motor neuron disease. Trans. Am. Neurol. Assoc. 91 (1966) 239–241.

GIMINEZ-ROLDAN, S. and N. A. ESTEBAN: Prognosis in hereditary amyotrophic lateral sclerosis. Arch. Neurol. 34 (1977) 706–708.

GUBBAY, S. S., E. KAHANA, N. ZILBER, G. COOPER, S. PINTOV and Y. LIEBOWITZ: Amyotrophic lateral sclerosis. A study of its presentation and prognosis. J. Neurol. 232 (1985) 295–300.

HASSIN, G. B.: Amyotrophic lateral sclerosis complicated by subacute combined degeneration of the cord. Arch. Neurol. Psychiatry 29 (1933) 125–138.

HAWKES, C. H., J. B. CAVANAGH, S. MOWBRAY and E. A. PAUL: Familial motor neurone disease: report of a family with five post-mortem studies. In: F. C. Rose (Ed.), Research Progress in Motor Neurone Disease. London, Pitman (1984) 70–98.

HIRANO, A., L. T. KURLAND, R. S. KROOTH and S. LESSELL: Parkinsonism-dementia complex, an endemic disease on the island of Guam. I. Clinical features. Brain 84 (1961a) 642–661.

HIRANO, A., N. MALAMUD and L. T. KURLAND: Parkinsonism-dementia complex, an endemic disease on the island of Guam. II — Pathological features. Brain 84 (1961b) 662–679.

HIRANO, A., L. T. KURLAND and G. P. SAYRE: Familial amyotrophic lateral sclerosis. Arch. Neurol. 16 (1967) 232–243.

HORTON, W. A., E. ROSWELL and J. A. BRODY: Familial motor neuron disease. Neurology 26 (1976) 460–465.

HUDSON, A. J.: Amyotrophic lateral sclerosis and its association with dementia, parkinsonism and other neurological disorders: A review. Brain 104 (1981) 217–247.

HUSQUINET, H. and G. FRANCK: Hereditary amyotrophic lateral sclerosis transmitted for five generations. Clin. Genet. 18 (1980) 109–115.

JOKELAINEN, M., J. PALO and J. LOKKI: Monozygous twins discordant for amyotrophic lateral sclerosis. Eur. Neurol. 17 (1978) 296–299.

JUERGENS, S. M. and L. T. KURLAND: Epidemiology. In: D. W. Mulder (Ed.), The Diagnosis and Treatment of Amyotrophic Lateral Sclerosis. Boston, Houghton Mifflin Professional Publishers Medical Division (1979) 35–51.

KEITH, R. A., J. VAN LOON, L. F. WUSSOW and R. M. WEINSHILBOUM: Thiol methylation pharmacogenetics: heritability of human erythrocyte thiol methyltransferase activity. Clin. Pharmacol. Ther. 34 (1983) 521–528.

KONNO, H., T. YAMAMOTO, Y. IWASAKI and H. IIZUKA: Shy-Drager syndrome and amyotrophic lateral sclerosis. Cytoarchitectonic and morphometric studies of sacral autonomic neurons. J. Neurol. Sci. 73 (1986) 193–194.

KOZLOWSKI, M. A., C. WILLIAMS, D. R. HINTON and C. A. MILLER: Heterotopic neurons in spinal cord of patients with ALS. Neurology 39 (1989) 644–648.

KURENT, J. E., A. HIRANO and J. M. FOLEY: Familial amyotrophic lateral sclerosis with spinocerebellar degeneration and peripheral neuropathy. J. Neuropathol. Exp. Neurol. 34 (1975) 110.

KURLAND, L. T. and D. W. MULDER: Epidemiologic investigations of amyotrophic lateral sclerosis. 2. Familial aggregation indicative of dominant inheritance. Neurology 5 (1955) 182–196; 249–268.

LAWYER, T. and M. G. NETSKY: Amyotrophic lateral sclerosis: A clinicoanatomic study of fifty-three cases. Arch. Neurol. Psychiatry 69 (1953) 171–192.

LI, T.-M., E. ALBERMAN and M. SWASH: Comparison of sporadic and familial disease amongst 580 cases of motor neuron disease. J. Neurol., Neurosurg. Psychiatry 51 (1988) 778–784.

LILIENFELD, D. E., E. CHAN, J. EHLAND, J. GODBOLD, P. J. LANDRIGAN, G. MARSH and D. P. PERL: Rising mortality from motoneuron disease in the USA, 1962–84. Lancet 2 (1989) 710–712.

MANNEN, T., M. IWATA, Y. TOYOKURA and K. NAGASHIMA: Preservation of a certain motoneurone group of the sacral cord in amyotrophic lateral sclerosis: Its clinical significance. J. Neurol., Neurosurg. Psychiatry 40 (1977) 464–469.

METCALF, C. W. and A. HIRANO: Amyotrophic lateral sclerosis. Clinicopathological studies of a family. Arch. Neurol. 24 (1971) 518–523.

MULDER, D. W.: Clinical limits of amyotrophic lateral sclerosis. In: L. P. Rowland (Ed.), Human Motor Neuron Diseases. New York, Raven Press (1982) 15–22.

MULDER, D. W., L. T. KURLAND, K. P. OFFORD and C. M. BEARD: Familial adult motor neuron disease: Amyotrophic lateral sclerosis. Neurology 36 (1986) 511–517.

MYERS, R. H., D. S. SAX, M. SCHOENFELD, E. D. BIRD, P. A.-WOLF, J. P. VONSATTEL, R. F. WHITE and J. B. MARTIN: Late onset of Huntington's disease. J. Neurol., Neurosurg. Psychiatry 48 (1985) 530–534.

OJEDA, V. J., CRAINGER, K. M. R. and T. DAY: Familial motor neurone disease associated with non-specific organic dementia. A clinico-pathological study of a family. Med. J. Australia 141 (1984) 430–433.

PLAITAKIS, A., S. BERL and M. D. YAHR: Neurological disorders associated with deficiency of glutamate dehydrogenase. Ann. Neurol. 15 (1984) 144–153.

PLAITAKIS, A. and J. T. CAROSCIO: Abnormal glutamate metabolism in amyotrophic lateral sclerosis. Ann. Neurol. 22 (1987) 575–579.

POSER, C. M., M. JOHNSON and L. D. BUNCH: Familial amyotrophic lateral sclerosis. Dis. Nerv. Syst. 26 (1965) 697–702.

ROSEN, A. D.: Amyotrophic lateral sclerosis. Clinical features and prognosis. Arch. Neurol. 35 (1978) 638–642.

SCHMITT, H. P., W. EMSER and C. HEIMES: Familial occurrence of amyotrophic lateral sclerosis, parkinsonism, and dementia. Ann. Neurol. 16 (1984) 642–648.

SELBY, G.: Hereditary motor neuron disease. Clin. Exp. Neurol. 24 (1987) 145–151.

SIDDIQUE, T., M. A. PERICAK-VANCE, B. R. BROOKS, R. P. ROOS, W.-Y. HUNG, J. P. ANTEL JP, T. L. MUNSAT, K. PHILLIPS, K. WARNER, M. SPEER, W. B. BIAS, N. A. SIDDIQUE and A. D. ROSES: Linkage analysis in familial amyotrophic lateral sclerosis. Neurology 39 (1989) 919–925.

SIDDIQUE, T., M. A. PERICAK-VANCE, R. P. ROOS, P. WATKINS, W.-Y. HUNG, B. R. BROOKS, F. NOORE, R. TANDAN, G. NICHOLSON, D. WILLIAMS, J. BEBOUT, M. ZEIDMAN, J. ANTEL, T. MUNSAT and A. D. ROSES: Chromosome 21 markers in familial amyotrophic lateral sclerosis. Neurology 40 (Suppl. 1) (1990) 721P.

STEVENTON, G., A. C. WILLIAMS, R. WARING, H. S. PALL and D. ADAMS: Xenobiotic metabolism in motor neurone disease. Lancet 2 (1988) 644–647.

STEVENTON, G. B., M. T. E. HEAFIELD, R. H. WARING and A. C. WILLIAMS: Xenobiotic metabolism in Parkinson's disease. Neurology 39 (1989) 883–887.

TAKAHASHI, K., H. NAKAMURA and E. OKADA: Hereditary amyotrophic lateral sclerosis. Histochemical and electron microscopic study of hyaline inclusions in motor neurons. Arch. Neurol. 27 (1972) 292–299.

TANAKA, J., H. NAKAMURA, Y. TABUCHI and K. TAKAHASHI:

Familial amyotrophic lateral sclerosis: features of multisystem degeneration. Acta Neuropathol. (Berl.) 64 (1984) 22–29.

TANDAN, R. and W. G. BRADLEY: Amyotrophic lateral sclerosis: Part 1. Clinical features, pathology, and ethical issues in management. Ann. Neurol. 18 (1985) 271–280.

VELTEMA, A. N., R. A. C. ROOS and G. W. BRUYN: Autosomal dominant adult amyotrophic lateral sclerosis. J. Neurol. Sci. 97 (1990) 93–115.

WARING, R. H., G. B. STEVENTON, S. G. STURMAN, M. T. E. HEAFIELD, M. C. G. SMITH and A. C. WILLIAMS: S-methylation in motorneuron disease and Parkinson's disease. Lancet 2 (1989) 356–357.

WEINSHILBOUM, R.: Methyltransferase pharmacogenetics. Pharmacol. Ther. 43 (1989) 77–90.

WEINSHILBOUM, R.: Sulfotransferase pharmacogenetics. Pharmacol. Ther. (1990) in press.

WILLIAMS, D. B.: Genetic factors in motor neuron disease. Ph. D. Thesis. University of Sydney, 1989.

WILLIAMS, D. B., D. A. FLOATE and J. LEICESTER: Familial motor neuron disease: differing penetrance in large pedigrees. J. Neurolog. Sci. 86 (1988) 215–230.

WOLFENDEN, W. H., A. F. CALVERT, E. HIRST, W. EVANS and J. G. MCLEOD: Familial amyotrophic lateral sclerosis. Clin. Exp. Neurol. 9 (1973) 51–55.

Handbook of Clinical Neurology, Vol. 15 (59): Diseases of the Motor System
J.M.B.V. de Jong, editor
© Elsevier Science Publishers B.V., 1991

CHAPTER 15

Amyotrophic lateral sclerosis in the Mariana Islands

RALPH M. GARRUTO and RICHARD YANAGIHARA

Laboratory of Central Nervous System Studies, National Institute of Neurological Disorders and Stroke,
National Institutes of Health, Bethesda, MD, USA

The high incidence of amyotrophic lateral sclerosis (ALS) among the indigenous Chamorro people of Guam, which has been well documented since the end of World War II (Koerner 1952; Arnold et al. 1953; Tillema and Wijnberg 1953; Kurland and Mulder 1954), represents a naturally occurring paradigm of a late-onset neurodegenerative disorder. High-incidence foci of ALS and parkinsonism-dementia (PD) also occur among 2 other geographically and genetically distinct populations in the western Pacific: the Auyu and Jakai of southern West New Guinea (Gajdusek 1963), and the Japanese of the Kii Peninsula of Honshu Island (Miura 1911) (Fig. 1). All three foci bear striking similarities to each other, and the cross-comparisons between these foci have added immeasurably to our understanding of other neurodegenerative disorders, such as Alzheimer disease, and of normal neuronal aging.

As a result of the high incidence of ALS, and the subsequent discovery of parkinsonism-dementia (PD) (Hirano et al. 1961a, b), in the Mariana Islands, the National Institute of Neurological Disorders and Stroke (then known as the National Institute of Neurological Diseases and Blindness) established a field research center on Guam in 1956 (Fig. 1). For the past 3 decades, the continued epidemiological, clinical and neuropathological surveillance of ALS and PD in the western Pacific have provided insights into their

etiology and pathogenesis, which may not have otherwise be forthcoming from studies of large cosmopolitan Western communities.

Although both ALS and PD occur in equally high incidence in the Mariana Islands and are commonly found in the same families and occasionally in the same individual, the disorders are usually clinically distinct. Therefore, the emphasis here will be on ALS, and PD will be discussed only where pertinent to ALS. Those interested in PD are referred to an earlier volume of the Handbook (Chen and Chase 1986) and to other recent reviews (Garruto 1989a, b).

HISTORIC PERSPECTIVE

Comprised of 15 islands, the Mariana archipelago is arranged in an arcuate chain extending 672 km in the western Pacific Ocean (13°14′ and 20°33′ North latitude, 144°54′ and 156°05′ East longitude), with Farallon de Pajaros at the extreme northern end and Guam at the extreme southern end (Fig. 1). Guam, the largest of the Mariana Islands, lies approximately 1100 km south of the Tropic of Cancer at 13° North latitude and 145° East longitude. It is 50 km in length and 6–14 km in width. A territory of the United States since 1898, Guam, together with the Caroline, Marshall and Gilbert Islands, forms the geographic and ethnographic region of Micronesia. Although

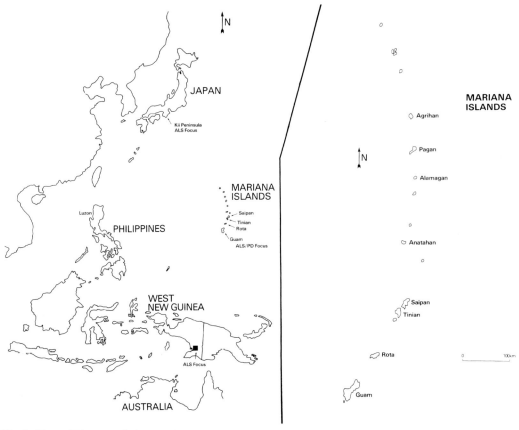

Fig. 1. Map of Guam and the Northern Mariana Islands where surveys for amyotrophic lateral sclerosis (ALS) and parkinsonism-dementia (PD) were conducted during the past 4 decades. The geographic location of the other 2 high-incidence foci of ALS and PD in the Kii Peninsula of Japan and West New Guinea are also shown.

previously part of the Trust Territory of the Pacific Islands, the inhabited islands north of Guam (Rota, Tinian, Saipan, Aguijan, Anatahan, Alamagan and Pagan) are now collectively referred to as the Commonwealth of the Northern Mariana Islands.

The native people of Guam and the Northern Mariana Islands are referred to as Chamorros. Although the precise origin of the aboriginal people of the Mariana Islands remains uncertain, it is believed that they migrated from the Indo-Malayan region circa 1530 BC (Thompson 1947; Spoehr 1957; Carano and Sanchez 1964). The first known contact between native Chamorros and Europeans occurred in 1521 with the landing of Ferninand Magellan near the southern village of Umatac on Guam. In 1565, Andres Miguel

Lopez de Legazpi claimed the islands (then known as the Ladrone Islands) for Spain, but no attempt was made to colonize them until 1668, when Father Diego Luis de Sanvitores, a Spanish Jesuit, arrived to found a mission on the Island. At the time of his arrival, an estimated high of 40 000 to 100 000 Chamorros lived throughout the islands, which he later renamed Mariana (Spoehr 1954; Underwood 1973). The ensuing 3 decades were marked by recurrent warfare with the Spanish. Epidemics of influenza, measles, and smallpox and destructive typhoons further reduced the native population such that in 1698, with the conquest of the Mariana Islands by Spain, the Chamorro population had been almost exterminated. The first census conducted by the Spanish in 1710 revealed a population of only

3539 Chamorros (Spoehr 1954; Underwood 1973). The Chamorro population continued to decline to less than 1500 during the latter half of the eighteenth century (Underwood 1973). It was during this period and the ensuing 50 years that matings flourished between Chamorros and members of other ethnic groups, particularly Filipinos and Spanish, resulting in the formation of a hybrid group of 'neo-Chamorros' which repopulated the Mariana Islands (Bowers 1951; Spoehr 1954; Plato et al. 1966; Lie-Injo 1967; Underwood 1976). Since the late eighteenth century, the population has remained genetically stable until the post-war population explosion of the 1950s, 60s, and 70s. Currently there are 106 000 people living on Guam (U.S. Bureau of Census, 1980). Of these, approximately 60 000 are Chamorros, 20 000 Filipinos and 26 000 other ethnic groups (mostly American military personnel and their dependents). Interestingly, the historical record does not indicate any unusual focus or mention of an ALS-like disease on Guam until the turn of the 20th century (Leach 1900).

CLINICAL FEATURES

Since the first systematic observations of ALS on Guam nearly 40 years ago (Koerner 1952; Arnold et al. 1953; Kurland and Mulder 1954; Tillema and Wijnberg 1953), approximately 400 patients from a mean Chamorro population base of 40 000 have been diagnosed as having ALS. All patients who developed disease were followed throughout their clinical course and approximately 75% have been autopsied.

ALS, known locally on Guam as *lytico* or *paralytico* does not differ clinically from ALS seen elsewhere, as first described by Aran (1850a, b), Duchenne (1855, 1860) and Charcot and Joffroy (1869) more than a century ago (Fig. 2). Typically, the onset of ALS is insidious, with progressive weakness, muscle atrophy, fasciculations and spasticity. The disease usually results in a flaccid paralysis several months to a year or so after onset. Atrophy of the small muscles of the hands is the presenting symptom in more than half of the patients, while spasticity appears first in about 15% of patients. Extrapyramidal signs, dementia and urinary or fecal

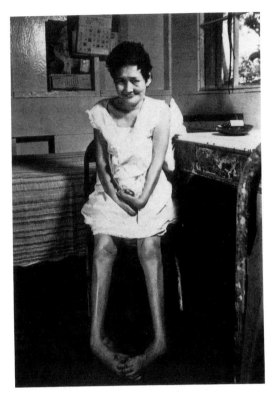

Fig. 2. A Chamorro patient with amyotrophic lateral sclerosis (ALS) of Guam. She shows characteristic muscle atrophy of the limbs and associated joint deformity. Onset of ALS occurred at age 39 with a rather long-term duration of 14 years. (Photograph courtesy of Dr. Kwang-Ming Chen)

incontinence occur rarely. Extraocular movements and pupillary responses are normal, and decubitus ulcers are not observed even in bedridden patients with long-standing ALS. In some patients there is a clinical overlap between ALS and PD, a topic extensively reviewed elsewhere (Chen 1979; Rodgers-Johnson et al. 1986).

No major temporal changes in clinical symptoms were found among 279 Chamorro patients who had onset of ALS between 1950 and 1979 (Table 1). However, the mean age at onset of ALS increased from 47.6 to 51.9 years in men and from 42.1 to 52.5 years in women, and the mean duration of disease decreased from 5.5 to 3.4 years in men and from 8.0 to 3.9 years in women during the 30-year period (Rodgers-Johnson et al. 1986). Approximately one-third of Guamanian patients with ALS live longer than 5

TABLE 1

Percent frequency of clinical features in amyotrophic lateral sclerosis of Guam between 1950 and 1979.

Clinical findings	1950–59 (%)	1960–69 (%)	1970–79 (%)
Lower motor neuron signs*	100	100	100
Atrophy	91	98	99
Fasciculations	80	92	89
Fasciculations and atrophy	73	92	88
Upper motor neuron signs*	97	99	95
Hyperreflexia and/or Babinski	79	82	79
Hyperreflexia and clonus	27	28	31
Babinski sign	45	63	45
Suck and/or snout	46	42	40
Jaw jerk	35	43	53
Bulbar/pseudobulbar palsy	70	64	67
Dysarthria	66	62	64
Extrapyramidal signs*	9	3	4
Tremor	7	3	4
Cogwheel rigidity	3	3	3
Bradykinesia	1	3	0
Masked facies	0	0	0
Dementia	6	3	5
Incontinence (fecal or urinary)	0	1	1

*Total number of patients with one or more physical sign showing involvement of that pathway.

years after the onset of disease (Mukai et al. 1982) without the use of mechanical ventilation, and nearly 15% live longer than 10 years (Uebayashi 1980). Women in this latter group usually exhibit a rapidly progressive course, which arrests in 5 years, with early confinement to a wheelchair or bed (Mukai et al. 1982). Men with long-duration ALS, on the other hand, tend to have a benign course and are usually ambulatory 10 years after onset of symptoms. In general, patients with younger age at onset and whose initial symptom is weakness in the upper extremities or spasticity of the lower extremities (common type) have a better prognosis than patients with bulbar palsy or distal muscular atrophy of the lower extremities (pseudopolyneuritic type) (Uebayashi 1980; Mukai et al. 1982).

NEUROPATHOLOGY

Widespread degeneration of anterior horn cells, accompanied by intracytoplasmic inclusions and neuroaxonal swellings, is the neuropathological hallmark of ALS worldwide, irrespective of whether the clinical variant is sporadic, familial or the Western Pacific form. Ultrastructurally, these inclusions consist of interwoven skeins or parallel arrays of phosphorylated neurofilament, a cytoskeletal protein that is a major determinent of axonal caliber and neuronal structure (Chou 1979; Averback 1981; Hirano et al. 1984a, b; Hoffman et al. 1987). Although the mechanism giving rise to this pathology are unknown, a fundamental abnormality of neurofilament biosynthesis or catabolism is likely.

The neuropathological findings of ALS on Guam, consisting of anterior horn cell loss and corticospinal tract degeneration, do not differ from changes seen in ALS patients elsewhere in the world. In addition to these classical changes, however, the neuropathology of ALS on Guam is characterized by neurofibrillary degeneration in the hippocampus, cranial nerve nuclei, substantia nigra, locus coeruleus and dentate nucleus of the cerebellum (Fig. 3 and Table 2). Neurofibrillary tangles (NFT), which occur in anterior horn cells in approximately 25% of patients, do not correspond to the severity of motor neuron

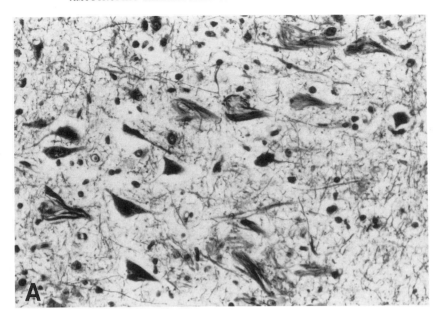

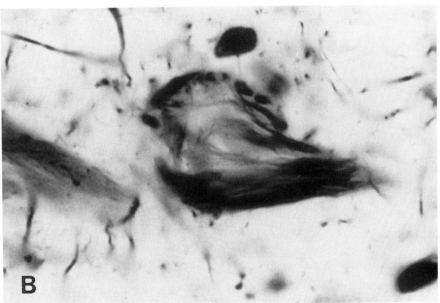

Fig. 3. Neurofibrillary tangle (NFT) bearing neurons in the hippocampus of a Guamanian patient with amyotrophic lateral sclerosis. (A) Large number of NFT-bearing neurons in a single field (Bodian silver, original magnification ×1000). Note the fibrillary nature of the proteins which accumulate to form neurofibrillary tangles. The same lesion is seen in brain tissues in Guamanian parkinsonism-dementia and is the neuropathological hallmark of these disorders (Bodian silver, original magnification ×5000).

damage or to the duration of disease. Granulo-vacuolar degeneration is commonly found in the pyramidal cells of Sommer's sector (Hirano et al. 1966). Occasionally, the extent of the neurofibrillary pathology and the neuronal loss and depigmentation in the substantia nigra in Guamanian ALS are indistinguishable from those observed in PD (Hirano et al. 1966). These lesions result from the abnormal accumulation of cytoskeletal proteins, particularly neurofila-

TABLE 2

Percent frequency of pathological findings in amyotrophic lateral sclerosis of Guam between 1950 and 1979.*

Pathological finding	1950–59 (%)	1960–69 (%)	1970–79 (%)
Brain			
Gross atrophy[a]	23	23	18
NFT	94	98	94
Localized changes[b]	33	47	62
Hirano bodies	2	6	26
Lewy bodies	2	6	3
Senile plaques	4	2	3
Cord			
NFT	16	29	21
Localized changes[c]	97	95	94

*Percentages are of the total number studied microscopically.
[a]Percentages are of number autopsied; all had macroscopic reports.
[b]Localized changes included neuronal loss, depigmentation of substantia nigra, and NFT formation in hippocampus, locus coeruleus and hypothalamus.
[c]Localized changes included anterior horn cell loss and demyelination of the corticospinal tracts.

ment and microtubule-associated protein tau, and the abnormal post-translational processing of a precursor protein into amyloid (Shankar et al. 1989; Garruto 1989b). These proteins accumulate and form argentophilic masses (NFT), that eventually lead to cellular dysfunction and death.

Interestingly, neurofibrillary degeneration, the common neuropathological denominator of ALS and PD of Guam, occurs at a much earlier age among neurologically normal Guamanians than reported for comparable populations in Japan and England (Anderson et al. 1979; Chen 1981). Approximately 70% of Chamorros, of either sex, older than 35 years of age have significant numbers of NFT, and in 15%, the severity and distribution of the neurofibrillary pathology is indistinguishable from that found in patients with PD. Quantitatively, NFT are distributed, in decreasing order of frequency, in the substantia innominata, hippocampal gyrus, amygdaloid nucleus, locus coeruleus, hypothalamus, and substantia nigra (Anderson et al. 1979). However, unlike ALS and PD, accompanying neuronal loss is not seen in such individuals. To what extent these NFT lesions represent a preclinical disease state is not known.

Ultrastructure

NFT in hippocampal neurons of Guamanian patients with ALS appear as paired helical filaments, not unlike those found in patients with Alzheimer disease (Wisniewski et al. 1970). By negative-stain electron microscopy, purified preparations of NFT from PD brain tissue appear as paired, helically coiled filaments having a diameter of 20 nm at the widest point but have an average crossover periodicity of 180 nm (Guiroy et al. 1987), unlike the 80-nm periodicity of paired helical filaments observed in patients with Alzheimer disease. Occasionally twisted filaments containing 3 filaments, each having diameters of 10 nm and morphologically similar to those seen in paired helical filaments, can be found (Guiroy et al. 1987). In addition, isolated, single filaments (diameter, 5–15 nm), composed of 3 or 4 subfilaments each measuring 2–5 nm in diameter, are rarely observed. Studies of purified preparations of NFT from brain tissues of Guamanian ALS patients currently in progress suggest that, unlike PD, the majority of filaments forming neurofilamentous bundles are single and unpaired rather than paired and helical.

Immunocytochemical characterization

Several cytoskeletal proteins, including neurofilament, microtubule-associated proteins 2 and tau, and ubiquitin have been implicated in the formation of NFT in Alzheimer disease. As determined by immunocytochemical techniques using monoclonal antibodies against phosphory-lated neurofilament, tau and paired helical fila-ments, the antigenic determinants of NFT in Guamanian ALS (and PD) resemble those re-ported in Alzheimer disease (Shankar et al. 1989). Specifically, these antibodies exhibit robust im-munoreactivity with tangle-bearing hippocampal neurons in patients with ALS (and PD) (Fig. 4), as well as in neurologically normal Guamanians with neurofibrillary degeneration. Thus, as in Alzheimer disease, tau and neurofilament appear to be 2 of the major antigenic components of NFT in ALS of Guam.

In Guamanian PD, amyloid β-protein, a 42-amino acid polypeptide forming paracrystalline

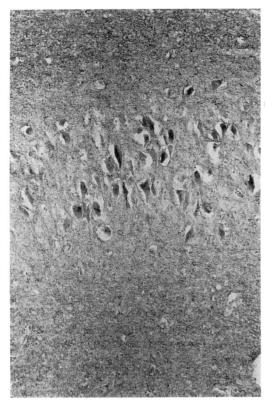

Fig. 4. Immunoperoxidase staining of cryostat-cut section of formalin-fixed hippocampus from a Guam-anian patient with amyotrophic lateral sclerosis showing intense immunoreactivity of neurofibrillary tangles with SMI-31, a monoclonal antibody directed against a phosphorylated epitope of the 200 kDa neu-rofilament protein (dilution, 1:5000). No staining was observed in brain sections of neurologically normal Guamanians without neurofibrillary degeneration.

arrays of β-pleated sheets, is a main constituent of the NFT. It exhibits green birefringence under polarized light after Congo red staining, and is highly resistant to proteases. Biochemically, amy-loid β-protein is composed of multimeric aggre-gates of a 42-amino-acid polypeptide with a relative molecular mass of 4.0–4.5 kDa, the same molecular mass reported for NFT, plaque and vascular amyloid in Alzheimer disease and Down syndrome (Guiroy et al. 1987). Studies are underway to sequence purified preparations of NFT isolated from Guamanian ALS tissue to determine if it is the same as that found in Guamanian PD and Alzheimer disease (D.C. Guiroy, pers. commun.).

The biochemical events leading to NFT forma-tion in Guamanian ALS are unknown, but aberrant phosphorylation of tau and neurofila-ment, and subsequent impairment of slow axonal transport, may serve as the common pathogenetic pathway in these disorders (Shankar et al. 1989). It is unlikely that NFT are composed of only one protein; rather, they are complex structures resulting from the abnormal post-translational processing and co-polymerization of several pro-teins, including tau, neurofilament, ubiquitin and probably amyloid β-protein.

EPIDEMIOLOGY

Temporal and geographic trends

Long before the earliest documentation of the phenomenal prevalence of ALS on Guam by Western physicians, the disease, called *lytico* or *paralytico* in the Chamorro language, was well recognized by the native Chamorros. The diagno-sis 'amyotrophic lateral sclerosis', however, did not appear on death certificates until 1931. In 1945, Zimmerman confirmed, at autopsy, 2 of 7 cases of ALS seen during a 1-month period at the Naval Hospital on Guam (Zimmerman 1945). Subsequent reports of an unusually high preva-lence of motor neuron disease among native Guamanians (Koerner 1952; Arnold et al. 1953; Tillema and Wijnberg 1953) prompted island-wide epidemiological surveys in 1953 and 1954 (Kurland and Mulder 1954). Incidence rates were more than 50 times the rate of 0.8 per 100 000 population in the continental United States, and ALS alone accounted for nearly 10% of all adult deaths on Guam.

During the house-to-house surveys for ALS conducted in the 1950s, several patients with parkinsonian features and dementia were identified, but a few years lapsed before investigators realized they were dealing, not with postencephalitic parkinsonism, but with a new entity, designated parkinsonism-dementia (PD). The clinical and neuropathological features of PD were originally described in detail in 1961 (Hirano et al. 1961a, b) and a review of this disorder appears in an earlier volume of the Handbook (Chen and Chase 1986). Surveys quickly established the high incidence of PD. PD, which like ALS is uniformly fatal, often occurred in families affected with ALS, and accounted for an equally high percentage of adult deaths on the island.

Incidence rates of ALS were highest in southern Guam, particularly among Chamorros living in the more traditional villages of Umatac and Merizo, and on the island of Rota, lying 70 km north of Guam (Kurland and Mulder, 1954; Reed and Brody 1975; Yanagihara et al. 1983; Garruto et al. 1985b). By contrast, Chamorros on the more northerly islands of Saipan and Tinian (Lessell et al. 1962) and the sparsely populated islands of Anatahan, Alamagan, Pagan and Agrihan, were affected less often, despite the fact that Chamorros inhabiting these islands are direct descendants of Chamorro migrants from Guam (Yanagihara et al. 1983).

Over the past 3 decades, however, the incidence rates of ALS (and PD) islandwide on Guam have declined dramatically (Figs. 5 and 6) (Yanagihara et al. 1983; Garruto et al. 1985b; Reed et al. 1987) such that the risk of developing ALS is only several-fold higher than that for non-Chamorros living on the United States mainland. Originally the male:female ratio was 2:1 for ALS in the 1940s and 1950s. Presently, however, the sex ratio approaches unity, with the most precipitous change occurring among males.

Geographically, the sharpest decline in ALS has occurred in southern Guam, where incidence rates have been highest, while the northern and central regions of the islands have had a less precipitous change (Fig. 7). The western region

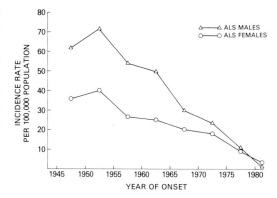

Fig. 5. Five-year average annual incidence rates of amyotrophic lateral sclerosis of Guam. Rates are for both males and females by year of onset of disease, age- and sex-adjusted to the 1960 Guamanian population.

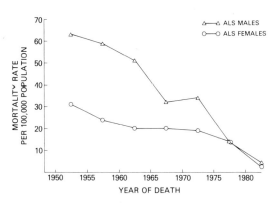

Fig. 6. Five-year average annual mortality rates of amyotrophic lateral sclerosis of Guam in males and females by year of death. Rates are age- and sex-adjusted to the 1960 Guamanian population.

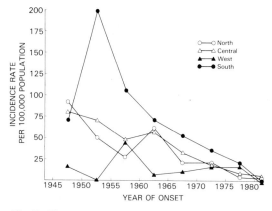

Fig. 7. Five-year average annual incidence rates of ALS among Chamorro males by geographic region. Rates for ALS females showed similar trends.

of the islands always had low rates, a trend which has continued throughout the 30-year period of surveillance.

On Guam, the disappearance of high-incidence ALS has occurred during the aggressive acculturation from a previously horticultural and fisherfolk subsistence economy, immediately following World War II, to an almost completely westernized culture having a cash economy today (Garruto et al. 1985b). Economic developments occurred more rapidly in the northern and central regions of the island, where large military installations, major harbor facilities, and the capital city of Agana are located. The southern villages were more impervious to change, but increasingly, men living in the south went to work in the more economically developed northern and central regions. Women in this oceanic Latino culture were slower to follow this trend and continued to remain at home. This may account for the earlier and sharper decline in disease incidence among men.

Equally important have been the declining incidence rates of ALS in the other 2 Pacific foci, again implying an environmental-etiologic factor that has decreased or disappeared with the advent of westernization and technology (Uebayashi 1980; Gajdusek 1984; Garruto and Yase 1986). Evidence from all 3 foci suggests a cohort effect and a geographic-residential distribution of disease involving exposure to environmental factors.

Migration studies

Between 1898 and 1950, strict military control of the island, geographic isolation, and lack of civilian transportation limited the opportunity of Guamanians to travel. With congressional passage of the Organic Act in 1950, Guamanians became citizens of the United States and a civilian government was established.

Epidemiological surveillance of Chamorro migrants established that ALS developed after long-term absences from Guam (Torres et al. 1957; Eldridge et al. 1969; Garruto et al. 1980). The minimum period of residence on Guam prior to onset of disease was from birth to age 18 years (Garruto et al. 1980). The longest period of

absence from Guam before onset of disease was over 3 decades. Based on the postwar estimates of the number of Chamorro migrants who left Guam, the crude average annual mortality rate for ALS was 5–7 per 100 000 population. The mortality rates for Chamorro migrants are one-third the rates for non-migrant Chamorros living on Guam. Thus, migrants appear less likely to develop disease than do those who remain on Guam, but their risk of developing ALS is still 5–10 times greater than that for the general population of the continental United States.

Although Filipinos and other ethnic groups migrated to Guam during the 18th century in an effort by the Spanish to bolster the decimated Chamorro population, it wasn't until after World War II that a resurgence in migration (particularly of Filipino males) to Guam occurred. Recent epidemiological studies of ALS and PD in Filipino migrants to Guam established that they migrated from the Ilocos region of northwestern Luzon (Garruto et al. 1981). All Filipino migrants who developed disease were men who had married Chamorro women and lived a traditional lifestyle in Chamorro villages. Their mean number of years of residence on Guam prior to onset of ALS was 17 and their mean age at migration was 38 years. PD has also developed in these migrants, but like Chamorro migrants, the number of cases is small (Garruto et al. 1981; Chen et al. 1982). Based on the surveillance data available, the average annual crude mortality rate for Filipino migrants is 8 per 100 000 population, a rate similar to that for Chamorro migrants (Garruto 1989a). A survey of ALS and PD in the Ilocos region of the Philippine Islands did not disclose any unusual prevalence rates in the parent population (R. Yanagihara, unpublished data). Thus the increased risk to Filipino migrants was likely acquired during their long-term residence on Guam.

The small endogamous Carolinian migrant population that remained genetically and culturally isolated in northern Guam has never developed a case of ALS (or PD) and only a single verified case of ALS in a Carolinian has been seen on Saipan (Elizan et al. 1966). Likewise, the much larger short-term transient American military population and their dependents living in confined military bases in northern and western Guam with their own foodstuffs and water supplies have no increased risk of developing ALS (Brody et al. 1978).

ETIOLOGICAL STUDIES

Genetics

On Guam, ALS occurred in high incidence, not only in the same villages, but in the same families, the same sibships and occasionally in the same individuals. In view of these observations, a prospective patient-control registry was established in 1958 to determine if first-degree (0.5) relatives and spouses of patients with ALS had a higher risk of developing disease than relatives of unaffected controls individually matched for age, sex and village (Plato et al. 1967). The Registry, which consisted of 3 separate categories (ALS patients, PD patients and matched controls), was initiated in 1958 and terminated 5 years later in 1963, at which time, it included 126 patients (77 with ALS and 42 with PD, and 7 with both ALS and PD) and an equal number of age-, sex- and village-matched controls; 994 living first-degree relatives of patients and 1218 of controls; and 88 living spouses of patients and 101 of controls. An analysis of this prospective registry, 25 years after its inception, disclosed a significantly increased risk of developing ALS or PD among parents, siblings and spouses of patients, but not among relatives of controls (Plato et al. 1986). Furthermore, offspring of both patients and controls showed no increased risk of disease. The increased risk among spouses of patients and its absence among their offspring further support the contention of a cohort effect and that exogenous, rather than genetic, factors are primarily important in the etiology of ALS in the western Pacific.

Numerous systematic genetic studies conducted on Guam over the past 3 decades, including pedigree analysis of residents of high-incidence villages, calculation of inbreeding coefficients for such villages, and identification of selected gene markers (such as HLA antigens at the A and B loci, blood group systems, red cell enzymes, immunoglobulin allotypes, serum proteins and dermatoglyphics), have not yielded a satisfactory genetic explanation (Plato and Cruz 1967; Plato et al. 1969; Reed et al. 1975; Hoffman et al. 1977; Blake et al. 1983; Garruto et al. 1983).

Clearly all Chamorros of the Mariana Islands derive from the same gene pool and there has been no genetic isolation between Chamorros on Guam and the other Mariana Islands during the past 400 years (Underwood 1976; Benfante et al. 1979; Yanagihara et al. 1983). Changes in the population gene pool would not account for the pattern of decline of ALS on Guam. Similarly, out-migration of high-risk individuals, abrupt changes in mating patterns (which remained unusually stable until 1965, when the decline was already apparent) or exhaustion of genetic susceptibles from the population also would not account for the declining incidence rates. In addition, the onset of both ALS and PD usually occurs during the post-reproductive years and therefore does not affect completed fertility. Finally, although ALS in the Kii Peninsula of Japan and West New Guinea are also familial, the data do not support a genetic etiology (Shiraki and Yase 1975; Gajdusek 1984; Garruto and Yase 1986).

Infectious agents

As early as 1962, Gibbs and Gajdusek attempted to experimentally transmit Guamanian ALS to several species of nonhuman primates, including chimpanzees, and to other laboratory animals, as they had with kuru and Creutzfeldt-Jakob disease (Gibbs and Gajdusek 1982). Several primates from those early experiments and from subsequent transmission attempts are still alive, and none has ever exhibited evidence of neurological disease. Attempts to isolate both conventional and unconventional viruses from cerebrospinal fluid and neural tissues (Gibbs and Gajdusek 1982) and to identify viral-specific proteins and nucleic acid sequences in brain and spinal cord tissues of Guamanian patients have been similarly unsuccessful (Viola et al. 1975, 1979; Kohne et al. 1981; Brahic et al. 1985; Mora et al. 1988).

A pattern of decline in ALS and PD affecting men more than women concomitant with a failure of the high incidence of these diseases to spread to the western part of Guam, to the Chamorros of Tinian, Saipan and more northerly islands, and from Chamorro migrants to non-Chamorro residents of the continental United States is inconsistent with an infectious etiology. The lack of geographical barriers delimiting affected from non-affected villages and the uniformity of ecology within each of the 3 high-incidence foci of ALS in the western Pacific also indicate that the etiology is not an infectious agent that depends on reservoir hosts or insect vectors.

Plant excitant neurotoxins

Neurotoxins have long been heralded as possible etiological agents in the Pacific foci of ALS. On Guam, tortillas prepared from the washed seeds of the false sago palm (*Cycas circinalis*) (called *federico* or *fadang* in the Chamorro language) are eaten occasionally (Whiting 1964). Cycad nuts, long known to contain neurotoxins which poduce an acute gait disturbance in foraging animals (Hall 1957; Mason and Whiting 1968; Hooper 1978), therefore, quickly became a major etiological candidate for these neurological disorders. However, β-N-methylamino-L-alanine (BMAA), a neurotoxin isolated from cycad seeds and shown to be toxic to chicks, rats and mice when administered in high doses, was thought to be present in too low concentrations in the seeds and leaves of *Cycas circinalis* to be a major factor in the development of neurological disease in man (Polsky et al. 1972). Other studies, including toxicity experiments in non-human primates (Yang et al. 1966; Hirano 1973; Sieber et al. 1980; Garruto et al. 1989b), were negative, although one rhesus monkey developed mild atrophy of 1 limb and degeneration of some motor neurons in the brain and spinal cord (Dastur 1964). This study, however, could not be confirmed; a decade later, none of 13 monkeys fed washed, unwashed or baked cycad for 1–5 years developed any neurological deficit (Dastur and Palekar 1974).

Recently, the cycad hypothesis has been resurrected (Spencer 1987; Spencer et al. 1987a, 1987b, 1987c). Cynomolgus monkeys (*Macaca fascicularis*), gavage fed synthetic BMAA for 2–12 weeks, developed neurological symptoms and some pathological changes in motor neurons of the brain and spinal cord (Spencer et al. 1987a). However, the doses administered were extremely high: based on a near maximum content of total BMAA in cycad seeds (approximately, 100 mg/100 g of fresh seed) (Duncan et al. 1988) and the dose given in these experiments (average, 250 mg of BMAA/kg body wt), each monkey would have had to ingest the equivalent of 0.5 kg of unwashed cycad per day or 42 kg over 12 weeks (Garruto et al. 1988). If humans are similarly susceptible to BMAA, an average 70 kg man would need to ingest 17.5 kg of unwashed cycad per day and approximately 1500 kg during a 12-week period to produce lesions or clinical symptoms. Washed cycad usually contains only trace amounts of toxin (Duncan et al. 1988, 1989a), so the amount of cycad necessary to produce clinical disease would have to be several orders of magnitude higher than 1500 kg. Even during World War II, when food was scarce on Guam, it is inconceivable that such quantities of cycad were ever consumed. Furthermore, BMAA does not easily cross the blood-brain barrier and millimolar quantities are necessary to kill neurons in cell culture (Duncan et al. 1989b; M.W. Duncan, pers. commun.).

Cycad is eaten and is used as a medicine (poultice) by inhabitants of regions bordering the high-incidence ALS foci, as well as by other populations within the Asia-Pacific Basin, including those in Malaysia, the Andaman Islands, South India, Sri Lanka, and the Philippine Islands (Burkhill 1935). Yet, an unusual prevalence of ALS has not been described in these regions, even when specifically looked for in cycad-eating populations (Kurland 1972; Gajdusek 1990). Furthermore, there is no precedent or evidence that plant excitant neurotoxins, including BMAA administered in sublethal doses, exert their effects years to decades after exposure, producing a cascading loss of neurons with the onset of clinical disease. Also, no cycad neurotoxin or its derivatives has been detected in brain or spinal cord tissues from patients with ALS or PD. It would appear, therefore, that cycad and its derivative BMAA are unlikely causes of ALS in these Pacific foci.

Metabolic defect

In the early 1960s, when the cycad hypothesis was first in vogue, Kimura and colleagues presented evidence indicating that toxic metals and essential minerals may be etiologically important in the ALS focus in the Kii Peninsula of Japan (Kimura et al. 1963). Over the next decade, Yase and others working in the Kii Peninsula and on Guam postulated that manganese and possibly calcium and aluminum were involved (Yase et al. 1968; Yase 1980). Subsequent studies demonstrated that all 3 Western Pacific foci displayed unusually low concentrations of calcium and magnesium in soil and drinking water coupled with high levels of aluminum and lead in soil (Tracey et al. 1959; Fosberg 1960; Shiraki and Yase 1975; Iwata et al. 1978; Yase 1978; Gajdusek and Salazar 1982; Garruto et al. 1984b).

That alkaline earth metals are involved in the pathogenesis of ALS (and PD) is supported by finding biochemical abnormalities of calcium and vitamin D metabolism in 53% and 46% of Guamanian and Japanese patients with ALS, respectively (Fig. 8) (Yase 1978; Yanagihara et al. 1984). Such abnormalities are also found in Caucasian patients with ALS in the continental United States (Mallette et al. 1977; Patten and Engel 1982; Yanagihara 1982; Glassberg et al. 1987a, 1987b). Furthermore, bone mass is decreased in neurologically normal Guamanian children and adults (Plato et al. 1982, 1984), compared to other ethnic groups in the continental United States. Analyses of hand-wrist X-rays indicate a significantly decreased bone mass in Guamanian patients with ALS, compared to neurologically normal Guamanian when factors such as age, sex, menopause status, disease duration and degree of disability are controlled (Garruto et al. 1989a).

The above data complement the recent demonstration of intraneuronal co-localization of calcium, aluminum and silicon in NFT-bearing hippocampal and spinal motor neurons of Guamanian patients with ALS (Garruto et al. 1985a, 1986; Piccardo et al. 1988). The localization and distribution of these metals in the central nervous system tissues in ALS patents have been reported by different laboratories using widely different

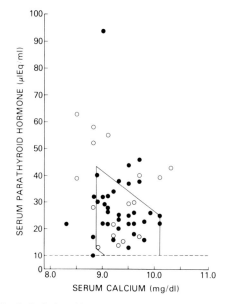

Fig. 8. Relationship between serum immunoreactive parathyroid hormone and serum calcium in Guamanian patients with amyotrophic lateral sclerosis (open circles) and parkinsonism-dementia (filled circles). The rhomboid represents the normal ranges as determined for 150 healthy whites. The broken line represents the limit of detection of the parathyroid hormone radioimmunoassay. All values for the 11 Chamorro controls (not shown) fell within the rhomboid. To convert serum calcium values to mmol/l multiply by 0.2495.

analytical techniques (Traub et al. 1980; Yase 1980; Yoshimasu et al. 1980; Perl et al. 1982; Linton et al. 1987; Yoshida et al. 1988). Using wavelength-dispersive spectrometry and computer-controlled electron beam X-ray microanalysis, chemical maps of the hippocampus and spinal cord were produced for Guamanian patients (Garruto et al. 1984a, 1985a, 1986; Linton et al. 1987; Piccardo et al. 1988).

The elemental images show the striking co-localization of calcium and aluminum (Fig. 9), and silicon in the perikarya and dendritic processes of the same neurons in brains and spinal cord tissues of ALS patients. No other element, except occasionally iron, was detected in abnormal concentrations in the neurons of these patients. Calcium and aluminum were also found in the walls of some cerebral blood vessels (Piccardo et al. 1988). In neurologically normal Guamanians without NFT formation, the abnor-

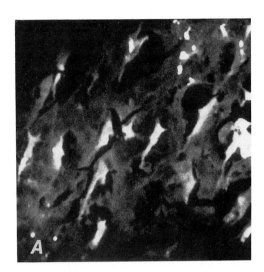 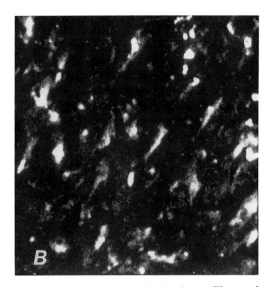

Fig. 9. Wavelength-dispersive spectrometry elemental images of (A) calcium and (B) aluminum. Elemental images show the striking co-localization of both elements within the cell body and dendritic processes of the same neurofibrillary tangle-bearing neurons in the hippocampus of a Guamanian patient with amyotrophic lateral sclerosis. The field size is 125 μm × 125 μm with an array size of 256 × 256 pixels. Image acquisition time was 4.5 h with a beam current of 0.2 μA and an accelerating voltage of 15 keV. Generally speaking, the brighter the image the higher the concentration of element. Semiquantitative estimates of calcium and aluminum for the brightest images in this field are 4300 and 700 ppm dry weight, respectively. No element could be imaged in neurologically normal Guamanians without neurofibrillary tangles.

mal deposition of calcium and aluminum did not occur and thus could not be imaged by X-ray microanalysis. Semiquantitative estimates of calcium, aluminum, and silicon concentrations (weight fraction) in NFT-bearing hippocampal neurons from Guamanian ALS patients exhibiting the brightest images were 4300, 700, and 3000 ppm dry weight, respectively. Aluminum, silicon and perhaps other elements in combination with calcium form mineral complexes in susceptible neurons, that is thought to lead to an alteration of cell function and NFT formation, the exact composition of which is dependent on the bioavailability of each cation species.

There is now compelling data to suggest a neurotoxic insult involving calcium, aluminum and silicon. We have proposed as a working hypothesis that a basic defect in mineral metabolism, associated with secondary hyperparathyroidism provoked by chronic nutritional deficiencies of calcium and magnesium, leads to increased intestinal absorption of toxic metals and deposition of calcium, aluminum, silicon and other metals as hydroxyapatites or aluminosili-

cates in neurons of patients with ALS. Intraneuronal deposition of these metals may interfere with slow axonal transport by disrupting the neuronal cytoskeleton and altering normal production or catabolism of cytoskeletal proteins. The metabolic defect with resultant deposition of calcium, aluminum, silicon and occasionally other metals such as iron may occur in utero, during infancy, childhood, adolescence or even later, but long before onset of clinical disease. Later intake of calcium, and subsequent correction of the secondary hyperparathyroidism is unlikely to prevent or reverse the neuronal damage and cell death that these deposits produce.

NEW EXPERIMENTAL MODELS

Based on the hallmark neurofibrillary changes in Guamanian ALS (and PD) and the mounting evidence to support a disturbance of calcium homeostasis coupled with the cellular desposition of calcium, aluminum and silicon, attempts have been made to experimentally model ALS and its pathologic changes. In 2 recent studies, in which aluminum was administered orally to nonhuman primates maintained on a low-calcium diet for 12–46 months, neuropathological changes compatible with early ALS (including chromatolysis, basophilic and hyaline-like inclusions and NFT) and the co-deposition of calcium and aluminum in spinal motor neurons were observed (Yano et al. 1987; Yase 1988; Garruto et al. 1989b; Yasui et al. 1990). These studies clearly support the conjecture that chronic dietary deficiencies of calcium and the increased uptake of aluminum leads to motor neuron pathology.

In a second experimental model of chronic aluminum intoxication, young adult New Zealand white rabbits, inoculated intracisternally once monthly with low doses of aluminum chloride, developed progressive hyperreflexia, hypertonia, gait impairment, weight loss, muscle wasting and abnormal righting reflexes over the course of 8 months (Strong et al. 1990b). Unlike the acute model of aluminum toxicity, no overt encephalopathic features were present. Spinal motor neuron perikarya, dendrites and axonal processes contained extensive argentophilic globular inclusions, and NFT-like inclusions were consistently present in brain stem nuclei. This chronic model of aluminum-induced spastic myelopathy mimics many of the clinical and neuropathological changes seen in human ALS and demonstrates a clinically and neuropathologically specific motor system degeneration.

In vitro experimental studies have been designed to complement and augment the in vivo studies (Strong et al. 1990a). Using dissociated fetal rabbit motor and hippocampal neurons in cell culture, neuron-specific thresholds of aluminum toxicity have been demonstrated. Motor neurons, exposed to different concentrations of aluminum in a chemically defined medium, exhibited a 10-fold greater sensitivity to aluminum intoxication than did hippocampal neurons, as measured by morphological criteria. This suggests that the striking disparity in sensitivity between motor neurons and hippocampal neurons represents neuron-specific thresholds of aluminum toxicity. Thus, these and other experimental studies, including a recently discovered model of plasticizer-induced spastic myelopathy (Strong et al. 1990c, 1991), are attempts to lead us closer to

a better understanding of the cellular and molecular mechanisms of motor system degeneration in human ALS.

CONCLUSION

The study of natural paradigms of environmentally induced neurodegenerative diseases occurring in high incidence in the western Pacific region has had a profound influence on formulating pathogenetic concepts of neurological disorders worldwide. In particular, the intensive multidisciplinary approach to the study of ALS on Guam for nearly 4 decades has had an immeasurable impact on our thinking about motor neuron disease. Although some of the thinking presented here is still speculative, it is based on decades of systematic epidemiological, clinical, neuropathological, etiological and experimental studies. The effort to study Pacific foci of ALS by numerous investigators worldwide has been enormous, and we hope we are nearer a solution to a problem that continues to plague people on a global scale.

REFERENCES

ANDERSON, F. H., E. P. RICHARDSON, JR., H. OKAZAKI and J. A. BRODY: Neurofibrillary degeneration on Guam: frequency in Chamorros and non-Chamorros with no known neurological disease. Brain 102 (1979) 65–77.

ARAN, F. A.: Recherches sur une maladie non encore décrite du système musculaire (atrophie musculaire progressive). Arch. Gén. Med. (Paris) 24 (1850a) 5–35.

ARAN, F. A.: Recherches sur une maladie non encore decrite du systeme musculaire (atrophie musculaire progressive). (2e article — Suite et fin.). Arch. Gén. Med. (Paris) 24 (1850b) 172–214.

ARNOLD, A., D. C. EDGREN and V. S. PALLADINO: Amyotrophic lateral sclerosis: fifty cases observed on Guam. J. Nerv. Ment. Dis. 117 (1953) 135–139.

AVERBACK, P.: Unusual particles in motor neuron disease. Arch. Pathol. Lab. Med. 105 (1981) 490–493.

BENFANTE, R. J., P. M. HOFFMAN, R. M. GARRUTO and D. C. GAJDUSEK: HLA antigens in the Chamorros of the Mariana Islands and comparisons with other Pacific populations. Hum. Biol. 51 (1979) 201–212.

BLAKE, N. M., R. L. KIRK, S. R. WILSON, R. M. GARRUTO, D. C. GAJDUSEK, C. J. GIBBS, JR. and P. HOFFMAN: Search for a red cell enzyme or serum protein marker in amyotrophic lateral sclerosis and parkinsonism-dementia of Guam. Am. J. Med. Genet. 14 (1983) 299–305.

BOWERS, N. M.: The Mariana, Volcano, and Bonin Islands. In: O. W. Freeman (ed.), Geography of the Pacific. New York, Wiley (1951) 205–235.

BRAHIC, M., R. A. SMITH, C. J. GIBBS, JR., R. M. GARRUTO, W. W. TOURTELLOTTE and E. CASH: Detection of picornavirus sequences by in situ hybridization in nervous system tissue of amyotrophic lateral sclerosis and control patients. Ann. Neurol. 18 (1985) 337–343.

BRODY, J. A., A. H. EDGAR and M. M. GILLESPIE: Amyotrophic lateral sclerosis. No increase among US construction workers in Guam. J. Am. Med. Assoc. 240 (1978) 551–552.

BURKILL, I. H.: A Dictionary of the Economic Products of the Malay Peninsula, Vol. 1 (A–H). Malaysia Ministry of Agriculture and Co-operatives Kuala Lumpur (1935) 729–731.

CARANO, P. and P. C. SANCHEZ: A Complete History of Guam. Rutland, VT, Charles E. Tuttle (1964).

CHARCOT, J. M. and A. JOFFROY: Deux cas d'atrophie musculaire progressive avec lésions de la substance grise et des faisceaux antérolateraux de la moëlle épinière. Arch. Physiol. 3 (1869) 629–744.

CHEN, K. M.: Motor neuron involvement in parkinsonism-dementia and its relationship to Guam ALS. In: Japan Medical Research Foundation, (Ed.) Amyotrophic Lateral Sclerosis. Proceedings of the International Symposium on Amyotrophic Lateral Sclerosis. Tokyo, University of Tokyo Press (1979) 319–344.

CHEN, K.-M. and T. N. CHASE: Parkinsonism-dementia. In: P. J. Vinken, G. W. Bruyn, and H. L. Klawans (Eds.), Handbook of Clinical Neurology, Vol 49, Extrapyramidal Disorders. Amsterdam, Elsevier Science Publishers (1986) 167–183.

CHEN, K.-M., T. MAKIFUCHI, R. M. GARRUTO and D. C. GAJDUSEK: Parkinsonian-dementia in a Filipino migrant: a clinicopathologic case report. Neurology 32 (1982) 1221–1226.

CHEN, L.: Neurofibrillary change on Guam. Arch. Neurol. 38 (1981) 16–18.

CHOU, S. M.: Pathognomy of intraneuronal inclusions in ALS. In: Japan Medical Research Foundation, Proceedings of the International Symposium on Amyotrophic Lateral Sclerosis. Tokyo, University of Tokyo Press (1979) 135–176.

DASTUR, D. K.: Cycad toxicity in monkeys: clinical, pathological and biochemical aspects. Fed. Proc. 23 (1964) 1368–1369.

DASTUR, D. K. and R. S. PALEKAR: The experimental pathology of cycad toxicity with special reference to oncogenic effects. Ind. J. Cancer (1974) 33–49.

DUCHENNE (DE BOULOGNE), G. B. A.: De l'atrophie musculaire avec transformation graisseuse. De l'électrisation localisée. Paris, Ed. Bailliere (1855) 622.

DUCHENNE (DE BOULOGNE), G.B.A.: Paralysie musculaire progressive de la langue, du voile du palais et des lèvres; affection non encore décrite comme espèce morbide distincte. Arch. Gén. Med. 16 (1860) 283–296.

DUNCAN, M. W., I. J. KOPIN, R. M. GARRUTO, L. LAVINE and S. P. MARKEY: 2-amino-3(methylamino)-propionic acid in cycad-derived foods is an unlikely cause of amyotrophic lateral sclerosis/parkinsonism. Lancet 2 (1988) 631–632.

DUNCAN, M. W., I. J. KOPIN, J. S. CROWLEY, S. M. JONES and S. P. MARKEY: Quantification of the putative neurotoxin 2-amino-3(methylamino)-propanoic acid (BMAA) in cycadales: Analysis of the seeds of some members of the family cycadaceae. J. Anal. Toxicol. 13 (1989a) A–G.

DUNCAN, M. W., A. M. MARINI, I. J. KOPIN and S. P. MARKEY: BMAA and ALS-PD: a causal link? International ALS-MND Update 2Q89 (1989b) 31.

ELDRIDGE, R., E. RYAN, J. ROSARIO and J. A. BRODY: Amyotrophic lateral sclerosis and parkinsonism-dementia in a migrant population from Guam. Neurology (Minneap.) 19 (1969) 1029–1037.

ELIZAN, T. S., K.-M. CHEN, K. V. MATHAI, D. DUNN and L. T. KURLAND: Amyotrophic lateral sclerosis and parkinsonism-dementia complex. A study in non-Chamorros of the Mariana and Caroline Islands. Arch. Neurol. 14 (1966) 347–355.

FOSBERG, F. R.: The vegetation of Micronesia. 1. General descriptions, the vegetation of the Mariana Islands, and a detailed consideration of the vegetation of Guam. Bull. Am. Mus. Nat. Hist. 119 (1960) 1–75.

GAJDUSEK, D. C.: Motor-neuron disease in natives of New Guinea. N. Engl. J. Med. 268 (1963) 474–476.

GAJDUSEK, D. C.: Environmental factors provoking physiological changes which induce motor neuron disease and early neuronal aging in high incidence foci in the western Pacific: In: F. C. Rose (Ed.), Progress in Motor Neuron Disease. Kent, Pitman Books, Ltd. (1984) 44–69.

GAJDUSEK, D. C.: Cycad toxicity not the cause of high incidence amyotrophic lateral sclerosis/parkinsonism-dementia on Guam, Kii Peninsula of Japan or in West New Guinea. In: A. J. Hudson (Ed.), Amyotrophic Lateral Sclerosis: Concepts in Pathogenesis and Etiology. Toronto, University of Toronto (1990) 317–325.

GAJDUSEK, D. C. and A. M. SALAZAR: Amyotrophic lateral sclerosis and parkinsonian syndromes in high incidence among the Auyu and Jakai people in West New Guinea. Neurology 32 (1982) 107–126.

GARRUTO, R. M.: Amyotrophic lateral sclerosis and parkinsonism-dementia of Guam: Clinical, epidemiological and genetic patterns. Am. J. Hum. Biol. 1 (1989a) 367–382.

GARRUTO, R. M.: Cellular and molecular mechanisms of neuronal degeneration: Amyotrophic lateral sclerosis, parkinsonism-dementia and Alzheimer disease. Am. J. Hum. Biol. 1 (1989b) 529–543.

GARRUTO, R. M. and Y. YASE: Neurodegenerative disorders of the Western Pacific: The search for mechanisms of pathogenesis. Trends Neurosci. 9 (1986) 368–374.

GARRUTO, R. M., D. C. GAJDUSEK and K.-M. CHEN: Amyotrophic lateral sclerosis among Chamorro migrants from Guam. Ann. Neurol. 8 (1980) 612–619.

GARRUTO, R. M., D. C. GAJDUSEK and K.-M. CHEN: Amyotrophic lateral sclerosis and parkinsonism-dementia among Filipino migrants to Guam. Ann. Neurol. 10 (1981) 341–350.

GARRUTO, R. M., C. C. PLATO, N. C. MYRIANTHOPOULOS, M. S. SCHANFIELD and D. C. GAJDUSEK: Blood groups, immunoglobulin allotypes and dermatoglyphic features of patients with amyotrophic lateral sclerosis and parkinsonism-dementia of Guam. Am. J. Med. Genet. 14 (1983) 289–298.

GARRUTO, R. M., R. FUKATSU, R. YANAGIHARA, D. C. GAJDUSEK, G. HOOK and C. E. FIORI: Imaging of calcium and aluminum in neurofibrillary tangle-bearing neurons in parkinsonism-dementia of Guam. Proc. Natl. Acad. Sci. USA 81 (1984a) 1875–1879.

GARRUTO, R. M., R. YANAGIHARA, D. C. GAJDUSEK and D. M. ARION: Concentrations of heavy metals and essential minerals in garden soil and drinking water in the Western Pacific. In: K.-M. Chen and Y. Yase (Eds.), Amyotrophic Lateral Sclerosis in Asia and Oceania. Taipei, National Taiwan University (1984b) 265–329.

GARRUTO, R. M., C. SWYT, C. E. FIORI, R. YANAGIHARA and D. C. GAJDUSEK: Intraneuronal deposition of calcium and aluminum in amyotrophic lateral sclerosis. Lancet 2 (1985a) 1353.

GARRUTO, R. M., R. YANAGIHARA and D. C. GAJDUSEK: Disappearance of high-incidence amyotrophic lateral sclerosis and parkinsonism-dementia on Guam. Neurology 35 (1985b) 193–198.

GARRUTO, R. M., C. SWYT, R. YANAGIHARA, C. E. FIORI and D. C. GAJDUSEK: Intraneuronal colocalization of silicon with calcium and aluminum in amyotrophic lateral sclerosis and parkinsonism-dementia of Guam. N. Engl. J. Med. 315 (1986) 711–712.

GARRUTO, R. M., R. YANAGIHARA and D. C. GAJDUSEK: Cycads and amyotrophic lateral sclerosis/parkinsonism-dementia. Lancet 2 (1988) 1079.

GARRUTO, R. M., C. C. PLATO, R. YANAGIHARA, K. FOX, J. DUTT, D. C. GAJDUSEK and J. TOBIN: Bone mass in patients with amyotrophic lateral sclerosis and parkinsonism-dementia of Guam. Am. J. Phys. Anthropol. 80 (1989a) 107–113.

GARRUTO, R. M., S. K. SHANKAR, R. YANAGIHARA, A. M. SALAZAR, H. L. AMYX and D. C. GAJDUSEK: Low-calcium, high-aluminum diet-induced motor neuron pathology in cynomolgus monkeys. Acta Neuropathol. (Berlin) 78 (1989b) 210–219.

GIBBS, C. J., JR. and D. C. GAJDUSEK: An update on long-term in vivo and in vitro studies designed to identify a virus as a cause of amyotrophic lateral sclerosis, parkinsonism-dementia and Parkinson disease. In: L. P. Rowland (Ed.), Advances in Neurology, Vol. 36, Human Motor Neuron Disease. New York, Raven Press (1982) 343–353.

GLASBERG, M., B. GROSS and M. KLEEREKOPER: Bone mineral content and osteopenia in amyotrophic lateral sclerosis. Abstract B2–13 in Abstracts of the International Conference on Amyotrophic Lateral Sclerosis, Kyoto (1987a) 117.

GLASBERG, M., M. KLEEREKOPER and J. GLASBERG: Parathyroid metabolism and calcium homeostasis in amyotrophic lateral sclerosis. Abstract No. A1–7 in Abstracts of the International Conference on Amyotrophic Lateral Sclerosis, Kyoto (1987b) 77.

GUIROY, D. C., M. MIYAZAKI, G. MULTHAUP, P. FISHER, R. M. GARRUTO, K. BEYREUTHER, C. MASTERS, G. SIMMS, C. J. GIBBS, JR. and D. C. GAJDUSEK: Amyloid of neurofibrillary tangles of Guamanian parkinsonism-dementia and Alzheimer disease share an identical amino acid sequence. Proc. Natl. Acad. Sci. USA 84 (1987) 2073–2077.

HALL, W. T. K.: Toxicity of the leaves of Macrozamia spp. for cattle. Queensland J. Agr. Sci. 14 (1957) 45–52.

HIRANO, A.: Progress in the pathology of motor neuron diseases. In: H. M. Zimmerman (Ed.), Progress in Neuropathology, Vol. II. New York, Grune and Stratton (1973) 181–215.

HIRANO, A.: Some current concepts of amyotrophic lateral sclerosis. Neurol. Med. (Tokyo) 4 (1976) 43–51.

HIRANO, A., L. T. KURLAND, R. S. KROOTH and S. LESSEL: Parkinsonism-dementia complex, an endemic disease on the island of Guam. I. Clinical features. Brain 84 (1961a) 642–661.

HIRANO, A., N. MALAMUD and L. T. KURLAND: Parkinsonism-dementia complex, an endemic disease on the island of Guam. II. Pathological features. Brain 84 (1961b) 662–679.

HIRANO, A., N. MALAMUD and L. T. KURLAND: Amyotrophic lateral sclerosis and parkinsonism-dementia complex on Guam. Further pathologic studies. Arch. Neurol. 15 (1966) 35–51.

HIRANO, A., H. DONNENFIELD, S. SASAKI and I. NAKANO: Fine structural observations of neurofilamentous changes in amyotrophic lateral sclerosis. J. Neuropathol. Exp. Neurol. 43 (1984a) 461–470.

HIRANO, A., I. NAKANO, L. T. KURLAND, D. W. MULDER, P. W. HOLLEY and G. SACCOMANNO: Fine structural study of neurofibrillary changes in a family with amyotrophic lateral sclerosis. J. Neuropathol. Exp. Neurol. 43 (1984b) 471–480.

HOFFMAN, P. M., D. S. ROBBINS, C. J. GIBBS, JR., D. C. GAJDUSEK, R. M. GARRUTO and P. I. TERASAKI: Histocompatability antigens in amyotrophic lateral sclerosis and parkinsonism-dementia on Guam. Lancet 2 (1977) 717.

HOFFMAN, P. N., D. W. CLEVELAND, J. W. GRIFFIN, P. W. LANDES, N. J. COWAN and D. L. PRICE: Neurofilament gene expression: A major determinant of axonal calibre. Proc. Natl. Acad. Sci. USA 84 (1987) 3472–3476.

HOOPER, P. T.: Cycad poisoning in Australia—etiology and pathology. In: R. F. Keeler, K. R. van Kampen, and L. F. James (Eds.), Effects of

Poisonous Plants on Livestock. New York, Academic Press (1978) 337–347.

IWATA, S., K. SASAJIMA, Y. YASE and K.-M. CHEN: Report of investigation of the environmental factors related to occurrence of amyotrophic lateral sclerosis in Guam Island. Tokyo, Japanese Ministry of Education (1978).

KIMURA, K., Y. YASE, Y. HIGASHI, S. UNO, K. YAMAMOTO, M. IWASAKI, I. TSUMOTO, M. SUGIURA, S. YOSHIMURA, K. NAMIKAWA, J. KUMURA, S. IWAMOTO, I. YAMAMOTO, Y. HANDA, M. YATA and Y. YATA: Epidemiological and geomedical studies on amyotrophic lateral sclerosis. Dis. Nerv. Sys. 24 (1963) 155–159.

KOERNER, D. R.: Amyotrophic lateral sclerosis on Guam: a clinical study and review of the literature. Ann. Intern. Med. 37 (1952) 1204–1220.

KOHNE, D. E., C. J. GIBBS, JR., L. WHITE, S. M. TRACY, W. MEINKE and R. A. SMITH: Virus detection by nucleic acid hybridization: Examination of normal and ALS tissues for the presence of poliovirus. J. Gen Virol. 56 (1981) 223.

KURLAND, L. T.: An appraisal of the neurotoxicity of cycad and the etiology of amyotrophic lateral sclerosis on Guam. Fed. Proc. 31 (1972) 1540–1542.

KURLAND, L. T. and D. W. MULDER: Epidemiologic investigations of amyotrophic lateral sclerosis. 1. Preliminary report on geographic distribution, with special reference to the Mariana Islands, including clinical and pathological observations. Neurology 4 (1954) 355–378, 438–448.

LEACH, P.: Sanitary report on Guam, L. I. In: Report of the surgeon general, U.S. Navy, Chief of the Bureau of Medicine and Surgery to the Secretary of the Navy. Washington, D.C., U.S. Government Printing Office (1900) 208–212.

LESSELL, S., A. HIRANO, J. TORRES and L. T. KURLAND: Parkinsonism-dementia complex. Epidemiological considerations in the Chamorros of the Mariana Islands and California. Arch. Neurol. 7 (1962) 377–385.

LIE-INJO, L. E.: Red cell carbonic anhydrase Ic in Filipinos. Am. J. Hum. Genet. 19 (1967) 130–133.

LINTON, R. W., S. R. BRYAN, D. P. GRIFFIS, J. D. SHELBURNE, C. E. FIORI and R. M. GARRUTO: Digital imaging studies of aluminum and calcium in neurofibrillary tangle-bearing neurons using SIMS secondary ion mass spectometry. Trace Elem. Med. 4 (1987) 99–104.

MALLETTE, L. E., B. M. PATTEN, J. D. COOK and W. K. ENGEL: Calcium metabolism in amyotrophic lateral sclerosis. Dis. Nerv. Sys. 38 (1977) 457–461.

MASON, M. M. and M. G. WHITING: Caudal motor weakness and ataxia in cattle in the Caribbean area following ingestion of cycads. Cornell Vet. 58 (1968) 541–554.

MIURA, K.: Amyotrophische Lateralsclerose unter dem Bilde von sog. Bulbarparalyse. Neurol. Jpn. 10 (1911) 366–369.

MORA, C., R. M. GARRUTO, P. BROWN, D. GUIROY, O. S. C. MORGAN, P. RODGERS-JOHNSON, M. CERONI, R. YANAGIHARA, L. G. GOLDFARB, C. J. GIBBS, JR. and D. C. GAJDU-

SEK: Seroprevalence of antibodies to HTLV-I in patients with chronic neurological disorders other than tropical spastic paraparesis. Ann. Neurol. 23 (Suppl.) (1988) S192–S195.

MUKAI, E., T. SAKAKIBARA, I. SOBUE and K.-M. CHEN: The prognosis of amyotrophic lateral sclerosis in Guam and Japan. Clin. Neurol. 22 (1982) 139–144.

PATTEN, B. M. and W. K. ENGEL: Phosphate and parathyroid disorders associated with the syndrome of amyotrophic lateral sclerosis. In: L. Rowland (Ed.), Advances in Neurology, Vol. 36. Human Motor Neuron Diseases. New York, Raven Press (1982) 181–200.

PERL, D. P., D. C. GAJDUSEK, R. M. GARRUTO, R. T. YANAGIHARA and C. J. GIBBS, JR.: Intraneuronal aluminum accumulation in amyotrophic lateral sclerosis and parkinsonism-dementia of Guam. Science 217 (1982) 1053–1055.

PICCARDO, P., R. YANAGIHARA, R. M. GARRUTO, C. J. GIBBS, JR. and D. C. GAJDUSEK: Histochemical and x-ray microanalytical localization of aluminum in amyotrophic lateral sclerosis and parkinsonism-dementia of Guam. Acta Neuropathol. (Berlin) 77 (1988) 1–4.

PLATO, C. C. and M. T. CRUZ: Blood group and haptoglobin frequencies of the Chamorros of Guam. Am. J. Hum. Genet. 19 (1967) 722–731.

PLATO, C. C., D. L. RUCKNAGEL and L. T. KURLAND: Blood group investigations in the Carolinians and Chamorros of Saipan. Am. J. Phys. Anthropol. 24 (1966) 147–154.

PLATO, C. C., D. M. REED, T. S. ELIZAN and L. T. KURLAND: Amyotrophic lateral sclerosis/parkinsonism-dementia complex of Guam. IV. Familial and genetic investigations. Am. J. Hum. Genet. 19 (1967) 617–632.

PLATO, C. C., M. T. CRUZ and L. T. KURLAND: Amyotrophic lateral sclerosis/parkinsonism-dementia complex of Guam: Further genetic investigations. Am. J. Hum. Genet. 21 (1969) 133–141.

PLATO, C. C., R. M. GARRUTO, R. YANAGIHARA, K.-M. CHEN, J. L. WOOD, D. C. GAJDUSEK and A. H. NORRIS: Cortical bone loss and measurements of the second metacarpal bone. I. Comparisons between adult Guamanian Chamorros and American Caucasians. Am. J. Phys. Anthropol. 59 (1982) 461–465.

PLATO, C. C., W. W. GREULICH, R. M. GARRUTO and R. YANAGIHARA: Cortical bone loss and measurements of the second metacarpal bone. II. Hypodense bone in post-war Guamanian children. Am. J. Phys. Anthropol. 63 (1984) 57–63.

PLATO, C. C., R. M. GARRUTO, K. M. FOX and D. C. GAJDUSEK: Amyotrophic lateral sclerosis and parkinsonism-dementia of Guam: a 25-year prospective case-control study. Am. J. Epidemiol. 124 (1986) 643–656.

POLSKY, R. I., P. B. NUNN and E. A. BELL: Distribution and toxicity of α-amino-β-methylaminopropionic acid. Fed. Proc. 31 (1972) 1473–1475.

REED, D. M. and J. A. BRODY: Amyotrophic sclerosis and parkinsonism-dementia on Guam, 1945–1972. I. Descriptive epidemiology. Am. J. Epidemiol. 101 (1975) 287–301.

REED, D. M., J. M. TORRES and J. A. BRODY: Amyotrophic lateral sclerosis and parkinsonism-dementia on Guam, 1945–1972. II. Familial and genetic studies. Am. J. Epidemiol. 101 (1975) 302–310.

REED, D. M., D. LABARTHE, K.-M. CHEN and R. STALLONES: A cohort study of amyotrophic lateral sclerosis and parkinsonism-dementia on Guam and Rota. Am. J. Epidemiol. 125 (1987) 92–100.

RODGERS-JOHNSON, P., R. M. GARRUTO, R. YANAGIHARA, K.-M. CHEN, D. C. GAJDUSEK and C. J. GIBBS, JR.: Amyotrophic lateral sclerosis and parkinsonism-dementia on Guam: a 30 year evaluation and clinical and neuropathological trends. Neurology 36 (1986) 7–13.

SHANKAR, S. K., R. YANAGIHARA, R. M. GARRUTO, I. GRUNDKE-IQBAL, K. S. KOSIK and D. C. GAJDUSEK: Immunocytochemical characterization of neurofibrillary tangles in amyotrophic lateral sclerosis and parkinsonism-dementia on Guam. Ann. Neurol. 25 (1989) 146–151.

SHIRAKI, H. and Y. YASE: Amyotrophic lateral sclerosis in Japan. In: P. J. Vinken and G. W. Bruyn (Eds.), Handbook of Clinical Neurology, Vol. 22. New York, Elsevier (1975) 353–419.

SIEBER, S. M., P. CORREA, D. W, DALGARD, K. R. MCINTIRE and R. H. ADAMSON: Carcinogenicity and hepatotoxicity of cycasin and its algycone methylazoxymethanol acetate in nonhuman primates. J. Natl. Cancer Inst. 65 (1980) 177–189.

SPENCER, P. S.: Guam ALS/Parkinsonism-dementia: a long-latency neurotoxic disorder caused by 'slow toxin(s)' in food? Can. J. Neurol. Sci. 14 (1987) 347–357.

SPENCER, P. S., P. B. NUNN, J. HUGON, A. C. LUDOLPH, S. M. ROSS, D. N. ROY and R. C. ROBERTSON: Guam amyotrophic lateral sclerosis-parkinsonism-dementia linked to a plant excitant neurotoxin. Science 237 (1987a) 517–522.

SPENCER, P. S., M. OHTA and V. S. PALMER: Cycad use and motor neuron disease in Kii Peninsula of Japan. Lancet 2 (1987b) 1462–1463.

SPENCER, P. S., V. S. PALMER, A. HERMAN and A. ASMEDI: Cycad use and motor neuron disease in Irian Jaya. Lancet 2 (1987c) 1273–1274.

SPOEHR, A.: Saipan. The Ethnology of a War-Devastated Island. Fieldiana: Anthropology. Anthropological Series of the Field Museum of Natural History, Vol. 41. Chicago, Natural History Museum Press (1954).

SPOEHR, A.: Marianas Prehistory; Archeological Survey and Excavations on Saipan, Tinian, and Rota. Fieldiana: Anthropology. Anthropological Series of the Field Museum of Natural History, Vol. 48. Chicago, Natural History Museum Press (1957).

STRONG, M. J., R. M. GARRUTO and R. YANAGIHARA: Aluminum-induced neurofilamentous lesions in dissociated motor neuron cultures. In: F. H. Norris and F. C. Rose (Eds), Amyotrophic Lateral Sclero-

sis — New Advances in Toxicology and Epidemiology. London, Smith-Gordon and Company, Ltd. (1990a) 175–180.

STRONG, M. J., R. YANAGIHARA, A. V. WOLFF, S. K. SHANKAR and R. M. GARRUTO: Experimental neurofilamentous aggregates: Acute and chronic models of aluminum-induced encephalomyelopathy in rabbits. In: F. H. Norris and F. C. Rose (Eds), Amyotrophic Lateral Sclerosis — New Advances in Toxicology and Epidemiology. London, Smith-Gordon and Company, Ltd. (1990b) 157–173.

STRONG, M. J., R. M. GARRUTO, A. V. WOLFF, R. YANAGIHARA, S. M. CHOU and S. D. FOX: N-Butylbenzenesulphonamide, a novel neurotoxic plasticising agent. Lancet 336 (1990c) 640.

STRONG. M. J., R. M. GARRUTO, A. V. WOLFF, S. M. CHOU, S. D. FOX and R. YANAGIHARA: N-Butylbenzenesulfonamide: a neurotoxic plasticizer inducing a progressive spastic myelopathy. Acta Neuropathol. (Berl.) 81 (1991) 235–241.

THOMPSON, L.: Guam and Its People. Princeton, Princeton University Press (1947).

TILLEMA, S. and C. J. WIJNBERG: 'Epidemic' amyotrophic lateral sclerosis on Guam: epidemiologic data Doc. Med. Geog. Trop. 5 (1953) 366–370.

TORRES, J., L. L. IRIARTE and L. T. KURLAND: Amyotrophic lateral sclerosis among Guamanians in California. Calif. Med. 86 (1957) 385–388.

TRACEY, J. I., JR., C. H. STENSLAND, D. B. DOAN, H. G. MAY, S. O. SCHLANGER and J. T. STARK: Military geology of Guam, Mariana Islands. Part I. Description of terrain and environment. Part II. Engineering aspects of geology and soils. Washington, D.C., Intelligence Division, Office of the Engineer headquarters, United States Army of the Pacific. Mimeographed (1959) 282 pp.

TRAUB, R. D., T. C. RAINS, R. M. GARRUTO, D. C. GAJDUSEK and C. J. GIBBS, JR.: Brain destruction alone does not elevate brain aluminum. Neurology 31 (1981) 986–990.

UEBAYASHI, Y.: Epidemiological investigation of motor neuron disease in the Kii Peninsula, Japan, and on Guam — the significance of long survival cases. Wakayama Med. Rep. 23 (1980) 13–27.

UNDERWOOD, J. H.: Population history of Guam: context of microevolution. Micronesica 9 (1973) 11–44.

UNDERWOOD, J. H.: the native origins of the neo-Chamorros of the Mariana Islands. Micronesica 12 (1976) 203–209.

U.S. BUREAU OF THE CENSUS: U.S. census of population: Guam, preliminary report, Washington, D.C., U.S. Government Printing Office (1980).

VIOLA, M. P., M. FRAZIER, L. WHITE, J. BRODY and S. SPIEGELMAN: RNA-instructed DNA polymerase activity in a cytoplasmic particulate fraction in brains from Guamanian patients. J. Exp. Med. 142 (1975) 483–494.

VIOLA, M. P., J. C. MYERS, K. L. GANN, C. J. GIBBS, JR. and R. P. ROOS: Failure to detect poliovirus genetic information in amyotrophic lateral sclerosis. Ann. Neurol. 5 (1979) 402–403.

WHITING, M. G.: Food practices in ALS foci in Japan, the Marianas and New Guinea. Fed. Proc. 23 (1964) 1343–1345.

WISNIEWSKI, H., R. D. TERRY and A. HIRANO: Neurofibrillary pathology. J. Neuropathol. Exp. Neurol. 29 (1970) 163–176.

YANAGIHARA, R.: Heavy metals and essential minerals in motor neuron disease. In: L. P. Rowland (Ed.), Advances in Neurology, Vol. 36. Human Motor Neuron Diseases. New York, Raven Press (1982) 233–247.

YANAGIHARA, R. T., R. M. GARRUTO and D. C. GAJDUSEK: Epidemiologic surveillance of amyotrophic lateral sclerosis and parkinsonism-dementia in the Commonwealth of the Northern Mariana Islands. Ann. Neurol. 13 (1983) 79–86.

YANAGIHARA, R., R. M. GARRUTO, D. C. GAJDUSEK, A. TOMITA, T. UCHIKAWA, Y. KONAGAYA, K.-M. CHEN, I. SOBUE, C. C. PLATO and C. J. GIBBS, JR.: Calcium and vitamin D metabolism in Guamanian Chamorros with amyotrophic lateral sclerosis and parkinsonism-dementia. Ann. Neurol. 15 (1984) 42–48.

YANG, M. G., O. MICKELSEN, M. E. CAMPBELL, G. L. LAQUEUR and J. C. KERESZTESY: Cycad flour used by Guamanians: effects produced in rats by long-term feeding. J. Nutr. 90 (1966) 153–156.

YANO, I., S. YOSHIDA, Y. UEBAYASHI, F. YOSHIMASU and Y. YASE.: Experimental study of degenerative CNS disease in monkeys. In: Abstracts of the International Conference of Amyotrophic Lateral Sclerosis, Kyoto (1987) 136.

YASE, Y.: The basic process of amyotrophic lateral sclerosis as reflected in the Kii Peninsula and Guam. In: W. A. den Hartog Jager, G. W. Bruyn and A. P. J. Heijstee (Eds.), Proceedings of the Eleventh World Congress of Neurology, Amsterdam, Netherlands, September 11–16, 1977, International Congress Series No. 434. Amsterdam, Excerpta Medica (1978) 413–427.

YASE, Y.: The role of aluminum in CNS degeneration with interaction of calcium. Neurotoxicology 1 (1980) 101–109.

YASE, Y.: Metal studies of ALS — Further development. In: T. Tsubaki and Y. Yase (Eds.), Amyotrophic Lateral Sclerosis. Amsterdam, Excerpta Medica (1988) 59–65.

YASE, Y., T. KUMAMOTO, F. YOSHIMASU and Y. SHINJO: Amyotrophic lateral sclerosis studies using neutron activation analysis studies. Neurology (India) 16 (1968) 46–50.

YASUI, M., Y. YASE and K. OTA.: Evaluation of magnesium, calcium, aluminum metabolism in rats and monkeys maintained on calcium-deficient diets. In: Abstracts of the Eighth International Neurotoxicology Conference, Little Rock, Arkansas (1990) 30.

YOSHIDA, S., Y. YASE, S. IWATA, Y. MIZUMOTO, K.-M. CHEN and D. C. GAJDUSEK: Comparative trace-elemental study on amyotrophic lateral sclerosis (ALS) and parkinsonism-dementia (PD) in the Kii Peninsula of Japan and Guam. Wakayama Med. Rep. 30 (1988) 41–53.

YOSHIMASU, F., M. YASUI, Y. YASE, S. IWATA, D. C. GAJDUSEK, C. J. GIBBS, JR. and K.-M. CHEN: Studies on amyotrophic lateral sclerosis by neutron activation analysis. 2. Comparative study of analytical results on Guam PD, Japanese ALS and Alzheimer disease cases. Folia Psychiatr. Neurol. Jpn. 34 (1980) 75–82.

ZIMMERMAN, H. M.: Monthly report to Medical Officer in Command, U.S. Naval Medical Officer in Command, U.S. Naval Medical Research Unit No. 2 (1945).

Handbook of Clinical Neurology, Vol. 15 (59): Diseases of the Motor System
J.M.B.V. de Jong, editor
© Elsevier Science Publishers B.V., 1991

Amyotrophic lateral sclerosis and parkinsonism-dementia in the Kii Peninsula: comparison with the same disorders in Guam and with Alzheimer's disease

HIROTSUGU SHIRAKI[1] and YOSHIRO YASE[2]

[1]*Shiraki Institute of Neuropathology, Tokyo, and* [2]*Division of Neurological Diseases, Wakayama Medical College, Wakayama, Japan*

Amyotrophic lateral sclerosis (ALS) remains an enigmatic disease despite intensive investigation. Since its recognition as a disease entity by Charcot in the period 1865 to 1869 (Charcot and Joffroy 1869), there has been continued worldwide interest in the disease.

A pioneer Japanese neurologist, Kinnosuke Miura, a pupil of Charcot gave the first detailed description of amyotrophic lateral sclerosis in Japan in 1911. He reported the clinical and the pathological findings, and also pointed out the apparently high frequency in the Kii Peninsula on the main island of Honshu. General interest in this disease in Japan has been growing due to numerous symposia on motor neuron disease and amyotrophic lateral sclerosis and a continued increase in reports on epidemiological, genetical, clinical and pathological studies, as well as related experimental studies.

A specific interest has centered around the cluster of ALS and the parkinsonism-dementia complex (PD) in the Western Pacific, including the Kii Peninsula of Japan, Guam in the Marianas and Western New Guinea. In the Kii Peninsula, PD is diagnosed clinically and no autopsy case reports are available. In New Guinea, autopsy reports have not been available for either ALS or PD (Fig. 1).

A recent steady decline in the incidence of ALS and PD in both the Kii Peninsula and Guam clearly indicates the possible contribution of environmental (exogenous) factors to the incidence of these diseases.

CHANGING EPIDEMIOLOGICAL PATTERN

As previously reported in Volume 22 of this Handbook series (Shiraki and Yase 1975), a high incidence of ALS was recognized in 2 districts in the southern part of the Kii Peninsula: Kozagawa (population 6200 in 1966) and Hobara (population 2100 in 1969). The annual incidence was 15 per 100 000 population in Kozagawa, and 55 per 100 000 in Hobara over a 20-year period. Compared with the average annual incidence of ALS of 0.3–0.4 per 100 000 population for the whole of Japan (Uebayashi 1988), these figures indicate an extremely high incidence of ALS in these areas.

Along with the steady decrease in incidence in the Kii Peninsula, no new cases have been seen in Kozagawa since 1980 and in Hobara since 1981. In Kozagawa the last ALS patient died in 1981, but the Hobara prevalence rate was still high in 1989. The dramatic disappearance from

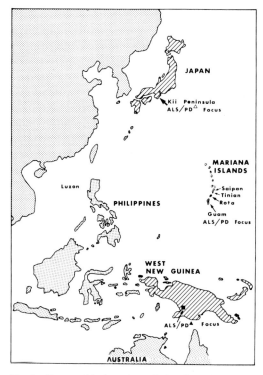

Fig. 1. Geographic location of the three high-incidence foci of ALS and PD in the western Pacific: Mariana Islands, Kii Peninsula of Japan and west New Guinea (Garruto and Yase 1986). △: *In Hobara*, ALS is clinically occasionally associated with PD, but no autopsy case of PD has been obtained as yet. The disease has reportedly developed in several patients who migrated to Hobara from other regions. ▲: *In west New Guinea*, parkinsonism with or without dementia occurs, but post-mortem examinations have not yet become available.

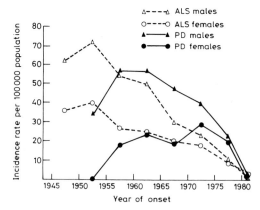

Fig. 2. Five-year average annual incidence rates in Guam of ALS and PD in males and females by years of onset of disease, age-adjusted to the 1960 Guam population (Garruto and Yase 1968).

Kozogawa has continued over the past 9 years of follow-up observation.

In addition, computation of the standardized mortality rate of ALS in Japan by sex, prefecture, and 3-year periods (Kondo and Minowa 1988) has shown that the Kii Peninsula has not been a high risk area since 1978.

A similar marked tendency towards decrease in incidence in both diseases was observed in Guam (Garruto and Yase 1986) (Fig. 2).

Although there was a familial occurrence of ALS cases in the Kii and Guam foci, hereditary trait transmission has not been demonstrated (Figs 3 and 4). Epidemiological trends rather suggest that these 2 disorders are clinical variants of a single disease entity due to an almost identical etiopathogenetic mechanism. In other words, this epidemiological trend of ALS and PD cases in the Kii and Guam foci indicates an environmental (exogenous) effect on residents rather than a genetic involvement in disease manifestation.

ANALYSIS OF ENVIRONMENTAL FACTORS

Environmental or exogenous factors include infectious or toxic agents with or without underlying individual vulnerability. Immunovirological studies using animal inoculation with CNS tissue from ALS cases have failed to produce animal models. Studies using neurotoxic substances are under way.

Based on the similarity of the neurotoxicity of BMAA (β-N-methylaminoalanine) contained in the cycad (*Cycas circinalis*), once widely distributed in Guam for dietary and/or medicinal use during and after World War II, and BOAA (β-N-oxalylaminoalanine) contained in *Lathyrus sativus* (the neurotoxic agent in lathyrism), the cycad had been suggested as a possible cause of ALS.

In addition, because of the abnormality in glutamate metabolism observed in ALS patients (Plaitakis and Caroscio 1987), chronic neurotoxic activation of glutamate receptors has been implicated since BMAA-induced excitotoxic changes

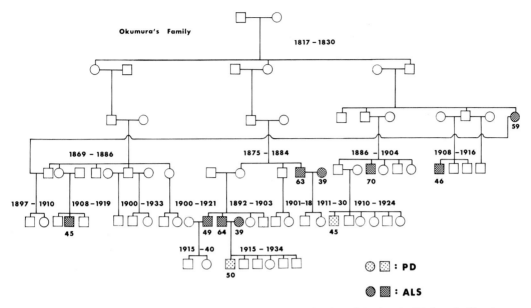

Fig. 3. A pedigree of ALS and PD from the Hobana area in the Kii Peninsula (Hajigami village).

are largely blocked by specific N-methyl-D-aspartate (NMDA) antagonists (Spencer et al. 1987). Weiss et al. (1989), on the contrary, reported that degeneration of cultured cortical neurons produced by both BMAA and BOAA may be predominantly mediated by the toxic activation of non-NMDA receptors. Clinical and experimental studies have led to elimination of cycad hypothesis as a factor in the etiopathogenesis of the disease (Garruto et al. 1989; Duncan et al. 1988). The participation of excitotoxins in neuronal degeneration is currently being studied in ALS clinical drug trials as well as in basic experimental studies.

As common environmental characteristics in the 3 foci of ALS in the Western Pacific, soil and water samples show a deficiency of minerals such as calcium and magnesium and excess of metals including aluminum and manganese (Yase 1972; Iwata et al. 1978; Garruto et al. 1984) (Figs 5 and 6). While soil concentrations of trace elements are not very good indicators of their influence in human nutrition, plants and animals are more likely to show evidence of trace element deficiencies or toxicities, because of their greater dependence on the soil for their supply of minerals.

The influence of the environment on plants

and animals has been studied in samples of rice and cattle hair, showing a corresponding deficiency or excess of these elements in the Kii Peninsula foci. There was a comparatively low content of calcium and high content of aluminum and manganese in rice, and an increase in manganese content in cattle hair (Yase 1972). Aluminum absorption into rice begins very early on, as a reflection of its presence in the environmental soil (Iwata et al. 1988) (Fig. 7). In humans, a comparative study of age-related changes in calcium metabolism was conducted in inhabitants of Kii and control areas (Fujita et al. 1977). Residents in the foci showed shorter stature, a higher prevalence of lumbago, a thinner clavicular cortex, lower serum levels of phosphorus, total protein and cholesterol, and a higher level of alkaline phosphatase than residents in the control areas.

In Guam, a survey of former manganese miners in the early 1960s suggested that manganese miners with a higher degree of exposure were more likely to develop ALS and/or PD than miners with a lesser exposure, other factors being equal (Yase 1972).

Although an ill-defined trend, a higher incidence of ALS was recognized in the middle and southern parts of Guam where the soil and rock

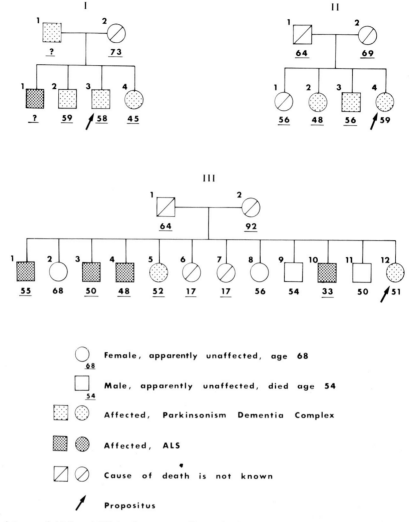

Fig. 4. Coexistence of ALS and PD in the same pedigrees in Guam. (Reproduced from Dr. A. Hirano et al. (1961a) by courtesy of the Editors of *Brain*.)

are of volcanic origin with a bauxite texture, whereas in the northern part, which consists of limestone formations, a lower incidence of ALS was found.

Although all water samples had an elevated manganese content, soil and water samples from the middle and southern parts of Guam showed a high content of manganese and aluminum, and absence or low content of calcium and magnesium (Iwata et al. 1978). It was noted that the calcium intake of Guam residents was exceedingly low, irrespective of age and sex (Fig. 8).

METAL/MINERAL ANALYSIS IN CNS AND BONE TISSUES IN ALS

Based on the biogeomedical nutritional chain, the metal and mineral content of CNS tissue of ALS cases was studied. Serial analytical studies of CNS tissue from autopsied ALS cases showed a high content of calcium and aluminum compared with controls, but there was no significant difference in manganese and copper content (Yase et al. 1974; Yoshimasu et al. 1976) (Fig. 9). The aluminum content of CNS tissue from PD cases

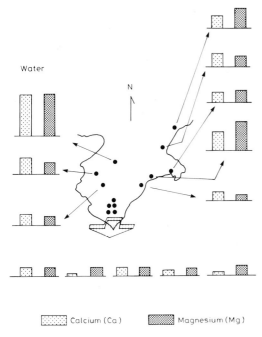

Fig. 5. Minerals and metals in river water in the Kii Peninsula.

in Guam was exceedingly high, as indicated in Table 1A (Yase et al. 1980; Yoshimasu et al. 1982). On the contrary, magnesium content in CNS tissue from ALS cases was significantly low compared with controls (Fig. 10). These 3 contradictory findings in CNS tissue of ALS/PD cases from Kii and Guam could be significant in formulating a hypothesis of the developmental mechanism of ALS/PD in both foci and will be discussed in detail below.

The calcium deposit, described as part of a metal-induced calcified degeneration, was found to be hydroxyapatite with the possible coexistence of additional trace elements other than aluminum, manganese and copper (Mizumoto et al. 1983). Using several methods, the finding of a 1.6 calcium to phosphorus ratio, similar to that of Ca-hydroxyapatite (Ca-Hap) was confirmed in the calcium deposits in CNS tissue of ALS cases (Mizumoto et al. 1983; Yoshida et al. 1987). In addition, X-ray microanalysis, infrared spectrometry, and X-ray diffraction studies have indicated a precipitation of Ca-Hap in the degenerative lesions of the cervical spinal cord and frontal cortex of ALS cases (Yoshida 1977, 1979; Iwata 1980; Yoshida et al. 1989) (Fig. 11).

Changes on the inner surface of the frontal bone were shown by X-ray. A marked decrease

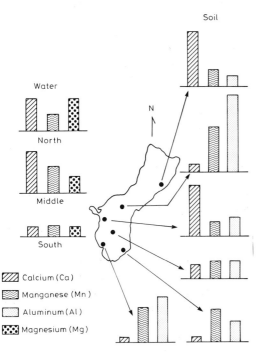

Fig. 6. Minerals and metals in water and soil in Guam.

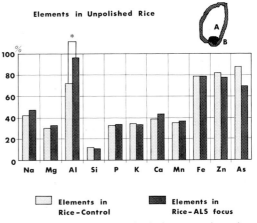

Fig. 7. Analysis of elements in the inner starchy endosperm (A) and other aleurone layer and germ (B) of unpolished rice. Samples from an ALS focus showed that aluminum is more evident in the starchy part (A) of rice than in samples from control areas, indicating early aluminum absorption in samples from the focus.

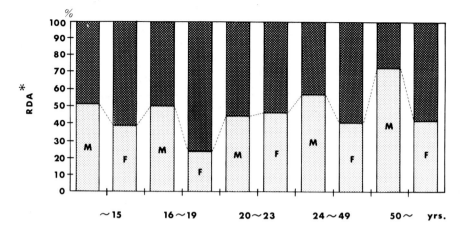

*RDA: Recommended Dietary Allowances established in 1980 by the U.S. National
Research Council: Average in all groups is 46.9% of RDA.

Fig. 8. Calcium intake of Guam residents is shown by sex and age group, indicating continuous generalized
calcium deficiency. This pilot study was done in 1984 by the Cooperation Nutrition Survey of the University
of Guam and the Western Human Research Center, University of California.

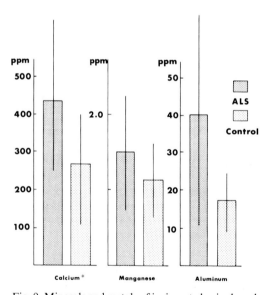

Fig. 9. Minerals and metals of incinerated spinal cord
specimens from ALS cases in the Kii Peninsula and
control cases (μg/g dry weight). *From computer-
controlled electron beam X-ray microanalysis and
wave length-dispersive spectrometry, Ca has now
been judged to be Ca-hydroxyapatite (see Fig. 11).

in calcium intensity was found in ALS cases and
a lesser decrease in PD cases (Mizumoto, Y.,
pers. commun.), while skull thickness and calcium
content in the frontal bone showed no differences
in the Guam ALS/PD cases and controls. (Murakami et al. 1984) (Fig. 12).

These findings, therefore, indicate a generalized
calcium and other metal/mineral dysmetabolism
in the degenerative process of ALS/PD under
prolonged exposure to a specific environment.

Reports on aluminum

Since Klatzo's work on the Al-induced neurofibrillary tangle (NFT) (Klatzo et al. 1965), reports
on the neurotoxic role of aluminum accumulated.
The high aluminum content of soil and water
samples in the Western Pacific foci suggests the
possible important role of aluminum in the CNS
degenerative process, since almost all autopsied
cases of ALS in Kii and Guam showed NFT
formation in CNS tissue.

A close relationship between aluminum and
NFTs has been indicated in cases of Alzheimer's
disease and ALS/PD in Guam (Nikaido et al.

TABLE 1A
Metals and/or minerals in CNS in ALS/PD and control cases in Guam.

	No. of cases	Cu	Al	Mn	Ca
ALS	7	26.8 ± 11.9	83.9 ± 53.3	0.98 ± 0.74	299 ± 118
PD	4	39.5 ± 18.3	_223.8 ± 130.0_	2.21 ± 1.03	233 ± 85
Cont.	3	28.8 ± 15.1	53.9 ± 38.6	1.37 ± 0.34	549 ± 155
				Mean ± S.D.	

TABLE 1B
Ultrastructural localization of aluminum (Al) within lumbar motor neurons of ALS and control cases.

Localization	Control case			ALS case			
	1	2	3	1	2	3	4
Nucleus							
*_nucleolus (r-RNA)_	0/5	0/5	0/5	$5^L/5$	$1^S/5$	$1^S/5$	$1^T/5$
euchromatin (RNA, DNA)	0/5	0/5	0/5	$4^S/5$	$1^S/5$	0/5	$1^T/5$
*_heterochromatin (DNA)_	0/5	0/5	0/5	$5^L/5$	0/5	0/5	$1^T/5$
Cytoplasm							
*_rough endoplasmic reticulum (rRNA)_	0/5	$1^T/5$	$1^T/5$	$5^L/5$	$3^T/5$	0/5	0/5
mitochondria	0/5	0/5	0/5	$3^S/5$	$1^T/5$	0/5	0/5
lipofuscin pigments	$1^T/5$	0/5	0/5	$3^T/5$	$1^T/5$	0/5	0/5
Bunina body (RNA positive)	–	–	–	–	$3^S/5$	–	–

Data are shown in fractional form as a number of detectable Al-L_1 edges per 5 probes. ^{L}: large Al-L_1 edge; ^{s}: small Al-L_1 edge; and ^{T}: trace of Al-L_1 edge.

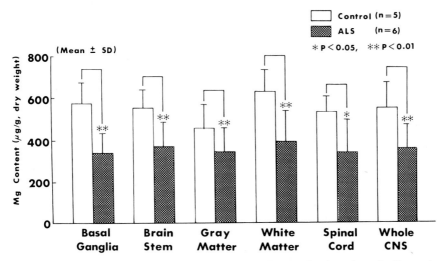

Fig. 10. Magnesium content of CNS tissue of ALS cases and controls; there is a significant decrease of magnesium content in ALS, as determined by inductively coupled plasma emission spectrometry (ICP). (By courtesy of Dr.M. Yasui.)

1972; Perl and Brody 1980; Perl et al. 1982). It has been reported that aluminum may be involved in the protein synthesis of the nuclear component of the neuron (DeBoni et al. 1976; Crapper and DeBoni 1977; Crapper et al. 1978, 1980; Lukiw et al. 1987; Wen and Wisniewski 1985). Al-induced neurofibrillary changes are composed of 10 nm normal neurofilaments, more closely re-

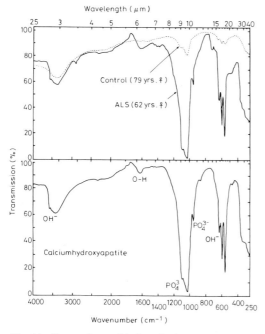

Fig. 11. Comparison of infrared absorption spectra of incinerated cervical cord specimens of ALS and control cases and Ca-hydroxyapatite (Yase, 1980).

	Ca	P	Fe	Cu	Zn	Sr	Pb
ALS (Japan)	↓↓	↓	—	—	↑	—	—
ALS (Guam)	↓↓	↓	↑	↑	↑	—	—
PD (Guam)	↓	↓	—	—	—	—	↑

Mizumoto et al : 1982

Fig. 12. Metal behavior in frontal bone of ALS/PD. The decreased content of calcium mainly affects the inner surface of the frontal bone (pers. commun. by late Dr. Mizutani).

sembling those of the early stage of ALS, whereas Alzheimer type NFTs show paired helical filaments (PHF).

Several reports of abnormal nucleic acid metabolism in the spinal motor neuron of ALS describe a condensation of the chromatin nucleoprotein with a decrease in nuclear and nucleolar volumes and in cytoplasmic rRNA content (Mann and Yates 1974; Davidson and Hartmann 1981a, b; Davidson et al. 1981), and with a possible defect in DNA repair (Bradley and Kasin 1982; Tandon et al. 1987).

Analytical electron microscopy combined with electron energy loss spectrometry (EELS) provides a useful and promising technique for microanalysis of biological thin specimens because of its high energy and spatial resolution (Johnson 1979). The details of EELS employed here were described more precisely by Yoshida et al. (1990).

Using EELS, aluminum was detected and found to be particularly high in the nuclear components, such as nucleoli and heterochromatin, but low in euchromatin in a single lumbar motor neuron of a sporadic ALS case, while negative in a control case (Figs 13 left and 14). In the cytoplasm of the same specimen, aluminum was also high in the rER cluster, low in the mitochondria and virtually absent in the lipofuscin granules (Yoshida et al. 1990) (Figs 13 right and 15). The detection of aluminum within DNA-containing chromatins and rRNA-containing cellular components strongly suggest a possible binding of aluminum to the phosphate groups of rRNA within the ribosomes (Shuma et al. 1982).

The Al/C ratios obtained from a lumbar motor neuron of the same ALS case in Figs 13 left, 14 and 15 are shown in Fig. 16. The same trend was found in the aluminum content of neuronal elements, while outside the motor neuron similar results were observed in astrocytes with significantly elevated Al/C ratios in the heterochromatin and glial fibers, but a rather low ratio in the euchromatin. In addition, the Al/C ratio in the nucleus (heterochromatin) of endothelial cells was significantly high, although no Al was detected in the basal lamina (Yoshida et al. 1990).

Table 1B shows the EELS data from a lumbar motor neuron in 4 ALS cases (all males, age at death 61–72 years; duration 1 year 8 months to 2 years 6 months) and 3 control cases (subdural hematoma, ischemic heart disease and Shy-Drager syndrome; all males, age at death 55–75 years). The data obtained from each of the cellular components are in fractional form as a number of detectable Al-L_1 edges per 5 probes with subscripts L, S and T, which indicate the relative size of the Al-L_1 edge (Yoshida et al. 1990).

Aluminum was also detected in the Bunina

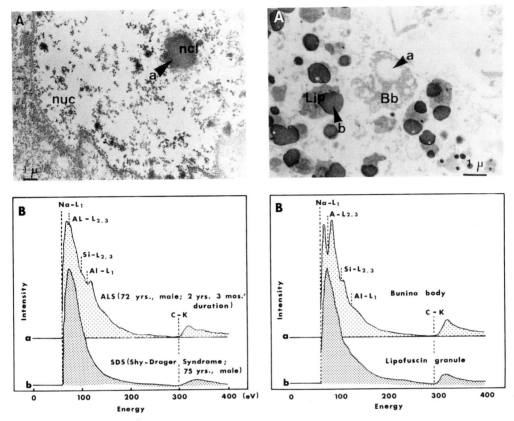

Fig. 13. *Left.* A. EM picture of nucleus (nuc) in a lumbar motor neuron (L_3) of an ALS case. B. EELS of nucleolus of ALS, as shown with *arrow-head* in (A;a) and in (B;a) as well as in control case 3 (B;b). In spectrum (B;a), Al-$L_{2,3}$ and -L_1 edges are clearly seen with a Si-$L_{2,3}$ edge, as compared with spectrum of control (B;b). *Right.* A. EM picture of a Bunina body (Bb) and lipofuscin in a lumber motor neuron (L_3) of another ALS case (62 year, male; 2 years and 3 months duration). B. EELS of Bunina body (Bb) and lipofuscin granules (Li) as shown with *arrow-head* in (A;a) and in (A;b), respectively. In contrast to lipofuscin granules (B;b) around the Bunina body, Al-$L_{2,3}$ and -L_1 edges are sharply visible in the Bunina body itself (B;a).

bodies, which are considered to be associated with ALS, and in spheroids with an accumulation of phosphorylated neurofilaments (Bizzini and Gambetti 1986; Manetto et al. 1988). In the Bunina bodies, silicon was also detected with or without aluminum (Okamoto et al. 1979; Yoshida et al. 1990) (Fig. 13 right).

Coexistence of aluminum and silicon has been reported in the nuclear region of NFT-bearing neurons in cases of Alzheimer's disease and Guam ALS/PD (Nikaido et al. 1972; Perl and Brody 1980; Perl et al. 1982), and in the CNS tissue of Japanese ALS (Yoshida et al. 1987). Aluminum or silicon or aluminosilicates, which are considered to form the interstitial environ-

ment of CNS, have been detected in the senile plaques (Candy et al. 1986). Furthermore, even in the aluminum-induced neurofibrillary tangles of experimental models, deposition of calcium with aluminum and silicon has been reported (Garruto et al. 1989).

DISCUSSION

Based on soil-plant-animal-human relationships, it is postulated that a low calcium and magnesium intake over a period of time induces a compensatory or secondary hyperparathyroidism with increasing bone resorption and an accumulation of calcium in target organs including nervous

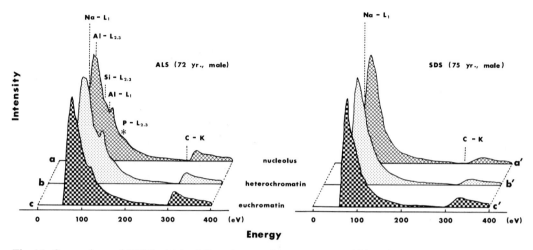

Fig. 14. Comparison of EELS obtained from intranuclear structures within a lumbar motor neuron of an ALS case (left) and a control case (right). In the ALS case, a sharp Al-L$_1$ edge is clearly visible with Si- and P-L$_{2,3}$ (*) edges in the nucleolus (spectrum a) and heterochromatin (spectrum b), but less marked in the euchromatin (spectrum c). In contrast, no Al-L$_1$ edges are seen in the control case (SDS: Shy-Drager syndrome; spectrum a′, b′ or c′).

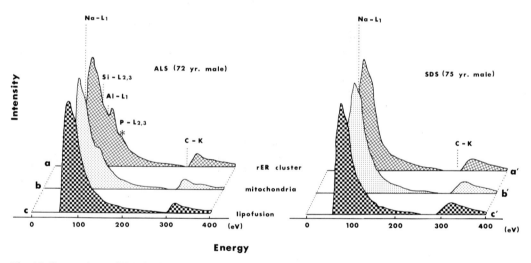

Fig. 15. Comparison of EELS obtained from cellular components in cytoplasm of an ALS case and a control case. In the ALS case, Al-L$_1$ edges are clearly visible in the r-ER cluster (spectrum a) and less marked in the mitochondria (spectrum b), but hardly visible in the lipofuscin granules (spectrum c). In the SDS case (Shy-Drager syndrome), no Al-L$_1$ edges are detected in any spectra (a′, b′ or c′).

tissue, kidney, and muscles. This situation may be worsened by trauma, infection, or overwork, which increase not only the intake of calcium but also of aluminum and other metals into CNS tissue (Yase 1977). This working hypothesis of metal/mineral dysmetabolism in ALS has been supported by experimental findings (Garruto and Yase 1986; Yano et al. 1989; Yasui et al. 1989; Kihira 1987) (Fig. 17).

DISTRIBUTION OF ALZHEIMER'S NEUROFIBRILLARY TANGLES IN ALS/PD AND DEMENTIA IN THE KII PENINSULA AND GUAM FOCI

The distribution of the neurofibrillary tangle (NFT) in a series of autopsied cases of typical PD in Guam summarized by Hirano et al. (1961a, b) is schematically illustrated in Fig. 18.

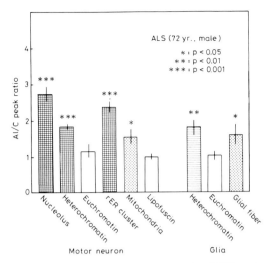

Fig. 16. Bar graphs of the Al-L/C-K edge (Al/C) peak ratio of each of the cellular components within motor neurons and glial cells. The results are standardized with the mean value of the Al/C peak ratio (1.00), which was obtained from the lipofuscin granules within the ALS motor neuron. The Al/C peak ratio is significantly increased in the nucleolus, heterochromatin, rER cluster, and mitochondria, but rather low in euchromatin, even in the nucleus. This tendency is the same as in the glial cells, i.e. the peak AL/C ratio is high in the heterochromatin and glial fibers but rather low in euchromatin.

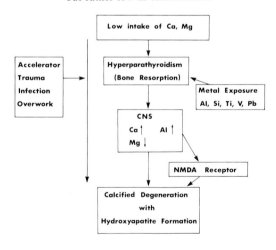

NMDA: N-Methyl-D-Aspartate

Fig. 17. Hypothesis for etiopathogenesis of ALS/PD in the Kii Peninsula and Guam Island.

We suggest that the severity of NFT, neuronal loss and/or gliosis in PD can be summarized in the following order: hippocampus, particularly Sommer's sector and presubiculum, substantia nigra, locus ceruleus ≥ hypothalamus and mammillary body, anterior perforate substance, substantia innominata, amygdaloid nucleus > periaqueductal region, area around third ventricle, third cranial nucleus > frontal cortex (particularly cingulate gyrus and frontal pole), temporal lobe (particularly its anterior portion) > lamina quadrigemina, tegmentum of pons and medulla oblongata (reticular and midline raphe nuclei) > dorsal motor nucleus of vagus, thalamus, corpus striatum, globus pallidus ≥ olfactory bulb and spinal cord.

In a Kii ALS case that was previously reported (Nakai 1969; Yase et al. 1972), the initial symptoms were atypical schizophrenic episodes with later on development of the typical ALS features. Typical NFT occurred mainly in the forebrain limbic system and progressed to the limbic-hypothalamic, midbrain-limbic system and other brain stem structures; the spinal cord was not examined.

A synthesis of the above-mentioned distributions of NFT in both Guam and Kii cases in general can be schematically illustrated in the following way: the distribution of NFT in both areas is predominantly in the forebrain limbic (medial and lateral), limbic midbrain, limbic-hypothalamic and adjacent structures, for example, substantia innominata, anterior perforate substance, etc. (Nieuwenhuys et al. 1988) (Fig. 19). This actually provides an answer to the question whether the mental disturbances noted in ALS and PD cases should really be called dementia or limbic dementia, or more simply psychotic state and/or mixed states. We are aware that the native Chamorro language may limit communication, and thus impede this kind of trial but the matter remains important for future investigation. Two points can be emphasized: ALS/PD in both foci may be closely related to 'subcortical dementia' since the fact that as far as the distribution is concerned, NFTs are mainly abundant in the limbic cortices and other limbic areas of the diencephalon and brain stem, and their presence in both compact and reticular zones of the substantia nigra may suggest the possibility that they play a significant role in the development of psychotic disturbances in ALS/PD.

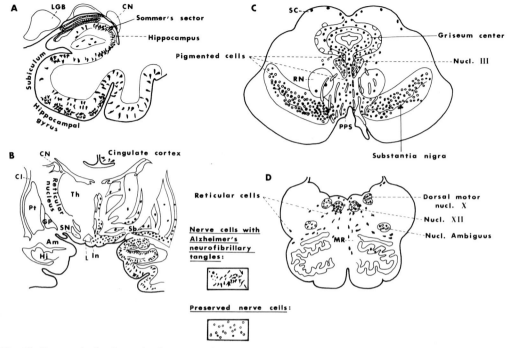

Fig. 18. Summarized schematic diagrams of the preferential sites of Alzheimer's neurofibrillary tangles, particularly in the limbic cortex, diencephalon, hypothalamus and brain stem of Chamorro people in Guam. (Reproduced from Drs. A. Hirano et al. (1961b) by courtesy of the Editors of *Brain*.)

ALS/PD IN KII PENINSULA AND GUAM: ETIOPATHOGENETIC RELATIONSHIP TO ALZHEIMER'S DISEASE

As previously discussed, a more detailed localization, particularly of aluminum and/or silicon is required because aluminum achieves its highest or a very high level in the nucleoli, rough endoplasmic reticulum and/or Bunina bodies, and occasionally in the nuclei and spheroids of the remaining anterior horn cells of the spinal cord from sporadic Japanese ALS cases, while being absent or merely traceable in control cases. It may be seriously asked whether these 2 disorders, i.e. ALS and PD in Kii and Guam and Alzheimer's disease, actually have a close relationship since a higher content of aluminum and/or silicon in the CNS of Alzheimer's disease has now gradually become well-known.

Crapper et al. (1973) emphasized that brain aluminum content was considerably increased in Alzheimer's disease (AD), particularly in the frontal, temporal and/or parahippocampal cortex. In a more extensive series (Crapper et al. 1976) they found that the increased Al content coincided with the distribution of NFTs, one of the more important characteristics of AD. McDermott et al. (1979), on the other hand, reported no significant difference in brain Al content between AD cases and age-matched controls. They concluded that an increased Al content paralleled the aging process but was not in itself specific for AD. Crapper et al. (1980) emphasized that, since the occurrence of NFT in AD as well as the Al content showed a topographical variation, the series studied by McDermott et al. (1979) was too large and this diluted the topographical difference of Al content, and further that NFTs were present in their controls.

Using neutron-capture radiochemical analysis, Markesberg et al. (1981) concluded that the findings supported the view of McDermott et al. (1979) mentioned above. Perl and Brody (1980), however, using electron microscopic X-ray spectrometric methods, found the Al content was frequently high in neurons and associated with NFT in AD brains. In 1982 Perl et al. found a remarkable accumulation of Al in the nuclei and

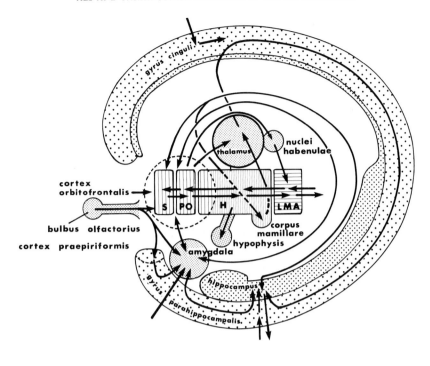

H: Hypothalamus

LMA: Limbic mid-brain area

PO: Preoptic area

S: Septum

Fig. 19. Schematic diagram of the limbic-hypothalamic system, divided into central and limbic circuits. (Modified figure from Dr. R. Nieuwenhuys (1988) with permission of Dr. Nieuwenhuys and Springer-Verlag.)

perinuclear areas of hippocampal neurons of ALS and PD patients in Guam. Garruto et al. (1984) also made the similar finding of a conspicuous deposit of Al in the NFT and in axons of the hippocampal neurons of Guam PD cases.

Birchall and Chapell (1988) claimed that Al is unquestionably neurotoxic since encephalopathy develops in some patients on long-term renal dialysis with an Al-rich dialysate. As has been pointed out by Hughes (1989), it is true that clinically or neuropathologically, AD is quite different from the Al neurotoxicity of dialysis encephalopathy. However, Birchall and Chapell (1988) emphasized that impairment of the intraneuronal inositol phosphate system may account for many of the changes observed in AD. The chemical considerations may suggest experimental

approaches to define the debated role of Al in the etiology of some forms of AD, for example, in studies of the effect of Al on the phosphatidyl-inositol-derived second messenger system.

An autopsied case of the juvenile type of Alzheimer's disease in a Japanese is of interest. Clinically, neurological disorders occurred at 28 years of age followed by cerebellar ataxia, spastic paraplegia, inability to walk, pyramidal signs and symptoms, and atrophy of the tongue. Near the terminal stage, a high-grade dementia, complete tetraplegia, dysphagia and dysarthria occurred in that order and the patient expired at 36 years of age. As shown in Fig. 20, her family history has a remarkable familial occurrence of neuropsychotic disorders. The present patient features as No. 11 in the third generation of her

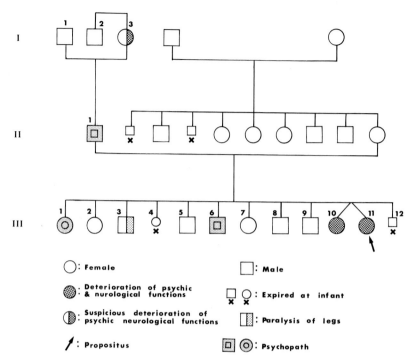

Fig. 20. Pedigree of juvenile type of Alzheimer's disease complicated by spastic tetraplegia in Japanese identical twins. (III: No. 10 and 11; propositus.)

siblings (Fig. 20), while her identical twin appears as No. 10. As indicated in Fig. 20, the clinical features are more or less similar to those of the propositus mentioned above but she was still alive at the time of last examination in 1982.

The present autopsy case showed a tremendous number of so-called primitive plaques without central core and Alzheimer's NFTs in both the cerebral cortices and subcortical grey matter (Figs 21A and B). Additionally, congophile angiopathy, occasionally combined with thrombus formation, had developed particularly in the occipital meningeal arteries. Consequently, the resulting small foci of softening were also disseminated in the subcortical white matter. The schematic distribution of these 3 lesions is shown in Fig. 22. Senile plaques were more widespread in the brain stem, cerebellum and spinal cord, while Alzheimer's NFTs were restricted to the midbrain (Fig. 23). The clear-cut distal dominant degeneration with or without gliosis, on the other hand, occurred symmetrically and bilaterally from the cerebral peduncle of the midbrain to

the pyramidal tract at all levels of the spinal cord (Figs. 22 and 23).

Intense gliosis also occurred symmetrically and bilaterally in the anterior horn at almost all levels of the spinal cord, while the anterior horn neurons were well-perserved, although vacuolar degeneration of the cytoplasm of a few motor neurons was occasionally encountered. The anterior nerve roots were occasionally and insidiously degenerated but definitive neurogenic muscular atrophy was insidious in both pathological and physiological senses. In addition, both spinocerebellar tracts were also involved (Figs 24A and B). The comparatively small number of references indicate that it is not uncommon for the juvenile type of AD to be combined not only with pyramidal tract degeneration but also with other systemic degenerations, although it is unfortunate that the aluminum content in this particular Japanese case (Matsuoka et al. 1967) was not examined.

Al and the pathogenesis of senile plaques in AD, Down's syndrome and chronic renal dialysis were described by Edwardson and Candy (1989).

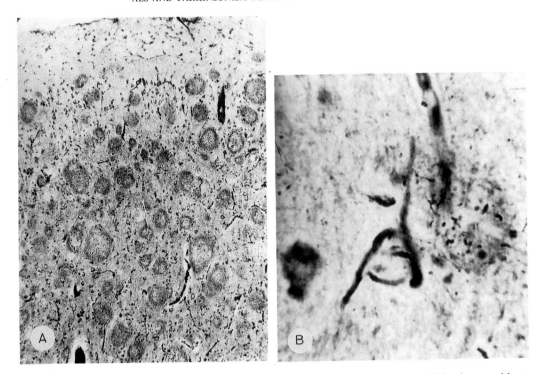

Fig. 21. A. Superficial cortical layers of the temporal tip. Round, variously sized, multiple plaques without central cores are widespread. B. Highly magnified nerve cell with Alzheimer's neurofibrillary tangles. (A and B: Hirano's simple silver.)

Using electron and proton-microprobe X-ray microanalysis, they showed that Al and Si were consistently present and confined locally to the central region of the senile plaque core in AD, Down's syndrome and 'normal' elderly cases. Using secondary ion mass spectrometry (SIMS), they found aluminosilicate deposits to be a specific and constant feature of the central region of the plaque core. In a preliminary study of 10 cases of dialysis encephalopathy using King's silver stain, on the other hand, they found high densities of immature plaques present in half of the patients. In addition, immunocytochemical staining disclosed the presence of A4 protein in the immature plaques. Imaging SIMS revealed a laminar distribution of what appeared to be an intraneuronal accumulation of Al in 8 of 10 dialysis cases, including some without senile plaques. This study adds weight to the view that Al, Si or aluminosilicates may have a role in the production of A4 amyloid protein and the development of senile plaques.

SIGNIFICANCE OF MULTIPLE FOAMY 'SPHEROID BODIES' PRESENT IN THE RETICULAR ZONE OF THE SUBSTANTIA NIGRA

We previously described 2 autopsy cases in Guam (Shiraki 1966; Shiraki and Yase 1975) which had been diagnosed clinically as ALS, one associated with severe deterioration of mental faculties and one with PD. Details of their clinical features are listed as Cases 13 and 14 in our previous paper (Shiraki and Yase 1975). Both cases showed quite unusual and similar neuropathological features, i.e. Hallervorden-Spatz disease-like features. As regards dark-brown pigmentary concretions of a cell-free nature and minute thioninophilic pigmentary granules in the cytoplasm and processes of astrocytes these examples of the Guam cases were quite similar to those of Hallervorden-Spatz disease. The structure of the spheroid bodies, however, was quite unusual, and disintegration of the compact zone of the substantia nigra had not been observed in Hal-

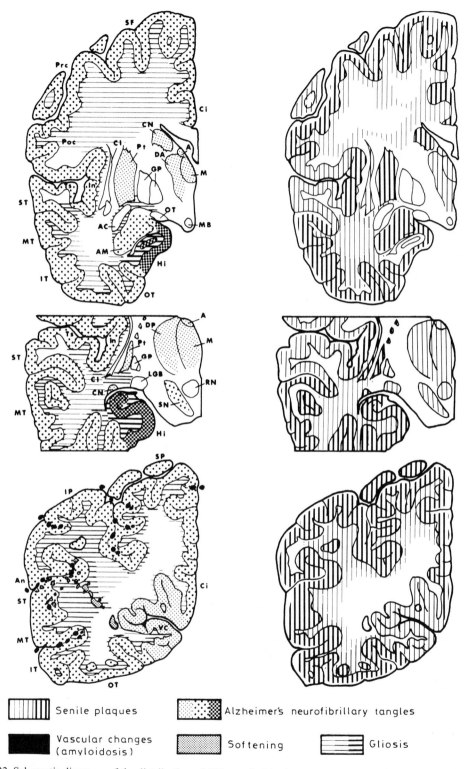

Fig. 22. Schematic diagrams of the distribution of different foci in the cerebrum of the same case as in Fig. 21.

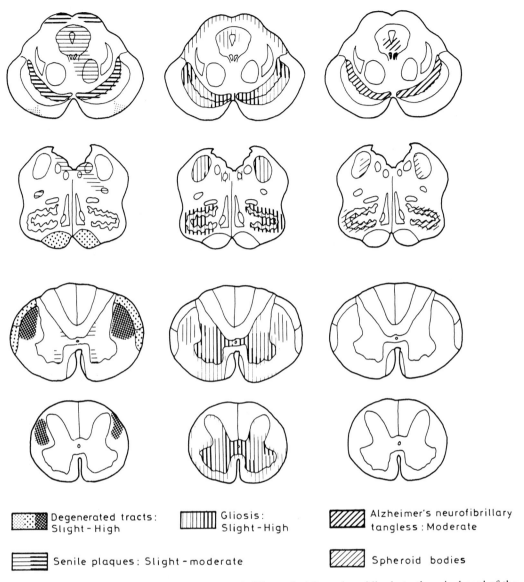

Degenerated tracts: Slight - High

Gliosis: Slight - High

Alzheimer's neurofibrillary tangless : Moderate

Senile plaques : Slight - moderate

Spheroid bodies

Fig. 23. Schematic diagrams of the distribution of different foci from the midbrain to the spinal cord of the same case as in Fig. 21 (see also Fig. 24).

lervorden-Spatz disease. In the present Guam ALS/PD cases, the depigmentation and disintegration of the pigmented cells of the compact zone occurred symmetrically and bilaterally (Figs 25A and B), while very many foamy spheroids and a number of dark-brown pigmentary concretions of a cell-free nature were visualized, mainly in the reticular zone where a great number of very small glycogen granules were accumulated in the spheroids (Figs 26A and B). Histochemical

examinations were made, as reported by Shiraki (1966). At the same time, careful examination for close spatial connections of these spheroids to nerve fiber structures showed no precise positive findings. It is unfortunate that electron microscopic examination was not made on the simple assumption that these foamy spheroids might be axon terminals of an axo-dendritic origin. In any case, the spheroids were extremely abundant in the reticular zone but few or even

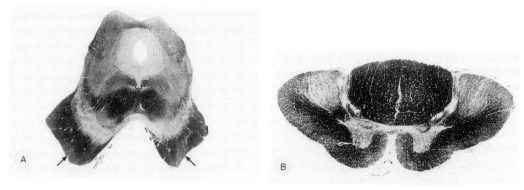

Fig. 24. Caudal part of the midbrain. Minimal demyelination occurs bilaterally and symmetrically in the corticospinal tracts of the cerebral peduncle (arrows). B. Middle thoracic cord. Demyelination is pronounced bilaterally and symmetrically in the lateral corticospinal tracts and less pronounced in the anterior pyramidal tracts. In addition, less obvious demyelination is also apparent in the dorsal and ventral spinocerebellar tracts on both sides. (A and B: Woelcke myelin.)

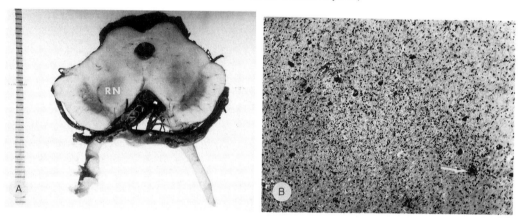

Fig. 25. A. Naked eye findings in the midbrain. Depigmentation in the compact zone of the substantia nigra and pronounced rusty brown discoloration in the reticular zone on both sides. The Sylvian aqueduct is conspicuously dilated. RN: red nucleus. B. Lateral part of compact zone. The pigmented cells are severely damaged and glial nuclei proliferate. *Arrow* indicates presumed neuronophagia of a degenerate pigment cell. (B: H.E.)

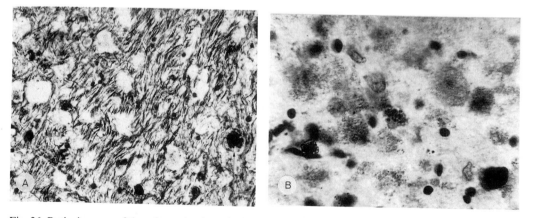

Fig. 26. Reticular zone of the substantia nigra. A. A majority of spheroid bodies develop a foamy appearance. *Arrow* indicates the only spheroid body laden with deeply argyrophilic granules centrally. B. Magnified foamy spheroid bodies fully laden with an abundance of tiny carmine-positive granules. (A: Hirano's simple silver, B: Best-carmine.)

absent in the globus pallidus. In general, the findings indicate that the present Guam ALS/PD examples cannot be regarded as cases of Hallervorden-Spatz disease, but should be regarded as Hallervorden-Spatz-like disease.

In this context, Arai et al. (1988) recently reported a case with interesting clinicopathological features. Spontaneous attacks of a peculiar twilight state with depersonalization began at the age of 30. These attacks were easily self-induced by focussing the eyes on a certain point. They could also be artificially induced by injection of atropine sulfate and immediately relieved by injection of adrenaline. They occurred frequently and usually lasted for 10 min to 3 h (Fig. 27A). Throughout the entire clinical course, there existed a slightly deformed and sluggish pupillary reflex to light on the left side, and persistent convergence palsy of both eyes, but it was noticeable that, except for the ocular disturbances mentioned, parkinsonian features did not develop. The patient died suddenly from unknown cause at 71 years of age, after a disease duration of 41 years. The EEG findings during spontaneous and artificially-induced attacks and attack-free periods are shown in Fig. 27B.

Autopsy showed pronounced depigmentation of the compact zone of the substantia nigra and rust-brown discoloration of the reticular zone (Fig. 28A). Disintegration of the pigmented cells of the compact zone and a corresponding gliosis occurred bilaterally and symmetrically (Fig. 28B). Again, a large number of the foamy spheroids were exclusively localized in the reticular zone of the substantia, as shown by H. E. and silver preparations (Fig. 29A). No glycogen granules were accumulated, as seen in the 2 Guam ALS/PD cases mentioned above.

Electron microscopic observation of these spheroids disclosed that no single unit membrane or myelin sheath lamella in the more distal portion surrounded the spheroids as a whole. Because of these characteristics alone, they were quite different from the spheroids of Hallervorden-Spatz disease. The septum-like membranes, which appeared to divide the foamy spheroids into several areas containing different-sized vacuoles, were scattered in the majority of these spheroids. Also, the outer margins of most of these 'spheroids' were surrounded by dense astrocytic glial fibers, which were also observed, together with the septum-like membranes, less obviously in the foamy spheroids (Fig. 29B). It is also important to note that neither neurofilaments nor intracytoplasmic organelle and/or membrano-lamellar structures were accumulated in the foamy spheroids in this particular case, and careful examination never disclosed synaptic junctions such as seen in Hallervorden-Spatz disease. Thus, the fundamental question is whether they can be regarded as spheroid bodies in the true sense of the word, or, in other words, of neuroaxonal origin, and the same question presumably can be applied to the 2 Guam cases mentioned above.

In this particular case, there was a clinicopathological dissociation, in that the disintegration of the compact zone of the substantia nigra was clear-cut but that parkinsonian features were completely lacking throughout the clinical course. We cannot yet account for this dissociation, though neuronal loss in the third cranial nerve nucleus closely corresponds to the clinical observations of the ocular disturbances. No 'foamy spheroids' nor other remarkable changes were encountered elsewhere in the CNS, such as the striopallidal, hypothalamic and other areas.

SPHEROID BODIES IN THE PALLIDO-NIGRAL SYSTEM IN MONKEYS

The veterinary pathologists Suzuki and Narama (1986, 1987) reported light and electron microscopic findings of spheroid bodies, similar to those observed in the human cases mentioned above, in the pallido-nigral system in monkeys. In 82 cases of 3–4 year old monkeys (*Macaca iris*) from the Philippines, spheroids were often seen in the pallidum, and less frequently in the reticular zone of the substantia nigra.

The first mentioned paper (Suzuki and Namara, 1986) is concerned only with these spheroids, the neuropathology of which closely resembles that of humans. As shown in Fig. 30A, multiple, large or small, mainly ovoid and/or rarely irregularly-shaped spheroids contained homogeneous or granular substances and/or fine vacuoles. They were weakly LFB- or Sudan Black B-positive and also

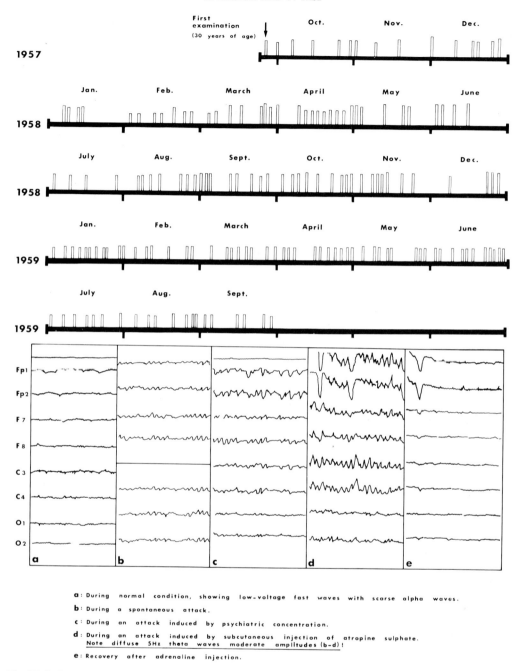

a: During normal condition, showing low-voltage fast waves with scarse alpha waves.

b: During a spontaneous attack.

c: During an attack induced by psychiatric concentration.

d: During an attack induced by subcutaneous injection of atropine sulphate.
Note diffuse 5Hz theta waves moderate ampiltudes (b-d)!

e: Recovery after adrenaline injection.

Fig. 27. A. Recurrent attacks of peculiar twilight state with depersonalization. The attacks occurred frequently in every month and year without seasonal differences. B. Electroencephalograms between and during attacks. (Reproduced from Dr. H. Honda (1960) by courtesy of the Editors of *Psychiatr. Neurol. Jpn.*)

weakly iron-positive, surrounded by a strongly GFAP-positive capsule which was continuous with strongly GFAP- and iron-positive, widely extended processes (Fig. 30B).

Electron microscopic examination of these spheroids showed that they exhibited no single unit external membrane or synaptic junctions and/or myelin sheath structures, while they were

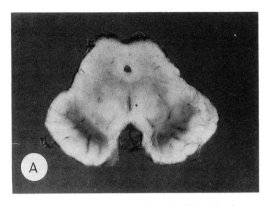
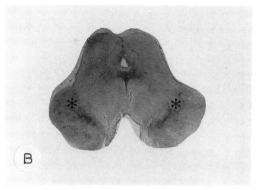

Fig. 28. Midbrain. A. Naked eye finding. Depigmentation in the compact zone of the substantia nigra and pronounced rusty-brown discoloration in the reticular zone on both sides. B. Gliosis is conspicuous in both the compact zone of the substantia nigra and oculomotor nucleus accompanied by neuronal loss. A small number of pigmented neurons are preserved in the lateral area (*asterisks*). (B: Holzer). (Reproduced from Dr. N. Arai et al. (1988) by courtesy of the Editors of *J. Neurol.* (Springer).)

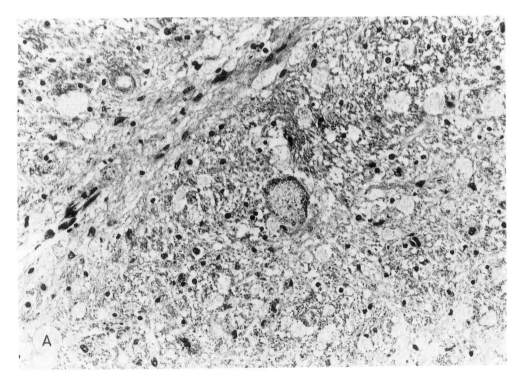

Fig. 29A.

surrounded by, but rarely contained, gliofilaments. The internal contents, on the other hand, were mainly highly electron-dense homogeneous, largish structures, and/or clusters of very fine, electron-dense granules with a diameter of 5 nm,

while others contained isolated or confluent vacuoles (Fig. 31). It was noticeable that there were no conspicuous organelles or structures as seen in neuroaxonal dystrophy.

It is true that the topography of the spheroids

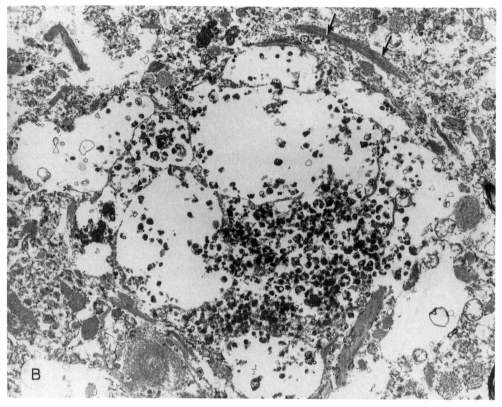

Fig. 29. Substantia nigra. A. Reticular zone. An abundance of variously-sized foamy spheroid bodies is observed. B. Electron microscopic picture of a spheroid body in Fig. 29A. The spheroid body has no unitary peripheral membrane but is surrounded by astrocytic glial fibers, one of which is indicated by *arrows*. The latter are also visible together with the septum-like membranes, which are divided into several structures. The latter contain vacuoles and a varying number of small electron-dense granules. (A: H.E.) (Reproduced from Dr. N. Arai et al. (1988) by courtesy of the Editors of *J. Neurol.* (Springer).)

and the nature of their constituents in these monkeys differed to some extent from those in the human cases. Also, no particular neurological disturbances were observed clinically.

MISCELLANEOUS NEUROPATHOLOGY IN GUAM AND KII ALS/PD CASES

Our previous papers on ALS/PD cases from both foci (Shiraki 1965; Shiraki and Yase 1975) noted that certain CNS aging processes of a physiological nature were accentuated in a majority of cases, such as spheroidal axonal bodies in Goll's nucleus, Marinesco bodies in the nuclei of the pigmented cells of the substantia nigra, while lipopigmentary granules were particularly noted

in the cytoplasmic processes of the astrocytes, all in increased number and intensity.

We also reported that multinucleate Purkinje cells were dislocated in the molecular layer of the cerebellar cortex in almost all cases of ALS/PD from both foci (Figs 32A and B), while multinucleate nerve cells were also sporadically encountered in the medial geniculate body, vestibular nuclei, and pigmented cells of the substantia nigra in a few cases. It may be postulated that the former finding suggests some precocious senility process, while the latter indicates a possible malformation of the CNS, presumably of prenatal origin.

Chen (1981), on the other hand, reported on the distribution and occurrence of NFTs in 28

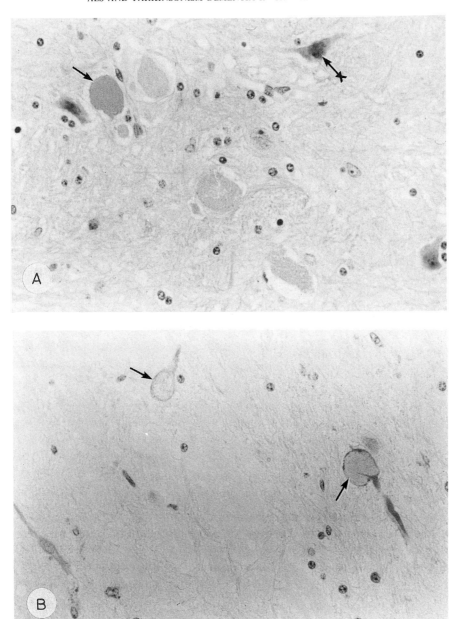

Fig. 30. Globus pallidus. A. Oval variably shaded, eosinophilic spheroid bodies, one of which is indicated by *arrow*, are widespread. Shrunken nerve cells, one of which is indicated by *arrow with cross*. B. Two faintly-stained spheroid bodies with marginal GFAP-positive astrocytic gliofibers and processes are indicated by *arrows*. (A: HE. B: GFAP.) (Reproduced from Dr. Y. Suzuki and Narama (1986) by courtesy of the Editor of *Neuropathology* (Kyoto).)

Guamanians with ALS/PD and compared these with 114 Guamanians without ALS/PD. He concluded that most Guamanians without ALS/PD during and after middle age had NFTs similar to those of the Guamanians with ALS/PD, but to a lesser degree and extent. In our opinion, such a controlled study is very important for a fundamental understanding of the develop-

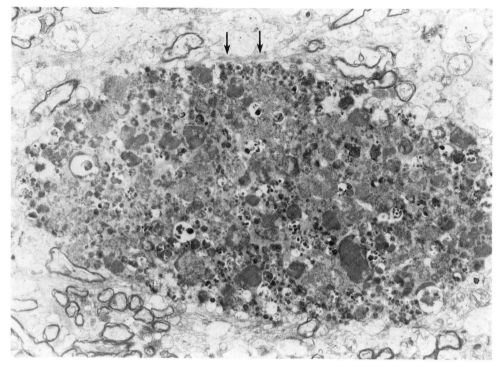

Fig. 31. Electron microscope study of spheroid. This spheroid may correspond with one of those illustrated in Fig. 30A. Moderately or weakly electron-dense, homogeneous, variably sized, multiple structures form the main components, while intervening clusters of very fine and more intensely electron-dense tiny granules with a diameter of 5 nm are as a rule surrounded by vacuoles. No single-unit membrane is visible outside this spheroid which is surrounded by astrocytic gliofibers, one of which is indicated by *arrows*. ×6250. (Reproduced from Dr. Y. Suzuki and Narama (1987) by courtesy of the Editors of *Neuropathology* (Kyoto).)

mental mechanism of ALS/PD in cases from both foci, but the possibilities of precocious senility and malformation of the CNS will be included in future examinations.

DISCUSSION

The fundamental epidemiological, clinical and pathological questions in relation to ALS and PD cases from both the Guam and Kii foci have to some extent been clarified, but many problems remain unsolved. For example, even if the two disorders in both foci are assumed to be clinical variants of a single entity due to an almost identical etiopathogenetic mechanism in which certain environmental (exogenous) factors such as metal/mineral dysmetabolism were well documented in a series of soil-plant-animal-human

analyses, the data for PD cases remain insufficient as compared with ALS cases. We still hesitate to conclude that the etiopathogenesis of PD belongs to the same category as that of ALS, since the clinicopathological features of PD alone correspond to the 'extrapyramidal system', although the pyramidal system is now postulated as the biggest output pathway of the extrapyramidal system. A clear-cut differentiation of the two systems is gradually evolving.

Acknowledgements
The authors are deeply grateful to their many closely-associated colleagues and to Professors R. Nieuwenhuys, K. Miyoshi and Y. Suzuki, to Dr. N. Arai and Dr. R.M. Garruto for use of their materials, and to D. Grier for her help with the proof-reading of the manuscript.

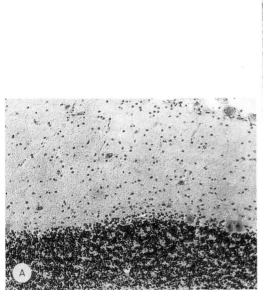
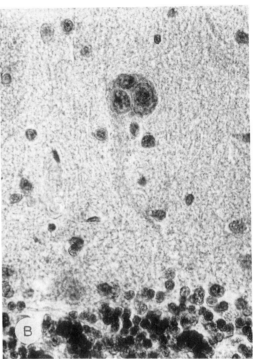

Fig. 32. Cerebellar cortex. A. Several Purkinje cells are dislocated in the molecular layer. B. A dislocated Purkinje cell in the molecular layer shows three nuclei at this higher magnification. (A and B: HE).

REFERENCES

ARAI, N., Y. HONDA, Y. AMANO and K. MISUGI: Foamy spheroid bodies in the substantia nigra. Report of an unusual case with recurrent attacks of peculiar twilight state. J. Neurol. (Springer) 235 (1988) 330–334.

BRADLEY, W. G. and F. KASIN: A new hypothesis of the etiology of amyotrophic lateral sclerosis. The DNA hypothesis. Arch. Neurol. 39 (1982) 677–680.

BIRCHALL, J. D. and J. S. CHAPPELL: Aluminum, chemical physiology and Alzheimer's disease. Lancet 2 (1988) 1008–1010.

BIZZINI, A. R. and P. GAMBETTI: Phosphorylation of neurofilaments in altered in aluminum. Acta Neuropathol. 71 (1986) 154–158.

CANDY, J. M., A. E. OAKIEY, J. KLIMOWSKI, T. A. CARPENTER, R. H. PERRY, J. R. ATACK, E. K. PERRY, G. BLESSED, A. FAIR-BRAIN and J. A. EDWARDSON: Aluminosilicates and senile plaque formation in Alzheimer's disease. Lancet 1 (1986) 354–356.

CHARCOT, J. M. and A. JOFFROY: Deux cas d'atrophie musculaire progressive avec lésion de la substance grise. Arch. Physiol. Norm. Pathol. 2 (1869) 354, 629, 744.

CHEN, L.: Neurofibrillary change on Guam. Arch. Neurol. (Chic.) 38 (1981) 16–18.

CRAPPER, D. R. and U. DEBONI: Aluminum and the Genetic Apparatus in Alzheimer Disease in the Aging Brain and Senile Dementia. New York, Plenum Press (1977) 229–246.

CRAPPER, D. R., S. S. KRISHNAN and A. J. DALTON: Brain aluminum distribution in Alzheimer's disease and experimental neurofibrillary degeneration. Science. 180 (1973) 511–513.

CRAPPER, D. R., S. S. KRISHNAN and S. QUITTKAT: Aluminum, neurofibrillary degeneration and Alzheimer's disease. Brain 99 (1976) 67–80.

CRAPPER, D. R., S. KARLIK and U. DEBONI: Aluminum and Other Metals in Senile (Alzheimer) Dementia in Alzheimer's Disease, Senile Dementia and Related Disorders. (Aging Vol. 7) New York, Raven Press (1978) 471–485.

CRAPPER, D. R., S. QUITTKAT, S. S. KIRSHNAN, A. J. DALTON and U. DEBONI: In nuclear aluminum content in Alzheimer's disease, dialysis encephalopathy, and experimental aluminum encephalopathy. Acta neuropathol. 50 (1980) 19–24.

DAVIDSON, T. J. and H. A. HARTMANN: RNA content and volume of motor neurons in amyotrophic lateral sclerosis. II. The number intumescence and nucleus dorsalis. J. Neuropathol. Exp. Neurol. 40 (1981a) 187–192.

DAVIDSON, T. J. and H. A. HARTMANN: Basic composition of RNA obtained from motor neurons in amyotrophic lateral sclerosis. J. Neuropathol. Exp. Neurol. 40 (1981b) 193–198.

DAVIDSON, T. J., H. A. HARTMANN and P. C. JOHNSON: RNA content and volume of motor neurons in amyotrophic lateral sclerosis. I. The cervical swelling. J. Neuropathol. Exp. Neurol. 40 (1981) 32–36.

DE BONI, U., A. OTROS, J. W. SCOTT and D. R. CRAPPER: Neurofibrillary degeneration induced by systemic aluminum. Acta Neuropathol. 35 (1976) 285–294.

DUNCAN, M. W., I. J. KOPIN, R. M. GARRUTO, L. LAVINE and S. P. MARKEY: 2-Amino-3 (methylamino)-propionic acid in cycad-derived foods is an unlikely cause of amyotrophic lateral sclerosis/parkinsonism. Lancet 2 (1988) 631–632.

EDWARDSON, J. A. and J. M. CANDY: Aluminum and the pathogenesis of senile plaques in Alzheimer's disease. Down's syndrome and chronic renal dialysis. Ann. Med. (Helsinki) 21 (1989) 95–97.

FUJITA, T., Y. OKAMOTO, T. TOMITA, Y. SAKAGUCHI, K. OTA and M. OHATA: Calcium metabolism in aging inhabitants of mountain versus seacoast communities in the Kii Peninsula. J. Am. Geriatrics Soc. 25 (1977) 254–258.

GARRUTO, R. M. and Y. YASE: Neurodegenerative disorders of the Western Pacific: the search for mechanisms of pathogenesis. TINS 9 (1986) 368–374.

GARRUTO, R. M., R. FUKATSU, R. YANAGIHARA, D. C. GAJDUSEK, G. HOOK and C. E. FIORI: Imaging of calcium and aluminum in neurofibrillary tangle-bearing neurons in parkinsonism-dementia of Guam. Proc. Natl. Acad. Sci. U.S.A. 81 (1984) 1875–1879.

GARRUTO, R. M., S. K. SHANKAR, R. YANAGIHARA, A. M. SALAZAR, H. L. AMYZ and D. C. GAJDUSEK: Low-calcium, high-aluminum diet-induced motor neuron pathology in cynomolgus monkeys. Acta Neuropathol. 78 (1989) 210–219.

HIRANO, A. L., T. KURLAND, R. S. KROOTH and S. LESSEL: Parkinsonism-dementia complex, an endemic disease on the island of Guam. I. Clinical features. Brain 84 (1961a) 642–661.

HIRANO, A., N. MALAMUD and L. T. KURLAND: Parkinsonism-dementia complex, an endemic disease on the island of Guam. II. Pathological features. Brain 84 (1961b) 662–679.

HONDA, Y.: Clinical studies on the diencephalon-related psychic symptoms. Psychiatr. Neurol. Jpn. 62 (1960) 297–325.

HUGHES, J. T.: Aluminum encephalopathy and Alzheimer's disease. Lancet (March 14, 1989) 490–491.

IWATA, S., K. SASAJIMA, Y. YASE and K-M CHEN: Report of investigation of the environmental factors related to occurrence of amyotrophic lateral sclerosis in Guam island. Overseas field research in 1976–77 supported by the Japanese Ministry of Education. (1978) 1–10.

IWATA, S.: Structural analysis of metal co-precipitated calcification products in the central nervous system with particular reference to ALS. Neurol. Med. 13 (1980) 103–109.

IWATA, S., K. SASAJIMA, Y. MIZUMOTO and T. KUSAKABE: Abnormal content and distribution of aluminum in a grain of rice planted on a district of high incidence of Japanese amyotrophic lateral sclerosis (ALS). In: Tsubaki and Yase (Eds), International Conference of Amyotrophic Lateral Sclerosis — Issue of Discussion and Poster Session. SIMUL International Inc., Osaka, Japan (1988) 237.

JOHNSON, D. E.: Energy Loss Spectrometry for Biological Research in Introduction of Analytical Electron Microscopy. New York, Plenum Press (1979) 245–258.

KIHIRA, T.: Morphological, morphometrical and metal analytical studies of oral aluminum neurotoxicity. Brain Nerve. (Tokyo) 39 (1987) 636–641.

KLATZO, I., H. WISNEIWSKI and E. STREICHER: Experimental production of neurofibrillary degeneration. 1. Light microscopic observations. J. Neuropathol. Exp. Neurol. 24 (1965) 187–199.

KONDO, K. and M. MINOWA: Epidemiology of motor neuron disease in Japan: Declining trends of the mortality rate. In: Tsubaki and Yase (Eds.), Proceedings of the International Conference of Amyotrophic Lateral Sclerosis. Amsterdam, Elsevier Science Publishers. (1988) 11–16.

LUKIW, W. J., T. P. A. KRUCK and D. R. CRAPPER: Alterations in human linker histone-DNA binding in the presence of aluminum salts in vitro and in Alzheimer's disease. Neurotoxicology 8 (1987) 291–302.

MANN, D. M. A. and P. O. YATES: Motor neuron disease: The nature of the pathogenic mechanism. J. Neurol. Neurosurg. Psychiatry 37 (1974) 1047–1052.

MANNETO, V., N. H. STERNBERGER, G. PERRY and L. A. STERNBERGER: Phosphorylation of neurofilaments is altered in amyotrophic lateral sclerosis. J. Neuropathol. Exp. Neurol. 47 (1988) 642–653.

MARKESBERG, W. R., W. D. EHEMANN, T. I. H. HOSSAIN, M. ALAUDIN and D. T. GOODIN: Instrumental neutron activation analysis of brain aluminum in Alzheimer disease and aging. Ann. Neurol. 10 (1981) 511–516.

MATSUOKA, T., K. MIYOSHI, T. SAKA, T. KAWAGOE, T. NISHIKIORI, M. HIRABAYASHI, T. SHISOZUKA, K. TSUKADA, A. AOKI, K. SHIMOKAWA and H. SHIRAKI: A case of encephalopathy with plaque-like bodies, neurofibrillary change, angiopathy and amyotrophic lateral sclerosis like lesions. Adv. Neurol. Sci. (Tokyo) 11 (1967) 801–811.

MCDERMOTT, J. R., A. I. SMITH, K. IQBAL and H. M. WISNIEWSKI: Brain aluminum in aging and alzheimer disease. Neurology 29 (1979) 809–814.

MIURA, K.: Amyotrophische Lateralsklerose unter dem Bilde von sog. Bulbärparalyse. Neurol. Jpn. 10 (1911) 366–369.

MIZUMOTO, Y., S. IWATA, K. SASAJIMA, S. YOSHIDA, F. YOSHIMASU and Y. YASE: Determination of Ca/P atomic ratio in spinal cord of amyotrophic lateral sclerosis and X-ray fluorescence analyses. Radioisotopes 32 (1983) 29–32.

MURAKAMI, N., F. YOSHIMASU, Y. YASE, S. IWATA and K-M. CHEN: Neutron activation analysis of the skull tissue in Japanese and Guamanian amyotrophic lateral sclerosis. Neurol. Med. (Tokyo) 20 (1984) 53–56.

NAKAI, Y.: Distribution of Alzheimer's neurofibrillary changes in amyotrophic lateral sclerosis. J. Wakayama Med. Assoc. 20 (1969) 355–370.

NIKAIDO, T., J. AUSTIN, L. TREUB and R. REIHART: Studies in aging of the brain. II. Microchemical analysis of the nervous system in Alzheimer patients. Arch. Neurol. 27 (1972) 549–554.

NIEUWENHUYS, R., J. J. VOOG and CHR. VAN HUIJZEN: The Human Central Nervous System. A Synopsis and Atlas, 3rd. revised ed. Berlin, Heidelberg, New York, London, Paris, Tokyo, Springer (1988).

OKAMOTO, K., M. MORIMATSU, S. HIRAI, E. NOGI and Y. ISHIDA: Intracytoplasmic inclusions (Bunina bodies) observed in a case of amyotrophic lateral sclerosis. Clin. Neurol. 19 (1979) 174–182.

PLAITAKIS, A. and J. T. CAROSCIO: Abnormal Glutamate Metabolism in Amyotrophic Lateral Sclerosis. Ann. Neurol. 22 (1987) 575–579.

PERL, D. P. and A. R. BRODY: Alzheimer's disease; X-ray spectrometric evidence of aluminum accumulation in neurofibrillary tangle-bearing neurons. Science 208 (1980) 297–299.

PERL, D. P., D. C. GAJDUSEK, R. M. GARRUTO, R. T. YANAGIHARA and GIBBS, C. J. JR: Intraneuronal aluminum accumulation in amyotrophic lateral sclerosis and parkinsonism-dementia of Guam. Science 217 (1982) 1053–1055.

SHIRAKI, H.: The Neuropathology of Amyotrophic Lateral Sclerosis (ALS) in the Kii Peninsula and other Areas of Japan in Motor Neuron Diseases: Research on Amyotrophic Lateral Sclerosis and Related Disorders. New York and London, Grune and Stratton (1965) 80–84.

SHIRAKI, H.: Some unusual neuropathologic features in Guam cases in comparison with those in the Japanese. With special reference to Hallervorden-spatz diseae-like lesions. Proc. 5th Internat. Congr. Neuropathol. Zurich, August 31–Sept. 3, 1965 Internat. Congr. Series No. 100. Excerpta Med. Found. Amsterdam/New York, London, Milano, Tokyo, Buenos Aires (1966) 201–207.

SHIRAKI, H. and Y. YASE: Amyotrophic lateral sclerosis in Japan. In: Handbook of Clinical Neurology, Vol. 22, System Disorders and Atrophies, Part II. North-Holland Pub. Comp. Amsterdam, Oxford, American Elsevier Pub. Co. Inc. New York (1975) 354–419.

SHUMA, H., A. P. SOMLYO, T. FREY and D. SAFER: Energy loss imaging in biology. Proc. 40th Ann. EMSA Meeting (1982) 416–417.

SPENCER, P. S., P. B. NUNN, J. HUGON, A. C. LUDOLPH, S. M. ROSS, D. N. ROY and R. C. ROBERTSON: Guam amyotrophic lateral sclerosis-parkinsonism-dementia linked to a plant excitant neurotoxin. Science 237 (1987) 517–522.

SUZUKI, Y. and I. NARAMA: Pigmentary deposition and spheroid formation in the pallidonigral system in monkeys (*Macaca iris*). Neuropathology (Kyoto) 7 (1986) 98.

SUZUKI, Y. and I. NARAMA: A question of whether spheroid bodies in the pallidum in monkeys (macaca iris) are of a dystrophic axonal nature or not. Neuropathology (Kyoto) 8 (1987) 90–91.

TANDON, R., S. H. ROBINSON, J. S. MUNZER and W. G. BRADLEY: Deficient DNA repair in amyotrophic lateral sclerosis cells. J. Neurol. Sci. 79 (1987) 189–203.

UEBAYASHI, Y.: Amyotrophic lateral sclerosis in the Western Pacific: A new aspect for the progressive disease process based on the changing epidemiological pattern. In: Tsubaki and Yase (Eds.), Proceedings of the International Conference of Amyotrophic Lateral Sclerosis. Amsterdam: Elsevier Science Publishers (1988) 17–23.

WEISS, J. H., J-Y. KOH and D. W. CHOI: Neurotoxicity of N-methylamino-L-alanine (BMAA) and N-Oxalylamino-L-alanine (BOAA) on cultured cortical neurons. Brain Res. 497 (1989) 64–71.

WEN, G. Y. and H. M. WISNIEWSKI: Histochemical localization of aluminum in the rabbit CNS. Acta Neuropathol. 68 (1985) 175–184.

YANO, I., S. YOSHIDA, Y. UEBAYASHI, F. YOSHIMASU and Y. YASE: Degenerative changes in the central nervous system of Japanese monkeys induced by oral administration of aluminum salt. Biomed. Res. 10 (1989) 33–41.

YASE, Y.: The pathogenesis of amyotrophic lateral sclerosis. Lancet 2 (1972) 292–296.

YASE, Y.: Amyotrophic lateral sclerosis in the Kii Peninsula, Japan. Neurol. Med. (Tokyo) 2 (1975) 17–24.

YASE, Y.: The basic process of amyotrophic lateral sclerosis as reflected in Kii Peninsula and Guam. In: W. A. Den Hartog Jager, G. W. Bruyn, A. P. J. Hejistee (Eds.), Proceedings of 11th World Congress of Neurology, Amsterdam (1977) 413–427.

YASE, Y., N. MATSUMOTO, K. AZUMA, Y. NAKAI and H. SHIRAKI: A Japanese case of amyotrophic lateral sclerosis with schizophrenic symptoms and showing Alzheimer's tangles. Arch. Neurol. (Chic.) 27 (1972) 118–128.

YASE, Y., F. YOSHIMASU, Y. UEBAYASHI, S. IWATA and K. KIMURA: Amytrophic lateral sclerosis. Interaction of divalent metals in CNS tissue and soft tissue calcification. Proc. Jpn. Acad. 50 (1974) 401–406.

YASE, Y., F. YOSHIMASU, M. YASUI, Y. UEBAYASHI, S. TANAKA, S. IWATA, K. SASAJIMA, D. C. GAJDUSEK and C. R. GIBBS JR: Amyotrophic lateral sclerosis — neutron activation analysis on Guamanian ALS and PD cases and their Chamorro controls — Annual Report of the Research Committee of Degenerative CNS Disease. The Ministry of Health and Welfare of Japan (1980) 296–302.

YASUI, M., T. KIHIRA, Y. YASE, F. YOSHIMASU and H. YOSHIDA: Magnesium study on amyotrophic lateral sclerosis. Magnesium Res. 2 (1989) 66.

YOSHIDA, S.: X-ray microanalytic studies on amyotrophic lateral sclerosis. I. Metal distribution compared with neuropathological findings in cervical spinal cord. Clin. Neurol. 17 (1977) 299–309.

YOSHIDA, S.: X-ray microanalytical studies on amyotrophic lateral sclerosis. II. The interrelationships of intraspinal blood supply metal deposition and degenerative changes. Clin. Neurol. 19 (1979) 283–291.

YOSHIDA, S., Y. YASE, S. IWATA, H. YOSHIDA and Y. MIZUMOTO: Trace-metals and its relationship to early pathological changes of motor neurons in amyotrophic lateral sclerosis. Clin. Neurol. 27 (1987) 518–527.

YOSHIDA, S., Y. YASE, Y. MIZUMOTO and S. IWATA: Aluminum deposition and Ca-hydroxyapatite formation in frontal cortex issue of amyotrophic lateral sclerosis. Clin. Neurol. 29 (1989) 421–426.

YOSHIDA, S., T. KIHIRA, K. MITANI, I. WAKAYAMA, Y. YASE, H. YOSHIDA and S. IWATA: Intraneuronal localization of aluminum: possible interaction with nucleic acids and pathogenetic role in amyotrophic lateral sclerosis. ALS. In: F. Clifford Rose and F. Norris (Eds.), New Advances in Toxicology and Epidemiology. London, Smith-Gordon (1990) 211–223.

YOSHIMASU, F., Y. UEBAYASHI, Y. YASE, S. IWATA and K. SASAJIMA: Studies on amyotrophic lateral sclerosis by neutron activation analysis. Folia Psychiatr. Neurol. Jpn. 30 (1976) 49–55.

YOSHIMASU, F., M. YASUI, Y. YASE, Y. UEBAYASHI, S. TANAKA, S. IWATA, K. SASAJIMA, D. C. GAJDUSEK, C. J. GIBBS and K-M. CHEN: Studies on amyotrophic lateral sclerosis by neutron activation analysis — 3. Systematic analysis of metals on Guamanian ALS and PD cases. Folia Psychiatr. Neurol. Jpn. 36 (1982) 173–179.

Handbook of Clinical Neurology, Vol. 15 (59): Diseases of the Motor System
J.M.B.V. de Jong, editor
© Elsevier Science Publishers B.V., 1991

CHAPTER 17

Hereditary spastic paraparesis (Strümpell-Lorrain)

R. P. M. BRUYN[1] and PH. SCHELTENS[2]

[1] *Department of Neurology, Oudenrijn Hospital, Utrecht, and* [2] *Department of Neurology, Free University Hospital, Amsterdam*

Hereditary spastic paraparesis in its pure form is a genetically heterogeneous disorder of the CNS of which slowly increasing spastic paraparesis is the pivotal clinical hallmark.

HISTORICAL SYNOPSIS

Close study of the original sources makes it unequivocally clear, that in spite of the fact that Erb (1875) described a pure motor disorder, consisting of a progressive paraparesis — probably a primary lateral sclerosis — it had nothing to do with a hereditary disease. Nor did the observation of Seeligmüller (1876) who reported 7 sibs, 4 of whom had progressive muscular atrophy as well as a bulbar paralysis, born from a consanguinous marriage.

Strümpell (Fig. 1), (1880, 1886, 1893 and 1904) and Lorrain (1898), in elaborating the features and emphasizing the heredofamiliar nature of the disorder, earned the eponymous credit in subsequent reports by others.

In his original paper Strümpell (1880) mentioned 3 brothers, 2 of whom developed a spastic gait at the ages of 37 and 56 years, respectively; the third brother suffered from severe arthritis deformans. Their father — and not the mother, as Harding (1981), and Boustany et al. (1987) posited — was said to have been 'a little lame'. Apart from the spastic paraparesis, they showed no neurological abnormalities other than increased tendon reflexes in the upper limbs. The first neuropathological description concerned the youngest brother (Strümpell 1886). In another report, Strümpell (1904) described the clinical follow-up and detailed the neuropathological changes in a man, on whom he had reported in 1893, who had developed the first symptoms at the age of 34 and who died 27 years later. His grandfather, father, two paternal uncles and one brother were also affected. The sole clinical sign was a slowly progressive spastic paraparesis, with relative sparing of muscle strength, as shown by his ability to walk (if crutch-aided) until the day he died. No sensory abormalities were detected. The neuropathological hallmark consisted of degeneration of the lateral pyramidal tracts extending from the lumbar to the upper cervical region. Remarkably, although all sensory modalities were intact, degeneration of Goll's tract increased upon ascent in the spinal cord.

The contribution of Lorrain to the knowledge of hereditary spastic paraparesis consisted of a thesis, in which he reviewed the literature and reported 3 personal observations (Lorrain 1898). The first contained the history of an 18-year-old woman, whose father was an alcoholic and whose mother suffered from agoraphobia. From his description it is clear that she suffered from a gradually progressive spastic paraparesis, without

Fig. 1. Prof. Dr. Adolf Strümpell.

(1916), Paskind and Stone (1933), Kahlstorf (1937), and Price (1939). In their exhaustive review, Bell and Carmichael (1939) studied 74 pedigrees selected on the criterion of 'pure' spastic paraparesis.

The term 'pure', seductively simple as it sounds, produced semantic confusion; many authors tried to adhere to the concept of pure spastic paraparesis, even though several of the families they reported clearly did not display it. This difficulty was well recognized by others who reported new case series, such as Landau and Gitt (1951), Schwarz (1952), Schwarz and Liu (1956), Behan and Maia (1974), Sutherland (1975) and Holmes and Shaywitz (1977).

Though it is the merit of Harding (1981, 1983 and 1984) to make an orderly division between pure and complicated forms, as well as between 2 types within the group showing dominant inheritance, only advanced molecular-biological techniques pinpointing the mutant gene/alleles and chemical identification of the gene product, will provide a definite nosology for Strümpell-Lorrain disease.

EPIDEMIOLOGY

sensory impairment. Intelligence was normal, but a horizontal nystagmus was noted in both directions together with bilateral optic atrophy. The onset was at the age of 9 years. The rest of the family, including a younger brother and sister, was reported to be normal. The second and third observation comprised the history of a 16-year-old girl and an unrelated 10-year-old boy, respectively, both with pure spastic paraparesis, and born of healthy parents. Again, the rest of the family was reported normal.

Around the turn of the century, a vast multitude of authors, stirred by Strümpell's reports and Lorrain's monograph studied this 'new' disease in increasing detail (Bayley 1897; Raymond and Souques 1896; Bernhardt 1891; Newmark 1893, 1904, 1906 and 1911; von Krafft-Ebing 1900; Kühn 1902).

The accumulated body of knowledge on a disorder which, as far as the neuropathological substrate is concerned, seemed similar to subacute combined degeneration, was reviewed by Rhein

Very few epidemiological studies have been undertaken, and those studies that have been done lack uniformity concerning diagnostic criteria, age of onset, and mode of inheritance. Skre's (1974) investigations in Western Norway led to an estimated prevalence of 12.1 : 100 000 and 1.9 : 100 000 for autosomal dominant and autosomal recessive forms respectively. No information is given about the number of pure cases.

The same holds true for the study of Werdelin (1986), who examined 23 patients with hereditary spastic paraparesis (4 families with autosomal dominant, 5 with autosomal recessive inheritance) in Denmark; most patients exhibited other neurological signs as well, such as ataxia, cranial nerve paresis and dementia. She found a prevalence ratio of 0.8 : 100 000 and 0.1 : 100 000 respectively for autosomal dominant and recessive types among 10–50-year-old men and women.

Until a consensus is reached concerning nosology, meaningful epidemiological studies will not be feasible.

NEUROPATHOLOGY

Contrasting with the wealth of clinical reports is the paucity of neuropathological descriptions, probably because of the slowly progressive course of the disease, the scarcely reduced life expectancy, and the fact that the patients usually die at home or in nursing homes.

In the first neuropathological description Strümpell (1886) found degeneration of the spinal pyramidal tracts decreasing from lower lumbar to upper cervical level, as well as an increasing degeneration of Goll's tract and, in a lesser degree of the spinocerebellar tracts ascending the spinal cord. Strümpell (1904), on studying the pathological changes in the spinal cord of a patient whom he had followed up for 15 years, found degeneration of the pyramidal tracts from the lower lumbar to upper thoracic levels. In the thoraco-cervical transition zone, degeneration of Goll's tract was apparent. No changes were found in the Goll and Burdach nuclei, nor in the medullary pyramids. In the same year, Newmark reported on 2 families with 1 autopsy. Degeneration of the pyramidal tract at the lumbar-sacral level reached its maximum at the upper lumbar level and was traceable, but, again, gradually diminishing up to the lower cervical segments, whereas degeneration of the dorsal columns began at the midlumbar level, but became most marked at the cervical level. Newmark's patient had minor sensory deficit below the knees, whereas Strümpell's 'pure' case did not.

A brother of Newmark's autopsied patient died at the age of 20 from tuberculosis, and was mentioned in another report (1906). He had minor spastic paraparesis (that manifested from the age of 8 onward), but did not have any sensory impairment, and could therefore be considered to be an example of the 'pure' type. At autopsy, in curious contrast to the former 2 cases, degeneration of Goll's tracts was much more pronounced than that of the pyramidal tracts. In the latter no abnormalities were discerned rostrally to the midlumbar region. Another brother came to autopsy (Newmark 1911). He had slight and varying sensory impairment below the knees and also died from TBC.

Examination disclosed bilateral pyramidal tract degeneration at the sacral level, decreasing rostrally whereas the dorsal column degeneration appeared slight at the lumbo-sacral level, but more and more pronounced at increasingly rostral levels. Raymond and Rose (1909) reported an autopsy in a woman, whose signs had begun at the age of 12. She died from tuberculosis at the age of 27. Surprisingly, they found no abnormalities at all in the spinal cord; possibly, inappropriate staining techniques played a role in this unsuspected negative finding.

Clearly, there is no clinicopathological mutual correlation between dorsal column degeneration and sensory impairment. Kahlstorf (1937) examined a member of the family reported by Specht (1925), in whom no sensory abnormalities had been detected. Again, besides degeneration of the crossed and uncrossed pyramidal tracts and a reduced number of Betz' cells, the dorsal columns had degenerated; most remarkably, the spinocerebellar tracts also showed degeneration, the cerebellum was hypoplastic, and the basal ganglia had atrophied; all clinically silent.

In Farago's (1947) report, the disease had occurred in 3 generations. The third generation included one monozygotic and one dizygotic pair of diseased twins. The male from the dizygotic and one male of the monozygotic pair committed suicide. Besides the paraparesis, the other monozygotic male had micturition and defaecation problems as well as hypalgesia below the navel. After his death from septic shock, histologic examination again revealed severe degeneration of the pyramidal tracts and a reduced number of Betz cells, as well as a rostrally increasing degeneration of Goll's tracts. Schwarz (1952) reviewed all clinical and neuropathological data up to 1950; the material considered left him no other alternative than the conclusion that many authors had mixed together pure and non-pure forms. From 24 reports with neuropathologic verification he considered only 7 to have dealt with the 'pure' form of hereditary muscle paraparesis: Strümpell (1886, 1904); Newmark (1893, 1904, 1906 and 1911); Jakob (1909); Kahlstorf (1937); Bischoff (1902); Raymond and Rose (1909) and Farago (1947). He added another 4 families with 32 affected persons, examples of the pure form. He gave a detailed description of

the neuropathological findings in one patient, a
descendent from a family, earlier described by
Bayley (1897), Spiller (1902) and again by
Rechtman and Alpers (1934). Again, bilateral
demyelination of the lumbar lateral corticospinal
tracts was noted, with pronounced demyelination
in the lateral and ventral columns at the lower
cervical level. He observed minor changes in the
spinocerebellar tracts and slight diffuse thinning
of the gracile fasciculi.

Four years later, Schwarz and Liu (1956)
reported another family, as well as a detailed
neuropathological description of a member of
one of the families that Schwarz had described
in 1952. They found significant demyelination of
the crossed corticospinal tracts, especially between midcervical and lower thoracic level, together with demyelination of the spinocerebellar
tracts, Goll's columns (especially at cervical level),
and loss of neurons in Clarke's nuclei.

The neuropathological data of 2 patients given
by Behan and Maia (1974), and Sack et al.
(1978) show essentially the same alterations as
those in the reports mentioned above, viz. cau-
dally increasing degeneration of the corticospinal
tracts and rostrally increasing degeneration of
the dorsal columns.

Kramer (1977) gave the neuropathological data
of a male patient with spastic paraparesis belonging to a family, earlier reported on by Bruyn
and Mechelse (1962). Both anterior and lateral
corticospinal tracts were degenerated more markedly at the lumbar level than at cervical level,
whereas the degeneration of the gracile fasciculi
was more pronounced in the cervical cord. The
lateral cerebellar tracts had also degenerated
(Figs 2 and 3). To our awareness no further
neuropathological 'pure' reports have been published, since 1977. The neuropathological data
are summarized in Table 1.

CLINICAL FEATURES

For practical reasons a distinction is made
between the pure and complicated forms of the
disorder. The complicated forms will not be dealt
with *in extenso*, as such falls beyond the scope
of this chapter.

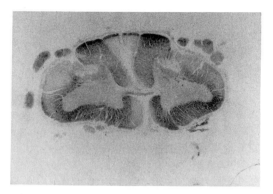

Fig. 2. Autopsy specimen from a member of the family reported by Bruyn and Mechelse (1962). The neuropathological findings were presented by Kramer
(1977). Transverse section through the cervical cord.
Demyelination of the anterior and lateral pyramidal
tracts, spinocerebellar tracts and Goll's fasciculi.

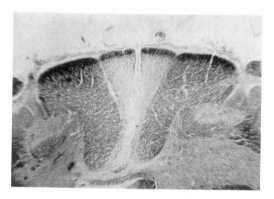

Fig. 3. Magnification of Fig. 2, clearly showing the
degeneration of the gracile fasciculi.

Pure spastic paraplegia

The essential clinical feature is a spastic weakness
of the lower limbs. Spasticity is more prominent
than weakness if the onset is in adult life. The
patients complain of stiffness or dragging of the
legs, rather than not being able to stand on their
legs. Presenting symptoms include a tendency to
trip on irregular surfaces or difficulty in running.
The affection is usually bilateral and, in the early
stages, often asymmetrical.

If the onset is in early life, motor delay is the
presenting symptom. Affected children may never
have walked, or do not walk until the age of 3
years (Harding 1981). Clearly, this feature holds
the danger of including cases of infantile cerebral

TABLE 1
Neuropathological findings.

Author	Pyramidal tr.			Dors. col.			Spinoc.	Betz
(year)	C	Th	L/S	C	Th	L/S		
Strümpell (1886)	+	+ +	+ + +	+ + +	+	+	+	n
Strümpell (1904)	−	+ +	+ + +	+ +	+	−	−	n
Newmark (1904)	+	+ +	+ + +	+ + +	+ +	+	−	n
Newmark (1906)	−	−	+	+ + +	+ +	+	−	n
Newmark (1911)	+	+ +	+ + +	+ + +	+ +	+	−	n
Raymond and Rose (1909)	−	−	−	−	−	−	−	−
Kahlstorf (1937)	+	?	?	+ + +	?	?	+	↓
Farago (1947)	+	+ +	+ + +	+ + +	+ +	+	+	↓
Schwarz (1952)	+ + +	?	+ +	+	?	?	+	?
Schwarz and Liu (1956)	+ +	+ + +	+	+ + +	+ +	+	+	n
Behan and Maia (1974)	+	+ +	+ + +	+ + +	+ +	+	−	n
Kramer (1977)	+	+ +	+ + +	+ + +	+ +	+	+	n
Sack et al. (1978)	+	+ +	+ + +	+ + +	+ +	+	−	n

C, Cervical; Th, Thoracic; L/S, Lumbosacral; +, mild; + +, moderate; + + +, severe degeneration; ?, not mentioned; Betz, number of Betz cells; ↓, reduced; n, normal; −, no degeneration.

diplegia. Children often show a tendency to walk on their toes at the time of their first steps. If the disease develops later in childhood, the patient may have walked normally, before a spastic, scissors gait develops. Inability to run is often mentioned. Several authors (Bickerstaff 1950; Sutherland 1975) have stressed the relative preservation of muscle strength in the presence of marked spasticity. Despite the absence of weakness, some patients need crutches to walk or become chairbound with progression of the disease (Behan and Maia 1974).

The patients' legs show marked spasticity and occasionally weakness. Weakness and spasticity of the arms is very rare and occurs late in the course of the illness, if at all. The tendon reflexes are invariably brisk, often clonic. Abdominal responses are usually preserved for a very long time (Sutherland 1975). A major diagnostic crite-

rion is the presence of extensor plantar responses. Ataxia is infrequently encountered, although it may be difficult to assess in the presence of severe spasticity.

Sensory impairment was found even in 'pure' cases by Bickerstaff (1950), Schwarz and Liu (1956), Behan and Maia (1974) and Schady and Sheard (1989, 1990). As mentioned earlier, there is such a striking discrepancy between the almost obligate degeneration of the posterior columns and the mild or absent clinical sensory impairment, that, today, it is widely accepted that mild sensory impairment, such as lack of vibration sense, does not rule out the presence of 'pure' hereditary spastic paraparesis.

Examination of cranial nerves is invariably normal and there are no cognitive changes. Sphincter dysfunction is not uncommon in the pure form (Wilson 1944; Farago 1947; Sutherland

1975). Cartlidge and Bone (1973) described 3 male patients in whom urgency and frequency of micturition were prominent. Opjordsmoen and Nyberg-Hansen (1980) have reported 9 patients from one family with spastic paraparesis, who suffered from neurogenic bladder disturbances. Some of them had sphincter symptoms before they manifested symptoms referable to the legs. In addition they also had syndactyly of the 4th and 5th finger.

Pes cavus, mentioned by Tyrer and Sutherland (1961) and by Garland and Astley (1950) as being a constant finding in spastic paraparesis, is in our view not pathognomonic for Strümpell's disease, but rather the result of the marked and prolonged spasticity. It is frequently encountered in various types of hereditary ataxias (Harding 1984b). Diagnostic criteria for pure spastic paraparesis are given in Table 2. Associated clinical features not permitted in the diagnosis of 'pure' spastic paraparesis are referred to as 'complicated' forms and will be shown in Table 3.

'COMPLICATED' FORMS OF SPASTIC PARAPARESIS

A distinction must be made between associated clinical features that do not interfere with the concept of 'pure' spastic paraparesis, such as mild sensory abnormalities, sphincter disturbances, pes cavus, and contractures of tendons, and syndromes in which spastic paraparesis is merely a part of a variety of clinical signs and symptoms, in general defined as 'complicated' forms.

Numerous families with such variants of hereditary spastic paraparesis or 'plus' forms have been reported. The range of additional clinical features associated with spastic parapare-

TABLE 2
Diagnostic criteria (1–5 are obligatory).

1. Spasticity in lower extremities
2. Paresis of lower extremities, usually not as pronounced as the spasticity
3. Hyperreflexia
4. Babinski's sign
5. Positive family history
6. Disturbances of sphincter function
7. Mild sensory impairment
8. Hyperreflexia and weakness of upper extremities.

sis is enormous. Many of these syndromes are extremely rare and in some instances unique to a particular family (Harding 1984a). Often the question can be raised whether the association is not merely coincidental.

In Table 3 an attempt is made to present an exhaustive summary of the conditions associated with spastic paraparesis, thus forming clinical entities clearly distinct from the pure form.

DIFFERENTIAL DIAGNOSIS

Differential diagnostic difficulties will arise when a patient with a spastic paraparesis has a positive family history. If so, then a scrupulous neurological examination should be done to classify the affected person either as 'pure' or 'complicated'. The list of complicated forms is exhaustive and 'new' complicated forms are frequently reported (Table 3).

If the family history is negative, then the patient is either the victim of a spontaneous mutation or other causes of paraparesis should be looked for, such as cerebral palsy, multiple sclerosis, amyotrophic lateral sclerosis, cervical myelopathy, spinal cord compression by e.g. solitary, or metastatic neoplasm, syringomyelia, neurosyphilis, neurolathyrism and subacute combined degeneration. Rare causes, such as a severe segmental narrowing of the spinal cord (Ramesh et al. 1989) or spinal epidural lipomatosis as a complication of iatrogenic Cushing's syndrome (Kaplan et al. 1989) should be included as well. Tropical spastic paraparesis, a myelopathy caused by the human T-cell lymphotropic type I retrovirus (HTLV-1), does not only occur in the tropics (Cruickshank et al. 1989; Power et al. 1989). Recently, Salazar-Grueso et al. described a family with 4 affected individuals in 3 generations, in whom the disease was associated with HTLV-1 infection (Salazar-Grueso et al. 1990).

Outbreaks of acute spastic paraparesis due to cassava toxicity have been reported in Tanzania under the name of Konzo (Howlett et al. 1990).

Careful history taking, neurological examination, laboratory tests and neuro-imaging techniques, such as MRI or myelography, should enable the clinician to establish the correct diagnosis.

TABLE 3
Conditions associated with spastic paraparesis.

Condition (inheritance)	Author (year)
Amyotrophy (AD or AR) (182700)	Seeligmüller (1876); Gee (1889); Ormerod (1904); Holmes (1905); Garland and Astley (1950); Refsum and Skillicorn (1954); Gilman and Horenstein (1964); Silver (1966a and 1966b); Gilman and Romanul (1975); Danadoost et al. (1977); van Gent et al. (1985); Gemignani (1986); Serena et al. (1990).
Cardiac defects (AR?)	Sutherland (1957); Tyrer and Sutherland (1961).
Cerebellar signs (AR or AD)	Roe (1963); Skre (1974); Scholtz and Swash (1985).
Deafness (XL) (312910)	Gemignani (1986); Wells and Jankovic (1986).
Epilepsy (AD)	Bruyn and Mechelse (1962); Kuroda et al. (1985).
Extrapyramidal signs (AD) (182800)	Osterreicher (1936); Dick and Stevenson (1953); Gilman and Horenstein (1964); Gilman and Romanul (1975). Bonduelle et al. (1953); Frank and Vuia (1973); Costell et al. (1989).
Fabry's disease (α-galactosidase deficiency) (XL)	Pierides et al. (1976).
Ferguson Critchley syndrome (with visual and sensory disturbances, cerebellar and extrapyramidal involvement) (AD)	Ferguson and Critchley (1929); Mahloudji (1963); Brown and Coleman (1966); Brown (1975); Harding (1982).
Glaucoma (with mental retardation) (AR) (270850)	Heijbell and Jagell (1981); Chenevix-Trench et al. (1986).
Hematological disorders (AD)	Chouza et al. (1984); Fujita et al. (1990).
Isaac's syndrome (AR)	Yokota et al. (1989).
Kallmann's syndrome (anosmia and hypogonadotrophic hypogonadism) (XL?) (308750)	Tuck et al. (1983).
Kjellin's syndrome (with amyotrophy, dementia and retinal degeneration) (AR)	Kjellin (1959 and 1981); Farmer et al. (1985).
Mast syndrome (with dysarthria, dementia and athetosis). (AR) (248900)	Cross and McKusick (1967b).
Mental retardation and dementia (AD or AR or XL)	Lobstein (1923); Appel and Van Bogaert (1951 and 1952); Sutherland (1957); Baar and Gabriel (1966); Mahloudji (1963); Gilman and Horenstein (1964); Ozsváth (1968); Allport (1971); Gilman and Romanul (1975); Rothner et al. (1976); Sjaastad et al. (1976); Katz et al. (1984); Young et al. (1984); Kenwrick et al. (1986).
Myotonia congenita (AD?)	Wessel et al. (1988).
Optic atrophy; retinal and macula degeneration (AD or AR) (182830;270700)	Louis-Bar and Pirot (1945); Bickerstaff (1950); Johnston and McKusik (1962); Bruyn and Went (1964); Mahloudji and Chuke (1968); Ozsváth (1968); Macrae et al. (1974); Rothner et al. (1976); Pagès and Pagès (1983); Katz et al. (1984).
Peripheral nerve involvement. (AD or AR)	Khalifeh and Zellweger (1963); Dyck and Lambert (1968); Koenig and Spiro (1970); Zellweger (1981); Cavanagh et al. (1979); Tredici and Minoli (1979); Abdallat et al. (1980); Schady and Sheard (1989 and 1990).
Precocious puberty (and mental retardation). (AD) (182820)	Raphaelson et al. (1983).
Psychiatric manifestations and euphoria. (AD or AR)	Bickerstaff (1950); Sutherland (1957); Jansen et al. (1988).
Skeletal abnormalities. (270710)	Fitzsimmons and Guilbert (1987).

TABLE 3 (continued)

Condition (inheritance)	Author (year)
Skin abnormalities (AR or AD) (270680;270750)	McNamara et al. (1975); Abdallat et al. (1980); Lison et al. (1981); Stewart et al. (1981); Bahemuka and Brown (1982); Powell et al. (1983); Mukamel et al. (1985); Dyck et al. (1988).
Troyer syndrome (with short stature, mental retardation, amyotrophy and athetosis). (AR) (275900)	Cross and McKusick (1967a); Neuhäuser et al. (1976).

AD, autosomal dominant; AR, autosomal recessive; XL, sex-linked.
The numbers between brackets correspond with the numbers in McKusick (1990).

Also some familial metabolic disorders should be taken into consideration as explanation for the paraparesis, e.g. mannosidosis, a glycoprotein lysosomal storage disease caused by deficiency of α-D-mannosidase (Mitchell et al. 1981; Kawai et al. 1985), clinically characterized by coarse facies, dementia, deafness, spastic paraparesis and hepatomegaly; homocarnosinosis, neurologically characterized by spastic paraparesis, mental deterioration and retinal pigmentation, and biochemically by an increased CSF concentration of homocarnosine (a brain specific dipeptide of GABA), and a reduced activity of the catabolizing enzyme homocarnosinase (Gjessing and Sjaastad 1974; Sjaastad et al. 1976, 1977; Perry et al. 1979); and Nasu-Hakola disease, a membranous lipodystrophy featuring a spastic paraparesis, dementia and peripheral neuropathy (Kitajima et al. 1989). Finally, adrenoleukodystrophy, an X-linked hereditary disorder comprising spastic paraparesis, mild distal polyneuropathy, hypogonadism and adrenal insufficiency due to accumulation of saturated, unbranched very long-chain fatty acids, and the spinal form of adrenoleukodystrophy, namely adrenomyeloneuropathy, an autosomal recessive disorder, exhibiting the same signs as the former, should be mentioned (O'Neill et al. 1985; Toifl et al. 1981; H. W. Moser, this volume). In this context, the report by Gutmann et al. (1990) should be noted also, in which 5 black brothers are described presenting with a spastic paraparesis, mental retardation, dysarthria, ataxia and tremor. Clinical, evoked potentials and MRI studies suggested white matter disease.

GENETIC ASPECTS

Mode of inheritance

In spite of the remark on the father of the Gaum brothers (Strümpell 1880) hereditary spastic paraparesis was initially thought to be 'familial' (i.e. autosomal recessive). In 1893, Strümpell described a patient ('Polster') with a pure spastic paraparesis, whose brother, father, 2 paternal uncles, as well as his paternal grandfather, displayed the same spastic gait. His 2 sisters and his mother were said to be completely healthy. Subsequent reports on the 'pure' form of the disorder have provided more evidence of an autosomal dominant mode of inheritance. However, autosomal recessive and sex-linked modes of inheritance have also been reported in the 'pure' form, but are more frequently encountered in the complicated forms, as is shown in Table 3.

In the following we will discuss the mode of inheritance in 'pure' Strümpell-Lorrain disease. A summary is given in Table 4, for easy reference.

Autosomal dominant

As shown in Table 4 this is the most frequent mode. It is characterized by occurrence of the disorder in every generation, and male to male and female inheritance. As a rule males and females are equally affected.

Dominant pedigrees of hereditary spastic paraparesis sometimes show reduced (or even non-) penetrance. An example of this phenomenon was reported by Bone et al. (1976). They described a

TABLE 4

Autosomal dominant pure pedigrees

Spiller (1902); Specht (1925); Farago (1947); Philip (1949); Bickerstaff (1950); Schwarz and Liu (1956); Aagenaes (1959); Carte Jr. (1963); Roe (1963); Fontaine et al. (1968); Thurmon and Walker (1971); Behan and Maia (1974); Marchau et al. (1974); Bone et al. (1976); Holmes and Shaywitz (1977); Kramer (1977); McLeod et al. (1977); Sack et al. (1978); Buge et al. (1979); Marchau and De Keyser (1979); Happel et al. (1980); Opjordsmoen and Nyberg-Hansen (1980); Burdick (1981); Harding (1981); Livingstone (1983); Boustany et al. (1985); Werdelin (1986); Boustany et al. (1987); Cooley et al. (1990a, 1990b); Scheltens et al. (1990).

Autosomal recessive pure pedigrees

Jones (1907); Bell and Carmichael (1939); Freund (1949); Bruins and Simons (1958); Ozsváth (1968); Meyer and Hopf (1971); Vernea and Symington (1977); McLeod et al. (1977); Rothschild et al. (1979); Harding (1981); De Coo et al. (1982); Werdelin (1986).

Sex-linked spastic paraparesis (pure?)

Johnston and McKusick (1962); Ford (1973); Raggio et al. (1973); Zatz et al. (1976); Keppen et al. (1987).

monozygotic twin pair, the offspring of an affected father, of which only one member was affected, for more than 10 years. However, the authors' conclusion of 'non-penetrance' might have been premature, since the age of the unaffected twin sister was 33 at the time of examination, and clinical symptoms can still develop at a later age. Harding (1981) distinguished 2 types of dominant 'pure' spastic paraparesis: type I, with an age of onset below 35 years, and type II with onset usually over 35 years. This phenomenon is sometimes called 'age related penetrance', and must be taken into account in genetic counseling. For instance a patient in a family with type I starts with a risk of 1 in 2, which becomes 1 in 4.6 at the age of 25 and diminishes to 1 in 11 at the age of 45 (Harding 1981; Baraitser 1985).

It should be stressed, however, that the age of onset bears only a tenuous relationship to the onset of symptoms, in that many of those with an early onset do not develop substantial mobility

problems until late in life. In this fashion, patients in pedigrees are said not to be affected, but on examination they may show mild spasticity and bilateral extensor responses (Baraitser 1985).

Autosomal recessive

Very few pure cases of hereditary spastic paraparesis with autosomal recessive inheritance have been reported. It is very likely, for reasons explained above, that subsequent reports of recessive pure hereditary spastic paraparesis, in fact describe unrecognised dominant cases. Parents and/or children have rarely been examined, and only stated to be unaffected by hearsay (Harding 1981).

In recessive inheritance, both parents, clinically unaffected, produce affected offspring, with male to female ratio of 1:1. As a rule only one generation is affected and very often there is consanguinity.

Jones (1907) reported pure hereditary spastic paraparesis in a family that consisted of 9 children from 2 unrelated parents. Only the boys (8) were affected, ranging in age from 1 to 17 years. Because of the suspicion of hereditary transmission through the females, he carefully examined the relatives of the mother, but could not find any affected member. Bell and Carmichael (1939) suggested that 49 out of 74 surveyed pedigrees might exhibit recessive inheritance, but the relatively low rate of first cousin marriages amongst the parents of the recessive patients would be in favour of dominant inheritance.

In 1958 Bruins and Simons reported a family with 179 members in 4 generations. They established the presence of hereditary spastic paraparesis in 15 members. Because of this small number and the presence of consanguinous marriages they assumed an autosomal recessive trait. It must be added, however, that some of the described cases had amyotrophy as well.

More pedigrees with recessive inheritance were reported by Ozsváth (1968), although most of them were complicated forms. Vernea and Symington (1977) described 2 families with 5 affected members. Onset was in middle or late life and affection was asymmetric in all members. The parents of the patients in one family had died in their 60s and were said to be unaffected.

Harding (1981) in her survey of 22 families found autosomal recessive inheritance in 5 members of 3 families. The clinical features were similar to those of the dominant cases. De Coo et al. (1982) reported 2 brothers with pure spastic paraparesis, with clinical onset in the first decade. Neurological examination of the parents revealed no abnormalities, and the rest of the family (54 members) was said not to be affected.

X-linked

The predominant affection of male patients, as already suggested by Strümpell (1880, 1893), has led to the presumption of sex linkage. In 1941 Haldane studied the inheritance of spastic paraplegia and concluded that the disorder was inherited in an autosomal dominant way, but a small fraction of all cases was probably inherited in a partially sex-linked recessive mode. Another study in which autosomal dominant inheritance with partial sex-linkage was noted, was the one by Landau and Gitt (1951), who studied 7 generations of a family having 238 members. In their case histories, however, cerebellar, lower motor neuron and extrapyramidal lesions were admixed, more an argument in favour of the possibility of coexistence of hereditary ataxia and hereditary spastic paraplegia in the same family, than an argument in favour of sex-linked hereditary spastic paraparesis. Later Johnston and McKusick (1962) described sex-linked paraparesis in 17 males in 5 generations. The disease began as spastic paraparesis, but over the years the patients developed signs in the upper limbs, and evidence of brain stem and cognitive dysfunction. An often cited study of Baar and Gabriel (1966) showing sex-linked spastic paraplegia in 13 males of a family of 54 members in 5 generations, is very poorly detailed with respect to the clinical histories and examination of family members. All affected members were born crippled, and suffered from a progressive tetraplegia and bulbar paralysis. Besides that, they were severely mentally retarded and showed several other associated features, including ataxia, nystagmus and microcephaly.

The pedigree of Thurmon et al. (1971), earlier described by Johnston and McKusick (1962), is X-linked, but is complicated with optic atrophy and amyotrophy. Raggio et al. (1973) reported X-linked spastic paraplegia but, surprisingly, their pedigree showed male to male transmission.

Ford (1973) described a 5-year-old boy who had been never able to walk, and who developed paraplegia in extension, followed by paraplegia in flexion at the age of 7 years. Later in the course of the disease he developed generalized muscle atrophy. Mental capacities were definitely below par. The patient's uncle was examined at the age of 20 years, and showed paraplegia, mild atrophy of the lower limbs, and absent ankle jerks and plantar responses. Among the siblings of the maternal great-grandmother, 4 brothers were affected, and 3 sisters were normal. The first sister gave birth to 2 afflicted sons, the second sisters (the great-grandmother of the patient) had 2 affected sons and 1 healthy daughter, the third sister had 5 sons, 4 of them affected. The healthy grandmother gave birth to 9 children, 4 of these were sons and all were affected. The 5 daughters, siblings of the patient's mother, were healthy and had, in all, 13 children, all normal except the patient. Clearly the mode of inheritance in this family was X-linked, but the severity of symptoms of the patient and the clinical signs observed in the uncle make a diagnosis of pure hereditary spastic paraparesis somewhat unlikely.

A family of 142 members with 24 affected males was reported by Zatz et al. (1976). The case reports of 12 affected members showed the pure form of the disease, although plantar responses are reported normal in 11, and clinical information regarding the other 12 is withheld.

In 1987 Keppen et al. described a large family having 12 males affected. Intelligence was normal in all cases, but some displayed atrophy of the lower limbs; abnormal plantar responses were absent in 4 patients.

Linkage studies

In search of the gene responsible for pure autosomal dominant hereditary spastic paraparesis, linkage analysis is a useful tool. Thus far these studies have failed to identify the chromosome on which this gene is located. Pedersen et al. (1980) found no linkage to HLA antigens in 2 autosomal recessive and 2 autosomal dominant familial spastic paraplegia families. Recently Kolodny et al. (1989) reported their genetic studies on 33 affected members in a family, characterized by late onset and slow progression. No statistically significant linkages could be found. However, lod scores were positive with loci of GC (vitamin D binding globulin) located on chromosome 4 and Rh located on chromosome 1.

ELECTROPHYSIOLOGICAL STUDIES

Neurophysiological studies on hereditary spastic paraparesis are scarcely reported, and those few reports that have been documented, often lack a uniformity with regard to recording techniques, criteria of abnormality, clinical criteria ('pure' vs. 'complicated'), and severity and duration of the disease.

Motor and sensory conduction velocities were determined in the median, ulnar and lateral popliteal nerves in 10 patients from 3 families with type I hereditary spastic paraparesis, of whom 2 showed an autosomal dominant inheritance and probably recessive in the third (McLeod et al. 1977). No abnormalities were detected, in consistency with the normal anatomy of posterior and anterior roots and anterior horn cells as well as peripheral nerves in hereditary spastic paraparesis. In a poorly detailed short report, however, Schady and Sheard (1989) reported 19 patients from 10 families with hereditary spastic paraparesis, of whom 6 (including 4 with a sensory polyneuropathy and one with distal amyotrophy, therefore not 'pure') had abnormal nerve conduction studies. The same authors (Schady and Sheard 1990) studied 23 patients from 14 kinships with hereditary spastic paraparesis, one of which also had upper limb amyotrophy, one had perceptive deafness, sensory loss and wasting of the extensor digitorum brevis muscle and one had optic atrophy. Four patients, members of the same family, with the pure form had reduced or absent sensory action potentials in arms and legs. Sensory conduction velocity, however, was normal. Motor conduction abnor-

malities were found in 2 of these patients. The amplitude of cortical SSEPs upon tibial nerve stimulation, recorded in 10 patients, was significantly reduced, but the latency of the P40 peak was normal in all cases. Thermal threshold in the hand was increased in 8 patients and in the foot in 17 patients. Heat pain thresholds were normal. Vibration thresholds were elevated in the hand in 1 patient, but in the foot in 13 out of 20 patients.

In a VEP study, Happel et al. (1980) reported on 5 members from one family with type II autosomal dominant inherited hereditary spastic paraparesis and 6 members from one family with type II autosomal recessive hereditary spastic paraparesis, earlier described by Rothschild et al. (1979). Although none of the patients complained of visual dysfunction, 9 of 11 patients had greatly increased latencies and changes in the wave form, suggesting a process of demyelination.

On the other hand, Livingstone et al. (1981) found normal pattern VEPs in 13 cases from 9 families with autosomal dominant hereditary spastic paraparesis and abnormal responses, uni- or bilaterally delayed P2 latency or increased interocular latency difference, in 3 of 7 patients with presumably recessively inherited hereditary spastic paraparesis, although the possibility of multiple sclerosis in these cases could not be excluded.

Pedersen and Trojaborg (1981) found increased latency of the pattern-evoked VEP in only 3 of 13 patients from 7 families with hereditary spastic paraparesis (autosomal dominant inheritance in 9, and recessive in 4, although their Table 2 showed autosomal dominant inheritance in 8, and recessive in 5). Some of the recessive cases also showed nystagmus and dysarthria. They also performed BEP studies which were normal in 11 patients, abnormal in one and not performed in one. Only 2 patients showed abnormal cortical SEPs following median nerve stimulation, whereas the cortical SEP was delayed in another third following tibial nerve stimulation, and 2 had both SEP and VEP abnormalities.

Thomas et al. (1981) reported 18 patients from 12 families with types I and II hereditary spastic paraparesis (autosomal dominant in 7 and probably recessive in 4). However, 2 patients also showed disordered eye movements, one had distal amyotrophy in the limbs and one patient was included, despite a negative family history. Motor and sensory conduction studies over the forearm were all within normal limits. No cervical SEP was detectable in 6 patients, and when obtainable the amplitude was reduced but the latency of the N13 peak was normal; the latter usually reflects a pre- and postsynaptic potential arising in the dorsal horn (Gasser and Graham 1933; Austin and McCough 1955).

Sixteen members of one family with type II autosomal dominant hereditary spastic paraparesis were examined by Dimitrijevic et al. (1982), of whom 4 were affected and 2 were probably affected (no complaints, but showing pyramidal signs). They performed SEP studies and found a poor definition of the components as most marked finding, besides a delay in the peak of the P1 deflection, suggesting an impairment of spinal afferent pathway. Nerve conduction studies were normal in all patients.

A peroneal H-reflex study was performed by Owens et al. (1982) in a family with autosomal dominant hereditary spastic paraparesis. A peroneal H-reflex, as indicator of upper motor neuron disease was found in 4 patients with definite hereditary spastic paraparesis, but also in 4 of 6 patients diagnosed as probably affected. The test does not have, however, any predictive value. Uncini et al. (1987) obtained SEPs and sural nerve conduction velocity from 3 patients from 1 family with autosomal dominant hereditary spastic paraparesis (type I) and found normal conduction velocities but reduced amplitudes and poorly defined components from peroneal nerve stimulation, suggestive of an axonal degeneration of the posterior columns rather than demyelination, which was earlier suggested by Thomas et al. (1981), Thomas (1982) and Dimitrijevic et al. (1982) and which is in accordance with the neuropathological findings (Behan and Maia 1974). Rossini and Gracco (1987), found no abnormalities in BEP and SEP studies in 3 patients with hereditary spastic paraparesis.

Where Strümpell-Lorrain's disease is a disorder, in which the neuropathological abnormalities are confined to the spinal cord, one may ask oneself the question about the usefulness of VEP and BEP studies. The latter usually yield normal results, whereas the former give contradictory results. SEP studies, on the other hand, give neurophysiological evidence of — even subclinical — impaired dorsal column function.

The introduction of a new technique, the transcranial magnetic stimulation of the motor cortex, in combination with magnetic stimulation over the spine, enables us to estimate the central motor conduction time (CMCT). CMCT was normal in the arms in 8 of 10 patients with hereditary spastic paraparesis (Claus et al. 1990) studied. Unfortunately, CMCT in the legs was examined in only 4 patients, but was abnormal in all.

OTHER LABORATORY STUDIES

Serological tests should rule out curable diseases, which can give rise to a similar neurological syndrome, such as neurosyphilis and subacute combined degeneration. In some cases myelography and MRI of the spinal cord should be done to exclude a spinal tumour or cervical myelopathy. In essence all these investigations are normal or negative. The CSF yields normal results, although some authors found an increased protein content in otherwise complicated forms of hereditary spastic paraparesis (Tyrer and Sutherland 1961; Silver 1966). Kjellin and Stibler (1975), using isoelectric focusing technique, however, found a double fraction in region '5' (pH 5.9–6.1) in 3 of 4 patients. The identities of these proteins remained unknown and no clinical data were given. An increased concentration of CSF homocarnosine has been demonstrated in families with hereditary spastic paraparesis, complicated by mental retardation and retinal pigmentation (Gjessing and Sjaastad 1974; Sjaastad et al. 1976, 1977; Perry et al. 1979).

THERAPY

Since the cause of hereditary spastic paraparesis is unknown, therapy can only be symptomatic. In view of the marked spasticity, drugs like baclofen and dantrolene sodium may help considerably, provided that they are given in small doses, to prevent loss of muscle power. Recently a new spasmolytic drug, tizanidine, has become available, which seems to be very efficacious, particularly in subjects experiencing muscle weakness on baclofen (Smolenski 1984). Diazepam, or other benzodiazepines are less useful because of the sedative effect. A beneficial effect of progabide, a GABA-agonist has been mentioned by Mondrup and Pedersen (1984).

L-Threonine, which inhibits spinal reflexes by increasing glycine levels, has been studied in patients with hereditary spastic paraparesis by Kolodny et al. (1989). Penn et al. (1989), and Young (1989), have advocated the use of intrathecal baclofen for severe spasticity. However, they reserve the intrathecal administration for patients not responding to oral baclofen. It is not yet possible to judge whether this therapy will be of use in patients with hereditary spastic paraparesis.

Probably the most effective way of training motor function and of reducing spasticity is physiotherapy. It has to be emphasized, however, despite obvious rivalry between 'schools' using their specific methods, a relative superiority of one method over another has never been demonstrated. It is most reasonable to take a pragmatic approach, tailoring the treatment to the needs of an individual patient in order to achieve maximum benefit concentrating on improvement of activities of daily living.

The role of surgery is restricted to correction of severe foot deformity or shortening of the Achilles tendons.

Acknowledgement
We wish to thank Mrs. E. M. van Deventer for expert librarian assistance.

REFERENCES

AAGENAES, O.: Hereditary Spastic Paraplegia. Acta Psychiatr. Scand. 34 (1959) 489–494.

ABDALLAT, A., S. M. DAVIS, J. FARRAGE and W. I. MCDONALD: Disordered pigmentation, spastic paraparesis and peripheral neuropathy in three siblings: a new neurocutaneous syndrome. J. Neurol. Neurosurg. Psychiatry 43 (1980) 962–966.

ALLPORT, R. B.: Mental retardation and spastic parapa-resis in four of eight siblings. Lancet 2 (1971) 1089.

APPEL, L. and L. VAN BOGAERT: Etudes sur la paraplégie spasmodique familiale. III La famille Tib. Acta Neurol. Psychiatr. Belg. 51 (1951) 716–730.

APPEL, L. and L. VAN BOGAERT: Etudes sur la paraplégie spasmodique familiale. IV La famille Fev: forme très tardive. Acta Neurol. Psychiatr. Belg. 52 (1952) 129–142.

AUSTIN, G. M. and G. P. MCCOUGH: Presynaptic component of intermediary cord potential. J. Neurophysiol. 18 (1955) 441–451.

BAAR, H. S. and A. M. GABRIEL: Sex-linked spastic paraplegia. Am. J. Ment. Defic. 71 (1966) 13–18.

BAHEMUKA, M. and J. D. BROWN: Heredofamilial syndrome of spastic paraplegia, dysarthria and cutaneous lesions in five siblings. Dev. Med. Child Neurol. 24 (1982) 513–524.

BARAITSER, M.: The Genetics of Neurological Disorders. Oxford, Oxford University Press. (1985) 201–212.

BAYLEY, W. D.: Hereditary spastic paraplegia. J. Nerv. Ment. Dis. 24 (1897) 697–701.

BEHAN, W. M. H. and M. MAIA: Strumpell's familial spastic paraplegia: genetics and neuropathology. J. Neurol. Neurosurg. Psychiatry 37 (1974) 8–20.

BELL, J. and E. A. CARMICHAEL: On hereditary ataxia and spastic paraplegia. Treas. Hum. Inherit. 4 (1939) 141–281.

BERNHARDT, M.: Beitrag zur Lehre von den familiaren Erkrankungen des Zentralnervensystems. Virchows Arch. Pathol. Anat. 126 (1891) 59–71.

BICKERSTAFF, E. R.: Hereditary spastic paraplegia. J. Neurol. Neurosurg. Psychiatry 13 (1950) 134–145.

BISCHOFF, E.: Pathologisch-anatomischer Befund bei familiärer infantiler spastischer Spinalparalyse. J. Psychiatr. Neurol. 22 (1902) 109–127.

BONDUELLE, M., J. GRUNER and P. BOUYGUES: Chorée de Huntington avec paraplégie spasmodique. Deux cas familiaux. Etude anatomique. Remarque sur les relations de la surdité et des lésions de l'olive superieur. Rev. Neurol. (Paris) 88 (1953) 126–131.

BONE, I., R. H. JOHNSON and M. A. FERGUSON-SMITH: Occurrence of familial spastic paraplegia in only one of monozygous twins. J. Neurol. Neurosurg. Psychiatry 39 (1976) 1129–1133.

BOUSTANY, R. M., E. FLEISHNICK, M. MARAZITA, M. A. SPENCE, J. B. MARTIN and E. H. KOLODNY: A pedigree with the autosomal dominant form of pure familial spastic paraplegia: linkage analysis of 49 individuals. Neurology 35 (Suppl.) (1985) 310.

BOUSTANY, R. M. N., E. FLEISCHNICK, C. A. ALPER, M. L. MARAZITA, M. A. SPENCE, J. B. MARTIN and E. H. KOLODNY: The autosomal dominant form of 'pure' familial spastic paraplegia: Clinical findings and linkage analysis of a large pedigree. Neurology 37 (1987) 910–915.

BROWN, J. W. and R. COLEMAN: Hereditary spastic paraplegia with ocular and extrapyramidal signs. Bull. Los Angeles Neurol. Soc. 31 (1966) 21–34.

BRUINS, J. W. and C. H. SIMONS: Paraplegia spastica hereditaria. Ned. Tijdschr. Geneeskd. 45 (1958) 2210–2218.

BRUYN, G. W. and K. MECHELSE: The association of familial spastic paraplegia and epilepsy in one family. Psychiatr. Neurol. Neurochir. (Amst.) 65 (1962) 280–292.

BRUYN, G. W. and L. N. WENT: A sex linked heredodegenerative neurological disorder, associated with Leber's optic atrophy. Part 1. Clinical studies. J. Neurol. Sci. 1 (1964) 59–80.

BUGE, A., R. ESCOUROLLE, G. RANCUREL, F. GRAY and B. F. PERTUISET: La paraplégie spasmodique familiale de Strümpell-Lorrain. Rev. Neurol. (Paris) 135 (1979) 329–337.

BURDICK, A. B., L. A. OWENS and CH. R. PETERSON: Slowly progressive autosomal dominant spastic paraplegia with late onset, variable expression and reduced penetrance: a basis for diagnosis and counseling. Clin. Genet. 19 (1981) 1–7.

CARTE JR, E. T.: Beitrag zur Kenntnis der dominanten vererbten, spastischen Spinalparalyse. Untersuchung an 179 Probanden einer schweizerischen Sippe. Thesis. Zurich, Art. Institut Orell Fussli, Zurich (1963).

CARTLIDGE, N. E. F. and G. BONE: Sphincter involvement in hereditary spastic paraplegia. Neurology (Minneap.) 23 (1973) 1160–1163.

CAVANAGH, N. P. C., R. A. EAMES, R. J. GALVIN, E. M. BRETT and R. E. KELLY: Hereditary sensory neuropathy with spastic paraplegia. Brain 102 (1979) 79–94.

CHENEVIX-TRENCH, G., R. LESHNER and P. MAMUNES: Spastic paresis, glaucoma and mental retardation. A probable autosomal recessive syndrome? Clin. Genet. 30 (1986) 416–421.

CHOUZA, C., J. L. CAAMANO, O. DE MEDINA, J. BOGACZ, C. OEHNINGER, R. DE VIGNALE, G. DE ANDA, E. NOVOA, R. DE BELLIS, H. CARDOZO, B. CRIPINO, S. ROMERO, H. CORREA and S. FERES: Familial spastic ataxia associated with Ehlers-Danlos syndrome with platelet dysfunction. Can. J. Neurol. Sci. 11 (1984) 541–549.

CLAUS, D., H. M. WADDY, A. E. HARDING, N. M. F. MURRAY and P. K. THOMAS: Hereditary motor and sensory neuropathies and hereditary spastic paraplegia: a magnetic stimulation study. Ann. Neurol. 28 (1990) 43–49.

COOLEY, W. C., G. MELKONIAN, C. MOSES and J. B. MOESCHLER: Autosomal dominant familial spastic paraplegia: description of a large New England family and a study of management. Dev. Ment. Child Neurol. 32 (1990a) 1087–1104.

COOLEY, W. C., E. RAWNSLEY, G. MELKONIAN, C. MOSES, D. MCCANN, B. VIRGIN, J. COUGHLAN and J. B. MOESCHLER: Autosomal dominant familial spastic paraplegia: report of a large New England family. Clin. Genet. 38 (1990b) 57–68.

COSTEFF, H., N. GADOTH, N. APTER, M. PRIALNIC and H. SAVIR: A familial syndrome of infantile optic atrophy, movement disorder, and spastic paraplegia. Neurology 39 (1989) 595–597.

CROSS, H. E. and V. A. MCKUSICK: The Troyer syndrome. A recessive form of spastic paraplegia with distal muscle wasting. Arch. Neurol. 16 (1967a) 473–485.

CROSS, H. E. and V. A. MCKUSICK: The Mast syndrome. A recessively inherited form of presenile dementia with motor disturbances. Arch. Neurol. 16 (1967b) 1–13.

CRUICKSHANK, J. K., P. RUDGE, A. G. DALGLEISH, M. NEWTON, B. N. MCLEAN, R. O. BARNARD, B. E. KENDALL and D. H. MILLER: Tropical spastic paraparesis and human T cell lymphotropic virus type I in the United Kingdom. Brain 112 (1989) 1057–1090.

DANADOOST, D. M., C. H. E. JACKSON and R. D. TEASDALL: The clinical variations of hereditary spastic paraplegia in four families. Henry Ford Hosp. Med. J. 25 (1977) 3–12.

DE COO, I. F. M., F. J. M. GABREELS, W. O. RENIER, E. J. COLON and B. G. A. TER HAAR: Recessively inherited pure spastic paraplegia: case study. Clin. Neurol. Neurosurg. 84 (1982) 247–253.

DICK, A. P. and C. J. STEVENSON: Hereditary spastic paraplegia. Report of a family with associated extrapyramidal signs. Lancet 1 (1953) 921–923.

DIMITRIJEVIC, M. R., J. A. R. LENMAN, T. PREVEC and K. WHEATLY: A study of posterior column function in familial spastic paraplegia. J. Neurol. Neurosurg. Psychiatry 45 (1982) 46–49.

DYCK, P. J. and E. H. LAMBERT: Lower motor and primary sensory neuron disease with peroneal muscular atrophy. II. Neurologic, genetic and electrophysiologic findings in various neuronal degenerations. Arch. Neurol. 18 (1968) 619–625.

DYCK, P. J., W. J. LITCHY and S. GOSSELIN: Dominantly inherited spastic paraplegia and multifocal palmoplantar hyperkeratosis. Rev. Neurol. (Paris) 144 (1988) 421–424.

ERB, W.: Ueber einen wenig bekannten spinalen Symptomencomplex. Berl. Klin. Wochenschr. 26 (1875) 357–359.

FARAGO, I.: Beitrag zur Vererbung und Pathohistologie der spastischen Spinalparalyse. Mschr. Psychiatr. Neurol. 114 (1947) 161–178.

FARMER, S. G., W. T. LONGSTRETH, R. E. KALINA and A. B. TODOROV: Fleck retina in Kjellin's syndrome. Am. J. Ophthalmol. 99 (1985) 45–50.

FERGUSON, F. R. and M. CRITCHLEY: A clinical study of an heredo-familial disease resembling disseminated sclerosis. Brain 52 (1929) 203–225.

FITZSIMMONS, J. S. and P. R. GUILBERT: Spastic paraplegia associated with brachydactyly and cone shaped epiphyses. J. Med. Genet. 24 (1987) 702–705.

FONTAINE, G., B. DUBOIS, J. P. FARRIAUX and E. MAILLARD: Une observation familiale de maladie de Strümpell-Lorrain. J. Neurol. Sci. 8 (1968) 183–187.

FORD, F. R.: Diseases of the Nervous System in Infancy, Childhood and Adolescence. Springfield, Charles C. Thomas, 5th Ed. (1973) 353–356.

FRANK, G. and O. VUIA: Chorea Huntington — amyotrofische Lateralsklerose — spastische Spinalparalyse. Zur Kombination von Systemerkrankungen. Z. Neurol. 205 (1973) 207–220.

FREUND, J.: Ueber spastische Spinalparalyse (Strumpell) bei einem Bruderpaar. Dtsch. Z. Nervenheilk. 161 (1949) 337–358.

FUJITA, Y., T. FUJI, A. NISHIO, K. TUBOI, K. TSUJI and M. NAKAMURA: Familial case of May-Hegglin anomaly associated with familial spastic paraplegia. Am. J. Hematol. 35 (1990) 219 221.

GARLAND, H. G. and C. E. ASTLEY: Hereditary spastic paraplegia with amyotrophy and pes cavus. J. Neurol. Neurosurg. Psychiatry 13 (1950) 130–133.

GASSER, H. S. and H. T. GRAHAM: Potentials recorded in the spinal cord by stimulation of the dorsal roots. Am. J. Physiol. 103 (1933) 303–320.

GEE, S.: Hereditary infantile spastic paraplegia. St. Bart. Hosp. Rep. 25 (1889) 81–83.

GEMIGNANI, F.: Spinocerebellar ataxia with localized amyotrophy of the hands, sensorineural deafness and spastic paraparesis in two brothers. J. Neurogenet. 3 (1986) 125–133.

GILMAN, S. and S. HORENSTEIN: Familial amyotrophic dystonic paraplegia. Brain 87 (1964) 51–66.

GILMAN, S. and F. C. A. ROMANUL: Hereditary dystonic paraplegia with amyotrophy and mental deficiency: clinical and neuropathological characteristics. In: P. J. Vinken and G. W. Bruyn (Eds.), Handbook of Clinical Neurology, Vol. 22. Amsterdam, North-Holland Publ. Co. (1975) 445–465.

GJESSING, L. R. and O. SJAASTAD: Homocarnosinosis: a new metabolic disorder associated with spasticity and mental retardation. Lancet 2 (1974) 1028.

GUTMANN, D. H., K. H. FISCHBECK and J. KAMHOLZ: Complicated hereditary spastic paraparesis with cerebral white matter lesions. Am. J. Med. Genet. 36 (1990) 251–257.

HALDANE, J. B. S.: The partial sex-linkage of recessive spastic paraplegia. J. Genet. 41 (1941) 141–147.

HAPPEL, L. T., H. ROTHSCHILD and C. GARCIA: Visual evoked potentials in two forms of hereditary spastic paraplegia. Electroencephalogr. clin. Neurophysiol. 48 (1980) 233–236.

HARDING, A. E.: Hereditary 'pure' spastic paraplegia: a clinical and genetic study of 22 families. J. Neurol. Neurosurg. Psychiatry 44 (1981) 871–883.

HARDING, A. E.: The clinical features and classification of the late onset autosomal dominant cerebellar ataxias. A study of 11 families, including descendants of 'the Drew family of Walworth'. Brain 105 (1982) 1–28.

HARDING, A. E.: Classification of the hereditary ataxias and paraplegias. Lancet 1 (1983) 1151–1155.

HARDING, A. E.: Complicated forms of hereditary spastic paraplegia. In: A. E. Harding. The Hereditary Ataxias and Related Disorders. Edinburgh, Churchill Livingstone (1984a) 191–205.

HARDING, A. E.: Hereditary 'pure' spastic paraplegia. In: A. E. Harding (Ed.), The Hereditary Ataxias and Related Disorders. Edinburgh, Churchill Livingstone (1984b) 174–190.

HEIJBELL, J. and S. JAGELL: Spastic paraplegia, glaucoma and mental retardation in three siblings. A new genetic syndrome. Hereditas 94 (1981) 203–207.

HOLMES, G.: Family spastic paralysis associated with amyotrophy. Rev. Neurol. Psychiatry 3 (1905) 1001–1002.

HOLMES, G. L. and A. SHAYWITZ: Strümpell's pure familial spastic paraplegia: case study and review of the literature. J. Neurol. Neurosurg. Psychiatry 40 (1977) 1003–1008.

HOWLETT, W. P., G. R. BRUBAKER, N. MLINGI and H. ROSLING: Konzo, an epidemic upper motor neuron disease studied in Tanzania. Brain 113 (1990) 223–235.

JAKOB, C.: Sobre un caso de paraplegia espasmodica familiar progressiva con examen histopatologico completo. Rev. Soc. Med. Argentine 17 (1909) 665–703.

JANSEN, P. H. P., A. KEYSER and B. C. M. RAES: Hypomanic behaviour associated with familial spastic paraplegia. Eur. Arch. Psychiatry Neurol. Sci. 238 (1988) 28–30.

JOHNSTON, A. W. and V. A. MCKUSICK: A sex-linked recessive form of spastic paraplegia. Am. J. Hum. Genet. 14 (1962) 83–94.

JONES, E.: Eight cases of hereditary spastic paraplegia. Rev. Neurol. Psychiatry 5 (1907) 98–106.

KAHLSTORF, A.: Klinischer und histopathologischer Beitrag zur hereditären spastischen Spinalparalyse. Z. Ges. Neurol. Psychiatr. 159 (1937) 774–780.

KAPLAN, J. G., E. BARASCH, A. HIRSCHFELD, L. ROSS, K. EINBERG and M. GORDON: Spinal epidural lipomatosis: a serious complication of iatrogenic Cushing's syndrome. Neurology 39 (1989) 1031–1034.

KATZ, D. A., A. NASEEM, D. S. HOROUPIAN, A. D. ROTHNER and P. DAVIES: Familial multisystem atrophy with possible thalamic dementia. Neurology 34 (1984) 1213–1217.

KAWAI, H., H. NISHINO, Y. NISHIDA, K. YONEDA, Y. YOSHIDA, T. INUI, K. MASUDA and S. SAITO: Skeletal muscle pathology of mannosidosis in two siblings with spastic paraplegia. Acta Neuropathol. (Berl.) 68 (1985) 201–204.

KENWRICK, S., V. IONASESCU, G. IONASESCU, CH. SEARBY, A. KING, M. DUBOWITZ and K. E. DAVIES: Linkage studies of X-linked recessive spastic paraplegia using DNA probes. Hum. Genet. 73 (1986) 264–266.

KEPPEN, L. D., M. F. LEPPERT, P. O'CONNELL, Y. NAKAMURA, D. STAUFFER, M. LATHROP, J. LALOUEL and R. WHITE: Etiological Heterogeneity in X-linked Spastic paraplegia. Am. J. Hum. Genet. 41 (1987) 933–943.

KHALIFEH, R. R. and H. ZELLWEGER: Hereditary sensory neuropathy with spinal cord disease. Neurology 13 (1963) 405–411.

KITAJIMA, I., M. KURIYAMA, F. USUKI, S. IZUMO, M. OSAME, T. SUGANUMA, F. MURATA and K. NAGAMATSU: Nasu-Hakola disease (membranous lipodystrophy). Clinical, histopathological and biochemical studies of three cases. J. Neurol. Sci. 91 (1989) 35–52.

KJELLIN, K. G.: Familial spastic paraplegia with amyotrophy, oligophrenia and central retinal degeneration. Arch. Neurol. 1 (1959) 133–140.

KJELLIN, K. G.: Spastic paraplegia with amyotrophy, mental deficiency and retinal degeneration (Kjellin syndrome). In: P. J. Vinken and G. W. Bruyn (Eds.), Handbook of Clinical Neurology, Vol. 42. Amsterdam, North-Holland Publ. Co. (1981) 173–175.

KJELLIN, K. G. and H. STIBLER: Protein patterns of cerebral fluid in hereditary ataxias and spastic paraplegia. J. Neurol. Sci. 25 (1975) 65–74.

KOENIG, R. H. and A. J. SPIRO: Hereditary spastic paraparesis with sensory neuropathy. Dev. Med. Child Neurol. 12 (1970) 576–581.

KOLODNY, E. H.,   R. M. N. BOUSTANY,   G. A. ROULEAU, J. H. GROWDEN and J. B. MARTIN: Familial spastic paraplegia: clinical observations and genetic studies. In: Progress in Clinical and Biological Research, Vol. 306. New York, Alan R. Liss. (1989) 205–211.

KRAMER, W.: Hereditary spinal spastic paraplegia (Strümpell-Lorrain's disease). Neuropathol. Appl. Neurobiol. 8 (1977) 488–489.

KÜHN, H.: Klinische Beitrage zur Kenntniss der hereditären und familiären spastischen Spinalparalyse. Dtsch. Z. Nervenheilk. 22 (1902) 132–152.

KURODA, S., Y. KAZAHAYA, S. OTSUKI and S. TAKAHASHI: Familial spastic paraplegia with epilepsy. Acta Med. Okayama 39 (1985) 113–117.

LANDAU, W. M. and J. J. GITT: Hereditary spastic paraplegia and hereditary ataxia. A family demonstrating a variety of phenotypic manifestations. Arch. Neurol. Psychiatry (Chic.) 66 (1951) 346–354.

LISON, M., B. KORNBRUT, A. FEINSTEIN, Y. HISS, H. BOICHIS and M. GOODMAN: Progressive spastic paraparesis, vitiligo, premature graying and distinct facial appearance: a new genetic syndrome in 3 sibs. Am. J. Med. Genet. 9 (1981) 331–337.

LIVINGSTONE, I. R., F. L. MASTAGLIA, R. EDIS and J. W. HOWE: Visual involvement in Friedreich's ataxia and hereditary spastic ataxia. A clinical and visual evoked response study. Arch. Neurol. 38 (1981) 75–79.

LIVINGSTONE, I. R. and D. F. ROBERTS: Hereditary spastic paraplegia: a clinical and genetic study of cases in the North-East of England. J. Genet. Hum. 31 (1983) 295–305.

LOBSTEIN, J.: Over den familiairen vorm der spastische diplegie. Psych. Neurol. Bladen 27 (1923) 52–71.

LORRAIN, M.: Contribution a l'étude de la paraplégie spasmodique familiale. Thesis. Steinheil, Paris (1898).

LOUIS-BAR, D. and G. PIROT: Sur une paraplégie spasmodique avec dégenérescence maculaire chez deux frères. Ophthalmologica 109 (1945) 1–43.

MACRAE, W., J. STIEFFEL and A. B. TODOROV: Recessive familial spastic paraplegia with retinal degeneration. Acta. Genet. Med. Gemellol. (Roma) 23 (1973) 249–252.

MAHLOUDJI, M.: Hereditary spastic ataxia simulating disseminated sclerosis. J. Neurol. Neurosurg. Psychiatry 26 (1963) 511–513.

MAHLOUDJI, M. and P. O. CHUKE: Familial spastic paraplegia with retinal degeneration. John Hopkins Med. J. 122 (1968) 142–144.

MARCHAU, M. M. B. and G. DE KEYSER: Familie onderzoek bij de ziekte van Strümpell-Lorrain. Tijdschr. Geneeskd. 35 (1979) 399–403.

MARCHAU, M. M. B., F. MORTIER and J. ROMBOUTS: De conclusie van een familie onderzoek: de ziekte van

Strümpell-Lorrain, een dominant autosomaal ziektebeeld. Tijdschr. Geneeskd. 30 (1974) 707–713.

MARSHALL, J.: Spastic paraplegia of middle age. Lancet 1 (1955) 643–646.

MCKUSICK, V. A., C. A. FRANCOMANO and S. E. ANTONARAKIS: Mendelian Inheritance of Man: Catalogs of Autosomal Dominant, Autosomal Recessive, and X-Linked Phenotypes, 9th Ed. Baltimore, The Johns Hopkins University Press (1990).

MCLEOD, J. G., J. A. MORGAN and C. REYE: Electrophysiological studies in familial spastic paraplegia. J. Neurol. Neurosurg. Psychiatry 40 (1977) 611–615.

MCNAMARA, J. O., J. R. CURRAN and H. H. ITABASHI: Congenital ichthyosis with spastic paraplegia of adult onset. Arch. Neurol. 32 (1975) 699–701.

MEYER, D. W. and H. C. HOPF: Beobachtungen über die adulte Form der recessiv erblichen spastischen Spinalparalyse. Z. Neurol. 199 (1971) 256–258.

MITCHELL, M. L., R. P. ERICKSON, D. SCHMID, V. HIEBER, A. K. POZNANSKI and S. P. HICKS: Mannosidosis: two brothers with different degrees of disease severity. Clin. Genet. 20 (1981) 191–202.

MONDRUP, K. and E. PEDERSEN: The clinical effect of the GABA-agonist, progabide, on spasticity. Acta Neurol. Scand. 69 (1984) 200–206.

MUKAMEL, M., R. WEITZ, A. METZKER and I. VARSANO: Spastic paraparesis, mental retardation, and cutaneous pigmentation disorder. A new syndrome. Am. J. Dis. Child. 139 (1985) 1090–1092.

NEUHÄUSER, G., C. WIFFLER and J. M. OPITZ: Familial spastic paraplegia with distal muscle wasting in the Old Order Amish; atypical Troyer syndrome or 'new' syndrome. Clin. Genet. 9 (1976) 315–323.

NEWMARK, L.: A contribution to the study of the family form of spastic paraplegia. Am. J. Med. Sci. 105 (1893) 432–440.

NEWMARK, L.: Ueber die familiäre spastische Paraplegie. Dtsch. Z. Nervenheilk. 27 (1904) 1–23.

NEWMARK, L.: Pathologisch-anatomischer Befund in einem weiteren Falle von familiärer spastischer Paraplegie. Dtsch. Z. Nervenheilk. 31 (1906) 224–230.

NEWMARK, L.: Klinischer Bericht über den siebenten Fall von spastischer Paraplegie in einer Familie und Ergebnis der dritten Autopsie aus derselben Familie. Dtsch. Z. Nervenheilk. 42 (1911) 419–431.

O'NEILL, B. P., J. W. SWANSON, F. R. BROWN, J. W. GRIFFIN and H. W. MOSER: Familial spastic paraparesis: an adrenoleukodystrophy phenotype? Neurology 35 (1985) 1233–1235.

OPJORDSMOEN, S. and R. NYBERG-HANSEN: Hereditary spastic paraplegia with neurogenic bladder disturbances and syndactylia. Acta Neurol. Scand 61 (1980) 35–41.

ORMEROD, J. A.: An unusual form of family paralysis. Lancet 1 (1904) 17–18.

ÖSTERREICHER, W.: Heredofamiliäres Syndrom. Morbus Parkinson juveniles mit spastische Spinalparalyse bei Inzucht in einer psychopathischen Familie. Med. Klin. 32 (1936) 1494–1496.

OWENS, L. A., CH. R. PETERSON and A. B. BURDICK: Familial

spastic paraplegia: a clinical and electrodiagnostic evaluation. Arch. Phys. Med. Rehabil. 63 (1982) 357–361.

OZSVÁTH, K.: Paralysis spinalis spastica familiaris. Dtsch. Z. Nervenheilk. 193 (1968) 287–323.

PAGÈS, M. and A.-M. PAGÈS: Leber's disease with spastic paraplegia and peripheral neuropathy. Eur. Neurol. 22 (1983) 181–185.

PASKIND, H. A. and T. T. STONE: Family spastic paralysis: report of three cases in one family and observations at necropsy. Arch. Neurol. Psychiatry (Chic.) 30 (1933) 481–500.

PEDERSEN, L., P. PLATZ, L. U. LAMM and J. DISSING: A linkage study of hereditary ataxia and related disorders. Hum. Genet. 54 (1980) 371–383.

PEDERSEN, L. and W. TROJABORG: Visual, auditory and somatosensory pathway involvement in hereditary cerebellar ataxia, Friedreich's ataxia and familial spastic paraplegia. Electroencephalogr. clin. Neurophysiol. 52 (1981) 283–297.

PENN, R. D., S. M. SAVOY, D. CORCOS, M. LATASH, G. GOTTLIEB, B. PARKE and J. S. KROIN: Intrathecal baclofen for severe spinal spasticity. N. Engl. J. Med. 320 (1989) 1517–1521.

PERRY, T. L., S. J. KISH, O. SJAASTAD, L. R. GJESSING, R. NESBAKKEN, H. SCHRADER and A. LOKEN: Homocarnosinosis: increased content of homocarnosine and deficiency of homocarnosinase in brain. J. Neurochem. 32 (1979) 1637–1640.

PHILIP, E.: Hereditary spastic paraplegia: report of six cases in one family. NZ. Med. J. 48 (1949) 22–25.

PIERIDES, A. M., G. HOLTI, A. L. CROMBIE, D. F. ROBERTS, S. E. GARDINER, A. COLLING and J. ANDERSON: Study on a family with Anderson-Fabry's disease and associated familial spastic paraplegia. J. Med. Gen. 13 (1976) 455–461.

POWELL, F. C., P. Y. VENECIE, H. GORDON and R. K. WINKELMANN: Keratoderma and spastic paralysis. Br. J. Dermatol. 109 (1983) 589–596.

POWER, CH., B. G. WEINSHENKER, G. A. DEKABAN, G. C. EBERS, C. S. FRANCIS and G. P. A. RICE: HTLV-I associated myelopathy in Canada. Can. J. Neurol. Sci. 16 (1989) 330–335.

PRICE, G. E.: Familial lateral sclerosis. J. Nerv. Ment. Dis. 90 (1939) 51–55.

RAGGIO, J. F., T. F. THURMON and E. ANDERSON: X-linked hereditary spastic paraplegia. J. LA State. Med. Soc. 125 (1973) 4–7.

RAMESH, V., D. GARDNER-MEDWIN and I. COLQUHOUN: Severe segmental narrowing of the spinal cord: an unusual finding in congenital spastic paraparesis. Dev. Med. Child Neurol. 31 (1989) 670–681.

RAPHAELSON, M. I., J. C. STEVENS, J. ELDERS, F. COMITE and W. H. THEODORE: Familial spastic paraplegia, mental retardation, and precocious puberty. Arch. Neurol. 40 (1983) 809–810.

RAYMOND, F. and F. ROSE: Autopsie d'une malade atteinte de paraplégie spastique familiale. Rev. Neurol. (Paris) 17 (1909) 781–785.

RAYMOND, F. and A. SOUQUES: Paraplégie spasmodique familiale. Sem. Med. (Paris) 8 (1896) 315.

RECHTMAN, A. M. and B. J. ALPERS: Hereditary spastic paraplegia: occurrence of the condition in five generations with presentation of two cases. Arch. Neurol. Psychiatry (Chic.) 32 (1934) 248–250.

REFSUM, S. and S. A. SKILLICORN: Amyotrophic familial spastic paraplegia. Neurology (Minneap.) 4 (1954) 40–47.

RHEIN, J. H. W.: Family spastic paralysis. J. Nerv. Ment. Dis. 44 (1916) 115–144.

ROE, P. F.: Hereditary spastic paraplegia. J. Neurol. Neurosurg. Psychiatry 26 (1963) 516–519.

ROSSINI, P. M. and J. B. CRACCO: Somatosensory and brainstem auditory evoked potentials in neurodegenerative system disorders. Eur. Neurol. 26 (1987) 176–188.

ROTHNER, A. D., F. YAHR and M. D. YAHR: Familial spastic paraparesis, optic atrophy, and dementia. NY State J. Med. 76 (1976) 756–758.

ROTHSCHILD, H., L. HAPPEL, D. RAMPP and E. HACKETT: Autosomal recessive spastic paraplegia: evidence for demyelination. Clin. Genet. 15 (1979) 356–360.

SACK, JR G. H., C. A. HUETHER and N. GARG: Familial spastic paraplegia. Clinical and pathological studies in large kindred. Johns Hopkins Med. J. 143 (1978) 117–121.

SALAZAR-GRUESO, E. F., T. J. HOLZER, R. A. GUTIERREZ, J. M. CASEY, S. M. DESAI, S. G. DEVARE, G. DAWSON and R. P. ROOS: Familial spastic paraparesis syndrome associated with HTLV-1 infection. N. Engl. J. Med. 323 (1990) 732–737.

SCHADY, W. and A. SHEARD: Sensory abnormalities in hereditary spastic paraplegia. J. Neurol. Neurosurg. Psychiatry 52 (1989) 418.

SCHADY, W. and A. SHEARD: A quantitative study of sensory function in hereditary spastic paraplegia. Brain 113 (1990) 709–720.

SCHELTENS, PH., R. P. M. BRUYN and G. J. HAZENBERG: A Dutch family with autosomal dominant pure spastic paraparesis (Strümpell-Lorrain). Acta Neurol. Scand. 82 (1990) 169–173.

SCHOLTZ, C. L. and M. SWASH: Cerebellar degeneration in dominantly inherited spastic paraplegia. J. Neurol. Neurosurg. Psychiatry 48 (1985) 145–149.

SCHWARZ, G. A.: Hereditary (familial) spastic paraplegia. Arch. Neurol. Psychiatry (Chic.) 68 (1952) 655–682.

SCHWARZ, G. A. and C. N. LIU: Hereditary (familial) spastic paraplegia. Arch. Neurol. Psychiatry (Chic.) 75 (1956) 144–162.

SEELIGMÜLLER, A.: Einige seltenere Formen von Affectionen des Rückenmarks. Dtsch. Med. Wochenschr. 2 (1876) 185–186; 197–198.

SERENA, M., N. RIZZUTO, G. MORETTO and G. ARRIGONI: Familial spastic paraplegia with peroneal amyotrophy. A family with hypersensitivity to pyrexia. Ital. J. Neurol. Sci. 11 (1990) 583–588.

SILVER, J. R.: Familial spastic paraplegia with amyotrophy of the hands. Ann. Hum. Genet. 30 (1966a) 69–75.

SILVER, J. R.: Familial spastic paraplegia with amyotrophy of the hands. J. Neurol. Neurosurg. Psychiatry 29 (1966b) 135–144.

SJAASTAD, O., J. BERSTAD, P. GJESDAHL and L. GJESSING: Homocarnosinosis 2. A familial metabolic disorder associated with spastic paraplegia, progressive mental deficiency and retinal pigmentation. Acta Neurol. Scand. 53 (1976) 275–290.

SJAASTAD, O., L. GJESSING, J. R. BERSTAD and P. GJESDAHL: Homocarnosinosis 3. Spinal fluid amino acids in familial spastic paraplegia. Acta Neurol. Scand. 55 (1977) 158–162.

SKRE, H.: Hereditary spastic paraplegia in Western Norway. Clin. Genet. 6 (1974) 165–183.

SMOLENSKI, C.: Vergleichsuntersuchung von Tizanidin und Baclofen bei der Behandlung der Spastizität. In: B. Conrad, R. Benecke and H. J. Bauer (Eds.). Die Klinische Wertung der Spastizitat. Stuttgart, Schattauer Press. (1984) 101–110.

SPECHT, R.: Ein Beitrag zur Lehre von der hereditären spastischen Spinalparalyse. Z. Ges. Neurol. Psychiatr. 99 (1925) 32–42.

SPILLER, W. G.: Fourteen cases of spastic spinal paralysis occurring in one family. Phil. Med. J. (1902) 1129–1131.

STEWART, M., G. TUNELL and A. EHLE: Familial spastic paraplegia, peroneal neuropathy, and crural hypopigmentation: a new neurocutaneous syndrome. Neurology 31 (1981) 754–757.

STRÜMPELL, A.: Beitrage zur Pathologie des Rückenmarks. Arch. Psychiatr. Nervenkr. 10 (1880) 676–717.

STRÜMPELL, A.: Ueber eine bestimmte Form der primären combinierten Systemerkrankung des Rückenmarks. Arch. Psychiatr. Nervenkr. 17 (1886) 217–238.

STRÜMPELL, A.: Ueber die hereditäre spastische Spinalparalyse. Dtsch. Z. Nervenheilk. 4 (1893) 173–188.

STRÜMPELL, A.: Die primäre Seitenstrangsklerose (spastische Spinalparalyse). Dtsch. Z. Nervenheilk. 27 (1904) 291–339.

SUTHERLAND, J. M.: Familial spastic paraplegia. Its relation to mental and cardiac abnormalities. Lancet 1 (1957) 169–170.

SUTHERLAND, J. M.: Familial spastic paraplegia. In: P. J. Vinken and G. W. Bruyn (Eds.), Handbook of Clinical Neurology, Vol. 22. Amsterdam, North-Holland Publ. Co. (1975) 421–431.

THOMAS, P. K.: Selective vulnerability of the centrifugal and centripetal axons of primary sensory neurons. Muscle Nerve 5 (1982) 117–121.

THOMAS, P. K., J. G. R. JEFFERYS, I. S. SMITH and D. LOULAKIS: Spinal somatosensory evoked potentials in hereditary spastic paraplegia. J. Neurol. Neurosurg. Psychiatry 44 (1981) 243–246.

THURMON, T. F. and B. A. WALKER: Two distinct types of autosomal dominant spastic paraplegia. Birth Defects 7 (1971) 216–218.

THURMON, T. F., B. A. WALKER, C. H. I. SCOTT and M. H. ABBOTT: Two kindreds with a sex linked recessive form of spastic paraplegia. Birth Defects 7 (1971) 219–221.

TYRER, J. H. and J. M. SUTHERLAND: The primary spinocerebellar atrophies and their associated defects, with a study of the foot deformity. Brain 84 (1961) 289–300.

TOIFL, K., B. MAMOLI and F. WALDHAUSER: A combination of spastic paraparesis, polyneuropathy and adrenocortical insufficiency. A childhood form of adrenomyeloneuropathy? J. Neurol. 225 (1981) 47–55.

TREDICI, G. and G. MINOLI: Peripheral nerve involvement in familial spastic paraplegia. Arch. Neurol. 36 (1979) 236–239.

TUCK, R. R., B. P. O'NEILL, H. GHARIB and D. W. MULDER: Familial spastic paraplegia with Kallmann's syndrome. J. Neurol. Neurosurg. Psychiatry 46 (1983) 671–674.

UNCINI, A., M. TREVISO, M. BASCIANI and D. GAMBI: Strümpell's familial spastic paraplegia: an electrophysiological demonstration of selective central distal axonopathy. Electroencephalogr. clin. Neurophysiol. 66 (1987) 132–136.

VAN GENT, E. M., R. A. HOOGLAND and F. G. I. JENNEKENS: Distal amyotrophy of predominantly the upper limbs with pyramidal features in a large kinship. J. Neurol. Neurosurg. Psychiatry 48 (1985) 266–269.

VERNEA, J. and G. R. SYMINGTON: The late form of pure familial spastic paraplegia. Clin. Exp. Neurol. 14 (1977) 37–41.

VON KRAFFT-EBING, R.: Ueber infantile familiäre spastische Spinalparalyse. Dtsch. Z. Nervenheilk. 17 (1900) 87–98.

WELLS, C. R. AND J. JANKOVIC: Familial spastic paraparesis and deafness. A new X-linked neurodegenerative disorder. Arch. Neurol. 43 (1986) 943–946.

WERDELIN, L.: Hereditary ataxias. Occurrence and clinical features. Acta Neurol. Scand. 73 (Suppl 106) (1986) 124 pp.

WESSEL, K., CH. KESSLER, A. ROSENGART and D. KOMPF: Myotonia congenita mit familiärer spastische Paraparese. Nervenarzt 59 (1988) 675–678.

WILSON, S. A. K.: Hereditary (family) spastic paraplegia or diplegia. In: A. N. Bruce (Ed.), Neurology. London, Edward Arnold & Co. (1944), 781.

YOKOTA, T., T. MATSUNAGA, T. FERRUKAWA and H. TSUKAGOSHI: Familial spastic paraplegia with syndrome of continuous muscle fiber activity (Isaacs). No to Shinkei 41 (1989) 589–592.

YOUNG, I. D., I. F. PYE and J. R. MOORE: Manifesting heterozygosity in sex-linked spastic paraplegia? J. Neurol. Neurosurg. Psychiatry 47 (1984) 311–313.

YOUNG, R. R.: Treatment of spastic paresis. N. Engl. J. Med. 320 (1989) 1553–1555.

ZATZ, M., C. PENHA-SERRANO and P. A. OTTO: X-Linked recessive type of pure spastic paraplegia in a large pedigree: absence of detectable linkage with Xg. J. Med. Genet. 13 (1976) 217–222.

ZELLWEGER, H.: Spastic paraplegia with sensory neuropathy. In: P. J. Vinken and G. W. Bruyn (Eds.) Handbook of Clinical Neurology, Vol. 42. Amsterdam, North-Holland Publ. Co. (1981) 179–180.

Handbook of Clinical Neurology, Vol. 15 (59): Diseases of the Motor System
J.M.B.V. de Jong, editor
© Elsevier Science Publishers B.V., 1991

CHAPTER 18

Ferguson-Critchley syndrome

R. P. M. BRUYN[1] and G. W. BRUYN[2]

[1] *Department of Neurology, Oudenrijn Hospital, Utrecht, and* [2] *Department of Neurology, Academic Hospital,
State University, Leiden, The Netherlands*

Just as the inclusion of this syndrome in a volume on Motor System Diseases seemed quite natural in the seventies (Brown 1975), there is scarcely an argument to maintain a chapter on this syndrome in the present updated volume 15 years later. What appeared to be at least a singular and perhaps new syndrome in 1929, and a nosologically challenging one in 1975, today has come to be regarded as a variant of late onset heredoataxias. This conceptual swing toward a variety of olivopontocerebellar degenerations has been brought about by the data that came forward on the 'Azorean' heredoataxias as well as by Harding's work (1982).

Only a few arguments might be marshalled to review Ferguson-Critchley syndrome here: traditional respect for the princeps observation; the syndrome's exemplary history bringing out in sharp educational profile the weakness of clinical and clinicopathological identification, and, finally, the diagnostic inadequacy of clinical parameters in hereditary disease, today being replaced by precise molecular-genetic data.

THE PRINCEPS REPORT ON THE DREW FAMILY OF WALWORTH (FIG. 1)

The clinical features originally outlined by Ferguson and Critchley in 1929 and later summarised by Critchley (1981) include: (1) Age at onset between 30 and 45 years in 11, and between 20 and 30 years in 2 individuals. (2) With respect to the sex ratio, males are affected about twice as often as females. (3) Duration of disease lasts from 5 to at least 16 years. (4) Initial and progressive spastic to spastic-ataxic gait is subsequently associated with dysarthria, external ophthalmoplegia with optic atrophy, palpebral retraction (producing a fixed, staring and blank facial expression) and sometimes nystagmus, light touch and vibration hypaesthesia with dysaesthesia in the legs, urge incontinence, and conspicuous euphoria with pathological laughter but no dementia. Pyramidal dysfunction was betrayed by proximal hypertonic paraparesis, brisk muscle stretch reflexes and extensor plantar reflexes, while hypokinesia together with the immobile face with perioral tremor and tiny myoclonic twitches indicated extrapyramidal system involvement. In about half the cases, cerebellar incoordination of some degree was noticed.

Because of the 'multisystemic' phenomenology and of the non-prominent ataxia, the *auctores intellectuales* interpreted this syndrome rather as one resembling multiple sclerosis than as one resembling autosomal dominant spastic ataxia (Brown 1892), because in the latter dementia is prominent and because euphoria and the ophthalmoplegia may be conspicuous in the first.

Work-up by Harding (1982) of 4 affected

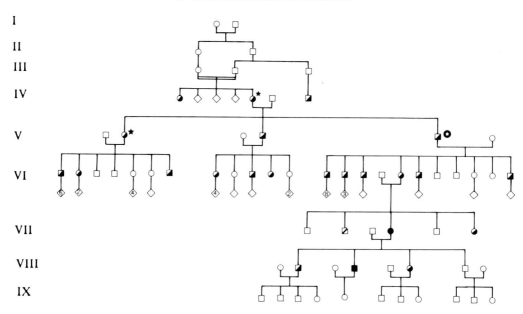

Fig. 1. The Drew family of Walworth.

persons (2 male, 2 female) in 2 subsequent generations of this family confirmed all features outlined above, with one conspicuous difference; the ataxia-component was appreciated as severe and the pyramidal component as mild, while mild distal muscle wasting indicative of lower motor neuron involvement was noted in one individual as a 'new' feature. On this basis, Harding categorised the Drew family as an example of autosomal dominant late onset cerebellar ataxia.

The 'weighing' of certain dysfunctions within a syndrome, i.e. a composite of (system) signs and symptoms, may remain a physician's matter of taste or preference until an objective test to determine the cause unequivocally replaces arbitrary clinical judgement. With respect to the Drew family, Ferguson and Critchley's preferential interpretation of the evidence resulted in a proposed resemblance of multiple sclerosis, just as Harding's resulted in a diagnosis of heredoataxia, in spite of a combination of components that militate against both: for the one, heredity, late onset, supranuclear ophthalmoplegia, obtainable abdominal reflexes and extrapyramidal symptoms, for the other the ophthalmoplegia, the prominent mental symptoms, the incontinence, to some degree the pyramidal, sensory,

and lower motor neuron signs (even Stephens et al. in 1958 pointed out the association of ataxia, amyotrophy, sensory impairment and supranuclear ophthalmoplegia), and the 2:1 sex ratio which is scarcely reconcilable with autosomal dominant Mendelian segregation. Neither of the 2 reports in 1929 and 1982 provide neuropathological, neurophysiological, CT or NMR data, and only in one case (who had syphilis) chemical data on the CSF was provided. The reader might as well weigh the evidence in favour of multisystem disease, or a new disorder. One will have to wait for hard data before a definitive diagnosis (and classification) is reached of the disease running in the Drew family.

A somewhat similar progressive disorder was reported from Iran in 1963 by Mahloudji. In a five-generation family with 18 affected persons (11 male, 7 female) dysfunction manifested between the ages 30 and 40, with initial gait impairment, followed by dysarthria, supranuclear ophthalmoplegia, moderate ataxia, pyramidal symptoms (hypertonia, brisk reflexes, extensor plantars), diminished vibration sense, all in variable degree, and with either present or absent abdominal cutaneous reflexes. No extrapyramidal symptoms were noted. Affected individuals were found in each generation, and both inside and outside a first cousin marriage (both of them affected) in the third generation. The conspicuous ocular and parkinsonian features were much less prominent in this family.

Partial autopsy material came forward in the Brown and Coleman report (1966) and in Brown's subsequent review (1975) in which a fourth family was introduced. Curiously, the affected male-female ratio inverted from 2:1 to 1:2 in the third and fourth families. Brown stipulated the 'parkinsonian' features, such as the immobile face with bulging eyes and staring gaze, tremulousness and perioral twitches, the festinating gait, and the pseudoophthalmoplegia with intact compensatory reflex gaze-movements. No clues indicative for multiple sclerosis were found in the CSF of two patients. The neuropathological findings included: (1) patchy, and incomplete demyelination in cerebral white matter without metachromasia. (2) deranged cytoarchitecture in cerebral cortex and patchy cell loss of Betz cells and third layer neurons, with PAS positive foci. (3) mild cell-degeneration in the dentate nucleus and cerebellar granular layer. (4) brainstem (olives) normal, but only one section available. (5) throughout the length of the cord, demyelination of the posterior funiculi (particularly the gracile fasciculus), the ventral and (less so) dorsal spinocerebellar tracts, mildly in the pyramidal tracts and anterior commissure; some cell loss in ventral horns and Clarke's columns.

On this not over-detailed histological basis, Brown (1975) preferred to interpret the cases known up to that date as a variant of Strümpell-Lorrain's hereditary spastic paraplegia, largely because the spinal cord changes were compatible with those described in the last-named disease.

The present authors found another five-generation pedigree, reported by Pitrangeli et al. (1981), in which a progressive disorder manifested certainly in 12 persons and possibly in another 11 (with a male/female ratio of 3:11), closely resembling the one originally described in 1929. Onset was again in the fourth decade with impaired gait and the symptoms included dysphagia, dysarthria, optic atrophy, supranuclear ophthalmoplegia, marked spastic (pyramidal) paraparesis, extrapyramidal signs such as rigidity and resting tremor, ataxia, dysaesthesias with vibration sense loss in the legs and as another marked feature muscle weakness and wasting with fasciculations of the distal limbs and bulbar province. EMG-examination confirmed the neurogenic origin by revealing denervation signs. A sural nerve biopsy uncovered the presence of segmental demyelination, and a CT scan exhibited atrophy of pons and cerebellum.

The sporadic mild distal muscle wasting noted by Harding (1982) in the Drew family was prominent in this Pitrangeli et al. pedigree, as were the spastic paretic, ocular and 'parkinsonian' components. There must have been also ventral horn cell degeneration with primary axonal dying-back neuronopathy in view of the clinical and EMG signs. Apparently, on weighing the evidence for diagnosis and in the absence of neuropathological data on the CNS, Pitrangeli et al. choose to attribute crucial value to the objective CT data and preferred not to consider the other objective procedures (EMG, nerve biopsy) of equivalent portent. They posed the diagnosis of 'atypical hereditary spinocerebellar degeneration'.

Arguments to coalesce syndromes such as the one *sub iudice* have grown in frequency and amplitude over the past 20 years. The condition reported as autosomal dominant spino-dentate degeneration by Boller and Segarra (1969, 1975), Taniguchi and Konigsmark (1971), and Pogacar et al. (1978) showed absence of dementia (except in some), normal or mildly affected inferior olives, and showed a combination of ataxic, pyramidal and sensory symptoms, some of them associated with intrinsic hand muscle wasting, nystagmus, optic atrophy, and diplopia with gaze paresis. The neuropathological findings showed that no clear distinction could be made with OPCA.

Also, the Woods-Schaumburg syndrome of nigrospinodentate degeneration essentially is an autosomal dominant spinocerebellar syndrome in a pedigree originating from the Azores islands. Ataxia, bulging eyes, prominent motor symptoms (pyramidal and extrapyramidal, and lower motor neurons) were present with prominent nuclear and some supranuclear ophthalmoplegia (Woods and Schaumburg 1972, 1975). Neuronal loss was noted in the substantia nigra, oculomotor, vestibular, pontine and dentate nuclei, spinal ventral horns and Clarke's columns, degeneration of spinocerebellar tracts, brachium pontis et conjunctivum, with intact cerebral and cerebellar cortex, inferior olives, pyramidal tracts and posterior funiculi. The involvement of the substantia nigra and the remarkable indemnity of the cerebellar cortex and inferior olives, on weighing the evidence, mitigate against the diagnosis of OPCA. However, because of the degeneration of the pontine nuclei, the Azorean origin of the pedigree, as well as the clinicopathological features that are compatible with a diagnosis of OPCA, recent reviewers of the heredoataxias were moved to wrap up the Woods-Schaumburg syndrome within the Azorean-Machado-Joseph condominium of OPCAs.

The same fate befell Striatonigral Degeneration (Adams 1968; Adams et al. 1961, 1964), today classified as a type of Multiple System Atrophy, because one of the original patients may have shown cerebellar ataxia and/or intention tremor at an early stage, and whose pontocerebellar, olivocerebellar, and ventral spinocerebellar fibres showed the pallor that is noted in OPCA, and one, who had shown urinary incontinence, was later considered to have had Shy-Drager syndrome.

In a judicious and cogent review, Adams (1986) exposed the close relationship between striatonigral degeneration, Shy-Drager syndrome and OPCA, and emphasised that multiplicity of system involvement in such cases is largely dependent on the patients' longevity, making *duration of disease a main determinant* for one system after another to become involved. This point is demonstrated *ad oculos* by the Swier family from Kampen, the Netherlands, that emigrated to the USA in 1860, and in which the increase as well as the shifting of symptoms over a period of 25 years emanates clearly from the reports by Gray and Oliver (1941), Schut (1950), Schut and Haymaker (1951), and Landis et al. (1974).

On shallower grounds of resemblance (such as ataxia, perioral twitching, bulging staring eyes, lower motor neuron deficit, (extra-) pyramidal signs, depletion of Purkinje cells or substantia nigra neurons), the heredoataxia category lumpingwise might easily accommodate also DRPLA

(Iizuka and Hirayama 1986), familial ALS (Veltema et al. 1990), the 'all-systems-atrophy' reported by Moffie (1961), the 'cerebello-pallido-luyso-nigral atrophy' reported by Morin et al. (1980), the autosomal dominant late onset condition simulating multiple sclerosis (!) and showing leukodystrophy on CTs, as reported by Eldridge et al. (1984), or even the fascinating Ferguson-Critchley-like syndrome without any ataxia at all as reported by Staal et al. (1983).

Harding (1981) perspicaciously argued that the quite unsatisfactory classification of (autosomal dominant late onset) heredoataxias on clinical, neuropathological and clinicopathological matrices, derives from the large inter- and intrafamilial variation of associated clinical/neuropathological features that may be mutually discordant, adding a second parameter to the factor time [as suggested by Koeppen et al. (1977), and eloquently stressed by Adams (1986)]. On genetic considerations, she tentatively outlined a group of kindreds with ophthalmoplegia, optic atrophy, extrapyramidal features, amyotrophy and dementia as probably genetically homogeneous, a group of cerebellar ataxia with pigmentary retinal dystrophy, a group in which ataxia, myoclonus and deafness are associated, and a group of a pure senile cerebellar syndrome. All the above would, to the present authors' understanding, imply that any classification based on the presence or absence of muscle stretch reflexes is as fatuous as the arbitrary parameter of age at onset (see e.g. Amit et al. 1986). But even the rather homogeneous clinical features of the first group, into which Ferguson-Critchley syndrome might ultimately well be proven to fit and belong to in the present writer's opinion, can be mimicked by late onset autosomal *recessive* 'multisystem disorder' or hexosaminidase A deficiency, as convincingly shown by Harding herself (Harding et al. 1984, 1987).

The arbitrary, if *bona fide*, preference for one feature such as ataxia, or spasticity, over any other from a clinical and/or neurological composite to serve as a criterion for either diagnosing or classifying a disorder, while a different selection probably would have been made, had the patient died sooner (or later), or suffered (or not) from an extraneous factor (such as alcohol abuse, or diabetes, or malnutrition, or infectious disease) that may accelerate selective neuronal death from endogenous factors (such as a gene-product or the lack of it), has contributed to the accumulation of a mass of system-atrophies, bewildering in their variety. Add to this the incompleteness of clinical or genealogical presentation of many reports, or the below-par neuropathological presentation and wrong conclusion [such as e.g. in Rosenberg et al. (1976)], and one may truly posit that optimal conditions have been set and met for chaos and confusion.

It will be clear, that enlarging the heredoataxia category by having it assimilate disorders that ultimately may be proven to be different entities, serves little purpose other than replacing one nuisance by another.

Unequivocal determination of genes and their products in the near future will rid neurologists of a quandary of which the 'Ferguson-Critchley syndrome' serves as an exemplary facet.

REFERENCES

ADAMS, R. D.: The striatonigral degenerations. In: P. J. Vinken and G. W. Bruyn (Eds), Handbook of Clinical Neurology, Vol. 6, Ch. 26. Amsterdam, North Holland Publ. Co. (1968a) 694–702.

ADAMS, R. D.: The striatonigral degenerations. In: P. J. Vinken and G. W. Bruyn (Eds), Handbook of Clinical Neurology, Vol. 49, Ch. 10. Amsterdam, Elsevier (1986b) 205–212.

ADAMS, R. D., L. VAN BOGAERT and H. VAN DER EECKEN: Dégénérescences nigrostriées et cérébello-nigrostriées. Psychiatr. Neurol. 142 (1961a) 219—259.

ADAMS, R. D., L. VAN BOGAERT and H. VAN DER EECKEN: Striatonigral degeneration. J. Neuropathol. Exp. Neurol. 23 (1964b) 384–608.

AMIT, R., G. GRANIT and Y. SHAPIRA: Familial ataxia with extreme difference in age of clinical onset. Neuropediatrics 17 (1986) 165–167.

BOLLER, F. and J. M. SEGARRA: Spinopontine degeneration. Eur. Neurol. 2 (1969a) 356–373.

BOLLER, F. and J. M. SEGARRA: Spino-pontine degeneration. In: P. J. Vinken and G. W. Bruyn, (Eds.), Handbook of Clinical Neurology, Vol. 21. Amsterdam, North Holland Publ. Co. (1975b) 389–416.

BROWN, J. W.: Hereditary spastic paraplegia with ocular and extrapyramidal symptoms (Ferguson-Critchley syndrome). In: P. J. Vinken and G. W. Bruyn (Eds.), Handbook of Clinical Neurology, Vol. 22, Ch. 18. Amsterdam, North Holland Publ. Co. (1975) 433–443.

BROWN, J. W. and R. F. COLEMAN: Hereditary spastic paraplegia with ocular and extrapyramidal signs. Bull. Los Angeles Neurol. Soc. 31 (1966) 21–34.

BROWN, S.: On hereditary ataxia, with a series of 21 cases. Brain 15 (1892) 250–282.

CRITCHLEY, M.: Ferguson-Critchley syndrome (Heredo-familial disease resembling disseminated sclerosis). In: P. J. Vinken and G. W. Bruyn (Eds.), Handbook of Clinical Neurology, Vol. 42. Amsterdam, North-Holland Publ. Co. (1981) 142–143.

ELDRIDGE, R., C. P. ANAYIOTOS, S. SCHLESINGER, D. COWEN, C. BEVER, N. PATRONAS and H. MCFARLAND: Hereditary adult-onset leukodystrophy simulating chronic progressive multiple sclerosis. N. Engl. J. Med. 311 (1984) 948–953.

FERGUSON, F. R. and MACDONALD CRITCHLEY: A clinical study of an heredofamilial disease resembling disseminated sclerosis. Brain 52 (1929) 203–225.

GRAY, R. C. and C. P. OLIVER: Marie's hereditary cerebellar ataxia (OPCA). Minn. Med. 24 (1941) 327–335.

HARDING, A. E.: 'Idiopathic' late onset cerebellar ataxia. A clinical and genetic study of 36 cases. J. Neurol. Sci. 51 (1981a) 259–271.

HARDING, A. E.: Genetic aspects of autosomal dominant late onset cerebellar ataxia. J. Med. Genet. 18 (1981b) 436–441.

HARDING, A. E.: The clinical features and classification of the late onset autosomal dominant cerebellar ataxias. A study of 11 families, including descendants of 'the Drew family of Walworth'. Brain 105 (1982) 1–28.

HARDING, A. E., J. V. DIENGDOH and A. J. LEES: Autosomal recessive late onset multisystem disorder with cerebellar cortical atrophy at necropsy: report of a family. J. Neurol. Neurosurg. Psychiatry 47 (1984) 853–856.

HARDING, A. E., E. P. YOUNG and F. SCHON: Adult onset supranuclear opthalmoplegia, cerebellar ataxia, and neurogenic proximal muscle weakness in a brother and sister: another hexosaminidase A deficiency syndrome. J. Neurol. Neurosurg. Psychiatry 50 (1987) 687–690.

IIZUKA, R. and K. HIRAYAMA: Dentatorubropallidoluysian atrophy. In: P. J. Vinken, G. W. Bruyn and H. L. Klawans (Eds.), Handbook of Clinical Neurology, Vol. 49, Ch. 23. Amsterdam, Elsevier (1986) 437–444.

KOEPPEN, A. H., M. B. HANS, D. I. SHEPHERD and P. V. BEST: Adult-onset hereditary ataxia in Scotland. Arch. Neurol. 34 (1977) 611–618.

LANDIS, D. M. D., R. N. ROSENBERG, S. C. LANDIS, L. SCHUT and W. L. NYHAN: Olivopontocerebellar degeneration. Arch. Neurol. 31 (1974) 195–207.

MAHLOUDJI, M.: Hereditary spastic ataxia simulating disseminated sclerosis. J. Neurol. Neurosurg. Psychiatry 26 (1963) 511–513.

MOFFIE, D.: Familial occurrence of neural muscle atrophy (Tooth-Marie-Charcot) combined with cerebral atrophy and parkinsonism. Psychiatr. Neurol. Neurochir. 64 (1961) 381–391.

MORIN, P., LECHEVALIER, B. and C. BIANCO: Atrophie

cérébelleuse et lésions pallido-luyso-nigriques avec corps de Lewy. Rev. Neurol. (Paris) 136 (1980) 381–390.

PITRANGELI, A., L. GESSINI, B. JANDOLO and E. OCCHIPINTI: Atypical hereditary spinocerebellar degeneration. Description of a case with family study. Ital. J. Neurol. Sci. 4 (1981) 387–390.

POGACAR, S., M. AMBLER, W. J. CONKLIN, W. A. O'NEIL and H. Y. LEE: Dominant spinopontine atrophy. Report of two additional members of family W. Arch. Neurol. 35 (1978) 156–162.

ROSENBERG, R. N., W. L. NYHAN, C. BAY and P. SHORE: Autosomal dominant striatonigral degeneration. Neurology 26 (1976) 703–714.

SCHUT, J. W.: Hereditary ataxia. Arch. Neurol. Psychiatr. 63 (1950) 535–569.

SCHUT, J. W. and W. HAYMAKER: Hereditary ataxia. J. Neuropathol. Clin. Neurol. 1 (1951) 183–213.

STAAL, A., S. Z. STEFANKO, F. G. I. JENNEKENS, L. H. PENNING DE VRIES-BOS and J. VAN GIJN: Autosomal recessive spino-olivo-cerebellar degeneration without ataxia. J. Neurol. Neurosurg. Psychiatry 46 (1983) 648–652.

STEPHENS, J., HOOVER, M. L. and J. DENST: On familial ataxia, neural amyotrophy and progressive external ophthalmoplegia. Brain 81 (1958) 556–566.

TANIGUCHI, R. and B. W. KONIGSMARK: Dominant spinopontine atrophy. Report of a family through three generations. Brain 94 (1971) 349–358.

VELTEMA, A. N., R. A. C. ROOS and G. W. BRUYN: Autosomal dominant adult amyotrophic lateral sclerosis. A six generation Dutch family. J. Neurol. Sci. 96 (1990) 93–115.

WOODS, B. T. and H. H. SCHAUMBURG: Nigrospinodentatal degeneration with nuclear ophthalmoplegia. J. Neurol. Sci. 17 (1972) 149–161.

WOODS, B. T. and H. H. SCHAUMBURG: Nigrospinodentatal degeneration with nuclear ophthalmoplegia. In: P. J. Vinken and G. W. Bruyn (Eds), Handbook of Clinical Neurology, Vol. 22, Ch. 17. Amsterdam, North Holland Publ. Co. (1975) 157–176.

Handbook of Clinical Neurology, Vol. 15 (59): Diseases of the Motor System
J.M.B.V. de Jong, editor
© Elsevier Science Publishers B.V., 1991

CHAPTER 19

Hereditary spastic paraplegia with retinal disease

CRAIG G. WELLS[1], W. T. LONGSTRETH, JR[2] and CRAIG H. SMITH[3]

[1]Department of Ophthalmology, RJ-10, University of Washington, Seattle, WA; [2]Harborview Medical Center, Division of Neurology, Seattle, WA, and [3]University of Washington, Seattle, WA, USA

Spastic paraplegia is a common feature of many neurologic conditions. Disease processes responsible include a variety of inherited and acquired conditions. Once the clinician has excluded acquired conditions with the appropriate evaluation, he or she is left with the possibility of a hereditary cause. Typically, hereditary spastic paraplegia is a mild, slowly-progressive disorder affecting mainly the lower limbs, with spasticity more prominent than paralysis. Little, if any, ataxia is present. Those patients with significant ataxia usually have vestibular and sensory abnormalities, less lower limb spasticity, and are best not considered part of the spectrum of hereditary spastic paraplegia. The earliest description (Strümpell 1880) was of autosomal dominant spastic paraplegia. Pedigrees supporting recessive inheritance were provided by Freud (1893), Jones (1907), and Johnston and McKusick (1962). Rarely, X-linked recessive inheritance is a cause. Initially it was believed that two-thirds of autosomal inheritance was recessive (Bell and Carmichael 1939). Recently, however, Harding (1981) has found most pedigrees support autosomal dominant inheritance. Many of these reports have demonstrated hereditary spastic paraplegia in association with other findings including amyotrophy (Silver 1966), extrapyramidal signs (Dick and Stevenson 1953), and sensory neuropathy (Cavanagh et al. 1979). Rowe (1963) has reviewed

findings in 215 families with hereditary spastic paraplegia.

Ophthalmic examination of a patient with hereditary spastic paraplegia may allow a more specific diagnosis to be made. When retinal findings occur in the setting of a hereditary spastic paraplegia, a number of rare conditions need to be considered, with diagnosis of these diseases requiring the combined effort of the neurologist and ophthalmologist. Associated features that are helpful in differentiating the disorders are outlined in Tables 1–4 and include the presence of mental retardation or dementia, retinal abnormalities, optic atrophy, age at presentation, neuro-imaging, electrophysiology, metabolic abnormalities, obesity, polydactyly, hypogonadism, skin changes, and dwarfism. We will also discuss briefly other disorders including those of Barnard and Scholz (1944), Behr (Landrigan et al. 1973), mitochondrial myopathy (Mullie et al. 1985), and hereditary spastic paraplegia with optic atrophy (Nyberg-Hansen and Refsum 1972).

Retinal alterations and optic atrophy are non-specific manifestations of many inherited and acquired diseases. Pagon (1988a) recently reviewed retinitis pigmentosa, the classic group of inherited disorders with retinal dystrophic changes and optic atrophy. Autosomal dominant, autosomal recessive, and X-linked recessive are

TABLE 1
General examination.

Syndrome	Age at presentation	Progressive	Other findings	Inheritance
Kjellin	20–40 yrs of age	Yes, slowly	–	Recessive
Sjögren-Larsson	Congenital	Minimal	Congenital ichthyosis convulsion; dental and osseous dysplasia; defective sweating (Gilbert et al., 1968)	Recessive
Laurence-Moon	Childhood	Yes	–	Recessive
Cockayne	Infancy	Yes	–	Recessive
Gordon-Capute-Konigsmark	Early childhood	Yes	–	Recessive
Hallervorden-Spatz	Infancy	–	Dwarfism, hearing loss, ataxia, bird-like facies	Recessive
Homocarnosinosis	Childhood to early adulthood	Yes	Microcephalic	Recessive

TABLE 2
Neurologic examination.

Syndrome	Spastic paraparesis	Mental retardation or dementia	Other neurologic findings
Kjellin	Yes	Mild	Weakness, muscle wasting
Sjögren-Larsson	Yes, diplegia or quadriplegia	Mental retardation, moderate, nonprogressive	Ataxia (rare)
Laurence-Moon	Yes	Mild mental retardation, not constant	Ataxia in some
Cockayne	Yes	Progressive mental retardation	Normal pressure hydrocephalus-ataxia
Gordon-Capute-Konigsmark	Yes, quadriplegia	Progressive, severe mental retardation	Wasting
Hallervorden-Sptaz	Yes	Progressive dementia	Basal ganglia signs
Homocarnosinosis	Yes	Yes, variable	–

TABLE 3
Ophthalmologic examination.

Syndrome	Retinal abnormality	Visual acuity	Optic atrophy
Kjellin	Yellow flecks in macula block fluorescence on angiography	20/20 to 20/100	No
Sjögren-Larsson	Glistening dots or crystals surround central yellowish deposit in 20% of patients. Late staining of central lesion on angiography.	?	No
Laurence-Moon	Atypical or typical retinitis pigmentosa	Early, severe loss of vision	Yes
Cockayne	Salt and pepper retinopathy (atypical retinitis pigmentosa)		Yes
Gordon-Capute-Konigsmark	Atypical retinitis pigmentosa		Yes
Hallervorden-Spatz	25% (Dooling et al. 1974) vascular attenuation discrete peripheral yellow flecks (Newell et al. 1979)		No
Homocarnosinosis	Macular gray pigment		

TABLE 4
Laboratory findings.

Syndrome	Electrophysiology	Pathology defect	Neuro-imaging
Kjelling	EEG, mild, diffuse, slowing,? EMG, mild polyneuropathy; ERG normal, EOG normal.	Mild atrophy on CT	
Sjögren-Larsson		Defective fatty alcohol oxidation(Rizzo et al., 1988)	–
Laurence-Moon	ERG, usually flat, rarely present, with progressive decline.	?	Cerebellar atrophy on CT (Rizzo et al., 1986)
Cockayne	–	?	–
Gordon-Capute-Konigsmark	Abnormal ERG, recordable; EEG normal (early) to diffusely slow.		MRI, characteristic basal ganglia changes
Hallervorden-Spatz	ERG, flat; EEG, normal, nerve coordination normal	Iron deposits in basal ganglia	–
Homocarnosinosis	EEG, normal; ERG, normal; EMG, normal	Elevated homocarnosine in CSF (?)	–

recognized. Inherited metabolic disorders such as gyrate atrophy (Simell and Takki 1973), abetalipoproteinemia (Bassen and Kornzweig 1950), and Refsum's disease (Refsum 1981) have retinal changes as well. Infections, such as tertiary syphilis (Folk et al. 1983), rubella (Krill 1967), and toxicity such as that associated with thioridazine (Miller et al. 1982) may produce a similar ocular appearance.

KJELLIN SYNDROME

Kjellin (1959) reported 2 families each with 2 members in the same generation affected by dementia, spastic paraplegia, and yellow retinal flecks in the macula. A similar case was seen by Schaffer (1922), who described ocular lesions in 1 of 2 brothers with spastic paraplegia. The ocular lesion was 'a pigmented focus about one quarter the diameter of the papilla, the center of which is somewhat lighter; around the focus there are some punctiform pigmentations'. Kjellin noted 'small atrophic foci and pigment displacement in the macula' that were progressive. Also noted was mental retardation, spastic paraplegia appearing at age 25, and amyotrophy associated with peripheral neuropathy and affecting mainly the small muscles of the hand. Mahloudji and Chuke (1968), MacCrae et al. (1974), and Farmer

et al. (1985) have reported additional cases. The latter reports details of 2 independent pedigrees with well documented neurologic, ocular and electrophysiologic findings.

Neurologic examination

The disorder presents in early adulthood with spastic paraparesis that progresses to paraplegia. The pronounced spasticity is far worse in the lower extremities than the upper extremities. Voluntary motor activity is ultimately lost in the lower extremities. The upper extremities may be normal in strength and coordination. Cranial nerve function, coordination testing in the upper extremities, and somatic sensation are normal.

Muscle wasting is unusual in hereditary spastic paraplegia. In Kjellin's families, amyotrophic changes noticed as fasciculations of the small muscles of the hand were present in 3 of the 4 affected individuals. Two individuals also showed atrophy in the leg muscles. The cases reported by Farmer et al. were affected with amyotrophy also, primarily in the lower extremities.

All reported patients have shown subnormal intelligence. Impairment has ranged from mild to moderately severe. Most individuals have been able to complete schooling before cognitive functions including memory and visual perceptual skills are affected.

Ophthalmologic examination

The retinal changes are unusual, constant and relatively unique. Visual acuity is normal or slightly reduced, however those patients with reduced acuity have not been aware of it, perhaps owing to their decreased cognitive function. The macula shows circular and dumbbell-shaped, yellow-white lesions surrounded by pigment arranged in a circinate pattern (Fig. 1). When observed with stereopsis these lesions appear to be at the level of the retinal pigment epithelium. The lesions may be difficult for the inexperienced observer to detect with the direct ophthalmoscope. This, in combination with the lack of symptoms, may mean that the association is underreported. The neurologist suspicious of this or the other diagnosis listed in Table 1 should request a full ophthalmological examination.

Intravenous fluorescein angiography provides information concerning the status of the retinal circulation, the underlying choroidal circulation, and the pigment-laden retinal pigment epithelium which lies between them. Fluorescein angiography (Fig. 2) shows that the yellow lesions block transmission of back-ground choroidal fluorescence centrally but transmit fluorescence around the periphery of the abnormality. The blocked fluorescence is consistent with lipofuscin accumulation in the retinal pigment epithelium, a common response of the retina to a variety of insults such as pattern dystrophy (Pautler 1985), fundus flavimaculatus (Eagle et al. 1980; Farmer 1985; Traboulski 1985), and Best's disease. The hyperfluorescence corresponds to the punctate pigmentation and depigmentation seen with the ophthalmoscope around the periphery of the yellow lesion. No leakage or staining with fluorescein dye is observed. The peripheral fundus is normal.

The most common disorder producing a similar ocular appearance would be autosomal dominant drusen (Deutman and Jansen 1970), which is easily distinguished on fluorescein angiography as the yellow lesion shows transmission hyperfluorescence. Fundus flavimaculatus is an autoso-

mal recessive ocular disorder which may appear clinically similar, but most cases will have diffuse blockage of background choroidal fluorescence (silent choroid) (Ernest and Krill 1966; Newell et al. 1972). Pattern dystrophies of the retinal pigment epithelium rarely may have similar ocular and angiographic findings, but are inherited in an autosomal dominant manner and do not have associated systemic findings (Watzke et al. 1982; Farmer 1985).

Laboratory findings

Pattern reversal visual-evoked potentials have shown mildly increased latencies bilaterally. The electroencephalogram shows mild diffuse slowing. Electroretinography and electrooculography are normal. Electromyogram and nerve conduction velocity are consistent with a mild polyneuropathy affecting the lower more than the upper extremity. Blood cell counts, syphilis serologies, folate, Vitamin B_{12}, amino acid screen, heavy metal screen, arylsulfatase A activity in leukocytes, rectal biopsy, cerebral spinal fluid examination, ceruloplasmin, and serum enzyme levels have not shown any informative abnormality.

Pattern of inheritance

Kjellin's 2 pedigrees and the 2 pedigrees of Farmer et al. (1985) are consistent with an autosomal recessive transmission.

SJÖGREN-LARSSON SYNDROME

Sjögren-Larsson syndrome is an autosomal recessive disorder distinguished by congenital ichthyiosis, mental retardation, spastic paraparesis, and retinal abnormalities (Jagell et al. 1980, 1981; Sjögren-Larsson 1957) (270200). Rizzo et al. (1988) has recently characterized an abnormality of fatty alcohol oxidation which appears to be responsible for this syndrome.

General and neurologic findings

The syndrome is usually recognized at birth as a result of the congenital ichthyosis. The skin appears dry, fissured, and hyperkeratotic. Skin

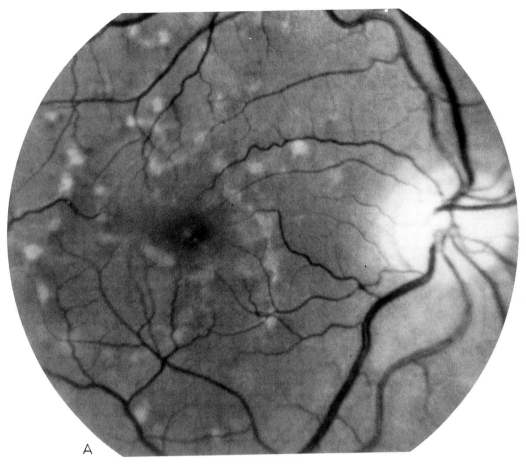

Fig. 1. Case 1. Lesions are located mostly within the vascular arcades. The remainder of the fundus is normal. Subtle hyperpigmentation surrounds some lesions. (A) Right eye. (B) left eye. (Reproduced from Farmer et al. (1985) by courtesy of the Editors of *Am. J. Ophthalmol.* © The Ophthalmic Publishing Company.)

biopsy findings are consistent with ichthyosiform erythroderma (Selmanowitz 1966). Other findings have included convulsions, dental and osseous dysplasia, and defective sweating. The spastic paraparesis may be a classic spastic paraplegia affecting the lower extremities or, in some individuals, may involve the upper extremities as well. Moderate, non-progressive mental retardation is present. Other neurologic findings such as ataxia are unusual.

Ophthalmologic examination

Glistening dots or crystals laying on or near the retinal surface surround a central yellowish deposit. This abnormality is present in between 20 and 100% of patients. Late staining of the central lesion is apparent on fluorescein angiography (Gilbert et al. 1968). It has been difficult to obtain an estimate of visual acuity in view of the mental retardation present. Optic atrophy is not present.

Pathophysiology

Rizzo et al. (1987, 1988, 1989) have recently described an inherited defect in the fatty alcohol cycle of affected individuals. Fatty alcohol: nicotinamide-adenine dinucleotide oxidoreductase (FAO) activity is deficient in cultured skin fibroblasts from affected individuals, and intermediate FAO activity is present in obligate heterozygotes.

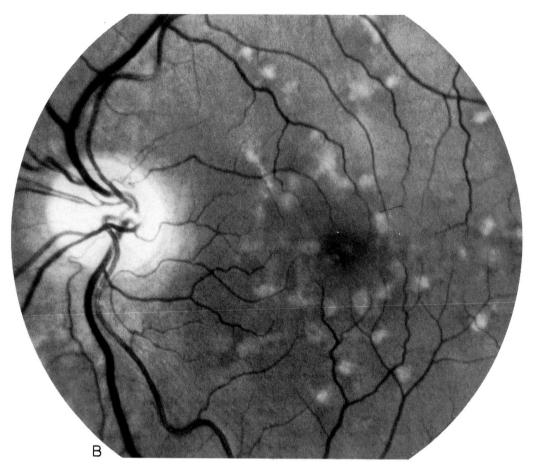

Fig. 1B.

Elevated fatty alcohol concentrations are present in the plasma. Rizzo suggests that accumulation of fatty alcohol or a metabolic product may be important in the pathogenesis of the disorder. The effect of dietary modification on disease progression is the subject of current investigation.

LAURENCE-MOON-BARDET-BIEDL SYNDROME

Laurence and Moon (1866) reported 4 siblings with mental retardation, hypogenitalism, ataxia, nystagmus, and retinal changes (245800). Two of the 4 showed severe choroidal atrophy, while the other 2 showed a pigmentary retinopathy. Hutchinson (1900) observed spastic paraplegia to develop at a later date in all 4 individuals. Bardet (1920) and Biedl (1922) independently described patients with retinitis pigmentosa, mental retardation, obesity, polydactyly, and hypogenitalism. Subsequent investigators (Solis-Cohen and Wiess 1925) concluded that those cases reported by Laurence and Moon, Biedl and Bardet represented a single disorder. This position has been held by most authorities until recently.

With additional experience, it has become apparent that the syndrome is more likely 2 separate disorders with many shared features (Schachat and Maumenee 1982). Laurence-Moon syndrome is much less common, with the natural history dominated by spastic paraparesis and severe visual loss. Polydactyly is rare. Modern descriptions are extremely rare, but Rizzo et al. (1986) report a case with the phenotype in which neurologic findings included paraparesis and

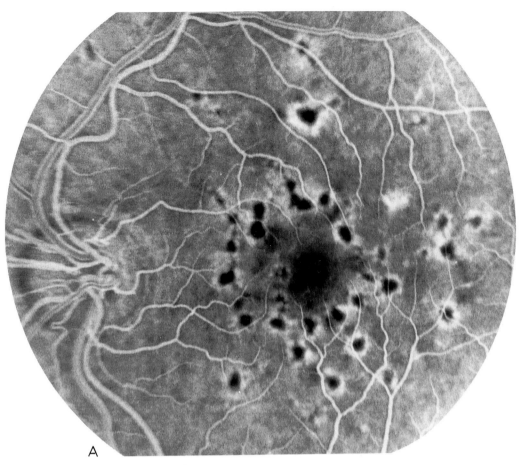

A

Fig. 2. Case 1. Midvenous and recirculation-phase fluorescein angiograms differ little in the appearance of the lesions. The area of hyperfluorescence surrounding the blocking lesions is more striking than the corresponding ophthalmoscopic change. The choroid and retinal pigment epithelium between the lesions appear normal. (A) Right eye. (B) Left eye. (Reproduced from Farmer et al. (1985) by courtesy of the Editors of *Am. J. Ophthalmol.* © The Ophthalmic Publishing Company.)

ataxia. In the more common Bardet-Biedl syndrome, patients have a severe retinal dystrophy, polydactyly, syndactyly or brachydactyly, obesity, and renal abnormalities which may progress to renal failure (Churchill et al. 1981; Leys et al. 1981; Campo and Aaberg 1982; Linné et al. 1986). Male patients have hypogenitalism (hypogonadism in female patients is not necessarily present). Delayed cognitive development and spastic paraparesis are not found (Green 1989).

General and neurologic examination

Although modern descriptions of Laurence-Moon patients are rare, findings in patients with the Bardet-Biedl syndrome are well-described [see above]. Almost 90% of patients exceed the 90th percentile of weight for height. Hutchinson originally described a slowly progressive paraplegia. Rizzo et al. (1986) describes impairment of gait starting at age 19. Gait was broad-based and stiff-legged. Terminal dysmetria and nystagmus were present. Deep tendon reflexes were symmetrically brisk with a very brisk jaw jerk, but no clonus or extensor-plantar responses were present. Delayed cognitive development is uncommon if assessed quantitatively. The disease presents in childhood and the pedigree is usually consistent with autosomal recessive inheritance. Considerable intrafamilial and interfamilial variation is present.

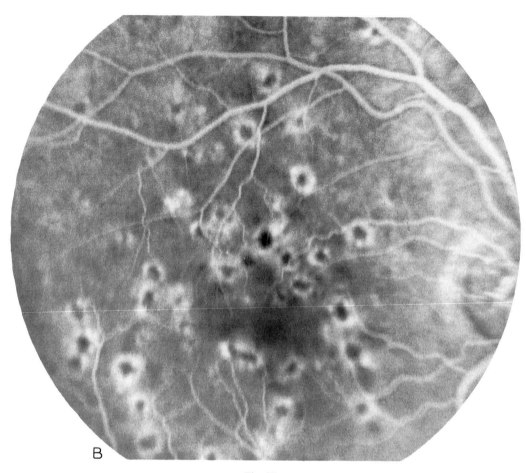

Fig. 2B.

Ophthalmologic examination

Atypical retinitis pigmentosa with early severe loss of vision (cone-rod dystrophy) is characteristically present. The changes cause pallor of the optic nerve, vascular attenuation, and salt-and-pepper pigmentary changes which may progress to complete choroidal atrophy in the affected fundus. Myopia, nystagmus, and cataracts are seen in the majority of patients. Fluorescein angiography shows widespread retinal pigment epithelial window defects, through which fluorescein dye in the choroid may be observed.

Laboratory findings

The electroretinogram is usually non-recordable. Cases with an initially recordable electroretino-gram have been reported. Rizzo et al. (1986) reports cerebellar atrophy on computed tomography (CT) and magnetic resonance imaging (MRI).

COCKAYNE SYNDROME

Cockayne's syndrome (Cockayne 1946) (216400) is a rare autosomal recessive form of cachectic dwarfism. Characteristic features include late infantile-onset growth failure, spastic paraparesis, mental retardation with progressive mental deterioration, normal pressure hydrocephalus, ataxia, sensory neural hearing loss, intention tremor, nystagmus, muscle rigidity, and incontinence (Pearce 1972). Premature aging with a characteristic 'bird-like facies' is present. More than 70 cases have been reported (Levin et al. 1983). These case reports show considerable clinical variation. Among the most constant findings is a 'salt-and-pepper' retinal dystrophy usually seen in conjunction with waxy pallor of the optic nerve and retinal vascular attenuation. Additional ocular findings include cataracts in about one third of reported cases, corneal ulcers or opacities in about 20%, nystagmus in 18%, and pupillary unresponsiveness to light in 18%, and normal pressure hydrocephalus (Brumback et al. 1978). Pathologic findings are detailed elsewhere (Soffer et al. 1979).

PROGRESSIVE QUADRIPARESIS, MENTAL RETARDATION, RETINITIS PIGMENTOSA, AND HEARING LOSS

Gordon (1976) reported 2 brothers with what appears to be a unique syndrome (270950). Neurologic findings include drooling and an expressionless face. At 5 years of age speech and word comprehension were lost. Extensive spasticity was present. By 9 years of age, the upper extremities were spastic and the lower extremities were rigid with extensive muscle wasting. The younger brother's course was similar with the exception of a profound hearing impairment which was noted early on. Ocular examination showed diffuse salt-and-pepper retinopathy, small, pale optic nerve heads, and retinal vascular attenuation.

A recordable but severely abnormal electroretinogram was present in 1 patient. The earliest electroencephalogram was read as normal; however, with the progression of neurologic difficulty, the electroencephalogram showed diffuse slowing.

HALLERVORDEN-SPATZ SYNDROME

Hallervorden and Spatz (1922) (234200) reported 5 siblings with progressive motor abnormalities, dementia, and speech problems. Pathologic examination showed a specific rusty brown discoloration of the globus pallidus and substantia nigra. More than 60 cases of a similar progressive disorder have now been described. Dooling et al. (1974) reviewed the 64 autopsy-proven cases reported at that time and concluded that 42 of these patients had a distinct entity.

Examination

Neurologic abnormalities include muscle tone abnormalities, dysarthria, choreoathetosis, a Parkinson-like tremor, progressive dementia, distal wasting, and immobility. Sensation is unaffected. The disease presents in infancy or childhood with progression to death in childhood.

Ocular findings include vascular attenuation, bone spicule pigmentation of typical retinitis pigmentosa, peripheral yellow flecks, and a bull's eye maculopathy (Luckenbach et al. 1983; Roth et al. 1971). These findings may be present in up to 25% of reported patients. Newell (1979) states that those patients with retinal degeneration had earlier onset and more rapid progression.

Laboratory studies

An electroretinogram performed in one family showed no response. The EEG and nerve-conduction studies done in the same family were normal. Magnetic resonance imaging (Schaffert et al. 1989; Sethi et al. 1988) may permit ante mortem diagnosis. The characteristic abnormality is a striking change in the globus pallidus, termed the 'eye of the tiger' sign.

HOMOCARNOSINOSIS

Sjaastad et al. (1976) reported a familial metabolic disorder characterized by spastic paraplegia, progressive mental deficiency, and retinal pigmentary changes (236130). The disorder was associated with elevated homocarnosine levels in the cerebral spinal fluid which were 20 times control present in 2 brothers and 1 sister. The clinically unaffected mother had similar cerebrospinal fluid findings, and the relation between the elevation and the clinical findings is uncertain.

Examination

Neurologic symptoms were the presenting sign and became apparent in childhood to early adulthood in the reported family. The major neurologic symptom was spastic paraplegia with predominant involvement of the legs. The most severely affected individual eventually developed spasticity in the upper extremities. Extensor-plantar responses, clonus of lower extremity reflexes, and very brisk upper extremity reflexes were characteristic at presentation. A cognitive deficit was present with IQs ranging from 60 to 90. Ocular findings were described as a 'diffuse gray pigmentation in both maculas'.

Laboratory studies

Laboratory investigations, consisting of electro-encephalogram, electromyogram, and electroretinogram were all normal. Of major interest was the presence of homocarnosine in the cerebral spinal fluid. Homocarnosine is a brain-specific imidazoledipeptide (Perry et al. 1971). The role of homocarnosine in cerebrospinal fluid metabolism is uncertain. While the clinical findings suggest a recessive disorder, the biochemical findings appear dominantly inherited. The exact nature of the relationship between the homocarnosine elevation and the pathologic process in the affected children is unclear.

RELATED SYNDROMES

Spastic paraplegia or quadriplegia and retinal degeneration associated with ophthalmoplegia is extremely rare. Four rather heterogeneous non-familial cases were reported by Barnard and Scholz in 1944, and other possible cases have been reported (Walsh 1947; Alfano and Berger 1957). Potentially confounding disorders including syphilis or Graves ophthalmopathy were present in all of the Barnard and Scholz patients, limiting the usefulness of their observations.

The association of retinal degeneration and ophthalmoplegia is unusual, most frequently being associated with the Kearn-Sayres syndrome (1958). This syndrome is one of the mitochondrial myopathies — a group of clinically heterogeneous disorders defined by abnormalities of muscle mitochondria (DiMauro 1985; Lombes et al. 1989). It is characterized by progressive external ophthalmoplegia with onset prior to age 15 and retinal pigmentary changes. Heart block, cerebrospinal fluid protein greater than 100 mg/dl, or ataxia are present in most patients. Weakness, dementia, short stature and sensori-

neural hearing loss have been reported. An unusual case with progressive macular changes, and spinocerebellar degeneration was associated with mitochondrial changes on muscle biopsy (Peterson et al. 1985).

Behr (1909) reported the association of infantile optic atrophy, spasticity, mental retardation, and spinal sensory loss (210000). An autosomal recessive inheritance was presumed. Landrigan et al. (1973) observed the additional sign of peripheral neuropathy in 2 siblings. The neurologic findings are variably progressive. The ocular findings are confined to the optic nerve, which is normal in size, but chalk white in color. The sensory deficit is severe, and the onset of secondary nystagmus has been reported as early as age 2 months.

Leber's syndrome (308900) of optic neuropathy with dramatic young adult-onset is sometimes associated with spasticity, muscle weakness, ataxia, and epilepsy. The inheritance is thought to be related to mitochondrial DNA (Scriver et al. 1989). This syndrome is discussed in detail in chapter 17 of this volume.

Refsum's disease (Refsum and Skillicorn 1954) (266500) is a metabolic abnormality characterized by the accumulation of phytanic acid in plasma. It is one of several syndromes related to defective beta oxidation of fatty acids in peroxisomes, a cellular organelle present in all cells. Many of these disorders have an associated retinal dystrophy (Moser 1986). The disease presents with defective dark adaptation and visual field constriction during teenage years. Although fundus changes may be minimal in the beginning, they evolve to more typical retinal dystrophy changes. The neurologic findings include hypertrophic sensory motor peripheral neuropathy with cranial nerve deficits and cerebellar signs.

Neuronal ceroid lipofuscinosis is a group of disorders, 3 of which are associated with a severe early-onset retinal dystrophy, followed shortly by or concomitant with neurologic deterioration. In the infantile (Scintaruori) (250730), late infantile (Jansky-Bielschowsky) (204500), and juvenile forms (Batten, Vogt-spielmeyer) (204200), patients usually present to ophthalmologists with severe visual loss (Hittner and Zeller 1975). Neurologic deterioration seizures, dementia, and death follow. Diagnosis is made by electron microscopy of lymphocytes or biopsy of skin, conjunctiva, or brain for specific inclusions (Dolman 1984).

Search for the cause of spastic paraplegia often leads to treatable causes. When the search does not and when the differential includes a hereditary spastic paraplegia, then a full ophthalmologic examination may lead to more specific diagnosis such as listed in Table 1. The retinal abnormalities are difficult to see on direct ophthalmoscopy so the neurologist should not rely on his or her examination of the retina to exclude the abnormality. Consultation with an ophthalmologist is necessary. With such an approach, possibly more of the otherwise non-specific hereditary spastic paraplegias could be more accurately characterized.

Acknowledgement
We gratefully acknowledge Roberta A. Pagon, M.D., for critical review of our manuscript.

REFERENCES

ALFANO, J. E. and J. P. BERGER: Retinitis pigmentosa, ophthalmoplegia, and spastic quadriplegia. Am. J. Ophthalmol. 43 (1957) 231–240.

BARDET, F.: Sur un syndrome d'obésité congénitale avec polydactylie et rétinite pigmentaire (contribution à l'étude des formes cliniques de l'obésité hypophysaire). Thèse, Paris 170 (1970) 107.

BARNARD, R. I. and R. O. SCHOLZ: Ophthalmoplegia and retinal degeneration. Am. J. Ophthalmol. 27 (1944) 621–624.

BASSEN, F. A. and A. L. KORNZWEIG: Malformation of the erythrocytes in a case of atypical retinitis pigmentosa. Blood 5 (1950) 381–387.

BEHR, C.: Die komplizierte, hereditär-familiäre Optikusatrophie des Kindesalters — Ein bisher nicht beschriebener Symptomkomplex. Klin. Monatsbl. Augenheilkd. 47 (1909) 138–160.

BELL, J. and E. A. CARMICHAEL: On Hereditary Ataxia and Spastic Paraplegia. Treasury of Human Inheritance, Vol. V, Pt. 3. London, Cambridge University Press (1939).

BIEDL, A.: Ein Geschwisterpaar mit adiposo-genitaler Dystrophie. Dtsch. Med. Wochenschr. 48 (1922) 1630.

BRUMBACK, R. A., F. W. YODER, A. D. ANDREWS ET AL.: Normal pressure hydrocephalus: recognition and relationship to neurological abnormalities in Cockayne's syndrome. Arch. Neurol. 35 (1978) 337–345.

CAMPO, R. V. and T. M. AABERG: Ocular and systemic manifestations of the Bardet-Biedl syndrome. Am. J. Ophthalmol. 94 (1982) 750–756.

CAVANAGH, N. P. C., R. A. EAMES, R. J. GALVIN, E. M. BRETT and R. E. KELLY: Hereditary sensory neuropathy with spastic paraplegia. Brain 102 (1979) 79–94.

CHURCHILL, D. N., P. MCMANAMON and R. M. HURLEY: Renal disease — a sixth cardinal feature of the Laurence-Moon-Biedl syndrome. Clin. Nephrol. 16 (1981) 151–154.

COCKAYNE, E. A.: Case reports: Dwarfism with retinal atrophy and deafness. Arch. Dis. Child. 21 (1946) 52–54.

DEUTMAN, A. F. and L. M. A. A. JANSEN: Dominantly inherited drusen of Bruch's membrane. Br. J. Ophthalmol. 54 (1970) 373.

DICK, A. P. and C. J. STEVENSON: Hereditary spastic paraplegia: report of a family with associated extrapyramidal signs. Lancet 1 (1953) 921–923.

DIMAURO, S., E. BONILLA, M. ZEVIANI, M. NAKAGAWA and D. C. DEVIVO: Mitochondrial myopathies. Ann. Neurol. 17 (1985) 521–538.

DOLMAN, C. L.: Diagnosis of neurometabolic disorders by examination of skin biopsies and lymphocytes. Semin. Diagn. Pathol. 1 (1984) 82–97.

DOOLING, E. C., W. C. SCHOENE and E. P. RICHARDSON: Hallervorden-Spatz syndrome. Arch. Neurol. 30 (1974) 70.

EAGLE, R. C. JR., A. C. LUCIER, V. B. BERNARDINO JR. and M. YANOFF: Retinal pigment epithelial abnormalities in fundus flavimaculatus. A light and electron microscopic study. Ophthalmology 87 (1980) 1189–1200.

ERNEST, T. J. and A. E. KRILL: Fluorescein studies in fundus flavimaculatus and drusen. Am. J. Ophthalmol. 62 (1966) 1.

FARMER, S. G.: Fleck retina in Kjellin's syndrome (Reply to Letter). Am. J. Ophthalmol. 99 (1985) 75.

FARMER, S. G., W. T. LONGSTRETH, JR., R. E. KALINA and A. B. TODOROV: Fleck retina in Kjellin's syndrome. Am. J. Ophthalmol. 99 (1985) 45–50.

FOLK, J. C., T. A. WEINGEIST, J. J. CORBETT ET AL.: Symphilitic neuroretinitis. Am. J. Ophthalmol. 95 (1983) 480–486.

FREUD, S.: Ueber familiaere Formen von cerebralen Diplegien. Neurol. Centralbl. (Mendel) 12 (1893) 512–515 and 542–547.

GILBERT, W. R., JR., J. L. SMITH and W. L. NYHAN: The Sjögren-Larsson syndrome. Arch. Ophthalmol. 80 (1968) 308–316.

GORDON, A. M., A. J. CAPUTE and B. W. KONIGSMARK: Progressive quadriparesis, mental retardation, retinitis pigmentosa, and hearing loss: report of two sibs. Johns Hopkins Med. J. 138 (1976) 142–145.

GREEN, J. S., P. S. PARFREY, J. D. HARNETT ET AL.: The cardinal manifestations of Bardet-Biedl syndrome, a form of Laurence-Moon-Biedl syndrome. N. Engl. J. Med. 321 (1989) 1002–1009.

HALLERVORDEN J. and H. SPATZ: Eigenartige Erkrankung in extrapyramidalen System mit besonderer Beteiligung des Globus pallidus und der der

Substantia nigra. Z. Neurol. Psychiatr. 79 (1922) 254.

HARDING, A. E.: Hereditary 'pure' spastic paraplegia: a clinical and genetic study of 22 families. J. Neurol. Neurosurg. Psychiatry 44 (1981) 871–883.

HITTNER, H. M. and ZELLER, R. S.: Ceroid-lipofuscinosis (Batten's disease). Arch. Ophthalmol. 93 (1975) 178–183.

HUTCHINSON, J.: Slowly progressive paraplegia and disease of the choroids with defective intellect and arrested sexual development in several brothers and a sister. Arch. Surg. (Lond.) 11 (1900) 118–122.

JAGELL, S., W. POLLAND and O. SANDGREN: Specific changes in the fundus typical for the Sjogren-Larsson syndrome. An ophthalmological study. Acta. Ophthal. Scand. 58 (1980) 321–330.

JAGELL, S., K.-H. GUSTAVSON and G. HOLMGREN: Sjogren-Larsson syndrome in Sweden. A clinical, genetic and epidemiological study. Clin. Genet. 19 (1981) 233–256.

JOHNSTON, A. W. and V. A. MCKUSICK: A sex-linked recessive form of spastic paraplegia. Am. J. Hum. Genet. 14 (1962) 83–94.

JONES, E.: Eight cases of hereditary spastic paraplegia. Rev. Neurol. 5 (1907) 98–106.

KEARNS, T. P. and G. P. SAYRE: Retinitis pigmentosa, external ophthalmoplegia, and complete heart block: unusual syndrome with histologic study in one of two cases. Arch. Ophthalmol. 60 (1958) 280–289.

KJELLIN, K.: Familial spastic paraplegia with amyotrophy, oligophrenia, and central retinal degeneration. Arch. Neurol. 1 (1959) 133.

KRILL, A. E.: The retinal disease of rubella. Arch. Ophthalmol. 77 (1967) 445–449.

LANDRIGAN, P. J., W. BERENBERG and M. BRESNAN: Behr's syndrome: Familial optic atrophy, spastic diplegia and ataxia. Dev. Med. Child Neurol. 15 (1973) 41–47.

LAURENCE, J. Z. and R. C. MOON: Four cases of retinitis pigmentosa occurring in the same family and accompanied by general imperfection of development. Ophthal. Rev. 2 (1866) 32–41.

LEVIN, P. S., W. R. GREEN, D. I. VICTOR and A. L. MACLEAN: Histopathology of the eye in Cockayne's syndrome. Arch. Ophthal. 101 (1983) 1093–1097.

LEYS, M. J., L. A. SCHREINER, R. M. HANSEN, D. L. MAYER and A. B. FULTON: Visual acuities and dark-adapted thresholds of children with Bardet-Biedl syndrome. Am. J. Ophthalmol. 106 (1988) 561–569.

LINNÉ, T., I. WIKSTAD and R. ZETTERSTRÖM: Renal involvement in the Laurence-Moon-Biedl syndrome. Acta Paediatr. Scand. 75 (1986) 240–244.

LOMBES, A., E. BONILLA and S. DIMAURO: Mitochondrial encephalomyopathies. Rev. Neurol. (Paris) 145 (1989) 671–689. Encephalomyopathies. Rev. Neurol. (Paris) 145 (1989) 671–689.

LUCKENBACH, M. W., W. R. GREEN, N. R. MILLER ET AL.: Ocular clinicopathologic correlation of Hallervorden-Spatz syndrome with acanthosis and pigmentary

retinopathy. Am. J. Ophthalmol. 95 (1983) 369–382.

MACRAE, W., J. STIEFFEL and A. B. TODORAV: Recessive familial spastic paraplegia with retinal degeneration. Acta Genet. Med. Gemellol. 23 (1974) 249.

MAHLOUDJI, M. and P. CHUKE: Familial spastic paraplegia with retinal degeneration. Johns Hopkins Med. J. 123 (1968) 142–144.

MILLER, F. S. III, A. H. BUNT-MILAM and R. E. KALINA: Clinical-ultrastructural study of thioridazine retinopathy. Ophthalmology 89 (1982) 1478–1488.

MOSER, H. W.: Peroxisomal disorders. J. Pediatr. 108 (1986) 89–91.

MULLIE, M. A., A. E. HARDING, R. K. H. PETTY, H. IKEDA, J. A. MORGAN-HUGHES and M. D. SANDERS: The retinal manifestations of mitochondrial myopathy. A study of 22 cases. Arch. Ophthalmol. 103 (1985) 1825–1830.

NEWELL, F. W., A. E. KRILL and T. B. FARKAS: Drusen and fundus flavimaculatus. Clinical, functional, and histologic characteristics. Trans. Am. Acad. Ophthalmol. Otolaryngol. 76 (1972) 88.

NEWELL, F. W., R. O. JOHNSON II and P. R. HUTTENLOCHER: Pigmentary degeneration of the retina in the Hallervorden-Spatz syndrome. Am. J. Ophthal. 88 (1979) 467–471.

NYBERG-HANSEN, R. and S. REFSUM: Spastic paraparesis associated with optic atrophy in monozygotic twins. Acta. Neurol. Scand. (Suppl) 48 (1972) 261–263.

PAGON, R. A.: Retinitis pigmentosa. Surv. Ophthalmol. 33 (1988a) 137–177.

PAGON, R. A.: Laurence-Moon-Biedl syndrome? (Letter to Editor). Mayo Clin. Proc. 63 (1988a) 209.

PAUTLER, S. E.: Fleck retina in Kjellin's syndrome (Letter to Editor). Am. J. Ophthalmol. 99 (1985) 618–619.

PEARCE, W. G.: Ocular and genetic features of Cockayne's syndrome. Can. J. Ophthalmol. 7 (1972) 435–444.

PERRY, T. L., S. HANSEN, K. BERRY, C. MOK and D. LESK: Free amino acids and related compounds in biopsies of human brain. J. Neurochem. 18 (1971) 521–528.

PETERSON, P. L., M. E. MARTENS, C.-P. LEE, J. S. HATFIELD, G. L. KLEPACH and J. GILROY: Mitochondrial dysfunction in spinocerebellar and macular degeneration. Ann. Neurol. 18 (1985) 146.

REFSUM, S.: Heredopathia atactia polyneuritiformis. Phytanic-acid storage disease. Refsum's disease: A biochemically well-defined disease with a specific dietary treatment. Arch. Neurol. 38 (1981) 605–606.

REFSUM, S. and S. A. SKILLICORN: Amyotrophic familial spastic paraplegia. Neurology (Minneap.) 4 (1954) 40–47.

RIZZO, J. F. III, E. L. BERSON and S. LESSELL: Retinal and neurologic findings in the Laurence-Moon-Biedl phenotype. Ophthalmology 93 (1986) 1452–1456.

RIZZO, W. B., D. A. CRAFF, A. L. DAMMANN and M. W. PHILLIPS: Fatty alcohol metabolism in cultured human fibroblasts: evidence for a fatty alcohol cycle. J. Biol. Chem. 262 (1987) 17412–17419.

RIZZO, W. B., A. L. DAMMANN and D. A. CRAFT: Sjogren-Larsson syndrome: Impaired fatty alcohol oxidation in cultured fibroblasts due to deficient fatty alcohol: nicotinamide adenine dinucleotide oxidoreductase activity. J. Clin. Invest. 81 (1988) 738–744.

RIZZO, W. B., A. L. DAMMANN, D. A. CRAFT ET AL.: Sjogren-Larsson syndrome: Inherited defect in the fatty alcohol cycle. J. Pediatr. 115 (1989) 228–234.

ROTH, A. M., R. S. HEPLER, M. MUKOYAMA ET AL.: Pigmentary retinal dystrophy in Hallervorden-Spatz disease: clinicopathological report of a case. Surv. Ophthalmol. 16 (1971) 24–35.

ROWE, P. F.: Hereditary spastic paraplegia. J. Neurol. Neurosurg. Psychiatry 26 (1963) 516–519.

SCHACHAT, A. P. and I. H. MAUMENEE: Bardet-Biedl syndrome and related disorders. Arch. Ophthalmol. 100 (1982) 285–288.

SCHAFFER, K.: Zur Pathologie und pathologischen Histologie der spastischen Heredodegeneration (hereditäre spastische Spinalparalyse). Dtsch. Z. Nervenheilkd. 73 (1922) 101–128.

SCHAFFERT, D. A., S. D. JOHNSEN, P. C. JOHNSON and B. P. DRAYER: Magnetic resonance imaging in pathologically proven Hallervorden-Spatz disease. Neurology 39 (1989) 440–442.

SCRIVER, C. R., A. L. BEAUDET, W. S. SLY and D. VALLE (Eds.): The Metabolic Basis of Inherited Disease. 6th ed. New York, McGraw-Hill Information Services Co. (1989) 920–922.

SELMANOWITZ, V.: Sjögren-Larsson syndrome. Arch. Dermatol. 93 (1966) 772–774.

SETHI, K. D., R. J. ADAMS, D. W. LORING and T. E. GAMMAL: Hallervorden-Spatz syndrome: clinical and magnetic resonance imaging correlations. Ann. Neurol. 24 (1988) 692–694.

SILVER, J. R.: Familial spastic paraplegia with amyotrophy of the hands. J. Neurol. Neurosurg. Psychiatry 29 (1966) 135.

SIMELL, O. and K. TAKKI: Raised plasma-ornithine and gyrate atrophy of the choroid and retina. Lancet 1 (1973) 1031–1033.

SJAASTAD, O. J. BERSTAD, P. GJESDAHL and L. GJESSING: Homocarnosinosis. 2. A familial metabolic disorder associated with spastic paraplegia, progressive mental deficiency, and retinal pigmentation. Acta Neurol. Scand. 53 (1976) 275–290.

SJÖGREN, T. and T. LARSSON: Oligophrenia in combination with congenital ichthyosis and spastic disorders. Acta Psychiatr. Scand. (Suppl) 32 (1957) 1–113.

SOFFER, D., H. W. GROTSKY, I. RAPIN ET AL.: Cockayne syndrome: Unusual neuropathological findings and review of the literature. Ann. Neurol. 6 (1979) 340–348.

SOLIS-COHEN, S. and E. WEISS: Dystrophia adopogenitalis with atypical retinitis pigmentosa and mental deficiency: the Laurence-Biedl syndrome: a report of four cases in one family. Am. J. Med. Sci. 169 (1925) 489–505.

STRÜMPELL, A.: Beiträge zur Pathologie des Rückenmarks. Arch. Psychiatr. 10 (1880) 676–717.

TRABOULSKI, E. I.: Fleck retina in Kjellin syndrome (Letter to Editor). Am. J. Ophthalmol. 99 (1985) 738–739.

WALSH, F. B.: Clinical Neuro-Ophthalmology. Baltimore, Williams and Wilkins Co. (1934, 1947).

WATZKE, R. C., J. C. FOLK and R. M. LANG: Pattern dystrophy of the retinal pigment epithelium. Ophthalmology 89 (1982) 1400–1406.

Handbook of Clinical Neurology, Vol. 15 (59): Diseases of the Motor System
J.M.B.V. de Jong, editor
© Elsevier Science Publishers B.V., 1991

CHAPTER 20

Hereditary secondary dystonias

EDWARD J. NOVOTNY JR

Department of Pediatrics, Yale University, School of Medicine, New Haven, CT, USA

The term dystonia refers both to a neurologic disorder and specific types of abnormal involuntary movements. The definition of dystonia adopted by the Dystonia Medical Research Foundation is: a syndrome of sustained muscle contractions, frequently causing twisting and repetitive movements, or abnormal postures (Fahn 1988). This syndrome can be classified into two major divisions according to etiology, either primary or secondary. The primary or idiopathic dystonias have this movement disorder as the only manifestation of neurological dysfunction. The secondary or symptomatic dystonias are heterogeneous in regard to etiology and frequently have other evidence of neurological or systemic dysfunction (Calne 1988). This chapter describes two types of secondary dystonias which have unique clinical and genetic characteristics.

Leber's disease and dystonia (308900)

Several families with both optic neuropathy and dystonia associated with striatal lesions have been reported (Bruyn and Went 1964; Miyoshi et al. 1969; Marsden et al. 1986). Numerous case reports of sporadic and familial dystonia with striatal lesions have also been described (Röyttä et al. 1981; Rondot et al. 1982; Mito et al. 1986). A family comprised of 5 generations in which 14

individuals had dystonia and 9 members had a hereditary optic neuropathy was recently reported (Novotny et al. 1986). The clinical and genetic features, its relation to other diseases and a discussion of the pathophysiology and investigation of this disorder is presented.

CLINICAL CHARACTERISTICS

The pedigree of a family in which several members had either a hereditary neuroretinopathy or dystonia is depicted in Fig. 1. The clinical characteristics of the ophthalmological disease are similar to those observed in patients with Leber's disease. The family members with the dystonia have variable degrees of intellectual impairment and muscle weakness and atrophy in addition to the movement disorder.

Case reports

Case 1 (III 10). This woman whose past medical history was unremarkable noted a decrease in the visual acuity of her right eye at 16 years of age. She had no associated ocular pain or headache. In 3 weeks she experienced loss of vision in her left eye. She also noted that her color perception was abnormal. Several weeks later she was seen by an ophthalmologist who noted bilateral optic nerve pallor and no retinal

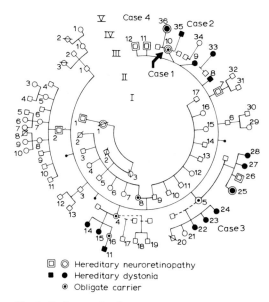

□ ◎ Hereditary neuroretinopathy
■ ● Hereditary dystonia
◉ Obligate carrier

Fig. 1. Pedigree of a family with both a hereditary dystonia and neuroretinopathy depicting the maternal inheritance pattern of the disorders. Cases are described in text.

vasculature abnormalities. Bilateral enlarged central scotomata were noted on perimetry. The diagnosis of Leber's optic neuropathy was made after learning that other family members had a similar eye disorder. At 35 years of age her neurological examination was normal except for the optic nerve and vision abnormalities.

Case 2 (IV 35). At 2.5 years of age, this boy developed a 'wide-based gait' and was observed to fall frequently. He began walking at 11 months of age and his neonatal and infantile history were unremarkable. Examination at age 3 years revealed generalized hypotonia, primarily involving the legs, an 'ataxic' gait, and hyporeflexia. There were no other abnormalities.

Evaluation included normal electromyography, skull and total spine roentgenograms, radionuclide brain scan, and a myelogram. Laboratory studies were remarkable for an elevated creatine kinase and aldolase. Muscle biopsy reportedly showed non-specific abnormalities.

Later increasing dysarthria was observed. He could not run without falling and had difficulty climbing stairs. He rode a tricycle at age 5 years when his parents also noted that he had marked

lability of affect. Psychometric testing performed at age 10 years revealed that he had a full scale IQ of 67 with a verbal IQ of 67 and performance IQ of 70 on the Wechsler Intelligence scale for Children-Revised (WISC-R). He scored at the tenth percentile on the Peabody Picture Vocabulary Test in both receptive and expressive language. At 12 years of age, his height was less than the 3rd percentile for age and his weight was at the fifth percentile. He was able to print his name and was observed to have a labile affect. His speech was dysarthric and he had bilateral facial weakness. He exhibited frequent episodes of blepharospasm and eye blinking. His voluntary tongue movements were slowed and dyskinetic. Muscle bulk and tone were decreased in all 4 extremities. Athetotic movements and dystonia were observed when he reached for objects. On ambulation his base was 18 inches and he kept his legs internally rotated at the hips and his feet were inverted. Festination of gait and poor postural reflexes were elicited. CT scan revealed well-demarcated, symmetric lucencies in the putamina and caudate nuclei.

Case 3 (IV 23). At 7 years of age this boy was observed to have a gait disturbance characterized by rigidity and dystonia of the left leg. Over the next several months he became dysarthric and his school performance declined. Left arm athetosis and dystonia was noticed at 8 years of age. At 10 years of age he visited a neurologist. He was able to run, but could not ride a bicycle. He could not read and his speech was severely dysarthric. Bilateral facial paresis and asymmetric (left greater than right) increased tone in his legs was noted. The tone in his right arm was normal and his left arm had increased tone and exhibited dystonic posturing with flexion at the elbow. He had muscle wasting in the distal left lower extremity. Psychometric testing at 11 years of age demonstrated that his full scale IQ was 67 on the WISC-R with a performance IQ of 64 and verbal IQ of 73.

He was able to read only simple sentences at 20 years of age and continued to have difficulty with his speech because of marked dysarthria. He had frequent oral-buccal dyskinesias on examination and facial weakness bilaterally. A left hemiatrophy was noted and he had increased tone in all 4 extremities with the left side and lower extremities affected greater than the right arm. His left arm remained flexed at the elbow and supinated and his left leg was extended at the knee with inversion and dorsiflexion of the left foot. His movements were athetotic and he had both truncal and appendicular dystonia. His movement disorder was accentuated when he attempted to perform complex motor tasks. Neuro-ophthalmological examination including direct and indirect ophthalmoscopy and slit lamp exam was normal. Visual acuity was 20/20 and J_1 in both eyes. A protanopsia was discovered on Farnsworth D_{15} color testing. Goldmann perimetry, fluorescein angiography, and fundus photography were normal.

Case 4 (IV 36). This girl was born at term via normal spontaneous vaginal delivery. The neonatal course was uncomplicated and developmental milestones were attained as expected during the first year of life. She walked at 12 months of age and 6 months later she began falling frequently. Her gait was described as 'wide-based' and she had poor postural reflexes. At 24 months of age she developed rigidity of both lower extremities and dysarthria. She could only walk with assistance at this time. By 5 years of age she had generalized dystonia, was wheelchair dependent and had difficulty feeding herself. Neurological examination at 7 years of age revealed a young girl with severe dysarthria, generalized dystonia affecting the left side greater than the right, and muscle wasting of the distal portion of the legs. She could not walk or stand without support and she had bilateral ankle clonus and extensor plantar responses. Diazepam was prescribed for her movement disorder. At age 10 years she had a symmetric, severe generalized dystonia and wasting of the distal musculature of all 4 extremities. Her deep tendon reflexes were increased symmetrically in all 4 extremities. Visual acuity and fundoscopic examination were normal. Three years later, she had surgery for release of flexion contractures of both knees. Bilateral hip dislocations and a mild thoracolumbar scoliosis were noted at this time.

A subacute loss of visual acuity was noted to begin at 16 years of age. This progressed over 4-6 months and was uncorrectable according to the parents who had taken her to an ophthalmologist. No complaints of ocular pain or signs of trauma were noted by the family.

At 17 years of age her height and weight were well below the third percentile for age. Her head circumference was at the fiftieth percentile for age. Several musculoskeletal deformities were noted. These included a right pes cavus, hind foot valgus and pes planus on the left, and a thoracolumbar scoliosis associated with a prominent lumbar lordosis. She was alert and had severe dysarthria. Bilateral optic atrophy was noted on fundoscopic examination. Frequent episodes of oral-buccal dyskinesias and blepharospasm were observed. Her voluntary tongue movement was slow and dyskinetic. Her gag reflex was hyperactive. Muscle tone was increased in all 4 extremities and generalized muscle wasting which was most prominent distally in the legs was noted. At rest she kept her arms flexed at the elbows and her legs extended at the knees and hips. Torticollis and increased dystonia was noted when she attempted to reach for objects. Deep tendon reflexes were brisk and symmetrical and she exhibited bilateral extensor plantar responses. CT scan demonstrated symmetrical lucencies in both striata.

Summary

The clinical characteristics and intrafamilial phenotypic variability of this disorder are demonstrated by the cases described above. Of the 55 individuals in this family which spans 5 generations, 21 suffer from a central nervous system disorder. Seven members are affected with a hereditary optic neuropathy and have no other neurological signs or symptoms. Fourteen individuals have a neurological disorder where dystonia is the most prominent feature. The dystonia often is the only neurological sign early in the course of the illness. Occasionally the dystonia is asymmetric, but it is typically generalized. Two of the 14 people also have an optic neuropathy. Four have documented intellectual impairment and 8 have short stature. Ten of the 14 also had evidence of muscle atrophy which was greatest in the distal lower extremities. Evidence of corticospinal tract involvement was frequently present late in the disease course.

LABORATORY INVESTIGATIONS

The following investigations were normal in at least 1 member of this pedigree with neurological disease: electromyography and nerve conduction studies, brain stem and median and posterior tibial somatosensory evoked potentials, routine hematology and serum biochemistry, serum copper and ceruloplasmin, serum lactate and pyruvate, serum creatine kinase and hepatic enzymes, thyroid function tests, quantitative plasma and urinary amino acids, qualitative urinary organic acid screen, leukocyte hexosaminidase A and B, β-galactosidase, β-glucuronidase, α-L-fucosidase, α-mannosidase, urinary arylsulfatase, urinary mucopolysaccharides and oligosaccharides by thin-layer chromatography and cerebrospinal fluid cytology, protein, and glucose. A muscle biopsy from the left gracilis of one member with the neurological disease showed excess fiber size variability and increased central nuclei on hematoxylin and eosin stain. Histochemical stains with NADH-tetrazolium reductase, acid phosphatase, pan esterase, ATPase (at pH 9.4 and 4.6), modified Gomori trichrome, phosphorylase, and oil red O were normal except for the fiber size variation. Electron microscopy was also normal.

Seven of the 14 members with the neurological disorder had neuroimaging studies (either CT scan or magnetic resonance imaging) and demonstrated abnormalities of the striatum which were most pronounced in the putamen (Seidenwurm et al. 1986). One woman with optic neuropathy had no neurological signs or symptoms and had a normal CT scan.

Four members in this family had an analysis of the restriction fragment patterns of their mitochondrial DNA (mtDNA). The 4 were the husband and wife (Case 1) and their 2 children who both had the neurological disorder. The husband had a normal neurological examination and the mother had optic atrophy without other neurological findings. The restriction fragment pattern analysis revealed polymorphisms which differentiated these mtDNAs from other individ-

uals and demonstrated the expected maternal inheritance of mtDNA. No insertion-deletion mutations were detected and none of these polymorphisms related to the disease. The restriction fragment patterns were similar to those found in individuals of European or Asian ancestry (Johnson et al. 1983). Further analysis of the mtDNAs has revealed that the point mutation at nucleotide 11778 that was recently discovered to be associated with Leber's optic atrophy was not present (Wallace et al. 1988; Singh et al. 1989).

GENETICS

The inheritance pattern of both the optic neuropathy and neurological disorder in this family is consistent with mitochondrial or maternal transmission. All affected individuals are from the maternal lineage and none of the 24 members from the paternal lineage are affected. This is the pattern of inheritance expected if the disorder is due to a mutation of mtDNA.

The phenotypic variability observed in this family also suggests that a defect of mitochondrial function mediated by a mtDNA mutation is responsible for the disorder. The molecular biology of mitochondria explains the observed diversity in phenotype. These molecular biological properties include heteroplasmic segregation of mitochondria during mitosis, threshold expression of the mutation, and tissue-specific isoenzymes. Heteroplasmic segregation refers to the fact that during development mitochondria are segregated to the cells of various tissues and organs randomly by mechanisms that have not been clearly defined. Each cell has 10–1000 mitochondria and each mitochondrion has 3–7 genomic DNA molecules that are normal or may have a certain mutation (Borst and Kroon 1969; Bogenhagen and Clayton 1974). The cell does not express the effects of that mutation until a specific number of mitochondria with the mutation are present. This is the property of threshold expression. Tissue-specific isoenzymes have also been identified and this may explain why certain organs are affected to a greater degree than others (Rizzuto et al. 1989).

RELATIONS TO OTHER DISORDERS

The family described here resembles 5 other families described by Bruyn and Went (1964), Miyoshi et al. (1969), and Marsden et al. (1986). Generalized dystonia was observed in all individuals with neurological symptoms and striatal lesions were discovered in neurologically impaired members who were examined by neuroimaging or neuropathologically. Eighteen of the 446 members of the pedigree reported by Bruyn and coworkers had similar clinical features. Six had both optic neuropathy and neurological symptoms. Six had either the neurological symptoms or the optic neuropathy alone. The 4 pedigrees described by both Miyoshi and Marsden had affected individuals in only 1 generation. Mental retardation was not present in the 6 cases in 2 families described by Marsden. However, Marsden and his colleagues stated that the inheritance pattern and clinical features of the families they described suggested a disorder of mitochondrial function similar to Leigh's disease.

Subacute necrotizing encephalomyelopathy or Leigh's disease is best described as a syndrome since both the clinical features and recent biochemical studies of the disorder have shown that it is an extremely heterogeneous condition. The clinical characteristics range from acute onset in the neonatal period with early death to chronic juvenile and adult variants (Kalimo et al. 1979). The neuropathology is variable as described by the review by Montpetit et al. (1971). At least 5 enzyme defects have been identified to cause this syndrome. Defects of pyruvate carboxylase (Hommes et al. 1968), pyruvate dehydrogenase complex (DeVivo et al. 1979), cytochrome c oxidase (Koga et al. 1990), complex I (NADH-coenzyme Q reductase) (Fujii et al. 1990), and succinate oxidation (Martin et al. 1988) have all been described. A common feature of these biochemical aberrations is that they all directly or indirectly affect mitochondrial oxidative metabolism. These oxidative defects may produce similar neuropathological findings (Walter et al. 1986). The pattern of neuropathological involvement observed in the family described here is analogous to other cases of Leigh's syndrome (Bargeton-Farkas et al. 1964; Rondot et al. 1982).

A definitive biochemical or molecular defect remains to be determined, though a disorder of mitochondrial oxidative metabolism is highly probable.

Hereditary putaminal necrosis (Druschky 1986) or infantile bilateral striatal necrosis (Mito et al. 1986), a term introduced by Friede (1975), has many similar features to the disorder described here in regard to the distribution of pathology. The families described by Miyoshi (1969) which are similar clinically to the pedigree reported here are often categorized as having infantile bilateral striatal necrosis (Mito et al. 1986). A major distinction is that the majority of cases have onset during the first year of life and are left with more severe neurological sequelae. This is undoubtedly due to the more extensive neuropathology documented in several cases (Paterson and Carmichael 1924; Erdohazi and Marshall 1979). Infantile bilateral striatal necrosis may represent one end of the spectrum of defective oxidative metabolism with Leber's optic neuropathy and asymptomatic carriers at the other end.

PATHOPHYSIOLOGY

Many aspects of this disorder suggest that it is due to a defect of oxidative metabolism, specifically dysfunction of the enzyme complexes of oxidative phosphorylation. It is also highly probable that this disorder is due to mutation of mtDNA. MtDNA is the only maternally inherited component of the human genome (Giles et al. 1980; Merrill and Harrington 1985). Each mtDNA codes for a large and small ribosomal RNA, 22 transfer RNAs, and 13 polypeptides that function in the enzyme complexes of oxidative phosphorylation (see Fig. 2). These polypeptides include 7 subunits of the 26 peptides of Complex I (NADH dehydrogenase complex), 1 subunit of the 10 peptides of Complex III (ubiquinol-cytochrome c oxidoreductase), 3 subunits of the 8 peptides of Complex IV (cytochrome c oxidase), and 2 subunits of the 12 peptides of Complex V (ATP synthase) (Wallace 1987). The exact nucleotide sequence of this 16 569 base pair portion of the human genome was determined by Anderson et al. (1981). From

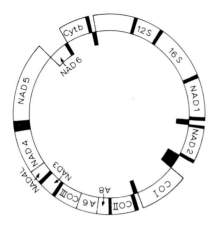

Fig. 2. Human mitochondrial DNA genomic map showing the location of the genes coding for specific mitochondrial proteins and RNAs. The coding regions on the outside of the map are on the heavy chain of the double-stranded mtDNA. The tRNAs are depicted by the solid rectangle (■); NAD, NADH dehydrogenase subunits; CO, cytochrome oxidase subunits; A, ATPase; Cytb, cytochrome *b*; and 12S and 16S, ribosomal RNAs.

analysis of the mtDNAs of affected family members of this pedigree it can be concluded that an insertion-deletion mutation is not responsible for this disease as has been described in other mitochondrial disorders (Holt et al. 1988; Zeviani et al. 1988). This suggests that the defect is a point mutation which will require sequencing to determine its location. This technique was recently applied to determine that Leber's hereditary optic neuropathy results from a point mutation of mtDNA (Wallace et al. 1988). This mutation alters the polypeptide of subunit 4 of NADH dehydrogenase. The activity of this enzyme complex was discovered to be defective in individuals with Leber's in an independent laboratory by Parker et al. (1989).

Although in this family and the pedigrees described by Bruyn and Went (1964) and Marsden et al. (1986), the maternal inheritance pattern suggest a mutation of mtDNA as the underlying molecular defect, other families with striatal necrosis may be the result of a nuclear DNA mutation since the majority of mitochondrial structural and functional proteins are coded for by the nuclear genome (Rizzuto et al. 1989; Wallace 1987). An autosomal recessive or dominant inheritance pattern would be observed in these families. However, many of the sporadic cases of striatal necrosis may be the result of mtDNA mutations since it is known that the spontaneous mutation rate of mammalian mtDNA is 5–10 times greater than that of the nuclear genes (Brown et al. 1979). Further investigation into the molecular biology of these disorders is needed to clarify these issues.

A clinical aspect of these disorders that requires explanation is the frequent observation that the affected individuals often have a fluctuating course with episodes of rapid deterioration in association with concurrent illness. This may be explained by environmental influences that augment the expression of the underlying defect. Several exogenous factors including hypoxia and ischemia may reduce the respiratory capacity of these cells with the mtDNA defect and lead to deterioration (Jellinger 1986). On the other hand, metabolic therapies directed toward increasing the reducing capacity may potentially decrease the rate and degree of deterioration (Eleff et al. 1984).

INVESTIGATION

Neuroimaging, preferably magnetic resonance imaging (MRI), should be obtained on anyone with dystonia to determine if a secondary etiology such as striatal necrosis is responsible for the disorder. The pattern of lesions identified on neuroimaging may correlate with specific enzyme defects (Medina et al. 1990). After exogenous factors have been excluded, 2 approaches to investigation of individuals with striatal necrosis may be undertaken.

First, investigation of the biochemical and enzymological aspects of the disorder is required. Measurements of bood lactate, pyruvate and amino acids may indicate the presence of an oxidative defect. Additional metabolic diseases that cause dystonia (Calne and Lang 1988) should be excluded by appropriate biochemical and enzyme assays as indicated by associated clinical features. Elevated cerebrospinal fluid lactate and pyruvate levels demonstrate the presence of central nervous system disease. Measurements of blood lactate and pyruvate with glucose infusion and exercise protocols increase the yield of determining the presence of a systemic or skeletal muscle defect. Investigation of skeletal muscle histochemistry and enzymology should be considered, but a limited number of laboratories perform extensive analyses of oxidative metabolism.

Another non-invasive technique that identifies biochemical defects in these individuals is nuclear magnetic resonance spectroscopy (NMRS). NMRS utilizing the phosphorus, ^{31}P, nucleus is able to measure intracellular pH and high energy metabolites such as ATP noninvasively. Studies on patients with disorders of mitochondrial function have already demonstrated specific changes in muscle using this technique (Radda et al. 1982; Arnold et al. 1985). NMRS investigations utilizing the proton nucleus currently permits measurement of cerebral levels of lactate and specific amino acids (Hanstock et al. 1988; Avison et al. 1990). Application of this technique will undoubtedly identify patients with defects of oxidative metabolism with elevated tissue levels of lactate that are not reflected in the blood.

Second, a molecular biological approach should be taken especially if there is evidence from the family history that the disorder is maternally inherited. Investigation of mtDNA is readily accomplished by obtaining blood samples from which the mtDNA is either extracted from platelets or from transformed lymphocytes. Digestion of the mitochondrial genome by site-specific restriction enzymes would detect the majority of insertion or deletion mutations. However, lesions smaller than 50 base pairs may not be detected. Alternatively, a region or all of the mtDNA genome could be sequenced and compared to the known nucleotide sequence. Single point mutations can then be identified. This is the most thorough method, but the procedure is extremely labor intensive. Finally, the gene products of the mtDNA can be studied. The various RNAs and their translation products are routinely investigated by either isoelectric focusing or polyacrylamide gel electrophoresis. As specific mutations are discovered and molecular probes for investigation of the nuclear genome are developed, these techniques will be routinely applied to the investigation of disorders

such as the one described here in which there is evidence of a defect of mitochondrial oxidative metabolism or mtDNA mutation (Wallace et al. 1986).

Hereditary dystonic paraplegia with amyotrophy and mental deficiency (105300)

CLINICAL AND NEUROPATHOLOGICAL CHARACTERISTICS

The clinical characteristics of this disorder were first reported by Gilman and Horenstein (1964). A family encompassing 3 generations in which 12 of 27 had neurological findings on examination was described to have variable expression of the neurological disorder. Two of the most severely affected members had a spastic paraparesis and intellectual impairment early in their course during the first decade of life. Muscular cramps and weakness involving the distal lower extremities later developed. Generalized dystonia with prominent involvement of the bulbar musculature began in adolescence and progressed through early adulthood. Their gait gradually deteriorated and bowel and bladder dysfunction was observed late in the illness. The 2 most severely affected individuals died in the fifth decade of life from respiratory complications of their neurological illness.

The least affected members with the disorder were asymptomatic and their ages ranged from 18 months to 40 years old. Corticospinal tract findings were the earliest signs observed in all members with neurological impairment. Two other boys had complaints of fatigue and leg cramps during the first decade of life. The disease was slowly progressive or remained static in the majority of affected members.

The neuropathology of this disorder as described by Gilman and Romanul (1975) demonstrated multisystem involvement in the 2 individuals who clinically had the most severe disease. Pronounced ventricular enlargement and atrophy of the caudate were present on gross examination. Gliosis and neuronal loss were revealed throughout the neocortex and were most prominent in the thalamic, subthalamic and red nuclei. The neostriatum and globus pallidus were affected to a moderate degree. Neuronal loss was observed in the substantia nigra and locus ceruleus. Anterior horn cell loss was noted at multiple levels throughout the spinal cord. Degeneration of both the lateral and anterior corticospinal tracts of the spinal cord was demonstrated. The large myelinated fibers of the ventral roots and peripheral nerves were decreased in number. Neuronal loss and gliosis was seen in the dorsal root ganglia. Denervation atrophy was demonstrated in all muscle groups examined.

GENETICS

The members with neurological impairment in this pedigree were in two generations. The most severely affected individuals, a woman and man well into their reproductive age, had no children. The 10 other less affected individuals were the progeny of 4 asymptomatic family members 2 of whom had evidence of corticospinal tract dysfunction on exam. Eight of these 10 were less than 12 years of age. The inheritance pattern is consistent with autosomal dominant transmission with incomplete penetrance.

DISCUSSION

The early corticospinal tract abnormalities which persist throughout the course of the illness indicate that this disorder is related to the hereditary spastic paraplegias. This group of neurogenetic diseases frequently have other associated neurological findings (Bundey 1985). Autosomal dominant (Boustany et al. 1987), autosomal recessive (Harding 1981) and X-linked (Keppen et al. 1987) inheritance patterns have been observed in both the 'pure' and complex variants of this group of diseases. Spastic paraplegia has also been associated with Leber's optic atrophy (Wilson 1963; Pages and Pages 1983). Maternal inheritance must be considered in these families. The family described here has many similarities with other families with spastic paraplegia described in the literature as related by Gilman and Romanul (1975). They related that the cases described by Miyoshi et al. (1969), Brown and Coleman (1966), and Dick and Stevenson (1953) were particularly analogous to this family.

Modern molecular genetic techniques have led to extraordinary advances in the diagnosis and understanding of many neurological illnesses (Botstein et al. 1980; Rosenberg 1984). The past use of enzyme and antigen markers present on specific chromosomes have had limited success in determining the molecular basis of hereditary neurological diseases (Frydman et al. 1985; Keppen et al. 1987). Modern recombinant DNA techniques provide many new chromosome markers to allow more detailed analysis of the human genome for evidence of linkage to specific regions of certain chromosomes (Botstein et al. 1980). These techniques have permitted the assignment of specific neurological diseases to portions of the human genome and ultimately to the discovery of the gene itself.

Application of these techniques to hereditary movement disorders has been particularly successful (Kurlan et al. 1989). The gene for Huntington's disease has been localized to chromosome 4 (Gusella et al. 1983). The gene for the autosomal dominantly inherited primary dystonias has recently been localized to chromosome 9 (Kramer et al. 1990). As more genes are discovered to be associated with clinically specific and neuropathologically distinct disorders, their function and interaction in producing complex neurological diseases such as the one described here will require further investigations.

REFERENCES

ANDERSON, S., A. T. BANKIER, B. G. BARRELL, M. H. L. DE BRUIJN, A. R. COULSON, J. DROVIN, I. C. EPERON, D. P. NIERLICH, B. A. ROE, F. SANGER, P. H. SCHRIER, A. J. H. SMITH, R. STADEN and I. G. YOUNG: Sequence and organization of the human mitochondrial genome. Nature 290 (1981) 457–465.

ARNOLD, D. L., D. J. TAYLOR and G. K. RADDA: Investigations of human mitochondrial myopathies by phosphorus magnetic resonance spectroscopy. Ann. Neurol. 18 (1985) 189–196.

AVISON, M., N. HERSCHKOWITZ, E. J. NOVOTNY, O. A. C. PETROFF, D. L. ROTHMAN, J. P. COLOMBO, C. BACHMANN, R. G. SHULMAN and J. W. PRICHARD: Proton NMR observation of phenylalanine and an aromatic metabolite in the rabbit brain in vivo. Pediatr. Res. 27 (1990) 566–570.

BARGETON-FARKAS, E., A. M. COCHARD, H. E. BRISSAUD, O. ROBAIN and J. C. LE BALLE: Encéphalopathie infantile familiale avec nécrose bilatérale et symétrique des corps striés. J. Neurol. Sci. 1 (1964) 429–445.

BOGENHAGEN, W. and D. A. CLAYTON: The number of mitochondrial deoxyribonucleic acid genomes in normal and human HeLa cells. J. Biol. Chem. 249 (1974) 7991–7995.

BORST, P. and A. M. KROON: Mitochondrial DNA: Physicochemical properties, replication, and genetic function. Int. Rev. Cytol. 26 (1969) 107–190.

BOTSTEIN, D., R. L. WHITE, M. SKOLNICK and R. W. DAVIS: Construction of a genetic linkage map in man using restriction fragment length polymorphisms. Am. J. Hum. Genet. 32 (1980) 314–331.

BOUSTANY, R. M., E. FLEISCHNICK, C. A. ALPER, M. L. MARAZITA, M. A. SPENCE, J. B. MARTIN and E. H. KOLODNY: The autosomal dominant form of 'pure familial spastic paraplegia': clinical findings and linkage analysis of a large pedigree. Neurology 37 (1987) 910–915.

BROWN, J. W. and R. F. COLEMAN: Hereditary spastic paraplegia with ocular and extra-pyramidal signs. Bull. Los Angeles Neurol. Soc. 31 (1966) 21–34.

BROWN, W. M., M. GEORGE, JR. and A. C. WILSON: Rapid evolution of animal mitochondrial DNA. Proc. Natl Acad. Sci. U.S.A. 76 (1979) 1967–1971.

BRUYN, G. W. and L. N. WENT: A Sex-linked heredodegenerative neurological disorder, associated with Leber's optic atrophy. Part 1. clinical studies. J. Neurol. Sci. 1 (1964) 59–80.

BUNDEY, S.: Spastic paraplegias. In: S. Bundey (Ed.), Genetics and Neurology. London, Churchill Livingstone (1985) 241–261.

CALNE, D. B. and A. E. LANG: Secondary dystonia. In: S. Fahn, C.D. Marsden and D. B. Calne (Eds.), Advances in Neurology, Vol. 50, Dystonia 2. New York, Raven Press (1988) 9–33.

DEVIVO, D. C., M. W. HAYMOND, K. A. OBERT, J. S. NELSON and A. S. PAGLIARA: Defective activation of the pyruvate dehydrogenase complex in subacute necrotizing encephalomyelopathy (Leigh disease). Ann. Neurol. 6 (1979) 483–494.

DICK, A. and C. STEVENSON: Hereditary spastic paraplegia. Report on a family with associated extrapyramidal signs. Lancet 264 (1953) 921–923.

DRUSCHKY, K-F.: Hereditary putaminal necrosis. In: P. J. Vinken, G. W. Bruyn and H. L. Klawans (Eds.), Handbook of Clinical Neurology, Vol. 5 (49): Extrapyramidal Disorders. Amsterdam, Elsevier (1986) 493–498.

ELEFF, S., N. G. KENNAWAY, N. R. M. BUIST, V. M. DARLEY-USMAR, R. A. CAPALDI, W. J. BANK and B. CHANCE: ^{31}P NMR study of improvement in oxidative phosphorylation by vitamins K_3 and C in a patient with a defect in electron transport at complex III in skeletal muscle. Proc. Natl. Acad. Sci. U.S.A. 81 (1984) 3529–3533.

ERDOHAZI, M. and P. MARSHALL: Striatal degeneration in childhood. Arch. Dis. Child. 54 (1979) 85–91.

FAHN, S.: Concept and classification of dystonia. In: S. Fahn, C. D. Marsden and D. B. Calne (Eds.), Advances in Neurology, Vol. 50, Dystonia 2. New York, Raven Press (1988) 1–8.

FRIEDE, R.: Various topographic patterns of postnatal neuron loss. In: Developmental Neuropathology. Wien/New York, Springer-Verlag (1975) 88–89.

FRYDMAN, M., B. BONNE-TAMIN, L. A. FARRER, P. N. CONNEALLY, A. MAGAZANIK, S. ASHBEL and Z. GOLDWITCH: Assignment of the gene for Wilson's disease to chromosome 13: linkage to esterase-D locus. Proc. Natl. Acad. Sci. U.S.A. 82 (1985) 1819–1821.

FUJII, T., M. ITO, T. OKUNO, K. MUTOH, R. NISHIKOMORI and H. MIKAWA: Complex-I (reduced nicotinamide-adenine dinucleotide coenzyme-Q reductase) deficiency in 2 patients with probable Leigh syndrome. J. Pediatr. 116 (1990) 84–87.

GILES, R. E., H. BLANC, H. M. CANN and D. C. WALLACE: Maternal inheritance of human mitochondrial DNA. Proc. Natl. Acad. Sci. U.S.A. 77 (1980) 6715–6719.

GILMAN, S. and S. HORENSTEIN: Familial amyotrophic dystonic paraplegia. Brain 87 (1964) 51–66.

GILMAN, S. and F. C. A. ROMANUL: Hereditary dystonic paraplegia with amyotrophy and mental deficiency: clinical and neuropathological characteristics. In: P. J. Vinken and G. W. Bruyn (Eds), Handbook of Clinical Neurology, Vol. 22: Disease of the Motor System. Amsterdam, North-Holland Publ. Co. (1975) 445–465.

GUSELLA, J. F., N. S. WEXLER, P. M. CONNEALLY, S. L. NAYLOR, M. A. ANDERSON, R. E. TANZI, P. C. WATKINS, K. OTTINA, M. R. WALLACE, A. Y. SAKAGUCHI, A. B. YOUNG, I. SHOULSON, E. BONILLA and J. B. MARTIN: A polymorphic DNA marker genetically linked to Huntington's disease. Nature 306 (1983) 234–238.

HANSTOCK, C. C., D. L. ROTHMAN, J. W. PRICHARD, T. JUE and R. G. SHULMAN: Spatially localized ^{1}H NMR spectra of metabolites in human brain. Proc. Natl. Acad. Sci. U.S.A. 85 (1988) 1821–1825.

HARDING, A. E.: Hereditary 'pure' spastic paraplegia: a clinical and genetic study of 22 families. J. Neurol. Neurosurg. Psychiatry 44 (1981) 871–883.

HOLT, I. J., A. E. HARDING and J. A. MORGAN-HUGHES: Deletions of muscle mitochondrial DNA in patients with mitochondrial myopathies. Nature 331 (1988) 717–719.

HOMMES, F. A., H. A. POLMAN and J. D. REERINK: Leigh's encephalomyelopathy: an inborn error of gluconeogenesis. Arch. Dis. Child. 43 (1968) 423–426.

JELLINGER, K.: (Exogenous) striatal necrosis. In: P. J. Vinken, G. W. Bruyn and H. L. Klawans (Eds), Handbook of Clinical Neurology, Vol. 5 (49): Extrapyramidal Disorders. Amsterdam, Elsevier (1986) 499–518.

JOHNSON, M. J., D. C. WALLACE, S. D. FERRIS, M. C. RATAZZI, L. L. CAVALLI-SFORZA: Radiation of human mitochondrial DNA types analyzed by restriction endonuclease cleavage patterns. J. Mol. Evol. 19 (1983) 255–271.

KALIMO, H., P. O. LUNDBERG and Y. OLSSON: Familial sub-

acute necrotizing encephalomyelopathy of the
adult form (adult Leigh syndrome). Ann. Neurol.
6 (1979) 200–206.

KEPPEN, L. D., M. F. LEPPERT, P. O'CONNELL, Y. NAKAMURA,
D. STAUFFER, M. LATHROP, J. M. LALOUEL and R. WHITE:
Etiological heterogeneity in X-linked spastic
paraplegia. Am. J. Med. Genet. 41 (1987)
933–943.

KOGA, Y., I. NONAKA, M. NAKAO, M. YOSHINO, M. TANAKA,
T. OZAWA, H. NAKASE and S. DIMAURO: Progressive
cytochrome-C oxidase deficiency in a case of
Leigh encephalomyelopathy. J. Neurol. Sci. 95
(1990) 63–76.

KRAMER, P. L., D. DE LEON, L. OZELIUS, N. RISCH, S. B. BRESS-
MAN, M. F. BRIN, D. E. SCHUBACK, R. E. BURKE, D. J. KWIAT-
KOWSKI, H. SHALE, J. F. GUSELLA, X. O. BREAKEFIELD and
S. FAHN: Dystonia gene in ashkenazi jewish popula-
tion is located on chromosome 9q32–34. Ann.
Neurol. 27 (1990) 114–120.

KURLAN, R., K. K. KIDD and D. PAULS: Linkage analysis
approach to hereditary movement disorders. J.
Neurogenet. 5 (1989) 161–171.

MARSDEN, C. D., A. E. LANG, N. P. QUINN, W. I. MCDONALD,
A. ABDALLAT and S. NIMIRI: Familial dystonia and
visual failure with striatal CT lucencies. J. Neurol.,
Neurosurg. Psychiatry 49 (1986) 500–509.

MARTIN, J. J., F. L. VAN DE VYVER and H. R. SCHOLTE: Defect
in succinate oxidation by isolated muscle mitochondria in a patient with symmetrical lesions in
the basal ganglia. J. Neurol. Sci. 84 (1988)
189–200.

MEDINA, L., T. L. CHI, D. C. DEVIVO and S. K. HILAL: MR
findings in patients with subacute necrotizing
encephalomyelopathy (Leigh syndrome): Correlation with biochemical defect. Am. J. Neuroradiol.
11 (1990) 379–384.

MERRILL, C. R. and M. G. HARRINGTON: The search for
mitochondrial inheritance of human diseases.
Trends Genet. (1985) 140–144.

MITO, T., T. TANAKA, L. E. BECKER, S. TAKASHIMA and
J. TANAKA: Infantile bilateral striatal necrosis:
clinicopathological classification. Arch. Neurol.
(Chic.) 43 (1986) 677–680.

MIYOSHI, K., T. MATSUOKA and S. MIZUSHIMA: Familial
holoptistic striatal necrosis. Acta Neuropathol.
(Berl.) 13 (1969) 240–249.

MONTPETIT, V. J. A., F. ANDERMANN, S. CARPENTER, J. S. FAW-
CETT, D. ZBOROWSKA-SLUIS and H. R. GIBERSON: Subacute necrotizing encephalomyelopathy. A review
and study of two families. Brain 94 (1971)
1–30.

NOVOTNY, E. J., JR., G. SINGH, D. C. WALLACE, L. J. DORF-
MAN, A. LOUIS, R. L. SOGG and L. STEINMAN: Leber's
disease and dystonia: A mitochondrial disease.
Neurology 36 (1986) 1053–1060.

PAGES, M. and A.-M. PAGES: Leber's disease with spastic
paraplegia and peripheral neuropathy. Eur. Neurol. 22 (1983) 181–185.

PARKER, W. D., JR., C. A. OLEY and J. K. PARKS: A defect
in mitochondrial electron-transport activity
(NADH-Coenzyme Q oxidoreductase) in Leber's

hereditary optic neuropathy. N. Engl. J. Med.
320 (1989) 1331–1333.

PATERSON, D. and E. A. CARMICHAEL: A form of familial
cerebral degeneration chiefly affecting the lenticu-
lar nuclei. Brain 47 (1924) 207–231.

RIZZUTO, R., H. NAKASE, B. DARRAS, U. FRANCKE, G. M.
FABRIZI, T. MENGEL, F. WALSH, B. KADENBACH, S. DI-
MAURO and E. A. SCHON: A gene specifying subunit
VIII of human cytochrome c oxidase is localized
to chromosome 11 and is expressed in both
muscle and non-muscle tissues. J. Biochem. 264
(1989) 10596–10600.

RADDA, G. K., P. J. BORE, D. G. GADIAN, B. D. ROSS, P. STYLES,
D. J. TAYLOR and J. A. MORGAN-HUGHES: ^{31}P NMR
examination of two patients with NADH-CoQ
reductase deficiency. Nature 295 (1982) 608–609.

RONDOT, P., J. DE RECONDO, P. DAVOUS, D. FREDY and
F.-X. ROUX: Rigidité extra-pyramidale avec dystonie,
atrophie optique et atteinte bilatérale du putamen
chez un adolescent. Forme juvénile de la maladie
de Leigh. Rev. Neurol. (Paris) 138 (1982) 143–148.

ROSENBERG, R. N.: Molecular genetics, recombinant
DNA techniques and genetic neurological disease.
Ann. Neurol. 15 (1984) 511–520.

RÖYTTÄ, M., I. OLSSON, P. SOURANDER and
P. SVENDSEN: Infantile bilateral striatal necrosis.
Clinical and morphological report of a case and
a review of the literature. Acta Neuropathol.
(Berl.) 55 (1981) 97–103.

SEIDENWURM, D., E. J. NOVOTNY, JR., W. MARSHALL and
D. ENZMANN: MR and CT in cytoplasmically inherited striatal degeneration. Am. J. Neuroradiol. 7
(1986) 629–632.

SINGH, G., M. T. LOTT and D. C. WALLACE: A mitochondrial DNA mutation as a cause of Leber's
hereditary optic neuropathy. N. Engl. J. Med.
320 (1989) 1300–1305.

VAN ERVEN, P. M. M., W. RUITENBEEK, F. J. M. GABREELS,
W. O. RENIER, J. C. FISCHER and A. J. M. JANSSEN: Disturbed oxidative metabolism in subacute necrotizing encephalomyelopathy (Leigh syndrome).
Neuropediatrics 17 (1986) 28–32.

WALLACE, D. C.: Maternal genes: mitochondrial diseases. In: V. A. McKusick, T. H. Roderick, J.
Mori and N. W. Paul (Eds.), Birth defects original
article series, Vol. 23, No. 3: Medical and Experimental Genetics: A Perspective. New York, Alan
R. Liss (1987) 137–190.

WALLACE, D. C., G. SINGH, L. C. HOPKINS and E. J. NOVOTNY,
JR.: Maternally diseases of man. In: E. Quagliar-
iello, E. C. Slater, F. Palmieri, C. Saccone and A.
M. Kroon (Eds.), Achievements and Perspectives
of Mitochondrial Research, Vol. 2: Biogenesis.
Amsterdam, Elsevier (1986) 427–436.

WALLACE, D. C., G. SINGH, M. T. LOTT, J. A. HODGE,
T. G. SCHURR, A. M. S. LEZZA, L. J. ELSAS II and E. K.
NIKOSKELAINEN: Mitochondrial DNA mutation as-
sociated with Leber's hereditary optic neuropathy.
Science 242 (1988) 1427–1430.

WALTER, G. F., J. M. BRUCHER, J. J. MARTIN, C. CEUTERICK,

P. PILZ and M. FREUND: Leigh's disease — several nosological entities with an identical histopathological complex? Neuropathol. Appl. Neurobiol. 12 (1986) 95–107.

WILSON, J.: Leber's hereditary optic atrophy. Some clinical and aetiological considerations. Brain 86 (1963) 347–362.

ZEVIANI, M., C. T. MORAES and S. DIMAURO: Deletions of mitochondrial DNA in Kearns-Sayre syndrome. Neurology 38 (1988) 1339–1346.

Handbook of Clinical Neurology, Vol. 15 (59): Diseases of the Motor System
J.M.B.V. de Jong, editor
© Elsevier Science Publishers B.V., 1991

CHAPTER 21

Spastic paraparesis due to metabolic disorders

HUGO W. MOSER[1], ANN BERGIN[2] and SAKKUBAI NAIDU[3]

[1]*Center for Research on Mental Retardation and Related Aspects of Human Development and* [2,3]*Department of Neurogenetics, The Kennedy Institute, Baltimore, MD, USA*

In this chapter we attempt to apply to the clinical problem of spastic paraparesis a set of advances in genetics and biochemistry that have implications for all of neurology. Thanks to the development of non-invasive biochemical assays it is now possible to achieve the precise and early diagnosis of a group of genetically determined disorders. In the past this often was possible only by biopsy or postmortem examination, and in some instances even not then. These are important advances. All of the metabolic disorders can be diagnosed prenatally, and carriers can be identified in most instances. Diagnosis of the affected person thus permits genetic counseling. Even more encouraging is the fact that specific therapies are being developed for several of these diseases.

CAUSES OF SPASTIC PARAPARESIS

A survey of 672 patients with spastic paraparesis admitted to Columbia-Presbyterian Medical Center in New York between 1960 and 1976 revealed that spinal cord trauma was the most common cause (27%) followed by tumors, multiple sclerosis and congenital anomalies (Ungar-Sargon et al. 1980). Spastic paraparesis is probably more common in developing countries, with tuberculous spinal osteitis, syphilis, schistosomiasis, epidural abscess, HTLV-1 infection (Roman et al.

1987) and chemical agents (Editorial 1984 and Osuntokun 1968) contributing to this higher incidence. In the New York series (Ungar-Sargon et al. 1980), 2.3% of the spastic paraparesis cases were familial and in 6.5% no etiology could be identified. The causes to be discussed in this chapter probably represent a subgroup of the 'familial' and 'unknown cause' cases, which in this New York series comprised 8.8% of the total.

FAMILIAL SPASTIC PARAPARESIS

While familial spastic paraparesis accounts for only a small proportion of the total, it is heterogeneous. It includes 'pure' forms in which clinical and pathological abnormalities are limited to the corticospinal tracts (Younger et al. 1988) and the more common types in which other parts of the nervous system (Behan and Maia 1974) or other organs are also involved. The most common pattern of inheritance is autosomal dominant (Sack et al. 1978; Harding 1981; Boustany et al 1987), autosomal recessive (Harding 1981), and more rarely X-linked recessive (Johnston and McKusick 1961; Kenwrick et al. 1962; Young et al. 1984; Keppen et al. 1987; Goldblatt et al. 1989). Even the X-linked recessive forms appear to be heterogeneous, since in one kindred it was mapped to Xq28 (Kenwrick et al. 1962), and in another to Xq21–22 (Keppen et al. 1987).

GENERAL ISSUES RELATING TO THE USE OF NEWLY DEVELOPED BIOCHEMICAL DIAGNOSTIC ASSAYS

Experience gained through the application of these new diagnostic tests has pointed out the need to re-examine disease classifications, and also has raised questions that are still unresolved. The need to re-examine diagnostic formulations stems from the probably universal finding that the range of phenotypic variability of these disorders is greater than had been realized. The usual sequence is that the biochemical basis of a disorder and a diagnostic assay are established by the study of severely affected individuals with a 'classical' phenotype, e.g. the well known Tay Sachs disease that causes death in early childhood. Tay Sachs disease was shown to be a GM_2 gangliosidosis. Then, as the diagnostic assay for GM_2 gangliosidosis was applied more widely, it was recognized that there are milder phenotypes, e.g. the adult form of GM_2 gangliosidosis. This finding necessitated re-examination of the clinical classification both of Tay Sachs disease and the GM_2 gangliosidosis. The same sequence applies to other disorders described in this chapter, and probably to those that are not yet recognized. The challenge is to correlate these new biochemical advances with classical neurology. The biochemical tests do not replace clinical judgement and experience. Anyone who thought that they might, has been chastened by the demonstration of 'pseudodeficiency states' in which an apparent biochemical defect is clinically insignificant, e.g. arylsulfatase A pseudodeficiency (Gieselman et al. 1989). The goal is to correlate biochemical and neurological findings. This process is still evolving. The intent is to re-examine, refine and modify clinical classifications and diagnostic criteria so that the newly developed diagnostic assays can be applied precisely and in a cost-efficient manner.

The long-range goal to apply the new diagnostic procedures in a precise and cost-effective manner has not yet been achieved, both because of insufficient knowledge and operational prob-

lems. The insufficiency of knowledge is due to the fact that we still do not have complete information about the full range of phenotypic variability. For example, it has been recognized only during the last few years that globoid leukodystrophy may present as spastic paraparesis in adulthood and without abnormalities in MRI, cerebrospinal protein levels or peripheral nerve function. Prior to this recognition the performance of this diagnostic assay in this setting would have been considered financially irresponsible. Since the chance for therapy and timely genetic counseling is greatest when the diagnosis is made early in the course of the disease, the tests may be most valuable when signs and symptoms are subtle. These considerations argue for a low 'threshold' for the performance of the assays.

Operational issues relate to the fact that several of the assays are performed only in a few specialized laboratories and that quality control and reimbursement mechanisms are in need of improvement. A network of regional laboratories is desirable but not yet generally available. Until this is changed the clinician must consult knowledgeable colleagues and utilize personal judgement and initiative to ensure that the appropriate tests are performed in a reliable manner.

SPASTIC PARAPARESIS DUE TO DEFINED METABOLIC DISORDERS

The metabolic disorders associated with spastic paraparesis to be discussed in this chapter all are genetically determined, and include both autosomal recessive and X-linked recessive modes of inheritance. As far as we know none of the familial spastic paraparesis cases cited above have as yet been linked to a specific biochemical defect. This probably reflects the genetic heterogeneity of spastic paraparesis and highlights the need for continued genetic and biochemical investigations.

Table 1 lists the metabolic disorders that may be associated with spastic paraparesis.

ADRENOMYELONEUROPATHY (300100)

Adrenomyeloneuropathy (AMN) is possibly the most common metabolic cause of spastic parapa-

TABLE 1

Disease entity	Mode of inheritance	Basic defect	Additional clinical features	Diagnostic test	Therapy
Adrenomyeloneuropathy	X-Linked recessive	Lignoceroyl CoA ligase	60% have Addison disease	Very long chain fatty acids	?Diet
Adrenoleukodystrophy Symptomatic heterozygotes	X-Linked recessive	Lignoceroyl CoA ligase	Adrenal function normal; misdiagnosed as multiple sclerosis	Very long chain fatty acids 15% false negatives adrenal function normal	?Diet
Globoid leukodystrophy	Autosomal recessive	Galactocerebrosidase	Often begins in adolescence with hemiparesis: intellect often normal	Enzyme assay	?Bone marrow transplant
Metachromatic leukodystrophy	Autosomal recessive	Sulfatidase	Preceded by psychosis, dementia; peripheral nerve involved	Enzyme assay rule out pseudodeficiency	?Bone marrow transplant
Pelizaeus Merzbacher disease	X-Linked recessive	Impaired synthesis of myelin proteolipid	Nystagmus	Mutations involving proteolipid gene	None
Cerebrotendinous xanthomatosis	Autosomal recessive	Impaired bile acid synthesis	Tendon xanthoma; cataract; mental defect	Increased serum cholestanol increased urine bile acid alcohols	Chenodeoxycholic acid
Sjögren-Larsson syndrome	Autosomal recessive	Long chain fatty alcohol reductase	Ichthyosis; mental retardation	enzyme assay	Low fat diet medium chain triglycerides
Arginase deficiency	Autosomal recessive	Arginase	Intermittent hyperammonemia; mental retardation	Plasma arginine increased; red blood cell enzyme assay; intermittent hyperammonemia	Low protein diet sodium benzoate
Homocarnosinosis	?	Carnosinase	Relation to phenotype uncertain; may be harmless	CSF homocarnosine	None may be needed
Cobalamin metabolism	Autosomal recessive	Various defects of cobalamin metabolism	Megaloblastic anemia sometimes; cerebellar deficit, dementia	Homocystine in plasma and urine; methylmalonic in urine specialized assays needed to define defect exactly	Cobalamin
Lesch Nyhan syndrome	Autosomal recessive	Hypoxanthine phosphoribosyl transferase	Self mutilation; athetosis; mental retardation	Uric acid/creatinine ratio in urine: enzyme assay in RBC	Allopurinol for kidney; none for nervous system
Salla disease	Autosomal recessive	Sialic acid transport	Dysmorphism; ataxia; mental retardation	Increased urine sialic acid; lysosomal storage	None

resis in adults. It was first described by Budka et al. (1976) and shortly thereafter by Griffin et al. (1977). It is now clear that AMN is the adult variant of adrenoleukodystrophy (ALD) (Moser et al. 1984). It is an X-linked recessive disorder. The gene has been mapped to Xq28, the terminal segment of the long arm of the X-chromosome (Migeon et al. 1981), and it is genetically linked to a DNA probe for that portion of the chromosome (Aubourg et al. 1987). The main biochemical abnormality is the accumulation of saturated very long chain fatty acids, particularly hexacosanoic acid (C26:0) in the nervous system white matter, adrenal cortex and testis (Igarashi et al. 1976; Moser et al. 1984). This accumulation is due to the impaired capacity to degrade these fatty acids (Singh et al. 1984a), a reaction that normally takes place in the peroxisome (Singh et al. 1984b). The underlying defect probably involves the impaired capacity to form the Coenzyme A derivative of very long chain fatty acids (Lazo et al. 1988; Wanders et al. 1988).

X-linked ALD shows great phenotypic variability. The most common and serious variant is childhood-ALD, which presents most commonly between 4 and 8 years, and often leads to a vegetative state within 2 years. Another variant, which is more common than had been realized, is Addison disease without neurological involvement. AMN is the phenotype in approximately 25% of persons with the ALD genotype. We have identified more than 300 AMN patients. The various phenotypes occur frequently within the same family.

The initial symptoms of AMN most commonly are noted in the early or mid-twenties as an impairment to engage in sports or run, or maneuver stairs. The deficit then progresses to increasing difficulty in walking due to spasticity, and after a decade or more assistive devices or wheelchair may be required. Bladder control problems are invariably present. Impaired sexual function develops later, although many of the AMN patients have fathered children. Impaired vibration sense in the distal aspects of the lower extremities is a universal finding, and some patients have a variable sensory level up to the thoracic region. The upper extremities may show increased deep tendon reflexes, but are functionally impaired only rarely. Pathologically the main abnormality is an axonopathy that involves the lumbar corticospinal, cervical gracile and the dorsal spinocerebellar tracts (Powers 1985). The peripheral nerve pathology also is mainly an axonopathy. Somatosensory evoked responses and brain stem auditory evoked responses are almost invariably abnormal. MRI often shows thoracic cord atrophy. Approximately 40% of patients also show brain white matter involvement by MRI, and approximately an equal number show some degree of dementia and depression. The cerebral MRI changes and the intellectual alterations may not always correlate.

Approximately 60% of AMN patients have Addison disease or impaired adrenocortical function. The Addison disease may have been present since childhood, with neurological disability supervening years or even decades later. Conversely, approximately 30% of AMN patients with established neurological deficits have normal adrenal function, including normal ACTH levels and cortisol response to ACTH.

Specific diagnosis of AMN depends upon demonstration of abnormally increased very long chain fatty acid levels in plasma (Moser et al. 1981), cultured skin fibroblasts (Moser et al. 1980) or red blood cells (Tsuji et al. 1981). Provided that the test is performed in a laboratory that has experience with these assays, this test is definitive. It is essential to perform this test in every patient with Addison disease. However, since a significant proportion of AMN patients have normal adrenal function, we believe that very long chain fatty acid analyses should be performed in all patients with unexplained progressive paraparesis. Since the disorder is X-linked recessive, male offspring of AMN patients will not be affected, but all daughters are obligate heterozygotes. Because of the importance of genetic counseling all sibs and other at-risk relatives should be tested for ALD.

It is now possible to normalize the plasma very long chain fatty acid levels by a dietary regimen that combines the dietary restriction of very long chain fatty acids with the administration of glycerol trioleate (Rizzo et al. 1987b; Moser et al. 1987) and glycerol trierucate (Rizzo et al. 1989a). Studies are in progress to determine whether this can prevent or ameliorate neurological disability. One patient with the childhood form of ALD benefited greatly from a bone marrow transplant (Aubourg et al. 1990). While methods to ameliorate the neurological deficit of AMN are still under investigation, it is essential that adrenal insufficiency be treated by appropriate steroid replacement therapy.

NEUROLOGICALLY SYMPTOMATIC FEMALE ADRENOLEUKODYSTROPHY HETEROZYGOTES

Approximately 15% of women who are heterozygous for ALD develop some degree of neurological disability (Pilz and Schiener 1973; O'Neill et al. 1982; Budka et al. 1983; O'Neill et al. 1984; Naidu et al. 1986; Powers et al. 1987; Noetzel et al. 1987; Naidu et al. 1989). Usually this resembles AMN, but is milder and of later onset with the mean age of onset of symptoms being 37 years. The affected women show spastic paraparesis combined with sphincter disturbances and impaired vibration sense in the distal lower extremities, and occasionally dysesthesias and numbness. Dementia is observed occasionally (Morariu et al. 1982). Adrenal insufficiency has been reported (Heffungs et al. 1980; Pilz and Schiener 1973), but it is rare. We have demonstrated normal adrenal function in more than 95% of our patients. Carriers can be identified by biochemical assays of very long chain fatty acids in plasma and cultured skin fibroblasts (Moser et al. 1983), but false negative results occur in about 15% of obligate carriers. In informative families the accuracy of carrier identification can be improved by DNA linkage studies (Aubourg et al. 1987). The diagnosis of neurologically involved ALD carriers is made more difficult because adrenal insufficiency, an important clue to the diagnosis in males, is almost never present. Furthermore the clinical manifestations often resemble those of multiple sclerosis (Dooley and Wright 1985). Up to this time almost all neurologically symptomatic ALD carriers who have been identified are relatives of male ALD patients. Clearly there must be symptomatic heterozygotes without a known affected male ALD relative. The diagnosis in these patients is complicated by the resemblance of the phenotype to multiple sclerosis and the fact that the present diagnostic assays for ALD heterozygotes (Moser et al. 1983) are not as reliable as those for affected males. Up to this time we have identified only 5 women in whom we felt confident to attribute neurological disability to the ALD heterozygote state in the absence of affected male relatives, compared to more than 150 women who were relatives of male ALD patients.

We consider it likely that male AMN patients and neurologically symptomatic ALD heterozygotes account for a larger proportion of spastic paraparesis than is currently recognized. More widespread application of diagnostic assays appears warranted, and there is also need for an improved method of heterozygote identification.

THE ADULT FORM OF GLOBOID LEUKODYSTROPHY (245200)

The globoid (Krabbe) form of leukodystrophy most commonly presents in infancy at 3–4 months of age and is rapidly progressive leading to death during the first or second year (Suzuki and Suzuki 1989). There is severe involvement of central nervous and peripheral nerve myelin with the characteristic accumulation of multinucleate (globoid) cells and the protein level in cerebrospinal fluid is markedly increased. Deficient function of the lysosomal enzyme galactosylceramidase is the basic defect, and this deficiency can be demonstrated in serum, leukocytes and in cultured skin fibroblasts and amniocytes.

Rarer late infantile and juvenile forms of the disease also occur (Hagberg 1984; Thomas et al. 1984; Fluharty et al. 1986; Fiumara et al. 1990; Goebel et al. 1990). These cases present at 1.5–15 years of age in a variable fashion with spastic hemiparesis or with visual failure, or cerebellar ataxia or with dementia-psychosis. The diagnosis of globoid leukodystrophy was often delayed because about half of these patients did not show increased cerebrospinal fluid protein levels or evidence of peripheral nerve involvement, abnormalities that in the past were considered obligatory. MRI is of diagnostic value. Initially there may be hyperdense plaque-like lesions followed later by reduction in gray matter and white matter mass with evidence of loss of myelin

(Baram et al. 1986; Cavanagh and Kendall 1986; Darras et al. 1986).

Still more recently it has become evident that globoid leukodystrophy may be seen in adults even as late as the eighth decade (Kolodny E., pers. commun.). These patients may present with slowly progressive hemiparesis in early adulthood. In others the hemiparesis begins in late childhood or adolescence and progresses slowly. Intellectual function may remain fully intact. Globoid leukodystrophy must therefore be included in the differential diagnosis of progressive hemiparesis or spastic paraparesis in adults. Diagnostic leads may be provided by white matter abnormalities demonstrable by MRI, but these may not be characteristic, and definitive diagnosis depends upon demonstration of galactocerebrosidase deficiency (Suzuki and Suzuki 1989). Pseudodeficiency of the enzyme, a finding that occurs commonly in metachromatic leukodystrophy, has not been documented for globoid leukodystrophy, although one report (Wenger and Riccardi 1976) suggests that it may exist. Diagnosis based upon enzyme deficiency may need to be buttressed by clinical findings and by sural nerve biopsy.

METACHROMATIC LEUKODYSTROPHY (250100)

Metachromatic leukodystrophy is probably the most common leukodystrophy, with an estimated incidence of 1:40 000 (Kolodny 1989). It presents most commonly in late infancy with gait disturbance and peripheral neuropathy and later loss of cognitive function. There is abnormal accumulation of the myelin lipid sulfatide and the urine also contains greatly increased levels of this substance. The sulfatide accumulation results from the deficient activity of arylsulfatase A, the enzyme that catalyzes the removal of the sulfate group. The arylsulfatase A gene has been cloned (Stein et al. 1989) and the molecular basis of the common subtypes has been determined (Polten et al. 1991). Laboratory diagnosis requires demonstration of greatly diminished arylsulfatase activity in leukocytes, serum or cultured skin fibroblasts, but it is of great importance to emphasize that this is not sufficient. This is due to the existence of what is called 'pseudoarylsulfa-

tase deficiency' (250100). In persons with pseudoarylsulfatase deficiency the enzyme activity may be as low as in patients with metachromatic leukodystrophy, but the person with pseudoarylsulfatase deficiency does not store sulfatide and appears free of neurological disability. The person with pseudodeficiency does not excrete abnormal amounts of sulfatide in the urine, and peripheral nerve function and sural nerve biopsy are normal. The laboratory confirmation of metachromatic leukodystrophy thus requires demonstration of enzyme deficiency and of sulfatide storage, such as decisively increased levels of urinary sulfatides. The 2 entities can be distinguished by studies in which cultured skin fibroblasts are presented with a sulfatide load. Cells from pseudodeficient patients can metabolize this load whereas those from metachromatic leukodystrophy patients cannot (Fluharty et al. 1978). Other methods of distinction are described in the review article by Kolodny (1989). The molecular basis of arylsulfatase A deficiency has recently been shown to be due to the loss of a polyadenylation signal and an N-glycosylation site (Gieselmann et al. 1989). It is thus likely that the pseudo and 'real' deficiency states will be distinguishable by DNA analysis. Finally, in rare circumstances metachromatic leukodystrophy can result from the deficiency of an activator protein (Stevens et al. 1981) (249900). These patients exhibit the clinical manifestations of metachromatic leukodystrophy and there is sulfatide storage, but arylsulfatase A activity (as tested in vitro) is normal. In vivo the enzyme fails to function because of the deficiency of the activator protein.

Metachromatic leukodystrophy may present as spastic quadriparesis in adults under 2 sets of circumstances. One is a long-surviving patient with the juvenile form of the disease. Here symptoms would have become evident toward the end of the first decade in adolescence, with progression to a vegetative or near-vegetative state in adulthood. MRI shows widespread cerebral demyelination and peripheral nerve conduction velocity is impaired. The second presentation is in the adult-onset form of the disease. This form of metachromatic leukodystrophy presents as a psychotic and dementing disorder in the second third or fourth decade and symptoms in the mental or psychological sphere dominate the clinical picture for one or more decades before motor or other clearly neurological manifestations become evident. It is not surprising that these patients were diagnosed as psychotic and that the leukodystrophy might not be identified until the terminal neurological stage of the illness or at the time of postmortem examination. Such diagnostic delays are now less common because of increased awareness of this type of disorder and the more widespread use of Magnetic Resonance Imaging studies. In fact, we must now also guard against over-diagnosis. As many as 7% of the general population may carry the pseudodeficiency gene, and it is important to utilize the steps described above to distinguish between pseudodeficiency of arylsulfatase A and metachromatic leukodystrophy.

PELIZAEUS MERZBACHER DISEASE (512080)

Pelizaeus Merzbacher (PM) disease is a slowly progressive genetically determined disorder characterized by early and prominent nystagmus and eye movement disorder, spastic quadriparesis and mental retardation. Its rate of progression is so slow that it may be mistaken for static cerebral palsy. Pathologically there is severe involvement of central nervous system myelin, with some areas of relatively preserved myelin, sometimes referred to as tigroid spots when the tissue is stained for myelin. The peripheral nervous system is spared (Seitelberger 1970). Several types and modes of inheritance have been observed.

While much remains to be learned about PM disease, recent biochemical and genetic studies provide important new information. Koeppen et al. (1987) reported that proteolipid apoprotein was not demonstrable in the postmortem brain tissue of an 18-year-old patient, whereas other myelin constituents were present albeit in diminished concentration. This finding is of significance since proteolipid is normally present only in central nervous system but not in peripheral nerve myelin and a primary defect in the formation of this compound would be consistent with the normal appearance of peripheral myelin. More significant, it had been shown that proteolipid protein was deficient in the jimpy mouse

(Kerner and Carson 1982) an X-linked animal model of PM disease. The gene encoding the proteolipid protein has been mapped to Xq22.

Two recent studies of PM families report mutations affecting the proteolipid gene (Troffater et al. 1989; Gencic et al. 1989). The mutation cosegregated with the PM phenotype with a lod score of 4.62. These mutations appear to be kindred-specific. It is considered likely that mutations affecting the gene that controls formation of proteolipid protein will be found in the classical X-linked type 1 form of PM disease, as well as in the connatal form (Gencic et al. 1989). In conclusion, it is likely that failure to form proteolipid protein is the basic defect in classical X-linked PM disease. In families known to carry a demonstrable mutation it may now be possible to identify carriers and to offer prenatal diagnosis.

CEREBROTENDINOUS XANTHOMATOSIS (213700)

Cerebrotendinous xanthomatosis (CTX) is an outstanding example of the importance of identifying metabolic disorders, because it can be treated successfully by a simple and inexpensive regimen. CTX was first described in 1937 (van Bogaert et al. 1937). It is an autosomal recessive disorder associated with dementia, psychiatric disturbances, cerebellar dysfunction, pyramidal tract involvement and peripheral neuropathy. Other features are cataracts, osteoporosis and fractures, and of great diagnostic value, xanthomas in the Achilles tendon and other sites (Berginer et al. 1989). The tendon xanthomata resemble those seen in hypercholesterolemic patients, but the key point is that in CTX the blood cholesterol levels are normal. In 1968 Menkes et al. made the important observation that CTX patients stored the abnormal steroid cholestanol which accumulates in the nervous system and the tendons and is thought to be the cause of the progressive tissue damage. The basic defect in CTX is the impaired synthesis of bile acids. It involves one of the hydroxylation steps in the conversion of cholesterol to bile acids (Bjorkhem and Skrede 1989). Through mechanisms that are not completely understood this leads to an overproduction of cholestanol. In respect to therapy, the key observation is that this overproduction can be abolished by the administration of chenodeoxycholic acid, a bile acid that fails to form in CTX (Salen et al. 1975).

Biochemical diagnosis depends upon the demonstration of increased levels of cholestanol in plasma (Bouwes Bavinck et al. 1986; Kasama and Seyama 1986) and of increased bile alcohol levels in urine (Batta et al. 1987; Koopman et al. 1987). Since plasma cholestanol levels may also be increased in liver disease, demonstration of increased urinary bile alcohol levels is important for the substantiation of the diagnosis. Carriers do not show biochemical abnormalities. These may be brought out however, by administration of cholestyramine which stimulates bile acid synthesis. Under these circumstances carriers excrete higher levels of bile acid alcohols than do controls (Koopman et al. 1986). MRI shows cerebral, cerebellar and spinal cord atrophy and increased signal intensity in the cerebrum and cerebellum in T2 weighted images (Bencze et al. 1990).

Oral administration of chenodeoxycholic acid in a daily dosage of 750 mg/day is beneficial. It reduces the levels of cholestanol in plasma and of urinary bile alcohols (Tint et al. 1989), improves intellectual function, cerebellar and pyramidal deficits, electroencephalogram, CT (Berginer et al. 1984), and visual and brain stem auditory evoked responses (Pedley 1985).

SJÖGREN-LARSSON SYNDROME (270200)

The Sjögren-Larsson syndrome is an autosomal recessive disorder characterized by the triad of ichthyosis, mental deficiency and progressive spastic diplegia (Sjögren and Larsson 1957). The motor deficit is usually a symmetrical diplegia and 75% of patients are confined to a wheelchair and speech and bulbar muscles are also involved. While most patients are mentally retarded, some have functioned in the borderline normal range. Approximately 50% of patients have glistening white dots on the retina, and seizures, short stature and kyphosis also occur (Rizzo et al. 1989b).

Rizzo and associates have shown increased levels of long chain alcohols in the plasma of patients with Sjögren-Larsson syndrome (Rizzo et al. 1989b). They demonstrated that normally there exists a series of reactions that convert the CoA derivative of long chain fatty acids to fatty alcohols, which may then be utilized for the synthesis of wax esters or glycerol ethers, or depending upon metabolic demand, be converted back to the fatty acid (Rizzo et al. 1987a). The latter reaction is catalyzed by the enzyme fatty alcohol : NAD + oxidoreductase. This enzyme is deficient in cultured skin fibroblasts of Sjögren-Larsson syndrome patients (Rizzo et al. 1988). This leads to tissue accumulation of long chain alcohols, and presumably this is the cause of ichthyosis and of the neurological and eye manifestations. These new findings raise the hope of dietary therapy by reduction of fat intake combined with the administration of medium chain triglycerides, and emphasize the importance of early diagnosis (Rizzo et al. 1989b).

ARGINASE DEFICIENCY (207800)

Familial hyperargininemia was first described by Terheggen et al. in 1969. The children are normal at birth, develop tetraplegia and spasticity beginning in the first few years which becomes more severe during late childhood and adolescence, and is associated with mild or moderately severe mental retardation. The disorder may be mistaken for a static form of cerebral palsy. An important diagnostic lead is the occurrence of intermittent vomiting, stupor or convulsions associated with moderately severe hyperammonemia. Patients may self-select a protein-poor diet (Bernar et al. 1986). The basic defect is the deficient activity of arginase, the last enzyme of the Krebs-Henseleit urea cycle. This enzyme splits arginine into ornithine and urea.

The diagnosis of arginase deficiency may be missed because the episodes of hyperammonemia are not as severe or prominent as in the other urea cycle disorders. Urinary amino acids may be entirely normal (Bernar et al. 1986). Plasma amino acid determinations show elevated levels of arginine and often of glutamine. Arginase deficiency is demonstrable in multiple tissues including erythrocytes (Michels and Beudet 1978).

Stabilization of the neurological deficit and

clinical improvement, has been obtained by restriction of arginine intake (Snyderman et al. 1977) which can be combined advantageously with the administration of sodium benzoate (Qureshi et al. 1984). It is encouraging that treatment begun at the time of birth appears to have prevented neurological deficits in one child with arginase deficiency (Snyderman et al. 1979).

HOMOCARNOSINOSIS (236130)

Homocarnosinosis is a rare metabolic disorder in which there is accumulation of homocarnosine, a brain-specific dipeptide of gamma-aminobutyric acid and histidine. Sjaastad et al. (1976) reported one Norwegian family in which 3 sibs with progressive paraparesis, dementia and retinal degeneration had elevated levels of homocarnosine in the cerebrospinal fluid. The association between the biochemical abnormality and the neurological syndrome is uncertain, since the homocarnosine level in cerebrospinal fluid level was also increased in the neurologically intact mother. Scriver and Perry (1989) propose that the homocarnosine accumulation results from a defective function of serum carnosinase. Because the biochemical abnormality has been observed in other asymptomatic persons, there is increasing doubt about the cause and effect relationship to the neurological abnormality. Scriver and Perry postulate that this along with other abnormalities of the omega amino acids 'impose no biological cost on the variant individual' (Scriver and Perry 1989).

DEFECTS IN COBALAMIN METABOLISM (277400; 277410)

Ten different inherited disorders of cobalamin (vitamin B_{12}) metabolism are known. Three involve absorption and transport; seven involve cellular utilization and coenzyme production (Fenton and Rosenberg 1989). Of pertinence to this chapter are rare instances of neurological dysfunction in adolescents and young adults that have been shown to be caused by genetically determined abnormalities of cobalamin metabolism. Identification as a cobalamin-related disorder is hampered by the fact that hematological changes are subtle or absent, and serum cobalamin levels and Schilling test are normal. Instead diagnosis depends upon demonstration of abnormalities involving the 2 enzymes known to depend on cobalamin coenzymes, namely methylmalonyl-CoA mutase and N^5-methyltetrahydrofolate:homocysteine methyltransferase. Defective function of these 2 enzymes results in abnormally high levels of methylmalonic acid and homocystine in the urine. The plasma level of homocystine is increased and that of methionine is reduced. These abnormalities are detected by measuring levels of amino acids in plasma and urine (Slocum and Cummings 1991), and of organic acids in urine (Sweetman 1991). Depending upon the exact nature of the cobalamin defect, the metabolic disturbance may involve either methylmalonic acid or homocystine metabolism or both (Fenton and Rosenberg 1989).

Examples of these neurological defects are 3 sibs described by Shinnar and Singer (1984) (277400) and a young woman described by Carmel et al. (1988) (277410). The proband in the first report was a 14-year-old girl who presented with dementia and myelopathy of 1 year's duration. Levels of homocystine in blood and urine and of methylmalonic acid in urine were increased, and specialized complementation studies identified a cobalamin C mutation. The same biochemical abnormality was demonstrated in an asymptomatic 12-year-old sister and 8-month-old brother. The proband improved remarkably after the administration of large doses of hydroxycobalamin. The second report described a 27-year-old woman who had a 5-year history of progressive ataxia, impaired vibration and position sense and hyperreflexia that had been diagnosed as multiple sclerosis. She had moderate megaloblastic anemia, but normal cobalamin levels and Schilling test. She had increased levels of homocystine in blood and urine, and decreased levels of plasma methionine, while levels of urinary methylmalonic acid were normal. It was concluded therefore that the metabolic defect involved methylcobalamin, which is a coenzyme for N^5-methyltetrahydrofolate:homocysteine methyl transferase. Weekly injections of 1 mg of hydroxycobalamin and the daily oral administration of 6 g of betaine resulted in a decrease of homocystine excretion and an improvement in the megaloblastic anemia. Fifteen months after initiation of therapy neurological function was improved, visual evoked responses had become normal and somatosensory evoked responses had improved.

These 2 reports have several implications. First, defects of cobalamin metabolism may present as neurological deficits in adolescents and adults and without the stigmata usually thought to be associated with cobalamin deficiency states. They can be diagnosed most readily by quantitative study of blood and urine amino acids and urinary organic acids. The exact definition of the cobalamin defect is a specialized procedure that can follow the initial identification. Most important, these disorders do respond to replacement therapy. The report by Shinnar and Singer suggests that presymptomatic identification and preventive therapy is feasible. It is likely that these disorders are more frequent than has been recognized so far, and in our view adds to the rationale for amino acid and organic acid studies in patients with undiagnosed progressive neurological disorders.

LESCH-NYHAN SYNDROME (308000)

Lesch-Nyhan syndrome is an X-linked disorder characterized by choreoathetosis, spastic quadriparesis, compulsive self mutilation and gouty arthritis (Stout and Caskey 1989). it will be discussed briefly here since spasticity persisting to adulthood is one of the clinical features (Mizuno 1986). The basic defect is the complete deficiency of the purine salvage enzyme hypoxanthine phosphoribosyltransferase (HPRT). The gene has been cloned and the molecular lesions are heterogeneous, since 11 separate types have been identified so far (Stout and Caskey 1989).

The diagnosis of Lesch Nyhan syndrome should be considered in patients with choreoathetoid cerebral palsy, with the striking self mutilation and hyperuricemia providing crucial leads. Most patients are moderately retarded.

Measurement of the urinary uric acid to creatinine ratio can be used as an initial screening test (Kaufman et al. 1968), but this test may give false positive results. Definitive diagnosis depends upon the measurement of HPRT activity in red blood cells. This assay can also be used for prenatal diagnosis. In families in which the mutation is defined at the DNA level, this provides a precise means of carrier detection and prenatal diagnosis.

The renal complications of Lesch-Nyhan syndrome can be treated with allopurinol. No specific therapy exists as yet for the neurological disorder (Stout and Caskey 1989).

SALLA DISEASE (268740)

Salla disease was first reported in 1979 by Aula et al. (1979) and takes its name from the area in Finland from which the first cluster of patients came. Four adult mentally retarded patients were described, 3 brothers and their female third cousin. All had psychomotor delay from infancy and all had coarsened facies. Ataxic gait was described in the 3 brothers in association with brisk reflexes and extensor plantar responses. Their cousin was wheelchair bound from childhood. The presence of vacuolated lymphocytes in the peripheral blood of the first patient suggested a storage disorder and this finding was consistent in all 4 of these cases. Increased urinary sialic acid was noted in 3 of the 4.

Subsequent reports (Renlund et al. 1983; Renlund 1984; Baumkotter et al. 1985; Wolburg-Buchholz et al. 1985; Ylitalo et al. 1986; Echenne et al. 1986) have confirmed the clinical findings as initially described. Other findings include exotropia (50%) and a transient nystagmus (50%) (Renlund 1984). The motor abnormalities may include an initial hypotonia in infancy later progressing to ataxia which is often associated with spasticity or rigidity (Renlund 1984). Athetosis has been reported in a significant proportion (Renlund 1984; Ylitalo et al. 1986). Speech development is grossly abnormal although language comprehension is said to be relatively preserved (Renlund et al. 1983). Seizures may also occur.

Somatic examination has revealed mild dysmorphism with hypertelorism and mild prognathism in some patients (Wolburg-Buchholz et al. 1985). Hepatosplenomegaly although uncommon may be seen (Baumkotter et al 1985). Radiological studies have shown thickened calvarium and some thoracic vertebral abnormalities (Wolburg-Buchholz et al. 1985). CT scan has shown basal, cortical and cerebellar atrophy (Gahl et al. 1989).

The course of the disease is one of fairly stable condition during the first 2 decades after which a slow deterioration in motor function and mental abilities is seen (Gahl et al. 1989). Life span may not be appreciably shortened. Post mortem studies have been carried out on 2 patients and revealed macroscopically evident loss of cerebral white matter which correlated with histological finding of loss of axons and myelin sheaths. This was associated with astrocytic proliferation. Findings were suggestive of primary axonal degeneration (Autio-Harmainen et al. 1988).

In addition to the more common form described above, a severe infantile form of the disease has also been described known as Infantile Sialic Acid Storage disease (Tondeur et al. 1982; Stevenson et al. 1983; Gillan et al. 1984). It is associated with dysmorphism, hepato-splenomegaly, severe neurological disability and death is in the first 5 years.

The diagnosis of this disorder depends on the finding in an appropriate clinical setting of increased urinary sialic acid (identifiable as *N*-acetylneuraminic acid) in association with the presence of lysosomal storage and the finding of free sialic acid intracellularly (Gahl et al. 1989). Evidence of lysosomal storage may be found in the peripheral blood but is not present in all cases. In these instances it may be found in the conjunctival tissue or in the capillary endothelium.

A defect in lysosomal transport of free sialic acid is the proposed underlying problem in both the sialic acid storage diseases. Loading studies have confirmed this phenomenon (Gahl 1987). Lack of egress of sialic acid from loaded fibroblast lysosomes occurred in Salla and Infantile Sialic Acid Storage disease (Gahl 1987). The precise pathogenesis is not described.

The disorder is inherited as an autosomal recessive trait. It has most frequently been described in those of Finnish extraction although there are reported cases from other ethnic groups (Echenne et al. 1986; Baumkotter et al. 1985;

Wolburg-Buchholz et al. 1985). Prenatal diagnosis is possible based on the finding of increased free sialic acid in amniocytes (Renlund and Aula 1987). No specific treatment of this disorder exists at this time.

CONCLUDING REMARKS

In this chapter we have discussed 12 genetic-metabolic entities that can cause spastic paraparesis in adults. In some of these disorders, such as adrenomyeloneuropathy, spastic paraparesis is the main and the most disabling disability; in others it is only part of a broader range of neurological disabilities. The disorders that have been presented represent a wide range of metabolic disturbances. They include lysosomal and peroxisomal disorders, and disorders of amino acid and of purine and pyrimidine metabolism. The selection was somewhat arbitrary and we have omitted disorders in which the metabolic defect is not yet defined. At this time the disorders with known metabolic defects account only for a small proportion of genetically determined spastic paraparesis. No doubt this percentage will increase. The wide range of metabolic disturbances and the wide range of phenotypic variation present challenges both in respect to the understanding, retention and improvement of available knowledge, and the development of a network of reliable, accessible and affordable laboratory services. These challenges can be met only by communication and cooperation between the clinical neurologist and the biochemist and geneticist. However, as shown in Table 1, actual or potential therapy is becoming available for 9 of the 12 disorders. This makes it worthwhile to meet these challenges.

REFERENCES

AUBOURG, P., G. H. SACK, D. A. MEYERS, J. J. LEASE and H. W. MOSER: Linkage of adrenoleukodystrophy to a polymorphic DNA probe. Ann. Neurol. 21 (1987) 240–249.

AUBOURG, P., S. BLANCHE, I. JAMBAQUE, F. ROCCHICCIOLI, G. KALIFA, C. NAUD-SAUDREAU, M.-O. ROLLAND, M. DEBRE, J. L. CHAUSSAIN, C. GRISCELLI, A. FISCHER and P.-F. BOUGNERES: Reversal of early neurologic and neuroradiologic manifestations of X-linked adrenoleukodystrophy by bone marrow transplantation. N. Engl. J. Med. 322 (1990) 1860–1866.

AULA, P., S. AUTIO, K. O. RAIVIO, J. RAPOLA, C. J. THODEN, S. L. KOSKELA and I. YAMASHINA: Salla disease. A new lysosomal storage disease. Arch. Neurol. 36 (1979) 88–94.

AUTIO-HARMAINEN, H., A. OLDFORS, P. SOURANDER, M. RENLUND, K. DAMMERT and S. SIMILA: Neuropathology of Salla disease. Acta Neuropathol. (Berl.) 75 (1988) 481–490.

BARAM, T., A. GOLDMAN and A. PERCY: Krabbe disease: specific MRI and CT findings. Neurology 36 (1986) 111–115.

BATTA, A. K., G. SALEN, S. SHEFER, G. S. TINT and M. BATTA: Increased plasma bile alcohol glucuronides in patients with cerebrotendinous xanthomatosis: effect of chenodeoxycholic acid. J. Lipid Res. 28 (1987) 1006–1012.

BAUMKOTTER, J., M. CANTZ, K. MENDLA, W. BAUMANN, H. FRIEBOLIN, J. GEHLER and J. SPRANGER: N-Acetyl-neuraminic acid storage disease. Hum. Genet. 71 (1985) 155–159.

BEHAN, W. M. H. and M. MAIA: Strumpell's familial spastic paraplegia: genetics and neuropathology. J. Neurol. Neurosurg. Psychiatry 37 (1974) 8–20.

BENCZE, K. S., D. R. VANDE POLDER and L. D. PROCKOP: Magnetic resonance imaging of the brain and spinal cord in cerebrotendinous xanthomatosis. J. Neurol. Neurosurg. Psychiatry 53 (1990) 166–167.

BERGINER, V., G. SALEN and S. SHEFER: Cerebrotendinous xanthomatosis. Neurol. Clin. 7 (1989) 55–74.

BERGINER, V. M., G. SALEN and S. SHEFER: Long-term treatment of cerebrotendinous xanthomatosis with chenodeoxycholic acid. N. Engl. J. Med. 311 (1984) 1649–1652.

BERNAR, J., R. A. HANSON, R. KERN, B. PHOENIX, K. N. F. SHAW and S. D. CEDERBAUM: Arginase deficiency in a 12 year old boy with mild impairment of intellectual function. J. Pediatr. 108 (1986) 432–435.

BJORKHEM, I. and S. SKREDE: Familial diseases with storage of sterols other than cholesterol: Cerebrotendinous xanthomatosis and phytosterolemia. In: C. R. Scriver, A. L. Beaudet, W. S. Sly, and D. Valle (Eds), Metabolic Basis of Inherited Disease. New York, McGraw Hill (1989) 1283–1302.

BOUSTANY, R. N., E. FLEISCHNICK, C. A. ALPER, M. L. MARAZITA, M. A. SPENCE, J. B. MARTIN and E. H. KOLODNY: The autosomal dominant form of 'pure' familial spastic paraplegia: clinical findings and linkage analysis of a large pedigree. Neurology 37 (1987) 910–915.

BOUWES BAVINCK, J. N., B. J. VERMEER, J. A. GEVERS LEUVEN, B. J. KOOPMAN and B. G. WOLTHERS: Capillary gas chromatography of urine samples in diagnosing cerebrotendinous xanthomatosis. Arch. Dermatol. 122 (1986) 1269–1272.

BUDKA, H., B. MOLZER, H. BERNHEIMER, H. LASSMANN, P. PILZ and K. TOIFI: Clinical, morphological and neurochemical findings in adrenoleukodystrophy and its variants. Neuropathology (Tokyo) 1 (1983) 209–224.

BUDKA, H., E. SLUGA and W. D. HEISS: Spastic paraplegia associated with Addison's disease: adult variant of adrenoleukodystrophy. J. Neurol. 213 (1976) 237–250.

CARMEL, R., D. WATKINS, S. I. GOODMAN and D. S. ROSENBLATT: Hereditary defect of cobalamin metabolism (cblG mutation) presenting as a neurologic disorder in adulthood. N. Engl. J. Med. 318 (1988) 1738–1741.

CAVANAGH, N. and B. KENDALL: High density on computed tomography in infantile Krabbe's disease: a case report. Dev. Med. Child. Neurol. 28 (1986) 799–802.

DARRAS, B., E. KWAN, H. GILMORE, B. EHRENBERG and E. RABE: Globoid cell leukodystrophy: cranial computed tomography and evoked potentials. J. Child. Neurol. 1 (1986) 126–130.

DOOLEY, J. M. and B. A. WRIGHT: Adrenoleukodystrophy mimicking multiple sclerosis. J. Can. Sci. Neurol. 12 (1985) 73–74.

ECHENNE, B., M. VIDAL, I. MAIRE, J. C. MICHALSKI, P. BALDET and J. ASTRUC: Salla disease in one non-Finnish patient. Eur. J. Pediatr. 145 (1986) 320–322.

EDITORIAL: Epidemic spastic paraparesis. Lancet 2 (1984) 904–905.

FENTON, W. A. and L. E. ROSENBERG: Inherited disorders of cobalamin transport and metabolism. In: C. R. Scriver, A. L. Beaudet, W. S. Sly, and D. Valle (Eds.), Metabolic Basis of Inherited Disease. New York, McGraw Hill (1989) 2065–2082.

FIUMARA, A., L. PAVONE, L. SICILIANO, A. TINE, E. PARANO and G. INNICO: Late-onset globoid cell leukodystrophy. Child. Nerv. Syst. 6 (1990) 194–197.

FLUHARTY, A. L., R. L. STEVENS and H. KIHARA: Cerebroside sulfate hydrolysis by fibroblasts from a metachromatic leukodystrophy parent with deficient arylsulfatase A. J. Pediatr. 92 (1978) 782–784.

FLUHARTY, A. L., L. NEIDENGARD, D. HOLTZMAN and H. KIHARA: Late-onset Krabbe disease initially diagnosed as cerebroside sulfatase activator deficiency. Metab. Brain Dis. 1 (1986) 187–195.

GAHL, W. A.: Disorders of lysosomal membrane transport. Cystinosis and Salla disease. Enzyme 38 (1987) 154–160.

GAHL, W. A., M. RENLUND and J. G. THOENE: Lysosomal transport disorders. In: C. R. Scriver, A. L. Beaudet, W. S. Sly and D. Valle (Eds.), Metabolic Basis of Inherited Disease. New York, McGraw Hill (1989) 2619–2647.

GENCIC, S., D. ABUELO, M. AMBLER and L. D. HUDSON: Pelizaeus-Merzbacher disease: an X-linked neurologic disorder of myelin metabolism with a novel mutation in the gene encoding proteolipid protein. Am. J. Hum. Genet. 45 (1989) 435–442.

GIESELMANN, V., A. POLTEN, J. KREYSING and K. VON FIGURA: Arylsulfatase A pseudodeficiency: loss of a polyadenylylation signal and N-glycosylation site. Proc. Nat. Acad. Sci. 86 (1989) 9436–9440.

GILLAN, J. E., J. A. LOWDEN, K. GASKIN and E. CUTZ: Congenital ascites as a presenting sign of lysosomal storage disease. J. Pediatr. 104 (1984) 225–231.

GOEBEL, H. H., K. HARZER, J. P. ERNST, J. BOHL and H. KLEIN: Late-onset globoid cell leukodystrophy: unusual ultrastructural pathology and subtotal beta-galactocerebrosidase deficiency. J. Child. Neurol. 5 (1990) 299–307.

GOLDBLATT, J., R. BALLO, B. SACHS and A. MOOSA: X-linked spastic paraplegia: evidence for homogeneity with a variable phenotype. Clin. Genet. 35 (1989) 116–120.

GRIFFIN, J. W., E. GOREN, H. SCHAUMBURG, W. K. ENGEL and L. LORIAUX: Adrenomyeloneuropathy: a probable variant of adrenoleukodystrophy. Neurology 27 (1977) 1107–1113.

HAGBERG, B.: Krabbe's disease: clinical presentation of neurological variants. Neuropediatrics 15 (1984) 11–15.

HARDING, A. E.: Hereditary 'pure' spastic paraplegia: A clinical and genetic study of 22 families. J. Neurol. Neurosurg. Psychiatry 44 (1981) 871–883.

HEFFUNGS, W., H. HAMEISIER and H. H. ROPERS: Addison Disease and Cerebral Sclerosis in an Apparently Heterozygous Girl: evidence of inactivation of the adrenoleukodystrophy locus. Clin. Genet. 18 (1980) 184–188.

IGARASHI, M., H. H. SCHAUMBURG, J. POWERS, Y. KISHIMOTO, E. KOLODNY and K. SUZUKI: Fatty acid abnormality in adrenoleukodystrophy. J. Neurochem. 26 (1976) 851–860.

JOHNSTON, A. W. and V. A. MCKUSICK: A sex-linked recessive form of spastic paraplegia. Am. J. Hum. Genet. 14 (1961) 83–93.

KASAMA, T. and Y. SEYAMA: Biochemical diagnosis of cerebrotendinous xanthomatosis using reversed phase thin layer chromatography. J. Biochem. 99 (1986) 771–775.

KAUFMAN, J. M., M. L. GREENE and J. E. SEEGMILLER: Urine uric acid to creatinine ratio: screening test for disorders of purine metabolism. J. Pediatr. 73 (1968) 583.

KENWRICK, S. V., V. IONASECU, G. IONASECU, C. SEARBY, A. KING, M. DUBOWITZ and K. DAVIES: Linkage studies of the X-linked recessive spastic paraplegia. Am. J. Hum. Genet. 14 (1962) 83–94.

KEPPEN, L. D., M. F. LEPPERT, O. O'CONNEL, Y. NAKAMURA, D. STAUFFER, M. LATHROP, J.-M. LALOUEL and R. WHITE: Etiological heterogeneity in X-linked spastic paraplegia. Am. J. Hum. Genet. 41 (1987) 933–943.

KERNER, A. L. and J. H. CARSON: Effect of the jimpy mutation on expression of myelin proteins in heterozygotes and hemizygous mouse brain. J. Neurochem. 43 (1982) 1017–1027.

KOEPPEN, A. H., N. A. RONCA, E. A. GREENFIELD and M. B. HANS: Defective biosynthesis of proteolipid protein in Pelizaeus-Merzbacher disease. Ann. Neurol. 21 (1987) 159–170.

KOLODNY, E.: Metachromatic leukodystrophy and multiple sulfatase deficiency. In: C. R. Scriver, A. L. Beaudet, W. S. Sly and D. Valle (Eds), Metabolic Basis of Inherited Disease. New York, McGraw Hill (1989) 1721–1760.

KOOPMAN, B. J., R. J. WATERREUS, H. W. C. VAN DEN BREKEL and B. G. WOLTHERS: Detection of carriers of cerebrotendinous xanthomatosis. Clin. Chim. Acta 158 (1986) 179–186.

KOOPMAN, B. J., B. G. WOLTHERS, J. C. VAN DER MOLEN, G. T. NAGEL and W. KRUIZINGA: Abnormal urinary bile acids in a patient suffering from cerebrotendinous xanthomatosis during oral administration of ursodeoxycholic acid. Biochim. Biophys. Acta 917 (1987) 238–246.

LAZO, O., M. CONTRERAS, M. HASHMI, W. STANLEY, C. IRAZU and I. SINGH: Peroxisomal lignoceroyl-CoA ligase deficiency in childhood adrenoleukodystrophy and adrenomyeloneuropathy. Proc. Nat. Acad. Sci. USA 85 (1988) 7647–7651.

MENKES, J. H., J. R. SCHIMSCHOCK and P. D. SWANSON: Cerebrotendinous xanthomatosis. Arch. Neurol. 19 (1968) 47–53.

MICHELS, V. V. and A. L. BEAUDET: Arginase deficiency in multiple tissues in argininemia. Clin. Genet. 13 (1978) 61–67.

MIGEON, B. R., H. W. MOSER, A. B. MOSER, J. AXELMAN, D. SILLENCE and R. A. NORUM: Adrenoleukodystrophy: Evidence for X-linkage, inactivation and selection favoring the mutant allele in heterozygous cells. Proc. Natl. Acad. Sci. USA 78 (1981) 5066–5070.

MIZUNO, T.: Long term follow-up of ten patients with Lesch-Nyhan syndrome. Neuropediatrics 17 (1986) 158.

MORARIU, M. A., J. L. CHASAN, R. A. NORUM, H. W. MOSER and B. R. MIGEON: Adrenoleukodystrophy variant in a heterozygous female. Neurology 32 (1982) 81.

MOSER, A. B., J. BOREL, A. ODONE, S. NAIDU, D. CORNBLATH, D. B. SANDERS and H. W. MOSER: A new dietary therapy for adrenoleukodystrophy: biochemical and preliminary clinical results in 36 patients. Ann. Neurol. 21 (1987) 240–249.

MOSER, H. W., A. B. MOSER, N. KAWAMURA, J. MURPHY, A. MILUNSKY, K. SUZUKI, H. SCHAUMBURG and Y. KISHIMOTO: Adrenoleukodystrophy: Elevated C-26 fatty acid in cultured skin fibroblasts. Ann. Neurol. 7 (1980) 542–549.

MOSER, H. W., A. B. MOSER, K. K. FRAYER, W. W. CHEN, J. D. SCHULMAN, B. P. O'NEILL and Y. KISHIMOTO: Adrenoleukodystrophy: increased plasma content of saturated very long chain fatty acids. Neurology 31 (1981) 1241–1249.

MOSER, H. W., A. B. MOSER, J. E. TROJAK and S. W. SUPPLEE: The identification of female carriers for adrenoleukodystrophy. J. Pediatr. 103 (1983) 54–59.

MOSER, H. W., A. B. MOSER, I. SINGH and B. R. O'NEILL: Adrenoleukodystrophy: Survey of 303 cases: biochemistry, diagnosis and therapy. Ann. Neurol. 16 (1984) 628–641.

NAIDU, S., D. CORNBLATH, A. MOSER and H. MOSER: Neurological abnormalities in ALD heterozygotes. Muscle Nerve 9 (1986) 129.

NAIDU, S., A. MOSER and H. MOSER: Phenotype of women heterozygous for adrenoleukodystrophy. Neurology 39 (1989) 370.

NOETZEL, M. J., W. M. LANDAU and H. W. MOSER: Adrenoleukodystrophy carrier state presenting as a chronic nonprogressive spinal cord disorder. Arch. Neurol. 44 (1987) 566–567.

O'NEILL, B. P., H. W. MOSER, K. M. SAXENA and L. C. MAR-MION: Adrenoleukodystrophy (ALD): Neurological disease in carriers and correlation with very long chain fatty acids (VLCFA) concentrations in plasma and cultured skin fibroblasts. Neurology 32 (1982) 216.

O'NEILL, B. P., H. W. MOSER, K. M. SAXENA and L. C. MARMION: Adrenoleukodystrophy: clinical and biochemical manifestations in carriers. Neurology 34 (1984) 789–801.

OSUNTOKUN, B. O.: An ataxic neuropathy in Nigeria: a clinical, biochemical and elecrophysiological study. Brain 91 (1968) 215–248.

PEDLEY, T. A., R. G. EMERSON, C. L. WARNER, L. P. ROWLAND and G. SALEN: Treatment of cerebrotendinous xanthomatosis with chenodeoxycholic acid. Ann. Neurol. 18 (1985) 517–518.

PILZ, P. and P. SCHIENER: Kombination von Morbus Addison und Morbus Schilder bei einer 43 Jahrigen Frau. Acta Neuropathol. 26 (1973) 357–360.

POLTEN, A., A. L. FLUHARTY, C. B. FLUHARTY, J. KAPPLER, K. VON FIGURA and V. GIESELMANN: Metachromatic leukodystrophy: the molecular basis for the common clinical subtypes. N. Engl. J. Med. (1991) in press.

POWERS, J. M.: Adrenoleukodystrophy (adreno-testiculo-leuko-myelo-neuropathic-complex). Clin. Neuropathol. 4 (1985) 181–199.

POWERS, J. M., H. W. MOSER, A. B. MOSER, K. A. CHAN, S. B. ELIAS and R. A. NORUM: Pathologic findings in adrenoleukodystrophy. Arch. Pathol. Lab. Med. 111 (1987) 151–153.

QURESHI, I. A., J. LETARTE, R. QUELLET, M. L. BATSHAW and S. BRUSKOW: Treatment of hyperargininemia with sodium benzoate and arginine-restricted diet. J. Pediatr. 104 (1984) 473–476.

RENLUND, M.: Clinical and laboratory diagnosis of Salla disease in infancy and childhood. J. Pediatr. 104 (1984) 232–236.

RENLUND, M. and P. AULA: Prenatal detection of Salla disease based on increased free sialic acid in amniocytes. Am. J. Med. Genet. 28 (1987) 377–384.

RENLUND, M., P. AULA, K. O. RAIVIO, S. AUTIO, K. SAINIO, J. RAPOLA and S. L. KOSKELA: Salla disease: a new lysosomal storage disorder with disturbed sialic acid metabolism. Neurology 33 (1983) 57–66.

RIZZO, W. B., D. A. CRAFT, A. L. DAMMANN and M. W. PHILLIPS: Fatty alcohol metabolism in cultured human fibroblasts. J. Biol. Chem. 262 (1987a) 17412–17419.

RIZZO, W. B., M. W. PHILLIPS, A. L. DAMMANN, R. Y. LESHNER, S. S. JENNINGS, J. AVIGAN and V. K. PROUD: Adrenoleukodystrophy: dietary oleic acid lowers hexacosanoate levels. Ann. Neurol. 21 (1987b) 232–239.

RIZZO, W. B., A. L. DAMMANN and D. A CRAFT: Sjögren-Larsson syndrome. J. Clin. Invest. 81 (1988) 738–744.

RIZZO, W. B., R. T. LESHNER, A. ODONE, A. L. DAMMANN, D. A. CRAFT, M. E. JENSONE, S. S. JENNINGS, S. DAVIS, R. JAITLY and J. A. SGRO: Dietary erucic acid therapy for X-linked adrenoleukodystrophy. Neurology 30 (1989a) 1415–1422.

RIZZO, W. B., A. DAMMANN, D. A. CRAFT, S. H. BLACK, A. H. TILTON, D. AFRICK, E. CHAVES-CARBALLO, G. HOLMGREN and S. JAGELL: Sjögren-Larsson syndrome: inherited defect in the fatty alcohol cycle. J. Pediatr. 115 (1989b) 228–234.

ROMÁN, G. C., B. S. SCHOENBERG, D. L. MADDEN, J. L. SEVER, J. HUGON, A. LUDOLPH and P. S. SPENCER: Human T-lymphotropic virus type I antibodies in the serum of patients with tropical spastic paraparesis in the Seychelles. Ann. Neurol. 44 (1987) 605–607.

SACK, G. H., C. A. HUETHER and N. GARG: Familial spastic paraplegia — clinical and pathologic studies in a large kindred. Johns Hopkins Med. J. 143 (1978) 117–121.

SALEN, G., T. W. MERRIWETHER and G. NICOLAU: Chenodeoxycholic acid inhibits increased cholesterol and cholestanol synthesis in cerebrotendinous xanthomatosis. Biochem. Med. 14 (1975) 57–74.

SCRIVER, C. R. and T. L. PERRY: Disorders of omega-amino acids in free and peptide-linked forms. In: C. R. Scriver, A. L. Beaudet, W. S. Sly and D. Valle (Eds.), Metabolic Basis of Inherited Disease. New York, McGraw Hill (1989) 755–771.

SEITELBERGER, F.: Pelizaeus-Merzbacher disease. In: P. J. Vinken and G. W. Bruyn (Eds.), Handbook of Clinical Neurology, Vol. 10: Leukodystrophies and Poliodystrophies. Amsterdam, North-Holland Publishing Co. (1970) 150–202.

SHINNAR, S. and H. S. SINGER: Cobalamin C mutation (methylmalonic aciduria and homocystinuria) in adolescence. N. Engl. J. Med. 311 (1984) 451–454.

SINGH, I., A. B. MOSER, H. W. MOSER and Y. KISHIMOTO: Adrenoleukodystrophy: Impaired oxidation of very long chain fatty acids in white blood cells, cultured skin fibroblasts and amniocytes. Pediatr. Res. 18 (1984a) 286–290.

SINGH, I., A. B. MOSER, S. GOLDFISCHER and H. W. MOSER: Lignoceric acid is oxidized in the peroxisomes: implications for the Zellweger cerebro-hepato-renal syndrome and adrenoleukodystrophy. Proc. Natl. Acad. Sci. U.S.A. 81 (1984b) 4203–4207.

SJAASTAD, O., J. BERSTAD, P. GJESDAHL and L. GJESSING: Homocarnosinosis. 2. A familial metabolic disorder associated with spastic paraplegia, progressive mental deficiency, and retinal pigmentation. Acta Neurol. Scand. 53 (1976) 275–290.

SJÖGREN, Y. and T. LARSSON: Oligophrenia in combination with congential ichthyiosis and spastic disorders: A clinical and genetic study. Acta Psychiatr. Neurol. Scand. 32 (Suppl. 113) (1957) 1–112.

SLOCUM, R. H. and J. G. CUMMINGS: Amino acid analysis of physiological samples. In: F. A. Hommes (Ed.), Techniques in Diagnostic Human Biochemical Genetics. Wiley-Liss, Inc. (1991) 87–126.

SNYDERMAN, S. E., C. SANSARICQ, W. J. CHEN, P. M. NORTON and S. V. PHANSALKAR: Argininemia. J. Pediatr. 90 (1977) 563–568.

SNYDERMAN, S. E., C. SANSARICQ, P. M. NORTON and F. GOLDSTEIN: Argininemia treated from birth. J. Pediatr. 95 (1979) 61–63.

STEIN, C. G., VOLKMAR, J. KREYSING, B, SCHMIDT, R. POHLMANN, A. WAHEED, H. E. MEYER, J. S. O'BRIEN and K. VON FIGURA: Cloning and expression of human arylsulfatase A. J. Biol. Chem. 264 (1989) 1252–1259.

STEVENS, R. L., A. L. FLUHARTY, H. KIHARA, M. M. SHAPIRO, L. J. KABACK, B. MARSH, K. SANDHOFF and G. FISSHER: Cerebroside sulfatase activator deficiency induced metachromatic leukodystrophy. Am. J. Hum. Genet. 33 (1981) 900.

STEVENSON, R. E., M. LUBINSKY, H. A. TAYLOR, D. A. WENGER, R. J. SCHROER and P. M. OLMSTEAD: Sialic acid storage disease with sialuria: clinical and biochemical features in the severe infantile type. Pediatrics 72 (1983) 441–449.

STOUT, J. T. and C. T. CASKEY: Hypoxanthine phosphoribosyltransferase deficiency: The Lesch-Nyhan syndrome and gouty arthritis. In: C. R. Scriver, A. L. Beaudet, W. S. Sly, and D. Valle (Eds.), Metabolic Basis of Inherited Disease. New York, McGraw Hill (1989) 1007–1028.

SUZUKI, K. and Y. SUZUKI: Galactosylceramide lipidosis: Globoid-cell leukodystrophy (Krabbe disease). In: C. R. Scriver, A. L. Beaudet, W. S. Sly and D. Valle (Eds.), Metabolic Basis of Inherited Disease. New York, McGraw Hill (1989) 1699–1720.

SWEETMAN, L: Organic acid analysis. In: F. A. Hommes (Ed.), Techniques in Diagnostic Human Biochemical Genetics. Wiley-Liss, Inc. (1991) 143–176.

TERHEGGEN, H. G., A. SCHWENK, A. LOWENTHAL, M. VAN SANDE and J. P. COLOMBO: Argininaemia with arginase deficiency. Lancet 2 (1969) 748.

THOMAS, P. K., J. P. HALPERN, R. H. M. KING and D. PATRICK: Galactosylceramide lipidosis: novel presentation as a slowly progressive spinocerebellar degeneration. Ann. Neurol. 16 (1984) 618–620.

TINT, G. S., H. GINSBERG, G. SALEN, N.-A. LE and S. SHEFER: Chenodeoxycholic acid normalizes elevated lipoprotein secretion and catabolism in cerebrotendinous xanthomatosis. J. Lipid Res. 30 (1989) 633–640.

TONDEUR, M., J. LIBERT, E. VAMOS, F. VAN HOOF, G. H. THOMAS and G. STRECKER: Infantile form of sialic acid storage disorder: clinical, ultrastructural, and biochemical studies in two siblings. Eur. J. Pediatr. 139 (1982) 142–147.

TROFATTER, J. A., S. R. DLOUHY, W. DEMYER, P. M. CONNEALLY and M. E. HODES: Pelizaeus-Merzbacher disease: tight linkage to proteolipid protein gene exon variant. Proc. Natl. Acad. Sci. 86 (1989) 9427–9430.

TSUJI, S., M. SUZUKI, T. ARIGA, M. SEKINE, M. KURIYAMA and T. MIYATAKE: Abnormality of long-chain fatty acids in erythrocyte membrane sphingomyelin from patients with adrenoleukodystrophy. J. Neurochem. 36 (1981) 1046–1049.

UNGAR-SARGON, J. Y., R. E. LOVELACE and J. C. M. BRUST: Spastic paraplegia-paraparesis. J. Neurol. Sci. 46 (1980) 1–12.

VAN BOGAERT, L., H.J. SCHERER and E. EPSTEIN: Une forme cérébrale de cholesterinose généralisée. Masson et Cie (Paris, 1937).

WANDERS, R. J. A., C. W. T. VAN ROERMUND, M. J. A. VAN WIJLAND, R. B. H. SCHUTGENS, H. VAN DEN BOSCH, A. W. SCHRAM and J. M. TAGER: Direct evidence that the deficient oxidation of very long chain fatty acids in X-linked adrenoleukodystrophy is due to an impaired ability of peroxisomes to activate very long chain fatty acids. Biochem. Biophys. Res. Commun. 153 (1988) 618–624.

WENGER, D. A. and V. M. RICCARDI: Possible misdiagnosis of Krabbe disease. J. Pediatr. 88 (1976) 76.

WOLBUG-BUCHHOLZ, K., W. SCHLOTE, J. BAUMKOTTER, M. CANTZ, H. HOLDER and K. HARZER: Familial lysosomal storage disease with generalized vacuolization and sialic aciduria. Sporadic Salla disease. Neuropediatrics 16 (1985) 67–75.

YLITALO, V., B. HAGBERG, J. RAPOLA, J. E. MANSSON, L. SVENNERHOLM, G. SANNER and B. TONNBY: Salla disease variants: Sialoylaciduric encephalopathy with increased dialidase activity in two non Finnish children. Neuropediatrics 17 (1986) 44–47.

YOUNG, I. D., I. F. PYE and J. R. MOORE: Manifesting heterozygosity in sex-linked spastic paraplegia? J. Neurol. Neurosurg. Psychiatry 47 (1984) 311–313.

YOUNGER, D. S., S. CHOU, A. P. HAYS, D. J. LANGE, R. EMERSON, M. BRIN, H. THOMPSON and L. P. ROLAND: Primary lateral sclerosis: a clinical diagnosis reemerges. Arch. Neurol. 45 (1988) 1304–1307.

Handbook of Clinical Neurology, Vol. 15 (59): Diseases of the Motor System
J.M.B.V. de Jong, editor
© Elsevier Science Publishers B.V., 1991

Differential diagnosis of spinal muscular atrophies and other disorders of motor neurons with infantile or juvenile onset

MARIANNE DE VISSER, PIETER A. BOLHUIS and PETER G. BARTH

Department of Neurology, Academic Medical Centre, University of Amsterdam, Amsterdam, The Netherlands

Spinal muscular atrophies comprise a heterogeneous group of disorders in which there is degeneration of the anterior horn cells of the spinal cord and sometimes the bulbar motor nuclei, but no evidence of pyramidal tract involvement or sensory disturbances. The main clinical features are wasting and weakness of muscles supplied by the affected anterior horn cells.

The differential diagnosis of spinal muscular atrophies in adults is dealt with extensively in Chapter 23. In this chapter only the infantile and juvenile spinal muscular atrophies will be considered. By far the greater majority of infantile and juvenile spinal muscular atrophies are inherited. Classification of spinal muscular atrophies, which can be divided according to at least 3 different sets of criteria (pattern of inheritance; distribution of weakness; or age of onset), is tentative until the primary biochemical defects are identified. However, a classification based on clinical and genetic differences is valid and still useful (Table 1), although rapid advance of molecular genetics undoubtedly will lead to another classification in the near future.

A. PROXIMAL SPINAL MUSCULAR ATROPHY

The more common chronic form of spinal muscular atrophies usually involves proximal muscles (Werdnig 1891, 1894; Hoffmann 1893, cited by Dubowitz 1975; Wohlfart et al. 1955; Kugelberg and Welander 1956), and follows an autosomal recessive trait. The clinical picture is dominated

TABLE 1

Classification of infantile/juvenile spinal muscular atrophies.

A. *Proximal spinal muscular atrophy*
   I. Infantile (severe)
      Autosomal recessive
      X-chromosomal recessive
   II. Intermediate
      Autosomal recessive
   III. Juvenile (mild)
      Autosomal recessive
      Autosomal dominant

B. *Distal spinal muscular atrophy*
      Autosomal recessive
      Autosomal dominant

C. *Progressive bulbar palsy*
      Autosomal recessive

D. *Scapuloperoneal and facioscapulohumeral spinal muscular atrophy*
      Autosomal dominant
      Autosomal recessive
      X-chromosomal recessive

E. *Monomelic amyotrophy*
      ? Autosomal dominant
      ? X-chromosomal recessive

by progressive symmetrical weakness of predominantly the proximal muscles, initially of the legs but later spreading to the arms and the trunk muscles. Extraocular muscle involvement is absent, and there is no marked facial weakness. Fasciculations are present in less than half of the patients. Conspicuously absent are sensory disturbances, CNS dysfunction, arthrogryposis, and other organ involvement, such as hearing loss, visual disturbances, and cardiomyopathy. The reader is referred to Chapters 5 and 6 of this volume for detailed descriptions of many aspects of the infantile and juvenile forms of proximal spinal muscular atrophy.

Among various proposed classification systems, that of Dubowitz (1978) is the most widely used, subdividing spinal muscular atrophy (SMA) in severe, intermediate and mild forms on the basis of the child's ability to sit unaided and to stand and walk unaided.

1. Severe SMA (Werdnig-Hoffmann disease): unable to sit unsupported;

2. Intermediate SMA: able to sit unsupported; unable to stand or walk unaided;

3. Mild SMA (Kugelberg-Welander disease): able to stand and walk unaided.

Recently, a gene for SMA has been localized on the long arm of chromosome 5 (Brzustowicz et al. 1990; Melki et al. 1990a). Mutations in the 5q12–5q13 region seem to be responsible for the majority of cases of the 3 autosomal recessive SMA forms, suggesting that the different clinical forms of proximal SMA are allelic and caused by different mutations of the same gene (Gilliam et al. 1990; Melki et al. 1990b). However, in a few SMA-families, no linkage with DNA-markers on chromosome 5 is found, which may indicate genetic heterogeneity. Clarification of this problem is important in view of the possibility to carry out prenatal diagnosis by means of DNA markers.

The onset of *severe SMA (Werdnig-Hoffmann disease)* (253300) is in utero or within the first 2 or 3 months of life. The infant shows generalized hypotonia, and there is marked weakness of the limbs, the axial, neck and intercostal musculature. These children are never able to raise their heads, to roll over or to sit. They are prone to recurrent respiratory infections and as a result rarely survive the first year of life.

Into the category of *intermediate SMA* (253550) fall the cases who progress quite normally in the first 6 months of life and usually achieve the ability to sit unaided, but are never able to stand or walk because of severe proximal weakness of the legs. Although some degree of weakness is usually present in the upper limbs, the ability to abduct the shoulders is relatively preserved. Most cases have a benign course with survival into adolescence or adulthood, but few patients with striking intercostal involvement and progressive scoliosis show a downhill course.

The major problem in diagnosis is to distinguish infantile SMA from an array of other diseases that cause hypotonia ('floppy infant') and delayed motor development in the neonate and infant (Table 2).

Once a neurogenic origin of proximal muscle

TABLE 2
Causes of hypotonia at birth and in infancy.

Congenital muscular dystrophy
Congenital myopathies
Congenital myotonic dystrophy
Congenital inflammatory myopathy
Metabolic myopathies
  lipid storage disease
  glycogen storage disease
  mitochondrial myopathy
Neuromuscular transmission defects
  myasthenia gravis
  congenital myasthenic syndromes
  neonatal transient myasthenia
  botulism
Fetal akinesia deformation sequence
Arthrogryposis multiplex congenita
Hereditary motor and sensory neuropathies
Congenital amyelinating neuropathies
Acute/chronic inflammatory polyneuropathy
X-linked recessive infantile SMA
Poliomyelitis

Rare disorders presenting with hypotonia due to anterior horn cell involvement: infantile neuronal degeneration, incontinentia pigmenti, infantile neuroaxonal dystrophy, a defect of methylcrotonyl-CoA carboxylase, De Sanctis Cacchione syndrome, and Moebius syndrome associated with amyotrophy of the limbs.
Other causes: cervical cord injury, cervical spinal cord atrophy, spinal cord infarction, spinal dysraphism, meningomyelocele, spinal cord aplasia, leukodystrophies, Down syndrome, Prader Willi syndrome, amino acidurias, organic acidurias, defect of generalised β-oxidation, glycogen storage diseases.

involvement has been established by means of electromyography and muscle biopsy, the differential diagnosis is narrowed down to the Fetal Akinesia Deformation Sequence (208150, 253310), due to a defect of anterior horn cells, neurogenic arthrogryposis multiplex congenita (AMC) (208100) due to anterior horn cell degeneration or peripheral neuropathies, hereditary motor and sensory neuropathies (HMSN) (18080, 11821), congenital amyelinating neuropathies (CAN), poliomyelitis, and acute (Guillain-Barré syndrome) or chronic inflammatory polyneuropathy.

The fetal akinesia deformation sequence and neurogenic AMC are described in Chapter 5 of this volume. These disorders are easily distinguishable from SMA because of the presence of joint contractures at birth.

Greenberg et al. (1988) described 4 male infants from 3 sibships in an extended family who were noted to have hypotonia, areflexia, and congenital joint contractures. Three also had fractures at birth. The findings of EMG and muscle biopsy were consistent with spinal muscular atrophy. Pedigree analysis strongly suggested that this disorder represents an X-linked recessive form of SMA.

The demyelinating forms of HMSN (Types I and III) in infancy and early childhood are usually characterised by distal weakness and wasting, foot deformities, and hypo- or areflexia. However, generalized hypotonia and/or delayed motor milestones in the first year of life is noticed in the majority of cases of HMSN type III, and, rarely, in autosomal recessive type I and autosomal dominant HMSN I (Harding and Thomas 1980a; Vanasse and Dubowitz 1981; Ouvrier et al. 1987). The diagnosis is made on the basis of markedly decreased motor nerve conduction velocities but one should keep in mind that in HMSN nerve conduction is usually normal at birth and evolves to maximal slowing by 3–5 years of life (Gutmann et al. 1983). In doubtful cases or in the absence of a positive family history, a clinical and electrophysiological study of the parents should be conducted to determine the type of inheritance and distinguish between types I, II and III (Vanasse and Dubowitz 1981; Rossi et al. 1983).

Early-onset neuronal HMSN manifesting with delayed walking, 'limp' feet or foot deformities, is described by Ouvrier et al. (1981), Julien et al. (1988), and Gabreëls-Festen (in press), amongst others. Most cases are autosomal recessively inherited and differ in clinical, genetic and morphological aspects from autosomal dominant HMSN type II. They show an earlier onset and greater clinical severity than usually seen in dominant type II with usually wheelchair dependency already in puberty or in adult life. Since motor nerve conduction velocities may be substantially reduced, a clear distinction between demyelinating and neuronal early-onset HMSN can only reliably be made by means of a nerve biopsy. Loss of large diameter fibers, a small total transverse fascicular area, without appreciable signs of fiber de- or regeneration, segmental demyelination or onion bulbs are the histological characteristics in early-onset neuronal HMSN.

Congenital amyelinating neuropathies may be accompanied by AMC, and are hereditary in about one third of the described cases (Pleasure et al. 1986; Charnas et al. 1988). The cases of CAN without AMC are often misdiagnosed as Werdnig-Hoffmann disease during life. A nerve biopsy revealing complete absence of myelin and onion-bulbs provides the key to a correct diagnosis, but this diagnostic technique has only in a few instances been employed during life. Since peripheral nerve system myelination occurs at 18 weeks, serial ultrasonography in order to assess fetal movement could provide an antenatal diagnosis in families with CAN and AMC.

Respiratory distress due to diaphragmatic involvement as the initial manifestation in neonatal spinal muscular atrophy has been reported by Mellins et al. (1974), McWilliam et al. (1985), and Schapira and Swash (1985). Since diaphragmatic movements are typically spared until the late stages of Werdnig-Hoffmann disease, prominent and early respiratory failure may represent a clinical variant.

The reader is referred to Chapter 5 of this volume for descriptions of other 'atypical' Werdnig-Hoffmann cases, e.g. pontocerebellar hypoplasia with infantile motor neuron disease (Norman's disease), also termed infantile neuronal degeneration (Steiman et al. 1980). Autopsy of these patients, who present with a clinical picture resembling Werdnig-Hoffmann disease, reveals a variety of central and peripheral nervous system lesions in addition to the characteristic chromatolysis and loss of motor neurons in anterior horn cells and cranial nerve nuclei.

Other rare disorders presenting with hypotonia in infancy, are an inborn error of leucine catabolism (β-hydroxyisovaleric aciduria and β-methylcrotonylglycinuria), xeroderma pigmentosum associated with neurological features (De Sanctis Cacchione syndrome) (278800), including choreoathetosis, mental deficiency, epilepsy, cerebellar ataxia, microcephaly, spasticity, deafness, and generalized hypotonia, due to either axonal polyneuropathy or degeneration of anterior horn cells and sensory ganglia (Trush et al. 1972), incontinentia pigmenti (Bloch-Sulzberger syndrome) (308300) associated with anterior horn cell degeneration, Moebius syndrome associated with progressive weakness and amyotrophy of the limb muscles (Tridon et al. 1971), and infantile or connatal neuroaxonal dystrophy (Seitelberger disease) (256600, 234200). Huttenlocher and Gilles (1967) have stressed the reduction in anterior horn cells at the caudal level of the spinal cord, found in a girl who had suffered from infantile neuroaxonal dystrophy as did 2 sisters. The inborn error of leucine catabolism, incontinentia pigmenti and infantile neuroaxonal dystrophy are described in Chapters 7 and 5, respectively, of this volume.

Poliomyelitis usually involves spinal motor neurons leading to asymmetrical paralysis of lower and upper extremities, and in about 10–15% of cases muscles supplied by certain of the brain stem nuclei will be affected. Details of this disease are given by Price and Plum (1978) and Wood and Anderson (1988).

In an immunodeficient child, a chronic neurological disease manifesting with myoclonic activity, muscle wasting, hypotonia and loss of tendon reflexes developed after she had received live oral poliovirus vaccine (Davis et al. 1977). CSF examination was normal; EMG showed fibrillation potentials in the leg muscles. At autopsy, focal loss of neurons was found in the anterior horn cells of the spinal cord, and inflammatory foci in the pons, thalamus, globus pallidus and premotor cortex. Poliovirus was isolated from throat and stool during life and from several sites within the brain at autopsy.

In a study of South African patients with suspected poliomyelitis, but from whom poliovirus was not isolated, a variety of causes of the paralysis was found (Gear 1984). Injury of the spinal cord complicated by periostitis or osteomyelitis was quite common. So were Guillain-Barré syndrome and non-polio enterovirus infections, especially various Coxsackie viruses. Exotic causes included respiratory paralysis due to snake bite, black widow spider bite, scorpion sting, ascending symmetrical paralysis with loss of reflexes and difficulty in speaking and breathing, known as tick paralysis, and paralysis caused by rabies, mumps, schistosomiasis and other parasitic worms such as *Cysticercus cellulosae*, *Echinococcus granulosus*, and *Coenurus cerebralis*. Of the many chemicals known to cause paralysis, arsenic was responsible for a poliolike paralytic syndrome in several members of a family. In one case, acute paralysis simulating paralytic poliomyelitis appeared to be due to intermittent porphyria. Dark urine containing excessive porphyrins revealed the true cause of the paralysis.

Guillain-Barré syndrome is a rare cause of the floppy infant (Gilmartin and Ch'ien 1977). Studies of sensory nerve action potentials are helpful in establishing the early differential diagnosis between Werdnig-Hoffmann disease and Guillain-Barré syndrome (Jones 1990).

Pasternak et al. (1982) described a 7-week-old infant presenting with an acute flaccid paraparesis. A diagnosis of chronic relapsing polyneuropathy was made, based on the acute onset, moderately reduced motor nerve conduction velocities on electroneurography, and elevated CSF protein. Despite corticosteroid treatment, there was relentless progression of the disease, and the child died at the age of 4 years. Postmortem examination confirmed the clinical diagnosis.

Hypotonia and paralysis of arms and/or legs are caused by cervical cord injury associated with breech delivery, ischemic infarction subsequent to umbilical artery catheterisation (Haldeman et al. 1983), congenital cervical spinal atrophy (Darwish et al. 1981), spinal dysraphism, meningomyelocele, and aplasia of the spinal cord.

Diseases such as Down syndrome, Prader Willi syndrome, leukodystrophies, organic acidurias, amino acidurias, defects of generalized β-oxidation and glycogen storage diseases may also present with hypotonia at birth, but it falls outside the scope of this chapter to elaborate on these disorders.

Mild SMA (Kugelberg-Welander disease) (253400) is characterised by progressive weakness of the proximal limb muscles in children who have normal milestones in the first year of life, and have subsequently achieved the ability to walk at a normal age or somewhat later.

Differential diagnosis is extensive, as it includes the heterogeneous group of the 'limb-girdle syndromes' (Table 3) (Panegyres et al. 1990). In most cases auxiliary diagnostic techniques, such as estimation of serum CK activity, EMG, muscle biopsy, and to some extent, scanning of muscles by computed tomography, real-time ultrasound imaging or Magnetic Resonance Imaging (MRI) are helpful in differentiating Kugelberg-Welander disease from the other limb-girdle syndromes, which are usually myopathic in origin. Very high serum CK activities seem to be exclusive to Duchenne muscular dystrophy, polymyositis and

TABLE 3

Differential diagnosis of juvenile limb-girdle syndromes.

Duchenne/Becker muscular dystrophy
Manifesting carriers of the gene for
  Duchenne/Becker muscular dystrophy
Limb-girdle muscular dystrophy
  autosomal recessive
  autosomal dominant
Emery-Dreifuss muscular dystrophy
Metabolic myopathies
  glycogen storage diseases
  lipid storage diseases
  mitochondrial myopathies
Congenital myopathies
Bethlem myopathy
Endocrine myopathies
Inflammatory myopathies
Sarcoidosis
Myoglobinuria
Neuromuscular transmission disorders
  myasthenia gravis
  congenital myasthenic syndromes
Proximal spinal muscular atrophy
  autosomal recessive
  autosomal dominant

---

dermatomyositis, but an up to 10-fold rise or even more can be found in Kugelberg-Welander disease (Bouwsma and Van Wijngaarden 1980).

EMG and muscle biopsy findings in Kugelberg-Welander disease were amply reviewed in Chapter 6 of this volume.

A skeletal muscle CT study on patients with Becker muscular dystrophy and Kugelberg-Welander disease revealed the presence of low attenuation areas throughout the leg muscles in the latter disease, whereas in Becker muscular dystrophy (part of) the muscle is replaced by fat in selected muscles with preservation or even hypertrophy of other muscles (De Visser et al. 1985; Sambrook et al. 1988). MRI showed similar patterns (Murphy et al. 1986), but it has the advantage of having no ionising radiation and no known adverse biological effects. Real-time ultrasound scanning produced a brightly speckled pattern of increased echo from the muscle in muscular dystrophy, whereas spinal muscular atrophy showed a moderate increase in muscle echo, associated with muscle atrophy (Heckmatt et al. 1988) or a moderate increase in echogenicity with a distinct heterogeneous 'moth-eaten' pattern (Fischer et al. 1988).

In a few myogenic disorders, e.g. Becker muscular dystrophy, polymyositis, and acid maltase deficiency, EMG may demonstrate fibrillation potentials and positive sharp waves (Swash and Schwartz 1988). Even histological examination of a muscle biopsy specimen in patients with Becker muscular dystrophy may mimic a neurogenic disease by showing small groups of atrophic muscle fibres (Bradley et al. 1978; Goebel et al. 1979; Ten Houten and De Visser 1984). Therefore, DNA studies and/or biochemical and immunohistochemical analysis of dystrophin should be included in the routine investigation of any male with either a chronic limb-girdle syndrome, especially if there is calf hypertrophy and/or a raised serum CK (Clarke et al. 1989; Lunt et al. 1989; Topaloglu et al. 1989; Zerres et al. 1990), or with the so-called 'quadriceps syndrome' (Sunohara et al. 1990).

A diagnosis of polymyositis is easily made on the basis of CK estimation and muscle biopsy. The same applies to acid maltase deficiency, and in addition, assay of acid maltase in muscle tissue, leukocytes, urine or fibroblasts elucidates the nature of this disease.

A wide variety of additional clinical manifestations has been described in Kugelberg-Welander disease, such as calf hypertrophy (Namba et al. 1970; Bouwsma and Van Wijngaarden 1978), extensor plantar responses (Namba et al. 1970), cranial nerve involvement (Namba et al. 1970; Gruber et al. 1983; Barois et al. 1989), sensory involvement (Winder and Auer 1989) and cardiac involvement (Sterz et al. 1971; Sugimura et al. 1973; Tanaka et al. 1976).

Since mental retardation is not a specific feature of Kugelberg-Welander, the family with chronic progressive spinal muscular atrophy of childhood, microcephaly and mental subnormality, described by Spiro et al. (1967) (271110) and reviewed in Chapter 5 of this volume, and the family with non-progressive autosomal recessive spinal muscular atrophy and non-progressive marked mental retardation (Staal et al. 1975) are considered variants of the disease.

Whether an autosomal recessive spinal muscular atrophy found in the Ryukyu islands (Kondo et al. 1970) (271200), manifesting with slowly progressive symmetrical proximal weakness with onset in early infancy, kyphoscoliosis, fasciculations and pes cavus, is a distinct clinical entity remains to be proven.

A rare disorder, clinically undistinguishable from Kugelberg-Welander disease is juvenile spinal muscular atrophy caused by hexosaminidase A deficiency (Johnson et al. 1982) (272800). Since this report many additional cases have been described.

Rarely, motor neuron disease presenting with pseudobulbar paresis, fasciculations, and diffuse leg weakness, is encountered in Sandhoff disease (hexosaminidase A and hexosaminidase B deficiency).

Juvenile-onset proximal spinal muscular atrophy resembling Kugelberg-Welander disease in conjunction with hyperlipoproteinemia and hypertriglyceridemia has been described. It is of interest that clinical features such as extensor plantar responses, talipes cavus, and hypertrophic calves were relatively common in these patients.

The reader is referred to Chapter 7 of this volume for detailed descriptions of the cases with hexosaminidase deficiency and lipid disturbances.

Kanda et al. (1990) described the pathology of the peripheral nervous system in 2 autopsied cases of group A xeroderma pigmentosum (De Sanctis-Cacchione syndrome). Clinical features included xeroderma pigmentosum, cerebellar ataxia, contractures of all joints, severe muscle atrophy, tendon areflexia, tongue fasciculations, dementia, dysarthria, impairment of perception of visual, auditory and tactile stimuli, and pyramidal features. Both patients died in early adulthood. Motor nerves including those of the oculomotor systems were severely affected, but sensory involvement was even more marked. Morphometric data suggest that the underlying pathogenetic mechanism is that of a neuronopathy.

Motor neuron degeneration, in addition to depletion of axons and myelin in the posterior columns and corticospinal tracts was also found at autopsy in a 31-year-old woman who had acquired developmental delay and hemiparesis during the perinatal period, but subsequently developed another clinical picture consisting of seizures, dysplastic facial changes and amyotrophy between the age of 8 and 10 (Landau et al. 1976). Eventually, she became severely retarded and totally bedridden with flaccid paralysis of the lower extremities, marked weakness of the muscles of the neck and the upper extremities, and aphagia.

Rapidly progressive weakness of the proximal limb muscles, and wasting and fasciculations of the tongue, in addition to slowly progressive ataxia, spasticity, choreo-athetosis, and early-onset seizures since early childhood, associated with sea-blue histiocytosis were described in a 15-year-old girl by Ashwal et al. (1984).

Autosomal dominant proximal spinal muscular atrophy (158500) with onset in infancy or early childhood is an uncommon disease (Pearn 1978), presenting with generalized weakness, in contrast to the proximal selectivity which is such a distinctive feature of Kugelberg-Welander disease. The disease runs a protracted course. Similar cases or families reported by Magee and De Jong (1960), Armstrong et al. (1966), Garvie and Woolf (1966), Lugaresi et al. (1966), Zellweger et al. (1969, 1972), and Emery (1971) are all cited by Pearn (1978). Tsugkagoshi et al. (1966)

described a family with what they called 'Kugelberg-Welander syndrome with dominant inheritance'. A characteristic feature of this family was the presence of bulbar palsy of very mild degree. Involvement of pontine nuclei, i.e. facial weakness was reported in a family with autosomal dominantly inherited juvenile proximal spinal muscular atrophy (Cao et al. 1976).

Differential diagnosis does not essentially differ from that of autosomal recessive Kugelberg-Welander disease (Table 3).

B. DISTAL SPINAL MUSCULAR ATROPHY

Hereditary distal spinal muscular atrophy (progressive spinal muscular atrophy of Charcot-Marie-Tooth type) (271200, 182960) has been described with autosomal dominant or recessive inheritance but also sporadic cases occur frequently. Weakness and wasting predominantly involve the lower legs (Dyck and Lambert 1968; McLeod and Prineas 1971; Harding and Thomas 1980b) and, rarely, the hands and/or forearms are primarily or exclusively affected (Meadows and Marsden 1969; O'Sullivan and McLeod 1978; Harding et al. 1983; Peiris et al. 1989). Onset is usually gradual and ranges from the first to the sixth decade but the majority develop symptoms in the first decade of life. The course is slowly progressive in most cases.

The differential diagnosis of distal spinal muscular atrophy is chiefly from the hypertrophic and neuronal forms of Charcot-Marie-Tooth syndrome (HMSN types I and II). The clinical distinction between the 3 disorders can be impossible to make, as a considerable proportion of patients with HMSN, particularly type II, have no clinical sensory loss. In type I, motor nerve conduction velocity (MNCV) is markedly decreased, and sensory conduction is abnormal. In type II HMSN, MNCV is normal or only slightly reduced and may, therefore, not be diagnostically helpful in individual cases. Thus, the essential distinguishing feature between distal spinal muscular atrophy and HMSN type II is the presence or absence of normal sensory action potentials.

Lander et al. (1976) described a kinship in which a purely motor, chronic progressive peripheral neuropathy predominantly affected the upper limbs. The authors mentioned an autosomal dominant inheritance pattern, although no male-to-male transmission was observed in the pedigree of the family.

Other, rare, hereditary neuropathies with or without specific biochemical abnormalities fall outside the scope of this chapter and will, therefore, not be discussed.

However, distal spinal muscular atrophy should be differentiated from autosomal dominant and recessive distal myopathies (Magee and DeJong 1965; Lapresle et al. 1972; Bautista 1978; Scoppetta et al. 1984), myotonic dystrophy, particularly in combination with polyneuropathy (Spaans et al. 1986; Brunner et al. 1991), and distal arthrogryposis. Since serum CK activity may be elevated in distal spinal muscular atrophy (Pearn and Hudgson 1979), differentiation from distal myopathies necessitates electromyography and histological examination of a muscle biopsy. Distal arthrogryposis is extensively described in Chapter 5 of this volume.

Kuskokwim disease (208200), an arthrogryposis-like autosomal recessive disorder, presenting with multiple joint contractures predominantly affecting the knees and ankles inducing distal atrophy or hypertrophy, has only been encountered in the Eskimo (Petajan et al. 1969). The normal electrophysiological and morphological findings indicate that the primary pathological condition does not lie in the motor unit.

In sporadic cases, causative factors such as cervical spine abnormalities with brachial radiculopathy, syringomyelia and poliomyelitis should be considered.

Additional features encountered in distal spinal muscular atrophy are the following: unilateral paralysis of diaphragm (Bertini et al. 1989), external ophthalmoparesis and cataracts (Dubrovsky et al. 1981), bilateral optic atrophy and sensorineural deafness (Iwashita et al. 1970; Chalmers and Mitchell 1987), partial optic atrophy, chorea and mental deficiency (Asano et al. 1960, see Chapter 5), vocal cord paralysis (Young and Harper 1980), and vocal cord paralysis and sensorineural hearing loss (Boltshauser et al. 1989). The presence of slight abnormalities in sensory conduction studies and on sural nerve biopsy in 2 out of 3 members of the family described in the latter report, raises the question whether we are actually dealing with a distal spinal muscular atrophy.

Distal muscular atrophy of predominantly the upper limbs with pyramidal features of the legs have been described by Silver (1966) (182700), and Van Gent et al. (1985). Since slight sensory abnormalities were found on clinical, electrophysiological and morphological examination in the latter report, and no electrophysiological or morphological study of the sensory nerves was undertaken in the former, it is doubtful whether amyotrophy of the hands is due to anterior horn cell or peripheral nerve lesions. Garland and Astley (1950) described autosomal dominantly inherited spastic paraplegia with amyotrophy of the lower legs and pes cavus. However, no ancillary investigations were carried out and it is not clear from their report whether the amyotrophy had juvenile onset as did the spastic paraplegia.

An autosomal recessive disorder, designated the Troyer syndrome (275900), was found exclusively among the inbred Amish of Holmes County, Ohio, and comprises a unique combination of progressive spastic paraplegia, distal muscle wasting due to lower motor involvement, growth retardation, delayed speech development with dysarthria and drooling, and cerebellar signs (Cross and McKusick 1967). A non-progressive variant with mental retardation has been reported by Neuhaüser et al. (1976).

Rare disorders presenting with distal neurogenic muscular atrophy are an X-linked disorder constituting Charcot-Marie-Tooth peroneal muscular atrophy and Friedreich ataxia (Van Bogaert and Moreau 1939) (302900), early-onset distal neurogenic muscular atrophy without sensory disturbances showing generalised neuroaxonal dystrophy in both the central and peripheral nervous system in combination with myriad Rosenthal fibers at autopsy (Ule 1972) (256600), progressive neural muscular atrophy with phenylketonuria (Meier et al. 1975) (16260), distal neurogenic muscular atrophy, mental retardation, Marfan-like features, and multiple epiphyseal dysplasia with lysosomal enzyme deficiencies (Goto et al. 1983), and an autosomal recessively inherited neuroacanthocytosis syndrome, including motor neuron disease, initially manifesting in the distal leg muscles, acanthocytosis, and a movement disorder (Spitz et al. 1985) (105400). In the case described by Meier et al., distal muscular atrophy was ascribed to a hypertrophic neuropathy, and the authors assumed that the combination of phenylketonuria and muscular atrophy was coincidental.

Ten members of 4 generations of a kinship with an autosomal dominant disorder with clinical resemblance to Refsum's disease (distal muscular atrophy, ataxia, retinitis pigmentosa, and diabetes mellitus) but without accumulation of phytanic acid or elevated protein levels of the cerebrospinal fluid, were reported by Furakawa et al. (1968) (158500). Distal muscle atrophy was a prominent feature in 8 of the 10 patients, and accompanied by mild hypesthesia. EMG showed signs of reinnervation and slightly reduced motor nerve conduction velocities. Since the propositus of the family was found to have slight distal sensory disturbances not associated with diabetes mellitus, distal muscular atrophy in this family should probably be considered a polyneuropathy.

c. PROGRESSIVE BULBAR PALSY

Progressive bulbar paralysis of childhood (Londe's disease) (211500) is a very rare, autosomal recessively inherited, disease of the motor nuclei of the brain stem. Onset is in early childhood, and the disease is fatal within 2 years after onset.

The controversy whether this clinical entity is part of the autosomal recessive proximal spinal atrophies, or the same entity as familial chronic progressive bulbopontine paralysis with deafness (Vialetto-Van Laere syndrome) (211530) will only be settled by identification of the gene products. Londe's disease and Vialetto-Van Laere syndrome are set out in detail in Chapter 5 of volume 21 and Chapter 9 of this volume.

Juvenile progressive bulbar palsy presenting with facial paralysis, and followed by dysphagia and dysarthria, impaired mastication, and weakness of neck and shoulder girdle muscles was described by Markand and Daly (1971). Since widespread denervation was found on EMG with normal motor conduction, and a muscle biopsy was characteristic of neurogenic atrophy, the authors felt that the primary site of the lesion was the anterior horn cell.

Two additional cases of juvenile progressive bulbar palsy were reported by Albers et al. (1983). Their patients also had ptosis in addition to facial diplegia and bulbar paresis, and therefore, were believed to have myasthenia gravis. The second patient also developed weakness of the neck and distal limb muscles. By means of electrophysiological studies including single fiber EMG, and the absence of acetylcholine receptor antibodies a diagnosis of myasthenia gravis was ruled out.

Dobkin and Verity (1976) described a family characterised by autosomal dominant, slowly progressive cranial and axial muscle wasting of juvenile onset. Autopsy findings in one family member showed cervicothoracic anterior horn cell loss. Cranial motor neuron loss could not be demonstrated. The disorder seemed unique, because it was associated with a cardiac conduction defect and a mitochondrial abnormality in skeletal muscle.

In children, bulbar palsy may be non-progressive and secondary to a congenital defect of cranial nerves (Moebius syndrome). The association of the Moebius syndrome with a variety of other developmental somatic anomalies has been widely recognised. Its co-existence with progressive peripheral neuropathy and hypogonadotrophic hypogonadism seems to be more than coincidental (Rubinstein et al. 1975; Abid et al. 1978). Tridon et al. (1971) described progressive weakness and amyotrophy of the limb muscles and Moebius syndrome in 2 infants.

Congenital suprabulbar paresis (Worster-Drought syndrome), originally described by Klippel and Pierre-Weil (1909) is characterised by selective weakness and impairment of movement of the orbicularis oris muscle, the tongue and the soft palate leading to dysarthria and profuse dribbling. In severe cases there is involvement of the pharyngeal and laryngeal muscles in early life. Associated features are epilepsy and mental retardation. Patton et al. (1986) suggested after having reviewed the cases described by Worster-Drought (Worster-Drought 1974, cited by Patton et al.), that some cases of congenital suprabulbar paresis are inherited as an autosomal dominant trait.

Cranial nerve paralysis of abrupt onset is usually caused by inflammatory diseases, e.g. poliomyelitis, Guillain-Barré syndrome, botulism, tick paralysis, diphtheritic polyneuritis, and acute disseminated encephalomyelitis. By far the most common cause of progressive multiple chronic paralysis of cranial nerves is a brain stem neoplasm. Congenital myopathies, oculopharyngeal muscular dystrophy, myotonic dystrophy, mitochondrial myopathies, myasthenia gravis and congenital myasthenic syndromes may mimic a disease of the brain stem nuclei. The same holds true for syringobulbia, vascular malformations, congenital anomalies, and cyst formations in the posterior fossa. Bulbar paralysis may develop late in the course of biochemically defined encephalopathies, e.g. infantile Gaucher's disease, Krabbe's disease, and metachromatic leukodystrophy.

D. SCAPULOPERONEAL AND FACIO-SCAPULO-HUMERAL SPINAL MUSCULAR ATROPHY

These disorders are amply reviewed in Chapter 4 of this volume. On the basis of this review and recently conducted linkage studies it is suggested that autosomal dominant facioscapulohumeral SMA does not exist at all.

Recently, Palmucci et al. (1991) described 3 familial cases of associated neurogenic facioscapuloperoneal and rigid spine syndrome, with autosomal recessive inheritance.

From the rare reports of scapuloperoneal SMA (181400) with juvenile onset, conflicting data about the pattern of heredity and the primary site of the lesion emerged. The occurrence of cardiomyopathy, muscle contractures of elbows and ankles, and rigid spine does not differentiate between the myogenic form of the scapuloperoneal syndrome (Emery-Dreifuss muscular dystrophy) and its neurogenic counterpart since these features have been observed in both conditions (Miller et al. 1985) (181350).

An unusual type of neural muscular atrophy and possible X-chromosomal inheritance pattern with clinical pictures ranging from facioscapulohumeral-peroneal distribution of weakness to generalised involvement identical to intermediate SMA was described by Skre (1978).

E. MONOMELIC SPINAL MUSCULAR ATROPHY

Monomelic spinal muscular atrophy is character-ised by insiduous onset of wasting and weakness confined to one limb, occurrence predominantly in males of 15–25 years old, and usually sporadic, initial slow progression followed by a stationary course, and lack of involvement of cranial nerves, brainstem, corticospinal tracts and sensory sys-tem. By far the majority of cases are reported from Japan (Hirayama et al. 1963; Sobue et al. 1978), India (Singh et al. 1980; Prabhakar et al. 1981; Gourie-Devi et al. 1984; Virmani and Mohan 1985), Sri Lanka (Peiris et al. 1989) and Malaysia (Tan 1985), and about two thirds of these patients show a uniform clinical profile of predominantly unilateral hand and forearm wasting, the so-called juvenile type spinal muscu-lar atrophy of the distal upper limb. This clinical entity has been described in detail in Chapter 8 of this volume.

Rarely, familial cases have been reported. Both Sobue et al. (1978) and Schlegel et al. (1987) described a father and a son with asymmetric atrophy and weakness of the hands and forearms. An X-linked recessive inheritance was presumed by Nedelec et al. (1987) who described unilateral lower leg atrophy and weakness in 2 half-brothers.

Unilateral weakness and wasting are also observed with scapulohumeral distribution (Kaeser et al. 1983; Gourie-Devi 1984; Virmani and Mohan 1985; De Visser et al. 1988), of the quadriceps muscle (Prabhakar et al. 1981; Gourie-Devi et al. 1983; Virmani and Mohan 1985; De Visser et al. 1988), and of the anterior crural muscles and/or the calf (Prabhakar et al. 1981; Gourie-Devi et al. 1983; Virmani and Mohan 1985; Nedelec et al. 1987; De Visser et al. 1988). In some patients the entire upper and/or lower limb (Meadows and Marsden 1969; Compernolle 1973; Prabhakar et al. 1981; Thijsse and Spaans 1983) was affected.

EMG and muscle biopsy findings had already suggested that the anterior horn cell was the primary site of the lesion, and this could be confirmed by an MR imaging study (Biondi et al. 1989) in patients with juvenile muscular atrophy of unilateral upper extremity, and in particular, by a necropsy study (Hirayama et al. 1987) demonstrating lesions only in the anterior horns of the spinal cord at C5–T1. The etiology of this condition remains an enigma, but some circulatory insufficiency in the anterior territory of the spinal cord resulting from repeated flexion of the neck is presumed to be a major pathoge-netic mechanism.

For the differential diagnosis with regard to the unilateral spinal muscular atrophy of the distal upper limb, the reader is again referred to Chapter 8 of this volume. Poliomyelitis virus or other neurotoxic enteroviruses, syringomyelia, and entrapment neuropathies (in the costo-clavic-ular region, at the shoulder, in the upper arm, in the forearm, at the elbow, at the wrist, at the pelvic exit, in the groin or upper thigh, at the knee, at the foot) may cause localised amyotro-phy. Other disorders which should be considered are space-occupying lesions, e.g. ependymoma of the filum terminale (Bourque and Dyck 1990), traumatic myelopathy, radiculopathy or plexo-pathy due to birth injury or spinal fracture, traumatic or spontaneous hydromyelia, and neu-ralgic amyotrophic plexopathy.

A rare condition, called hypertrophic mono-neuropathy should also be taken into consider-ation. This condition occurs particularly in children and young adults and has a tendency to affect the radial nerve (Simpson and Fowler 1966; Hawkes et al. 1974).

In the initial stage, distal spinal muscular atrophy confined to the hands may be unilateral and thus indistinguishable from juvenile muscular atrophy of the upper extremity (O'Sullivan and McLeod 1978; Peiris et al. 1989).

Unilateral scapulohumeral muscular atrophy particularly affecting the pectoralis muscle must be distinguished from the Poland anomaly con-sisting of congenital unilateral absence of the sternocostal part of this muscle in addition to ipsilateral syndactyly (David 1972).

MOTOR UNIT HYPERACTIVITY STATE

Muscular hyperactivity may arise from a variety of distinctive disorders of the motor unit. Disorders having the spinal cord as their site of origin include tetanus, strychnine poisoning, familial congenital stiff-man syndrome (Layzer 1979), Satoyoshi syndrome (Satoyoshi et al. 1972), and rabies (Hemachudha 1989).

INFANTILE AND JUVENILE AMYOTROPHIC
LATERAL SCLEROSIS

The existence of slowly progressive, benign infantile and juvenile motor neuron disease (infantile and juvenile ALS) (105400) is indisputable. These rare conditions seem to be more frequent in Tunisia, where the disease is found to be most often familial, following an autosomal recessive trait. Ben Hamida et al. (1990) described 43 patients belonging to 17 families with a bilateral pyramidal syndrome, weakness with atrophy and fasciculations of the hands and/or legs, with or without a (pseudo)bulbar syndrome and without sensory disturbances. The anterior horn cell was considered the primary site of involvement as far as the amyotrophy was concerned on the basis of normal motor and sensory conduction velocities, normal sensory action potential amplitudes, and minor changes on nerve biopsy. The authors subdivided their patients into 3 groups: (1) upper limb and sometimes bulbar amyotrophy with a bilateral pyramidal syndrome; (2) spastic paraplegia with peroneal muscular atrophy; and (3) a spastic pseudobulbar form.

Silver (1966) and Van Gent et al. (1985) reported cases with autosomal dominant transmission showing distal amyotrophy of predominantly the upper limbs and a pyramidal syndrome of the legs. The question whether amyotrophy is due to anterior horn cell involvement in these 2 reports has already been raised (see under 'distal spinal muscular atrophy').

The 3 siblings with early-onset spastic paresis initially in the legs, gradually extending to the upper extremities, progressive generalised muscular wasting and weakness associated with fasciculations, and marked skeletal deformities described by Refsum and Skillicorn (1954), resemble the cases of group 1 although the former are affected more severely.

The cases of group 2 are different from the patients with peroneal muscular atrophy and pyramidal features described by Dyck and Lambert (1968) and Harding and Thomas (1984), since the majority of their patients showed sensory disturbances both clinically and electrophysiologically. However, in 5 of the 25 cases of Harding and Thomas involvement of the primary sensory neuron was not present. Garland and Astley (1950) have made a similar observation of peroneal atrophy of the distal SMA type associated with spastic paraplegia.

A family with a unique autosomal dominant disorder was described by Gilman and Horenstein (1964) (105300) showing severe dystonic disabling paraplegia complicated by pseudobulbar palsy and distal wasting of all 4 limbs with fasciculations in the severely affected members.

Yokochi et al. (1989) described the results of neuropathological examination of a 5-year-old boy in whom symptoms of both the upper and lower motor neuron had occurred. Combined degenerative processes in the upper and lower motor neurons, the spinocerebellar and olivocerebellar systems, and the ventral thalamic nuclei were found. Lewy-body-like intraneuronal hyaline inclusions which showed trilaminar membranous profiles at the ultrastructural level were detected in the spinal anterior horn, Clarke's dorsal nucleus, facial nerve nucleus, inferior olivary nucleus, and substantia nigra. These structures differ from the basophilic intracytoplasmic lesions of sporadic juvenile ALS (Berry et al. 1969; Nelson and Prensky 1972; Oda et al. 1978). They differ also from the eosinophilic intranuclear inclusions of neuronal intranuclear inclusion disease, a rare degenerative disease which is characterised by the association of a central nervous system disorder with extrapyramidal signs and anterior horn cell dysfunction (Goutières et al. 1990).

REFERENCES

ABID, F., R. HALL, P. HUDGSON and R. WEISER: Moebius syndrome, peripheral neuropathy and hypogonadotrophic hypogonadism. J. Neurol. Sci. 35 (1978) 309–315.

ALBERS, J. W., S. ZIMNOWODZKI, C. M. LOWREY and B. MILLER: Juvenile progressive bulbar palsy. Clinical and electrodiagnostic findings. Arch. Neurol. 40 (1983) 351–353.

ASHWAL, S., T. V. THRASHER, D. R. RICE and D. A. WENGER: A new form of sea-blue histiocytosis associated with progressive anterior horn cell and axonal degeneration. Ann. Neurol. 16 (1984) 184–192.

BAROIS, A., B. ESTOURNET, G. DUVAL-BEAUPÈRE, J. BATAILLE and D. LECLAIR-RICHARD: Amyotrophie spinale infantile. Rev. Neurol. 145 (1989) 299–304.

BAUTISTA, J., E. RAFEL, J. M. CASTILLA and R. ALBERCA: Hereditary distal myopathy with onset in early infancy: observation of a family J. Neurol. Sci. 37 (1978) 149–158.

BEN HAMIDA, M., F. HENTATI and C. BEN HAMIDA: Hereditary motor system diseases (chronic juvenile amyotrophic lateral sclerosis). Conditions combining a bilateral pyramidal syndrome with limb and bulbar amyotrophy. Brain 113 (1990) 347–363.

BERRY, R. G., R. A. CHAMBERS, S. DUCKETT and R. TERRERO: Clinico-pathological study of juvenile amyotrophic lateral sclerosis. Neurology 19 (1969) 312.

BERTINI, E., J. D. GADISSEUX, G. PALMIERI, E. RICCI, M. DI CAPUA, G. FERRIERE and G. LYON: Distal infantile spinal muscular atrophy associated with paralysis of the diaphragm: a variant of infantile spinal muscular atrophy. Am. J. Med. Genet. 33 (1989) 328–335.

BIONDI, A., D. DORMONT, I. WEITZNER, JR., P. BOUCHE, P. CHAINE and J. BORIES: MR imaging of the cervical cord in juvenile amyotrophy of distal upper extremity. Am. J. Neuroradiol. 10 (1989) 263–268.

BOGAERT, L. VAN and M. MOREAU: Combinaison de l'amyotrophie de Charcot-Marie-Tooth et de la maladie de Friedreich. Encéphale 34 (1939) 312–322.

BOLTSHAUSER, E., W. LANG, T. SPILLMANN and E. HOF: Hereditary distal muscular atrophy with vocal cord paralysis and sensorineural hearing loss: a dominant form of spinal muscular atrophy? J. Med. Genet. 26 (1989) 105–108.

BOURQUE, P. R. and P. J. DYCK: Selective calf weakness suggests intraspinal pathology, not peripheral neuropathy. Arch. Neurol. 47 (1990) 79–80.

BOUWSMA, G. and G. K. VAN WIJNGAARDEN: Spinal muscular atrophy and hypertrophy of the calves. J. Neurol. Sci. 44 (1978) 275–279.

BRADLEY, W. G., M. Z. JONES, J.-M. MUSSINI and P. R. W. FAWCETT: Becker-type muscular dystrophy. Muscle Nerve 1 (1978) 111–132.

BRUNNER, H. G., F. SPAANS, H. J. M. SMEETS, M. COERWINKEL-DRIESSEN, T. HULSEBOS, B. WIERINGA and H.-H. ROPERS: Genetic linkage with chromosome 19 but not chromosome 17 in a family with myotonic dystrophy associated with hereditary motor and sensory neuropathy. Neurology 41 (1990) 80–84.

BRZUSTOWICZ, L. M., T. LEHNER, L. H. CASTILLA, G. K. PENCHASZADEH, K. C. WILHELMSEN, R. DANIELS, K. E. DAVIES, M. LEPPERT, F. ZITER, D. WOOD, V. DUBOWITZ, K. ZERRES, I. HAUSMANOWA-PETRUSEWICZ, J. OTT, T. L. MUNSAT and T. C. TILLAM: Genetic mapping of chronic childhood-onset spinal muscular atrophy to chromosome 5q11.2–13.3. Nature 344 (1990) 540–541.

CAO, A., C. CIANCHETTI, L. CALISTI and W. TANGHERONI: A family of juvenile proximal spinal muscular atrophy with dominant inheritance. J. Med. Gen. 13 (1976) 131–135.

CHALMERS, N. and J. D. MITCHELL: Optico-acoustic atrophy in distal spinal muscular atrophy. J. Neurol., Neurosurg. Psychiatry 50 (1987) 238–239.

CHARNAS, L., B. TRAPP and J. GRIFFIN: Congenital absence of peripheral myelin: abnormal Schwann cell development causes lethal arthrogryposis multiplex congenita. Neurology 38 (1988) 966–974.

CLARKE, A., K. E. DAVIES, D. GARDNER-MEDWIN, J. BURN and P. HUDGSON: Xp21 DNA probe in diagnosis of muscular dystrophy and spinal muscular atrophy. Lancet 1 (1989) 443.

COMPERNOLLE, T.: A case of juvenile muscular atrophy confined to one upper limb. Eur. Neurol. 10 (1973) 237–242.

CROSS, H. E. and V. A. MCKUSICK: The Troyer syndrome. A recessive form of spastic paraplegia with distal muscle wasting. Arch. Neurol. 16 (1967) 473–485.

DARWISH, H., H. SARNAT, C. ARCHER, K. BROWNELL and S. KOTAGAL: Congenital cervical spinal atrophy. Muscle Nerve 4 (1981) 106–110.

DAVID, T. J.: Nature and etiology of the Poland anomaly. N. Engl. J. Med. 287 (1972) 487–489.

DAVIS, L. E., D. BODIAN, D. PRICE, I. J. BUTLER and J. H. VICKERS: Chronic progressive poliomyelitis secondary to vaccination of an immunodeficient child. N. Engl. J. Med. 297 (1977) 241–245.

DOBKIN, B. H. and M. A. VERITY: Familial progressive bulbar and spinal muscular atrophy. Juvenile onset and late morbidity with ragged-red fibers. Neurology 26 (1976) 754–763.

DUBOWITZ, V.: Infantile spinal muscular atrophy. In: P. J. Vinken and G. W. Bruyn (Eds), Handbook of Clinical Neurology, Vol 22, System Disorders and Atrophies, Part II. Amsterdam, North-Holland Publ. Co. (1975) 81–101.

DUBOWITZ, V.: Muscle Disorders in Childhood. London, W. B. Saunders (1978) 149–157.

DUBROVSKY, A., A. L. TARATUTO and R. MARTINO: Distal spinal muscular atrophy and ophthalmoparesis. A case with selective type 2 fiber hypotrophy. Arch. Neurol. 38 (1981) 594–596.

DYCK, P. J. and E. H. LAMBERT: Lower motor and primary sensory neuron diseases with peroneal muscular atrophy. Arch. Neurol. 18 (1968) 619–625.

FISCHER, A. Q., D. W. CARPENTER, P. L. HARTLAGE, J. E. CARROLL and S. STEPHENS: Muscle-imaging in neuromuscular disease using computerized real-time sonography. Muscle Nerve 11 (1988) 270–275.

FURUKAWA, T., A. TAKAGI, K. NAKAO, H. SUGITA, H. TSUKAGOSHI and T. TSUBAKI: Hereditary muscular atrophy with ataxia, retinitis pigmentosa, and diabetes mellitus. A clinical report of a family. Neurology 18 (1968) 942–947.

GABREËLS-FESTEN, A. A. W. M., E. M. G. JOOSTEN, F. J. M. GABREËLS, F. G. I. JENNEKENS, R. H. J. M. GOOSKENS and D. F. STEGEMAN: Hereditary motor and sensory neuropathy of neuronal type with onset in early childhood. Brain (1991) in press.

GARLAND, H. G. and C. E. ASTLEY: Hereditary spastic paraplegia with amyotrophy and pes cavus. J. Neurol., Neurosurg. Psychiatry 13 (1950) 130–133.

GEAR, J. H. S.: Nonpolio causes of polio-like paralytic syndromes. Rev. Infect. Dis. 6 (Suppl. 2) (1984) S379–S384.

GENT, E. M. VAN, R. A. HOOGLAND and F. G. I. JENNEKENS: Distal amyotrophy of predominantly the upper limbs with pyramidal features in a large kinship. J. Neurol., Neurosurg. Psychiatry 48 (1985) 266–269.

GILLIAM, T. C., L. M. BRZUSTOWICZ, L. H. CASTILLA, T. LEHNER, G. K. PENCHASZADEH, R. J. DANIELS, B. C. BYTH, J. KNOWLES, J. E. HISLOP, Y. SHAPIRA, V. DUBOWITZ, T. L. MUNSAT, J. OTT and K. E. DAVIES: Genetic homogeneity between acute and chronic forms of spinal muscular atrophy. Nature 345 (1990) 823–825.

GILMAN, S. and S. HORENSTEIN: Familial amyotrophic dystonic paraplegia. Brain 87 (1964) 51–66.

GILMARTIN, R. C. and L. T. CH'IEN: Guillain-Barre syndrome with hydrocephalus in early infancy. Arch. Neurol. 34 (1977) 567–569.

GOEBEL, H. H., H. PRANGE, F. GULLOTTA, H. KIEFER and M. Z. JONES: Becker's X-linked muscular dystrophy. Histological, enzyme-histochemical, and ultrastructural studies of two cases, originally reported by Becker. Acta Neuropathol. 46 (1979) 69–77.

GOTO, I., H. NAKAI, T. TABIRA, Y. TANAKA, H. SHIBASAKI and Y. KUROIWA: Juvenile neurogenic muscle atrophy with lysosomal enzyme deficiencies: new disease or variant of mucopolysaccharidosis? J. Neurol. 229 (1983) 45–54.

GOURIE-DEVI, M., T. G. SURESH and S. K. SHANKAR: Monomelic amyotrophy. Arch. Neurol. 41 (1984) 388–394.

GOUTIÈRES, F., J. MIKOL and J. AICARDI: Neuronal intranuclear inclusion disease in a child: diagnosis by rectal biopsy. Ann. Neurol. 27 (1990) 103–106.

GREENBERG, F., K. R. FENOLIO, J. F. HEJTMANCIK, D. ARMSTRONG, J. K. WILLIS, E. SHAPIRA, H. W. HUNTINGTON and R. L. HAUN: X-linked infantile spinal muscular atrophy. Am. J. Dis. Child. 142 (1988) 217–219.

GRUBER, H., J. ZEITLHOFER, J. PRAGER and P. PILS: Complex oculomotor dysfunctions in Kugelberg-Welander disease. Neuroophthalmology 3 (1983) 125–128.

GUTMANN, L., A. FAKADEJ and J. E. RIGGS: Evolution of nerve conduction abnormalities in children with dominant hypertrophic neuropathy of the Charcot-Marie-Tooth type. Muscle Nerve 6 (1983) 515–519.

HALDEMAN, S., G. W. FOWLER, S. ASHWAL and S. SCHNEIDER: Acute flaccid neonatal paraplegia: a case report. Neurology 33 (1983) 93–95.

HARDING, A. E. and P. K. THOMAS: Hereditary distal spinal muscular atrophy. A report on 34 cases and a review of the literature. J. Neurol. Sci. 45 (1980a) 337–348.

HARDING, A. E. and P. K. THOMAS: Autosomal recessive

forms of hereditary motor and sensory neuropathy. J. Neurol., Neurosurg. Psychiatry 43 (1980b) 669–678.

HARDING, A. E. and P. K. THOMAS: Peroneal muscular atrophy with pyramidal features. J. Neurol. Neurosurg. Psychiatry 47 (1984) 168–172.

HARDING, A. E., P. G. BRADBURY and N. M. F. MURRAY: Chronic asymmetrical spinal muscular atrophy. J. Neurol. Sci. 59 (1983) 69–83.

HAWKES, C. H., J. M. JEFFERSON, E. L. JONES and W. T. SMITH: Hypertrophic mononeuropathy. J. Neurol. Neurosurg. Psychiatry 37 (1974) 76–81.

HECKMATT, J. Z. and V. DUBOWITZ: Real-time ultrasound imaging of muscles. Muscle Nerve 11 (1988) 56–65.

HEMACHUDHA, T. Rabies. In: P. J. Vinken and G. W. Bruyn (Eds), Handbook of Clinical Neurology, Vol. 56. Amsterdam, Elsevier (1989) 383–404.

HIRAYAMA, K., T. TSUBAKI, Y. TOYOKURA and S. OKINAKA: Juvenile muscular atrophy of unilateral upper extremity. Neurology 13 (1963) 373–380.

HIRAYAMA, K., M. TOMONAGA, K. KITANO, T. YAMADA, S. KOJIMA and K. ARAI: Focal cervical poliopathy causing juvenile muscular atrophy of distal upper extremity: a pathological study. J. Neurol. Neurosurg. Psychiatry 50 (1987) 285–290.

HOUTEN, R. TEN and M. DE VISSER: Histopathological findings in Becker-type muscular dystrophy. Arch. Neurol. 41 (1984) 729–733.

HUTTENLOCHER, P. R. and F. H. GILLES: Infantile neuroaxonal dystrophy. Clinical, pathologic, and histochemical findings in a family with 3 affected siblings. Neurology 17 (1967) 1174–1184.

IWASHITA, H., N. INOUE, S. ARAKI and Y. KUROIWA: Optic atrophy, neural deafness, and distal neurogenic amyotrophy. Arch. Neurol. 22 (1970) 357–364.

JOHNSON, W. G., H. J. WIGGER, H. R. KARP, L. A. GLAUBIGER and L. P. ROWLAND: Juvenile spinal muscular atrophy: a new hexosaminidase deficiency phenotype. Ann. Neurol. 11 (1982) 11–16.

JONES, H. R.: EMG evaluation of the floppy infant: differential diagnosis and technical aspects. Muscle Nerve 13 (1990) 338–347.

JULIEN, J., C. VITAL, A. LAGUENY and X. FERRER: Hereditary motor and sensory neuropathy type II with axonal lesions. J. Neurol. 235 (1988) 254–255.

KAESER, H. E., R. FEINSTEIN and W. TACKMANN: Unilateral scapulohumeral muscular atrophy. Eur. Neurol. 22 (1983) 70–77.

KANDA, T., M. ODA, M. YONEZAWA, K. TAMAGAWA, F. ISA, R. HANAKAGO and H. TSUKAGOSHI: Peripheral neuropathy in xeroderma pigmentosum. Brain 113 (1990) 1025–1044.

KLIPPEL, M. and PIERRE-WEIL: Syndrome labioglossolaryngé pseudo-bulbaire héréditaire et familial. Rev. Neurol. 17 (1909) 102–103.

KONDO, K., T. TSUBAKI and F. SAKAMOTO: The Ryukyuan muscular atrophy. An obscure heritable neuromuscular disease found in the islands of Southern Japan. J. Neurol. Sci. 11 (1970) 359–382.

KUGELBERG, E. and L. WELANDER: Heredofamilial juvenile muscular atrophy simulating muscular dystrophy. Arch. Neurol. Psychiatry 75 (1956) 500–509.

LANDAU, W. M., R. M. TORACK and M. A. GUGGENHEIM: Congenital retardation and central motor defect with later evolution of seizure disorder, orofacial dysplasia, and amyotrophy. Neurology 26 (1976) 869–873.

LANDER, C. M., M. J. EADIE and J. H. TYRER: Hereditary motor peripheral neuropathy predominantly affecting the arms. J. Neurol. Sci. 28 (1976) 389–394.

LAPRESLE, J., M. FARDEAU and J. GODET-GUILLAIN: Myopathie distale et congénitale, avec hypertrophie des mollets. J. Neurol. Sci. 17 (1972) 87–102.

LAYZER, R. B.: Motor unit hyperactivity states. In: P. J. Vinken and G. W. Bruyn (Eds), Handbook of Clinical Neurology, Vol. 41: Diseases of Muscle, Part II. Amsterdam, North-Holland Publ. Co. (1979) 295–316.

LUNT, P. W., W. J. K. CUMMING, H. KINGSTON, A. P. READ, R. C. MOUNTFORD, M. MAHON and R. HARRIS: DNA probes in differential diagnosis of Becker muscular dystrophy and spinal muscular atrophy. Lancet 1 (1989) 46–47.

MAGEE, K. R. and R. N. DEJONG: Hereditary distal myopathy with onset in infancy. Arch. Neurol. 13 (1965) 387–390.

MARKAND, O. N. and D. D. DALY: Juvenile type of slowly progressive bulbar palsy: report of a case. Neurology 21 (1971) 753–758.

MCLEOD, J. G. and J. W. PRINEAS: Distal type of chronic spinal muscular atrophy. Clinical, electrophysiological and pathological studies. Brain 94 (1971) 703–714.

MCWILLIAM, R. C., D. GARDNER-MEDWIN, D. DOYLE and J. B. P. STEPHENSON: Diaphragmatic paralysis due to spinal muscular atrophy. An unrecognised cause of respiratory failure in infancy? Arch. Dis. Child. 60 (1985) 145–149.

MEADOWS, J. C. and C. D. MARSDEN: A distal form of chronic spinal muscular atrophy. Neurology 19 (1969) 53–58.

MEIER, C., J. LÜTSCHG, F. VASSELLA and A. BISSCHOFF: Progressive neural muscular atrophy in a case of phenylketonuria. Dev. Med. Child Neurol. 17 (1975) 625–630.

MELKI, J., S. ABDELHAK, P. SHETH, M.-F. BACHELOT, P. BURLET, A. MARCADET, J. AICARDI, J. P. CARRIERE, M. FARDEAU, D. FONTAN, G. PONSOT, T. BILLETTE, C. ANGELINI, C. BARBOSA, G. FERRIERE, G. I. LANZILL, A. OTTOLINI, M. C. BABRON, D. COHEN, A. HANAUER, F. CLERGET-DARPOUX, M. LATHROP, A. MUNNICH and J. FREZAL: Gene for chronic proximal spinal muscular atrophies maps to chromosome 5q. Nature 344 (1990a) 767–768.

MELKI, J., P. SHETH, S. ABDELHAK, P. BURLET, M.-F. BACHELOT, M. G. LATHROP, J. FREZAL, A. MUNNICH and THE FRENCH SPINAL MUSCULAR ATROPHY INVESTIGATORS: Mapping of acute (type I) spinal muscular atrophy to chromosome 5q12–q14. Lancet 2 (1990b) 271–273.

MELLINS, R. B., A. P. HAYS, A. P. GOLD, W. E. BERDON and J. D. BOWDLER: Respiratory distress as the initial manifestation of Werdnig-Hoffmann disease. Pediatrics 53 (1974) 33–40.

MILLER, R. G., R. B. LAYZER, M. A. MELLENTHIN, M. GOLABI, R. A. FRANCOZ and J. A. MALL: Emery-Dreifuss muscular dystrophy with autosomal dominant transmission. Neurology 35 (1985) 1230–1233.

MURPHY, W. A., W. G. TOTTY and J. E. CARROLL: MRI of normal and pathologic skeletal muscle. Am. J. Radiol. 146 (1986) 565–574.

NAMBA, T., D. C. ABERFELD and D. GROB: Chronic proximal spinal muscular atrophy. J. Neurol. Sci. 11 (1970) 401–423.

NEDELEC, C., F. DUBAS, J. L. TRUELLE, F. POUPLARD, F. DELESTRE and I. PENISSON-BESNIER: Amyotrophie spinale progressive distale et asymétrique des membres inférieurs à caractère familial. Rev. Neurol. 143 (1987) 765–767.

NELSON, J. S. and A. L. PRENSKY: Sporadic juvenile amyotrophic lateral sclerosis. A clinicopathological study of a case with neuronal cytoplasmic inclusions containing RNA. Arch. Neurol. 27 (1972) 300–306.

NEUHAÜSER, G., C. WIFFLER and J. M. OPITZ: Familial spastic paraplegia with distal muscle wasting in the Old Order Amish; atypical Troyer syndrome or 'new' syndrome. Clin. Genet. 9 (1976) 315–323.

ODA, M., N. AKAGAWA, Y. TABUCHI and H. TANABE: A sporadic juvenile case of the amyotrophic lateral sclerosis with neuronal intracytoplasmic inclusions. Acta Neuropathol. 44 (1978) 211–216.

O'SULLIVAN, D. J. and J. G. MCLEOD: Distal chronic spinal muscular atrophy involving the hands. J. Neurol, Neurosurg. Psychiatry 41 (1978) 653–658.

OUVRIER, R. A., J. G. MCLEOD, G. J. MORGAN, G. A. WISE and T. E. CONCHIN: Hereditary motor and sensory neuropathy of neuronal type with onset in early childhood. J. Neurol. Sci. 51 (1981) 181–197.

OUVRIER, R. A., J. G. MCLEOD and T. E. CONCHIN: The hypertrophic forms of hereditary motor and sensory neuropathy. Brain 110 (1987) 121–148.

PALMUCCI, L., T. MONGINI, C. DORIGUZZI, M. MANISCALCO and D. SCHIFFER: Familial autosomal recessive rigid spine syndrome with neurogenic facio-scapulo-peroneal muscle atrophy. J. Neurol., Neurosurg. Psychiatry 54 (1991) 42–45.

PANEGYRES, P. K., F. L. MASTAGLIA and B. A. KAKULAS: Limb-girdle syndromes. Clinical, morphological and electrophysiological studies. J. Neurol. Sci. 95 (1990) 201–218.

PASTERNAK, J. F., K. FULLING, J. NELSON and A. L. PRENSKY: An infant with chronic, relapsing polyneuropathy responsive to steroids. Dev. Med. Child. Neurol. 24 (1982) 504–524.

PATTON, M. A., M. BARAITSER and E. M. BRETT: A family with congenital suprabulbar palsy (Worster Drought syndrome). Clin. Genet. 29 (1986) 147–150.

PEARN, J.: Autosomal dominant spinal muscular atrophy. A clinical and genetic study. J. Neurol. Sci. 38 (1978) 263–275.

PEARN, J. and P. HUDGSON: Distal spinal muscular atrophy. A clinical and genetic study of 8 kindreds. J. Neurol. Sci. 43 (1979) 183–191.

PEIRIS, J. B., K. N. SENEVIRATNE, H. R. WICKREMASINGHE, S. B. GUNATILAKE and R. GAMAGE: Non familial juvenile distal spinal muscular atrophy of upper extremity. J. Neurol., Neurosurg. Psychiatry 52 (1989) 314–319.

PETAJAN, J. H., G. L. MOMBERGER, J. AASE and D. G. WRIGHT: Arthrogryposis syndrome (Kuskokwim disease) in the Eskimo. J. Am. Med. Assoc. 209 (1969) 1481–1486.

PLEASURE, J. R., S. SHUMAN and L. B. RORKE: Congenital absence of peripheral nervous system (PNS) myelin presenting as arthrogryposis multiplex congenita. Pediatr. Res. 20 (part 2) (1986) 465A.

PRABHAKAR, S., J. S. CHOPRA, A. K. BANERJEE and P. V. S. RANA: Wasted leg syndrome: a clinical, electrophysiological and histopathological study. Clin. Neurol. Neurosurg. 83(1) (1981) 19–28.

PRICE, R. W. and F. PLUM: Poliomyelitis. In: P. J. Vinken and G. W. Bruyn (Eds), Infections of the Nervous System, Vol. 34. Amsterdam, North-Holland Publ. Co. (1978) 93–132.

REFSUM, S. and S. A. SKILLICORN: Amyotrophic familial spastic paraplegia. Neurology 4 (1954) 40–47.

ROSSI, L. N., J. LÜTSCHG, C. MEIER and F. VASSELLA: Hereditary motor sensory neuropathies in childhood. Dev. Med. Child Neurol. 25 (1983) 19–31.

RUBINSTEIN, A. E., R. E. LOVELACE, M. M. BEHRENS and L. A. WEISBERG: Moebius syndrome in association with peripheral neuropathy and Kallmann syndrome. Arch. Neurol. 32 (1975) 480–482.

SAMBROOK, P., D. RICKARDS and W. J. K. CUMMING CT muscle scanning in the evaluation of patients with spinal muscular atrophy (SMA). Neuroradiology 30 (1988) 487–496.

SATOYOSHI, E. and K. YAMADA: Recurrent muscle spasms of central origin. Arch. Neurol. 16 (1967) 254–264.

SCHAPIRA, D. and M. SWASH: Neonatal spinal muscular atrophy presenting as respiratory distress: a clinical variant. Muscle Nerve 8 (1985) 661–663.

SCHLEGEL, U., F. JERUSALEM, W. TACKMANN, A. CORDT and Y. TSUDA: Benign juvenile focal muscular atrophy of upper extremities — a familial case. J. Neurol. Sci 80 (1987) 351–353.

SCOPPETTA, C., M. L. VACCARIO, C. CASALI, G. DI TRAPANI and G. MENNUNI: Distal muscular atrophy with autosomal recessive inheritance. Muscle Nerve 7 (1984) 478–481.

SILVER, J. R.: Familial spastic paraplegia with amyotrophy of the hands. J. Neurol., Neurosurg. Psychiatry 29 (1966) 135–144.

SIMPSON, D. A. and M. FOWLER: Two cases of localized hypertrophic neurofibrosis. J. Neurol. Neurosurg. Psychiatry 29 (1966) 80–84.

SINGH, N., K. P. SACHDEV and A. K. SUSHEELA: Juvenile muscular atrophy localized to the arms. Arch. Neurol. 37 (1980) 297–299.

SKRE, H., S. I. MELLGREN, P. BERGSHOLM and J. E. SLAGSVOLD: Unusual type of neural muscular atrophy with a possible X-chromosomal inheritance pattern. Acta Neurol. Scand. 58 (1978) 249–260.

SOBUE, I., N. SAITO, M. IIDA and K. ANDO: Juvenile type of distal and segmental muscular atrophy of upper extremities. Ann. Neurol. 3 (1978) 429–432.

SPAANS, F., F. G. I. JENNEKENS, J. F. MIRANDOLLE, J. B. BIJLSMA and G. C. DE GAST: Myotonic dystrophy associated with hereditary motor and sensory neuropathy. Brain 109 (1986) 1149–1168.

SPIRO, A. J., M. H. FOGELSON and A. C. GOLDBERG: Microcephaly and mental subnormality in chronic progressive spinal muscular atrophy of childhood. Dev. Med. Child Neurol. 9 (1967) 594–601.

SPITZ, M. C., J. JANKOVIC and J. M. KILLIAN: Familial tic disorder, parkinsonism, motor neuron disease, and acanthocytosis: a new syndrome. Neurology 35 (1985) 366–371.

STAAL, A., L. N. WENT and H. F. M. BUSCH: An unusual form of spinal muscular atrophy with mental retardation occurring in an inbred population. J. Neurol. Sci. 25 (1975) 57–64.

STEIMAN, G. S., L. B. RORKE and M. J. BROWN: Infantile neuronal degeneration masquerading as Werdnig-Hoffmann disease. Ann. Neurol. 8 (1980) 317–324.

STERZ, H., G. HARRER, H. MARCHET, H. P. KASERER, H. SCHLAMBERGER, Hj. SAMEC and U. STARK: Primäre und neurogene Skelettmuskelerkrankungen bzw. -paralysen mit schweren kardialen Rhythmusstörungen. Z. Kreislaufforsch. 60 (1971) 1–13.

SUGIMURA, F., M. IIJIMA, Y. OZAWA, Y. OOKI and S. WATANABE: Two cases of Kugelberg-Welander's disease with cardiopathy. Clin. Neurol. (Tokyo) 13 (1973) 79–86.

SUNOHARA, N., K. ARAHATA, E. P. HOFFMAN, H. YAMADA, J. NISHIMIYA, E. ARIKAWA, M. KAIDO, I. NONAKA and H. SUGITA: Quadriceps myopathy: forme fruste of Becker muscular dystrophy. Ann. Neurol. 28 (1990) 634–639.

SWASH, M. and M. S. SCHWARTZ: Neuromuscular Diseases. A Practical Approach to Diagnosis and Management, 2nd ed. London, Springer-Verlag (1988) 1–316.

TAN, C. T.: Juvenile muscular atrophy of distal upper extremities. J. Neurol. Neurosurg. Psychiatry 48 (1985) 285–286.

TANAKA, H., N. UEMURA, Y. TOYAMA, A. KUDO, Y. OHKATSU and T. KANEHISA: Cardiac involvement in the Kugelberg-Welander syndrome. Am. J. Cardiol. 38 (1976) 528–532.

THRUSH, D. C., G. HOLTI, W. G. BRADLEY, M. J. CAMPBELL and J. N. WALTON: Neurological manifestations of xeroderma pigmentosum in two siblings. J. Neurol. Sci. 22 (1974) 91–104.

THIJSSE, W. J. and F. SPAANS: Unilateral spinal muscular atrophy. Clin. Neurol. Neurosurg. 85 (1983) 117–121.

TOPALOGLU, H., Y. RENDA, G. KALE and K. GUCUYENER: Muscular dystrophy or spinal muscular atrophy? Lancet 1 (1989) 960–961.

TRIDON, P., J.-M. ANDRÉ, M. ANDRÉ, B. BRICHET and

G. ARNOULD: Syndrome de Moebius et amyotrophies des membres (A propos de 5 cas). Rev. Neurol. 124 (1971) 367–378.

TSUKAGOSHI, H., H. SIGUTA, T. FURUKAWA, T. TSUBAKI and E. ONO: Kugelberg-Welander syndrome with dominant inheritance. Arch. Neurol. 14 (1966) 378–381.

ULE, G.: Progressive neurogene Muskelatrophie bei neuroaxonaler Dystrophie mit Rosenthalschen Fasern. Acta Neuropathol. (Berl.) 21 (1972) 332–339.

VANASSE, M. and V. DUBOWITZ: Dominantly inherited peroneal muscular atrophy (Hereditary motor and sensory neuropathy Type I) in infancy and childhood. Muscle Nerve 4 (1981) 26–30.

VIRMANI, V. and P. K. MOHAN: Non-familial, spinal segmental muscular atrophy in juvenile and young subjects. Acta Neurol. Scand. 72 (1985) 336–340.

VISSER, M. DE and B. J. VERBEETEN JR: Computed tomography of the skeletal musculature in Becker-type muscular dystrophy and benign infantile spinal muscular atrophy. Muscle Nerve 8 (1985) 435–444.

VISSER, M. DE, B. W. ONGERBOER DE VISSER and B. VERBEETEN JR.: Electromyographic and computed tomographic findings in five patients with monomelic spinal muscular atrophy. Eur. Neurol. 28 (1988) 135–138.

WINDER, T. R. and R. N. AUER: Sensory neuron degeneration in Kugelberg-Welander disease. J. Can. Sci. Neurol. 16 (1989) 67-70.

WOHLFART, G., J. FEX and S. ELIASSON: Hereditary proximal spinal muscular atrophy: a clinical entity simulating progressive muscular dystrophy. Acta Psychiatr. Neurol. Scand. 30 (1955) 395–406.

WOOD, M. and M. ANDERSON: Neurological infections. In: Sir J. Walton (Ed.), Major Problems in Neurology. Philadelphia, W. B. Saunders (1988) 487–502.

YOKOCHI, K., M. ODA, J. SATOH and Y. MORIMATSU: An autopsy case of atypical infantile motor neuron disease with hyaline intraneuronal inclusions. Arch. Neurol. 46 (1989) 103–107.

YOUNG, I. D. and P. S. HARPER: Hereditary distal spinal muscular atrophy with vocal cord paralysis. J. Neurol., Neurosurg. Psychiatry 43 (1980) 413–418.

ZERRES, K., S. RUDNIK-SCHÖNEBORN and M. RIETSCHEL: Heterogeneity in proximal spinal muscular atrophy. Lancet 2 (1990) 749–750.

Handbook of Clinical Neurology, Vol. 15 (59): Diseases of the Motor System
J.M.B.V. de Jong, editor
© Elsevier Science Publishers B.V., 1991

Differential diagnosis of sporadic amyotrophic lateral sclerosis, progressive spinal muscular atrophy and progressive bulbar palsy in adults

ELISABETH S. LOUWERSE, PETER A. E. SILLEVIS SMITT and
J. M. B. VIANNEY DE JONG

Department of Neurology, Academic Medical Centre, University of Amsterdam, Amsterdam, The Netherlands

Progressive weakness is one of the most alarming symptoms in neurology and the ultimate diagnosis of motor neuron disease (MND) has serious consequences for the patient and his family. Therefore, every effort must be made to exclude disorders with a better prognosis or treatable cause. The aim of this chapter is to provide a rational aid.

There are no signs, symptoms, or laboratory tests which are pathognomonic for MND. The diagnosis rests mainly on the characteristic clinical picture and course of the disease, after other conditions have been excluded (Rowland 1980; Mulder 1982).

Li et al. (1986) have demonstrated, that the combination of progressive weakness, muscle wasting and fasciculation of the limbs or tongue, hyperactive tendon jerks and the absence of sensory loss has a 98% sensitivity and an 86% specificity to discriminate amyotrophic lateral sclerosis (ALS) from other neurological diseases, especially multiple sclerosis, cervical myelopathy, and stroke.

Although many neurologists consider the diagnosis of ALS an easy task (Adams and Victor 1985), the literature contains many reports of delay and actual error in diagnosis (Mulder 1980; O'Reilly et al. 1982; Belsh and Schiffman 1990). In a series of 46 patients, referred because of questionable ALS, only 4 cases appeared to have classical ALS (Mulder 1980). The most frequent misdiagnoses were cervical myelopathy (12 cases), inflammatory neuropathy (5 cases), myopathy, multiple sclerosis, syringomyelia, basilar invagination, foramen magnum tumor, brain tumor, lacunar brain infarction, brachial plexus neuropathy, arsenic myelopathy, Shy-Drager syndrome, and meningioma of the cervical cord. Thus, most mistakes are prevented by EMG and CSF examination, and by imaging of the cervical cord, craniovertebral junction and posterior fossa. On the other hand, O'Reilly et al. (1982) reported a high rate of initial misdiagnosis in patients who ultimately proved to suffer from ALS. In patients presenting with unilateral or pseudo-polyneuritic forms, initial diagnosis was correct in only 38%. In a similar retrospective study of 33 ALS patients, 14 were initially misdiagnosed (Belsh and Schiffman 1990). Of the 14 patients, 3 underwent spinal laminectomy because of suspected radiculopathy. Two patients were diagnosed only after they had developed respiratory failure.

NOMENCLATURE

Motor neuron disease (MND) consists of the following syndromes:

1. *amyotrophic lateral sclerosis (ALS)*: a progressive disease of the anterior horn cells and pyramidal tracts, with or without concomitant involvement of the bulbar motor nuclei and corticobulbar tracts,
2. *progressive spinal muscular atrophy (SMA)*: a progressive disease of the anterior horn cells, with or without concomitant involvement of the bulbar motor nuclei,
3. *progressive bulbar palsy (PBP)*: a progressive disease of the bulbar motor nuclei, with or without concomitant involvement of the corticobulbar tracts,
4. *primary lateral sclerosis (PLS)*: a progressive disease of the pyramidal tracts, with or without concomitant involvement of the corticobulbar tracts.

TABLE 1

Diagnostic criteria of motor neuron disease (MND).

1. *Lower motor neuron signs:*
   weakness of bulbar and/or spinal muscles
   muscle atrophy of bulbar and/or spinal muscles
   fasciculation of bulbar and/or spinal muscles

2. *Upper motor neuron signs are absent in SMA and present in ALS, and may manifest as:*
   increased tendon reflexes
   reflex spread
   clonus
   extensor plantar responses
   Hoffmann signs
   spasticity
   increased jaw jerk
   increased gag reflex
   positive corneomandibular responses
   positive snout response
   uncontrolled laughing, crying or yawning
   decreased tongue alternating movements, finger tapping or foot tapping

3. *Signs and symptoms, that are usually absent in MND:*
   sensory disturbances
   sphincter disturbances
   significant autonomic nervous system disturbances
   dementia
   parkinsonism
   cerebellar signs
   anterior visual system abnormalities
   involvement of the extraocular muscles

4. *Insidious onset of motor deficits within weeks to months*

5. *Progression of motor deficits within weeks to years, manifested as:*
   worsening of focal signs
   spread to other parts of the body

6. *Secondary or symptomatic forms of the disease have to be excluded by:*
   history taking
   physical examination
   neurophysiological tests (EMG, nerve conduction studies)
   CSF, blood and urine tests
   neuroimaging

DIAGNOSIS (Tables 1–3)

Diagnostic criteria of MND are summarized in Table 1. The clinical syndrome in adults has been described in more detail in Chapter 2 (SMA), Chapter 11 (ALS) and Chapter 12 (PBP) of this volume.

Suggested guidelines for *additional investigations* are given in Table 2. They are used primarily to identify treatable causes of MND. Careful history taking and clinical examination will help to guide the diagnostic investigations in the right direction.

Usually, *laboratory evaluations* will not reveal any abnormalities except for a slightly raised creatine kinase (Williams and Bruford 1970). CSF studies show normal cell count and protein content, no oligoclonal bands and negative Borrelia burgdorferi and syphilis serology. However, in some patients with otherwise classical ALS or SMA, CSF protein content exceeds 50 mg/dl (25% of cases) and oligoclonal bands are present (12% of cases) (Younger et al. 1990). A CSF protein content of over 0.75 g/l is frequently observed in patients with MND associated with paraproteinemia or malignant tumors, in particular lymphoma (Younger et al. 1990). CSF examination is considered essential for excluding treatable causes of the syndrome.

Electrophysiological studies are of utmost importance for the diagnosis of MND (Table 3) (Bradley 1987). Needle EMG will disclose signs of denervation and reinnervation. The presence of fasciculation is not required, but does support the diagnosis. Sensory nerve conduction velocity (NCV) studies are nearly always normal. Motor NCV is normal, unless there is severe atrophy. Conduction studies should exclude the presence of multifocal conduction blocks (Feasby et al. 1985; Parry and Clarke 1985, 1988). H-reflex and

TABLE 2

Recommended investigations to exclude secondary or symptomatic motor neuron disease (MND) in adults.

1. *Laboratory studies*
   a. *Cerebrospinal fluid (CSF):*
      cell count, protein, oligoclonal bands, *Borrelia burgdorferi* and syphilis serology
   b. *Blood*:
      hematological studies: hemoglobin, white blood cell count and differentiation, erythrocyte sedimentation rate (ESR)
      electrolytes: sodium, potassium, calcium
      metabolic and endocrine screening tests: glucose, renal function, lactate, thyroid function
      immunological studies: serum protein electrophoresis or, more sensitive, immunofixation electrophoresis
      infections: *Borrelia burgdorferi* and syphilis serology
      muscle enzymes: creatine kinase (CK)
   c. *Urine*
      protein excretion

2. *Electrophysiology*
   EMG of at least two limbs
   motor and sensory nerve conduction studies

3. *Neuroimaging*
   neuroimaging of the posterior fossa and craniovertebral junction, if bulbar symptoms are present
   neuroimaging of the spinal cord and craniovertebral junction, if there are only spinal symptoms

4. *Special investigations are performed, when a specific disease is suspected:*
   laboratory tests: parathyroid function, hexosaminidase A and B, ganglioside antibodies, antinuclear antibody (ANA), rheumatoid factor analysis, lysozyme, angiotensin converting enzyme (ACE), HIV-1, HTLV-1, toxoplasmosis, amino acid spectrum of serum and urine, toxicological studies (esp. metals), acanthocytes, fatty acids (C26: 0 to C22: 0), lipoproteins
   neurophysiological examinations: H-reflex, M-wave, multifocal proximal conduction blocks, repetitive stimulation at low and high rate, EMG of bulbar muscles
   neuropathological examinations: muscle and nerve biopsy

TABLE 3

Electromyography (EMG) and nerve conduction studies (NCV) in the differential diagnosis of motor neuron disease (MND).

Disorder	Activity at rest	Form of motor unit potential (MUP)
Anterior horn cell disease	Fibrillations, fasciculations, positive sharp waves	High amplitude, long duration, polyphasic potentials
Demyelinating neuropathy	Occasional fibrillation	Normal amplitude, long duration, polyphasic potentials
Axonal neuropathy	Fibrillations, fasciculations, positive sharp waves	High amplitude, long duration, polyphasic potentials
Disorders of neuromuscular transmission	None	Normal or myopathic in severe cases
Myopathic disorders	Usually none, fibrillations may be seen	Small amplitude, brief duration, polyphasic potentials

Disorder	Weak contraction	Maximal effort
Anterior horn cell disease	Normal MUPs	High amplitude, reduced number of MUPs activated, that fire repetitively, polyphasic potentials
Demyelinating neuropathy	Normal MUPs	High amplitude, reduced number of MUPs activated, that fire repetitively, polyphasic potentials
Axonal neuropathy	Normal MUPs	High amplitude, reduced number of MUPs activated, that fire repetitively, polyphasic potentials
Disorders of neuromuscular transmission	Normal or myopathic in severe cases	In myasthenia gravis, initial compound MUP is nearly normal, the amplitude declines at a slow rate of stimulation (2–3 Hz); in the Lambert-Eaton myasthenic syndrome, the amplitude of initial compound MUP is small and increases at a fast rate of stimulation (10–30 Hz)
Myopathic disorders	Increased number of MUPs, that fire repetitively	Low amplitude, short duration, increased number of MUP's activated, polyphasic potentials, full pattern in weak muscles

Disorder	Motor NCV	Sensory NCV	Motor distal latency	Amplitude	H-reflex and F-response
Anterior horn cell disease	Normal, some degree of reduction is acceptable in weak and wasted muscles	Normal	Normal	High	Normal or slightly delayed
Demyelinating neuropathy	Slowed	Slowed	Prolonged	Normal or low	Delayed
Axonal neuropathy	Normal or slightly reduced, lack of impulse dispersion after proximal stimulation	Normal or slightly reduced	Normal	Absent or low	Delayed
Disorders of neuromuscular transmission	Normal	Normal	Normal	Normal	Normal
Myopathic disorders	Normal	Normal	Normal	Low	Normal

F-wave studies are normal, unless there is marked axonal motor fiber loss.

The most common structural lesions mimicking MND should be excluded by *neuroimaging*, especially posterior fossa tumors, vascular pathology, craniovertebral junction abnormalities, cervical radiculo-myelopathy, and cervical or high thoracic cord tumors.

Muscle and nerve biopsy studies are only required, if the clinical syndrome is aspecific, in order to exclude myopathic disorder, neuropathies, inflammation or vasculitis. Muscle biopsy should be performed in a muscle showing clinical evidence of lower motor neuron involvement. Neurogenic atrophy with type grouping and target formation is required for the diagnosis of MND. More advanced cases of the disease exhibit secondary myopathic changes (Achari and Anderson 1974). Significant infiltration with mononuclear inflammatory cells, vasculitis, necrosis, ragged red fibers or rimmed vacuoles rule out the diagnosis or suggest additional disease. Sensory nerve biopsy should be normal.

DIFFERENTIAL DIAGNOSIS (Table 4)

The differential diagnosis will follow the first section of the classification of the World Federation of Neurology (Chapter 1 of this volume) and is listed in Table 4.

A. SPORADIC, SECONDARY OR SYMPTOMATIC MOTOR NEURON DISEASES

A clinical syndrome indistinguishable from classical ALS, SMA or PBP has been described in association with tumors, structural abnormalities, time-linked exposure to toxins, physical or infectious agents, and specific, acquired or genetic, laboratory abnormalities. Prognosis and treatment may be different and therefore mandate correct diagnosis.

In 'secondary' or 'symptomatic' MND, the definition of causation requires resolution of symptoms or cessation of progression after treatment. Otherwise, it is better to speak of an 'association' between MND and a specific abnormality or disease.

In addition, of some of the discussed causes of MND-mimic syndromes, it is still unknown whether they cause a pure motor neuropathy or an anterior horn cell disease.

1. PHYSICAL CAUSES

a. Trauma

Although trauma resulting in direct injury to the spinal cord is the main cause of spastic paraplegia in the world, trauma is hardly of differential diagnostic consideration in MND due to the acute nature of the neurological deficit and preceding trauma. Moreover, the clinical picture is of a more or less complete cord transection.

Many cases of ALS and SMA have been reported to occur shortly after a significant trauma (Norris, Chapter 2, present volume). Case-control studies in this subject are inconclusive (Kondo and Tsubaki 1981; Deapen and Henderson 1986). The same goes for the association of MND following the use of pneumatic tools (Gallagher and Sanders 1983). The cases developing a benign amyotrophy restricted to the affected limb, following a soft tissue injury with massive ecchymosis deserve special consideration (Norris 1975).

b. Spinal cord or brain stem compression

Spondylotic myeloradiculopathy. Differentiation between spondylotic *myeloradiculopathy* and SMA and ALS can be challenging, especially in cases of spondylosis with long-standing neural encroachment at multiple levels but with minimal or no sensory symptoms. Signs and symptoms of *radiculopathy* are usually confined to the C6 and C7 roots and consist of pain and sensory disturbances in the corresponding dermatomes. Less often the complaint is of actual weakness. On examination, there may be weakness, atrophy and hypo- or areflexia conform with the offended root, resembling the clinical syndrome of SMA. Neck muscle spasm is responsible for local tender areas.

The clinical picture of cervical or thoracic *myelopathy* may mimic ALS with spastic tetraparesis and superimposed lower motor neuron signs. In cervical myelopathy, there may be moderate

TABLE 4

Differential diagnosis of sporadic motor neuron disease (MND) in adults.

A. SECONDARY OR SYMPTOMATIC MND

1. *Physical causes*
   Trauma
   Spinal cord and brainstem compression: spondylotic myeloradiculopathy, anomalies of the craniocervical junction, syringomyelia and syringobulbia, cerebellar ectopia without syringomyelia, spinal tumors, intracranial neoplasms
   Vascular causes: acute ischemic myelopathy, progressive vascular myelopathy, spinal arteriovenous malformation (AVM)
   Electrical injury
   Radiotherapy: post-irradiation myelo- and radiculopathy, radiogenic lesions of cranial and peripheral nerves

2. *Toxins*
   Bacterial exotoxins: tetanus, botulism, diphtheria
   Metals and trace elements: lead, mercury, aluminum, manganese, selenium
   Industrial toxins, solvents and pesticides: toluene, hexacarbons, organophosphates, organic pesticides
   Excitotoxic amino acids: domoic acid, glutamate, β-N-oxalylamino-L-alanine (BOAA), β-N-methyl-amino-L-alanine (BMAA)
   Drugs: dapsone, phenytoin
   Strychnine

3. *Infections*
   Viral: poliovirus, HIV-1, HTLV-1, herpes zoster, encephalitis lethargica, other viruses affecting motor neurons
   Bacterial: syphilis, borreliosis (Lyme disease)
   Protozoan: toxoplasmosis
   Infectious agent suspected: poliomyelitis-like syndrome in bronchial asthma, Creutzfeldt-Jakob disease

4. *Immune disorders*
   Paraproteinemia
   Anti-GM_1 and anti-GD_1 ganglioside antibodies
   Paraneoplastic disorders

5. *Endocrine and electrolyte disorders*
   Hypoglycemic hyperinsulinism
   Diabetic amyotrophy
   Hyperthyroidism
   Hyperparathyroidism, osteomalacia and hypercalcemia

or severe wasting of one or both hands with or without wasting of other arm muscles (Wilkinson 1969). Extensive fasciculation, usually confined to the wasted muscles, is sometimes observed (Kasdon 1977). Sensory signs, if present, may have a 'glove' distribution in the arms, but in the legs a definite sensory level cannot usually be found. Bladder symptoms occur late and only in the most affected patients.

Differentiation of cervical or thoracic spondylotic *myeloradiculopathy* from MND can be remarkably difficult in the absence of sensory symptoms and in the presence of cord or root compression, or both, as documented by myelography or MRI. The presence of bulbar and pseudobulbar motor neuron lesions strongly suggests that problems related to spondylosis are coincidental. Another clue is the clinically and electromyographically more widespread lower motor neuron involvement in MND in the absence of concomitant spondylotic compression. Nerve conduction velocity (NCV) studies seem to contribute as well (Thacker et al. 1988). However, Kasdon (1977) decribed EMG-confirmed fasciculations in the lower extremities that disappeared after surgical decompression in 3 of the 4 reported cases. Electromyographically, there were no fibrillations in the legs.

Generally, surgical treatment of spondylotic radiculomyelopathy is reserved for patients with

Table 4 continued

B. HEREDITARY MND WITH KNOWN ENZYME DEFICIENCY

Hexosaminidase deficiency

Adrenoleukodystrophy and -myeloneuropathy

Neuroacanthocytosis, with or without an associated disturbance of lipid metabolism

Multiple lysosomal enzyme deficiency

C. MND WITH A MORE BENIGN PROGNOSIS

Monomelic motor neuron disease

Slowly progressive, proximal bulbospinal spinal muscular atrophy (SMA)

D. NEUROLOGIC DISORDERS MIMICKING MND, BUT NOT AFFECTING MOTOR NEURONS

1. *Peripheral nerve and nerve roots*

   Mononeuropathy and plexus lesions

   Motor neuropathy
2. *Neuromuscular junction*

   Myasthenia gravis

   Lambert-Eaton syndrome
3. *Myopathic disorders*, including dystrophia myotonica
4. *Multiple sclerosis*

E. MND PRESENTING AS PRIMARY LATERAL SCLEROSIS OR PROGRESSIVE BULBAR PALSY

(differential diagnosis discussed in text)

F. MND WITH ADDITIONAL SIGNS AND SYMPTOMS, INCLUDING DEMENTIA, PARKINSON-ISM, ATAXIA, DYSAUTONOMIA, OPHTHALMOPLEGIA, SENSORY DISTURBANCES

(differential diagnosis discussed in text)

G. FASCICULATIONS OR MUSCLE CRAMPS WITHOUT WEAKNESS, ATROPHY AND DENER-VATION

(differential diagnosis discussed in text)

---

severe pain or progressive neurological deficit caused by documented root or cord compression. In some of these patients with rapidly progressive neurological deficit, surgical decompression may be the only way to ensure accurate differentiation (Norris et al. 1980). Another reason for spinal surgery may be the maintenance of a good general state of health in a patient with MND and debilitating signs and symptoms felt to be caused by cord or root compression (Norris et al. 1980).

Anomalies of the craniocervical junction. Abnormalities at the craniovertebral junction give rise to a combination of spinal, lower cranial nerve, and cerebellar signs, sometimes resembling the syndrome of PBP, ALS or SMA. In the series of Mulder (1980), a case with basilar impression was mistaken for ALS, exemplifying the need for appropriate neuro-radiological imaging of the craniovertebral junction. MRI has replaced all other procedures for diagnosis.

The observed anomalies may consist of platybasia, basilar invagination, junction of the atlas and the foramen, atlantoaxial dislocation or complete separation of the odontoid from the axis, and Arnold-Chiari deformation. Congenital junction of the atlas and the foramen is the most common of these (McCrae 1953). The abnormalities are congenital or acquired. Atlantoaxial disclocation can be caused by rheumatoid arthritis or trauma. In all the congenital anomalies of the foramen magnum there is a high incidence of syringomyelia. Hydrocephalus is frequently present in Arnold-Chiari malformation.

Syringomyelia and syringobulbia. Exceptionally, sensory disturbances are lacking in early stages of syringomyelia. The appearance of muscle wasting and weakness sometimes leads to the erroneous diagnosis of SMA (Siriaungkul and Shuangshoti 1988). The development of amyotrophy and spastic signs prior to the onset of dissociated sensory loss may lead to the diagnosis of ALS (Rafalowska and Wasowics 1968; Schliep 1978; Petit et al. 1984). Isolated syringobulbia is very rare, but involvement of the lower cranial nerves is present in one half of all patients with syringomyelia. Palatal weakness, hemiatrophy and fasciculation of the tongue are common. Differentiation of these disorders from MND is usually easy due to the characteristic dissociated sensory loss in combination with segmental weakness and atrophy of the hands and arms, and loss of tendon reflexes in syringomyelia. Other aids in the differentiation are the frequent occurrence of trophic skin disorders, scoliosis and kyphoscoliosis, neurogenic arthropathy, Horner's syndrome, trigeminal sensory disturbance, nystagmus and bladder disturbances. The correct diagnosis is made by myelography or, preferably, by MRI.

Cerebellar ectopia without syringomyelia. Alani (1985) reported a 19-year-old patient with a SMA-like syndrome probably due to primary cerebellar ectopia without syringomyelia. Reflexes were preserved and there were no objective sensory abnormalities.

Spinal tumors. Pain is the commonest symptom in patients with spinal tumors, followed by progressive paresis, sensory disturbances and urinary retention (Guidetti and Fortuna 1975).

Malignant *extradural tumors* are characterized by severe radicular pain and rapidly progressive neurological deficit. However, motor disturbances can be the presenting symptom in a minority of non-malignant cases and high cervical cord compression has long been associated with hand muscle wasting attributed to anterior horn cell damage (Symonds and Meadows 1937; Stark et al. 1981; Levison et al. 1973; Alani 1985).

Tumors of the foramen magnum, usually meningiomas or neurinomas, can produce spastic tetraparesis, more severe in the arms than in the legs and atrophy of the small hand muscles. Most often, these signs are accompanied by nuchal pain, paresthesia of the arms, disturbance of position sense or by involvement of the accessory nerve (Love et al. 1954; Dodge et al. 1956; Stein et al. 1963).

A major problem might occur with multiple sites of compression as in *neurofibromata*, especially if the patient lacks other signs of von Recklinghausen's disease and the family history is negative or incomplete (McKusick et al. 1990: 162200; Norris et al. 1980). Another is *hereditary multiple exostoses* causing cervical cord compression and femoral neuropathy (McKusick et al. 1990: 133700).

Diffuse cerebrospinal gliomatosis can present clinically and electromyographically with symptoms and signs of the upper and lower motor neuron only (Schmidbauer et al. 1989).

Some tumors can already be diagnosed on plain films of the spine. Osteolytic and osteoblastic lesions, particularly with irregular or rough borders, strongly suggest malignancy. Neurinomas characteristically widen the intervertebral foramina. Expansion of the spinal canal extending over 2 or more segments is strongly suggestive of intramedullary tumor. The presence of calcifications in a tumor suggests meningioma. Myelography can document cord compression and differentiate intramedullary, extramedullary and extradural tumors. MRI is superior in demonstrating spinal cord tumors.

Intracranial neoplasms. Symptoms of intracranial neoplasms can mimic PBP and ALS. *Parasagittal* and *foramen magnum meningioma, brainstem astrocytoma* and *ependymoma* deserve special attention in this regard (Norris et al. 1980). The presence of atrophy of the hand muscles in posterior fossa tumors has been described by Symonds and Meadows (1937) and can be especially misleading in the differentiation from MND.

c. Vascular causes of motor neuron degeneration

Acute ischemic myelopathy. Selective anterior horn cell degeneration following ischemia has been described after aortic clamping in repair of coarctation (Dodson and Landau 1973), after cardiac arrest (Gilles and Nag 1971) and due to spinal cord infarction in dissecting aortic aneurysm and atheromatous emboli (Herrick and Mills 1971). However, these cases are more an illustration of the sensitivity of motor neurons to ischemia than that they are a differential diagnostic consideration in MND.

Progressive vascular myelopathy. Jellinger and Neumayer (1962) described 21 cases of a syndrome that they considered to result from chronic vascular insufficiency of the spinal cord. Most patients were in their sixties and suffered from slowly progressive weakness. Signs of lower and upper motor neuron pathology were present mimicking the clinical picture of ALS. In 4 of the 21 reported cases, sensory disturbances or pain were noted and in 3 cases there were sphincter problems. Bulbar symptoms developed in 10 cases and death due to bulbar palsy and respiratory insufficiency occurred in 7 of these. The neuropathological picture consisted of hyaline changes of intramedullary vessels, varying white matter changes, and neurogenic muscle atrophy, but without necrosis of the anterior horns. Clinical differentiation from ALS can be made by the late onset, the relatively mild bulbar involvement, and if present, sensory and sphincter disturbances.

Spinal arteriovenous malformations (AVM). Simultaneous involvement of the upper and lower motor neuron in arteriovenous malformation (AVM) can lead to an incorrect diagnosis or MND. Patients are usually middle-aged men with a steadily progressive syndrome of gradual onset consisting of pain, weakness, sensory disturbances in the legs, and severe disturbance of micturition. Symptoms are often exacerbated by exercise or specific postures, an association providing strong support for the diagnosis of spinal angioma (Aminoff and Logue 1974a, b). Some authors report an association with skin angiomata (Djindjian 1978). In most cases of AVM, CSF examination is abnormal with an elevated protein and mild pleocytosis.

Both ALS and SMA can be simulated in patients with spinal angioma without sensory disturbances, but bladder involvement early in the course of the disease, pain, the characteristic exacerbation by exercise and certain postures, and the usual restriction of signs to the lower part of the body, will indicate the need for myelography or MRI (Rosenblum et al. 1987).

d. Electrical injury

Electrical injury is caused by either lightning stroke or technical electrical accidents. Panse's (1975) review of the literature includes 12 cases in which a progressive pure motor syndrome with signs and symptoms of lower and usually upper motor neuron involvement developed after electrical trauma. Clinically and pathologically these cases were indistinguishable from ALS. A case-control study of 518 ALS patients and 518 matched controls showed an increased incidence in ALS of occupations at risk of electrical exposure and of electrical shocks producing unconsciousness (Deapen and Henderson 1986). However, the case-control study of MND in Japan did not show such an association (Kondo and Tsubaki 1981; Norris, Chapter 2, present volume).

e. Radiotherapy

Post-radiation myelopathy and radiculopathy. The most common type of radiation injury to the spinal cord is a delayed, progressive and chronic myelopathy, occurring 3 months to 6 years (commonly 6–12 months) after radiation therapy (Reagan et al. 1968). Differentiation of chronic radiation myelopathy with pyramidal tract signs from ALS is usually easy since the clinical picture is primarily that of a progressively declining motor, sensory and sphincter function (Pallis et al. 1961; Reagan et al. 1968; Jellinger and Sturm 1971).

The rare syndrome of selective lower motor neuron damage following radiotherapy may mimic SMA. This condition was first described in 3 patients by Greenfield and Stark (1948) and subsequent cases have been reported by Maier et al. (1969), Sadowsky et al. (1976), Kristensen et al. (1977), Rowland (1980), Horowitz and Stewart (1983), and Lagueny et al. (1985). This type of myelopathy or radiculopathy is most likely to occur in patients undergoing abdomino-pelvic irradiation or after complete neuraxis radiation. Symptoms usually appear from 3 to 23 months after irradiation, although presentations after 13 and 14 years have been described (Maier et al. 1969; Lagueny et al. 1985). The clinical picture invariably consists of lower limb weakness, atrophy, diminished or absent reflexes, in the absence of sensory or sphincter disturbances. The course is progressive over some months to 2 years followed by stabilization without remission. One patient in the series of Kristenson et al. (1977) worsened progressively.

Electrophysiological studies reveal signs of denervation and normal motor and sensory NCV (Sadowsky et al. 1976; Lagueny et al. 1985). Detailed electrophysiological studies indicate that the site of damage is either the anterior horn cells of the lower spinal cord or the lumbar and sacral motor nerve roots, rather than the lumbosacral plexus (Horowitz and Stewart 1983). The CSF protein can be substantially raised (Sadowsky et al. 1976; Lagueny et al. 1985). Myelography is always normal as is CT of pelvis and abdomen.

Differential diagnosis is from spinal metastases or diffuse infiltrations. Treatment is remedial (Henson and Urich 1982).

Radiogenic lesions of cranial and peripheral nerves are comparatively rare due to their low radiosensitivity. The lesions are usually of the late type occurring months to years after radiotherapy and mainly have to be differentiated from neoplastic, paraneoplastic, and drug induced lesions.

Most frequent is *brachial plexopathy* which seems to be due more often to neoplastic infiltration than to radiation injury (Bagley et al. 1978; Kori et al. 1981). Latencies in both entities vary from 3 months to 25 years (Kori et al. 1981). Severe pain and Horner syndrome are more common in tumor infiltration, whereas lymphedema and involvement of the upper plexus are more typical of radiation injury (Kori et al. 1981). MND will usually not be a serious differential diagnostic consideration, since in brachial plexopathy sensory disturbances and pain are present in an early stage (Henson and Urich 1982).

2. TOXINS

a. Bacterial exotoxins

Tetanus. The clinical syndrome of tetanus with palsy of cranial nerves, dysphagia, respiratory problems, spasticity and increased tendon reflexes bears resemblance with ALS. It can be distinguished by the subacute onset of symptoms, the characteristic clinical features with painful muscle spasms, the absence of muscle weakness or atrophy, and typical EMG-findings (Edmondson and Flowers 1979; Trujillo et al. 1980).

Botulism. The clinical picture of botulism may have the appearance of rapidly progressive SMA. Differential diagnosis is based on the acute onset of botulism, rapidly progressive course and spontaneous recovery of the motor syndrome, the presence of characteristic ocular signs, and enlarged pupils, that are unresponsive to light. In addition, isolation of the organism or toxin and EMG recordings may support the diagnosis (Merson et al. 1974; Gutman and Pratt 1976; Arnon 1980).

Diphtheria. The combination of bulbar signs and subsequent evolvement of weakness of the limbs, may initially resemble PBP or SMA. Diphtheria is suspected by the history of an upper respiratory infection or wound infection, characteristic course of the disease with onset of bulbar symptoms, followed by progressive weakness of the limbs and spontaneous stabilization and recovery of symptoms, the absence of atrophy and fasciculation, the presence of sensory impairment and ocular symptoms, electrophysiological findings consistent with sensorimotor demyelinating neuropathy and, occasionally, raised CSF protein and cells (McDonald and Kocen, 1984).

b. Metals and trace elements

Lead. Chronic lead intoxication may cause a predominantly lower motor neuron syndrome due to a peripheral neuropathy or possibly, an anterior horn cell disorder (Campbell et al. 1968, 1970; Boothby et al. 1974). Upper motor neuron signs are sometimes observed, leading to diagnostic confusion with ALS (Wilson 1907; Aub et al. 1925; Livesley and Sissons 1968).

Suggestive of lead poisoning are blue lines at the gumtooth margin, anemia with basophilic stippling and abdominal colics. At this moment, a history of increased lead exposure is more important than any laboratory value. Urinary levels of lead may be normal despite an elevated body burden (Campbell et al. 1970; Barry 1975). Bone lead content probably reflects the total body burden of lead more reliably.

Some patients with ALS or SMA and a history of lead exposure may benefit from treatment with chelating agents (Simpson et al. 1964; Livesley and Sissons 1968; Campbell et al. 1970). Patients with upper motor neuron signs are less likely to respond to treatment (Currier and Haerer 1968; Campbell et al. 1970).

Mercury. Mercury poisoning may cause encephalopathy, ataxia, rapid tremors, peripheral neuropathy and perhaps MND (Rustam et al. 1975). Brown (1954) described a patient with progressive muscle weakness and bulbar palsy in association with mercury exposure indistinguishable from SMA. Urinary levels of mercury were raised. Kantarjian (1961) studied 114 patients with neurologic symptoms after exposure to organic mercury compounds in Iraq. Six developed an ALS-like syndrome and 5 exhibited symptoms and signs simulating SMA. Barber (1978) and Adams et al. (1983) reported a resolution of symptoms and signs after withdrawal from exposure in 2 patients with predominantly lower motor neuron signs exposed to elemental mercury and mercuric oxide. The lower motor neuron signs in the reported patients might have been caused by a peripheral neuropathy or anterior horn cell lesion. No pathologically proven cases of MND that responded convincingly to therapy have been reported.

Aluminum. Aluminum intoxication has been named as a possible cause of ALS on Guam (Yoshimasu et al. 1980; Garruto et al. 1989; see also Chapters 15 and 16, this volume). Outside the endemic areas, Patten and Brown (1990) reported 2 patients with ALS working at the same aluminum canning company. Urinary and serum levels of aluminum were increased. The diagnosis of ALS was confirmed at autopsy in one patient. The causal relation between the high aluminum exposure and subsequent evolvement of ALS is uncertain, because both patients died

a few weeks after institution of desferrioxamine treatment due to cardiac infarction and aspiration.

Manganese. Manganese intoxication may be seen in manganese miners and manufacturers of batteries. The clinical syndrome comprises confusion, hallucinations and extrapyramidal features. Progressive weakness, fatigability, corticospinal and corticobulbar signs mimicking ALS may be added (Mena 1967). Usually, the history of manganese poisoning and the extrapyramidal and psychiatric features facilitate differential diagnosis from ALS.

Manganese intoxication has been proposed as a causative factor of ALS in the Western Pacific high incidence foci (see for further details Chapter 15 and 16, this volume).

Selenium. A relation between selenium and MND is suggested by the report of a cluster of 4 cases from an area of South Dakota with a high soil content of selenium (Killness and Hochberg 1977). Only 1 of the 4 reported patients, however, had high urinary levels. Also, the increased spinal cord selenium content in ALS patients without exposure to selenium probably represents an epiphenomenon of the disease without known causal relation (Kurlander and Patten 1979; Mitchell et al. 1986).

c. Industrial toxins, solvents and pesticides

The list of pesticides and industrial toxins, known to be harmful to the nervous system is ever expanding.

Toluene and hexacarbons. The most toxic industrial compounds are probably toluene and hexacarbons, which are used as glue solvents and in rubber, enamels and paints. The neurologic sequelae are various. Most toxins cause a mixed sensorimotor neuropathy (Anonymous 1979). Rarely, concomitant corticospinal tract signs are observed (Hormes et al. 1986). The clinical syndrome may resemble ALS or SMA, but the presence of sensory disturbances and neurophysiologic findings will point to a neuropathy.

There are conflicting reports on the relation

between solvent exposure and subsequent evolvement of MND (Gunnarson and Lindberg 1989). Two studies showed an increase of MND in workers in the leather industry (Hawkes and Fox 1981; Buckley et al. 1983; Hawkes et al. 1989), but other careful studies failed to confirm these findings (Deapen and Henderson 1986; Martyn 1989).

Organophosphates and tri-ortho-cresyl phosphate (TOCP). Organic phosphorous compounds are widely used as insecticides and as softeners in plastic industry. The best known is tri-ortho-cresyl phosphate (TOCP), which caused an outbreak of jake paralysis after drinking Jamaica ginger (Charlin and Brunschwig 1948) and after the use of cooking oil containing TOCP (Smith and Spalding 1959).

Acute poisoning manifests with gastrointestinal complaints, headache, sweating, salivation, bronchial spasm, and miosis (Senanayake and Karalliedde 1987).

Chronic effects of poisoning consist of a distal weakness, wasting, occasional fasciculation, pyramidal signs and cranial nerve involvement. Nerve conduction velocities are normal or only slightly slowed, pointing to an anterior horn cell lesion or an axonal degeneration with some segmental demyelination (Waida et al. 1974).

The clinical picture of chronic poisoning bears resemblances with ALS, PBP or SMA (Charlin and Brunschwig 1948; Smith and Spalding 1959; Namba et al. 1971; Waida et al. 1974).

Organic pesticides. A case with rapidly progressive bulbar ALS associated with overexposure to the organic pesticides, pyrethrin, organochlorine derivatives and the propellant freon has been reported by Pall et al. (1987). The diagnosis of ALS has been confirmed by autopsy. It is unclear, whether the exposure and subsequent development of the disease are causally related or represent a chance association.

d. Excitotoxic amino acids

Domoic acid and glutamate. Domoic acid is an excitotoxic amino acid structurally related to glutamate, an excess of which is known to be

harmful to the central nervous system (Choi 1988). Late in 1987, consumption of mussels contaminated with domoic acid resulted in an outbreak of gastrointestinal and neurological symptoms in Canada (Perl et al. 1990). The acute phase of the intoxication developed within 48 h after ingestion of the mussels, with generalized weakness, occasionally accompanied by fasciculation and muscle atrophy. Deep tendon reflexes were either increased with extensor plantar signs or diminished. Up to 1 year after poisoning, a consistent part of the patient population still exhibited mild weakness with atrophy of the distal muscles. EMG findings were consistent with an anterior horn cell disease or axonal neuropathy (Teitelbaum et al. 1990).

The chronic sequelae of domoic acid poisoning closely resemble MND, but the acute onset of the disease distinguishes both disorders.

β-N-oxalylamino-L-alanine (BOAA). BOAA is an excitotoxic amino acid derived from the chickling pea. Poisoning is linked to human lathyrism (Ludolph et al. 1987).

β-N-methylamino-L-alanine (BMAA). BMAA is an excitotoxic amino acid present in the seed of the false sago palm *Cycas circinalis*. BMAA has been implicated as a causative factor in the development of the ALS-parkinsonism-dementia syndrome in the high incidence foci (Spencer et al. 1988).

e. Drugs

Dapsone. Dapsone may produce an SMA-like pure motor syndrome with weakness and atrophy, especially of the hands, and normal or diminished tendon jerks. Motor NCV are normal or only slightly reduced in the presence of marked denervation. The disorder probably represents an axonal neuropathy rather than MND (Gutmann et al. 1976; Homeida et al. 1980).

Phenytoin (diphenylhydantoin). Direkze and Fernando (1977) reported a 29-year-old man with profuse fasciculations, but without weakness. The calf muscles were flabby and tendon reflexes brisk. EMG showed large motor unit action potentials with normal sensory and motor nerve conduction velocities. Six months after cessation of phenytoin therapy, the clinical and neurophysiological abnormalities had completely resolved. In spite of the absence of weakness and denervation, this case may be considered as a reversible form of MND or axonal neuropathy.

f. Strychnine

Strychnine poisoning is a disorder of the motor neurons characterized by muscle cramps, twitching and hyperreflexia. It is differentiated from MND by the acute onset of the disease, the absence of weakness and atrophy, and characteristic EMG recordings (O'Callaghan et al. 1982; Boyd et al. 1983).

Cumulative toxicity has not been reported due to the rapid clearance of the drug (Goodman et al. 1980).

3. INFECTIONS

a. Viral infections

Poliovirus: acute anterior poliomyelitis. Several days after a poliovirus infection, paralysis may develop in a minority of the infected patients. The legs are most affected, but the disease may also attain the bulbar and arm muscles. Reflexes are diminished or abolished. Atrophy can be detected within 3 weeks of onset of paralysis. In the meningeal phase of the disease, the CSF initially shows a lymphocytic pleocytosis and later a slightly raised protein.

The acute onset of the disease and subsequent complete or partial recovery marks off acute poliomyelitis from more chronic and progressive MND.

Late post-poliomyelitis spinal muscular atrophy (SMA). Patients with prior paralytic poliomyelitis may complain of a decrease in functional capacity many years after recovery from the acute infection. Only in a minority of the patients progressive disability is caused by a distinct syndrome: post-poliomyelitis SMA (Alter et al. 1982). It has the hallmarks of idiopathic SMA, like excessive fatigue, progressive muscle weakness and atrophy. The onset is insidious, usually involving the orginally affected limb. Deep tendon reflexes are normal or absent. In some cases, there may be evidence of upper motor neuron involvement at bulbar or spinal levels (Zilkha 1962; Campbell et al. 1969; Mulder et al. 1972; Salazar-Gruesco et al. 1987). The rate of progression of the disease is remarkably slow (Dalakas and Hallett 1988a).

The diagnosis of postpoliomyelitis SMA is based on repetitive examinations showing progressive weakness and atrophy of muscles (Mulder, Chapter 3, present volume). EMG, muscle biopsies and laboratory evaluations do not distinguish newly symptomatic and asymptomatic patients with prior acute poliomyelitis (Hayward and Seaton 1979; Cashman et al. 1987).

Persistent infection by poliovirus or echovirus in agammaglobulinemia. Chronic viral infection of the CNS may cause a progressive pure motor syndrome in immunodeficient patients. Progressive weakness with wasting of the limbs and, incidentally, pyramidal tract signs may lead to an incorrect diagnosis of ALS or SMA. Davis et al. (1977) reported the syndrome in an immunodeficient child after poliovirus vaccination and Mease et al. (1981) in a patient with agammaglobulinemia and persistent echovirus infection. The clinical symptoms in the latter patient resolved after treatment with immunoglobulins.

Human immunodeficiency virus (HIV). There has been only 1 reported case of a 26-year-old man who developed rapidly progressive ALS associated with HIV-1 infection (Hoffman et al. 1985). However, both the central and peripheral nervous system are frequently involved in HIV-1 infection (McArthur 1987; Fischer and Enzensberger 1987; Lange et al. 1988; Dalakas and Pezeshkpour 1988).

Motor disturbances are an important feature of the *AIDS dementia complex.* Deterioration of cognitive and behavioral functions, unsteady gait, weakness of the legs, sometimes accompanied by tremors, loss of fine-motor coordination, hyperreflexia, and, occasionally, extensor plantar responses are prominent. MRI or CT-scan and lumbar puncture are required to exclude other treatable causes of the dementia and motor syndrome.

Vacuolar myelopathy is another directly HIV-1 related disorder and is often associated with the AIDS dementia complex. The syndrome may mimic ALS with progressive spastic paraparesis, spinal ataxia, hyperreflexia, extensor plantar reflexes, and only mild sensory disturbances. In later stages an ascending paresis (tetraparesis) and micturition disturbances can be present. Myelography and MRI appear normal, whereas somatosensory evoked potentials may be slowed at the medullary level (Petito et al. 1985; Helweg-Larsen et al. 1988). HIV-1 p24Ag can often be demonstrated in CSF. CSF changes are otherwise aspecific. Differential diagnosis in AIDS myelitis includes herpes simplex, varicella zoster and cytomegalovirus infection (Britton et al. 1985; Tucker et al. 1985).

Peripheral nervous system complications occur in at least 15% of patients with AIDS and can be classified as *peripheral neuropathies*, *polymyositis* and other disorders, including the Guillain-Barré syndrome and chronic inflammatory demyelinating polyneuropathy (Dalakas and Pezeshkpour 1988). The clinical syndrome with progressive motor deficits may be indistinguishable from SMA. The presence of sensory symptoms, elevation of CSF protein, pleocytosis, electrophysiology, characteristic findings in nerve biopsies, and positive HIV-1 serology will lead to the right diagnosis.

HTLV-1. HTLV-1-associated ALS and tropical spastic paraparesis are extensively reviewed by Román et al. (Chapter 23, present volume).

Foci of high incidence include Japan, the Caribbean, and the Seychelles, but cases have also been reported from the United Kingdom (Cruickshank et al. 1989) and Canada (Power et al. 1989). All British patients and 5 of the 6 Canadian patients had migrated from an area endemic for HTLV-1 infection.

The clinical syndrome is characterised by a slowly progressive spastic paraparesis with back pain, impotence, spastic bladder, and minimal sensory symptoms. In up to 25% of patients, peripheral nerve involvement leads to loss of ankle reflexes and diminished vibration percep-

tion in the toes. In some patients there is also involvement of the lower motor neuron with marked muscle wasting and fasciculation leading to a syndrome called 'pseudo'-ALS (Evans et al. 1989; Vernant et al. 1989). Lower extremity symptoms precede weakness of the upper extremities. Sometimes there is bulbar involvement with fasciculation of the tongue and dystonia, but dysphagia and dysarthria are commonly absent. Back pain, sensory symptoms and bladder dysfunction differentiate this 'pseudo'-ALS syndrome from classical ALS. Moreover, in ALS, absence of dysarthria and dysphagia is atypical in the advanced stage. Other associated neurological and systemic symptoms can help to differentiate (Table 1, Chapter 23, present volume). CSF may show mild lymphocytic pleocytosis and mild to moderate increase in protein and oligoclonal bands. The diagnosis can be further confirmed by the presence of HTLV-1 antibodies in blood and CSF and testing for HTLV-1 DNA in white blood cells. Neurophysiologic studies can be confusing in showing changes consistent with an anterior horn cell lesion.

Herpes zoster. Within 2–3 weeks after cutaneous herpes zoster, segmental paresis may occur in about 5% of patients (Thomas and Howard 1972). The facial muscles are often affected. The prognosis is good with 55% of patients showing full recovery and 30% significant improvement. Mainly in immunocompromised hosts, dissemination of the infection with zoster encephalomyelitis is a dreaded complication. Differentiation from SMA and PBP is mainly by the typical cutaneous lesions of the disease and severe pains. Furthermore, the (sub)acute onset of the disease, spontaneous recovery of the symptoms and segmental paresis preclude confusion with MND.

Encephalitis lethargica. In survivors of the 1912–1926 encephalitis lethargica epidemic, a progressive lower and upper motor neuron syndrome may develop after a latency of months to years (Greenfield and Matthews 1954). The muscles of the arms are usually affected first. Amyotrophy may precede or be superimposed on other sequelae of the disease, like parkinsonism, oculogyric crises, narcolepsy, and myoclonus. The course is usually slowly progressive, but more rapid forms with death within 1 or 2 years have also been described. The frequency with which amyotrophy has been reported makes it unlikely that it is only a chance association of 2 separate diseases, but it is probably very rare at present.

Other viruses affecting motor neurons. Several viruses other than the polioviruses may cause an acute, more or less selective infection of the anterior horn cells, including Coxsackie viruses, echoviruses, enteroviruses-70 and -71, mumps, Herpes simplex, adenovirus type 7 and arboviruses (Lenette et al. 1960; Magoffin et al. 1961; Grist et al. 1978; Kaufman et al. 1980). Enterovirus-70 is probably one of the viruses responsible for the *acute hemorrhagic conjunctivitis* associated with a severe polio-like illness (Bharucha and Mondkar, 1972; Hung and Kuno 1979; Hatch et al. 1981).

However, a more chronic anterior horn cell disease like MND has not been reported in relation to these viruses.

b. Bacterial infections

Syphilis. Syphilitic amyotrophy, has been observed by neurologists in the first half of this century (Nonne 1921; Ostheimer et al. 1924; Martin 1925; Mackay and Hall 1933), but has also been noted more recently (Hooshmand et al. 1972; Luxon et al. 1979; El Alaoui-Faris et al. 1990).

Clinically, flaccid weakness and wasting usually begin in the distal muscles of the arms (syphilitic wrist drop) and spread in months or years to the muscles of shoulder-girdle, trunk and limbs. The onset is often accompanied by pain in the neck or shoulder-girdle. Tendon reflexes are diminished or absent. Rarely, reflexes are pathologically brisk, with extensor plantar signs (Nonne 1921; Heathfield and Turner 1951; El Alaoui-Faris et al. 1990). Bulbar signs may be present (Cook 1953; El Alaoui-Faris et al. 1990).

Four of the 5 patients reported by El Alaoui-Faris et al. (1990) improved after antibiotic treatment and in 1 patient the disease stabilized with a follow-up period of 5–13 years.

Thus, neurosyphilis may cause a clinical syndrome consistent with the diagnosis SMA, ALS, or PBP, which may be cured or stopped by antibiotic treatment. Therefore, the CSF of each patient presenting with an MND-like syndrome should be tested for syphilis.

Borreliosis (Lyme disease). Within 2 months after an acute infection with the tick-borne spirochete *Borrelia burgdorferi*, a minority of the patients will exhibit signs of meningitis, radiculoneuritis, and cranial neuritis, particularly facial palsies (Pachner and Steere 1985). Months to years after the infection focal demyelinating encephalomyelitis and neuropathy may develop (Reik et al. 1985). Chronic radiculitis and neuritis are often accompanied by pain and sensory symptoms, but may manifest as a pure motor syndrome without pain resembling SMA (Wokke et al. 1987a, b; Ackermann et al. 1988). Coincidence of motor radiculoneuropathy and encephalomyelitis may cause an ALS-like clinical picture (Ackermann et al. 1985; Waisbren et al. 1987; Fredrikson and Link 1988; Halperin et al. 1989, 1990).

Clues for differential diagnosis are a history of tick-bite, influenza-like illness, erythema chronicum migrans, and large joint oligo-arthritis. About 30% of the patients with chronic neuroborreliosis have early neurologic abnormalities (Logigian et al. 1990). Antibodies to *Borrelia burgdorferi* in serum and CSF are often raised. EMG findings are consistent with radiculopathy, demyelinating or axonal motor neuropathy (Cherington and Snyder 1968).

Pitfalls comprise the absence of a history of tick bite or erythema chronicum migrans (Reik et al. 1986; Halperin et al. 1990). Moreover, patients with lower motor neuron involvement may fail to show raised intrathecal antibody production (Halperin et al. 1990). *Borrelia burgdorferi* antibodies cross-react to *Treponema* (syphilis), *Mycobacterium* (tuberculosis), *Leptospira* and *Toxoplasma gondii.*

Recognition of the syndrome is important, because some patients with predominantly lower motor neuron involvement responded to antibiotic treatment, if treated early in the course of the disease (Halperin et al. 1990; Logigian et al. 1990).

c. Protozoan

Toxoplasmosis (Toxoplasma gondii). Toxoplasma gondii, an obligate intracellular protozoan parasite, is the most commonly encountered cause of neurological opportunistic infection. Cerebral toxoplasmosis occurs in 28% of AIDS patients (Cohn et al. 1989). The clinical spectre of infections include acute myositis (Rowland and Greer 1961), polyradiculoneuritis (Lavaud et al. 1979; Bouchez et al. 1985), encephalitis (Bach and Armstrong 1983) or, less commonly, myelitis (Leys et al. 1984; Mehren et al. 1988). Infection of anterior horn cells is probably rare. A syndrome resembling rapidly progressive ALS, that responded to treatment, has been reported in an immunologically non-compromised man by Dubois et al. (1989b).

d. Infectious agent suspected

Poliomyelitis-like syndrome in bronchial asthma. A poliomyelitis-like syndrome has been reported in children recovering from an acute attack of bronchial asthma (Hopkins 1974; Danta 1975; Wheeler and Ochoa 1980; Liebeschuetz 1981). The patients exhibited acute flaccid paresis, muscle tenderness and paresthesia, without fever or meningeal signs. Residual weakness and atrophy of an upper or lower extremity may develop, the proximal arm muscles being most affected. CSF showed mononuclear pleocytosis and elevated protein concentration in some cases.

Viral anterior poliomyelitis is assumed to be the cause of the symptoms, although a virus has not yet been identified. Manson and Thong (1980) suggested an immunological pathogenesis. Wheeler and Ochoa (1980) found bursts of motor unit potential synchronously with deep inspiration, which may point to a lesion or compression of the brachial plexus or cervical nerve roots.

Creutzfeldt-Jakob disease (CJD). Clinical features of the amyotrophic form of CJD may include pyramidal and lower motor neuron signs (Allen et al. 1971). The development of dementia, myoclonus, and, occasionally, extrapyramidal and cerebellar signs, seizures, rapid down-hill course and characteristic EEG abnormalities are clues for the diagnosis of CJD. Most cases of ALS and dementia, previously considered amyotrophic CJD, appeared non-transmittable to primates and lacked the characteristic autopsy features of CJD (Salazar et al. 1983).

4. IMMUNE DISORDERS

a. Paraproteinemia

In a small percentage of the normal population, monoclonal paraproteins are present, especially in the elderly. In MND, the prevalence of benign and malignant paraproteinemia seems to be inordinately high as compared to the general population (Shy et al. 1986; Younger et al. 1990). Paraproteins associated with MND are either of the IgM or, less frequently, of the IgA or IgG type. In some cases, the immunoglobulins reacted with GM_1 and GD_{1b} ganglioside (Freddo et al. 1986) and cross-reacted with glycoproteins of the peripheral nerve and spinal cord (Latov et al. 1988). Usually, paraproteins are benign, but they may also occur in association with plasma cell malignancies, like sclerotic myeloma and multiple myeloma, Waldenström's macroglobulinemia and chronic lymphatic leukemia.

The neurologic syndromes are those of a mixed sensorimotor neuropathy, motor neuropathy, SMA (Bauer et al. 1977; Latov et al. 1980; Latov 1982; Shy et al. 1986), ALS or PBP (Brownell et al. 1970; Chazot et al. 1976; Younger et al. 1990). The clinical features of lower motor neuron involvement alone may be caused by an anterior horn cell disease, peripheral neuropathy, or both. Loss of motor neurons has been demonstrated at autopsy (Brownell et al. 1970; Bauer et al. 1977). Neurophysiologically, multifocal conduction blocks may be demonstrated (Pestronk et al. 1988a).

A patient with SMA, ALS or PBP is more likely to have paraproteinemia, if CSF protein is elevated and CSF oligoclonal bands are present (Younger et al. 1990). Immunofixation electrophoresis on agarose is more sensitive in detecting paraproteins than conventional electrophoresis on cellulose acetate gels (Younger et al. 1990). The effect of immunosuppressive therapy and plasmapheresis in patients with MND or motor neuropathy is uncertain (Patten 1984; Shy et al. 1986). All reported cases that responded to treatment had a predominantly lower motor neuron syndrome (SMA or motor neuropathy) and paraproteinemia due to benign plasma cell dyscrasia (Patten 1984; Shy et al. 1986), Waldenström's macroglobulinemia (Peters and Clotonoff 1968; Rowland et al. 1982), multiple myeloma or sclerotic plasmacytoma (Driedger and Pruzanski 1980). The effect of treatment in patients with upper motor neuron signs has yet to be determined.

b. Anti-GM₁ and anti-GD₁ ganglioside antibodies

Some patients with otherwise classical ALS or SMA appear to have high serum titers of anti-GM₁ anti-GD₁ ganglioside antibodies. Antibodies are of the IgM or IgG type. The first reports were in patients with plasma cell dyscrasia (Freddo et al. 1986), but they also occur in patients without paraproteinemia (Pestronk et al. 1990). The clinical syndromes associated with ganglioside antibodies are motor neuropathy, SMA, and less commonly, ALS (Latov et al. 1988; Pestronk et al. 1988a, b; Shy et al. 1988; Sadiq et al. 1990).

Of interest is that plasmapheresis and immunosuppressive treatment had been followed by improvement in some cases with predominantly lower motor neuron signs and high titers of anti-GM₁ IgM antibodies (Latov et al. 1988; Pestronk et al. 1988b; Shy et al. 1988). Like in paraproteinemia, some of the reported cases responding to treatment might be considered as having motor neuropathy rather than MND and it is unknown whether treatment is effective in patients with upper motor neuron signs (ALS).

In view of the incidental response to treatment, perhaps each patient with predominantly lower motor neuron signs should be tested for anti-GM₁ and anti-GD₁ antibodies, especially if CSF protein is elevated (Sadiq et al. 1990), CSF oligoclonal bands are present (Sadiq et al. 1990), polyclonal or monoclonal serum IgM immunoglobulins are increased, and, electrophysiologically, proximal multifocal conduction blocks are demonstrated (Pestronk et al. 1988a).

c. Paraneoplastic disorders

The large studies in the literature are in favor of a chance association in older age groups of MND and malignancy (Henson and Urich 1982), probably with the exception of lymphoma (Younger et al. 1991). On the other hand, progressive weakness in cancer patients may be caused by other paraneoplastic disorders as encephalomyelitis, necrotizing myelopathy, the Guillain-Barré syndrome, carcinomatous neuromyopathy, polymyopathy, and endocrine myopathy, that can all be confused with ALS or SMA (Patchell and Posner 1985).

Association of Hodgkin's disease and non-Hodgkin lymphoma with a mild form of paraneoplastic *motor neuropathy* or *lower motor neuron syndrome* is often emphasized (Schold et al. 1978). Younger et al. (1991) reviewed 25 patients and reported 9 new cases with the combination of lymphoma and motor neuron disease (MND). In 14 of these 34 patients signs of *upper motor neuron* involvement were present. In one third, MND preceded the diagnosis of the tumor. Clinical clues to the diagnosis of lymphoma were monoclonal paraproteinemia, increased CSF protein content (especially when > 0.75 g/l), and the presence of oligoclonal bands in CSF. Younger et al. (1991) detected paraproteinemia in 3 of 7 patients and this discovery led in 2 patients to detection of asymptomatic non-Hodgkin lymphoma. Two patients had multifocal conduction blocks on nerve conduction studies that disappeared after treatment of the lymphoma by chemotherapy. Previous radiotherapy does not explain this association of lymphoma with MND. No clear relationship exists between therapy of lymphoma and improvement of MND, since only 3 of 16 patients treated with chemotherapy responded.

Paraneoplastic *encephalomyelitis* and *subacute necrotizing myelopathy* may present as a subacute ALS-like picture associated with fasciculations, atrophy, profound weakness in a patchy distribution, diminished or absent tendon jerks, and pyramidal tract signs (Brain et al. 1965).

Carcinomatous *neuromyopathy* is usually associated with carcinoma of the lung or ovary. Clinically, it is characterized by proximal weakness and wasting, depressed reflexes and absence of sensory symptoms, which render it clinically indistinguishable from SMA (Croft and Wilkinson, 1965; Campbell and Paty, 1974). Besides signs of denervation, EMG shows small polyphasic motor unit potentials (Campbell and Paty, 1974). Muscle biopsy demonstrates increase of small angular fibers and atrophy of type 2 fibers (Barron and Hefner, 1978). A beneficial response in one patient to cyclophosphamide treatment has been reported by Bruyland et al. (1984).

The pure motor deficits due to paraneoplastic *polymyopathies* and *endocrine myopathic disorders* can be differentiated from MND by the absence of fasciculation and reflex abnormalities, and myopathic EMG and muscle biopsy.

The paraneoplastic *Guillain-Barré (GBS)* and *Lambert-Eaton myasthenic syndromes* (LEMS) can easily be confused with SMA. The Guillain-Barré syndrome has primarily been identified in association with Hodgkin's disease. Both disorders are discussed in sections D-1 and D-2 of this chapter.

Since paraneoplastic syndromes are defined as nervous system disorders occurring exclusively or in higher incidence in patients with cancer that are not caused by metastasis or other directly to the tumor related factors as chemotherapy, radiation, infection, etc., all of these factors have to be excluded (Henson and Urich 1982; Patchell and Posner 1985).

5. ENDOCRINE AND ELECTROLYTE DISORDERS

a. Hypoglycemic hyperinsulinism

An MND-like syndrome with progressive distal weakness of the arms, wasting and occasional fasciculation is a highly exotic finding in patients with pancreatic islet cell tumors (Tom and Richardson, 1951; Mulder et al. 1956). Some patients complain of numbness or paresthesia, but few of them have objective sensory signs at examination.

Electrophysiologically, signs of denervation and reinnervation with normal motor NCV in the acute stage and slightly reduced NCV in the later stages have been found (Lambert 1960; Danta 1969; Harrison 1976).

Hypoglycemic amyotrophy is differentiated from idiopathic SMA by the associated symptoms of hypoglycemia that may be provoked by exercise or fasting. It is unknown whether the syndrome is caused by a motor neuropathy or by an anterior horn cell lesion (Moersch and Kernohan 1938). More importantly, symptoms have been arrested or improved after removal of the tumor (Silfverskiöld 1946; Barris 1953; Harrison 1976; Jaspan et al. 1982).

b. Diabetic amyotrophy

Diabetic amyotrophy may be observed in elderly patients with unknown or untreated diabetes mellitus. The prognosis is good, if hyperglycemia is adequately controlled.

Clinically, the patients develop asymmetrical progressive weakness and wasting of the thigh, with diminished or absent tendon jerks (Garland 1959; Bruyn and Garland 1970). The scapulohumeral muscles are only rarely involved and bulbar muscles are invariably spared. Rarely, an isolated pyramidal tract sign supervenes, often caused by concomitant disorders as cervical spondylosis. Symptoms develop abruptly or subacutely and patients often complain of severe pain in the lower back, hip, or thigh. Sensation is normal or only mildly disturbed. The EMG shows denervation and motor NCV are slightly reduced with prolonged motor latencies of the femoral nerves (Subramony and Wilbourn 1982; Lamontagne and Buchtal 1970).

Differentiation of diabetic amyotrophy from SMA can be made clinically by the limitation of weakness to the proximal muscles of the lower limbs, the presence of pain, areflexia, and by the typical neurophysiological findings, in combination with an increased serum glucose.

c. Hyperthyroidism

Hyperthyroidism may lead to brisk reflexes, myopathy, peripheral neuropathy, spastic paraparesis, myasthenic syndromes, muscle cramps and ocular symptoms (Bradley and Walton 1971; Feibel and Campa 1976; Greene 1976). However, the association of hyperthyroidism and ALS is still controversial (Rosati et al. 1980).

Thyrotoxic *myopathy* usually presents with wasting and proximal weakness of the limbs and rarely involves the bulbar muscles (Ramsay 1966; Kammer and Hamilton 1974). Fasciculation has been reported by Harman and Richardson (1954). Serum CK activity is usually normal. The clinical syndrome resolves during treatment. Differentiation from SMA or, if reflexes are brisk, from ALS, is by EMG, which discloses myopathic changes (Ramsay 1965).

Thyrotoxic sensorimotor or *purely motor neuropathy* may be demyelinating or axonal, and predominantly affects the legs (Feibel and Campa 1976). Especially, pure motor axonopathy may clinically and electrophysiologically resemble SMA.

Also, *spastic paraparesis* and *pseudobulbar signs* improving after treatment, have been described in hyperthyroidism (Ravera et al. 1960; Melamed et al. 1975; Garcia and Fleming 1977; Bulens 1981; Shaw et al. 1989). Simultaneous corticospinal and corticobulbar tract involvement and lower motor neuron signs caused by thyrotoxic myopathy or neuropathy may initially be diagnosed as ALS (Melamed et al. 1975; Mottier et al. 1981; Fisher et al. 1985; Serradel et al. 1990). However, EMG will reveal myopathic or (axonal) polyneuropathic changes rather than MND. The signs of thyrotoxicosis itself also give a clue to the underlying diagnosis, but they may be absent in elderly patients or masked by the use of beta-blockers (Thomas et al. 1970). Therefore, thyroid function should be evaluated in each patient with progressive weakness of unknown cause with or without pyramidal signs.

d. Hyperparathyroidism, osteomalacia and hypercalcemia

Vicale (1949) first noted progressive proximal weakness of the legs with wasting, hypotonia and occasional fasciculation in hyperparathyroidism, osteomalacia and hypercalcemia. Reflexes are normal or brisk. The clinical features may lead to the erroneous diagnosis of SMA or ALS (Dubois et al. 1989; Norris, Chapter 2, present volume). However, EMG and muscle biopsy findings are almost always consistent with myopathy or neuropathy. It is important to recognize

the syndrome, because treatment resolves the symptoms (Patten 1984). A pitfall is, that serum creatine kinase activity is usually normal and calcium may be normal in hyperparathyroidism and osteomalacia (Smith and Stern 1967, 1969; Frame 1976).

The association between a true anterior horn cell disease and hyperparathyroidism is generally considered a chance association. However, Murphy et al. (1960) have reported a patient with hyperparathyroidism and an ALS-like syndrome, that completely disappeared following parathyroid surgery.

B. HEREDITARY MOTOR NEURON DISEASES WITH KNOWN ENZYME DEFICIENCY

In 5–10% of the patients with MND a family aggregation is present. In additon, amyotrophy may be part of the clinical syndrome in several other inherited neurologic disorders as Huntington's disease and the heredoataxias. Hereditary forms of MND in adults are discussed in Chapter 14 of this volume (Familial ALS), Chapter 2 (Adult hereditary SMA), Chapter 12 (Progressive bulbar palsy) and Chapters 17–19 (Hereditary spastic paraparesis and paraplegia). Hereditary metabolic derangements in infantile and juvenile SMA are reviewed by Troost in Chapter 7 of this volume.

For the differential diagnosis of sporadic, adult cases of MND, we will discuss the hereditary forms of the disease with known biochemical abnormalities. We wish to emphasize, that especially in recessive disorders, family history may be negative.

1. HEXOSAMINIDASE DEFICIENCY

Reduced activity of hexosaminidase (Hex) A or Hex A and B results in deposition of GM_2-gangliosides in the nervous system and viscera. Gangliosidosis may lead to multisystem degeneration of the nervous system, but can also cause pure motor syndromes, superficially mimicking ALS or SMA. The mode of inheritance is autosomal recessive with different genotypes.

Within the same family, the expression of the disease may be heterogeneous. Two genotypes are associated with MND:

1. An α-subunit defect, located on chromosome 15, leads to deficiency of the isoenzyme Hex A (McKusick et al. 1990: 272800). Hex A deficiency is associated with Tay-Sachs disease in children, but also with SMA (Johnson 1982; Mitsumoto et al. 1985; Parnes et al. 1985) and ALS (Yaffe et al. 1979; Mitsumoto et al. 1985).

2. A β-subunit defect causes a deficiency of Hex A and B (McKusick et al. 1990: 268800). It may manifest as Sandhoff disease, Ramsay Hunt syndrome, but also as juvenile ALS with pyramidal and bulbar signs (Cashman et al. 1986; Rubin et al. 1988). Cerebellar atrophy has been demonstrated by head scans (Mitsumoto et al. 1985).

Patients with MND and Hex deficiency differ from idiopathic cases by the younger age of onset (before 40 years), the more protracted course, cerebellar or extrapyramidal signs, and occasional seizures.

Still, Hex deficiency is rare. Gudesblatt et al. (1988) failed to find a deficiency in 50 sporadic ALS patients and in 52 ALS patients with a positive family history, early onset, or protracted course.

The reader is referred to Chapter 7 of this volume for further details on this subject.

2. ADRENOLEUKODYSTROPHY AND -MYELONEUROPATHY

Adrenoleukodystrophy and -myeloneuropathy are X-linked peroxisomal diseases with accumulation of very long-chain saturated fatty acids (VLCFA) in tissue and body fluids (McKusick et al. 1990: 300100).

Of differential diagnostic importance in MND is adrenomyeloneuropathy, that may begin in adulthood. The combination of pyramidal tract involvement and peripheral neuropathy may cause a clinical picture resembling ALS. Additional signs are slurred speech, ataxia, mental deterioration, hypogonadism, and adrenal insufficiency (Moser et al. 1984, 1987).

The presence of sensory disturbances, slowed nerve conduction velocities, adrenal insufficiency, and X-linked inheritance usually distinguishes adrenomyeloneuropathy from ALS. In plasma the ratio of VLCFA to long-chain fatty acids (C26:0 to C22:0) is elevated. VLCFA can also be demonstrated in skin fibroblasts and white blood cells. Patients may benefit from oleic acid diet with restriction of VLCFA (Moser et al. 1987).

Adrenoleukodystrophy and -myeloneuropathy is reviewed by Moser in Chapter 21 of this volume.

3. NEUROACANTHOCYTOSIS, WITH OR WITHOUT AN ASSOCIATED DISTURBANCE OF LIPID METABOLISM

Acanthocytes are abnormally shaped red blood cells with spiny projections. They occur in association with disturbances of lipid metabolism and neurological diseases.

Abetalipoproteinemic acanthocytosis is an autosomal recessive inherited disease, that manifests in childhood with steatorrhoea, weakness of the shoulders and legs, areflexia, extensor plantar responses, sensory loss, athetoid movements, visual field defects, ataxia and retinitis pigmentosa (McKusick et al. 1990: 200100). Loss of anterior horn cells and demyelination of the CNS and peripheral nerves have been found at autopsy (Sobrevilla et al. 1964).

In adults, *normolipoproteinemic acanthocytosis* (McKusick et al. 1990: 100500) and *hypolipoproteinemic acanthocytosis* (McKusick et al. 1990: 107730) may lead to diagnostic confusion with SMA or ALS due to a motor neuropathy with or without CNS involvement. The disease is inherited in an autosomal dominant manner and is characterized by progressive proximal weakness and wasting of the distal muscles, fasciculation, hypotonia, and areflexia. Reported additional clinical features comprise dystonic or choreiform movements, dysarthria and dysphagia, pyramidal signs, ataxia, epileptic seizures, dementia, pes cavus, and sphincter disturbances (Mars et al. 1969; Aminoff 1972; Bruyn 1986).

Cholesterol and triglyceride levels, and screening for circulating acanthocytes in freshly made blood films should be performed in patients suspected of MND with one of the above mentioned additional signs or in case of an autosomal dominant inheritance.

The subject is discussed by Troost in Chapter 7 of this volume.

4. MULTIPLE LYSOSOMAL ENZYME DEFICIENCY

Spinal muscular atrophy and hyperreflexia in association with childhood onset mental retardation, congenital malformations and Marfan-like posture have been described in children with lysosomal enzyme deficiencies (Goto et al. 1983). A lower motor neuron syndrome has been reported by Dalakas et al. (1989) in 3 young patients with an intralysosomal accumulation of cystine and nephropathic cystinosis (McKusick 1990: 219900).

For further details the reader is referred to Troost in Chapter 7 of this volume.

C. MOTOR NEURON DISEASE WITH A MORE BENIGN PROGNOSIS

1. MONOMELIC MOTOR NEURON DISEASE (MND)

A mild form of MND with a more benign prognosis has been reported by several authors, especially from Japan, India, Sri Lanka and Malaysia (Sobue et al. 1978). Age of onset in predominantly male patients was in 90% between 18 and 22 years. The course was progressive for a few years followed by stabilization. In most cases the distal arm muscles were affected with atrophy and fasciculation. Tendon reflexes were diminished, normal or increased, but no patient exhibited pathological reflexes. Other reported features were exacerbation by cold and hyperhidrosis. No bulbar signs or sensory loss were noted. Most importantly, in 70% of the patients, clinical involvement was restricted to one limb. EMG, however, was abnormal on the asymptomatic contralateral side in 90%.

The disorder may represent a forme fruste of autsomal recessive Kugelberg-Welander disease (Furakawa et al. 1977; McKusick et al. 1990: 158600), but most cases are sporadic (Gourie-Devi et al. 1984; Riggs et al. 1984; Serratrice et al. 1985; Hanson et al. 1986).

In an individual patient, it is impossible to predict the course of the disease. Favorable prognostic factors for monomelic MND are male sex, age of onset under 40 years, absence of bulbar or pyramidal tract signs, confinement of clinical and EMG abnormalities to 1 or 2 limbs only.

Especially in monomelic MND, other causes of focal weakness and atrophy such as mononeuro- or plexopathy, poliomyelitis and spinal cord trauma have to be excluded by careful clinical examination and EMG.

The subject is extensively reviewed by Hirayama in Chapter 8 and by De Visser et al. in Chapter 22 of this volume.

2. SLOWLY PROGRESSIVE, PROXIMAL BULBOSPINAL SMA

Slowly progressive, proximal bulbospinal SMA is an X-linked recessive disorder frequently being confused with sporadic SMA (McKusick et al. 1990: 313200). The disease begins in adulthood and has a remarkably benign prognosis with an average disease duration of 24.8 years (Kennedy et al. 1968; Evans 1987). There is a considerable number of single cases and new mutations. Six of the 10 cases described by Harding et al. (1982) had no affected relatives.

The diagnosis should be considered in a male patient, if the course of the disease is slow, onset is between the age of 30 and 50 years, preceded by years of muscle cramps, if weakness begins in the proximal muscles of the lower limbs, subsequently spreading to the shoulder girdle; if fasciculation of the lower face and tongue is prominent; if tremor of the outstretched hands, gynecomastia, testicular atrophy and late-onset diabetes are present, if dysarthria, dysphagia and upper motor neuron signs are absent, and if other male family members are affected.

For further details, the reader is referred to Padberg in Chapter 4 of present volume.

D. NEUROLOGIC DISORDERS MIMICKING THE CLINICAL PICTURE OF MOTOR NEURON DISEASE (MND), BUT NOT AFFECTING MOTOR NEURONS

A number of neurologic conditions can cause a similar clinical picture as SMA, ALS or PBP.

Usually, the history, clinical findings, and results of EMG and CSF tests are sufficiently characteristic to distinguish the disorder from MND. In other cases, it may be of added help to perform special diagnostic tests in order to determine the site of origin of the motor disturbances. Some neurologic syndromes deserve special attention, because they are sometimes mistaken for MND.

1. PERIPHERAL NERVES AND NERVE ROOTS

Clinically and electrophysiologically, it may be difficult to distinguish SMA from pure motor neuropathy or radiculopathy (Dyck 1982). Widespread fasciculation and atrophy are common in MND, but may also occur in peripheral root and nerve disorders. The characteristic EMG abnormalities of denervation and reinnervation are present in both disorders and nerve conduction velocities (NCV) may be only slightly decreased in axonal neuropathies. Conversely, in MND, NCV may be slowed in wasted and denervated muscles and in cold limbs.

a. Mononeuropathy and plexus lesions

Lesions of a single nerve or plexus present an important differential diagnostic consideration in SMA and in monomelic MND. Especially, at the onset of the disease, careful sensory testing and motor and sensory NCV will usually differentiate from ulnar palsy, carpal tunnel syndrome, peroneal nerve and brachial plexus lesions. Moreover, in SMA, EMG abnormalities can usually be found in clinically unaffected limbs.

b. Polyneuropathy

We will discuss some specific neuropathies that are frequently confused with MND:

Axonal neuropathies. Especially, axonal motor neuropathies may simulate SMA, because motor nerve conduction velocities (NCV) are normal or only slightly reduced as are the F response and H reflex.

Pure or predominantly motor neuropathies. Most neuropathies are easily differentiated from MND by the presence of sensory disturbances, but relatively pure motor neuropathy may occur (Table 5).

Neuropathies with bulbar symptoms. Diagnostic confusion with MND increases by the presence of bulbar symptoms due to cranial nerve involvement in some neuropathies (Table 5).

Neuropathies with extensor plantar responses. In rare cases of motor or sensorimotor neuropathy, plantar responses may be extensor from concomitant cervical spondylosis, cerebrovascular pathology, or CNS involvement (Table 5).

Chronic inflammatory demyelinating polyradiculoneuropathy. Chronic inflammatory demyelinating polyradiculoneuropathy (CIDP) may masquerade as SMA with or without bulbar deficits (Dyck et al. 1975a). However, wasting and fasciculation are far less pronounced than in MND. In 90% of the cases, an elevation of the CSF protein is found, sometimes with oligoclonal bands. Pleocytosis is rare, except in HIV-associated cases. Electrophysiological abnormalities comprise marked slowing of motor and sensory NVC, and prolonged latencies. Sural nerve biopsy shows segmental demyelination-remyelination. Diagnostic criteria for CIDP are listed by Barohn et al. (1989).

Demyelinating motor neuropathy with proximal multifocal conduction blocks. There have been several reports of patients with an initial diagnosis of MND due to a pure motor demyelinating neuropathy characterized by weakness, atrophy and fasciculation, and relatively preserved reflexes. Electrophysiologically, proximal multifocal conduction blocks have been demonstrated (Chad et al. 1986; Parry and Clarke 1985, 1988). Association with high serum titers of antibodies to gangliosides and response to immunosuppressive treatment have been reported (Pestron et al. 1988b).

The neurophysiologic phenomenon of proximal conduction block may also occur in patients with otherwise classical ALS. Four of the 120 ALS

TABLE 5

(Radiculo-) neuropathies with predominantly motor involvement, bulbar signs or pyramidal signs leading to diagnostic confusion with MND.

	Predominantly motor deficits	Bulbar symptoms	Pyramidal signs
Metabolic derangements			
diabetes mellitus	+	−	−
uremia	+	−	−
porphyria	+ +	+	+
Infectious states			
diphtheria	+ +	+ +	−
borreliosis	+ +	+	+
varicella zoster	+	+	−
leprosy	−	+	+
Inflammatory and granulomatous disorders			
combined peripheral (CIDP) and CNS inflammatory demyelinating syndrome	+ +	+	+
Guillain-Barré syndrome	+ +	+	−
systemic lupus erythematosus (SLE)	+ +	+	+
Behçet's disease	+	+	+
sarcoidosis	+ +	+	+
amyloidosis	+	−	+
Intoxications			
gold	+ +	−	−
lead	+ +	+	+
mercury	+ +	+	+
organophosphates	+ +	+	+
methyl bromide	+	−	+
triorthocresylphosphate (TOCP)	+	−	+
dapsone	+ +	−	−
Deficiencies			
vitamin B-12	−	+	+
Cancer			
paraneoplastic disorders	+	+	+
Hereditary disorders			
HMSN	+	+	+
Refsum disease	+	+	+
adrenomyeloneuropathy	+	+	+
metachromatic leukodystrophy	+	+	+
polyglucosan body disease	−	−	+
neurofibromatosus	+	+	+

+ + = frequently present; + = may be present; − = absent

patients (33%), examined by Younger et al. (1990), appeared to have multifocal proximal conduction blocks. The significance of this finding, especially in relation to immunosuppressive therapy, has yet to be determined.

Combined peripheral and CNS inflammatory demyelinating syndrome. In several patients, CIDP is accompanied by a relapsing multifocal CNS inflammatory demyelinating syndrome with white matter lesions on magnetic resonance imaging (MRI) and oligoclonal bands in the CSF (Mendell et al. 1987; Thomas et al. 1987). Combined peripheral and CNS demyelinating syndrome may cause an ALS-like clinical picture.

Hereditary motor and sensory neuropathies. The clinical picture of hereditary motor and sensory neuropathies (HMSN) may be similar to SMA with distal weakness and wasting, fasciculation, and muscle cramps (Norris, Chapter 2, present volume). Differentiation of SMA from HMSN type II (axonal degeneration) may be difficult, because motor NCV in HMSN type II may be normal or only slightly decreased whereas sensory disturbances may be absent. The presence of extensor plantar responses in HMSN with upper motor neuron involvement (McKusick et al. 1990: 162380) may lead to confusion with ALS. Usually, HMSN can be distinguished from MND by the slow rate of progression, heredity, predominance of distal weakness and wasting, absence of fasciculation, and the presence of sensory disturbances, high arches, hammer toes, abnormal motor and sensory NCV, and sural nerve biopsy abnormalities.

The Guillain-Barré syndrome. The Guillain-Barré syndrome (GBS) is characterized by rapidly progressive flaccid motor weakness of the limbs with normal to absent deep tendon reflexes and, occasionally bulbar palsy (Anderson and Sidén 1982). Important considerations in distinguishing GBS from rapidly progressive SMA and PBP include the course of the disease, additional symptoms as frequent sensory loss, ophthalmoplegia, and autonomic disturbances. CSF and EMG examination are critical for confirming the

diagnosis (McLeod 1981; Spaans 1985; Brown and Feasby 1984).

2. NEUROMUSCULAR JUNCTION

a. Myasthenia gravis (MG)

Signs and symptoms in myasthenia gravis (MG) may closely resemble those in PBP or SMA. The presence of fasciculation and muscle cramps in patients receiving anticholinesterase treatment may be misleading. On the other hand, some MND patients are very fatigable and benefit from anticholinesterase medication (Mulder et al. 1959).

MG usually presents with weakness and fatigability of the eyelids and extraocular muscles. However, dysarthria, dysphagia, weakness of the limbs and diminished vital capacity may be present without ocular symptoms. Muscle wasting is less frequently observed. If present, atrophy is confined to the tongue, face and shoulder girdle. Fasciculation is rare.

Marked fluctuations in muscle strength over time, and considerable response to anticholinesterase treatment are characteristic for MG. Another important diagnostic clue is the typical EMG finding of decrementing muscle action potentials during repetitive nerve stimulation. Antibodies against acetylcholine receptor are specific for the diagnosis, but may be absent (Lindstrom et al. 1976).

b. The Lambert-Eaton myasthenic syndrome (LEMS)

Differentiation between the Lambert-Eaton myasthenic syndrome (LEMS) and early SMA may be difficult. Characteristic features of LEMS are aching and weakness of the proximal leg and trunk muscles, diminished or absent tendon reflexes, and autonomic signs. In later stages of the disease, weakness of the arms supervenes. Involvement of the bulbar and extraocular muscles is rare. Muscle wasting is less pronounced than in SMA and fasciculations are absent. In half of the patients a neoplasm, especially an oat cell carcinoma of the lung, can be demonstrated (Newson-Davis and Murray 1984; O'Neill et al.

1988). The EMG abnormalities of LEMS are characteristic with low amplitudes at supramaximal stimulation, a decrease of amplitude at low-frequency stimulation, and an incremental response at high-frequency stimulation or upon voluntary muscle contraction.

3. MYOPATHIES

Myopathies, particularly polymyositis, can be confused with SMA and early stages of ALS. Progressive muscle weakness and wasting occur in both disorders. A confounding phenomenon is that, in MND, creatine kinase (CK) levels may be slightly elevated (Harrington et al. 1983) and muscle biopsies may reveal myopathic abnormalities (Achari and Anderson 1974). On the other hand, in myopathic disorders, bulbar symptoms and fibrillation potentials may be present.

A patient with proximal weakness of the limbs and normal or diminished reflexes is more likely to have a myopathic disorder than SMA, when fasciculations are absent, muscle wasting is less pronounced, weakness is confined to the proximal muscles and EMG shows myopathic changes. If EMG reveals aspecific changes, a muscle biopsy of a carefully selected muscle is recommended (Buchthal and Kaminiecka 1982). The combination of choking and difficulty in chewing, muscle wasting, especially of the facial muscles and hands in *myotonic dystrophy* (McKusick et al. 1990: 160900) may be confused with MND. However, the additional clinical features of myotonic dystrophy, like myotonia, typical facial appearance, ptosis, cataract, endocrine abnormalities, positive family history and EMG findings have little or nothing in common with MND.

4. MULTIPLE SCLEROSIS (MS)

Cases of definite or probable MS, according to the diagnostic criteria described by Poser et al. (1983), can hardly be confused with MND. Potential confusion may arise in case of the first attack of MS, when clinically evident symptoms outside the motor system are lacking or when the course is progressive without remissions or exacerbations.

Cases of MS simulating MND have already been reported in earlier literature (Charcot 1879). If muscle atrophy occurs in MS, especially the hand muscles are involved (Muller 1949; Abb and Schaltenbrand 1956; Kurtzke 1970; Fisher 1983; Paty and Poser, 1984). Muscle atrophy in MS may be explained by concomitant demyelinating polyneuropathy (Thomas et al. 1987), motor radiculopathy (Noseworthy et al. 1980), entrapment neuropathy, disuse atrophy, or focal destruction of anterior horn cells by demyelinating plaques. Focal anterior horn cell destruction in the lower cervical region has been found on autopsy (Davison 1934). Extensive and widespread muscle wasting is rare (Brauer 1898; Bau-Prussak 1930) and a more diffuse widespread disease of the motor neurons similar to MND is less likely, unless both diseases attack the same individual (Tyler 1982; Hader et al. 1986). Interesting in this respect, are the reports of Paty and Poser (1984) of widespread mild atrophy, fasciculation and EMG evidence of denervation during an exacerbation of MS with subsequent recovery indicating a reversible lesion of peripheral nerves or of anterior horn cells. However, EMG examinations in MS patients with muscle atrophy by Fisher et al. (1983) turned out to be normal.

Fasciculation is almost never mentioned by authors of extensive reviews on MS (Davison 1943; Muller 1949; Abb and Schaltenbrand 1956).

The presence of CSF oligoclonal bands cannot be used to reliably distinguish MS from MND, because they occur in both diseases (Younger et al. 1990). Also, MRI periventricular white matter lesions may be seen in normal population, especially in the elderly (Gerard and Weisberg 1986).

To summarize, the following clinical features differentiate MS from MND; the absence of fasciculation, the absence of EMG evidence of more widespread denervation, the absence of muscle atrophy except in the upper extremities, the presence of clinical or MRI evidence of lesions in the CNS outside the motor system (visual, cerebellar, sensory or sphincter disturbances), course of the disease in remissions and exacerbations, young age, long disease duration, the presence of increased IgG production and CSF oligoclonal bands, MRI white matter lesions consistent with demyelinating plaques, and abnormal visual or sensory evoked responses.

E. DIFFERENTIAL DIAGNOSIS OF MOTOR NEURON DISEASE (MND) PRESENTING AS PRIMARY LATERAL SCLEROSIS (PLS) OR PROGRESSIVE BULBAR PALSY (PBP)

1. PRIMARY LATERAL SCLEROSIS

According to neurologists with considerable reputation, ALS does not usually present with upper motor neuron signs without clinical or EMG signs of lower motor neuron involvement (Norris et al. 1980; Mulder et al. 1982). As distinct from other neurological disorders with pyramidal tract involvement, in ALS sphincter disturbances and extensor plantar responses may be absent, and abdominal reflexes normal or even brisk. The differential diagnosis of spastic paraplegia is reviewed by Bruyn in Chapter 24 of this volume.

2. PROGRESSIVE BULBAR PALSY (PBP)

In MND with bulbar onset, the patient presents with complaints of dysphagia or dysarthria. Eventually, the disease will spread to the trunk and limbs.

On examination, bulbar palsy with fasciculations and atrophy dominates the clinical picture and may be combined with positive pseudobulbar reflexes and occasional forced laughing, crying, or yawning. EMG discloses denervation in affected bulbar muscles.

According to Adams and Victor (1985), ALS with bulbar onset is the only common clinical disorder in which spastic and atrophic bulbar palsy co-exist. The differential diagnosis of dysarthria and dysphagia is listed below; many of the disorders have been described in previous sections of this chapter and by Bruyn in Chapter 12 of this volume.

a. Dysarthria and dysphagia from lesions of the bulbar motor nuclei and efferent fibers

Weakness of the muscles of the face, palate, pharynx, larynx, or tongue may result from lesions of the bulbar nuclei, efferent fibers or muscles. It gives rise to slurred speech, and deficient pronunciation of consonants. Main causes include the Guillain-Barré syndrome, diph-theria, sarcoidosis, syphilis, botulism, bulbar poliomyelitis, thrombosis of the posterior inferior cerebellar artery, fusiform aneurysm of the basilar artery, and posterior fossa tumor.

b. Dysarthria and dysphagia from bilateral lesions of the corticobulbar tracts

Dysarthria and dysphagia in pseudobulbar palsy may be caused by chronic progressive spinobulbar spasticity, a form of primary lateral sclerosis (Gastaut et al. 1988), multiple sclerosis, lacunar state, bilateral cerebrovascular accidents, falx meningioma, and Binswanger's disease.

c. Dysarthria and dysphagia from extrapyramidal lesions

Dysarthria in *Parkinson's disease* is characterized by monotonous, slurred speech due to reduced volume, diminished prosody and articulation. Dysphagia and dysarthria roughly parallel the severity of the parkinsonian motor disturbances. Fasciculations and atrophy of bulbar musculature and pseudobulbar signs and symptoms are absent.

In *progressive supranuclear palsy (PSP)*, dysarthria can be severe (Steel et al. 1972). Other features of PSP are increased jaw and facial jerks, exaggerated palatal and pharyngeal reflexes, and, occasionally, forced laughter and crying. The presence of volitional vertical gaze palsy, axial rigidity and development of profound bradykinesia, in the absence of fasciculation and atrophy of bulbar muscles, are clues for the diagnosis of PSP.

d. Dysarthria and dysphagia in myasthenia gravis (MG)

Presentation with bulbar symptoms is common in myasthenia gravis (MG). Dysarthria and dysphagia develop during conversation, eating and swallowing. Bulbar muscle atrophy is sometimes observed. The presence of ocular symptoms and a positive response to anticholinesterase treatment, the absence of fasciculation and pseudobulbar signs, and typical EMG findings distinguish MG from PBP or SMA.

e. Dysarthria due to lesions of the cerebellum and connections

Speech in cerebellar dysarthria may be slurred and slow, as in pseudobulbar palsy. However, attempts of the patients to circumvent the problem by enunciating words syllable by syllable, will give rise to the characteristic scanning speech. In addition to dysarthria, other cerebellar symptoms and signs are usually present. Causes of cerebellar dysarthria include multiple sclerosis, posterior fossa tumors, abnormalities at the craniovertebral junction, Friedreich's ataxia (McKusick et al. 1990: 229300), and vascular lesions of the hindbrain.

f. Dysarthria and dysphagia due to involuntary movements

In chorea and dystonia, involuntary movements of the tongue, lips, larynx, and respiratory muscles may interfere with speech. Speech is jerky, explosive and irregular.

g. Dysarthria and dysphagia in dystrophia myotonica and myopathies

Dysarthria in dystrophia myotonica and other myopathic disorders usually is a late event. Differentiation from PBP is by clinical signs of general myopathy, elevated CK, myopathic changes on EMG, muscle biopsy examination and absence of pyramidal tract signs.

h. Dysarthria and dysphagia from infections or mechanical obstruction of oropharynx and esophagus

The clinician should be aware that dysarthria and, in particular, dysphagia may also result from inflammation or mechanical obstruction of the oropharynx, larynx, or esophagus, such as retropharyngeal abscess, neoplasms, aneurysm of the thoracic aorta, mediastinal lymph nodes, aberrant subclavian arteria, iron deficiency anemia, enlarged thyroid, tuberculous laryngitis, and candidiasis, amongst others. As distinct from MND, signs of neurogenic bulbar weakness (atrophy, fasciculation, EMG abnormalities, pseudobulbar features) are lacking.

F. DIFFERENTIAL DIAGNOSIS OF MOTOR NEURON DISEASES WITH ADDITIONAL NEUROLOGIC SIGNS AND SYMPTOMS

In general, MND is a disease of motor neurons with sparing of the extraocular muscles, bladder, bowel and sensory systems. However, in a minority of patients with otherwise classical MND, careful testing will reveal mild eye movement disorders (Leveille et al. 1982), sensory loss (Dyck et al. 1975), mild autonomic disturbances (Chida et al. 1989), and neuropsychological changes (David and Gilham 1986). On the other hand, some patients show a combination of MND and overt dementia, parkinsonism (Hudson, Chapter 13, present volume), cerebellar signs, autonomic failure (Oppenheimer 1988), objective sensory loss (Norris et al. 1980) or ophthamoplegia (Hayashi et al. 1987). In these patients careful additional investigations are warranted to exclude other diseases.

1. DEMENTIA

When systematically tested, ALS patients without clinically evident mental deterioration may show slight cognitive impairment (David and Gillham 1986; Gallassi et al. 1989).

The association between ALS and dementia has increasingly been recognized over recent years (Hudson, 1981; Neary et al., 1990) and attains an incidence of about 3.5% in sporadic and 7% in familial cases (Hudson et al., 1986). Formerly identified as part of the *ALS-dementia-parkinsonism complex* of Guam, there have since been several reports of the association from Japan (Mitsuyama and Takamiya, 1979) and from Western countries (Finlayson et al., 1973; Pinsky et al., 1975; Wikström et al., 1982; Salazar et al. 1983). This dementia is generally considered a *frontal lobe dementia* closely related to MND (Hudson, Chapter 13, present volume; Neary et al., 1990).

The existence of an amyotrophic form of *Creutzfeldt-Jakob disease* is regarded as unproven by many authors because of negative transmission experiments (Salazar et al. 1983). Lower motor neuron deficit is regarded to occur late in the disease and is then accompanied by other cerebral and cerebellar signs (Salazar et al. 1983). However, Allen et al. (1971) have described a patient with the amyotrophic form of Creutzfeldt-Jakob disease and a positive transmission experiment (Connolly et al. 1988).

The association of ALS with *Alzheimer's* and *Pick's disease* is probably based on chance. *Diffuse Lewy body disease* is characterized by parkinsonian features and cortical dementia (Gibb et al. 1985). A disorder resembling diffuse Lewy body disease with distal arm amyotrophy and fasciculations, tetrapyramidal syndrome, dementia and rigidity was reported by Delisle et al. (1987).

The dementia in the *lacunar state* may combine with pseudobulbar palsy. Differentiation from PBP is by lack of signs of lower motor neuron involvement (EMG) and leukoencephalopathy and lacunes on CT or MRI.

Huntington's disease (McKusick et al. 1990: 143100) and *Wilson's disease* (McKusick et al. 1990: 277900) are other examples of dementias with prominent involvement of the motor system.

In *Hallervorden-Spatz disease* onset is in childhood or adolescence with speech and gait disturbance, dystonia, rigidity, tremor, delusions, corticospinal tract signs, and dementia (McKusick et al. 1990: 234200). In a later stage, progression to extreme stiffness, seizures, retinitis pigmentosa and optic atrophy clearly differentiate from ALS. Sometimes Hallervorden-Spatz disease presents as ALS-dementia complex (Bots and Staal 1973).

2. PARKINSONISM

ALS and parkinsonism do occur in the same patient, not only in the Western Pacific high incidence foci, but also, in about 1.5% of all cases in the western world (Hudson, Chapter 13, present volume).

Autosomal dominant *Joseph disease* (also Azorean neurologic disease or Machado-Joseph disease) is clinically manifested by weakness, fasciculation, distal wasting, hyper- or areflexia,

extensor plantar responses and bulbar signs, resembling the clinical syndrome of ALS or SMA (Fowler, 1984; McKusick et al. 1990: 109150). However, progressive gait ataxia, limitation of gaze and parkinsonian features clearly distinguish the disorder from MND. A sporadic form of this disorder has been reported (McQuinn and Kemper 1987).

Corticobasal degeneration combines pyramidal tract signs and dysarthria with parkinsonism, ataxia, supranuclear gaze palsy, parietal lobe signs, 'alien hand' sign, dementia, and focal myoclonus (Gibb et al. 1989).

In *diffuse Lewy body dementia* the following features are noted: age at onset 20–70 years, equal sex ratio, disease duration 1–25 years, severe dementia, unequivocal triad of parkinsonian rigidity/bradykinesia/tremor, with focal symptoms such as dysphasia, acalculia, etc. Lewy bodies abundantly occur in allo- and isocortex, locus coeruleus, substantia nigra and dorsal motor X nucleus (Gibb et al. 1985; Dickson et al. 1987; Burckhardt et al. 1988).

In contrast, mild extrapyramidal and no parkinsonian features are seen in 'senile dementia of the Lewy body type' which occurs in about 20% of all hospitalized demented aged. Onset is over the age of 70 with hallucinatory confusion. Lewy-bodies occur disseminated through the brain stem and allo- and isocortex.

A disorder resembling the above, with distal arm amyotrophy and fasciculations plus a tetrapyramidal syndrome, dementia, and rigidity was reported by Delisle et al. (1987). An autosomal dominant parkinson-dementia syndrome with non-Alzheimer-amyloid plaques was reported by Rosenberg et al. (1989).

3. ATAXIA

Pathologically, degeneration of spinocerebellar neurons seems common in ALS (Williams et al. 1990). However, if cerebellar signs are present in a patient with a clinical syndrome resembling ALS or SMA, other disorders should be considered.

Friedreich's ataxia is the commonest of progressive spinocerebellar degeneration (McKusick et al. 1990: 229300). The combination of weakness with wasting, areflexia, and extensor plantar responses in the legs may mimic ALS. However, distal loss of joint position and vibration sense, ataxia of gait and scoliosis distinguish Friedreich's ataxia from ALS. Other differentiating features comprise optic atrophy, nystagmus, abnormal extra-ocular movements, deafness, characteristic widespread T-wave inversion on ECG, and axonal sensory polyneuropathy findings on EMG.

The ataxia of *olivopontocerebellar atrophies (OPCA)* combines with many associated findings (McKusick et al. 1990: 164400, 164500, 164600, 164700). Also, pyramidal signs in the limbs, distal wasting and fasciculation of the face and tongue may occur (Harding 1991). Idiopathic degenerative late onset ataxia, labeled OPCA as well, can occur in combination with *striatonigral degeneration* and *progressive autonomic failure* (Shy-Drager syndrome; McKusick et al. 1990: 146500) giving rise to *multiple system atrophy*. In multiple system atrophy, pyramidal tract signs and lower motor neuron dysfunction combine with an assortment of neurological deficits including parkinsonism, cerebellar, cranial nerve and autonomic dysfunction.

In conclusion, for patients presenting with MND and associated neurological deficits — particularly broadbased gait — CT scan or MRI for detection of cerebellar or brainstem atrophy (Staal et al. 1990) and exclusion of other diseases is warranted.

4. DYSAUTONOMIA

In MND, the only clinical signs of (minor) autonomic disturbances are confined to paretic extremities and consist of cold, cyanotic, slightly edematous limbs (Norris et al. 1980). More advanced tests may demonstrate subclinical cardiovascular changes associated with sympathetic hyper- and parasympathetic hypofunction (Chida et al. 1989).

Progressive autonomic failure (McKusick et al. 1990: 146500) and *multiple system atrophy* combine autonomic failure, including urinary and rectal incontinence, with muscle wasting, fasciculation, pyramidal tract signs and other signs of neurological dysfunction (Oppenheimer 1988).

Differentiation is from polyneuropathies with motor and autonomic involvement, and from poisoning with organophosphate insecticides which act as anticholinesterase drugs.

In MND, even when the patient is tetraplegic, sphincter functions are preserved. When bowel and bladder disturbances are present in combination with pyramidal tract signs and progressive weakness, special attention is required to rule out spinal cord compression and multiple sclerosis.

5. OPHTHALMOPLEGIA

Clinical involvement of extraocular muscles in MND is very rare. Electro-oculography showed decreased saccadic or smooth pursuit velocities in 4 of 10 ALS patients studied by Leveille et al. (1982). Rare cases showing nystagmus have been reported by Kushner et al. (1984). However, oculomotor disturbances occur frequently in ALS patients who are respirator-supported (Hayashi et al. 1987) and, eventually, suffer a totally locked-in state (Hayashi and Kato 1989). The pupillary light reaction is spared (Harvey et al., 1979).

Differential diagnosis of ophthalmoplegia can be found in Daroff et al. (1990), Glaser and Bachinsky (1990), Leigh and Zee (1983), and Sergott et al. (1984).

When eye movement disturbances and clinically normal pupils occur together with involvement of other voluntary muscles, particularly myasthenia gravis must be excluded. Other differential diagnostic considerations include the Lambert-Eaton myasthenic syndrome, neuropathies, that may be drug-induced (D-penicillamine, aminoglycosides, vincristine or vinblastine) or toxin-induced (botulism [pupillary involvement], tetanus), dysthyroid myopathy, chronic progressive external ophthalmoplegia, oculopharyngeal dystrophy, myotonic dystrophy, polymyositis, primary or metastatic tumors at the brain base, inflammatory processes at the brain base (sarcoidosis, syphilis, neuroborreliosis, tuberculosis, mycosis, and cryptococcosis), the Guillain-Barré syndrome (Fisher's syndrome), vascular disease of the basilar artery system, demyelination (multiple sclerosis), diabetes, Wernicke's encepha-

lopathy, Whipple's disease, and progressive supranuclear palsy.

6. SENSORY LOSS

The absence of sensory symptoms and findings is
an important clue in the diagnosis of MND.
Sometimes, however, patients with MND complain about 'pin and needle' paresthesia (Norris
et al., 1980; Tyler and Shefner, Chapter 11, present
volume). Occasionally, patchy areas of diminished
pain and temperature sensitivity are observed
(Norris et al. 1980). More often, the only abnormality at neurological examination is blunting of
vibration perception in the toes, consistent with
age (Norris et al. 1980). When carefully tested,
reduced vibration and tactile sense in the lower
extremities may be present in 18% of patients
(Mulder et al. 1983; Radtke et al. 1986). If
prominent sensory abnormalities are present, other
conditions, like peripheral neuropathies, demyelinating disorders, spinal cord lesions and heredoataxias should be carefully excluded.

G. DIFFERENTIAL DIAGNOSIS OF FAS-CICULATIONS OR MUSCLE CRAMPS WITHOUT MUSCLE WEAKNESS

Fasciculations in the presence of weakness and
denervation are frequently encountered in MND,
but also in peripheral neuropathies, especially
in amyloid neuropathy, alcoholic neuropathy,
mononeuritis multiplex and recovery phase of
the Guillain-Barré syndrome (Layzer 1982).

Painful muscle cramps, weakness and occasional fasciculation may occur in myopathies,
particularly in McArdle's disease (McArdle 1951;
Pearson and Rimmer 1961; McKusick et al.
1990: 232600), the syndrome of delayed muscle
relaxation induced by exercise (Brody 1969),
myotonia (Stohr et al. 1975; Streib 1987), and
hypothyroidism (Norris and Penner 1960).

Syndromes, characterized by fasciculations,
myokymia or muscle cramps, but without muscle
weakness or EMG signs of denervation (fibrillation and positive sharp waves), should not be
classified as motor neuron disease. Fasciculations
are common in healthy individuals (Reed and

Kurland 1963). Actually, when a patient presents
with fasciculation or muscle cramps without
weakness, wasting, or denervation, it is unlikely
that he has or will ever develop MND (Fleet
and Watson, 1986). In that case, the following
causes may be considered:

– dehydration, electrolyte or metabolic disturbances due to vomiting, diarrhea, excessive
  perspiration, tetany, uremia, dialysis, diuretics,
  laxatives, hypothyroidism, hypoadrenalism,
  pregnancy
– drugs and toxins: cholinesterase inhibitors,
  bronchodilator agents like salbutamol and
  terbutaline, strychnine, black widow spider
  bite, tetanus
– benign fasciculation-cramp syndrome (syn.
  Denny-Brown and Foley syndrome) (Hudson
  et al. 1978)
– syndrome of continuous muscle-fibre activity
  (syn. Isaacs syndrome, neuromyotonia, arma-
  dillo syndrome) (Isaacs 1961; McKusick et al.
  1990: 121020)
– myokymia-hyperhidrosis-impaired muscle relaxation syndrome (Greenhouse et al. 1967;
  McKusick et al. 1990: 137200)
– painful legs-moving toes syndrome (Spillane
  et al. 1971)
– Satoyoshi syndrome (Satoyoshi and Yamada
  1967)
– Stiff-man syndrome (Moersch and Woltman
  1956; Olafson et al. 1964; McKusick et al.
  1990: 184850)
– myelopathy with rigidity, spasm or continuous
  motor unit activity (Whiteley et al. 1976;
  Howell et al. 1979)
– spinal myoclonus
– occupational cramps and writer's cramp
– restless legs syndrome (Ekbom 1970)
– claudicatio intermittens.

Excellent reviews have been presented by
Layzer (1979 and 1982) and Rowland (1985).

Acknowledgements
We are grateful to Mrs Cox Schouten for her
assistance in the preparation of the manuscript.

REFERENCES

ABB, L. and G. SCHALTENBRAND: Statistische Untersuchungen zum Problem der Multiplen Sklerose.

II. Mitteilung. Das Krankheitsbild der Multiplen Sklerose. Dtsch. Z. Nervenheilkd. 174 (1956) 199–218.

ACHARI, A. N. and M. S. ANDERSON: Myopathic changes in amyotrophic lateral sclerosis. Neurology 24 (1974) 477–481.

ACKERMANN, R., E. GOLLMER and B. RHESE-KÜPPER: Progressiven Borrelien-Encephalomyelitis. Dtsch. Med. Wochenschr. 110 (1985) 1039–1042.

ACKERMANN, R., B. RHESE-KÜPPER, E. GOLLMER and R. SCHMIDT: Chronic neurologic manifestations of erythema migrans borreliosis. Ann. N.Y. Acad. Sci. 539 (1988) 16–23.

ADAMS, C. R., D. K. ZIEGLER and J. T. LIN: Mercury intoxication simulating amyotrophic lateral sclerosis. J. Am. Med. Assoc. 250 (1983) 642–643.

ADAMS, R. D. and M. VICTOR: Principles of Neurology. New York, McGraw Hill (1985) 891.

ALANI, S. M.: Denervation in wasted hand muscles in a case of primary cerebellar ectopia without syringomyelia. J. Neurol., Neurosurg. Psychiatry 48 (1985) 84–85.

ALLEN, I. V., E. DERMOTT, H. H. CONNOLLY and L. J. HURWITZ: A study of a patient with the amyotrophic form of Creutzfeldt-Jakob disease. Brain 94 (1971) 715–724.

ALTER, M., L. T. KURLAND and C. A. MOLGAARD: Late progressive muscular atrophy and antecedent poliomyelitis. In: L. P. Rowland (Ed.), Human Motor Neuron Diseases, Advances in Neurology, Vol. 36. New York, Raven Press (1982) 303–308.

AMINOFF, M. J.: Acanthocytosis and neurological disease. Brain 95 (1972) 749–760.

AMINOFF, M. J. and V. LOGUE: Clinical features of spinal vascular malformations. Brain 97 (1974a) 197–210.

AMINOFF, M. J. and V. LOGUE: The prognosis of patients with spinal vascular malformations. Brain 97 (1974b) 211–218.

ANDERSON, T. and A. SIDEN: A clinical study of the Guillain-Barré syndrome. Acta Neurol. Scand. 66 (1982) 316–327.

ANONYMOUS: Editorial. Hexacarbon neuropathy. Lancet 2 (1979) 942.

ARNON, S. S.: Infant botulism. Ann. Rev. Med. 31 (1980) 541–560.

AUB, J. C., L. T. FAIRHALL, A. S. MINOT and P. REZNIKOFF: Lead poisoning. Medicine 4 (1925) 1–250.

BACH, M. C. and R. M. ARMSTRONG: Acute toxoplasmic encephalitis in a normal adult. Arch. Neurol. 40 (1983) 596–597.

BAGLEY, J. F. H., J. W. WALSH and B. CADY, et al.: Carcinomatosis versus radiation-induced brachial plexus neuropathy in breast cancer. Cancer 41 (1978) 2154–2157.

BARBER, T. E.: Inorganic mercury intoxication reminiscent of amyotrophic lateral sclerosis. J. Occup. Med. 20 (1978) 667–669.

BAROHN, R. J., J. T. KISSEL, J. R. WARMOLTS and J. R. MENDELL: Chronic inflammatory polyradiculoneuropathy: clinical characteristics, course, and recommendations for diagnostic criteria. Arch. Neurol. 46 (1989) 874–884.

BARRIS, R. W.: Pancreatic adenoma (hyperinsulinism) associated with neuromuscular disorders. Ann. Intern. Med. 38 (1953) 124–129.

BARRON, S. A. and R. R. HEFFNER: Weakness in malignancy; evidence for a remote effect of tumor on axons. Ann. Neurol. 4 (1978) 268–274.

BARRY, P. S. I.: A comparison of concentrations of lead in human tissues. Br. Ind. Med. 32 (1975) 119–139.

BAU-PRUSSAK, S.: Über amyotrophische Form der multiplen Sklerose. Zentralbl. Gesamte Neurol. Psychiatr. 54 (1930) 495.

BAUER, M., R. BERGSTROM, B. RITTER and Y. OLSSON: Macroglobulinemia Waldenstrom and motor neuron syndrome. Acta. Neurol. Scand. 55 (1977) 245–250.

BELSH, J. M. and P. L. SCHIFFMAN: Misdiagnosis in patients with amyotrophic lateral scleroris. Arch. Intern. Med. 150 (1990) 2301–2305.

BHARUCHA, E. P. and V. P. MONDKAR: Neurological complications of a new conjunctivitis. Lancet 2 (1972) 970.

BOOTHBY, J. A., P. V. DEJESUS and L. P. ROWLAND: Reversible forms of motor neuron disease (lead neuritis). Arch. Neurol. 31 (1974) 18–26.

BOTS, G. TH. and A. STAAL: Amyotrophic lateral sclerosis-dementia complex, neuroaxonal dystrophy, and Hallervorden-Spatz disease. Neurology 23 (1973) 35–39.

BOUCHEZ, B., J. POIRRIEZ, G. E. ARNOTT, F. FOURRIER, C. CHOPIN and M. BLONDEL: Acute polyradiculoneuritis during toxoplasmosis. J. Neurol. 231 (1985) 347.

BOYD, R. E., P. T. BRENNAN, J. F. DENG, D. F. ROCHESTER and D. A. SPYKER: Strychnine poisoning. Recovery from profound lactid acidosis, hyperthermia, rhabdomyolysis. Am. J. Med. 74 (1983) 507–512.

BRADLEY, W. G.: Recent views on amyotrophic lateral sclerosis with emphasis on electrophysiological studies. Muscle Nerve 10 (1987) 490–502.

BRADLEY, W. G. and J. N. WALTON: Neurologic manifestations of thyroid gland. Postgrad. Med. 50 (1971) 118–121.

BRAIN, R., P. B. CROFT and M. WILKINSON: Motor neurone disease as a manifestation of neoplasm. Brain 88 (1965) 479–500.

BRAUER, L.: Muskelatrophie bei Multiper Sklerose. Neurol. Centralbl. 17 (1898) 635–640.

BRITTON, C. B., R. MESA-TEJADA, C. M. FENOGLIO, A. P. HAYS, G. G. GARVEY and J. R. MILLER: A new complication of AIDS: thoracic myelitis caused by herpes simplex virus. Neurology 35 (1985) 1071–1074.

BRODY, T. A.: Muscle contracture induced by exercise. A syndrome attributable to decreased relaxing factor. N. Engl. J. Med. 290 (1969) 187–192.

BROWN, I. A.: Chronic mercurialism: a cause of the clinical syndrome of amyotrophic lateral sclerosis. Arch. Neurol. Psychiatry 72 (1954) 674–681.

BROWN, W. F. and T. E. FEASBY: Conduction block and

denervation in Guillain-Barré polyneuropathy. Brain 107 (984) 219–239.

BROWNELL, B., D. R. OPPENHEIMER and J. T. HUGHES: The central nervous system in motor neuron disease. J. Neurol., Neurosurg. Psychiatry 33 (1970) 338–357.

BRUYLAND, M., S. VAN BELLE, D. SCHALLIER, D. EBINGER and J. J. MARTIN: Good response of a paraneoplastic neuromyopathy to cyclophosphamide. Cancer Treat. Rep. 68 (1984) 787–789.

BRUYN, G. W. and H. GARLAND: Neuropathies of endocrine origin. In: P. J. Vinken and G. W. Bruyn (Eds), Handbook of Clinical Neurology, Vol. 8. Amsterdam, North-Holland Publishing Co. (1970) 29–71.

BRUYN, G. W.: Chorea-acanthocytosis. In: P. J. Vinken, G. W. Bruyn and H. L. Klawans (Eds), Handbook of Clinical Neurology, Vol. 5. Amsterdam, North-Holland Publishing Co. (1986) 327–334.

BUCHTHAL, F. and Z. KAMIENIECKA: The diagnostic yield of quantified electromyography and quantified muscle biopsy in neuromuscular disorders. Muscle Nerve 5 (1982) 265–280.

BUCKLEY, J., C. WARLOW, P. SMITH, D. HILTON-JONES, S. IRVIN and J. R. TEW: Motor neurone disease in England and Wales, 1959–1979. J. Neurol. Neurosurg. Psychiatry 46 (1983) 197–205.

BULENS, C.: Neurologic complications of hyperthyroidism. Remission of spastic paraplegia, dementia, and optic neuropathy. Arch. Neurol. 38 (1981) 669–670.

BURCKHARDT, C. R., C. M. FILLEY and B. K. KLEINSCHMIDT-DE MASTERS: Diffuse Lewy body disease and dementia. Neurology 38 (1980) 1520–1528.

CAMPBELL, A. M. G. and E. R. WILLIAMS: Chronic lead intoxication mimicking motor neurone disease. Br. Med. J. 4 (1968) 582.

CAMPBELL, A. M. G., E. R. WILLIAMS and J. PEARCE: Later motor neuron degeneration following poliomyelitis. Neurology 19 (1969) 1101–1106.

CAMPBELL, A. M. G., E. R. WILLIAMS and D. BARLTROP: Motor neurone disease and exposure to lead. J. Neurol. Neurosurg. Psychiatry 33 (1970) 877–885.

CAMPBELL, M. J. and D. W. PATY: Carcinomatous neuromyopathy: 1. Electrophysiological studies. J. Neurol. Neurosurg. Psychiatry 37 (1974) 131–141.

CASHMAN, N. R., J. P. ANTEL, L. W. HANCOCK et al.: N-acetyl-beta-hexosaminidase beta locus defect and juvenile motor neuron disease: A case study. Ann. Neurol. 19/6 (1986) 568–572.

CASHMAN, N. R., R. MASELLI, R. L. WOLLMANN, R. ROOS, E. NICHOLS, F. BROWN, R. SIMON and J. P. ANTEL: Electromyography and muscle biopsy do not distinguish newly symptomatic from asymptomatic patients with prior paralytic poliomyelitis (abstract). Neurology 37 (Suppl 1) (1987) 214.

CHAD, D. A., K. HAMMER and J. SARGENT: Slow resolution of multifocal weakness and fasciculation: A reversible motor neuron syndrome. Neurology 36 (1986) 1260–1263.

CHARCOT, J. M.: Lectures on diseases of nervous system. Translation by G. Sigerson, 2nd ed. Philadelphia, HG Lea (1879) 164.

CHARLIN, A. and R. BRUNSCHWIG: Pseudosyndrome de sclérose latérale amyotrophique consécutif à un intoxication par le triorthocrésylphosphate. Rev. Neurol. 80 (1948) 68–69.

CHAZOT, G., B. BERGER, H. CARRIER et al.: Manifestations neurologiques des gammapathies monoclonales. Rev. Neurol. 132 (1976) 195–212.

CHERINGTON, M. and R. SNYDER: Tick paralysis: neurophysiological studies. N. Engl. J. Med. 278 (1968) 95–97.

CHIDA, K., S. SAKAMAKI and T. TAKASU: Alterations in autonomic function and cardiovascular regulation in amyotrophic lateral sclerosis. J. Neurol. 236 (1989) 127–130.

CHOI, D. W.: Glutamate neurotoxicity and diseases of the nervous system. Neuron 1 (1988) 623–634.

COHN, J. A., A. MCMEEKING, W. COHEN, J. JACOBS and R. S. HOLZMAN: Evaluation of the policy of empiric treatment of the suspected Toxoplasma encephalitis in patients with Acquired Immunodeficiency Syndrome. Am. J. Med. 86 (1989) 521–527.

CONNOLLY, J. H., I. V. ALLEN and E. DERMOTT: Transmissible agent in the amyotrophic form of Creutzfeldt-Jakob disease. J. Neurol. Neurosurg. Psychiatry 51 (1988) 1459–1460.

COOK, R. E.: Progressive bulbar palsy due to syphilis. Am. J. Syph. 37 (1953) 161–164.

CROFT, P. B. and M. WILKINSON: The incidence of carcinomatous neuromyopathy in patients with various types of carcinoma. Brain (1965) 427–434.

CRUICKSHANK, J. K., P. RUDGE, A. GT. DALGLEISH, M. NEWTON, B. N. MCLEAN, R. O. BARNARD, B. E. KENDALL and D. H. MILLER: Tropical spastic paraparesis and human T cell lymphocytic virus type 1 in the United Kingdom. Brain 112 (1989) 1057–1090.

CURRIER, R. D. and A. F. HAERER: Amyotrophic lateral sclerosis and metalic toxins. Arch. Environ. Health 17 (1968) 712–719.

DALAKAS, M. C. and M. HALLETT: The post-polio syndrome. In: F. Plum (Ed.), Advances in Contemporary Neurology. Philadelphia, F. A. Davis Company (1988a) 51–94.

DALAKAS, M. C. and G. H. PEZESHKPOUR: Neuromuscular diseases associated with human immunodeficiency virus infection. Ann. Neurol. 23 (Suppl.) (1988b) 38–48.

DALAKAS, M. C., L. CHARNAS, G. H. PEZESHKPOUR, T. KUWABARA and W. A. GAHL: A motor neuron-like neuromuscular disease in patients with nephropathic cystinosis. Neurology 39 (Suppl 1) (1989) 399.

DANTA, G.: Hypoglycemic peripheral neuropathy. Arch. Neurol. (Chic.) 21 (1969) 121–132.

DANTA, G.: Electrophysiological study of amyotrophy associated with acute asthma (asthmatic amyotrophy). J. Neurol., Neurosurg. Psychiatry 38 (1975) 1016–1021.

DAROFF, R. B., B. T. TROOST and R. J. LEIGH: Supranuclear disorders of eye movements. In: J. S. Glaser (Ed.), Neuro-ophthalmology. Philadelphia, Lippincott (1990) 299–323.

DAVID, A. S. and R. A. GILLHAM: Neuropsychological study of motor neuron disease. Psychosomatics 27 (1986) 441–445.

DAVIS, L. E., D. BODIAN, D. PRICE et al.: Chronic progressive poliomyelitis secondary to vaccination of an immunodeficient child. N. Engl. J. Med. 297 (1977) 241–245.

DAVISON, C., S. P. GOODHART and J. LANDER: Multiple sclerosis and amyotrophies. Arch. Neurol. Psychiatry (Chic.) 31 (1934) 270–289.

DEAPEN, D. M. and B. E. HENDERSON: A case-control study of amyotrophic lateral sclerosis. Am. J. Epidemiol. 123 (1986) 790–799.

DELISLE, M. B., P. GORCE and E. HIRSCH: Motor neuron disease, parkinsonism and dementia. Report of a case with diffuse Lewy-body-like intracytoplasmic inclusions. Acta Neuropathol. (Berl.) 75 (1987) 104–108.

DELISLE, M. B., P. GORCE and E. HIRSCH: Motor neuron disease, parkinsonism and dementia. Report of a case with diffuse Lewy-body-like intracytoplasmic inclusion. Acta Neuropathol. (Berl.) 75 (1987) 104–108.

DENNY-BROWN, D. and J. M. FOLEY: Myokymia and the benign fasciculation of muscular cramps. Trans. Assoc. Am. Physicians 61 (1948) 88–96.

DIREKZE, M. and P. S. L. FERNANDO: Transient anterior horn cell dysfunction in diphenylhydantoin therapy. Eur. Neurol. 15 (1977) 131–134.

DICKSON, D. W., P. DAVIES and R. MAYEUX: Diffuse Lewy body disease. Acta Neuropathol. (Berl.) 75 (1987) 8–15.

DJINDJAN, R.: Angiomas of the spinal cord. In: P. J. Vinken, G. W. Bruyn, N. C. Myrianthopoulos and H. L. Klawans (Eds), Handbook of Clinical Neurology, Vol. 32. Amsterdam, North-Holland Publishing Co. (1978) 465–510.

DODGE, H. W., J. G. LOVE and C. M. GOTTLIEB: Benign tumors at the foramen magnum. J. Neurosurg. 13 (1956) 603–617.

DODSON, W. E. and W. M. LANDAU: Motor neuron loss due to aortic clamping in repair of coarctation. Neurology 23 (1973) 539–542.

DRIEDGER, H. and W. PRUZANSKI: Plasma cell neoplasia with peripheral neuropathy. A study of five cases and review of the literature. Medicine 59 (1980) 301–310.

DUBOIS, F., P. BERTRAND and J. EMILE: Amyotrophie spinale progressive et adénome parathyroidien. Rev. Neurol. 145 (1989a) 65–68.

DUBOIS, F., H. PETIT, O. GODEFROY, C. BONTE and J. D. GUIEU: Amyotrophie spinale aigue au cours d'une Toxoplasmose. Rev. Neurol. (Paris) 145 (12) (1989b) 857–859.

DYCK, P. J.: Are motor neuropathies and motor neuron diseases separable? In: L. P. Rowland (Ed.), Human Neuron Diseases, Advances in Neurology, Vol. 36. New York, Raven Press (1982) 105–114.

DYCK, P. J., A. C. LAIS, M. OHTA, J. A. BASTRON, J. OKAZAKI and R. V. GROOVER: Chronic inflammatory polyradiculoneuropathy. Mayo Clin. Proc. 50 (1975a) 621–637.

DYCK, P. J., J. C. STEVENS, D. W. MULDER and R. E. ESPINOSA: Frequency of nerve fiber degeneration of peripheral motor and sensory neurons in amyotrophic lateral sclerosis. Neurology (Minneap.) 25 (1975b) 781–785.

EDMONDSON, R. S. and M. W. FLOWERS: Intensive care in tetanus: Management, complications, and mortality in 100 cases. Br. Med. J. 1 (1979) 1401–1404.

EKBOM, K. A.: Restless leg syndrome. In: P. J. Vinken and G. W. Bruyn (Eds), Handbook of Clinical Neurology, Vol. 8. Amsterdam, North-Holland Publishing Co. (1970) 311–320.

EL ALAOUI-FARIS, M., A. MEDEJEL, K. AL ZEMMOURI, M. YAHYAOUI and T. CHKILI: Le syndrome de sclérose laterale amyotrophique d'origine syphilitique. Etude de 5 cas. Rev. Neurol. (Paris) 146 (1990) 41–44.

EVANS, R., W. R. KENNEDY and R. ROELOFS.: A 20-years follow-up of proximal bulbar-spinal muscular atrophy. Neurology 37 (Suppl 1) (1987) 214–215.

EVANS, B. K., I. GORE, L. E. HARRELL, T. ARNOLD and S. J. OH: HTLV-1-associated myelopathy and polymyositis in a US native. Neurology 39 (1989) 1572–1575.

FEASBY, T. E., W. F. BROWN, J. J. GILBERT and A. F. HAHN: The pathological basis of conduction block in human neuropathies. J. Neurol., Neurosurg. Psychiatry 48 (1985) 239–244.

FEIBEL, J. H. and J. F. CAMPA: Thyrotoxic neuropathy (Basedow's paraplegia). J. Neurol. Neurosurg. Psychiatry 39 (1976) 491–497.

FINLAYSON, M. H., A. GUBERMAN and J. B. MARTIN: Cerebral lesions in familial amyotrophic lateral sclerosis and dementia. Acta Neuropathol. (Berl.) 26 (1973) 237–246.

FISHER, M., R. R. LONG and D. A. DRACHMAN: Hand muscle atrophy in multiple sclerosis. Arch. Neurol. 40 (1983) 811–815.

FISHER, M., J. E. MATEER, I. ULLRICH and J. A. GUTRECHT: Pyramidal tract deficits and polyneuropathy in hyperthyroidism. Combination clinically mimicking amyotrophic lateral sclerosis. Am. J. Med. 78 (1985) 1041–1044.

FISCHER, P. A. and W. ENZENSBERGER: Neurological complications in AIDS. J. Neurol. 234 (1987) 269–279.

FLEET, W. S. and R. T. WATSON: From benign fasciculations and cramps to motor neuron disease. Neurology 36 (1986) 997–998.

FOWLER, H. L.: Machade-Joseph-Azorean disease. Arch. Neurol. 41 (1984) 921–925.

FRAME, B.: Neuromuscular manifestations of parathyroid disease. In: P. J. Vinken and G. W. Bruyn (Eds), Handbook of Clinical Neurology, Vol. 27. Amsterdam, North-Holland Publishing Co. (1976) 283–320.

FREDDO, L., R. K. YU, N. LATOV, P. D. DONOFRIO, A. P. HAYS, H. S. GREENBERG, J. W. ALBERS, A. G. ALLESSI and D. KEREN: Gangliosides GM_1 and GD_{1b} are antigens for IgM M-protein in a patient with motor neuron disease. Neurology 36 (1986) 454–458.

FREDRIKSON, S. and H. LINK: CNS-borreliosis selectively affecting central motor neurons. Acta Neurol. Scand. 78 (1988) 181–184.

FURAKAWA, T., N. AKAGAMI and S. MARUYAMA: Chronic neurogenic quadriceps amyotrophy. Ann. Neurol. 2 (1977) 528–530.

GALLAGHER, J. P. and M. SANDERS: Apparent motor neuron disease following the use of pneumatic tools. Ann. Neurol. 14 (1983) 694–695.

GALLASSI, R., P. MONTAGNA, A. MORREALE, S. LORUSSO, P. TINUPER, R. DAIDONE and E. LUGARESI: Neuropsychological, electroencephalogram and brain computed tomography findings in motor neuron disease. Eur. Neurol. 29 (1989) 115–120.

GARCIA, C. A. and H. FLEMMING: Reversible corticospinal tract disease due to hyperthyroidism. Arch. Neurol. 34 (1977) 647–648.

GARLAND, H.: Neurological complications of diabetic mellitus: clinical aspects. Proc. R. Soc. Med. 53 (1959) 137–141.

GARCIA, C. A. and R. H. FLEMMING: Reversible corticospinal tract disease due to hyperthyroidism. Arch. Neurol. 34 (1977) 647–648.

GARRUTO, R. M., S. K. SHANKAR, R. YANAGIHARA, A. M. SALAZAR, H. L. AMYX and D. C. GAJDUSEK: Low-calcium, high-aluminium diet induced motor neuron pathology in cynomolgus monkeys. Acta Neuropathol. 78 (1989) 210–219.

GASTAUT, J. L., B. MICHEL, D. FIGARELLA-BRANGER and H. SOMMA-MAUVAIS: Chronic progressive spinobulbar spasticity. Arch. Neurol. 45 (1988) 509–513.

GERARD, G. and L. A. WEISBERG: MRI periventricular lesions in adults. Neurology 36 (1986) 998–1001.

GIBB, W. R. G., M. M. ESIRI and A. J. LEES: Clinical and pathological features of diffuse cortical Lewy body disease (Lewy body dementia). Brain 110 (1985) 1131–1153.

GIBB, W. R. G., P. J. LUTHERT and C. D. MARSDEN: Corticobasal degeneration. Brain 112 (1989) 1171–1192.

GILLES, F. H. and D. NAG: Vulnerability of human spinal cord in transient cardiac arrest. Neurology 21 (1971) 833–839.

GLASER, J. S. and B. BACHYNSKI: Infranuclear disorders of eye movements. In: J. S. Glaser (Ed.), Neuro-ophthalmology. Philadelphia, Lippincott (1990) 361–418.

GOODMAN GILMAN, A., L. S. GOODMAN and A. GILMAN: Goodman's and Gilman's The Pharmacological Basis of Therapeutics. New York, Macmillan (1980) 585–587.

GOTO, I., H. NAKAI, T. TABIRA, N. SHINO, Y. TANAKA, H. SHIBASAKI and Y. KUROIWA: Juvenile neurogenic muscle atrophy with lysosomal enzyme deficiencies: new diseases or variant of mucopolysaccharidosis? J. Neurol. 229 (1983) 45–54.

GOURIE-DEVI, M., T. G. SURESH and S. K. SHANKAR: Monomelic amyotrophy. Arch. Neurol. 41 (1984) 388–394.

GREENE, R.: The thyroid gland: its relationship to neurology. In: P. J. Vinken and G. W. Bruyn (Eds), Handbook of Clinical Neurology, Vol. 27. Amsterdam, North-Holland Publishing Co. (1976) 255–277.

GREENFIELD, J. G. and W. B. MATTHEWS: Post-encephalitic parkinsonism and amyotrophy. J. Neurol. Neurosurg. Psychiatry 17 (1954) 50–56.

GREENFIELD, M. M. and F. M. STARK: Post-irradiation neuropathy. Am. J. Roentgen, Radium Ther. Nucl. Med. 60 (1948) 617–622.

GREENHOUSE, A. H., J. M. BICKNELL, R. N. PESCH and D. F. SEELINGER: Myotonia, myokymia, hyperhidrosis, and wasting of muscle. Neurology 17 (1967) 263–268.

GRIST, N. F., E. J. BELL and F. ASSAAD: Enterovirus in human disease. Prog. Med. Virol. 24 (1978) 114–157.

GUDESBLATT, M., M. D. LUDMAN, J. A. COHEN, R. J. DESNICK, S. CHESTER, G. A. GRABOWSKI and J. T. CAROSCIO: Hexosaminidase A activity and amyotrophic lateral sclerosis. Muscle Nerve 2 (1988) 227–230.

GUIDETTI, B. and A. FORTUNA: Differential diagnosis of intramedullary and extramedullary tumours. In: P. J. Vinken, G. W. Bruyn and H. L. Klawans (Eds), Handbook of Clinical Neurology, Vol. 19. Amsterdam, North-Holland Publishing Co. (1975) 51–75.

GUNNARSSON, L. G. and G. LINDBERG: Amyotrophic lateral sclerosis in Sweden 1970–1983 and solvent exposure. Lancet 1 (1989) 958–959.

GUTMANN, L. and L. PRATT: Pathophysiologic aspects of human botulism. Arch. Neurol. 33 (1976) 175–179.

GUTMANN, L., J. D. MARTIN and W. WELTON: Dapsone motor neuropathy and axonal disease. Neurology 26 (1976) 514–516.

HADER, W. J., B. ROZIDILSKY and C. P. NAIR: The concurrence of multiple sclerosis and amyotrophic lateral sclerosis. Can. J. Neurol. Sci. 13 (1986) 66–69.

HALPERIN, J. J., B. J. LUFT, A. K. ANAND, C. T. ROQUE, O. ALVAREZ, D. J. VOLKMAN and R. J. DATTWYLER: Lyme neuroborreliosis: central nervous system manifestations. Neurology 39 (1989) 753–759.

HALPERIN, J. J., G. P. KAPLAN, S. BRAZINSKY, T. F. TSAI, T. CHENG, A. IRONSIDE, P. WU, J. DELFINER, M. GOLIGHTLY, R. H. BROWN, R. J. DATTWYLER and B. J. LUFT: Immunologic reactivity against Borrelia burgdorferi in patients with motor neuron disease. Arch. Neurol. 47 (1990) 586–594.

HANSON, M. R., A. J. WILBOURN, R. LEDERMAN et al.: Focal motor neuron disease: Clinical and electromyographic features. Muscle Nerve 9 (1986) 654.

HARDING, A. E.: X-linked recessive bulbo-spinal muscular atrophy: a report of ten cases. J. Neurol. Neurosurg. Psychiatry 45 (1982) 1012.

HARDING, A. E.: Cerebellar and Spinocerebellar Disorders. In: Neurology in clinical practice Vol. II. Boston, Butterworth-Heinemann (1991) 1603–1624.

HARMAN, J. B. and A. T. RICHARDSON: Generalized myokymia in thyrotoxicosis. Lancet 2 (1954) 473–474.

HARRIMAN, D. G. F., D. TAVERNER and A. L. WOOLF: Ekbom's syndrome and burning paraesthesiae. Brain 93 (1970) 393–406.

HARRINGTON, T. M., M. D. COHEN, J. D. BARLESON and W. W. GINSBURG: Elevation of creatine kinase in amyotrophic lateral sclerosis. Arthritis Rheum. 26 (1983) 201–205.

HARRISON, M. J. G.: Muscle wasting after prolonged hypoglycaemic coma: case report with electrophysiological data. J. Neurol. Neurosurg. Psychiatry 39 (1976) 465–470.

HARVEY, D. G., R. M. TORACK and H. E. ROSENBAUM: Amyotrophic lateral sclerosis with ophthalmoplegia: A clinicopathologic study. Arch. Neurol. 36 (1979) 615–617.

HATCH, M. H., M. D. MALISON and E. L. PALMER: Isolation of enterovirus 70 from patients with acute hemorrhagic conjunctivitis in Key West, Florida. N. Engl. J. Med. 305 (1981) 1648–1649.

HAWKES, C. H. and A. J. FOX: Motor neurone disease in leather workers. Lancet 1 (1981) 507.

HAWKES, C. H., J. B. CAVANAGH and A. J. FOX: Motoneuron disease: a disorder secondary to solvent exposure? Lancet 1 (1989) 73–76.

HAYASHI, H., S. KATO, T. KAWADA and T. TSUBAKI: Amyotrophic lateral sclerosis: oculomotor function in patients in respirators. Neurology 37 (1987) 4131–1432.

HAYASHI, H. and S. KATO: Total manifestations of amyotrophic lateral sclerosis. ALS in the totally locked-in state. J. Neurol. Sci. 93 (1989) 19–35.

HAYWARD, M. and D. SEATON: Late sequelae of paralytic poliomyelitis: a clinical and electromyographic study. J. Neurol. Neurosurg. Psychiatry 442 (1979) 117–122.

HEATHFIELD, K. W. G. and J. W. A. TURNER: Syphilitic wrist-drop. Lancet 2 (1951) 566–569.

HELWEG-LARSEN, S., J. JAKOBSEN, F. BOESEN et al.: Myelopathy in AIDS. A clinical and electrophysiological study of 23 Danish patients. Acta Neurol. Scand. 77 (1988) 64–73.

HENSON, R. A. and H. URICH: Cancer and the Nervous System. Oxford, Blackwell (1982).

HERRICK, M. K. and P. E. MILLS: Infarction of spinal cord. Two cases of selective gray matter involvement secondary to asymptomatic aortic disease. Arch. Neurol. 24 (1971) 228–241.

HOFFMAN, P. M., B. W. FESTOFF, L. T. GIRON et al.: Isolation of LAV/HTLV-III from a patient with amyotrophic lateral sclerosis. N. Engl. J. Med. 312 (1985) 324.

HOMEIDA, M., A. BABIKR and T. K. DANESHMEND: Dapsone-induced optic atrophy and motor neuropathy. Br. Med. J. 281 (1980) 1180.

HOOSHMAND, H., M. R. ESCOBAR and S. W. KOPF: Neurosyphilis, a study of 241 patients. J. Am. Med. Assoc. 219 (1972) 726–729.

HOPKINS, I. J.: A new syndrome: poliomyelitis-like illness associated with acute asthma in childhood. Aust. Paediatr. J. 10 (1974) 273–276.

HORMES, J. T., C. M. FILLEY and N. L. ROSENBERG: Neurologic sequelae of chronic solvent vapor abuse. Neurology 36 (1986) 698.

HOROWITZ, S. L. and J. D. STEWART: Lower motor neuron syndrome following radiotherapy. Can. J. Neurol. Sci. 10 (1983) 56–58.

HOWELL, D. A., A. J. LEES and P. J. TOGHILL: Spinal internuncial neurones in progressive encephalomyelitis with rigidity. J. Neurol. Neurosurg. Psychiatry 42 (1979) 773–785.

HUDSON, A. J.: Amyotrophic lateral sclerosis and its association with dementia, Parkinsonism and other neurological disorders: a review. Brain 104 (1981) 217–247.

HUDSON, A. J., W. F. BROWN and J. J. GILBERT: The muscular pain-fasciculation syndrome. Neurology 28 (1978) 1105–1109.

HUDSON, A. J., A. DAVENPORT and W. J. HADER: The incidence of amyotrophic lateral sclerosis in southwestern Ontario, Canada. Neurology 36 (1986) 1524–1528.

HUNG, T. P. and R. KONO: Neurologic complications of acute haemorrhagic conjunctivitis (a polio-like syndrome in adults). In: P. J. Vinken and G. W. Bruyn (Eds), Handbook of Clinical Neurology, Vol. 38. Amsterdam, North-Holland Publishing Co. (1979) 595–623.

ISAACS, H.: A syndrome of continuous muscle-fiber activity. J. Neurol. Neurosurg. Psychiatry 24 (1961) 319–325.

JASPAN, J. B., R. L. WOLLMAN, L. BERNSTEIN et al.: Hypoglycemic peripheral neuropathy in assocation with insulinoma: Implication of glucopenia rather than hyperinsulinism. Case report and literature review. Medicine 61 (1982) 33–44.

JELLINGER, K. and E. NEUMAYER: Myelopahie progressive d'origine vasculaire. Acta Neurol. Psychiatr. Belg. 62 (1962) 944–956.

JELLINGER, K. and K. W. STURM: Delayed radiation myelopathy in man. Report of twelve necropsy cases. J. Neurol. Sci. 14 (1971) 389–408.

JOHNSON, W.: Hexosaminidase deficiencies and neuromuscular diseases. Muscle Nerve 5 (Suppl.) (1986) 70.

KAMMER, G. M. and C. R. HAMILTON: Acute bulbar muscle dysfunction and hyperthyroidism: A study of four cases and review of literature. Am. J. Med. 56 (1974) 464–470.

KANTARJIAN, A. D.: A syndrome clinically resembling amyotrophic lateral sclerosis following chronic mercurialism. Neurology 11 (1961) 639–644.

KASDON, D. L.: Cervical spondylotic myelopathy with reversible fasciculations in the lower extremities. Arch. Neurol. 34 (1977) 774–776.

KAUFMAN, J. M., T. L. KEMPER and S. PETERS: Coxsackie B2 infection simulating poliomyelitis pathologically. Neurology (NY) 30 (1980) 396.

KENNEDY, W. R., M. ALTER and J. H. SUNG: Progressive proximal spinal and bulbar muscular atrophy of late onset. A sex-linked recessive trait. Neurology 18 (1968) 671–680.

KILNESS, A. W. and F. H. HOCHBERG: Amyotrophic lateral sclerosis in a high selenium environment. J. Am. Med. Assoc. 237 (1977) 2843–2844.

KONDO, K. and T. TSUBAKI: Case-control studies of motor neuron disease. Association with mechanical injuries. Arch. Neurol. 38 (1981) 220–226.

KORI, S. H., K. M. FOLEY and J. B. POSNER: Brachial plexus lesions in patients with cancer: 100 cases. Neurology 31 (1981) 45–50.

KRISTENSON, O., B. MELGARD and A. V. SCHLT: Radiation myelopathy of the lumbo sacral spinal cord. Acta Neurol. Scand. 56 (1977) 217–222.

KURLANDER, H. M. and B. M. PATTEN: Metals in spinal cord tissue of patients dying of Motor Neuron Disease. Ann. Neurol. 6 (1979) 21–24.

KURTZKE, J. F.: Symptomatology of Multiple Sclerosis. In: P. J. Vinken and G. W. Bruyn (Eds), Handbook of Clinical Neurology, Vol. 9. Amsterdam, North-Holland Publishing Co. (1970) 179–209.

KUSHNER, M. J., M. PARRISH, A. BURKE, M. BEHRENS, A. P. HAYS, B. FRAME and L. P. ROWLAND: Nystagmus in motor neuron disease: Clinicopathological study of two cases. Ann. Neurol. 16 (1984) 71–77.

LAGUENY, A., M. AUPY, P. AUPY, X. FERRER, P. HENRY and J. JULIEN: Syndrome de la corne antérieure postradiothérapique. Rev. Neurol. (Paris) 141 (1985) 222–227.

LAMBERT, E. H., D. W. MULDER and J. A. BASTRON: Regeneration of peripheral nerves with hyperinsulin neuropathy. Neurology 10 (1960) 851–854.

LAMONTAGNE, A. and F. BUCHTHAL: Electrophysiological studies in diabetic neuropathy. J. Neurol. Neurosurg. Psychiatry 33 (1970) 442–452.

LANGE, D. J., C. B. BRITTON, D. S. YOUNGER and A. P. HAYS: The neuromuscular manifestations of human immunodeficiency virus infections. Arch. Neurol. 45 (1988) 1084–1088.

LATOV, N.: Plasma cell dyscrasia and motor neuron diseases. In: L. P. Rowland (Ed.), Human Motor Neuron Diseases, Advances in Neurology, Vol. 36 New York, Raven Press (1982) 273–280.

LATOV, N., W. H. SHERMAN, R. NEMNI et al.: Plasma-cell dyscrasia and peripheral neuropathy with a monoclonal antibody to peripheral-nerve myelin. N. Engl. J. Med. 303 (1980) 618–621.

LATOV, N., A. P. HAYS, P. D. DONOFRIO et al.: Monoclonal IgM with unique specificity to gangliosides GM1 and GD1b and to lacto-N-tetraose associated with human motor neuron disease. Neurology 38 (1988) 763–768.

LAVAUD, J., B. COCHOIS, H. P. DE LEERSNYDER and A. BAROIS: Le toxoplasme, nouvel agent possible de la polyradiculoneurite de Guillain-Barré. Nouv. Press. Méd. 8 (1970) 131.

LAYZER, R. B.: Motor units hyperactivity states. In: P. J. Vinken and G. W. Bruyn (Eds), Handbook of Clinical Neurology, Vol. 41. Amsterdam, North-Holland Publishing Co. (1979) 295–316.

LAYZER, R. B.: Diagnostic implications of clinical fasciculation and cramps. In: L. P. Rowland (Ed.), Human Motor Neuron Diseases. New York, Raven Press (1982) 23–29.

LEIGH, R. J. and D. S. ZEE: The Neurology of Eye Movement. Philadelphia, F. A. Davis (1983).

LENETTE, E. H., G. E. CAPLAN and R. L. MAGOFFIN: Mumps virus infection simulating paralytic poliomyelitis. Pediatrics 25 (1960) 788–797.

LEVEILLE, A., J. KIERNAN, J. A. GOODWIN and J. ANTEL: Eye movements in amyotrophic lateral sclerosis. Arch. Neurol. 39 (1982) 684–686.

LEVISON, J. A., J. RANSOHOFF and J. GOODGOLD: Electromyographic studies in a case of foramen magnum meningioma. J. Neurol. Neurosurg. Psychiatry 36 (1973) 561–564.

LEYS, D., M. PARENT, F. LESOIN, D. CAMUS and H. PETIT: Myelite pseudo-tumorale au cours d'une toxoplasmose. Sem. Hôp. Paris 60 (1984) 1437–1438.

LI, T. M., S. J. DAY, E. ALBERMAN and M. SWASH: Differential diagnosis neurone disease from other neurological conditions. Lancet 2 (1986) 731–733.

LIEBESCHUETZ, H. J.: Poliomyelitis-type illness associated with severe asthma in a child. J. R. Soc. Med. 74 (1981) 71–72.

LINDSTROM, J. M., M. E. SEYBOLD, V. A. LENNON, S. WHITTINGHAM and D. D. DUANE: Antibody to acetylcholine receptor in myasthenia gravis. Neurology 26 (1976) 1054–1059.

LIVESLEY, B. and C. E. SISSONS: Chronic lead intoxication mimicking motor neuron disease. Br. Med. J. 4 (1968) 387–388.

LOGIGIAN, E. L., R. F. KAPLAN and A. C. STEERE: Chronic neurologic manifestations of Lyme disease. N. Engl. J. Med. 323 (1990) 1438–1444.

LOVE, J. G., E. P. THELEN and H. W. DODGE: Tumors of the foramen magnum. J. Int. Coll. Surg. 22 (1954) 1–17.

LUDOLPH, A. C., J. HUGON, M. P. DWIVEDI, H. H. SCHAUMBURG and P. S. SPENCER: Studies on aetiology and pathogenesis of motor neuron diseases. 1. Lathyrism: Clinical findings in established cases. Brain 110 (1987) 149–165.

LUXON, L., A. J. LEES and R. J. GREENWOOD: Neurosyphilis today. Lancet 1 (1979) 90–93.

MACDERMOTT: Frontal lobe dementia and motor neuron disease. J. Neurol. Neurosurg. Psychiatry 53 (1990) 23–32.

MACKAY, R. P. and G. W. HALL: Syphilitic amyotrophy. Arch. Neurol. Psychiatry 29 (1933) 241–254.

MAGOFFIN, R. L., E. H. LENETTE, A. C. HOLLISTER JR. et al.: An etiologic study of clinical paralytic poliomyelitis. J. Am. Med. Assoc. 175 (1961) 269–278.

MAIER, J. G., R. H. PERRY, W. SAYLOR and M. H. SULAK: Radiation myelitis of the dorsolumbar spinal cord. Radiology 93 (1969) 153–160.

MANSON, J. I. and Y. H. THONG: Immunological abnormalities in the syndrome of poliomyelitis-like illness associated with acute bronchial asthma (Hopkin's syndrome). Arch. Dis. Child. 55 (1980) 26–32.

MARS, H., L. A. LEWIS, A. L. ROBERTSON, A. BUTKUS and G. H. WILLIAMS: Familial hypo-β-lipoproteinemia with neurologic manifestations. Neurology 17 (1976) 285.

MARTIN, J. P.: Amyotrophic meningo-myelitis (spinal progressive muscular atrophy of syphilitic origin). Brain 48 (1925) 153–182.

MARTYN, C. N.: Motoneuron disease and exposure to sovents. Lancet 1 (1989) 394.

MCARDLE, B.: Myopathy due to a defect in muscle glycogeen breakdown. Clin. Sci. 10 (1951) 13–33.

MCARTHUR, J. C.: Neurologic manifestations of AIDS. Medicine 66 (1987) 407–437.

MCCRAE, C. L.: Bony abnormalities in the region of the foramen magnum: Correlation of the anatomic and neurologic findings. Acta. Radiol. 40 (1953) 335.

MCDONALD, W. I. and R. S. KOCEN: Diphteric neuropathy. In: P. J. Dyck, P. K. Thomas, E. H. Lambert and R. Bunge (Eds), Peripheral Neuropathy, Vol. II. Philadelphia, W. B. Saunders Co. (1984) 2010–2017.

MCKUSICK, V. A., C. A. FRANCOMANO and I. E. ANTONORAKIS: Mendelian Inheritance in Man, 9th ed. Baltimore, Johns Hopkins University Press (1990).

MCLEOD, J. G.: Electrophysiological studies in the Guillain-Barre-syndrome. Ann. Neurol. 9 (Suppl.) (1981) 20–27.

MCQUINN, B. A. and T. L. KEMPER: Sporadic case resembling autosomal-dominant motor system degeneration (Azorean disease complex). Arch. Neurol. 44 (1987) 341–344.

MEASE, P. J., H. D. OCHS and R. J. WEDGWOOD: Successful treatment of echovirus meningoencephalitis and myositis-fasciitis with intravenous immune globulin therapy in a patient with X-linked agammaglobulinemia. N. Engl. J. Med. 304 (1981) 1278–1281.

MENDELL, J. R., S. KOLKIN, J. T. KISSEL, K. L. WEISS, D. W. CHAKERES and K. W. RAMMOHAN: Evidence for central nervous system demyelination in chronic inflammatory demyelinating polyradiculoneuropathy. Neurology 37 (1987) 1291–1294.

MEHREN, M., P. J. BURNS, D. O. BURNS, F. MAMANI, C. S. LEVY and R. LAURENO: Toxoplasmic myelitis mimicking intrameduallary spinal cord tumor. Neurology 38 (1988) 1648–1650.

MELAMED, E., M. BERMAN and S. LAVY: Posterolateral myelopathy associated with thyrotoxicosis. N. Engl. J. Med. 293 (1975) 778–779.

MENA, I., O. MARIN, S. FUENZALIDA and G. C. COTZIAS: Chronic manganese poisoning: clinical picture and manganese turnover. Neurology 17 (1967) 128.

MENDELL, J. R., S. KOLKIN, J. T. KISSEL, K. L. WEISS, D. W. CHAKERES and K. W. RAMMOHAN: Evidence for central nervous system demyelination in chronic inflammatory demyelinating polyradiculoneuropathy. Neurology 37 (1987) 1291–1294.

MERSON, M. H., J. M. HUGHES and V. R. DOWELL et al.: Current trends in botulism in the United States. J. Am. Med. Assoc. 229 (1974) 1305–1308.

MITCHELL, J. D., B. W. EAST, I. A. HARRIS, R. J. PRESCOTT and B. PENTLAND: Trace elements in the spinal cord and other tissues in motor neuron disease. J. Neurol. Neurosurg. Psychiatry 49 (1986) 211–215.

MITSUMOTO, H., R. J. SLIMAN, I. A. SCHAFER, C. S. STERNICK, B. KAUFMAN, A. WILBOURN and S. J. HORWITZ: Motor neuron disease and adult hexosaminidase A deficiency in two families: evidence for multisystem degeneration. Ann. Neurol. 17 (1985) 378–385.

MITSUYAMA, Y. and S. TAKAMIYA: Presenile dementia with motor neurone disease in Japan: A new entity? Arch. Neurol. 36 (1979) 592–593.

MOERSCH, F. P. and J. W. KERNOHAN: Hypoglycemia, neurologic and neuropathologic studies. Arch. Neurol. Psychiatry 39 (1938) 242–257.

MOERSCH, F. P. and H. W. WOLTMAN: Progressive fluctuating muscular rigidity and spasm ('Stiff-Man Syndrome'): Report of a case and some observations in 13 other cases. Proc. Mayo Clin. 31 (1956) 421–427.

MOSER, H. W., A. B. MOSER, I. SINGH and B. O. O'NEILL: Adrenoleukodystrophy: Survey of 303 cases: Biochemistry, diagnosis and therapathy. Ann. Neurol. 16 (1984) 628–641.

MOSER, A. B., J. BOREL, A. ODONE, S. NAIDU, D. CORNBLATH, D. B. SANDERS and H. W. MOSER: A new dietary therapy for adrenoleukodystrophy: Biochemical and preliminary clinical results in 36 patients. Ann. Neurol. 21 (1987) 240–249.

MOTTIER, D., G. BERGERET, M. F. PERREAUT, A. MISSOUM, J. BASTARD and D. MABIN: Myopathie thyroïdienne chronique simulant une sclérose latérale amyotrophique. Nouv. Presse Méd. 10 (1981) 1655.

MULDER, D. W., J. A. BASTOON and E. H. LAMBERT: Hyperinsulin neuropathy. Neurology 6 (1956) 627–635.

MULDER, D. W., E. H. LAMBERT and L. M. EATON: Myasthenic syndrome in patients with amyotrophic lateral sclerosis. Neurology 9 (1959) 627–631.

MULDER, D. W., R. A. ROSENBAUM and D. D. LAYTON, JR.: Late progression of poliomyelitis or forme fruste amyotrophic lateral sclerosis. Mayo Clin. Proc. 47 (1972) 756–761.

MULDER, D. W.: Commentary. Differential diagnosis of adult motor neuron diseases. In D. W. Mulder (Ed.), The Diagnosis and Treatment of Amyotrophic Lateral Sclerosis. Boston, Houghton Mifflin Prof. Publ. (1980) 79–83.

MULDER, D. W.: Clinical Limits of Amyotrophic Lateral Sclerosis. In: L. P. Rowland (Ed.), Human Motor Neuron Diseases, Advances in Neurology, Vol. 36. New York, Raven Press (1982) 15–22.

MULDER, D. W., W. BUSHEK, E. SPRING, J. KARNES and P. J. DYCK: Motor neuron disease (ALS): Evaluation of detection thresholds of cutaneous sensation. Neurology 33 (1983) 1625–1627.

MULLER, R.: Studies on disseminated sclerosis with special reference to symptomatology, course and prognosis. Acta. Med. Scand. 133 (Suppl. 222) (1949) 1–214.

MURPHY, T. R., W. H. REMINE and M. K. BURBANK: Hyperparathyroidism: Report of a case in which parathyroid adenoma presented primarily with profound muscular weakness. Proc. Mayo Clin. 35 (1960) 629–640.

NAMBA, I. T., C. T. NOLTE, J. JACKREL and D. GROB: Poisoning due to organophosphate insecticides: Acute and chronic manifestations. Am. J. Med. 50 (1971) 475.

NEARY, D., J. S. SNOWDEN, D. M. A. MANN, B. NORTHEN, P. J. GOULEING, J. NEWSON-DAVIS and M. F. MURRAY:

Plasma exchange and immunosuppressive drug treatment in the Lambert-Eaton myasthenic syndrome. Neurology 34 (1984) 480–485.

NONNE, M.: Syphilis un Nervensystem. Berlin, Verlag Karger (1921) 625–630.

NORRIS, F. H.: Adult spinal motor neuron disease: Progressive muscular atrophy (Aran's disease) in relation to amyotrophic lateral sclerosis. In: P. J. Vinken and G. W. Bruyn (Eds), Handbook of Clinical Neurology, Vol. 22, Part II. Amsterdam, North-Holland Publishing Co. (1975) 1–56.

NORRIS, F. H. and B. J. PENNER: Hypothyroid myopathy. Clinical, electromyographic and ultrastructural observations. Arch. Neurol. 14 (1966) 547–589.

NORRIS, F. H., E. H. DENYS and K. S. U: Differential diagnosis of adult motor neuron diseases. In: D. W. Mulder (Ed.), The Diagnosis and Treatment of Amyotrophic Lateral Sclerosis. Boston, Houghton Mifflin Prof. Publ. (1980) 53–87.

NOSEWORTHY, J. H. and L. P. HEFFERMAN: Motor radiculopathy — an unusual presentation of multiple sclerosis. Can. J. Neurol. Sci. 7 (1980) 207–209.

O'CALLAGHAN, W. G., N. JOYCE and H. E. COUNIHAN: Unusual strychnine poisoning and its treatment: Report of eight cases. Br. Med. J. (1982) 285–478.

OLAFSEN, R. A., D. W. MULDER and F. M. HOWARD: 'Stiff-Man' Syndrome: A review of literature, report of three additional cases and discussion of pathophysiology and therapy. Proc. Mayo Clin. 39 (1964) 131–144.

O'NEILL, J. H., N. M. F. MURRAY and J. NEWSOM-DAVIS: The Lambert-Eaton myasthenic syndrome. A review of 50 cases. Brain 111 (1988) 577–596.

OPPENHEIMER, D.: Neuropathology and neurochemistry of autonomic failure. In: R. Bannister (Ed.), Autonomic Failure, A Textbook of Clinical Disorders of the Autonomic Nervous System, 2nd ed. Oxford, Oxford University Press (1988) 451–483.

O'REILLY, D. F., P. W. BRAZIS and F. A. RUBINO: The misdiagnosis of unilateral presentations of amyotrophic lateral sclerosis. Muscle Nerve 5 (1982) 724–726.

OSTHEIMER, A. J., G. WILSON and N. W. WINKELMAN: Syphilis as the cause of muscular atrophy of spinal origin. Am. J. Med. Sci. 167 (1924) 835.

PACHNER, A. M. D. and A. C. STEERE: The triad of neurological manifestations of Lyme disease: meningitis, cranial neuritis, and radiculoneuritis. Neurology 35 (1985) 47–53.

PALL, H. S., A. C. WILLIAMS, R. H. WARING and E. ELIAS: Motor neurone disease as manifestation of pesticide toxicity. Lancet 2 (1987) 685.

PALLIS, C., A. M. JONES and J. D. SPILLANE: Cervical spondylosis. Incidence and implications. Brain 77 (1954) 274–289.

PANSE, F.: Electrical trauma. In: P. J. Vinken, G. W. Bruyn, R. Braakman and H. L. Klawans (Eds), Handbook of Clinical Neurology, Vol. 32 ch 34. Amsterdam, North-Holland Publishers Co. (1975) 683–729.

PARNES, S., G. KARPATI, S. CARPENTER, N. M. K. NG YING KIN, L. S. WOLFE and L. SURANYI: Hexosaminidase-A deficiency presenting as atypical juvenile-onset spinal muscular atrophy. Arch. Neurol. 42 (1985) 1176–1180.

PARRY, G. J. and S. CLARKE: Pure motor neuropathy with multifocal conduction block masquerading as motor neuron disease. Muscle Nerve 8 (1985) 617.

PARRY, G. J. and S. CLARKE: Multifocal acquired demyelinating neuropathy masquerating as motor neuron disease. Muscle Nerve 11 (1988) 103–107.

PATCHELL, R. A. and J. B. POSNER: Neurologic complications of systemic cancer. Neurol. Clin. 3 (4) (1985) 729–750.

PATTEN, B. M.: Neuropathy and motor neuron syndromes associated with plasma cell disease. Acta Neurol. Scand. 69 (1984) 47–61.

PATTEN, B. M. and S. BROWN: Amyotrophic lateral sclerosis associated with aluminum intoxication. In: F. C. Rose and F. H. Norris (Eds), Amyotrophic Lateral Sclerosis; New Advances in Toxicology and Epidemiology. London, Smith-Gorden (1990) 205–209.

PATY, D. W. and C. POSER: Peripheral nerve signs and amyotrophy. In: C. Poser (Ed.), The Diagnosis of Multiple Sclerosis. Stuttgart, New York, Thieme Stratton Inc., (1984) 39–40.

PEARSON, C. M., W. F. RIMMER and H. M. MOMMAERTS: A metabolic myopathy due to absence of muscle phosphorylase. Am. J. Med. 30 (1961) 502–517.

PERL, T. M., L. BEDARD, T. KOSATSKY et al.: An outbreak of toxic encephalopathy caused by eating mussels contaminated with domoic acid. N. Engl. J. Med. 322 (1990) 1775–1780.

PESTRONK, A., R. ADAMS, L. CLAWSON, D. R. CORNBLATH, D. B. DRACHMAN, D. GRIFFIN and R. W. KUNCL: Multifocal motor neuropathy: Clinical features of patients with anti-GM1 ganglioside antibodies. Neurology 38 (Suppl. 1) (1988a) 251.

PESTRONK, A., D. R. CORNBLATH, A. A. ILYAS et al.: A treatable multifocal motor neuropathy with antibodies to GM_1 ganglioside. Ann. Neurol. 24 (1988b) 73–78.

PESTRONK, A., V. CHAUDHRY, E. L. FELDMAN, J. W. GRIFFIN, D. R. CORNBLATH, E. H. DENYS, M. GLASBERG, R. W. KUNCL, R. K. OLNEY and W. C. YEE: Lower motor neuron syndromes defined by patterns of weakness, nerve conduction abnormalities, and high titers of antiglycolipid antibodies: The Am. Neurol. Assoc. (1990) 316–326.

PETERS, H. A. and D. V. CLATANOFF: Spinal muscular atrophy secondary to macroglobulinemia. Reversal of symptoms with chlorambucil therapy. Neurology 18 (1968) 101–107.

PETIT, H., M. ROUSSEAUX, G. GOZET and M. MAZINGUE: Amyotrophie spinale cervico-thoracique pure par hydromyélie. Stabilisation par dérivation ventriculaire. Rev. Neurol. (Paris) 140 (1984) 144–147.

PETITO, C. K., B. A. NAVIA, E. S. CHO, B. D. JORDAN, D. C. GEORGE and R. W. PRICE: Vacuolar myelopathy pathologically resembling subacute combined degeneration in patients with the acquired immunodeficiency syndrome. N. Engl. J. Med. 312 (1985) 874–879.

PINSKY, L., M. H. FINLAYSON, I. LIBMAN and B. H. SCOTT: Familial amyotrophic lateral sclerosis with dementia: a second Canadian family: Clin. Genet. 7 (1975) 186–191.

POSER, C. M., D. W. PATY, L. SCHRENBERG et al.: New diagnostic criteria for multiple sclerosis: guidelines for protocols. Ann. Neurol. 13 (1983) 227–231.

POWER, C., B. G. WEINSHENKER, G. A. DEKABAN, G. C. EBERS, G. S. FRANCIS and G. P. A. RICE: HTLV-1 associated myelopathy in Canada. Can. J. Neurol. Sci. 16 (1989) 330–335.

RADTKE, R. A., A. ERWIN and C. W. ERWIN: Abnormal sensory evoked potentials in amyotrophic lateral sclerosis. Neurology 36 (1986) 796–801.

RAFALOWSKA, J. and B. WASOWICZ: Syringomyelia simulating amytrophic lateral sclerosis. Pol. Med. J. 7 (1968) 1214–1218.

RAMSAY, I. D.: Electromyography in thyrotoxicosis. Q. J. Med. 34 (1965) 255–267.

RAMSAY, I. D.: Muscle dysfunction in hyperthyroidism. Lancet 2 (1966) 931–934.

RAVERA, J. J., J. M. CERVIÑO, G. FERNANDEZ et al.: Two cases of Grave's disease with signs of a pyramidal lesion: Improvement in neurologic signs during treatment with antithyroid drugs. J. Clin. Endocrinol. 20 (1960) 876–880.

REAGAN, T. J., J. E. THOMAS and M. Y. COLBY: Chronic progressive radiation myelopathy. Its clinical aspects and differential diagnosis. J. Am. Med. Assoc. 103 (1968) 106–110.

REED, D. M. and L. T. KURLAND: Muscle fasciculation in a healthy population. Arch. Neurol. 9 (1963) 363–367.

REIK, L., L. SMITH, A. KHAN and W. NELSON: Demyelinating encephalopathy in Lyme disease. Neurology 35 (1985) 267–269.

REIK, L., W. BURGDORFER and J. O. DONALDSON: Neurologic abnormalities in Lyme disease without erythema chronicum migrans. Am. J. Med. 81 (1986) 73–78.

RIGGS, J. E., S. S. SCHOCHET and L. GUTMANN: Benign focal amyotrophy. Variant of chronic spinal muscular atrophy. Arch. Neurol. 41 (1984) 678–679.

ROSATI, G., I. AIELLO, R. TOLA and E. GRANIERI: Amyotrophic lateral sclerosis associated with thyrotoxicosis. Arch. Neurol. 37 (1980) 530–531.

ROSENBERG, R. N., J. B. GREEN and C. L. WHITE: Dominantly inherited dementia and parkinsonism with non-Alzheimer amyloid plaques. Ann. Neurol. 25 (1989) 152–815.

ROSENBLUM, B., E. H. OLDFIELD, J. L. DOPPMAN and G. DI-CHIRO: Spinal arteriovenous malformations: a comparison of dural arteriovenous fistulas and intradural AVM's in 81 patients. J. Neurosurg. 67 (1987) 795–802.

ROWLAND, L. P.: Motor neuron diseases: The clinical syndromes. In: D. W. Mulder (Ed.), The Diagnosis and Treatment of Amyotrophic Lateral Sclerosis. Boston, Houghton Mifflin (1980) 7–27.

ROWLAND, L. P.: Cramps, spasms and muscle stiffness. Rev. Neurol. 141 (1985) 261–273.

ROWLAND, L. P. and M. GREER: Toxoplasmic polymyositis. Neurology 11 (1961) 367–370.

ROWLAND, L. P., R. DEFENDINI, W. SHERMAN et al.: Macroglobulinemia with peripheral neuropathy simulating motor neuron disease. Ann. Neurol. 11 (1982) 532–536.

RUBIN, M., G. KARPATI, L. W. WOLFE, S. CARPENTER, M. H. KLAVINS and D. J. MAHURAN: Adult onset motor neuronopathy in the juvenile type of hexosaminidase A and B deficiency. J. Neurol. Sci. 87 (1988) 103–119.

RUSTAM, H., R. VON BURG, L. AMIN-ZAKI and S. EL HASSANI: Evidence for neuromuscular disorder in methylmercury poisoning. Arch. Environ. Health 30 (1975) 190.

SADIQ, S. A., F. P. THOMAS, K. KILIREAS et al.: The spectrum of neurologic disease associated with anti-GM I antibodies. Neurology 40 (1990) 1067–1072.

SADOWSKY, C. H., E. SACHS and J. OCHOA: Postradiation motor neuron syndrome. Arch. Neurol. 33 (1976) 786–787.

SALAZAR, A. M., C. L. MASTERS, D. C. GAJDUSEK and C. J. GIBBS: Syndromes of amyotrophic lateral sclerosis and dementia: Relation to transmissible Creutzfeldt-Jakob disease. Ann. Neurol. 14 (1983) 17–26.

SALAZAR-GRUESO, E. F., N. R. CASHMAN, R. MASELLI and R. P. ROOS: Upper motor neuron (UMN) findings in patients with antecedent poliomyelitis. Neurology 37 (Suppl. 1) (1987) 214.

SATOYOSHI, E. and K. YAMADA: Recurrent muscle spasm of central origin. Arch. Neurol. 16 (1976) 254–263.

SCHLIEP, G.: Syringomyelia and syringobulbia. In: P. J. Vinken, G. W. Bruyn and H. L. Klawans (Eds), Handbook of Clinical Neurology, Vol. 32, Ch. 10. Amsterdam, North-Holland Publishing Co. (1978) 255–327.

SCHMIDBAUER, M., C. MÜLLER, I. PODREKA, B. MAMOLI, E. SLUGA and L. DEECKE: Diffuse cerebrospinal gliomatosis presenting as motor neuron disease for two years. J. Neurol. Neurosurg. Psychiatry 52 (1989) 275–278.

SCHOLD, S. C., E. S. CHO, M. SOMASUNDARAM et al.: Subacute motor neuronopathy: A remote effect of lymphoma. Ann. Neurol. 5 (1979) 271–287.

SENANAYAKE, N. and L. KARALLIEDDE: Neurotoxic effects of organophosphate insecticides. N. Engl. J. Med. 316 (1987) 761.

SERGOTT, R. C., J. S. GLASER and L. J. BERGER: Simultaneous, bilateral diabetic ophthalmoplegia. Report of two cases and discussion of differential diagnosis. Ophthalmology 91 (1984) 18–22.

SERRADELL, A. P., J. R. GONZALEZ, J. M. C. TORRES, J. L. TRULL, J. O. BIELSA and A. U. ELOLA: Syndrome de sclérose latérale amyotrophique et hyperthyroïdie guérison sous antithyroïdiens. Rev. Neurol. Paris 146 (3) (1990) 219–220.

SERRATRICE, G., A. POU-SERRADEL, J. F. PELLISSIER, H. ROUX,

J. LAMARCO-CIVRO and J. POUGET: Chronic neurogenic quadriceps amyotrophies. J. Neurol. 232 (1985) 150–153.

SHAW, P. J., D. BATES and P. KENDALL-TAYLOR: Hyperthyroidism presenting as a pyramidal tract disease. Br. Med. J. 49 (1989) 49–50.

SHY, M. E., L. P. ROWLAND, T. SMITH et al.: Motor neuron disease and plasma cell dyscrasia. Neurology 36 (1986) 1429–1436.

SILFVERSKIOLD, B. P.: Polyneuritis hypoglycemia. Later peripheral paresis after hypoglycemic attacks in two insuloma patients. Acta. Med. Scand. 125 (1946) 502–504.

SIRIAUNGKUL, S. and S. SHUANGSHOTI: Syringomyelia associated with developmental anomalies and clinically pure motor presentation. J. Med. Assoc. 71 (1988) 345–349.

SIMPSON, J. A., D. A. SEATON and J. F. ADAMS: Response to treatment with chelating agents of anaemia, chronic encephalopathy, and myelopathy due to lead poisoning. J. Neurol. Neurosurg. Psychiatry 27 (1964) 536–541.

SMITH, H. V. and J. M. K. SPALDING: Outbreak of paralysis in Morocco due to orthocresyl phosphate poisoning. Lancet 2 (1959) 1019.

SMITH, R. and G. STERN: Myopathy, osteomalacia and hyperparathyroidism. Brain 90 (1967) 593–602.

SMITH, R. and G. STERN: Muscular weakness in osteomalacia and hypoparathyroidism. J. Neurol. Sci. 8 (1969) 511–520.

SOBREVILLA, L. A., M. L. GOODMAN and C. A. KANE: Demyelinating central nervous system disease, macular atrophy and acanthocytosis (Bassen-Kornzweig syndrome). Am. J. Med. 37 (1964) 821–828.

SOBUE, I., N. SAITO, M. IIDA and N. ANDO: Juvenile type of distal and segmental muscular atrophy of upper extremities. Ann. Neurol. 3 (1978) 429–432.

SPAANS, F.: Guillain-Barré syndrome with exclusively motor involvement. Electroenceph. clin. Neurophysiol. 61 (1985) 16P.

SPENCER, P. S., P. B. NUNN, J. HUGON, A. C. LUDOLPH, S. M. ROSS, D. N. ROY and R. C. ROBERTSON: Guam amyotrophic lateral sclerosis-parkinsonism-dementia linked to a plant excitant neurotoxin. Science 237 (1988) 517–522.

SPILLANE, J. D., P. W. NATHAN, R. E. KELLY et al.: Painful legs and moving toes. Brain 94 (1971) 541–556.

STAAL, A., J. D. MEERWALDT, K. J. VAN DONGEN, P. G. H. MULDER and H. V. M. BUSCH: Non-familial degenerative disease and atrophy of brainstem and cerebellum. Clinical and CT data in 47 patients. J. Neurol. Sci. 95 (1990) 259–269.

STARK, R. J., C. KENNARD and M. SWASH: Hand wasting in spondylotic high cord compression: An electromyographic study. Ann. Neurol. 9 (1981) 58–62.

STEELE, J. C.: Progressive supranuclear palsy. Brain 95 (1972) 693–704.

STEIN, B. M., N. E. LEEDS, J. M. TAVERAS and J. L. POOL: Meningiomas of the foramen magnum. J. Neurosurg. 20 (1963) 740–751.

STOHR, M., W. SCHLOTE, H. D. BUNDSCHU and H. E. REICHEN-

MILLER: Myopathica myotonica. J. Neurol. 210 (1975) 41–66.

STREIB, E. W.: Differential diagnosis of myotonic syndromes. Muscle Nerve 10 (1987) 603–615.

SUBRAMONY, S. H. and A. J. WILBOURN: Diabetic proximal neuropathy. Clinical and electromyographic studies. J. Neurol. Sci. 53 (1982) 293–304.

SYMONDS, C. P. and S. P. MEADOWS: Compression of the spinal cord in the region of the foramen magnum. With a note on the surgical approach by Julian Taylor. Brain 60 (1937) 52–84.

TEITELBAUM, J. S., R. J. ZATORRE, S. CARPENTER, D. GENDRON, A. C. EVANS, A. GJEDDE and N. R. CASHMAN: Neurologic sequelae of domoic acid intoxication due to the ingestion of contaminated mussels. N. Engl. J. Med. 322 (1990) 1781–1787.

THACKER, J. E., S. MISRA and B. C. KAYIYAR: Nerve conduction studies in upper limbs of patients with cervical spondylosis and motor neurone disease. Acta Neurol. Scand. 78 (1988) 45–48.

THOMAS, F. B., E. L. MAZZAFERRI and T. G. SKILLMANN: Apathetic thyrotoxicosis: A distinctive clinical and laboratory entity. Ann. Intern. Med. 72 (1970) 679–685.

THOMAS, J. E. and F. M. HOWARD: Segmental zoster paresis — a disease profile. Neurology 22 (1972) 459–466.

THOMAS, P. K.: Metabolic neuropathy. J. R. Coll. Phys. (Lond.) 7 (1973) 154.

THOMAS, P. K., R. W. H. WALKER, P. RUDGE et al.: Chronic demyelinating peripheral neuropathy associated with multifocal central nervous system demyelination. Brain 110 (1987) 53–76.

TOM, M. I. and J. C. RICHARDSON: Hypoglycaemia from islet cell tumor of pancreas with amyotrophy and cerebrospinal nerve cell changes. J. Neuropathol. Exp. Neurol. 10 (1951) 57–66.

TRUJILLO, M. J., A. CASTILLO, J. V. ESPANA et al.: Tetanus in the adult: Intensive care and management experience with 233 cases. Crit. Care Med. 8 (1980) 419–423.

TUCKER, T., R. D. DIX, C. KATZEN, R. L. DAVIS and J. W. SCHMIDLEY: Cytomegalovirus and herpes simplex virus ascending myelitis in a patient with acquired immune deficiency syndrome. Ann. Neurol. 18 (1985) 74–79.

TYLER, H. R.: Nonfamilial amyotrophy with dementia or multisystem degeneration and other neurological disorders. In: L. P. Rowland (Ed.), Human Motor Neuron Disease. Advances in Neurology, Vol. 36. New York, Raven Press (1982) 173–179.

VERNANT, J. C., G. BUISSON, R. BELLANCE, M. A. FRANCOIS, O. MADKA and O. ZAVARO: Pseudo amyotrophic lateral sclerosis, peripheral neuropathy and chronic polyradiculoneuritis in patients with HTLV-1 associated paraplegia. In: G. C. Roman, J. C. Vernant and M. Osame (Eds), HTLV-1 and the Nervous System. New York, Alan R. Liss Publishers (1989) 361–365.

VICALE, C. T.: The diagnostic features of a muscular syndrome resulting from hyperparathyroidism, osteomalacia owing to renal tubular acidosis, and perhaps to related disorders of calcium metabolism. Trans. Am. Neurol. Assoc. 74 (1949) 143–147.

WAIDA, R. S., C. SADAGOPAN, R. B. AMIN and H. V. SARDESAI: Neurological manifestations of organophosphorous insecticide poisoning. J. Neurol. Neurosurg. Psychiatry 37 (1974) 841–847.

WAISBREN, B. A., N. CASHMAN, R. F. SCHELL and R. JOHNSON: Borrelia burgdorferi antibodies and amyotrophic lateral sclerosis. Lancet 2 (1987) 332–333.

WHEELER, S. D. and J. OCHOA: Poliomyelitis-like syndrome associated with asthma. A case report and review of literature. Arch. Neurol. 37 (1980) 52–33.

WHITELEY, A. M., M. SWASH and H. URICH: Progressive encephalomyelitis with rigidity. Its relation to subacute myoclonic spinal neuronitis and to the stiff man syndrome. Brain 99 (1976) 27–42.

WIKSTROM, J., A. PAETAU, J. PALO, R. SULKAVA and M. HALTIA: Classic amyotrophic lateral sclerosis with dementia. Arch. Neurol. 39 (1982) 681–683.

WILKINSON, M.: Motor Neuron Disease and Cervical Spondylosis. In: F. H. Norris and L. T. Kurland (Eds), Contemporary Neurology Symposia, Vol. II. Motor Neuron Diseases: Research on Amyotrophic Lateral Sclerosis and Related Disorders. New York, Grune and Stratton (1969) 130–134.

WILLIAMS, C., M. A. KOZLOWSKI, D. R. HINTON and C. A. MILLER: Degeneration of spinocerebellar neurons in amyotrophic lateral sclerosis. Ann. Neurol. 27 (1990) 215–225.

WILLIAMS, E. R. and A. BRUFORD: Creatine phosphokinase in motor neuron disease. Clin. Chim. Acta 27 (1970) 53–56.

WILSON, S. A. K.: The amyotrophy of chronic lead poisoning: Amyotrophic lateral sclerosis of toxic origin. Rev. Neurol. Psychiatry 5 (1907) 441–445.

WOKKE, J. H. J., J. VAN GIJN, A. ELDERSON and G. STANEK: Chronic forms of Borrelia burgdorferi infection of the nervous system. Neurology 37 (1987a) 1031–1034.

WOKKE, J. H. J., J. DE KONING, G. STANEK and F. G. I. JENNEKENS: Chronic muscle weakness caused by Borrelia burgdorferi meningoradiculitis. Ann. Neurol. 22 (1987b) 389–392.

YAFFE, M. G., M. KABACK, M. GOLDBERG, J. MILES, H. ITABASHI, H. MCINTYRE and T. MOHANDAS: An amyotrophic lateral sclerosis-like syndrome with hexosaminidase-A deficiency: A new type of GM_2 gangliosidosis. Neurology 29 (1979) 611.

YOSHIMASU, F., M. YASUI, Y. YASE, S. IWATA, D. C. GAJDUSEK, C. J. GIBBS JR. and K. M. CHEN: Studies on amyotrophic lateral sclerosis by neutron activation analysis. 2: Comparative study of analytic results on Guam PD, Japanese ALS and Alzheimer Disease cases. Folia Psychiatr. Neurol. (Jpn.) 34 (1980) 75–82.

YOUNGER, D. S., S. CHOU, A. P. HAYS et al.: Primary lateral sclerosis. Arch. Neurol. 45 (1988) 1304–1307.

YOUNGER, D. S., L. P. ROWLAND, N. LATOV et al.: Motor neuron disease and amyotrophic lateral sclerosis: Relation of high CSF protein context to paraproteinemie and clinical syndromes. Neurology 40 (1990) 595–599.

YOUNGER, D. S., L. P. ROWLAND, N. LATOV, A. P. HAYS, D. J. LANGE, W. SHERMAN, G. INGHIRAMI, M. A. PESCI, D. M. KNOWLES, J. POWERS, J. R. MILLER, M. R. FETELL and R. E. LOVELACE: Lymphoma, Motor neuron diseases, and amyotrophic lateral sclerosis. Ann. Neurol. 29 (1991) 78–86.

ZILKHA, K. J.: Discussion on motor neuron disease. Proc. R. Soc. Med. 55 (1962) 1028–1029.

Handbook of Clinical Neurology, Vol. 15 (59): Diseases of the Motor System
J.M.B.V. de Jong, editor
© Elsevier Science Publishers B.V., 1991

CHAPTER 24

Differential diagnostic work-up of spastic paratetraplegia

G.W.BRUYN

Department of Neurology, Academic Hospital, State University Leiden, The Netherlands

As with most other syndromes in neurology, the diagnostic management of a case of spastic paraplegia ideally runs a three-stage course:

— establishment of the nature of dysfunction, called *functional diagnosis* in this case meaning the verification whether the spastic paraplegia is a pure pyramidal one or associated with other dysfunctions, whether sensory, coordinative, autonomic, cranial nerves, and higher nervous systemic.
— establishment of the site of the lesion, called *anatomical diagnosis*.
— defining the cause of the lesion, called *aetiological* or *nosological diagnosis*.

Whether this diagnostic course is run speedingly or haltingly, is successfully concluded or ends in failure, depends on a great many factors, of which quite a few are inaccessible to quantification. The cooperation of the patient and the veracity of his statements, the extent of the diagnostician's theoretical and empirical knowledge or his inclination to use reasoning before comfortably resorting to auxiliary techniques, or his degree of satisfaction with low-sophistication pattern-recognition, are elements in his (differential) diagnostic process as multiple as they are elusive.

Given the fact that one is dealing with a case of spastic paraplegia, the decisional flow-chart of the physician's acts (both of commission and omission), is quite apart from verification whether the case is one of 'pure' spastic paraplegia or one with associated signs and symptoms of descending or ascending spinal tracts other than the pyramidal, determined also by parameters other than the functional deficit sensu strictiori:

Did the deficit develop acutely or insidiously? Is it present in an infant or in an aged person? Has the patient perhaps been vocationally or geographically exposed to factors known to be capable of causing paraplegia? Is the patient male or female? Does the patient have a history or exhibit symptoms indicative of systemic, metabolic, infectious or genetic disease into which the paraplegia might nicely fit? All these are considerations, which, if timely entertained, may spare the patient unnecessary ordeal of (invasive) diagnostic procedures, such as EMG, spinal angiography, or phlebography, and myelography. Then again, modern scanning techniques that have revolutionised classic clinical neurology, taught us that even the most astute physician of undisputed diagnostic acumen not seldom fails in defining the anatomical and aetiological diagnosis, whereas the lesion may be fairly readily visualised by an NMR- or CT-procedure. Accordingly, the enviromental setting of the problem (are all advanced chemical, and radiological techniques available at hand?) also exerts an as unquantifiable as undeniable influence on the diagnostic elucidation of a case of spastic paraplegia. Even so, in some cases a fairly extended 'longitudinal' observation may be required for definitive elucidation.

The following pages will attempt to provide a sketch of rational diagnostic management, starting with the symptoms at the bedside. For brevity's sake, the field of 'cerebral palsy', i.e. diplegia spastica infantilis due to prae- and perinatal adversities, and the multitude of inborn errors of metabolism manifesting in the earliest years, will not be considered. Excellent textbooks on the neurology of inborn metabolic diseases provide the necessary detail (Adams and Lyon 1982; Stanbury and Wijngaarden 1983). In practically all these instances the clinician will be led to the correct diagnosis by data from history, gestation, delivery, and the symptoms associated with the paraplegia, such as mental retardation, seizures, ataxia, dystonia, and in some cases even by the odour of the patient's urine, as e.g. in maple syrup urine disease, isovaleric acidemia and β-methylcrotonyl-CoA carboxylase deficiency (see also Table 1).

The traditional aetiological classification comes in only at the third stage of work-up, though it must be kept in mind that neurological nosology never provides sharp boundaries and remains arbitrary. Some nosological categories are straightforward, such as trauma (whether mechanical, electrical, dysbaric, thermal, radiation or chemical) and infection (whether virus, bacteria, fungal or other), neoplasm, and intoxication. Others, contrariwise remain confusing: a vascular lesion may be either a bleed, a clot, a stenosis, a vascular inflammation, or a space occupying angioma or AVM; a metabolic lesion may be carential, such as avitaminosis, or dysregulatory, such as myxoedema and diabetes mellitus, but still might have to be grouped with genetic causes. For example Wilson's disease is genetic, metabolic, and a cellular intoxication by copper deposition. A similar reasoning obtains for Refsum's disease and quite a few enzymopathies. Disorders that used to be considered as genetic (Kuru, tropical spastic paraparesis) now are interpreted as intoxicatory and viral, respectively. The diagnostician should be aware, on the basis

of epidemiological data of the area he works in, where the best chances for aetiological diagnosis lie. Also, the bed-side neurologist will profit from awareness of modern data on the pyramidal tract, as synoptically reviewed by Davidoff (1990). Strickly speaking, the concept that spastic para/tetraplegia with certainty denotes a pyramidal tract lesion because such a lesion produces: (a) loss of voluntary movement; (b) paresis/paralysis; (c) spasticity; and (d) increased myotatic reflexes and extensor plantars, appears fallacious and hardly tenable anymore. Loss of finesse, speed, precision and agility seem to be the only characters of movement lost by a pyramidal lesion.

FUNCTIONAL DIAGNOSIS

The neurological examination in the stage of 'functional diagnosis' will verify whether the case is the result of exclusive interruption of the corticospinal tracts or whether there are signs and symptoms of associated damage.

As early as this moment, the physician has taken the history and therefore possesses the valuable information as to whether the paraplegia was installed acutely or developed subacutely or even insidiously. That information contains cues for both anatomic and aetiologic diagnosis. The majority of acute instances are the result from exogenous trauma to spine and spinal cord, followed by permanent abolition of all motor, sensory and reflex function below a certain level (anatomical diagnosis) and a transient flaccid paraplegia (diaschizis or spinal shock) which reverts into spastic paraplegia in a matter of days. In a minority of instances no such history of sudden exogenous injury can be obtained, and diagnostic considerations should entertain such clinical entities as e.g. intraspinal bleed from an angioma or AVM or even such rare events as fibrocartilaginous nucleus pulposus embolism (Bots et al. 1981; Kestle et al. 1989), or epidural (or subdural) hematoma which presents a highly characteristic picture and usually is subacute but may be acute (Bruyn and Bosma 1976).

Chronic progressive spastic paraplegia offers a classic tableau: gait-difficulties ranging from fatigue on effort to frequent falls and stumbling to complete impossibility of ambulance, a bilateral

TABLE 1
Disorders that may include spastic para/tetraplegia in infancy and childhood.

Aminoacid metabolism
  Maple syrup urine disease
  Homocystinuria
  Argininaemia
  Pyroglutamic acidaemia
  Non-ketotic hyperglycinaemia
  Homocarnosinosis

Lipid metabolism
  Bassen-Kornzweig α-β-lipoproteinaemia
  Cerebrotendinous xanthomatosis (van Bogaert)
  GM_1 gangliosidosis (Tay-Sachs)
  GM_2 gangliosidosis (late infantile/juvenile)
  Krabbe globloid cell leukodystrophy
  Niemann-Pick A, C/D
  Farber lipogranulomatosis
  Metachromatic leukodystrophy
  Austin multiple sulfatase deficiency
  Sudanophilic leukodystrophy
  Pelizaeus-Merzbacher disease
  Adreno(myelo)leuko(neuro) dystrophy
  Cockayne disease
  Orthochromatic leukodystrophy
  Leptomeningeal angiomatosis
  Oculomeningeal angiomatosis
  Alexander's disease
  Neuronal ceroid lipofuscinosis
  Seitelberger neuroaxonal dystrophy

(Poly) Hexose metabolism, incl. mucopolysaccharidoses
  Hurler disease, Hunter disease
  Mannosidosis
  Fucosidosis
  Leigh's necrotic encephalomyelopathy
  Canavan-van Bogaert-Bertrand's disease?
  Lafora body disease? (progr. myoclonus epilepsy)
  Mucolipidosis V
  Sialuria (French type)
    (Finnish type, Salla disease)
  Sulfite oxidase deficiency
  Triosephosphate isomerase deficiency

Infectious
  Viral: tropical spastic paraparesis
    Behçet disease
    Vogt-Koyanagi disease

Organic acids
  glutaric aciduria I

Purine metabolism
  Lesch-Nyhan syndrome

Varia
  Hyperbilirubinaemia
  Pseudo(pseudo) hypoparathyroidism
  Xeroderma pigmentosum (de Sanctis-Cacchione)
  Neuro-amyloidosis, type 7 (Ohio oculo-
    leptomen ingeal type)
  T-cell deficient ataxic diplegia
  Sjögren-Larssen syndrome
  Cerebral palsy, spastic/athetoid
  Kallmann syndrome
  Hereditary spastic paraplegia
    Pubertas praecox
    Deafness
    Hyperkeratosis
    Amyotrophy
    Macular or retinal degeneration
    Syndactyly
    Sensory neuropathy
    Vitiligo + grey hair (Lison 1981)
    Lingua scrotalis, ophthalmoplegia (Levic 1975)
    Dementia
    Optic atrophy
    Renal dysfunction (Fanconi syndrome)
      (Troyer syndrome)
  Laubenthal syndrome (Troost 1984)
  Hallervorden-Spatz disease
  Dentatorubropallidonigral degeneration
  Striatopallidal degeneration
  Trichorhexis nodosa
  Charlevoix-Saguenay ataxia
  Sanger-Brown ataxia
  Boller-Segarra spinopontine ataxia, OPCA
    (Machado-Joseph type)
  Eales disease
  Cerebro-osteo-nephrodysplasia
  Oculocerebral syndrome (Cross et al. 1967; Balci
    1974)
  Oculorenal cerebellar syndrome
  Erythrophagocytic lymphohistiocytosis
  Toluene intoxication

Developmental anomalies and osseous disease
  Smith-Lemli-Opitz
  Corpus callosum agenesis
  Aqueductal stenosis
  Syringomyelia/hyperkeratosis
  Intradural enterogenous cyst
  Hereditary multiple exostoses
  Klippel-Feil syndrome
  Dandy Walker syndrome
  Spinal angioma and AVM
  Kjellin syndrome
  Rhizomelic chondrodysplasia punctata (Conrad)

circumductional bringing forward of the legs with the fibular side of the feet dragging along the floor, stiffness of the legs, muscular hypertonia with clasp-knife resistance, increased muscle-stretch reflexes of adductor, quadriceps and gastrocnemius with (sub-) clonus, crowned by the extensor plantar reflex and a more or less wide gamma of Babinski-variants, and not infrequently most painful muscle spasms ('brain stem fits') together constitute the basic elements of anatomical diagnosis. Strikingly, not so much the 'plegia', the loss of muscle strength or force is impaired, but its voluntary initiation and control. In minor degree paraplegia, signs and symptoms may become unequivocal after the patient has taken a hot bath or been put to the effort of exercise. In these chronic-stationary or slowly-progressive cases one does rarely observe symptoms of disinhibited autonomic reflex-activity such as are notorious in acute traumatic cases. Nonetheless, livid patterned discolouration of the feet, swelling of the lower legs and feet in depending positions of the legs are often observed. Slight difficulties with micturition and defaecation may be present. In a number of instances the muscle-stretch reflexes appear abolished, either due to extreme spasticity or contractures.

One should, particularly in the presence of sphincter problems, remain alert as to the possibility of a conus syndrome, or, in the presence of deranged position and vibration sense, consider the possibility of vitamin B-12 deficiency. Also in some cases the examination yields the picture of a combined upper and lower motor neuron deficit, in which not only reflexes are abolished but muscle atrophy and fasciculations occur and extensor plantar responses are unobtainable. Quite apart from the intriguing problem of 'central (cortical) amyotrophy' as seen in the interosseus muscles in cases of hemiplegia, such combinations are known in ALS, cervical spondylarthrotic myelopathy, adrenomyeloneuropathy, Pelizaeus-Merzbacher disease, syringomyelia and quite a few other entities.

Fortunately, the diagnostician's burden is alleviated somewhat by the fact the 'pure' pyramidal paraplegia constitute the minority group of instances. This considerably reduces the requisite

acts to be carried out in the second stage, the stage of anatomical diagnosis.

CAUSATIVE LESION

With respect to the site of the causative lesion, the entire length of the pyramidal tracts theoretically would qualify for consideration were it not for the presence of additional symptoms from adjacent structures, the metameric segmental structuring of the spinal cord, and intrinsic level-tied variations in the 'pyramidal symptom-complex'. Cortical stimulation and ablation experiments (Bucy 1957; Bucy and Keplinger 1961) have provided arguments that lesions of area 4 produce loss of fine movements, some contralateral weakness and hypotonia rather than spasticity. Lesions of the premotor area (area 6α) tend to render the pyramidal weakness spastic. Why capsular lesions have predominant spasticity and (pes) peduncular lesions are flaccid (Bucy and Keplinger 1961) has never been satisfactorily explained. Pertinent research over the last decades has shown that the traditional views on the pyramidal tract still taught unmodified in present-day curriculi have to be thoroughly revised (Davidoff 1990; Iwatsubo et al. 1990).

The elicitation of so-called regression-reflexes (grasping, groping, sucking, snouting, head retraction) will provide the diagnostician with a firm footing as to the telencephalic localisation of the cause of spastic para- or tetraplegia. Moreover, the concurrent presence of pseudobulbar palsy removes any further doubt as to the site of the lesion.

The often quoted example of a falcine meningioma compressing both premotor Brodmann area 4 and producing pure spastic paraplegia is one disorder the present writer has not come across. One might expect bilateral cortical spastic paraplegia to be observed frequently in thrombosis or thrombophlebitis of the superior sagittal sinus, developing either acutely or insidiously on the basis of predisposing or systemic illness, otorhinological infection, or the puerperium. However, the acute form mainly manifests with faciobrachial weakness, focal seizures, obtundation, the notorious 'oedème en casque' (caput Medusae), and sanguinous CSF, while the chronic form shows headache and papilledema. Bilateral subdural haematoma (De Preux and Stephano 1981) or bilateral paracentral angioma (Arsini and Decu 1979) may produce paraplegia but are exceptional and not without associated localising symptoms. Telencephalic bilateral corticospinal damage without involvement of adjacent connexions producing pseudo-bulbar palsy (see the 'opercular syndrome', Bruyn and Gathier 1969, Mao et al. 1989) must be exquisitely rare. However, in (obstructive) internal hydrocephalus progressive widening of the lateral ventricles will affect chiefly the ventricle-near corticospinal tract fibres coming from the leg-areas whereas those from the arm/face/hand areas largely escape. This change may well be the substrate of the gait-apraxia in normal presssure hydrocephalus, but the symptom does not exist in exclusivity: there are associated symptoms that aid the diagnostician.

In general, one safely courts the assumption that the pyramidal tract fans out to such a degree within the hemispheres, that telencephalic process to produce spastic paraplegia must of necessity be generalised and diffuse, and as a consequence involve other telencephalic structures, and most likely to be of vasuclar or metabolic nature. Where the pyramidal tract runs in much more 'condensed' bundles (capsula interna, pedunculi), a localised lesion limited-in-size, may be expected to knock out both of them but, then again, one has already left the wide telencephalic space for the crowdedness of the upper brainstem (diencephalon, mesencephalon) where even small intraneuraxial lesions damaging the corticospinal tracts will involve other tracts or nuclei that are in close proximity, and produce tell-tale symptoms as to the site of the lesion. Essentially the same pertains to the pontine and medulla oblongata course of the pyramidal fibres.

In short, cerebral paraplegia (tetraplegia) carries as many distinctive features as does spinal. Just like the familiar example of it at the initial extremity of life's scale, cerebral palsy, with its characteristic 'extra'-pyramidal overtones, a well-known (if insufficiently understood) example at the terminal end of life, senile cerebral 'paraplégie en flexion', reveals the message. Its tell-tale fetal curled-up position with ultimate contractures is heralded by a shuffling, wide-based, hesitant, and slow gait executed with tiny steps ('démarche à petits pas') not infrequently developing into the 'motor ritardando' of putaminal-lesioned cats, in neurology known as parkinsonian freezing, the body is moved like a solid block; turning about becomes next to impossible, posture becomes flexed, standing still ends with backward falling (Bruns' retropulsive 'frontal ataxia'), the examiner cannot persuade the patient to relax his muscles and is not surprised upon any passive manoeuvering by him of the patient's limbs to encounter paratonia ('Gegenhalten'). Earlier neurologists interpreted this disorder as 'frontal'. The striking parkinsonian, i.e. putaminal, components in this senile cerebral spastic paraplegia, ultimately frequently (but not obligatorily) associated with dementia, betrays the basal-ganglionic involvement. The present writer must confess at this point to be unable to offer a coherent anatomical or physiological explanation why certain cases develop into a paraplégie en flexion and others in paraplégie en extension; the first is observed particularly in cerebral and the last-named usually in below-pontine-level lesions. In both, the lateral and medial reticular formation may be involved, and therefore flexor- and extensor reflexes are released.

Bilateral capsular lesions are rare. I have observed them in the MERRF-syndrome, as well as in hereditary dystonic tetraplegia (Bruyn-Went's disease 1964), a condition also likely to be of mitochondrial dys-enzymatic origin; the CT-scan in these cases reveals typical slit-like hypodense lesions in both putamina. Curiously, in quite a few of the 'heredodegenerative' disorders, such as Hallervorden-Spatz disease, familial ALS, or Strümpell-Lorrain's hereditary paraplegia, the degeneration of the corticospinal tracts can be demonstrated to occur above the pontine level only exceptionally. The abundant associated symptoms in the neurolipidoses and leuko- and poliodystrophies (Krabbe's globoid cell leukodystrophy, Alexander's disease, Pelizaeus-Merzbacher's disease, Gaucher's disease, cerebrotendinous xanthomatosis, Bassen-Kornzweig's disease) leave no doubt as to the telencephalic sedes morbi.

Neurological tradition holds that in capsular lesions facial, lingual and occasionally brachial weakness predominates over crural paresis. Spasticity develops rapidly. Since Fisher's work, one accepts that pure motor hemiplegia is most often due to lacunar infarctions in the middle to posterior part of the internal capsule, but can scarcely be distinguished on clinical evidence from lacunar pontine pure motor hemiplegia (Orgogozo and Bogousolavsky 1989). Recent work has shown that the paralysis may be either evenly distributed over face, arm and leg ('proportional') or not ('non-proportional'). Bilateral status lacunaris in the capsulae internae is a major factor underlying pseudobulbar palsy.

The location of the pyramidal tract within the capsula interna varies individually and also at various levels above the 'Frankfurter Horizontale' (Hanaway and Young 1977; Ross 1980). Just below the cella media of the lateral ventricles the tract is located in the genu capsulae internae, while still lower, lateral to the mass of the thalamus, it runs in the middle to posterior part of the posterior crus. There it is accompanied by a mass of descending area 6 and parietal (area 3) fibres.

Bilateral peduncular lesions are definitely exceptional. One may come across them in chordoma of the clivus Blumenbachi, and in such instances basal cranial nerve deficit and increased intracranial pressure again will put the diagnostic process on the right track. Moveover, just as in the case of medulla oblongata lesions of the pyramidal tracts, the plegia is flaccid rather than spastic, as was pointed out by Bucy et al. (1966), Lawrence and Kuypers (1968), and Gilman and Marco (1971). Bilateral crural compression such as may ultimately occur in supratentorial mass shift with transtentorial (hiatus-) herniation resulting in mesencephalic (supra-rubral) 'functional transection' as evidenced by coma, extensor-pronation spasms, total oculomotor palsy, etc., the associated signs betray the level and the cause of the syndrome.

At a still more caudal level, pontine processes rarely are so symmetric as to produce para- or tetraplegia, and if they are, the accompanying symptomatology reveals the level. Central pontine myelinolysis produces flaccid (and seldom spastic) paraplegia plus pseudobulbar palsy, deranged consciousness and pupillary signs. The same obtains for the basilar artery syndrome and

locked-in-syndrome, in which sensory impairment is added to the semeiology. The supranuclear, facial and lingual paralysis predominates over the crural palsy in pontine lesions. Cranial nerve deficit is decidedly infrequent with this location. The absence of spastic paralysis in itself is no argument against a pontine lesion. In brainstem encephalitis, that used to be due to von Economo's encephalitis lethargica or bulbar polymyelitis, it is nowadays seen more often as a result of varicella-zoster, Epstein-Barr virus, Japanese B/St. Louis/tick-borne and parainfectious encephalitides or as a Guillain-Barré type (Miller Fisher syndrome) the symptoms of bulbar weakness, obtundation and ataxia with ophthalmoplegia and respiratory difficulties do rarely if ever include limb palsy (Al-Din et al. 1982). On the other hand, hepatic (porto-caval-shunt) encephalopathy that usually manifests with a typical picture of metabolic brain disease even to the degree of coma, may (if exceptional) predominantly manifest with progressive spastic paraplegia (Zieve et al. 1960).

The rapid development of coma, respiratory dysregulation, ocular bobbing, attacks of tonic extensor spasms ('brain stem fits') with fatal outcome within 48 h used to be a reliable set of data for pontine hemorrhage in the pre-CT scan era. Smaller pontine hemorrhages and infarctions with a much less grim prognosis (but also with less unequivocally clear symptoms) are nowadays diagnosed fairly rapidly by virtue of neuroradiological procedures. In (para-) median dorsal pontine lesions spastic paraplegia may occur. The cortico-spinal fibres for the upper limbs occupy a more lateroventral position within the basis pontis.

Spastic paraplegia as part of a symptomcomplex produced by intrinsic medulla oblongata lesions is decidedly rare. In the multitude of brain stem syndromes that proliferated about the turn of the 19th century, ranging from Weber's and Benedikt's to Dejerine's and Babinski-Nageotte's have been subjected to critical analysis in the post-World War II-decades and simplified to such a degree, i.e. median, paramedian, lateral and dorsal, that no qualified neurologist will consider their diagnosis as beset with unsurmountable difficulties. For an intrinsic lesion at this neuraxis-level to produce spastic paraplegia, a *bilateral* medial medullary lesion (Ho and Meyer 1981) should have to be present; such a rare lesion is almost always vascular and will extend to involve the medial lemniscus and hypoglossal nucleus and nerve. Hemiplegia à bascule developing into tetraplegia with lingual palsy and atrophy, associated with impairment of gnostic sensation are the classical features. Bilateral medial medullary syndrome includes non-explainable non-crossed facial palsy in 50% of instances, which can be a confusing, false-localising sign. In contrast to Wallenberg's lateral syndrome the medial syndrome when it occurs is quite often bilateral. Among the brain stem gliomas spastic para/tetraplegia is a common but fairly late symptom of the rapidly developing semeiology, that usually begins with facial palsy and spasms, ataxia, mental changes and other cranial nerve deficit in a youngster. The hereditary olivo- and spino-cerebellar diseases that involve the pyramidal tract are extensively dealt with in other chapters of this volume.

In lesions of the foramen magnum (mostly being meningiomas and neurilemmomas), nucho-occipital pains, radiating to the head, shoulders, arms and down the spine with progressive paresis of the arms and subsequently the legs, atrophy of muscles well caudal to the lowest level of the tumor (intrinsic muscles of the hands, intercostal, and even lumbar muscles), Lhermitte's sign, bizarre cold sensations in the neck and thorax and lower cranial nerve palsies, sometimes spontaneously but transiently remitting, point the way to correct localisation (Symonds and Meadows 1937; Cohen 1974). Quite a few cases are referred because of suspected 'multiple sclerosis' (Meyer et al. 1984).

In the localisation of a spinal cause for spastic para/tetraplegia, the physician is on a fairly paved road at the bedside, because of the metameric organisation of the final motor path.

It hardly needs emphasis that the lowest spinal level for a lesion to produce full spastic tetraplegia must be the C4–C5 level, the presence of diaphragmatic respiration (C4), and myotatic reflexes (biceps C5/6; triceps C7/8) being useful aids in examination. In a triplegia one should recall that the pyramidal decussation for the

upper limbs occurs slightly more rostral than the one for the legs at C1 level. In acute (traumatic) cases, during the stage of spinal diaschizis, the only real localisation difficulties constitute the conus terminalis syndrome and the cauda equina syndrome, the semeiology of which may closely mimic that of acute spinal paraplegia at higher levels.

Examination of the spine may be helpful, particularly in chronic progressive cases; a change in the perception of percussion of the spinous processes, a gibbus-formation, a kyphoscoliosis, or a slight torticollis may as well serve to define the lesion's level as does associated anaesthesia, and do better when no sensory deficit is found. Even the presence of scapular malformation (Sprengel's deformity) is helpful in e.g. Klippel-Feil syndrome with spastic paraplegia, quite apart from the associated mirror movements and syringomyelic symptoms (Critchley 1927; Du Toit 1931; Guillain and Mollaret 1931). Lesions involving thoracic levels can be defined by observing Beevor's sign, and by brisk myotatic abdominal reflexes in contradistinction to the extrinsic cutaneous abdominals that are abolished at the appropriate étage. A considerable variety of methods and manoeuvres have been developed in the pre-War era to enable the neurologist to define the location of a lesion at bedside-examination. Of the delightful standard-works on this art (Monrad-Krohn 1954; de Jong 1958; Wartenberg 1945,1952) modern neurology has become either ignorant or neglectful.

In subacute and chronic-progressive cases of spastic paraplegia, that are likely to be caused by neoplasm, infection, genetic-metabolic disease, or intoxication, the examiner will have asked — in taking the history — about the presence of radicular pains ('pseudo-neuralgic prelude') and verify their presence during examination. These neuralgic pains produced by compression (or rather stretch) of spinal nerve roots due to the growth of an extra- or intradural space-occupying process ultimately leading to compressio medullae, may be regarded as the best indicators of the lesion's spinal level. They are band-like unilateral pains irradiating from the spine around the trunk, and can be provoked by asking the patient to sneeze or cough, or exert effort with the glottis closed (pushing, lifting a weight, straining at stool) thereby raising the intraspinal pressure. Within the distribution of the revelant root, (discrete) muscle atrophy and fasciculation may develop, which may mislead the examiner. The pseudoneuralgic pains may increase in the recumbent position at night. Another sign which is frequently obtainable in cervical and cervico-thoracic lesions, of course, is Lhermitte's sign, in which electric-like paraesthesias irradiating from the spine caudad and towards the acra of the limbs are produced by flexion of the head.

In the majority of spinal spastic paraplegias, the associated sensory deficit reaching up to a definite 'transverse' line is the easiest help in establishing the lesion's site, whether in the form of a Brown-Séquard syndrome or a complete transection. In addition, the presence of a 'disso-ciated sensibility impairment' permits the exam-iner to localise the lesion within a transverse spinal cord plane.

A number of findings apparently discrepant with the established pyramidal lesion may be confusing such as:

— the presence of fasciculations and muscle atrophy. If these are occlusively observed in the intrinsic muscles of the hand, one should entertain the diagnosis of spondylarthrotic myelopathy or syringomyelia, or foramen magnum pathology.
— diffusely distributed fasciculations and the absence of unequivocally disinhibited myo-tatic reflexes or of extensor plantar reflexes, such as is likely to be found in amyotrophic lateral sclerosis.
— unobtainable myotatic reflexes in the legs (whether knee- or Achilles-jerk) in the absence of spasticity or massive pareses, yet combined with marked rectal- and urinary sphincter trouble, early decubitus ulcers, and occasion-ally extensor plantar reflexes. This indicates an epiconus, or a conus terminalis syndrome, usually of traumatic, vascular, or neoplastic (Norstrom et al. 1961; Nassar et al. 1968) nature and of extremely rare occurrence. The conus terminalis (ending at the L1–2 disc) is sheathlike enclosed by the L3,4,5,S1 spinal nerve root of the cauda equina. Associated

cauda equina deficit may further obscure the issue (see Table 2).

Auxiliary diagnostic techniques

The essentially relevant data stemming from the history, the functional diagnosis and the anatomi-cal diagnosis constitute the platform on which the clinician selects from the palette of auxiliary techniques that synchronously serve the purposes of defining to optimal detail the site and the nature of the causative lesion. In this selection the 'cost-benefit' principle is the prevailing axi-oma, 'cost' including the risk of the subsequent diagnostic procedure for the patient. That risk cannot be simplified as the decision remains as to whether or not to opt for an invasive or non-invasive technique. Any clinical elucidation of a suspected metabolic, systemic or intoxicatory disorder underlying the spastic para/tetraplegia will necessitate at least a vein-puncture to obtain blood for laboratory work-up, and may go as far as bone-marrow puncture, or biopsy of the skin, liver, and rectal mucosa. In addition, clinical neurophysiological methods, particularly EMG (to verify whether associated localised or general-ised neurogenic muscle involvement is present), and to a lesser extent evoked potential methods (VEP, BAEP, SSEP, etc) are either invasive or at least uncomfortable for the patient, while a contrast-enhanced CT-scan also requires vena-puncture.

Without advocating a generally valid diagnostic decisional flow-chart which can never satisfy all individual instances, the brief outline below should simply serve as a general sketch. Auxiliary procedures and techniques can be grosso modo distinguished into those of clinical neurophysiol-ogy, of the chemical laboratory (including CSF examination), and of roentgenology.

Inasmuch as the great majority of instances of spastic para/tetraplegia are caused by trauma, followed by space-occupying processes and vascu-lar disease one of the procedures to select first because of its high informational yield is plain roentgenograms (including plain tomography if necessary) of the spine and skull, according to the surmised lesion-level. These will identify e.g. systemic bone disease (such as Paget's, or

TABLE 2
Caudal cord syndromes.

	Epiconus (L/5–S2/3)	Conus* (S3–Co1)
micturition	retention	ischuria paradoxa
defaecation	retention	incontinence
potentia coeundi		
— erection	+	−
— ejaculation	−	−
	(dissociated impotence)	
sensory deficit	lateral footsole	saddle-type
(often dissociated)	laterocaudal lower leg	
	penis	
motor deficit	m. quadriceps femoris	m. flex. halluc. long
	mm. piriformis	m. flex. halluc. brevis
	mm. abductor.	m. flex. digit. long.
	hamstrings	
	mm. peronei	
Reflexes		
— Patellar	+	+
— Achilles	−	+
— Plantar	indifferent	flexor
— anal	±	−
— bulbocavernous	±	−

* The corticospinal tract does not directly innervate Onufrowicz' nucleus (Iwatsubo et al. 1990).

osteochondroplasia, or compressive spondylarthrosis), acquired bone disease (fracture, vertebral haemangioma, cervical or thoracic slipped disks, gibbus-formation, rheumatoid C1/2 luxation, discitis, spondyl(omyel)itis, vertebral metastases ('vertèbre d'ivoire' and vertebral collapse), or tumoral bone/peduncle erosion ('vertèbre borgne'), solitary myeloma, vertebral scalloping, enlarged intervertebral foramina, etc.), congenital and developmental anomalies (basilar impression, Klippel-Feil syndrome, metameric displacements, kyphoscolioses, diastematomyelia, canal stenosis, multiple exostoses, spina bifida), not to mention the variety of visualisable cranial lesions that may substantially serve to clinch the diagnosis.

The diagnostic-contributory value of CSF examination has lost much of its traditional glamour. In certain instances and disorder-categories it is even contraindicated (intracranial hypertension due to space-occupying processes) or may worsen the patient's deficit (spinal epi/subdural hematoma, subtransverse traumatic or neoplastic cord-lesion), and in others it is not so much contraindicated or hazardous as devoid of rational meaning as it will not contribute in the slightest way to diagnosis (such as in hereditary spastic paraplegia and its variants, certain types of spinocerebellar disease, metabolic/deficiency or intoxication disorders). The value of lumbar puncture, and in some way of cisterna magna puncture, lies only in its yield of physical and chemical-cytological data. It enables the neurologist to diagnose the pressure and pressure-gradient of the CSF by recording opening-pressure, pulsations and waves, closing-pressure, response to abdominal pressure and Queckenstedt-Stookey manoeuvre, including the Kaplan-variants of flexed and extended head of the patient. This is of decisive influence in proving the presence of an intrathecal block due to space-occupying lesions. It also is of considerable contributory value in the definition of infectious, demyelinating and immunologic-systemic disease by yielding information on the presence of microorganisms, the number and composition of cells, the protein content and its composition, and occasionally, glucose and chloride content. Finally, modern techniques such as culture of microorganisms, antigen-antibody titer assays, immunofluorescent staining, isoelectric focusing of protein fractions, and spectrophotometric assay of methaemoglobin, bilirubin and oxyhaemoglobin may adduce decisive diagnostic data.

Depending on the case at hand, one may often spare the patient at least one lumbar or cisterna magna puncture by using the in situ needle for both obtaining CSF samples and introducing contrast medium for (dynamic) myelo-radiculography. Myelography permits one to visualise vermicular non-filling indicative of an angioma, the parsimonious pockets of arachnoiditis, the swelling of the cord in intramedullary tumor or syringomyelia, the arrest or block of the intrathecal contrast-transit in both intra- and extramedullary space occupying lesions, with the qualification that establishment of one (cranial or caudal) pole of the lesion in such a case is effected, but definition of the other pole requires a second myelography. The availability of immediate neurosurgical intervention is one of the aspects that will influence the decision whether or not to carry out myelography.

In this respect again, modern methods have revolutionised old-fashioned neurology, inasmuch as the non-invasive method of NMR is diagnostically so superior to any other auxiliary technique, that the use of plain X-rays and NMR of the spine and contents will suffice in practically all instances. Developmental anomalies, cysts, granulomatous infections, space-occupying processes from malignancy to lipoma to slipped disc, tethered cord, demyelination, syringomyelia, constitute the wide range of lesions readily accessible for NMR-diagnosis. Only the precise definition of the feeders and drainers of an angioma or AVM as required by the neurosurgeon necessitates ultimate spinal angiography, again a technique that should — for experience's sake — remain restricted to a few neuroradiological centres.

Nosology

The aetiological stage has already been entered in the activation of auxiliary diagnostic techniques. There is scarcely sense in attempting to provide an exhaustive list of potential causes of spastic paraplegia. Even a limited list, with traditional nosological categories containing diseases ranked according to their frequency will not reflect reality always and anywhere: rare causes in one country or region may belong to the most frequent in others, while the apparent sharp lines dividing the categories are not all that absolute. In principle, any disease or combination of diseases may involve the CNS and produce spastic paraplegia.

Trauma

By far the greater majority of paraplegias in the world are due to trauma, whether sustained in or on the vehicles of transport in our modern world, in industry, or in sports, and, alas, in fights and wars. These are mostly (hyper-) acute, concern those belonging to the active younger age groups, offer the characteristic tableau of more or less complete cord transection (unless, and rarely, due to bilateral frontal impression fractures or bilateral subdural hematomas (Benvenuti et al. 1987)), and have been amply reviewed in Volumes 16–19 of this Handbook.

The topic of cranio-cerebrospinal injury is, of course, not exhausted with mechanical-ballistic accidents. Electro-trauma, both industrial and natural (lightning), is known to damage the CNS, as reviewed by Panse (1975). Electrotrauma of the spinal cord may produce not only the well-known ischaemic and transient keraunoparalysis but also tetra- or paraplegia (Kanitkar and Roberts 1988).

One should also keep in mind that ultra-short electromagnetic radiation exerts harmful effects on the nervous tissue in spite of its rather high resistance to it. Iatrogenic radiation myelopathy with spastic paraplegia (as usually in intramedullary lesions, associated with dissociated sensibility impairment) has been reviewed by Palmer (1976) and recently by Berlit et al (1987).

Another type of physical injury, dysbarism, or caisson disease, notorious for its symptoms called the 'bends' and the 'chokes', readily damages the spinal cord, often with the picture of combined degeneration, but occcasionally as a spastic paraplegia (Synek and Glasgow 1985; Calder et al. 1989).

In thermo-trauma (insolation, heat stroke) of the spinal cord, the paraplegia is flaccid, though in the company of extensor plantar signs (Yaqub 1987).

This nosological category falls outside the scope of this chapter of a volume on (heredo-) 'degenerative' systemic motor diseases.

Space-occupying lesions

The category of neoplastic processes, whether primary solitary or secondary multiple (metastatic) comes next in frequency, if one includes all other space-occupying processes, a sleight-of-hand natural to a discipline of which the object of study is enclosed in a rigid bony box and pipe putting constraints on their contents. This widened group ranges from solid tumors to fluid ones, such as hematomas, whether intraparenchymal, sub- or epidural. In the spinal tumors the presenting symptom most often is pain, gradually followed by progressive paresis and sensory impairment and urinary retention. Most of them are extramedullary and develop before the age of 70 (Fried et al. 1988), some, such as meningiomas, occur predominantly in females. Some may present rather suddenly, and these have a less favourable prognosis, e.g. sub- and epidural spinal hematoma (Bruyn and Bosma 1976; Servadei et al. 1987). Some of these lesions show quite an aberrant course, e.g. protracted over nearly 20 years in an intramedullary ganglioglioma (Pialat et al. 1987), protracted (20 years) and relapsing in extradural arachnoid cyst (Labauge et al. 1989) or in herniated thoracic disc (Besson et al. 1988), which mostly occurs below Th.8 and may be subacute after a fall (Hedge and Staas 1988) in which spinal cord ischemia plays a prominent part (Fisher 1986) and hyperacute in ependymal cyst (Wackym et al. 1988), or subacute in Ewing sarcoma (Ben-Meir et al. 1989), hyperparathyroid brown tumor (Yokota et al. 1989) or thyroid carcinoma (Ginsberg et al. 1987), or lymphocytic leukemia (Lustman et al. 1988). Clearly, the neurologist cannot afford to remain unaware of systemic causes in his aetiological considerations, as exemplified by spinal cord compression in extramedullary haematopoiesis in myelofibrosis, thalassemia, or sickle cell anemia (Ammoumi et al. 1975; Mann et al. 1987; Thompson 1988), corticosteroid therapy

management (in sarcoidosis; Leys et al. 1986) and its complications [spinal epidural lipomatosis (Cénac et al. 1987; Kaplan et al. 1989)]. Even if the patient is known to have malignant disease elsewhere, the spastic paraplegia need not to be caused by a spinal metastasis, but can result from paramalignant necrotising myelopathy (Misumi et al. 1989), or, worse, be iatrogenic as in intrathecal methotrexate treatment.

Delay of diagnosis occurs all too often in spite of the known beneficial effect of neurosurgical intervention (Livingstone and Perrin 1978; Dunn et al. 1980; Greenberg et al. 1980). The wide array of spinal space occupying lesions have been reviewed in Volumes 23–26 of this Handbook.

As with the nosological category of injuries, in the category of space-occupying process the spastic paraplegia is invariably associated with other long tract signs and sphincter disturbances compatible with the diagnosis of a (sub-) complete cord transection. As such, again, this category comes hardly into consideration within the frame of system atrophy pur sang to which this volume is devoted.

Infections

It needs no emphasis that one can hardly deny any microorganism the potential to invade the vertebral canal and its contents and thereby to produce either a granulomatous or purulent lesion compresssing the cord or producing a myelitis, whether bacterial, viral, fungal or helminthic. In the 19th century, Erb's spastic paraplegia concept was based on syphilitic cervical pachymeningitis; an ALS-like syndrome was reported in vascular-meningeal syphilis (Storm-Mathisen 1978). As to the other historical scourge in medicine, paraplegia due to spinal tuberculoma, or transverse myelitis (tuberculous radiculomyelitis) was a diagnosis early neurologists had always to keep in mind (Chortis 1958; Freilich and Swash 1979), and in certain regions of the world still do today (Scrimgeour et al. 1987). Tuberculous paraplegia may develop during pregnancy (Govender et al. 1989) or even after decompressive laminectomy (Rand and Smith 1989). Diagnostic vertebral body biopsy by percutaneous puncture for histology and antibiotic sensitivity assay, plus anterior vertebral fusion is the best management. In the Middle-East area, spastic paraplegia has a fair chance of being a symptom of brucellosis, which may even produce multiple sclerosis-like symptoms and is best routinely tested for (Al-Deeb et al. 1988). In Peru, Bartonella-infection is known to produce spastic paraplegia (Trelles and Trelles 1978).

Of the other bacteria apt to produce infections of the nervous system, staphylococci are those most frequently found in the epidural abscess underlying spastic paraplegia (d'Angelo and Whisler 1978; Mattle et al. 1986). The extensive tourism by air in recent times forces the diagnostician to be aware of such diseases that are regarded as exotic in his own country. Definitely rare are Neisseria meningitidis (Boothman et al. 1988), Whipple's disease (Koudouris et al. 1963) and pertussis encephalopathy (Miller 1955) as causes of spastic paraplegia.

Among the viruses as causative agents of spastic paraplegia, the leading role once played by measles-virus and other paramyoviruses producing SSPE (Swoveland and Johnson 1988) as well as by a variety of virus producing Reye-syndrome (Davis 1988), varicella zoster (Gilden and Vafai 1988), and Epstein-Barr virus (Gotlieb-Stematsky and Arlazoroff 1988), has been taken over by HTLV-III (Harris et al. 1988) and HTLV-I species (Román 1988), with the qualification that diagnostic awareness with respect to the last-named virus should be heightened in the Caribbean and Japanese islands. Much rarer causes to be listed are Behçet's disease (Alema 1978; Inaba 1989), Vogt-Koyanagi-Harada disease (Imomata and Kato 1988). Exceptional are poliomyelitis (Foley and Beresfort 1974) and progressive rubella panencephalitis (Wolinsky 1978). The cutaneous, ocular and mucosa-manifestations make diagnosis not too difficult, just as the skin lesions in verruga peruana (Bartonellosis) in Behçet's and Vogt-Koyanagi diseases. Instances of spastic paraplegia occur among the 'slow virus' diseases such as Creutzfeldt-Jakob disease, Gerstmann-Sträussler-Scheinker disease, etc.

With respect to fungal disease, the most widespread organism that affects the central nervous system is *Cryptococcus neoformans*. In-volvement of the spine, meninges and cord is no exception; diagnosis and treatment are well-established (Weenink and Bruyn 1978, 1988; Govender and Charles 1987). Less well recognised are the spinal aspergillomas (Kingsley et al. 1979; Sheth et al. 1985) and Candida granuloma (Kumar et al. 1979). The blastomycetes-group is prevalent in particular areas of the world and therefore varies in position on the diagnostic ranking list; spastic paraplegia is frequent in coccidioidomycosis (Winn and Hartstein 1988), but much less so in North American blastomycosis (Goneya 1978), paracoccidioidomycosis (Gonzalez 1988), and nocardiosis (Mandell and Neu 1988). Zygomycosis, previously called mucomycosis, is characterised by vessel wall infection and septic thrombosis; only a few cases of myelomalacia with paraplegia have been reported (Kalayjian et al. 1988).

In all these conditions, it is above all the (iatrogenically) immunodepressed or disease-induced immunocompromised patient that should be considered as an increased risk. Proper work-up includes careful examination of blood and CSF (including bacteriological and fungal techniques), attention to details of vocation and geographic origin of the patient as well as heed of the non-neurological symptoms.

Finally, with respect to paraplegias due to helminth infestation, these will remain a major challenge in the South-American, African and Asian areas as long as medical-hygiene per capita expenditures stay at the present minute levels and education of hygienic awareness seems ineffectual. Because untold hundreds of millions are infested and only an infinitesimal fraction of those who die come — for folkoristic or religious motives — to autopsy, any estimate of CNS involvement is unrealistic. Spastic paraplegia has been seen in schistosomiasis (Blansjaar 1988; Suchet et al. 1987), Trichinosis (Kramer and Aita 1988), Filariasis (Dumas and Avode 1988), Angiostrongyliasis (Hung and Chen 1988), Echinococcus (Kammer 1988) and Cysticercosis (Sotelo 1988) either as a myelitis or mass lesion (abscess). The Guinea-worm, *Dracunculus medinensis*, has infested an estimated half million people in Nigeria, and only 6 cases have been reported (Khawaja et al. 1976; Odaibo et al. 1986).

In most infectious lesions the abnormal findings on ESR, leucocyte count and differential, CSF-cell and -protein content will point in the right direction. The fairly rapid development of neurologic deficit including sensory and sphincter symptoms, as well as an NMR or myelography will clinch the diagnosis.

Chronic spinal arachnoiditis with ultimate cavitation of the cord is as rare as it is difficult to diagnose. Systemic signs are usually lacking; the history may include a variety of factors such as infection, iatrogenic intrathecal drug or oil contract medium administration or spinal surgical interventions (Shaw et al. 1978; Whisler 1978). This disorder is exceptionally rare after subarachnoid hemorrhage (Tjandra et al. 1989). It is not a '-itis' sensu strictiori.

Vascular lesions

This nosological category is not a causal but structural one, focussed on diseases of the vessels. Apart from those rare instances of bilateral frontal angioma and subdural hematoma mentioned previously and anterior communicating artery aneurysm (Maiuri et al. 1986), there are few intrinsic special vascular disorders that lead to spastic paraplegia. More often, disease of the aorta underlies the syndrome.

Angiodysgenetic necrotising myelopathy at the cervico-thoracic level (Foix-Alajouanine disease) is so exquisitely rare, that entertaining its possibility in one's mind for differential diagnosis can be considered as more than a luxury.

Spinal cord ischemia and myelomalacia, whether acute or chronic usually includes (dissociated) impairment of sensation and sphincter trouble in the semeiology. The sensory impairment often indicates the 2 watershed zones of approximately L I and approximately Th IV. In addition, the syndrome is usually acute. If not, the clinician should recognise the preceding stage of neurogenic intermittent claudication. In the chronic cases, ranging from spinal arterial steal ischemia in Paget's disease (well treatable by calcitonin, Porrini et al. 1987; Awwad and Sundoram 1987) to atherosclerotic stenosis of the major feeders of the spinal cord (thyreocervical trunk, intercostal and Adamkiewicz arteries) that join to form the radicular arteries and ultimately the anterior spinal artery.

In contradistinction to (compressive) mass lesions that invariably herald the deficit by pains, development of the vascular paraplegias not exceptionally is indolent, particularly in sudden aortic thrombosis. The confusing pleiomorphic presentation of spinal-radicular ischemia in (pseudo-) aneurysmatic dissection of the aorta invariably includes pains. In both groups of patients there is a history of vascular disease, hypertension, coronary infarction and they are of advanced age, while the paraplegia often is flaccid initially to become spastic later, and signs of other long tracts are associated. Of course, the arterial pulsation in the lower limbs are absent, though they may be felt as weak in insidious, subtotal, dissection or stenosis (Kochar et al. 1987; Bolduc et al. 1989; Chatlani et al. 1989; Zull and Cydulka 1988). Diagnostic pitfalls here form the younger age group, whose aortic thrombosis is caused by blunt trauma, and in whom the diagnostic search fails to reveal a vertebral fracture as cause for the paralysis (Sumpio and Gusberg 1987; Bednarski and Nayduck 1989). Still worse, a patient known with malignant disease and vertebral metastasis may develop sudden paraplegia due to aortic thrombosis (Moore et al. 1989). Again, palpation of leg-artery pulsations is a simple, diagnostically effective geste.

Iatrogenic paraplegia in (vascular) surgery, although fortunately rare, yet constitutes not such a negligible risk that one should omit to inform the patient before surgical intervention: intraaortic balloon pump (Scott and Gioti 1985; Seifert and Silverman 1986; Riggle and Oddi 1989), ductus arteriosus ligation (Beg et al. 1987), aortic coarctation operation (Brewer et al. 1972) and aortic segment resection (von Segesser et al. 1988) are life-saving procedures at which the price-tag of a wheel-chair bound life may be tied. The part played — whether etiological or pathogenetical — by vasospasm, ischaemia and edema in the spinal cord has not been clarified in other iatrogenic paraplegias that are initially flaccid and become spastic: paraplegia after coeliac plexus block (Woodham and Hanna 1989) and after sublaminar segmental spinal instrumentation (Johnson et al. 1986). In pulmonary surgery, the same factors seem to play a part or epidural hematoma was found in quite a number of cases (Batellier et al. 1989, Jöhr and Salathé 1988). In patients on anticoagulant treatment, or with coagulopathy, a simple lumbar puncture may provoke subarachnoid hemorrhage or an epi/subdural hematoma (Bruyn and Bosma 1976; Hanakita et al. 1987; Faillace et al. 1989). The iatrogenic sequelae of epidural anesthesia are so numerous as to require a chapter of its own (Usubiaga 1975; Tashiro et al. 1987; Gustafsson et al. 1988).

An underrated iatrogenic 'paraplegia' is the one seen after surgical interventions under general anaesthesia with NO_2 in those patients with marginally low stores of vitamin B12; the nitrous oxide inactivates what B12 is left and myelopathy develops (Schilling 1986; Holloway and Alberico 1990). The same mechanism operates in nitrous oxide abuse, as reported in nurses and dental surgeons (Layzer 1978; Blanco and Peters 1983). Nosologically, this group should be ranged with the intoxication-induced deficiencies.

A very rare iatrogenic spastic paraplegia may follow drastic oral treatment of malignant hypertension (Brown et al. 1987).

Spinal cord angioma or AVM, predominantly occurring in males, may present suddenly and with pain, with remitting-relapsing and with a chronic progressive course. In acute cases the paraplegia is flaccid, and often combined with haemorrhage and 'steal-ischemia', whereas in chronic cases the paraplegia is often spastic. Associated symptoms may be mild. Once-suspected, NMR or — better still — spinal angiography are the superior procedures for diagnosis (Hurt et al. 1978; Zentner et al. 1989). Vertebral hemangioma with its tell-tale vertebral body trabeculae or trellis-work on plain X-rays, may be another, if rare, cause of paraplegia, occasionally becoming manifest during pregnancy (Liu and Yang 1988).

Intoxications

Spastic paraplegia due to intoxications are as intriguing as, from a numerical point of view, they are exotic. No attention will be given to clioquinol-induced subacute myeloopticoneuropathy, as this is a thing of the past (Sobue 1979). The epidemic of tri-orthocresylphosphate poisoning in Algeria also lies behind us (Cavanagh and Koller 1979), and also the diftri-ethyltin poisoning by 'Stalinon' in France (Alajouanine et al. 1958).

In tropical countries there are still endemics which silently cripple thousands of people.

Particularly in certain geographical areas (Reddy 1979), but also elsewhere (Fisher et al. 1989) endemic fluorosis from groundwater sources, leading to skeletal overgrowth, ligamentous calcification and spinal cord compression, is a disease the diagnostician should suspect.

The same obtains for neuro-lathyrism, which may produce a clean Erb's spastic paraplegia often misdiagnosed as multiple scelerosis (Prasad and Sharan 1979).

In Africa and Mozambique, the staple diet of Cassava, if improperly prepared, repeatedly leads to endemics of spastic paraparesis, striking thousands of people. These are known as 'Mantakassa' (Cliff et al. 1984), 'Konzo' in Tanzania (Howlett et al. 1990), or 'Buka-buka' in Zaire (Carton et al. 1989) and essentially due to cyanide-intoxication. They should not be confused with tropical HLTV-I spastic paraplegia, nor with Jamaican vomiting illness due to unripe ackee-fruit ingestion with resultant hypoglycin A intoxication, nor with tropical ataxia neuropathy.

Another rare but epidemiologically interesting disorder is seen among the population of addicts in the Western metropoles: spongiform leukoencephalopathy with spastic para/tetraplegia, regression reflexes, extensor spasms and cerebellar ataxia after heroin-inhalation (Ell et al. 1981; Wolters et al. 1982).

Spastic paraplegia has also been observed in chronic alcohol-intoxication of the Marchiafava-Bignami type (Brion 1976).

Iatrogenic spastic paraplegia due to neurotoxicity is a well-known sequel of intrathecal methotrexate or cytosine-arabinoside administration, particularly during or after X-irradiation (Werner 1988). It is also seen after intrathecal mitozantrone (Lakhani et al. 1986), and even following intrathecal hypertonic saline (Kim et al. 1988).

Metabolic and deficiency disease (excluding inborn errors)

Spastic paraplegia within this nosological category has been noted to occur in a minority of pellagra-patients (Still 1976), exceptionally in Crohn's disease (Cooke 1976), in Leigh's necrotising encephalomyelopathy (David et al. 1976), and in a number of lysosomal diseases. Mannosidosis in the adult may present with spastic paraplegia (Kawai et al. 1985). Though an inborn metabolic error, argininemia should be mentioned here, because it manifests at a rather late stage (4th–6th year of life) with spastic para/tetraplegia (Terheggen et al. 1982).

For a survey of spastic paraplegia in neurolipidoses the reader is referred to Vol. 10 of this Handbook. Van Bogaert's cerebrotendinous xanthomatosis is one example of this group often displaying spastic paraplegia (Berginer et al. 1989; Fiorelli et al. 1990), and a hereditary disease closely mimicking it, viz. β-sitosterolemia with xanthomatosis may, exceptionally, produce paraplegia (Hatanaka et al. 1990).

Systemic disease

Worth mentioning here are sarcoidosis (Matthews 1979), Bechterew's ankylosing spondylitis (Whitfield 1979), Sjögren-Larsson's disease easily diagnosable by its ichthyosis and mental retardation, Eales' disease (Sawhney et al. 1986), dermatomyositis (Haguenau et al. 1989), rheumatoid arthritis either through C1/2 subluxation or granulomatous cord compression (Matsumine et al. 1988), Fanconi syndrome, subacute combined degeneration of the cord, and Hutchinson-Laurence-Moon (*not* Bardet-Biedl) syndrome (Green et al. 1989). If one considers Vaquez' polycythemia vera as a systemic disorder, the reports of paraplegia as a complication of Vaquez' disease by Grunberg et al. (1950) and Filho and Levy (case 7; 1955) should be mentioned. Adult Hallervorden-Spatz disease manifesting with Parkinsonian features and spastic paraparesis may be classified with this group (Alberca et al. 1987; Eidelberg et al. 1987) as well as a hereditary adult leukodystrophy simulating multiple sclerosis, Pelizaeus-Merzbacher disease and adreno-myeloneuropathy (El-

dridge et al. 1984). The same obtains for striopallidodentate calcification (Löwenthal and Bruyn, 1968). Sporadic cases of spastic paraplegia in a variety of diseases, too many to be listed here, are listed in Vols. 42 and 43 of this Handbook.

Hereditary disease

The variants of pure autosomal dominant and recessive spastic paraplegia are set out in detail in Chapters 17 and 18 of this volume. Diagnostic pitfalls may be encountered: in autosomal dominant spastic paralysis, associated with dementia and ataxia, as first reported by Worster-Drought (1933) and Worster-Drought et al. (1940, 1944), the underlying cause has been established as one of hereditary cerebral amyloid angiopathy ('English type', not the Icelandic or Dutch type) as recently reported by Plant et al. (1990).

It will be clear that spastic paraplegia as a symptom can be expected to occur in a multitude of disorders and diseases. An exhaustive aetiological review of those would require a monograph. The rather succinct and (by virtue of that) lacunar review presented above deals only with those that the clinical neurologist has a fair chance to be presented with.

REFERENCES

ADAMS, R. D. and G. LYON: Neurology of Hereditary Metabolic Diseases of Children. New York, McGraw Hill Cy (1982).

ALAJOUANINE, TH., L. DÉROBERT and S. THIEFFREY: Etude clinique d'ensemble 210 cas d'intoxication pour les sels organiques d'étain. Rev. Neurol. (Paris) 98 (1958) 85–96.

ALBERCA, R., E. RAFEL, I. CHINCHON, J. VADILLO and A. NAVARRO: Late onset Parkinsonian syndrome in Hallervorden-Spatz disease. J. Neurol. Neurosurg. Psychiatry 50 (1987) 1665–1668.

ALEMA, G.: Behçet's disease. In: P. J. Vinken and G. W. Bruyn (Eds.), Handbook of Clinical Neurology, Vol. 34: Infections of the Nervous System, Ch. 34. Amsterdam, North-Holland Publ. Co. (1978) 475–512.

AL-DEEB, S. M., B. A. YAQUB, H. S. SHARIF and S. M. AL-RAJEH: Neurobrucellosis. In: P. J. Vinken, G. W. Bruyn, H. L. Klawans and A. A. Harris (Eds.), Handbook of Clinical Neurology, Vol. 52, Microbial

Disease, Ch. 42. Amsterdam, Elsevier (1988) 581–600.

AL-DIN, A. N., M. ANDERSON, E. R. BICKERSTAFF and I. HARVEY: Brain stem encephalitis and the Miller-Fisher syndrome. Brain 105 (1982) 481–495.

AMMOUMI, A. A., J. H. SHER and D. SCHMELKA: Spinal cord compression by extra-medullary hemopoietic tissue in sickle cell anemia. J. Neurosurg. 43 (1975) 483–485.

ARSENI, C. and P. DECU: Paraplegia due to bilateral angioma of the paracentral area. Neurochirurgia 22 (1979) 194–196.

AWWAD, E. E. and M. SUNDARAM: Vertebral Paget's disease causing paraparesis. Orthopedics 10(3) (1987) 528 + 531–533.

BALCI, S., B. SAY and T. FIRAT: Corneal opacity, microphthalmia mental retardation, microcephaly and generalized muscular spasticity associated with hyperglycinemia. Clin. Genet. (Kbh.) 5 (1974) 36–39.

BATELLIER, J., J. M. WIHLM, G. MORAND and J. P. WITZ: Paraplégie par hématome extra-dural rachidien après lobectomie pulmonaire élargie pour cancer. Ann. Chir. 43 (1989) 210–214.

BEDNARSKI, J. J. and D. A. NAYDUCH: Thoracic aorta rupture as the cause of paraplegia: A diagnostic dilemma (case report). J. Trauma 29 (1989) 531–533.

BEG, M. H., D. M. R. D. EKRAMULLAH, S. H. AHMAD and M. D. RAYAZUDDIN: Paraplegia after ductus ligation accompanied by injury to intercostal artery. J. Thorac. Cardiovasc. Surg. 93 (1987) 934–941.

BEN-MEIR, P., E. PICARD, A. SAGI, B. GREBER, Y. HERTZANU, P. TIBERIN, D. M. FLISS and H. ZIRKIN: Sarcome d'Ewing extra-osseux épidural: le dixième cas. Rev. Neurol. (Paris) 145 (1989) 324–327.

BENVENUTI, D., F. MAIURI and A. LAVANO: Paraparesis due to bilateral subdural hematoma. Acta Neurol. (Napoli) 9 (1987) 288–290.

BERGINER, V. M., G. SALEN and S. SHEPER: Cerebrotendinous xanthomatosis. Neurol. Clin. 7 (1989) 55–74.

BERLIT, P., M. HÄRLE and A. JOHANN: Zervikale Strahlenmyelopathie mit spastischer Paraparese der Ärme. Nervenarzt 58 (1987) 40–46.

BESSON, I., J. C. MIETTE and G. ROUALDES: Hernie discale dorsale D8–D9 révélée par une compresssion medullaire. La Presse Méd. 17 (1988) 1813.

BLANCO, G. and H. A. PETERS: Myeloneuropathy and macrocytosis associated with nitrous oxide abuse. Arch. Neurol. 90 (1983) 416–418.

BLANSJAAR, B. A.: Schistosomiasis, In: P. J. Vinken, G. W. Bruyn, H. L. Klawans and A. A. Harris (Eds.), Handbook of Clinical Neurology, Vol. 52: Microbial Disease, Ch. 39. Amsterdam, Elsevier (1988) 535–543.

BOLDUC, M. E., S. CLAYSON and P. N. MADRAS: Acute aortic thrombosis presenting as painless paraplegia. J. Cardiovasc. Surg. 30 (1989) 506–508.

BOOTHMAN, B. R., J. M. BAMFORD and M. R. PARSONS: Paraplegia as a presenting feature of meningococcal meningitis. J. Neurol. Neurosurg. Psychiatry 51 (1988) 1241.

BOTS, G. T. A. M., A. R. WATTENDORF and O. J. S. BURUMA: Acute myelopathy caused by fibrocartilagenous emboli. Neurology 31 (1981) 1250–1256.

BREWER, L. A., R. G. FOSBURG, G. A. HULDE and J. J. VERSKA: Spinal cord complications following surgery for coarctation of the aorta. J. Thorac. Cardiovasc. Surg. 64 (1972) 368–381.

BRION, S.: Marchiafava-Bignami disease. In: P. J. Vinken and G. W. Bruyn (Eds.), Handbook of Clinical Neurology, Vol. 28: Metabolic and Deficiency Diseases of the Nervous System, Part II, Ch. 12. Amsterdam, North-Holland Publ. Co. (1976) 317–330.

BROWN, P., M. GROSS and M. HARRISON: Paraplegia following oral hypotensive treatment of malignant hypertension. J. Neurol. Neurosurg. Psychiatry. 50 (1987) 104–118.

BRUYN, G. W. and N. J. BOSMA: Spinal extradural hematoma. In: P. J. Vinken and G. W. Bruyn (Eds.), Handbook of Clinical Neurology, Vol. 26, Ch. 1. Amsterdam, North Holland Publ. Co. (1976) 1–30.

BRUYN, G. W. and J. C. GATHIER: The operculum syndrome. In: P. J. Vinken, and G. W. Bruyn (Eds.), Handbook of Clinical Neurology, Vol. 2. Amsterdam, North Holland Publ. Co. (1969) 776–783.

BRUYN, G. W. and L. N. WENT: A sex-linked heredodegenerative neurological disorder associated with Leber's optic atrophy. J. Neurol. Sci. 1 (1964) 59–80.

BUCY, P. C.: The Prefrontal Motor Cortex. Univ. Ill. Press, Urbana, Ill. (1944).

BUCY, P. C.: Is there a pyramidal tract? Brain 80 (1957) 376–392.

BUCY, P. C. and J. E. KEPLINGER: Section of the cerebral peduncles. Arch. Neurol. 5 (1961) 132–139.

BUCY. P. C., R. LADLI and A. EHRLICH: Destruction of the pyramidal tract in the monkey. J. Neurosurg. 25 (1966) 1–23.

CALDER, I. M., A. C. PALMER, J. T. HUGHES, J. F. BOLT and J. D. BUCHANAN: Spinal cord degeneration associated with type II decompression sickness: case report. Paraplegia 27 (1989) 51–57.

CARTON, H., K. KAYEMBE, KABEYA, ODIO, A. BILLIAU and K. MAERTENS: Epidemic spastic paraparesis in Bandundu (Zaire). J. Neurol. Neurosurg. Psychiatry 49 (1989) 620–627.

CAVANAGH, J. B. and W. C. KOLLER: Triorthocresylphosphate poisoning. In: P. J. Vinken and G. W. Bruyn (Eds.), Handbook of Clinical Neurology, Vol. 37: Intoxications of the Nervous System, Part. II, Ch. 15. Amsterdam, North-Holland Publ. Co. (1979) 471–477.

CÉNAC, A., J. AUDOIN, F. LAMOTHE., A. TOUTA and J. M. VETTER: Lipomatose extra-durale et paraplégie brusque. Guérison après laminectomie. Rev. Méd. Interne 8 (1987) 533–534.

CHATLANI, P. T., M. G. VAN DESSEL, K. R. KRISHNAN and G. A. MCLOUGHLIN: Abdominal aortic aneurysm presenting as paraplegia: case report. Paraplegia 27 (1989) 146–147.

CHORTIS, P.: Transverse myelitis occurring during tuberculous meningitis. Dis. Chest 33 (1958) 506–508.

CLIFF, J., A. MARTELLI, A. MOLIN and H. ROSLING: Mantakassa; epidemic of spastic paraparesis due to chronic cyanide intoxication. Bull. WHO 62 (1984) 477–492.

COHEN, L.: Tumors in the region of the foramen magnum. In: P. J. Vinken and G. W. Bruyn (Eds.), Handbook of Clinical Neurology, Vol. 17, Ch. 21. Amsterdam, North Holland Publ. Co. (1974) 719–730.

COOKE, W. T.: Neurological manifestations of malabsorption. In: P. J. Vinken and G. W. Bruyn, (Eds.), Handbook of Clinical Neurology, Vol. 28: Metabolic and Deficiency Diseases of the Nervous System, Part II, Ch. 8. Amsterdam, North-Holland Publ. Co. (1976) 236.

CRITCHLEY, M.: Sprengel's deformity with paraplegia. Br. J. Surg. 14 (1926/7) 243–249.

CROSS, H. E., V. A. MCKUSICK and W. BREEN: A new oculocerebral syndrome with hypopigmentation. J. Pediat. 70 (1967) 398–406.

D'ANGELO, C. M. and W. W. WHISLER: Bacterial infections of the spinal cord and its coverings. In: P. J. Vinken and G. W. Bruyn (Eds.), Handbook of Clinical Neurology, Vol. 33, Part I: Infections of the Nervous System, Ch. 11. Amsterdam, North-Holland Publ. Co. (1978) 187–194.

DAVID, R. B., P. MAMUNES and W. I. ROSENBLUM: Necrotising encephalomyelopathy (Leigh). In: P. J. Vinken and G. W. Bruyn, (Eds.), Handbook of Clinical Neurology, Vol. 28: Metabolic and Deficiency Diseases of the Nervous System, Part II, Ch. 14. Amsterdam, North-Holland Publ. Co. (1976) 349–364.

DAVIDOFF, R. A.: The pyramidal tract. Neurology 40 (1990) 332–339.

DAVIS, L. E.: Reye's syndrome. In: P. J. Vinken, G. W. Bruyn, H. L. Klawans and A. A. Harris (Eds.), Handbook of Clinical Neurology, Vol. 52: Microbial Disease, Ch. 9. Amsterdam, Elsevier (1988) 149–177.

DE JONG, R. N.: The Neurologic Examination, 2nd ed. London, Pitman Medical Publ. Co. (1958).

DE PREUX, J. and S. STEPHANOV: Paraparesis due to bilateral subdural hematoma. Surg. Neurol. 16 (1981) 346–348.

DUMAS, M. and G. AVODE: Filariasis. In: P. J. Vinken, G. W. Bruyn, H. L. Klawans and A. A. Harris (Eds.), Handbook of Clinical Neurology, Vol. 52: Microbial Disease, Ch. 36. Amsterdam, Elsevier (1988) 513–520.

DUNN, R. C., W. A. KELLY, R. N. WOHNS and J. F. HOWE: Spinal epidural neoplasm. A 15-year review. J. Neurosurg. 52 (1980) 47–51.

DU TOIT, F.: A case of congenital elevation of the scapula (Sprengel's deformity) with defect of the cervical spine associated with syringomyelia. Brain 54 (1931) 421–429.

EIDELBERG, D., A SOTREL, C. JOACHIM, D. SELKOE ET AL.: Adult onset Hallervorden-Spatz disease with neurofibrillary pathology. Brain 110 (1987) 993–1013.

ELDRIDGE, R., C. P. ANAYIOTOS, S. SCHLESINGER, D. COWEN ET AL.: Hereditary adult-onset leukodystrophy simulating chronic progressive multiple sclerosis. N. Engl. J. Med. 311 (1984) 948–953.

ELL, J. J., D. UTTLEY and J. R. SILVER: Acute myelopathy in association with heroin addiction. J. Neurol., Neurosurg. Psychiatry 44 (1981) 448–450.

FAILLACE, W. J., I. WARRIER and A. I. CANADY: Paraplegia after lumbar puncture in an infant with previously undiagnosed hemophilia A. Treatment and Perioperative Considerations. Clin. Pediatr. 28 (1989) 136–138.

FILHO, R. M. and J. A. LEVY: Aspectos neurológicos da policitemia vera. Arg. Neuropsiq. 13 (1955) 313–321.

FIORELLI, M., V. DI PIERO, S. BASTIANELLO ET AL: Cerebrotendinous xanthomatosis: Clinical and MRI study (a case report). J. Neurol. Neurosurg. Psychiatry 53 (1990) 76–78.

FISHER, R. G.: The ominous discoloration of the spinal cord due to thoracic disk protrusions: a historical note. J. Neurol. Neurosurg. Psychiatry 49 (1986) 844–846.

FISHER, R. L., T. W. MEDCALF and M. C. HENDERSON: Endemic fluorosis with spinal cord compression. Arch. Intern. Med. 149 (1989) 679–700.

FOLEY, K. M. and R. H. BERESFORD: Acute poliomyelitis begining as transverse myelopathy. Arch. Neurol. 30 (1974) 182–183.

FREILICH, D. and M. SWASH: Diagnosis and management of tuberulous paraplegia J. Neurol. Neurosurg. Psychiatry 42 (1979) 12–18.

FRIED, H., H. G. NIEBELING and D. HOHREIN: Die tumorbedingte medulläre Kompression—Erfahrungen an 570 Patienten. Zentralbl. Neurochir. 49 (1988) 270–272.

GILDEN, D. H., A. VAFAI: Varicella-zoster: In: P. J. Vinken, G. W. Bruyn, H. L. Klawans and A. A. Harris (Eds.), Handbook of Clinical Neurology, Vol. 52: Microbial Disease, Ch. 13. Amsterdam, Elsevier (1988) 229–247.

GILMAN, S. and L. A. MARCO: Effects of medullary pyramidotomy in the monkey. Brain 94 (1971) 495–514, 515–30.

GINSBERG, J., J. D. PEDERSEN, CHR. VON WESTARP and A. B. MCCARTEN: Cervical cord compression due to extension of a papillary thryoid carcinoma. Am. J. Med. 82 (1987) 156–8.

GONEYA, E. F.: The spectrum of primary blastomycotic meningitis: a review of CNS blastomycosis. Ann. Neurol. 3 (1978) 26–39.

GONZALEZ, G. T.: Paracoccidiodomycosis. In: P. J. Vinken, G. W. Bruyn, H. L. Klawans and A. A. Harris (Eds.), Handbook of Clinical Neurology, Vol. 52: Microbial Disease, Ch. 32. Amsterdam, Elsevier (1988) 455–465.

GOTLIEB-STEMATSKY, T. and A. ARLAZOROFF: Epstein-Barr virus. In: P. J. Vinken, G. W. Bruyn, H. L.

Klawans and R. R. McKendall (Eds.), Handbook of Clinical Neurology, Vol. 56: Viral Disease, Ch. 14. Amsterdam, Elsevier (1988) 249–261.

GOVENDER, S. and R. W. CHARLES: Cryptococcal infection of the spine. A case report. S. Afr. Med. J. 71 (1987) 782–783.

GOVENDER, S., S. C. MOODLEY and M. J. GROOTBOOM: Tuberculous paraplegia during pregnancy. A report of 4 cases. S. Afr. Med. J. 75 (1989) 190–192.

GREEN, J. S., P. S. PARFREY, J. D. HARNETT ET AL: The cardinal manifestation of Bardet-Beidl syndrome, a form of Laurence-Moon-Biedl syndrome. N. Engl. J. Med. 321 (1989) 1002–1009.

GREENBERG, H. S., J. H. KIM and J. B. POSNER: Epidural spinal cord compression from metastatic tumor. Results with a new treatment protocol. Ann. Neurol. 8 (1980) 361–366.

GRUNBERG, A., J. L. BLAIR and R. M. RAWCLIFFE: Unusual neurological symptoms in polycythaemia rubra vera. Edinburgh. Med. J. 57 (1950) 305–308.

GUILLAIN, G. and P. MOLLARET: Syndrome de Klippel-Feil avec quadriplégie spasmodique. Variété étiologique particulière de hémiplégie spinale ascendante chronique. Rev. Neurol. (Paris) 1 (1931) 436–444.

GUSTAFSSON, H., H. RUTBERG and M. BENGTSSON: Spinal hematoma following epidural analgesia. Report of a patient with ankylosing spondylitis and a bleeding diathesis. Anaesthesia 43 (1988) 220–222.

HAGUENAU, M., J. CH. PIETTE and G. ROBERT: Paraparésie spasmodique progressive et dermatomyosite. Rev. Neurol. (Paris) 145 (1989) 330–334.

HANAKITA, J., H. MIYAKE and T. TOYODA: Acute paraplegia due to spinal subarachnoid hematoma after lumbar puncture. Case report. Neurol. Med. Chir. (Tokyo) 27 (1987) 1005–1009.

HANAWAY, J., R. R. YOUNG: Localization of the pyramidal tract in the internal capsule of man. J. Neurol. Sci. 34 (1977) 63–70.

HARRIS, A. A., J. SEGRETI and H. A. KESSLER: The neurology of AIDS. In: P. J. Vinken, G. W. Bruyn, H. L. Klawans and R. R. McKendall (Eds.), Handbook of Clinical Neurology, Vol. 56: Viral Disease, Ch. 27 Amsterdam, Elsevier (1988) 489–506.

HATANAKA, I., H. YASUDA and H. IDAKA: Spinal cord compression with paraplegia in xanthomatosis due to normocholesterolamic sitosterolemia. Ann. Neurol. 28 (1990) 390–393.

HEGDE, S. and W. E. STAAS, JR.: Thoracic disc herniation and spinal cord injury. Am. J. Phys. Med. Rehabil. 67 (1988) 228–229.

HO, K. L. and K. R. MEYER: The medial medullary syndrome. Arch. Neurol. 38 (1981) 385–387.

HOLLOWAY, K. L. and A. M. ALBERICO: Postoperative myeloneuropathy: a preventable complication in patients with B_{12} deficiency. J. Neurosurg. 72 (1990) 732–736.

HOWLETT, W. P., G. R. BRUBAKER, N. MLINGI and H. ROSLING: Konzo, an epidemic upper motor neuron disease studied in Tanzania. Brain 113 (1990) 223–225.

HUNG, T.-P. and E.-R. CHEN: Angiostrongylosis. In: P. J. Vinken, G. W. Bruyn, H. L. Klawans and A. A. Harris (Eds.), Handbook of Clinical Neurology, Vol. 52: Microbial Disease, Ch. 40. Amsterdam, Elsevier (1988) 545–562.

HURT, M., R. HOUDART and R. DJINDJIAN: AVM of the spinal cord. A series of 150 cases. Prog. Neurol. Surg. 9 (1978) 238–266.

INABA, G.: Behçet's disease. In: P. J. Vinken, G. W. Bruyn, H. L. Klawans and R. R. McKendall (Eds.), Handbook of Clinical Neurology, Vol. 56: Viral Diseases, Ch. 32. Amsterdam, Elsevier (1989) 593–610.

INOMATA, H. and M. KATO: Vogt-Koyanagi-Harada disease. In: P. J. Vinken, G. W. Bruyn, H. L. Klawans and R. R. McKendall (Eds.), Handbook of Clinical Neurology, Vol. 56: Viral Disease, Ch. 33. Amsterdam, Elsevier (1988) 621–626.

IWATSUBO, T., S. KUZUHARA and A. KANEMITSU: Corticofugal projections to the motor nuclei of the brain and spinal cord. Neurology 40 (1990) 309–312.

JOHNSTON, C. E., L. T. HAPPEL and R. NORRIS: Delayed paraplegia complicating sublaminar segmental spinal instrumentation. J. Bone Jt. Surg. 68A (1986) 556–563.

JÖHR, M. and M. SALATHÉ: Paraplegia nach Pneumonektomie. Eine anästhesiologische oder eine chirurgische Komplikation. Schweiz. Med. Wschr. 118 (1988) 1412–1414.

KALAYJIAN, R. C., R. H. HERZIG, A. M. COHEN and M. C. HUTTON: Thrombosis of the aorta caused by mucormycosis. South Med. J. 81 (1988) 1180–1182.

KAMMER, W. S.: Echinococcosis. In: P. J. Vinken, G. W. Bruyn, H. L. Klawans and A. A. Harris (Eds.), Handbook of Clinical Neurology, Vol. 52: Microbial Disease, Ch. 37. Amsterdam, Elsevier (1988) 523–527.

KANITKAR, S. and A. H. N. ROBERTS: Paraplegia in an electrical burn: A case report. Burns 14 (1988) 49–50.

KAPLAN, J. G., E. BARASCH and A. HIRSCHFELD: Spinal epidural lipomatosis. A serious iatrogenic complication of Cushing's syndrome. Neurology 39 (1989) 1031–1034.

KAWAI, H., H. NISHINO and Y. NISHIDA: Skeletal muscle pathology of mannosidosis in two siblings with spastic paraplegia. Acta Neuropathol. (Berl.) 68 (1985) 201–204.

KESTLE, J. R. W., CH. H. TATOR and W. KUCHARCYK: Intervertebral disc embolization resulting in spinal cord infarction. J. Neurosurg. 71 (1989) 938–941.

KHAWAJI, M. S., J. F. B. DOSSETOR and J. H. LAWRIE: Extradural guinea worm abscess. J. Neurosurg. 43 (1976) 627–630.

KIM, R. C., R. W. PORTER, B. H. CHOI and S. W. KIM: Myelopathy after the intrathecal administration of hypertonic saline. Neurosurgery 22 (1988) 942–945.

KINGSLEY, D. P. E., E. WHITE, A. MARKS and A. COXON: Intradural extramedullary aspergilloma complicating chronis lymphatic leukemia. Br. J. Radiol. 52 (1979) 916–917.

KOCHAR, G., M. N. KOTLER, J. HARTMAN, S. E. GOLDBERG, W. PARRY, R. PARAMESWARAN and M. SCANLON: Thrombosed aorta resulting in spinal cord ischemia and paraplegia in ischemic cardiomyopathy. Am. Heart J. 113 (1987) 1510–1513.

KOUDOURIS, S. D., T. N. STERN and R. A. UTTERBACK: Involvement of the central nervous system in Whipple's disease. Neurology 13 (1963) 397–404.

KRAMER, M. D. and J. F. AITA: Trichinosis. In: P. J. Vinken, G. W. Bruyn, H. L. Klawans and A. A. Harris (Eds.), Handbook of Clinical Neurology, Vol. 52: Microbial Disease, Ch. 41. Amsterdam, Elsevier (1988) 563–579.

KUMAR, S., R. C. RAZA, A. TANDON and M. NAYAR: Subdural spinal granuloma due to candida tropicalis. J. Neurosurg. 50 (1979) 395–396.

LABAUGE, R., M. PAGÈS, D. TESTARD and J. M. PRIVAT: Kyste arachnoïdien extradural de la région dorsale d'évolution regressive et à rechute. Rev. Neurol. (Paris) 145 (1989) 405–407.

LAKHANI, A. K., A. G. ZUIABLE, C. M. POLLARD ET AL.: Paraplegia after intrathecal mitozantrone. Lancet 2 (1986) 1393.

LAWRENCE, D. G. and H. G. J. M. KUYPERS: The functional organisation of the motor system in the monkey. Brain 91 (1968) 1–14, 15–36.

LAYZER, R. B.: Myeloneuropathy after prolonged exposure to nitrous oxide. Lancet 2 (1978) 1227–30.

LEVIC, Z. M., B. S. STEFENOVIC, M. Z. NIKOLIC and D. T. PISTELJIC: Progressive nuclear ophthalmoplegia associated with mental deficiency, lingua scrotalis and other neurologic and ophthalmologic signs in a family. Neurology (Minneap.) 25 (1975) 68–71.

LEYS, D., M. PARENT and H. PETIT: Atteinte médullaire isolée. Récidive d'une sarcoidose. Rev. Neurol. (Paris), 142 (1986) 931–932.

LISON, M., KORNBRUT, B. and A. FEINSTEIN: Progressive spastic paresis, vitiligo, premature greying, and distinct facial appearance: a new genetic syndrome in 3 sibs. Amer. J. Med. Genet. 9 (1981) 351–357.

LIU, CH-L. and D-J. YANG: Paraplegia due to vertebral hemangioma during pregnancy. A case report. Spine 13 (1988) 107–108.

LIVINGSTONE, K. E. and R. G. PERRIN: The neurosurgical management of spinal metastases causing cord and cauda equina compression. J. Neurosurg. 49 (1978) 839–843.

LÖWENTHAL, A. and G. W. BRUYN: Calcification of the striatopallidodentate system. In: P. J. Vinken and G. W. Bruyn (Eds.), Handbook of Clinical Neurology, Vol. 6, Ch. 27. Amsterdam, North-Holland Publ. Co. (1968) 703–725.

LUSTMAN, F., J. FLAMENT-DURANT, H. COLLE, M. LAMBERT and B. SZTERN: Paraplegia due to epidural infiltration in a case of chronic lymphocytic leukemia. J. Neuro-Oncol. 6 (1988) 259–260.

MAIURI, F., M. GANGEMI, G. CORRIERO and F. D'ANDREA: Anterior communicating artery aneurysm presenting with sudden paraplegia. Surg. Neurol. 25 (1986) 397–398.

MANN, K. S., C. P. YUE, K. H. CHAN, L. T. MA and H. NGAN: Paraplegia due to extramedullary hematopoiesis in thalassemia. J. Neurosurg. 66 (1987) 938–940.

MANDELL, W. and H. C. NEU: Nocardial infections. In: P. J. Vinken, G. W. Bruyn, H. L. Klawans and A. A. Harris (Eds.), Handbook of Clinical Neurology, Vol. 52: Microbial Disease, Ch. 31. Amsterdam, Elsevier (1988) 445–453.

MAO, C. C., B. M. COULL and L. A. C. GOLPER: Anterior operculum syndrome. Neurology 39 (1989) 1169–1172.

MATSUMINE, A., K. SHICHIKAWA, K. YAMASHITA, A. UCHIDA and K. YONENOBU: Rheumatoid arthritis causing paraplegia. J. Bone Jt. Surg. 70A (1988) 1410–4111.

MATTHEWS, W. B.: Neurosarcoidosis. In: P. J. Vinken and G. W. Bruyn (Eds.), Handbook of Clinical Neurology, Vol. 38: Neurological Manifestations of Systemic Diseases, Part I, Ch. 21. Amsterdam, North-Holland Publ. Co. (1979) 521–542.

MATTLE, H., A. JASPERT, M. FORSTING, J. P. SIEB, P. HÄNNY and U. EBELING: Der akute spinale Epiduralabszess. Dtsch. Med. Wochenschr. 111 (1986) 1642–1646.

MEYER, F. B., M. J. EBERSOLD and D. F. REESE: Benign tumors of the foramen magnum. J. Neurosurg. 61 (1984) 136–142.

MILLER, H.: Neurological complications of the acute specific fevers. Proc. R. Soc. Med. 49 (1955) 139–146.

MISUMI, H., H. ISHIBASHI, K. KANAYAMA, W. KAYIYAMA, H. NOMURA, T. SUGIMOTO, K. HIROSHIGE and Y. NIHO: Necrotizing myelopathy associated with hepatocellular carcinoma. Jpn. J. Med. 27 (1988) 333–336.

MONRAD-KROHN, G. H.: Die klinische Untersuchung des Nervensystems. Stuttgart, Thieme. (1954).

MOORE, M. R., D. D. BLATTER, J. WEISSMANN and H. R. TYLER: Acute aortic thrombosis causing sudden paraplegia in a patient with known thoraco-lumbar spinal metastasis: The diagnostic usefulness of magnetic resonance imaging. Neurosurgury 25 (1989) 105–109.

NASSAR, S. I., J. W. CORREL and E. M. HOUSEPION: Intramedullary cystic lesions of the conus medullaris. J. Neurol. Neurosurg. Psychiatry. 31 (1968) 106–109.

NORSTROM, C. W., J. W. KERNOHAN and J. G. LOVE: One hundred primary caudal tumors. J. Am. Med. Assoc. 178 (1961) 1071–1077.

ODAIBO, S. K., I. A. AWOGUN and K. OSHAGBEMI: Paraplegia complicating dracontiasis. J. R. Coll. Surg. Edinburgh 31 (1986) 376–378.

ORGOGOZO, J. M., and J. BOGOUSSLAVSKY: Lacunar syndromes. In: P. J. Vinken, and G. W. Bruyn, H. L. Klawans and J. F. Toole (Eds.), Handbook of Clinical Neurology, Vol. 54: Vascular Diseases, Part II, Ch. 14. Amsterdam, Elsevier (1989) 235–269.

PALMER, J. J.: Radiation myelopathy. In: P. J. Vinken and G. W. Bruyn, (Eds.), Handbook of Clinical Neurology, Vol. 26: Injuries of the Spine and Spinal cord, Part II, Ch. 8. Amsterdam, North-Holland Publ. Co. (1976) 81–95.

PANSE, F.: Electrical trauma. In: P. J. Vinken and G. W, Bruyn (Eds.), Handbook of Clinical Neurology, Vol. 23: Injuries of the Brain and Skull, Part I, Ch. 34. Amsterdam, North-Holland Publ. Co. (1975) 683–729.

PIALAT, J., B. BANCEL, P. PIERLUCA, I. PELISSOU: Le gangliogliome intra-médullaire. Une observation, revue de la littérature. Ann. Pathol. 7 (1987) 41–46.

PLANT, G. T., T. RÉVÉSZ, R. O. BARNARD, A. E. HARDING and P. C. GAUTIER-SMITH: Familial cerebral amyloid angiopathy with nonneuritic amyloid plaque formation. Brain 113 (1990) 721–747.

PORRINI, A. A., J. A. MALDONADO COCCO and O. GARCIA-MORTEO: Spinal artery steal syndrome in Paget's disease of bone. Clin. Exp. Rheumatol. 5 (1987) 377–378.

PRASAD, L. S. and R. K. SHARAN: Lathyrism In: P. J. Vinken and G. W. Bruyn (Eds.), Handbook of Clinical Neurology, Vol. 36: Intoxications of the Nervous System, Part I, Ch. 19. Amsterdam, North-Holland Publ. Co. (1979) 505–514.

RAND, C. and M. A. SMITH: Anterior spinal tuberculosis; paraplegia following laminectomy. Ann. R. Coll. Surg. Engl. 71 (1989) 105–109.

REDDY, D. R.: Skeletal fluorosis. In: P. J. Vinken and G. W. Bruyn (Eds.), Handbook of Clinical Neurology, Vol. 36: Intoxications of the Nervous System, Part I, Ch. 18. Amsterdam, North-Holland Publ. Co. (1979) 465–504.

RIGGLE, P. and M. A. ODDI: Spinal cord necrosis and paraplegia as complication of the intraaortic balloon. Crit. Care Med. 17 (1989) 475–476.

ROMÁN, G. C.: Tropical spastic paraparesis and HTLV-1 myelitis. In: P. J. Vinken and G. W. Bruyn, H. L. Klawans and R. R. McKendall (Eds.), Handbook of Clinical Neurology, Vol. 56: Viral Disease, Ch. 29. Amsterdam, Elsevier (1988) 523–542.

ROSS, E. D.: Localisation of the pyramidal tract in the internal capsule by whole brain dissection. Neurology 30 (1980) 59–64.

SAWHNEY, I. M. S., J. S. CHOPRA, S. K. BANSAL and A. K.-GUPTA: Eales' disease with myelopathy. Clin. Neurol. Neurosurg. 88 (1986) 213–215.

SCHILLING, R. F.: Is nitrous oxide a dangerous anesthetic for vitamin B_{12}-deficient subjects? J. Am. Med. Assoc. 255 (1986) 1605–1606.

SCOTT, and GIOTI: Late paraplegia as a consequence of intraaortic balloon pump. Ann. Thorac. Surg. 40 (1985) 300–303.

SCRIMGEOUR, E. M., J. KAVEN and D. C. GAJDUSEK: Spinal tuberculosis — The commonest cause of non-traumatic paraplegia in Papua New Guinea. Tropic. Geographic. Med. 39 (1987) 218–221.

SEIFERT, P. E. and N. A. SILVERMAN: Late paraplegia resulting from intraaortic balloon pump. Ann. Thorac. Surg. 41 (1986) 700–702.

SERVADEI, F., C. TREVISAN, G. BIANCHEDI and R. PADOVANI: Spontaneous dorsal epidural haematoma: Usefulness of magnetic resonance imaging and importance of operative treatment even in cases with complete paraplegia. Acta Neurochir. (Wien) 89 (1987) 137–139.

SHAW, M. D. M., J. D. RUSSEL and K. W. GROSSART: The changing pattern of spinal arachnoiditis. J. Neurol. Neurosurg. Psychiatry 41 (1978) 97–107.

SHETH, N. K., B. VARKEY and D. K. WAGNER: Spinal cord aspergilloma. Am. J. Clin. Pathol. 84 (1985) 763–769.

SOBUE, I: Clinical aspects of SMON. In: P. J. Vinken, G. W. Bruyn, (Eds.), Handbook of Clinical Neurology, Vol. 37: Intoxications of the Nervous System, Part II, Ch. 5. Amsterdam, North-Holland Publ. Co. (1979) 115–140.

SOTELO, J.: Cysticercosis. In: P. J. Vinken, G. W. Bruyn, H. L. Klawans and A. A. Harris (Eds.), Handbook of Clinical Neurology Vol. 8(52) Microbial Disease, Ch. 38. Amsterdam, Elsevier (1988) 529–534.

STANBURY, J., J. B. WIJINGAARDEN and D. S. FREDERICKSON: The Metabolic Basis of Inherited Diseases, 2 Vols, 5th ed. New York, McGraw Hill Cy. (1983).

STILL, CH. N.: Nicotinic acid and nicotinamide deficiency. In: P. J. Vinken and G. W. Bruyn (Eds.), Handbook of Clinical Neurology, Vol. 28: Metabolic and Deficiency Diseases of the Nervous System, Part II, Ch. 4. Amsterdam, North-Holland Publ. Co. (1976) 49–58.

STORM-MATHISEN, A.: Syphilis. In: P. J. Vinken and G. W. Bruyn (Eds.), Handbook of Clinical Neurology, Vol. 33: Infections of the Nervous System, Part I, Ch. 17. Amsterdam, North-Holland Publ. Co. (1978) 337–394.

SUCHET, I., C. KLEIN, T. HORWITZ, S. LALLA and M. DOODHA: Spinal cord schistosomiasis; A case report and review of the literature. Paraplegia 25 (1987) 491–496.

SUMPIO, B. E. and R. J. GUSBERG: Aortic thrombosis with paraplegia: An unusual consequence of blunt abdominal trauma. J. Vasc. Surg. 6 (1987) 12–14.

SWOVELAND, P. T. and K. P. JOHNSON: SSPE and other paramyovirus infections. In: P. J. Vinken, G. W. Bruyn, H. L. Klawans and R. R. McKendall (Eds.), Handbook of Clinical Neurology, Vol. 56: Viral Disease, Ch. 23. Amsterdam, Elsevier (1988) 417–437.

SYMONDS, C. and S. P. MEADOWS: Compression of the spinal cord in the neighbourhood of the foramen magnum. Brain 60 (1937) 52–76.

SYNEK, V. M. and G. L. GLASGOW: Recovery from alpha coma after decompression sickness complicated by spinal cord lesions at cervical and midthoracic levels. Electroencephalogr. Clin. Neurophysiol. 60 (1985) 417–419.

TASHIRO, CH., M. IWASAKI, K. NAKAHARA and I. YOSHIYA: Postoperative paraplegia associated with epidural narcotic administration. Can. J. Anaesth. 34 (1987) 190–192.

TERHEGGEN, H. G., A. LOWENTHAL and J. P. COLOMBO: Clinical and biochemical findings in argininemia. Adv. Exp. Biol. Med. 153 (1952) 111–116.

THOMPSON, S. A.: Extra-medullary haemopoiesis causing paraplegia. Australas. Radiol. 32 (1988) 141–143.

TJANDRA, J. J., T. R. K. VARMA and R. D. W. WEEKS: Spinal arachnoiditis following subarachnoid haemorrhage. Aust. N.Z. J. Surg. 59. (1989) 84–87.

TRELLES, J. O. and L. TRELLES: Neurological manifestations of verruga peruana. In: P. J. Vinken and G. W. Bruyn (Eds.), Handbook of Clinical Neurology, Vol. 34: Part II, Infections of the Nervous System, Ch. 36. Amsterdam, North-Holland Publ. Co. (1978) 659–673.

TROOST, D. and A. VAN ROSSUM: Cerebral calcifications and cerebellar hypoplasia in two children: clinical, radiologic and neuropathological studies — A separate neurodevelopmental entity. Neuropediatrics 15 (1984) 102–109.

USUBIAGA, J. E.: Neurological complications following epidural anesthesia. Int. Anesthesiol. Clin. 13 (1975) 1–153.

VON SEGESSER, L. K., H. BURKI, K. SCHNEIDER, R. SIEBENMANN, E. R. SCHMID and M. TURINA: Die Chirurgie der Aneurysmen der deszendierenden thorakalen Aorta und Paraplegie. Helv. Chir. Acta 55 (1988) 503–508.

WACKYM, PH. A., T. FEUERMAN, G. F. GADE and T. J. DUBROW: Ependymal cyst of the spinal cord presenting with acute paraplegia. J. Neurol. Neurosurg. Psychiatry 51 (1988) 885.

WARTENBERG, R.: The Examination of Reflexes. Chicago, Year Book Publ. Inc. (1945).

WARTENBERG, R.: Diagnostic Tests in Neurology. Chicago, Year Book Publ. Inc. (1952).

WEENINK, H. R. and G. W. BRUYN: Cryptococcosis of the nervous system. In: P. J. Vinken and G. W. Bruyn (Eds.), Handbook of Clinical Neurology, Vol. 35: Infections of the Nervous System, Part III, Ch. 22. Amsterdam, North-Holland Publ. Co. (1978) 459–502.

WEENINK, H. R. and G. W. BRUYN: Cryptococcosis. In: P. J. Vinken, G. W. Bruyn, H. L. Klawans and A. A. Harris (Eds.), Handbook of Clinical Neurology, Vol. 52: Microbial disease, Ch. 29. Amsterdam, Elsevier (1988) 429–436.

WERNER, R. A.: Paraplegia and quadriplegia after intrathecal chemotherapy. Arch. Phys. Med. Rehabil. 69 (1988) 1054–1056.

WHISLER, W. W.: Chronic spinal arachnoiditis. In: P. J. Vinken and G. W. Bruyn (Eds.), Handbook of Clinical Neurology, Vol. 33: Infections of the Nervous System, Part I, Ch. 13. Amsterdam, North-Holland Publ. Co. (1978) 263–274.

WHITFIELD, A. G. W.: Neurological complications of ankylosing spondylitis. In: P. J. Vinken and G. W. Bruyn (Eds.), Handbook of Clinical Neurology, Vol. 38: Neurological Manifestations of Systemic Diseases, Part I, Ch. 20. Amsterdam, North-Holland Publ. Co. (1979) 505–520.

WINN, R. E. and A. I. HARTSTEIN: Coccidiodomycosis. In: P. J. Vinken, G. W. Bruyn, H. L. Klawans and A. A. Harris (Eds.), Handbook of Clinical Neurology, Vol. 52: Microbial Disease, Ch. 28. Amsterdam, Elsevier (1988) 409–427.

WOLINSKY, J. S.: Progressive rubella panencephalitis. In: P. J. Vinken, and G. W. Bruyn (Eds.), Handbook of Clinical Neurology, Vol. 34: Infections of the Nervous System, Part II, Ch. 17. Amsterdam, North-Holland Publ. Co. (1978) 331–341.

WOLTERS, E. CH., G. K. VAN WIJNGAARDEN and F. C. STAM: Leucoencephalopathy after inhaling heroin pyrolysate. Lancet 2 (1982) 1233–1237.

WOODHAM, M. J. and M. H. HANNA: Paraplegia after coeliac plexus block. Case report. Anaesthesia 44 (1989) 487–489.

WORSTER-DROUGHT, C., T. R. HILL, and W. H. MCMENEMEY: Familial cerebral amyloid angiopathy presenting as recurrent cerebral haemorrhage. J. Neurol. Psychopathol. 14 (1933) 27–34.

WORSTER-DROUGHT, C., J. G. GREENFIELD and W. H. MCMENEMEY: A form of familial presenile dementia with spastic paralysis: (including the pathological examination of a case). Brain 63 (1940) 237–254.

WORSTER-DROUGHT, C., J. G. GREENFIELD and W. H. MCMENEMEY: A form of familial presenile dementia with spastic paralysis. Brain. 67 (1944) 38–43.

YAQUB, B. A.: Neurologic manifestations of heatstroke at the Mecca pilgrimage. Neurology 37 (1987) 1004–1006.

YOKOTA, N., T. KURIBAYASHI, M. NAGAMINE, M. TANAKA, S. MATSUKURA and S. WAKISAKA: Paraplegia caused by brown tumor in primary hyperparathyroidism. J. Neurosurg. 71 (1989) 446–448.

ZENTNER, J., W. HASSLER, J. GAWEHN and G. SCHROTH: Intramedullary cavernous angiomas. Surg. Neurol. 31 (1989) 64–68.

ZIEVE, L., D. F. MENDELSOHN and M. GOEPFERT: Shunt-encephalopathy, occurrence of permanent myelopathy. Ann. Int. Med. 53 (1960) 53–62.

ZULL, D. N. and R. CYDULKA: Acute paraplegia: a presenting manifestation of aortic dissection. Am. J. Med. 84 (1988) 765–770.

Handbook of Clinical Neurology, Vol. 15 (59): Diseases of the Motor System
J.M.B.V. de Jong, editor
© Elsevier Science Publishers B.V., 1991

CHAPTER 25

HTLV-1-associated motor neuron disease

GUSTAVO C. ROMÁN[1], JEAN-CLAUDE VERNANT[2] and MITSUHIRO OSAME[3]

[1] *Neuroepidemiology Branch, National Institute of Neurological Disorders and Stroke, National Institutes of Health, Bethesda, MD, U.S.A.,* [2] *Department of Neurology, Hôpital P. Zobda-Quitman, Centre Hospitalier Régional et Universitaire de Fort-de-France, Martinique, French West Indies, and* [3] *Third Department of Internal Medicine, Kagoshima University School of Medicine, Kagoshima, Japan*

Motoneuronal involvement is a well-recognized complication of human enteroviral infections, ranging from the typical and rather selective lesion of lower motor neurons caused by polioviruses in acute paralytic poliomyelitis, to the less frequent occurrence of paralytic forms of infection with some strains of coxsackieviruses, human echoviruses, and other enteroviruses (Jubelt and Lipton 1989). In the 1970s, numerous cases of lower motor neuron lesions were described in several countries during the pandemic spread of acute hemorrhagic conjunctivitis caused by the human enterovirus EV 70 (Hung and Kono 1979; Vejjajiva 1989). EV 71 has also been reported as a cause of aseptic meningitis and epidemic outbreaks of paralytic disease mimicking poliomyelitis (Melnick 1984). Among other viruses, tick-borne encephalitis viruses, the Vilyuisk encephalitis agent, and the Zil'ber agent have also been implicated as possible causes of motor neuron disease (Johnson 1976). Interest in the possible role of animal and human retroviruses in motoneuronal disease has been reawakened by the recent description of clinical forms resembling amyotrophic lateral sclerosis (ALS) among patients with chronic myelopathies associated with infection by the human T-lymphotropic virus type 1 (HTLV-1). This clinical form is reviewed here under the name pseudo-ALS.

MURINE NEUROTROPIC RETROVIRUSES

Animal retroviral infections may produce selective involvement of lower motor neurons. Gardner et al. (1973) described a spontaneous disease occurring naturally in wild mice (*Mus musculus*), which was characterized clinically by the development of hind limb paralysis and muscle atrophy in aging mice, occurring after a long latent period. Neuropathological lesions (Fig. 1) consisted of spongiosis, gliosis, and vacuolar motoneuronal degeneration of anterior lateral horns of the lumbosacral spinal cord, without inflammation (Andrews and Gardner 1974). Electron microscopy studies showed abundant intracytoplasmic type C viral particles (Figs. 2 and 3) inside anterior horn motor neurons (Gardner et al. 1976).

The disease was found to be caused by an animal retrovirus, the Casitas strain of murine leukemia virus (Cas-MuLV). This RNA tumor virus also produces spontaneous lymphomas in aging mice. Cas-MuLV is exogenously acquired early in life via maternal milk, and the pattern of infection discloses geographic and familial clustering. Resistance to infection with the neurotropic strains of MuLV is genetically determined, and infection early in life is required for the development of paralysis (Hoffman and Panitch 1989). Infected animals develop life-long viremia and tolerance to the virus. Delayed infection

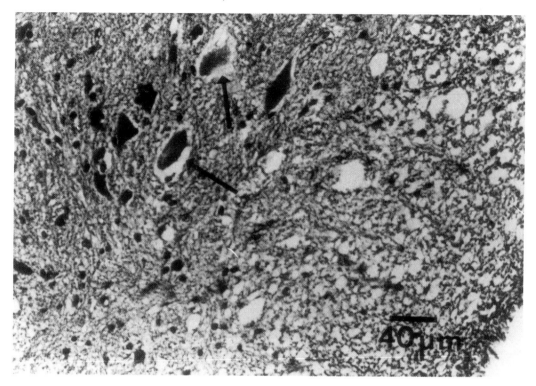

Fig. 1. Degenerating motor neurons (arrows) in anterior horn of lower spinal cord in LC wild mice. Prominent vacuolar changes are present in the neuropil. Notice absence of inflammatory cells. (Hematoxylin and eosin, ×250). Reproduced from Gardner et al. 1976 by courtesy of the Editors of *Amyotrophic Lateral Sclerosis: Recent Research Trends.*

beyond the 6th day of life confers resistance to paralysis (Robbins et al. 1989). Although the pathogenesis of the neural lesions is still unclear despite intensive research (Robbins et al. 1989), the causal etiology of the neurotropic MuLV has been demonstrated by the development of paralytic disease in NIH Swiss mice inoculated with virus passaged in tissue culture. Infectious virus was produced by injection of viral DNA genome into mouse fibroblasts in tissue culture (Openshaw 1989). Other MuLVs with neurotropic capacities include the WM 1504E strain of MuLV, temperature sensitive mutants of Moloney MuLV, and a rat-passaged strain of Friend MuLV (Hoffman and Panitch 1989).

HUMAN T-LYMPHOTROPIC VIRUS TYPE 1
(HTLV-1)

Prior to the description of chronic myelitides associated to the first human retrovirus, HTLV-

1, clinical cases with simultaneous and nearly symmetrical involvement of pyramidal tracts and lower motor neurons characteristic of classic ALS had not been observed in animal or human viral diseases. Consequently, the occurrence of forms of HTLV-1-associated myelopathy/tropical spastic paraparesis (HAM/TSP) mimicking ALS (ALS-like or pseudo-ALS forms) is of considerable importance. These cases have been well-documented (Vernant et al. 1989, 1990; Arimura et al. 1989; Evans et al. 1989) and will be reviewed here after a brief description of the general features of HAM/TSP. Detailed reviews of HAM/TSP have been recently published (Román 1989; Vernant and Román 1989; Román et al. 1990).

Clinical features

Following the isolation of HTLV-1 (Poiesz et al. 1980) and the demonstration of its etiologic role

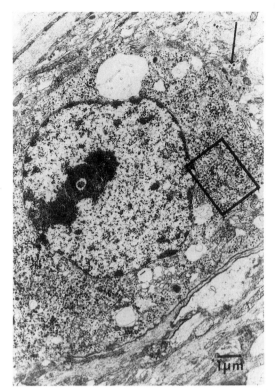

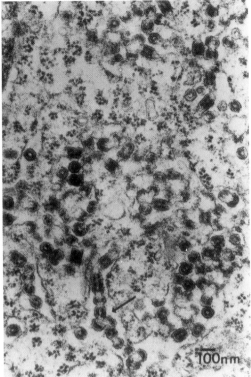

Fig. 2. Anterior horn motor neuron containing numerous intracytoplasmic aberrant type C virus particles. Extracellular type C virions are also present (arrow). The enclosed area is shown under higher magnification in Fig. 3. (Uranyl acetate and lead citrate, × 10 000.) Reproduced from Gardner et al. 1976 by courtesy of the Editors of *Amyotrophic Lateral Sclerosis: Recent Research Trends*.

Fig. 3. Aberrant type C virus particles budding into cisterns of rough endoplasmic reticulum of anterior horn neuron shown in Fig. 2. The particles have a cylindrical configuration, sometimes with multiple budding forms from a common stalk (arrow). (Uranyl acetate and lead citrate, × 60 000.) Reproduced from Gardner et al. 1976 by courtesy of the Editors of *Amyotrophic Lateral Sclerosis: Recent Research Trends*.

in adult T-cell leukemia (ATL) (Yoshida et al. 1982), high prevalence of HTLV-1 antibodies in the general population was documented in Jamaica and the French West Indies (Blattner et al. 1982; De Thé et al. 1983). In Martinique, a group of patients with tropical spastic paraparesis (TSP), diagnosed by Vernant (1989) over a period of 14 years, was found to have antibodies against HTLV-1 (Gessain et al. 1985). Simultaneously and independently, Osame et al. (1986) described in patients from Kyushu island, a region of Japan endemic for ATL, a chronic form of spastic paraparesis which they called HTLV-1-associated myelopathy (HAM).

Neuroepidemiologic studies of TSP by Román and collaborators (1985, 1987a, b) in the islands of Tumaco, off the Pacific coast of Colombia, and Mahé (Seychelles) in the Indian Ocean, demonstrated that despite the geographic distance, TSP in these areas presents with common clinical and epidemiologic features (Román and Román 1988).

These studies prepared the way for the demonstration of a common etiology. The association of HTLV-1 and TSP was soon confirmed by the presence of positive titers in serum and CSF of patients with TSP from Colombia and Jamaica (Rodgers-Johnson et al. 1985, 1988), as well as from the Seychelles islands (Román et al. 1987b). Finally, Román and Osame (1988) concluded that HTLV-1-associated TSP and HAM are essentially the same disease.

Clinical presentation of HAM/TSP. Onset of the disease occurs in late adult life, usually after 40 years of age. HAM/TSP tends to predominate in women, with a male-to-female ratio of 1:2.5 and a mean age of 53 years. In Japan, the geographic distribution of HAM cases parallels that of ATL (Osame et al. 1989). A ratio of 1 case of HAM per 3 600 carriers was observed and it is anticipated that the proportion will increase to 1 per 2 000 carriers. More than 700 cases of HAM have been reported in a nationwide survey in Japan as of March, 1989 (Osame et al. 1990).

HAM/TSP is characterized by a chronic and slowly progressive spastic paraparesis with back pain, impotence in males, and spastic bladder. There are minimal sensory signs and symptoms, mainly limited to loss of vibratory perception distally in the toes. In most patients, pyramidal tract signs, including symmetrically brisk knee reflexes, crossed adductor responses, clonus of ankles, and Babinski signs are found. Spasticity is moderate and affects the thigh adductors and, to a lesser extent thigh extensors and gastrocnemius muscles. Leg weakness involves the proximal muscle groups, mainly glutei and iliopsoas. This pattern of spasticity and weakness results in a typically slow scissoring gait, with conspicuous dragging and shuffling of the feet. The severe spasticity of lathyrism, with lurching gait on the toes is not seen in HAM/TSP. Peripheral nerve involvement is seen in up to 25% of HAM/TSP patients affecting predominantly the lower limbs. It is characterized mainly by loss of ankle reflexes, and decreased perception of vibration in the toes, without loss of proprioception.

Clinical neurophysiology. Somatosensory evoked responses are abnormal in almost 70% of HAM/TSP patients in a pattern suggestive of bilateral involvement of the fasciculus gracilis. Increased distal latencies with moderate slowing of distal and proximal (F-waves) motor conduction velocities in peroneal and tibial nerves has been demonstrated in some patients.

Other manifestations. An HTLV-1-associated inflammatory myopathy (polymyositis), with or without spastic paraparesis, has been documented (Goudreau et al. 1988; Tarras et al. 1989; Francis and Hughes 1989; Evans et al. 1989; Wiley et al. 1989; Masson et al. 1989; Morgan et al. 1989). Direct infection of muscle fibers by HTLV-1 with accumulation of the *tat* protein in the oxidative muscle fibers and inflammatory reaction against the virus appears to be the pathogenic mechanism (Wiley et al. 1989).

Among other non-neurologic manifestations of HTLV-1 infection, often seen in association with HAM/TSP, the following have been described (Román et al. 1989): pulmonary alveolitis is usually observed in close to 80% of the patients; and, less frequently, uveitis, arthropathy, Sjögren syndrome, vasculitis, xerosis and ichthyosis, cryoglobulinemia, and IgG monoclonal gammopathy. The dual occurrence of HAM/TSP and ATL or lymphoma remains exceptional (Kawai et al. 1989).

Laboratory examination. CSF examination is usually normal, although some patients may have lymphocytic pleocytosis (range: 10–200 mean: 9 cells/mm^3) during the acute stages. About 1% of the lymphocytes in the blood and CSF may show the typical 'flower' appearance of ATL. T-cell activation is reflected by the increase in OKT10 and OKIa1 positive cells; there is also an increase in the OKT4/OKT8 ratio with decrease of NK cells. Elevation of serum IgG (mean 1870 ± 524 mg/dl) and IgA (346 ± 218 mg/dl) are also common. Moderately increased CSF proteins (mean: 37 mg/dl S.D. = 18) with elevation of IgG, β_2-microglobulin, and oligoclonal bands occur in most patients. The blood-brain-barrier is intact and intrathecal synthesis of IgG antibodies against HTLV-1 is demonstrated by isoelectric focusing (Gessain et al. 1988). Polyclonal B cells activation is suggested by higher HTLV-1 antibody titers in serum (range: 512X–8192X) and CSF (range: 16X–1024X) of patients with myelopathy than in ATL patients or carriers (Kitajima et al. 1988).

HTLV-1 DNA can be detected in peripheral blood and CSF lymphocytes of patients with HAM/TSP by Southern blot hybridization or by enzymatic DNA amplification (Bhagavati et al. 1988; Yoshida et al. 1989). Isolation of HTLV-1 from patients with HAM/TSP has been reported by several groups (Hirose et al. 1986; Jacobson et al. 1988; Sarin et al. 1989; Gessain et al. 1989). Myelopathy viruses showed a 97–100%

homology with viruses isolated from ATL cases (Tsujimoto et al. 1988; Evangelista et al. 1990).

Diagnostic guidelines for HAM/TSP

Table 1 summarizes the clinical and laboratory guidelines for the diagnosis of HAM/TSP, based on the recommendations of the World Health Organization (WHO) Meeting of a Scientific Group on HTLV-1 Infections and Associated Diseases (Kagoshima, Japan, 10–15 December 1988). The use of the name HAM/TSP was recommended by this group (WHO 1989).

The clinical diagnosis of ALS is based on the simultaneous presence of usually symmetric signs of upper and lower motor neuron lesion indicative of disease of the voluntary motor system, along with absence of sensory and other neurologic signs, as well as sparing of extrinsic eye muscles and urinary sphincters. Although the presence of pyramidal signs, muscle atrophy, and prominent fasciculations has been reported in patients with HAM/TSP, these patients usually

TABLE 1
Clinical and laboratory diagnosis of HAM/TSP.

I. Clinical diagnosis

The florid clinical picture of chronic spastic paraparesis may not be present early in the disease. A single symptom or clinical sign may be found.

A. *Age and sex:* Mostly sporadic and adult onset with female preponderance, but sometimes familial and occasionally seen in childhood.

B. *Onset:* Usually insidious but may be sudden.

C. *Main neurologic manifestations:*
    (1) Chronic spastic paraparesis, usually slowly progressive, sometimes static after initial progression.
    (2) Weakness of lower limbs, more marked proximally.
    (3) Bladder disturbances usually an early feature; constipation usually occurs later, impotence or decreased libido are common.
    (4) Sensory symptoms, such as tingling, pins and needles, burning, etc. are more prominent than objective physical signs.
    (5) Low-lumbar pain with radiation to the legs is common.
    (6) Vibration sense is frequently impaired, proprioception less often affected.
    (7) Hyperreflexia of lower limbs, often with clonus and Babinski sign.
    (8) Hyperreflexia of upper limbs, positive Hoffmann and Trömner sign frequent, weakness may be absent.
    (9) Exaggerated jaw jerk in some patients.

D. *Less common neurological findings:* Cerebellar signs, optic atrophy, deafness, nystagmus, other cranial nerve deficits, tremor of fingers and hands, absent or depressed ankle jerks. Convulsions, cognitive impairment, confusion, dementia, or impaired consciousness are rare.

E. *Other associated neurological manifestations:* Muscular atrophy and fasciculations (Pseudo-ALS), polymyositis, peripheral neuropathy, polyradiculopathy, cranial neuropathy, meningitis, encephalopathy.

F. *Systemic manifestations which may be associated with HAM/TSP:* Pulmonary alveolitis, uveitis, Sjögren syndrome, arthropathy, vasculitis, ichthyosis, cryoglobulinemia, monoclonal gammopathy, adult T-cell leukemia/lymphoma.

II. Laboratory diagnosis

The diagnosis is usually confirmed by the following features.
A. Presence of HTLV-1 antibodies or antigens in blood and CSF.
B. CSF may show mild lymphocytic pleocytosis.
C. Lobulated lymphocytes may be present in blood and/or CSF.
D. Mild to moderate increase in CSF protein may be present.
E. Viral isolation from blood or CSF, when possible.

---

have other signs and symptoms, namely subtle
sensory deficits and bladder involvement, which
justify the use of the term 'pseudo-ALS' for this
clinical variant.

Clinical presentation

Based on reported cases (Vernant et al. 1989;
Arimura et al. 1989; Evans et al. 1989) the
following clinical features are usually present.

Age, sex and race.   Pseudo-ALS has been reported to occur at a mean age of onset older
than 60 years of age, with female preponderance,
affecting Black and Japanese patients. Age of
onset of pseudo-ALS appears to be significantly
later than in the general group of HAM/TSP
patients (mean age of onset: 53 years). Female
preponderance is common to both groups, and
Blacks and Japanese have been the racial groups
most commonly affected.

Initial symptoms.   Onset and progression of the
disease are usually slow and chronic, evolving
for a period of years. Initial symptoms were
those of upper motor neuron lesion affecting the
lower extremities, mainly difficulty walking, leg
weakness, back pain. Bladder problems, including
increased urinary frequency and incontinence,
and sensory symptoms were recorded in a third
of the patients. Abnormal sensations included
numbness and burning of the feet, and a band-
like feeling of tightness about the waist. The
above symptoms are practically constant in
HAM/TSP. Weight loss and wasting of the
muscles was reported in all patients. At the time
of diagnosis most patients complained of weakness of the upper extremities. The latter is an
uncommon complaint in patients with the classic
form of HAM/TSP.

Neurological signs.   Most patients presented
abnormally brisk reflexes in the 4 limbs, usually
with bilateral Babinski signs. Marked muscle
wasting and fasciculations were also constant,
usually involving small hand muscles (Fig. 4).
Atrophy of scapular and shoulder muscles and,
in one case, atrophy of the quadriceps, was also
reported. Prominent atrophy and fasciculations

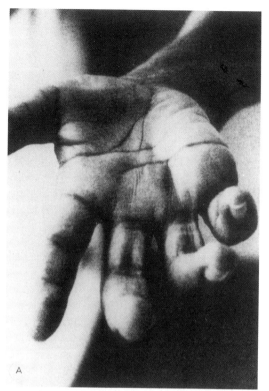

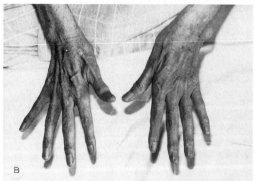

Fig. 4. Atrophy of small muscles of the hands in
patients with pseudo-ALS. A. Patient from Martinique with HTLV-1 infection and atrophy of median-
and ulnar-innervated hand muscles. Reproduced
from Vernant et al. 1989 by courtesy of the Editors
of *HTLV-I and the Nervous System*. B. Atrophy of
interossei hand muscles in a Japanese patient with
HTLV-1-associated pseudo-ALS. Reproduced from
Arimura et al. 1989 by courtesy of the Editors of
HTLV-I and the Nervous System.

of the tongue were present in 2 cases (Fig. 5)
and 1 patient had laryngeal paralysis with dysphonia. Dysphagia and dysarthria were not
commonly mentioned.

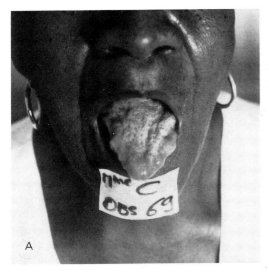

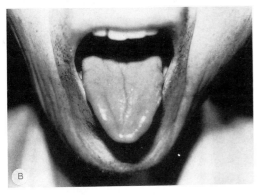

Fig. 5. Atrophy of the tongue in patients with pseudo-ALS. A. Black patient from Martinique with scalloping of the lateral aspects of the tongue resulting from prominent atrophy. Fasciculations were also present. Reproduced from Vernant et al. 1989 by courtesy of the Editors of *HTLV-I and the Nervous System*. B. Appearance of the tongue in Japanese patient in Fig. 4B. Atrophy and fasciculations of the tongue were present. Reproduced from Arimura et al. 1989 by courtesy of the Editors of *HTLV-I and the Nervous System*.

Clinical neurophysiology studies. In the few reported studies (Arimura et al. 1989; Evans et al. 1989) the sensory and motor nerve conduction velocities in the upper and lower limbs were usually normal, or minimally slow. F-waves were also normal. Needle EMG studies showed widespread denervation with high-amplitude motor unit potentials, fibrillation and fasciculation potentials. These changes were considered typical for ALS. Ludolph et al. (1988) observed changes

consistent with denervation in EMG studies in 5 of 19 patients with HAM/TSP from the Seychelles.

Evoked potentials were normal in the Japanese patient. In the black woman, these studies showed prolonged interpeak latencies and abnormal potentials consistent with a myelopathy.

Neuropathology studies. Muscle biopsies have been reported in 2 patients. In the patient from Japan (Arimura et al. 1989) biopsy showed small angulated fibers and group atrophy (Fig. 6) indicative of lower motor neuron disease. In a Black patient from the USA (Evans et al. 1989) the biopsy showed a combination of denervated angular fibers with type I fiber grouping, consistent with chronic denervation. In addition, this patient also showed a typical inflammatory myopathy.

No autopsy reports of the pseudo-ALS form of HAM/TSP are available. However, earlier description of neuropathologic changes in the Jamaican form of TSP indicated the presence of inflammation surrounding anterior horn cells, neuronophagia, as well as degeneration and loss of lower motor neurons (Robertson and Cruickshank 1972). Neuropathologic studies of patients with HAM/TSP in Japan have also shown central chromatolysis of anterior horn cells (Fig. 7) and mononuclear cell infiltration (Fig. 8).

Laboratory studies. HTLV-1 infection in these patients has been demonstrated by the presence of HTLV-1 antibodies in serum, confirmed by Western blotting, and, in a single case (Evans et al. 1989), also by polymerase chain reaction (PCR) in leukocytes. Of interest is the unpublished observation by one of the authors (J.-C.V.) of a 60-year-old woman with clinical ALS and negative HTLV-1 and HIV antibodies, who presented twice a positive PCR for a 210-base sequence of the HTLV-1 *gag* gene.

CONCLUSION

The occurrence of motor neuronal lesions in patients with HAM/TSP has been observed in areas of HTLV-1 endemia and appears to be well documented in the literature. The name 'pseudo-ALS' is suggested for these clinical forms with simultaneous and predominant involvement

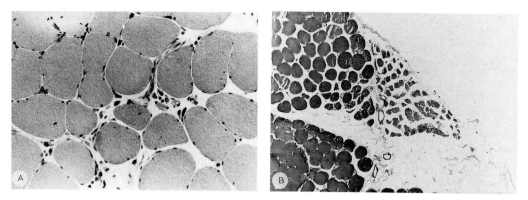

Fig. 6. Muscle biopsy (biceps brachii) from: (A) Japanese patient, and, (B) Martiniquaise woman with pseudo-ALS form of HAM/TSP. Notice small angulated fibers (arrowhead) and grouped fiber atrophy (long arrows) consistent with neurogenic atrophy. (Both, hematoxylin and eosin stain. Bar = 100 μm). (A) Reproduced from Arimura et al. (1989) by courtesy of the Editors of *HTLV-I and the Nervous System.*

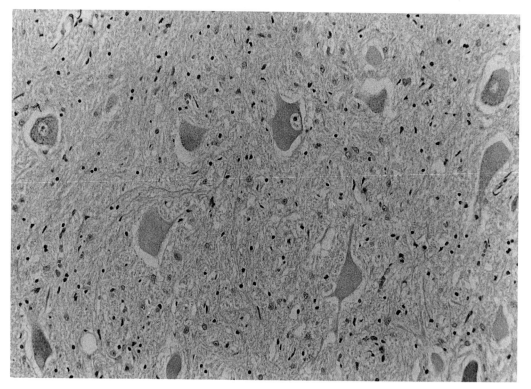

Fig. 7. Section of anterior horn (L$_4$ spinal cord level) of patient with HAM/TSP. Central chromatolysis of some large motor neurons is present. (Hematoxylin and eosin stain).

of lower motor neurons. The clinical progression appears to be more slow and benign than in classic bulbar ALS.

Determination of HTLV-1 antibodies in the serum and CSF of patients with ALS from endemic areas failed to demonstrate a significant difference in the number of positive cases when compared with patients with other neurologic diseases. However, the PCR technique should probably be used in these cases in view of the presence of positive PCR in seronegative patients. Nevertheless, it appears that only a minor propor-

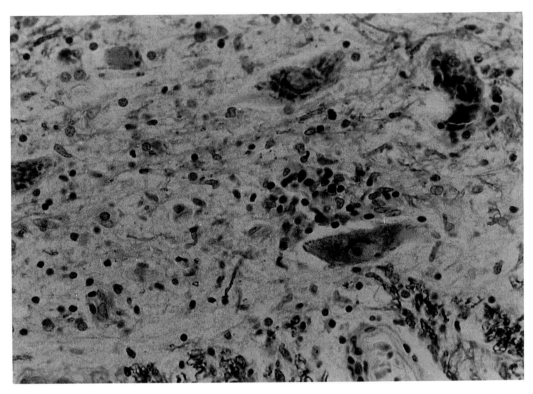

Fig. 8. Mononuclear cell infiltration surrounding motor neurons of anterior horn at the lumbar level in patient with HAM/TSP. (Klüver-Barrera stain).

tion of cases of HAM/TSP present with the pseudo-ALS form. The reasons for the selective lesion of anterior horn cells are unexplained at the present time. Further research in this field may offer clues for the causation of ALS.

REFERENCES

ANDREWS, J. M. and M. B. GARDNER: Lower motor neuron degeneration associated with type C RNA virus infection in mice: Neuropathologic features. J. Neuropathol. Exp. Neurol. 33 (1974) 285–307.

ARIMURA, K., H. NAKASHIMA, W. MATSUMOTO, H. OHSAKO, K. HASHIGUCHI, I. HIGUCHI and M. OSAME: HTLV-I-associated myelopathy (HAM) presenting with ALS-like features. In: G. C. Román, J.-C. Vernant and M. Osame (Eds.), HTLV-I and the Nervous System. New York, Alan R. Liss, Inc. (1989) 367–370.

BHAGAVATI, S., G. EHRLICH, R. W. KULA, S. KWOK, J. SNINSKY, V. UDANI and B. J. POIESZ: Detection of human T-cell lymphoma/leukemia virus type I DNA and antigen in spinal fluid and blood of patients with chronic progressive myelopathy. N. Engl. J. Med. 318 (1988) 1141–1147.

BLATTNER, W. A., V. S. KALYANARAMAN, M. ROBERT-GUROFF, T. A. LISTER, D. A. G. GALTON, P. S. SARIN, M. H. CRAWFORD, D. CATOVSKY, M. GREAVES and R. C. GALLO: The human type-C retrovirus, HTLV, in blacks from the Caribbean region, and relationship to adult T-cell leukemia/lymphoma. Int. J. Cancer 30 (1982) 257–264.

DE THÉ, G., M. F. MACLANE, N. MONPLAISIR, T. H. LEE and M. ESSEX: Présence d'anticorps antiHTLV (human T cell leukemia virus) chez les donneurs de sang de la Martinique. C. R. Acad. Sci. (Paris) 297 (1983) 195–197.

EVANGELISTA, A., S. MAROUSHEK, H. MINNIGAN, A. LARSON, E. RETZEL, A. HAASE, D. GONZALEZ-DUNIA, D. MCFARLIN, E. MINGIOLI, S. JACOBSON, M. OSAME and S. SONODA: Nucleotide sequence analysis of a provirus derived from an individual with tropical spastic paraparesis. Microb. Pathogen. 8 (1990) 259–278.

EVANS, B. K., I. GORE, L. E. HARRELL, T. ARNOLD and S. J. OH: HTLV-I-associated myelopathy and polymyositis in a US native. Neurology 39 (1989) 1572–1575.

FRANCIS, D. A. and R. A. C. HUGHES: Polymyositis and HTLV-I antibodies. Ann. Neurol. 25 (1989) 311.

GARDNER, M. B., B. E. HENDERSON, J. E. OFFICER, R. W. RONGEY, J. C. PARKER, C. OLIVER, J. D. ESTES and R. J. HUEBNER: A spontaneous lower motor neuron disease apparently caused by indigenous type-C

RNA virus in wild mice. J. Natl. Cancer Inst. 51 (1973) 1243–1254.

GARDNER, M. B., S. RASHEED, V. KLEMENT, R. W. RONGEY, J. C. BROWN, R. DWORSKY and B. E. HENDERSON: Lower motor neuron disease in wild mice caused by indigenous type C virus and search for a similar etiology in human amyotrophic lateral sclerosis. In: J. M. Andrews, R. T. Johnson and M. A. B. Brazier (Eds.), Amyotrophic Lateral Sclerosis: Recent Research Trends. New York, Academic Press (1976) 217–234.

GESSAIN, A., F. BARIN, J.-C. VERNANT, O. GOUT, L. MAURS, A. CALENDER and G. DE THÉ: Antibodies to human T-lymphotropic virus type-1 in patients with tropical spastic paraparesis. Lancet 2 (1985) 407–410.

GESSAIN, A., C. CAUDIE, O. GOUT, J.-C. VERNANT, L. MAURS, C. GIORDANO, G. MALONE, E. TOURNIER-LASSERVE, M. ESSEX and G. DE THÉ: Intrathecal synthesis of antibodies to human T lymphotropic virus type I and the presence of IgG oligoclonal bands in the cerebrospinal fluid of patients with endemic tropical spastic paraparesis. J. Infect. Dis. 157 (1988) 1226–1234.

GESSAIN, A., F. SAAL, V. MOROZOV ET AL.: Characterization of HTLV-I isolates and T lymphoid cell lines derived from French West Indian patients with tropical spastic paraparesis. Int. J. Cancer 43 (1989) 327–333.

GOUDREAU, G., G. KARPATI and S. CARPENTER: Inflammatory myopathy in association with chronic myelopathy in HTLV-I seropositive patients. (Abstract). Neurology 38 (Suppl. 1) (1988) 206.

HIROSE, S., Y. UEMURA, M. FUJISHITA, T. KITAGAWA, M. YAMASHITA, J. IMAMURA, Y. OHTSUKI, H. TAGUCHI and I. MIYOSHI: Isolation of HTLV-I from cerebrospinal fluid of a patient with myelopathy. Lancet 2 (1986) 397–398.

HOFFMAN, P. M. and H. S. PANITCH: Retrovirus infections including visna. In: P. J. Vinken, G. W. Bruyn and H. L. Klawans (Eds.), Handbook of Clinical Neurology, Vol. 56, R. R. McKendall (Ed.), Viral Disease. Amsterdam, Elsevier Science Publishers B. V. (1989) 453–466.

HUNG, T. and R. KONO: Neurological complications of acute haemorrhagic conjunctivitis (a polio-like syndrome in adults). In: P. J. Vinken, G. W. Bruyn and H. L. Klawans (Eds.), Handbook of Clinical Neurology, Vol. 38: Neurological Manifestations of Systemic Diseases, Part I. Amsterdam, North-Holland Publishing Co. (1979) 595–623.

JACOBSON, S., C. S. RAINE, E. S. MINGIOLI and D. E. MCFARLIN: Isolation of an HTLV-I-like retrovirus from patients with tropical spastic paraparesis. Nature (Lond.) 331 (1988) 340–343.

JOHNSON, R. T.: Virological studies of amyotrophic lateral sclerosis: an overview. In: J. M. Andrews, R. T. Johnson and M. A. B. Brazier (Eds.), Amyotrophic Lateral Sclerosis: Recent Research Trends. New York, Academic Press, (1976) 173–180.

JUBELT, B. and H. L. LIPTON: Enterovirus infections. In: P. J. Vinken, G. W. Bruyn and H. L. Klawans (Eds.), Handbook of Clinical Neurology, Vol. 56, R. R. McKendall (Ed.), Viral Disease. Amsterdam, Elsevier Science Publishers B. V. (1989) 307–347.

KAWAI, H., Y. NISHIDA, M. TAKAGI, K. NAKAMURA, K. MASUDA, S. SAITO and A. SHIRAKAMI: HTLV-I-associated myelopathy with adult T-cell leukemia. Neurology 39 (1989) 1129–1131.

KITAJIMA, I., M. OSAME, S. IZUMO and A. IGATA: Immunological studies of HTLV-I-associated myelopathy. Autoimmunity 1 (1988) 125–131.

LUDOLPH, A. C., J. HUGON, G. C. ROMAN, B. S. SCHOENBERG and P. S. SPENCER: A clinical neurophysiologic study of tropical spastic paraparesis. Muscle Nerve 11 (1988) 392–397.

MASSON, C., M. P. CHAUNU, D. HENIN, M. MASSON and J. CAMBIER: Myelopathie, polymyosite et manifestations sytemiques associees au virus HTLV-I. Rev. Neurol. (Paris) 149 (1989) 838–841.

MELNICK, J. L.: Enterovirus type 71 infections: a varied clinical pattern sometimes mimicking paralytic poliomyelitis. Rev. Infect. Dis. 6 (1984) S387–S390.

MORGAN, O. ST. C., P. RODGERS-JOHNSON, C. MORA and G. CHAR: HTLV-I in polymyositis in Jamaica. Lancet 2 (1989) 1184–1187.

OPENSHAW, H: Animal viral disease as models of CNS disease in man. In: P. J. Vinken, G. W. Bruyn and H. L. Klawans (Eds.), Handbook of Clinical Neurology, Vol. 56, R. R. McKendall (Ed.), Viral Disease. Amsterdam, Elsevier Science Publishers B. V. (1989) 581–592.

OSAME, M., K. USUKU, S. IZUMO, N. IJICHI, H. AMITANI, A. IGATA, M. MATSUMOTO and M. TARA: HTLV-I associated myelopathy, a new clinical entity. Lancet 1 (1986) 1031–1032.

OSAME, M., A. IGATA and M. MATSUMOTO: HTLV-I-associated myelopathy (HAM) revisited. In: G. C. Roman, J.-C. Vernant, M. Osame (Eds.), HTLV-I and the Nervous System. New York, Alan R. Liss, Inc. (1989) 213–223.

OSAME, M., R. JANSSEN, H. KUBOTA, H. NISHITANI, A. IGATA, S. NAGATAKI, M. MORI, I. GOTO, H. SHIMABUKURO, R. KHABBAZ and J. KAPLAN: Nationwide survey of HTLV-I-associated myelopathy in Japan: Association with blood transfusion. Ann. Neurol. 28 (1990) 50–56.

POIESZ, B. J., F. W. RUSCETTI, A. F. GAZDAR, P. A. BUNN, J. D. MINNA and R. C. GALLO: Detection and isolation of type-C retrovirus particles from fresh and cultured lymphocytes of a patient with cutaneous T-cell lymphoma. Proc. Natl. Acad. Sci. USA 77 (1980) 7415–7419.

ROBBINS, D. S., J. A. BILELLO and P. M. HOFFMAN: Pathogenesis and treatment of neurotropic murine leukemia virus infections. In: G. C. Román, J.-C. Vernant and M. Osame (Eds), HTLV-I and the Nervous System. New York, Alan R. Liss, Inc. (1989) 575–587.

ROBERTSON, W. G. and E. K. CRUICKSHANK: Jamaican (tropical) myeloneuropathy. In: J. Minckler (Ed.), Pathology of the Nervous System, Vol. 3. New York, McGraw-Hill Book Co. (1972) 2466–2476.

RODGERS-JOHNSON, P., D. C. GAJDUSEK, O. ST. C. MORGAN, V. ZANINOVIC, P. S. SARIN and D. S. GRAHAM: HTLV-I and HTLV-III antibodies and tropical spastic paraparesis. Lancet 2 (1985) 1247–1248.

RODGERS-JOHNSON, P., O. ST. C. MORGAN, C. MORA, P. SARIN, M. CERONI, P. PICCARDO, R. M. GARRUTO, C. J. GIBBS JR. and D. C. GAJDUSEK: The role of HTLV-I in tropical spastic paraparesis. Ann. Neurol. 23 (Suppl.) (1988) S121–S126.

ROMÁN, G. C.: Tropical spastic paraparesis and HTLV-I myelitis. In: P. J. Vinken, G. W. Bruyn and H. L. Klawans (Eds.), Handbook of Clinical Neurology, Vol. 56, R. R. McKendall (Ed.), Viral Disease. Amsterdam, Elsevier Science Publishers B. V. (1989) 525–542.

ROMÁN, G. C. and L. N. ROMÁN: Tropical spastic paraparesis: a clinical study of 50 patients from Tumaco (Colombia) and review of the worldwide features of the syndrome. J. Neurol. Sci. 87 (1988) 121–138.

ROMÁN, G. C. and M. OSAME: Identity of HTLV-I-associated tropical spastic paraparesis and HTLV-I-associated myelopathy. Lancet 1 (1988) 651.

ROMÁN, G. C., L. N. ROMÁN, P. S. SPENCER and B. S. SCHOENBERG: Tropical spastic paraparesis: a neuroepidemiological study in Colombia. Ann. Neurol. 17 (1985) 361–365.

ROMÁN, G. C., P. S. SPENCER, B. S. SCHOENBERG, J. HUGON, A. LUDOLPH, P. RODGERS-JOHNSON, B. O. OSUNTOKUN and C. F. SHAMLAYE: Tropical spastic paraparesis in the Seychelles islands. A clinical and case-control neuro-epidemiologic study. Neurology 37 (1987a) 1323–1328.

ROMÁN, G. C., B. S. SCHOENBERG, D. L. MADDEN, J. L. SEVER, J. HUGON, A. LUDOLPH and P. S. SPENCER: Human T-lymphotropic virus type I antibodies in the serum of patients with tropical spastic paraparesis in the Seychelles. Arch. Neurol. 44 (1987b) 605–607.

ROMÁN, G. C., J.-C. VERNANT and M. OSAME (Eds): HTLV-I and the Nervous System. New York, Alan R. Liss, Inc. (1989).

ROMÁN, G. C., L. N. ROMÁN and M. OSAME: Human T Lymphotropic Virus Type I Neurotropism. Prog. Med. Virol. 37 (1990) 190–210.

SARIN, P. S., P. RODGERS-JOHNSON, D. K. SUN ET AL.: Comparison of a human T-cell lymphotropic virus type I strain from cerebrospinal fluid of a Jamaican patient with tropical spastic paraparesis with a prototype human T-cell lymphotropic virus type I. Proc. Natl. Acad. Sci. U.S.A. 86 (1989) 2021–2025.

TARRAS, S., W. A. SHEREMATA, S. SNODGRASS and R. AYYAR: Polymyositis and chronic myelopathy associated with presence of serum and cerebrospinal fluid antibody to HTLV-I. In: G. C. Roman, J.-C. Vernant and M. Osame (Eds.), HTLV-I and the Nervous System. New York, Alan R. Liss, Inc. (1989) 435–441.

TSUJIMOTO, A., T. TERUUCHI, J. IMAMURA, K. SHIMOTOHNO, I. MIYOSHI and M. MIWA: Nucleotide sequence analysis of a provirus derived from HTLV-I-associated myelopathy (HAM). Mol. Biol. Med. 5 (1988) 29–42.

VEJJAJIVA, A.: Acute hemorrhagic conjunctivitis with nervous system complications. In: P. J. Vinken, G. W. Bruyn and H. L. Klawans (Eds.), Handbook of Clinical Neurology, Vol. 56, R. R. McKendall (Ed.), Viral Disease. Amsterdam, Elsevier Science Publishers B. V. (1989) 349–354.

VERNANT, J.-C.: From 'myélite Martiniquaise' to HTLV-I-associated paraplegia. In: G. C. Román, J.-C. Vernant and M. Osame (Eds.), HTLV-I and the Nervous System. New York, Alan R. Liss, Inc. (1989) xli–xliii.

VERNANT, J.-C. and G. C. ROMÁN: Les paraplégies associées au virus HTLV-I. Rev. Neurol. (Paris) 145 (1989) 260–266.

VERNANT, J.-C., G. BUISSON, R. BELLANCE, M. A. FRANÇOIS, O. MADKAUD and O. ZAVARO: Pseudo-amyotrophic lateral sclerosis, peripheral neuropathy and chronic polyradiculoneuritis in patients with HTLV-1-associated paraplegias. In: G. C. Román, J.-C. Vernant, M. Osame (Eds), HTLV-I and the Nervous System. New York, Alan R. Liss, Inc. (1989) 361–365.

VERNANT, J.-C., R. BELLANCE, G. G. BUISSON, S. HAVARD, J. MIKOL and G. ROMÁN: Peripheral neuropathies and myositis associated to HTLV-I infection in Martinique. In: W. A. Blattner (Ed.), Human Retrovirology: HTLV. New York, Raven Press, Ltd. (1990) 225–235.

WHO REGIONAL OFFICE FOR THE WESTERN PACIFIC: Report of the Scientific Group on HTLV-I Infections and Associated Diseases, Kagoshima, Japan, 10–15 December 1988. Document (WP) CDS (S)/ICP/PHC/014, Manila, Philippines, March, (1989). WHO Wkly. Epidem. Rec. 49 (1989) 382–383.

WILEY, C. A., M. NERENBERG, D. CROS and M. C. SOTO-AGUILAR: HTLV-I polymyositis in a patient also infected with the human immunodeficency virus. N. Engl. J. Med. 320 (1989) 992–995.

YOSHIDA, M., I. MIYOSHI and Y. HINUMA: Isolation and characterization of retrovirus from cell lines of human adult T cell leukemia and its implications in the disease. Proc. Natl. Acad. Sci. U.S.A. 79 (1982) 2031–2035.

YOSHIDA, M., M. OSAME, H. KAWAI, M. TOITA, N. KUWASAKI, Y. NISHIDA, Y. HIRAKI, K. TAKAHASHI, K. NOMURA, S. SONODA, N. EIRAKU, S. IJICHI and K. USUKU: Increased replication of HTLV-I in HTLV-I-associated myelopathy. Ann. Neurol. 26 (1989) 331–335.

Handbook of Clinical Neurology, Vol. 15 (59): Diseases of the Motor System
J.M.B.V. de Jong, editor
© Elsevier Science Publishers B.V., 1991

CHAPTER 26

Palliative treatment of motor neuron disease

RICHARD ALAN SMITH, ELLEN GILLIE and JONATHAN LICHT

Center for Neurologic Study, San Diego, CA, U.S.A.

Advances in the basic sciences make it certain that the cause of motor neuron disease (MND) will be understood and that a cure will be forthcoming. Sadly for those affected, no therapy currently available can arrest or modify the course of this dreaded disease. In short, MND remains one of the last scourges of humankind, not to have materially benefited from the medical and scientific progress which has characterized the twentieth century. In spite of this rather discouraging picture, the lives of individual patients have been enormously improved as a result of progress in patient care (Norris et al. 1985). Persons are now living and even working with MND for years as a result of medical interventions which support basic bodily functions and technological innovations which provide for higher human needs such as communication.

A number of important social factors have coalesced to bring us to this level of care. On the technical side, miniaturization has reduced the size of components which has favorably influenced not only the cost, but the reliability and transportability of medical devices. Use of a respirator, for example, no longer brings to mind confinement in an iron lung. Patients on ventilators are now free to move about their communities, and travel to distant places is possible. In addition to the availability of new technology,

the will to use it in support of the disabled is not thought extraordinary. Both patients' and physicians' attitudes towards prolongation of useful life has been transformed as a result of emphasis on the abilities rather than the disabilities of the handicapped (Smith 1987). While it has been argued that society may need to ration medical resources as a matter of efficiency there is almost universal agreement that these considerations are not germane in the context of the physician-patient relationship (Fried 1975). At this level of discussion the decision making process is rightfully shifting toward the patient at the expense of physician. This has had the effect of empowering patients — the consequence being that patients are now on a more equal footing with their care providers than they have been in the past (Cassel 1980).

For some curious reasons, patients with MND are invariably persons one would cherish as friends or family. As a group they make few demands on physicians who are committed to their care but they, like many persons with untreatable, chronic disease, resent caretakers who are indifferent or callous. As a rule MND patients have strong family ties. Recognizing this it is well advised to think of care in the context of the home. This necessitates involving family members as partners. Physicians may be humbled in this role since lay persons are often adept at picking up nuances of care that may be unfamiliar to trained professionals.

In general MND patients enjoy relatively sound health in contrast to numerous other diseases which threaten multiple systems, leading to organ failure, opportunistic infections, etc. Accordingly the health needs of patients with advanced MND may be relatively modest, allowing patients to live out their lives at home except on brief occasions when they require hospital care. This distinguishes MND from kidney failure, for example, which requires weekly medical visits for dialysis or AIDS which is associated with a variety of complications and necessitates frequent or prolonged hospitalization. Although not always feasible, home care is able to meet the daily needs of MND patients even when the disease is far advanced (Norris et al. 1987b).

For the physician and care providers who assume responsibility for MND patients it is important to organize a treatment plan in a systematic manner. To the degree possible it is essential to familiarize the patient and family with the treatment strategy. This requires an ongoing relationship which can take various forms. During an office visit it is necessary to set aside enough time to establish meaningful interaction. Unfortunately the economics of medical practice often intrude on this goal. Discounting this factor, it is certain that some physicians are more suited than others to the task of caring for the chronically ill (Gould 1980). Sensing this, some patients actively search out a compassionate caregiver, often consulting with a number of medical doctors at the onset of their illness until they find someone who is willing to make a commitment to them. Specialized centers have been developed in association with some hospitals (Hudson 1987). These take a multidisciplinary approach to care. In addition to medical services these centers often distribute educational materials and sponsor support groups which can put patients and family in touch with others who constitute a network of persons with kindred interests.

Once the diagnosis of MND has been established, emphasis for the patient switches to symptomatic and possibly experimental treatment. At the outset there are a number of non-specific recommendations that need to be touched upon. Many of these may reflect biases since they are often based on common sense or experience rather than medical fact, but they are nonetheless important. Patients with incurable illness understandably want to know whether factors such as mental attitude, nutrition, exercise, or physical therapy will favorably influence the course of their disease. Although the impact of these is uncertain, it cannot be excluded that non-specific factors may account for the long-term survival of some MND patients (Mulder and Howard 1976). There are patients who live with the disease for years, possibly because they have altered their lifestyle in some significant way.

Although one should avoid endorsement of a particular belief system, it is reasonable to suggest that mental attitude may influence the course of MND. Through discourse, reading or religious practice patients should be encouraged to explore their resolve to live. More tangible recommendations regarding activities of daily life are welcome. Considering that there may be a dietary link to MND, patients should review their eating habits (Spencer et al. 1987). Our own biases are towards a varied, high fiber, natural diet modestly supplemented with vitamins. At the minimum this dietary recommendation may favorably influence elimination which is commonly a problem. Consumption of highly processed, canned, artificially flavored and colored food is discouraged, but there is no perfect diet and in some areas and in some seasons strict adherence to these principles would be unrealistic due to expense or availability of fresh food stuffs. As for daily activities a regular program of exercise is recommended although patients are advised to avoid fatigue. Heat and massage may help alleviate cramps and muscular spasms. Although these and similar recommendations are not proven to be beneficial, at the minimum they provide comfort. In an age of computers and transplants we should not forget the possible therapeutic benefit of simple measures.

MENTATION AND MOOD

In interaction with MND patients most observers have the impression that they are stoical in attitude and generally in control of their circumstances. After study of a small group of patients it was concluded that this seemingly healthy adjustment was attained through the exercise of denial of depression and anxiety (Brown and Mueller 1970). Carrying this further it was suggested that there might be a premorbid MND personality which had etiologic implications. Fortunately, these assumptions have been challenged by several investigators. On a standard measure of personality (MMPI), MND patients were similar to persons with general medical illnesses (Peters et al. 1978). Studying a group of 40 patients, Houpt et al. concluded that the personality profile of MND patients was comparable to persons with other serious illnesses (Houpt et al. 1977). Using one test parameter, 32.5% of persons with MND were moderately depressed. Only 2.5% were severely so. A small percentage (10%) exhibited evidence for endogenous depression, suggesting that they might respond to drug therapy.

Although not unique to MND, pathologic laughter or tearfulness is almost pathognomonic for the disease. This may be exaggerated in some instances to the point of embarrassment. Considering that any behavioral idiosyncrasy is apt to impugn one's mental competence it is well advised to explain the basis for these symptoms to the patient and their acquaintances. The pathologic substrate for this behavior is generally assumed to result from bilateral involvement of corticobulbar pathways which is common in MND (Schiffer et al. 1985). The emotional lability which results from loss of cortical inhibition of the brainstem can be thought of as an example of a disconnection syndrome. On empirical grounds treatment with amitriptyline or levodopa can be considered (Udaka et al. 1984, Schiffer et al. 1985). These therapies have been reported to be effective in persons with cerebrovascular disease and multiple sclerosis and anecdotally in MND.

While isolated examples of dementia in association with MND have long been recognized, it is only recently that attention has focused on the impact of the disease on cognition. As a group Gallassi and his colleagues found slight cognitive impairment in a group of 22 patients in comparison with matched controls (Gallassi et al. 1985). This took the form of impaired reasoning, and abstraction and interference with planning and organization. Recall appeared to be spared. Another study comparing MND patients with a matched control group consisting of persons with non-dementing disease of the nervous system, failed to confirm these conclusions upholding the general impression that cognitive processes are spared in MND except in exceptional instances (Poloni et al. 1986). When dementia is associated with MND it is usually less severe than that associated with Alzheimer's disease. This might be expected from the pathology which, except for patients from the Western Pacific, is usually not associated with neurofibrillary change (Hudson 1987). When patients are linguistically or cognitively impaired this adds further hardship to an already difficult situation.

SPEECH

Depending on the mode of onset speech may be involved early or late in the course of MND. When it is lost it is one of the cruelest consequences of the disease. Under normal circumstances speech is made possible as a result of the modulation of air by the coordinated action of the vocal cords, palate, tongue, and lips which can be thought of as a series of musculoskeletal valves. As a consequence of weakness and spasticity, MND patients typically demonstrate a mixed dysarthria which is characterized by flaccid and spastic elements (Darley et al. 1969). With spasticity, hyperadduction of the vocal cords can result causing elevated laryngeal resistance to exhalation, and a characteristic asperity of voice. Weakness of one or both vocal cords, rare in MND, gives speech a breathy character associated with inhalatory stridor. Although vocal characteristics may vary depending on when in the course of the disease the patient is examined, typically patients with motor neuron disease exhibit hypernasality of speech due to escape of air into the nasal pharynx and difficulty in the pronunciation of consonants (Aronson 1980). Further, vocalization is usually strained, slow in rate, and even in pitch.

In general rehabilitative efforts to preserve

speech are unsatisfactory. A palatal lift has been reported to be useful in patients who exhibit velopharyngeal incompetence (Gonzalez and Aronson 1969). The use of silicone injection theoretically might benefit patients with weakness of the vocal cords (Smith et al. 1967). As a general measure attention should be directed to lessening secretions which can affect articulation. When inability to talk is due to breathlessness patients can regain the ability to speak after placement on a respirator. To achieve this, the tracheostomy tube must be uncuffed when the patient wishes to speak or a specially adapted tube with an independent air supply must be employed (Smith and Norris 1975).

NUTRITION

Most MND patients have inadequate caloric intake even when no or mild dysphagia is observed. This occurs when loss of appetite, difficulty with food preparation or feeding precede problems with chewing and swallowing. By the time patients exhibit difficulty with semi-solid and pureed foods they may have experienced severe weight loss (20–25%). Surprisingly most patients are at a loss to account for this, reporting that their diets are normal.

In normal fasting subjects, weight loss is accompanied by a state of negative nitrogen balance, indicating loss of protein as well as fat (Davidson 1987). Since there is reason to assume caloric restriction has the same effect on patients, it is likely that at least some of the muscle breakdown that occurs in MND is due to catabolism of muscle. Further there is evidence of insulin resistance in MND which would be expected to impair nutrition even with adequate caloric intake (Festoff 1987). Whatever the merits of these factors, it is well established that malnourished MND patients who undergo a feeding procedure stabilize or add to their weight. This would seem to argue for a dietary rather than metabolic cause for weight loss in MND.

All phases of deglutition can be affected in MND. Since the lower cranial nerves are disproportionately involved, chewing and swallowing are invariably affected in the course of the disease (Hillel and Miller 1989). On formal evaluation with cinefluorography MND patients show abnormalities in the first and second phases of swallowing. In the first stage there is slowness of the passage of the bolus from the mouth to the pharynx and poor nasopharyngeal occlusion causing nasal regurgitation. In the second, liquid pools in the vallecula and pyriform recesses and can enter the laryngeal vestibule before elevation of the larynx can protect the airway and prevent tracheal aspiration. Both bulbar and pseudobulbar disease can contribute to the problem.

A few simple principles can help patients maintain their weight and minimize the risk of aspiration. Liquids are usually harder to get down than solids. The taste and temperature of food can favorably influence swallowing. It is difficult for MND patients to eat with the head in an upright position. For a while attention to these details can help to offset a problem with mild dysphagia. More decisive action is dictated by 3 factors: evidence for aspiration, dehydration, or severe weight loss. When indicated supplemental feeding is easily undertaken. Accordingly, this should not be delayed until there is a crisis. Further it is not, in our judgement, appropriate to starve patients in the belief that this is a merciful, final act. For terminal patients a simple nasogastric feeding tube is appropriate. Newer versions are thin and flexible and thus well tolerated for several weeks. Their drawbacks include nasal and oropharyngeal irritation and for some a visible tube is offensive on aesthetic grounds. The position of the tube needs to be radiographically confirmed prior to institution of feeding to ensure correct placement in the small bowel. For long-term supplementation, percutaneous gastrostomy (PEG) is the procedure of choice (Larson et al. 1983). This ranks among the major advances in symptomatic care to have been made available to MND patients. This procedure, made possible as a result of technical innovation, has enjoyed wide acceptance, displacing cervical esophagostomy and cricopharyngeal myotomy, which deserve mention only in historical context. The appeal of PEG is its ease of implementation, minimal risk, and discomfort and avoidance of general anesthesia. Although it can be performed in an outpatient setting, brief hospitalization (1–2 days) is preferred (Fig. 1).

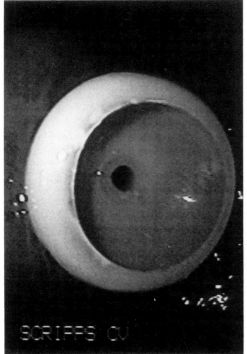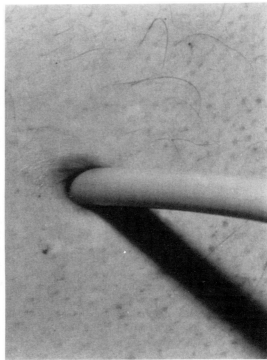

Fig. 1. Fiber optic view of the inside of the stomach following implantation of a percutaneous gastrostomy tube (a), exterior view of the tube as it exits the abdominal wall.

Ideally, the operation should be carried out as soon as indicated to minimize the risks due to dehydration and other factors. The puncture site is visualized either through an endoscope or in the radiology suite using fluoroscopy. Once the site of placement has been identified a stab wound is made through the abdominal wall with a large bore needle through which a guide wire and subsequently the tube is passed into the stomach. At the end of the procedure, air should be removed from the stomach to prevent abdominal distension which may compromise respiration. Within 12 hours water can be introduced into the tube. Thereafter, feeding with a hospital formula can begin initially at half strength and within 48 h at full strength. Feeding can be done by bolus, following the normal meal time pattern or by pump at a rate of 1 ml/min. This provides 1500–2000 calories per day. Complications include local and peritoneal infections, but serious problems are few and the wound site is usually well healed and easily maintained with minimal daily care. If aspiration is a concern due to loss

of protection of the pylorus a thinner tube can be passed into the jejunum. Continuous feeding will then be required.

RESPIRATION

The principal factors which can lead to respiratory complications in neuromuscular diseases are: (1) aspiration; (2) diaphragmatic failure; (3) impaired cough; (4) intercurrent respiratory infection; and (5) sleep apnea.

One of the main goals of regular follow up is to try to prevent potentially treatable problems such as aspiration from evolving into major, possibly fatal complications. Protection of the airway may involve a feeding procedure in which case a gastroenterologist may end up making a major contribution to the patient's pulmonary well being. Decisions of this nature can safely be made on clinical grounds, sometimes supplemented with tests such as a swallowing study which can demonstrate aspiration radiographically. To address the problem completely, attention must also be directed to control of secretions which can be aspirated independently of food and liquids. In the effort to prevent pneumococcal pneumonia and influenza, vaccination may be employed, or in the case of the flu, amantadine may be prescribed prophylactically if there is a known exposure. Use of IPPB (intermittent positive pressure breathing) or an incentive spirometer may prevent atelectasis which can further embarrass pulmonary function (Fig. 2). The benefits of these and a myriad of other recommendations of this sort have not specifically been tested in controlled studies. Diaphragmatic weakness is the principal cause for the inexorable deterioration of pulmonary function that is seen over time in patients with MND (Howard et al. 1989). At rest the diaphragm contributes the major force of respiration (Derenne et al. 1978). Since innervation of the diaphragm originates from the midcervical spinal cord it is unduly susceptible because the disease preferentially involves the cervical segments early in the course of motor neuron disease (Wohlfart and Swank 1941). Unfortunately the thoracic and accessory muscles

Fig. 2. Use of an incentive spirometer by a patient with MND.

of respiration, which are relatively spared due to their thoracic innervation, cannot easily compensate for diaphragmatic weakness. Two consequences result from this. There is progressive deterioration of pulmonary function, and clearing of secretions is made difficult due to the lack of a forceful cough.

Respiratory complaints which should alert the physician to diaphragmatic failure include the presence of breathlessness at rest, particularly while supine, poor cough, and diminution in the volume of speech. Orthopnea occurs because the abdominal contents fall upon the diaphragm when patients are flat in bed. If the diaphragm is weak this adds a notable burden, resulting in dyspnea. This can be formally documented by a change in the vital capacity when the patient goes from the upright to a supine position (Howard et al. 1989) (Fig. 3).

Pulmonary function studies can provide eloquent documentation of the respiratory status of patients with neuromuscular disease, although there are pitfalls to their use especially in persons who have trouble holding a mouthpiece firmly in place (Griggs et al. 1980). In a hospital setting elaborate studies can be undertaken, but in the office small, relatively inexpensive electronic spirometers can be employed. Diaphragmatic weakness should be assessed by determining the maximal inspiratory pressure (MIP). Respiratory fatigue is best measured by maximum voluntary ventilation (MVV), and evidence for restriction of flow by measurement of the forced expiratory volume in 1 s (FEV-1) obtained during a forced vital capacity (FVC).

Several detailed studies of pulmonary function have been undertaken in MND patients (Nakano et al. 1976; Fallat et al. 1979). Serial measurements in a group of patients demonstrated a gradual decline of FVC with more precipitous decline in the months before death. As a group MND patients show a decrease in total lung capacity and vital capacity and an increased residual volume. In some patients expiration may suddenly be interrupted, suggesting erratic closure of the glottis or vocal cords. This is best demonstrated in flow vs. volume studies (flow volume loops). Surprisingly patients with rather marked deterioration of ventilatory tests can

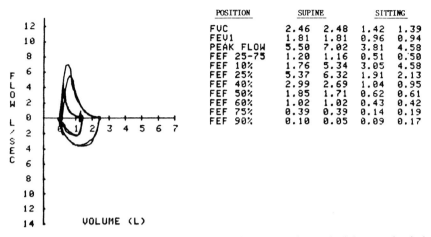

POSITION	SUPINE		SITTING	
FVC	2.46	2.48	1.42	1.39
FEV1	1.81	1.81	0.96	0.94
PEAK FLOW	5.50	7.02	3.81	4.58
FEF 25-75	1.20	1.16	0.51	0.50
FEF 10%	1.76	5.34	3.05	4.58
FEF 25%	5.37	6.32	1.91	2.13
FEF 40%	2.99	2.69	1.04	0.95
FEF 50%	1.85	1.71	0.62	0.61
FEF 60%	1.02	1.02	0.43	0.42
FEF 75%	0.39	0.39	0.14	0.19
FEF 90%	0.10	0.05	0.09	0.17

Fig. 3. Flow volume loop in sitting (a), and supine (b) positions. Note the marked decrease in vital capacity from 2.46 to 1.42 liters as well as the decrease in the flows and size of the loop.

maintain normal blood gases. Thus chronic hypoxemia and hypercapnia are not characteristic of MND.

Sleep studies are probably underutilized in the case of MND. On several bases MND patients ought to be unduly susceptible to disturbed sleep. Under normal circumstances, collapse of the upper airway on inspiration is prevented in part by muscular effort involving the tongue and pharyngeal muscles. Weakness of these muscles along with accumulation of secretions should lead to obstruction of the airway resulting in sleep apnea. Although expected, obstructive and central sleep apnea have only recently been described in MND (Carre et al. 1988, Howard et al. 1989). They are rarely seen as a presenting complaint. It is not certain that these problems occur as often as would be anticipated, possibly no more so than occurs in an older population. In patients who exhibit daytime somnolence or fatigue, treatment with continuous positive pressure ventilation (CPAP) should be considered at night with the goal of preventing hypoxemia and hypercapnia, sleep fragmentation and daytime symptoms (Howard et al. 1989).

Although respiratory failure is usually insidious in onset and evolution, it may occur suddenly, sometimes as the initial presenting symptom (Serpick et al. 1965). Respiratory distress is, of course, an ominous sign. In the majority of cases it signals a terminal stage of illness, but some patients may linger before succumbing to their illness. For a time it is possible to provide respiratory assistance in the form of negative pressure devices (cuirass, pneumobelt or tank), a rocking bed, or by mask using intermittent positive pressure, but these options are temporary solutions (Bach et al. 1987; Kerby et al. 1987). Over the short term, assistive devices and possibly medication may relieve muscular fatigue, which has been thought to complicate respiratory failure secondary to neuromuscular disease and carbon dioxide retention (Shiffman and Belsh 1989). Further, respiratory muscle fatigue can be exacerbated by intercurrent pulmonary infections which should be treated vigorously unless a decision has been made to defer further therapy. To initiate long-term ventilatory support it is necessary to proceed to tracheostomy and use of a positive pressure respirator. This can provide satisfactory ventilation for years.

A wise physician will anticipate the need for life support prior to the occurrence of respiratory arrest. Considering the tremendous financial and emotional demands, the decision making process should involve thoughtful deliberation rather than rash action. To facilitate this, the wishes of the patient and family about life support should be discussed at appropriate times during follow up and the patient should execute a directive to physicians. This insures compliance with the patient's wishes under most circumstances. Although there are no absolute guidelines, in our opinion, life support should be reserved for patients who can be taken care of at home (Norris et al. 1987a). Ideally the patient should maintain use of an arm or hand, although this is only a relative consideration since over time most patients become totally dependent. At the point where the patient feels that prolongation of life is without purpose, it is ethical to disconnect the respirator. Knowing that physicians are willing to do this should be a comfort to persons who elect to use a ventilator, but fear losing control over their destiny. The following case illustrates, from our point of view, a successful result.

The patient, a 66-year-old male with MND, dating from 1977, presented with weakness of the left wrist followed by progressive weakness of the limbs and neck musculature. As the disease progressed noted shortness of breath was noted, especially when supine. In April of 1980 the patient was hospitalized because of respiratory failure. The following day a tracheostomy was performed and the patient was placed on a Bennett MA-1 respirator. Postoperatively his course was complicated by aspiration pneumonia. With resolution of this problem attention was directed to returning the patient to his home. A portable, volume controlled respirator was obtained. The cuffed tracheostomy tube was replaced with a trachspeaking tube which allowed the patient to phonate without impeding ventilatory assistance. In the week before discharge the patient was placed on a general medical ward where his family and care providers were comprehensively trained in all aspects of his care. He was discharged after 3 weeks of hospitalization, after which he was cared for at home. A number of modifications were made there to facilitate bathing, transfer, and communication. A motorized van with a hydraulic lift was obtained to allow limited outings. Over the ensuing year speech was lost and in 1981 and 1982 he suffered bouts of pneumonia, secondary to aspiration. Due to persistent tracheal irritation the standard balloon cuffed tracheostomy tube which had been used since the patient lost the ability to speak was replaced with a model employing a foam cuff.

In spite of almost complete loss of motor function the patient learned to operate a personal computer using a microprocessor switch. A software package allowed him to interface with the computer using Morse code. In 1985 a percutaneous gastrostomy was performed when it became impossible for the patient to take food by mouth. This operation was complicated by cellulitis which responded to antibiotic therapy. In 1987 he developed cholecystitis. This was treated at home with intravenous antibiotics. Hospitalization for aspiration pneumonitis was again required in the summer of 1989. Daily bronchoscopy and antibiotics were successfully employed and the patient returned home.

This case, which is not unique in our experience represents the successful management of a patient with advanced MND over a period of 10 years. Although this has entailed the expenditure of enormous human and financial resources, hospitalization has only been required on a few occasions and this has never been prolonged. Ignoring his considerable handicap, the patient has managed a family business and maintained his role as head of his household.

SECRETIONS

Drooling is one of the more vexing problems to confront MND patients. In childhood salivary incontinence is acceptable, but for adults it is humiliating. On clinical grounds drooling appears to be secondary to muscular weakness which compromises the containment or transport of secretions (Smith and Goode 1970). Weakness of the lower lip permits overflow from the mouth in some instances, but generally it is difficult with swallowing that causes secretions to flood the oral cavity. Hypersalivation is unlikely to be the problem although this has not been carefully studied.

Drooling is usually subtle at the outset, first noted on the pillow in the morning. Later it is present while the patient is upright. Characteristically, patients will be seen to frequently dab themselves with tissue. This can go almost unnoticed or it can be extreme as a result of constant outpouring of saliva. Secretions may become extremely viscid, especially after consumption of milk products. This may be partially relieved with the use of potassium iodide (SSKI), n-acetylcysteine (mucomyst) or papase. Beyond the matter of inconvenience and embarrassment associated with drooling, aspiration of saliva poses a constant risk in some patients. For both these reasons considerable attention has been given to the management of this aspect of MND.

Approximately 1500 ml of saliva is produced daily, the bulk of it by the major salivary glands (Stuchell and Mandel 1988). Salivation is principally under the influence of the parasympathetic nervous system. On this basis the mainstay of therapy is the use of medication with anticholinergic effects. This can be administered orally or via a skin patch. Amitriptyline hydrochloride, given nightly at the outset and then during the day once patients have accommodated to its sedative effect, is generally beneficial. A scopolamine patch, applied behind the ear for 24–72 h, reduces salivary secretion more effectively than atropine (Talmi et al. 1988). Other drugs which may be employed include trihexyphenidyl hydrochloride, benzatropine mesylate, or propantheline bromide. It is important to recall that older patients medicated with anticholinergic drugs may develop untoward complications including confusion, hallucinations, and urinary retention.

For instances of intractable drooling a number of therapeutic strategies have been employed. These attempt to alter the functional competence of the salivary glands. Based on the fortuitous relation of the tympanic plexus and chorda tympani within the ear denervation of the salivary glands can be accomplished readily with tympanic neurectectomy (Goode and Smith 1970). Although technically feasible, glandular excision, salivary duct ligation or diversion, or radiotherapy are not advised.

WEAKNESS-SPASTICITY-CRAMPS

Although muscular weakness is the hallmark of MND, it is only recently that the pattern of weakness in MND has been characterized quantitatively. Munsat pioneered the longitudinal assessment of maximum voluntary isometric contraction as a means of documenting disease progression (Munsat et al. 1988). These studies indicated that the time of onset, degree of

weakness, and rate of deterioration were similar in adjacent anatomic regions, implying that MND exhibits a regional character. Over the course of the illness deterioration was found to be generally linear, although at the outset and end of the disease process patients could appear to be stable. Carrying this type of analysis further, Brooks discerned that spread of the disease is generally from closely approximated anatomic sites (Brooks et al. 1990). Thus spread is typically from one limb to its fellow.

With the exception of temporary facilitation of strength with administration of edrophonium chloride in the instance of a pseudomyasthenic syndrome which may occur in MND, there is little that can be currently done pharmacologically to alleviate weakness (Mulder et al. 1959). It is commonly asked whether exercise is advisable. One can infer from studies with training programs that this is beneficial. In MND both maximum oxygen consumption and work capacity drop in relation to loss of strength and the normal metabolic responses to prolonged submaximal exercise are impaired (Karpati et al. 1979; Sanjak et al. 1987). With resistive training endurance capacity has been shown to be considerably enhanced (McCartney et al. 1988). While medical professionals have historically been concerned about prescribing exercise for MND patients, due to the example of poliomyelitis in which overwork weakness has been described, it is reasonable to advise a program of exercise for muscles which retain some strength (Cashman et al. 1987; Milner-Brown and Miller 1988). Although unproven, this may retard muscular atrophy due to disuse, prevent contractures and minimize the risk of bedsores. When the patient is unable to move a limb, passive exercise is indicated. In the presence of proximal upper extremity weakness, range of motion exercises are essential to prevent the patient from developing frozen shoulders. This is a common, extremely painful condition which can be prevented with daily physical therapy.

Spasticity, leading to contractures and scissoring of the lower extremities, is a management problem in some instances. Generally, it is not prominent because lower motor neuron signs predominate. Classically, spasticity was thought to result from hyperactive gamma discharge. More current is the notion that supraspinal segments elicit presynaptic inhibition of fusimotor fibers (Ia terminals) through release of GABA from inhibitory interneurons (Young and Weighner 1987). In the case of a supraspinal lesion this inhibition is lost leading to hypertonus. Further, spastic patients may be more handicapped by reduction in reciprocal inhibition than by hyper-excitability of the stretch reflex itself.

The use of muscle relaxants in MND to deal with spasticity may be of value, but can be disadvantageous (Landau 1974). Hypertonus of the quadriceps and exaggeration of the extensor thrust reflex facilitate ambulation in paraparetic patients through a splinting effect. If medication is warranted, baclofen is probably the drug of choice. Prolonged administration of benzodiazepines can be associated with depression. Dantrium sodium, which acts peripherally by reducing the release of calcium from the sarcoplasmic reticulum during muscle contraction, is contraindicated (Rivera et al. 1975). In some patients it causes profound weakness. Temporary reduction of muscle tone may be obtained by cooling a limb with cold packs which reduces excitability of muscle spindles and the input from cutaneous pain receptors (Lenman 1985). In the rare instance of severe spasticity consideration can be given to the use of a motor point block to decrease spasticity of the hip adductors (Sinaki 1988).

Muscle cramps are an early, not generally persistent, complaint. They characteristically occur with stretching at which time they involve the calf, hands, abdominal or chest muscles. They are often preceded or accompanied by fasciculations in the same muscle. Electromyography (EMG) has demonstrated that cramps are usually confined to part of a particular muscle (Norris et al. 1957). They appear to be triggered by hyperexcitability of the intramuscular portion of motor nerve terminals, possibly as a result of distortion of the nerve terminals upon muscle shortening or as a consequence of metabolic changes in the extracellular space which attend muscle contraction. The immediate remedy for cramps entails static stretching of the involved muscle. This may activate the Golgi tendon

organs which subsequently inhibit the motor neuron supplying the affected muscles (Botte et al. 1988). Prophylactic treatment traditionally involves the administration of quinine sulfate or phenytoin which are membrane stabilizing agents. These are usually administered at night. Based on a recent report one might wonder whether these medications are as effective as generally thought (Warburton et al. 1987).

ATYPICAL SYMPTOMS

As a result of the predominance of motor symptoms seen in MND, there are a variety of symptoms which are not considered to be 'classical' features of the disease. The prototypic patient is thought to be one whose mind, senses and sphincters are spared. Further, the disease is generally pain-free except for the occurrence of cramps. Although this description usually holds, physicians who see large numbers of patients are aware of the occurrence of atypical symptoms. Patients may be demented, or have difficulties with language, and may exhibit extrapyramidal signs. Alteration of taste and smell, common late in the disease, may occur before the onset of paralysis (Steele et al. 1990). While sensory complaints are minimal, sophisticated testing of altered sensation can be demonstrated (Dyck et al. 1979). Severe pain, primarily involving the neck, back or limbs can bother patients at any time, even awakening them from sleep (Newrick and Langton-Hewer 1984). Loss of sphincter control may occur as a result of impairment of autonomic regulation at the spinal or supranuclear level (Mitsumoto et al. 1988). Breakdown of skin is not common to MND, but patients are not immune to this if they are in bed or in a wheelchair much of the time. Protective cushions and mattresses can minimize this problem.

When unusual symptoms occur, they obviously should bring to mind the possibility of complicating factors and in some instances an alternative diagnosis. As more is written about motor neuron disease, the full extent of the syndrome will obviously be better understood. It is possible that some of the unusual symptoms may be a clue to the etiology of the disease, but this remains to be proven. Many of the non-specific symptoms of MND are amenable to treatment, decubitus ulcers responding to cleansing and application of an artificial covering, incontinence to the use of oxybutynin chloride or a catheter and pain to the administration of analgesics.

ASSISTIVE DEVICES

The primary purpose of rehabilitation procedures in MND is to maintain optimal functioning in the course of daily living, but the relationship between functional ability and equipment use is complex (Lord et al. 1987). Initially, all that may be necessary are simple orthotics such as a wrist splint or a foot lift.

With progression of the disease, dependency on adaptive equipment increases. With loss of neck muscles, bracing is required to keep the head upright. The use of an underarm sling can minimize the effects of subluxation of the shoulders. For a time, specially designed utensils or dressing aids help the patient compensate for loss of hand function. While elaborate orthotic devices are sometimes prescribed, in practice these often prove to be cumbersome, expensive, and of limited utility (Newrick and Langton-Hewer 1984).

As ambulation becomes more difficult, a cane or a walker is necessary, but at some point a wheelchair is indicated. This becomes mandatory when the patient begins to fall. Contemporary light weight chairs are easier to use and more portable. Self-propelled vehicles are expensive, but if a patient is able to operate them, they offer a degree of independence. With the addition of a portable respirator even patients with advanced disease can be mobile using a van equipped with a hydraulic lift (Fig. 4).

Modifications in the home are encouraged since it is the ideal setting for long-term care. Barriers to the free passage of a wheelchair need to be eliminated, entrances ramped and bathrooms modified. Railings should be installed near the toilet and shower to provide support. A raised toilet seat and a chair in the shower can facilitate personal hygiene. A sliding board can be used to facilitate transfer from a wheelchair to a commode. If the patient is difficult to move, a hydraulic lift allows a lone care provider to

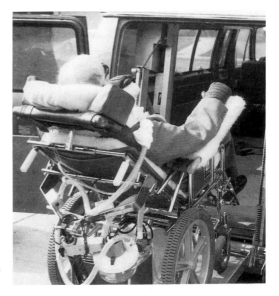

Fig. 4. Patient with advanced MND entering a van. Note motorized ramp and respirator attached to the wheelchair.

lower a patient in and out of a tub or from a bed to a wheelchair. An electric hospital type bed with overhead trapeze and an electric lift chair will fill out the furniture needs of most patients. A variety of mattresses are available which protect patients from developing decubitus ulcers.

The greatest single advance for handicapped patients has been the availability of a range of consumer electronics, including telephones with automatic dialing, intercoms, alarms, printers, speech synthesizers, and computers. Linking some of these permits patients to control lights and appliances, and to communicate. Unfortunately, no currently available equipment imitates the intimacy of normal speech. Further, the interface between patient and machine is not ideal especially for those who have limited use of their upper extremities. With even slight hand function, patients can input a computer using a microprocessor switch (Dymond et al. 1988). For completely paralyzed patients, an infra-red switch triggered by eye blink can be employed. Two basic computer strategies are utilized. The patient can address the computer directly using Morse code or assemble information from a menu which is scrolled automatically before him on a monitor

(Fig. 5). Irrespective of the system, communication is painstakingly slow. This can be augmented by triggering phrases that are stored in a memory bank. If patients choose, they can vocalize their thoughts using a speech synthesizer. In spite of these advances, it is still common to see patients with sophisticated equipment using simple means such as an alphabet or ETRAN board to get their ideas across (Fig. 6). Based on the rapid advances of technology, it must be presumed that innovation will greatly enhance the lives of persons with neuromuscular disease in the years ahead.

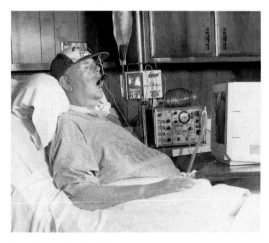

Fig. 5. Quadriplegic MND patient operating a word processor with a self designed oral switch. Note gastrostomy feeding bag and ventilator.

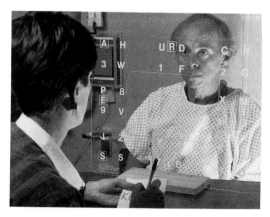

Fig. 6. MND patient with bulbar palsy using an ETRAN board to communicate.

TERMINAL CARE

Attitudes about death not only vary among individuals, but among cultures as well. Economic and other considerations may impinge on the decision to let a disease take its course, or to extend life as long as possible. A number of ethical and legal issues increasingly are being discussed for the purpose of establishing legal standards for care of the terminally ill (Beresford 1989). These have had the effect of protecting the patient's right to forgo treatment and the physician's right not to prolong life. The proper role for a physician in these matters is to serve the patient's interest rather than that of the state or some other party (Goldblatt 1984). In this capacity, physicians uphold the highest ideals of their profession and of society. In instances where the physician's values conflict with those of the patient, it is incumbent upon him to transfer responsibility to a colleague.

There is now legal precedent in some countries for the withdrawal of life support systems (Wanzer et al. 1989). It is important for patients to inform their caretakers as to their wishes about such matters. This should be documented in the form of a 'living will'. Except in the Netherlands, euthanasia does not enjoy legal sanction (Singer and Siegler 1990). However, in the final stages of illness, the compassionate use of sedative medications is justified to relieve pain or suffering. This can be done in the hospital, or in the home in conjunction with hospices which generally distinguish themselves in the care of the terminally ill.

CONCLUSION

Physicians caring for persons with motor neuron disease can see parallels between themselves and those who cared for persons with poliomyelitis in the recent past. In both instances, there is no cure for the disease. Rather, symptomatic measures are all that can be made available. Based on remarkable advances in technology, it is now possible to treat most of the symptoms of MND effectively. While theoretically the lives of patients can be extended for years, a number of human factors ultimately dictate whether the patient should embark on such a course. To resolve these decisions requires considerable time and insight, challenging physicians to realize the highest ideals of their profession. It can only be hoped that the intimate relationship between patient and physician in these circumstances continues to be respected by the many parties who are involved in the delivery of health care.

Acknowledgements
This chapter is dedicated to patients Charles Bordner, Fred Clark, William Doro, Warren Irwin, Arthur Levien, Curtis Robert, John Stockton, and James Waldal, and to friends and supporters: Cindy Beechner, Betty Bordner, Bruce Clark, William and Ruth Frank, Arlene and George Hecht, Jack and Marianne Olson, Francis, Patty, Gloria and Wanda Stockton, and George and Joan Thagard. Colleagues Benjamin Brooks, Forbes Norris, and Hiroshi Mitsumoto provided thoughtful comments, and Drs Robert Fallat and Gregory Weiner illustrations. These contributions are greatly appreciated.

REFERENCES

ARONSON, A. E.: Definition and scope of the communication disorder In: D. W. Mulder (Ed.), The Diagnosis and Treatment of Amyotrophic Lateral Sclerosis. Boston, Houghton Mifflin Professional Publishers (1980) 217–223.

BACH, J. R., A. S. ALBA, G. BOHATIUK, L. SAPORITO and M. LEE: Mouth intermittent positive pressure ventilation in the management of postpolio respiratory insufficiency. Chest 91 (1987) 859–864.

BERESFORD, H. R.: Legal aspects of termination of treatment decisions. Neurol. Clin. 7(4) (1989) 775–787.

BOTTE, M. J., V. L. NICKEL and W. H. AKESON: Spasticity and contracture/physiologic aspects of formation. Clin. Orthop. Relat. Res. 233 (1988) 7–18.

BROOKS, B. R., R. DEPAUL, Y. D. TAN, M. SANJAK, R. SUFIT and J. ROBBINS: Motor neuron disease. In: R. Porter and Bl Schoenberg (Eds.), Controlled Clinical Trials in Neurological Disease. Boston, Kluwer Academic Publishers (1990) 249–281.

BROWN, W. A. and P. S. MUELLER: Psychological function in individuals with amyotrophic lateral sclerosis (ALS). Psychosom. Med. 32(2) (1970) 141–152.

CARRE, P. C., A. P. DIDIER, Y. M. TIBERGE, L. J. ARBUS and P. J. LEOPHONTE: Amyotrophic lateral sclerosis presenting with sleep hypopnea syndrome. Chest 93(6) (1988) 1309–1312.

CASHMAN, N., R. MASELLI, R. WOLLMANN, R. ROOS, R. SIMON and J. ANTEL: Late denervation in patients with antecedent paralytic poliomyelitis. N. Engl. J. Med. 317(1) (1987) 7–12.

CASSEL, C.: Patient autonomy as therapy. In: D. W.

Mulder (Ed.), The Diagnosis and Treatment of Amyotrophic Lateral Sclerosis. Boston, Houghton Mifflin Professional Publishers (1980) 322–325.

DARLEY, F. L., A. E. ARONSON and J. R. BROWN: Differential diagnostic patterns of dysarthria. J. Speech Hearing Res. 12 (1969) 246–269.

DAVIDSON, M. B.: Effect of growth hormone on carbohydrate and lipid metabolism. Endocr. Rev. 8 (1987) 115–131.

DERENNE, J., P. T. MACKLEM and C. ROUSSOS: The respiratory muscles: mechanics, control and pathophysiology. Am. Rev. Respir. Dis. 118 (1978) 373–390.

DYCK, P. J., J. C. STEVENS, D. W. MULDER and R. ESPINOSA: Frequency of nerve fiber degeneration of peripheral motor and sensory neurons in amyotrophic lateral sclerosis: morphometry of deep and superficial peroneal nerves. Neurology 25 (1979) 781–785.

DYMOND, E. A., R. POTTER, P. A. GRIFFITHS and E. J. MCCLEMONT: A week in the life of Mary: the impact of microtechnology on a severely handicapped person. J. Biomed. Eng. 10 (1988) 483–490.

FALLAT, R. J., B. JEWITT, M. BASS, B. KAMM and F. H. NORRIS: Spirometry in amyotrophic lateral sclerosis. Arch. Neurol. 36 (1979) 74–80.

FESTOFF, B. W.: Amyotrophic lateral sclerosis. In: R. A. Davidoff (Ed.), Handbook of the Spinal Cord. New York, Marcel Dekker, Inc. (1987) 607–663.

FRIED, C.: Rights and health care, beyond equity and efficiency. N. Engl. J. Med. 293(5) (1975) 241–245.

GALLASSI, R., P. MONTAGNA, C. CIARDULLI, S. LORUSSO, V. MUSSUTO and A. STRACCIARI: Cognitive impairment in motor neuron disease. Acta Neurol. Scand. 71 (1985) 480–484.

GOLDBLATT, D.: Decisions about life support in amyotrophic lateral sclerosis. Semin. Neurol. 4(1) (1984) 104–110.

GONZALEZ, J. B. and A. E. ARONSON: Palatal lift prosthesis for treatment of anatomic and neurologic palatopharyngeal insufficiency. Cleft Palate J. 8 (1969) 91–104.

GOODE, R. L. and R. A. SMITH: The surgical management of sialorrhea. The Laryngoscope, LXXX(7) (1970) 1078–1089.

GOULD, B. S.: Psychiatric aspects. In: D. W. Mulder (Ed.), The Diagnosis and Treatment of Amyotrophic Lateral Sclerosis. Boston, Houghton Mifflin Professional Publishers (1980) 157–165.

GRIGGS, R. C., K. M. DONOHOE, M. J. UTELL and D. GOLDBLATT: Pulmonary function testing. In: D. W. Mulder (Ed.), The Diagnosis and Treatment of Amyotrophic Lateral Sclerosis. Boston, Houghton Mifflin Professional Publishers (1980) 291–297.

HILLEL, A. D. and R. MILLER: Bulbar amyotrophic lateral sclerosis: patterns of progression and clinical management. Head Neck 11 (1989) 51–59.

HOUPT, J. L., B. S. GOULD and F. H. NORRIS: Psychological characteristics of patients with amyotrophic lateral sclerosis (ALS). Psychosom. Med. 39(5) (1977) 299–303.

HOWARD, R. S., C. M. WILES and L. LOH: Respiratory complications and their management in motor neuron disease. Brain 112 (1989) 1155–1170.

HUDSON, A. J.: Amyotrophic lateral sclerosis and its association with dementia, parkinsonism and other neurological disorders: a review. Brain 104 (1981) 217–247.

HUDSON, A. J.: Outpatient management of amyotrophic lateral sclerosis. Semin. Neurol. 7(4) (1987) 344–351.

KARPATI, G., G. KLASSEN and P. TANSER: The effects of partial chronic denervation on forearm metabolism. Can. J. Neurol. Sci. 6(2) (1979) 105–112.

KERBY, G. R., L. S. MAYER and S. K. PINGLETON: Nocturnal positive pressure ventilation via nasal mask. Am. Rev. Resp. Dis. 135 (1987) 738–740.

LANDAU, W. M.: Spasticity: the fable of a neurological demon and the emperor's new therapy. Arch. Neurol. 31 (1974) 217–219.

LARSON, D. E., C. R. FLEMING, J. O. BEVERLY and K. W. SCHROEDER: Percutaneous Endoscopic Gastrostomy, simplified access for enteral nutrition. Mayo Clin. Proc. 58 (1983) 103–107.

LENMAN, J. A. R.: The pharmacology of spasticity. In J. Eccles and M. Dimitrijeuvic (Eds.), Achievements in Restorative Neurology Upper Motor Neuron Functions and Dysfunctions. New York, Karger (1985) 31–41.

LORD, J. P., J. S. LIEBERMAN, M. M. PORTWOOD, W. M. FOWLER and R. CARSON: Functional ability and equipment use among patients with neuromuscular disease. Arch. Phys. Med. Rehabil. 68 (1987) 348–352.

MCCARTNEY, N., D. MOROZ, H. GARNER and A. J. MCCOMAS: The effects of strength training in patients with selected neuromuscular disorders. Med. Sci. Sports Exercise 20(4) (1988) 362–368.

MILNER-BROWN, H. S. and R. G. MILLER: Muscle strengthening through high-resistance weight training in patients with neuromuscular disease. Arch. Phys. Med. Rehabil. 69 (1988) 14–19.

MITSUMOTO, H., R. HANSON and D. CHAD: Amyotrophic lateral sclerosis, recent advances in pathogenesis and therapeutic trials. Arch. Neurol. 45 (1988) 189–202.

MULDER, D. W. and F. M. HOWARD: Patient resistance and prognosis in amyotrophic lateral sclerosis. Mayo Clin. Proc. 51 (1976) 537–541.

MULDER, D. W., E. H. LAMBERT and L. M. EATON: Myasthenic syndrome in patients with amyotrophic lateral sclerosis. Neurology 9(10) (1959) 627–631.

MUNSAT, T. L., P. L. ANDRES, L. FINISON, T. CONLON and L. THIBODEAU: The natural history of motorneuron loss in amyotrophic lateral sclerosis. Neurology 38 (1988) 409–413.

NAKANO, K. K., H. BASS, H. R. TYLER and R. J. CARMEL: Amyotrophic lateral sclerosis: a study of pulmonary function. Dis. Nerv. Syst. 37 (1976) 32–35.

NEWRICK, P. G. and R. LANGTON-HEWER: Motor neurone disease: can we do better? — a study of 42 patients. Br. Med. J. 289 (1984) 539–542.

NORRIS, F. H., E. L. GASTEIGER and P. O. CHATFIELD: An electromyographic study of induced and spontaneous muscle cramps. Electroencephalogr. Clin. Neurophysiol. 9 (1957) 139.

NORRIS, F. H., R. A. SMITH and E. H. DENYS: Motor neuron disease: towards better care. Br. Med. J. 291 (1985) 259–262.

NORRIS, F. H., D. HOLDEN, K. KANDAL and E. STANLEY: Home nursing care by families for severely paralyzed patients. In: V. Cosi, A. C. Kato, W. Parlette, P. Pinelli and P. Poloni (Eds.), Amyotrophic Lateral Sclerosis, Therapeutic, Psychological and Research Aspects. New York, Plenum Publishing Corporation (1987a) 231–238.

NORRIS, F. H., R. A. SMITH, E. H. DENYS, D. HOLDEN, R. ELKIN, R. J. FALLAT, H. TUCKER and E. STANLEY: Home care of the paralyzed respirator patient with amyotrophic lateral sclerosis. In: L. I. Charash, R. E. Lovelace, S. G. Wolf, A. H. Kutscher, D. P. Royce and C. F. Leach (Eds.), Realities in Coping with Progressive Neuromuscular Diseases. Philadelphia, The Charles Press Publishers Inc. (1987b) 72–83.

PETERS, K. P., W. M. SWENSON and D. W. MULDER: Is there a characteristic personality profile in amyotrophic lateral sclerosis, a Minnesota multiphasic personality inventory study. Arch. Neurol. 35 (1978) 321–322.

POLONI, M., E. CAPITANI, L. MAZZINI and M. CERONI: Neuropsychological measures in amyotrophic lateral sclerosis and their relationship with CT scan-assessed cerebral atrophy. Acta Neurol. Scand. 74 (1986) 257–260.

RIVERA, V. M., W. B. BREITBACH and L. SWANKE: Dantrolene in amyotrophic lateral sclerosis. J. Am. Med. Assoc. 223 (1975) 863–864.

SANJAK, M., D. PAULSON, R. SAFIT, W. REDDAN, E. BEAULIEU, A. SHAG and B. R. BROOKS: Physiologic and metabolic response to progressive and prolonged exercise in amyotrophic lateral sclerosis. Neurology 37 (1987) 1217–1220.

SCHIFFER, R. B., R. M. HERNDON and R. A. RUDICK: Treatment of pathologic laughing and weeping with amitriptyline. N. Engl. J. Med. 312(23) (1985) 1480–1482.

SERPICK, A. A., E. L. BAKER and T. E. WOODWARD: Motor system disease: Review and discussion of a case presenting with alveolar hypoventilation. Arch. Intern. Med. 115 (1965) 192–197.

SHIFFMAN, P. L. and J. M. BELSH: Effect of inspiratory resistance and theophylline on respiratory muscle strength in patients with amyotrophic lateral sclerosis. Am. Rev. Respir. Dis. 139 (1989) 1418–1423.

SINAKI, M.: Exercise and rehabilitation measures in amyotrophic lateral sclerosis. In: T. Tsubaki and Y. Yase (Eds.), Amyotrophic Lateral Sclerosis. Amsterdam, Elsevier Science Publishers B.V. (1988) 343–367.

SINGER, P. and M. SIEGLER: Euthanasia — a critique. N. Engl. J. Med. 322(26) (1990) 1881–1883.

SMITH, R. A.: On behalf of the patient. In: V. Cosi, A. C. Kato, W. Parlette, P. Pinelli and M. Poloni (Eds.), Amyotrophic Lateral Sclerosis, Therapeutic, Psychological, and Research Aspects. New York, Plenum Publishing Corporation (1987) 319–322.

SMITH, R. A. and R. L. GOODE: Sialorrhea. N. Engl. J. Med. 283 (1970) 917–918.

SMITH, R. A. and F. NORRIS: Symptomatic care of patients with amyotrophic lateral sclerosis. J. Am. Med. Assoc. 234(7) (1975) 715–717.

SMITH, R. O., C. J. SANDS, N. M. GOLDBERG, R. U. MASSEY and J. R. GAY: Injection of silicone lateral to a vocal cord in a patient with progressive bulbar palsy. Neurology 17 (1967) 1217–1218.

SPENCER, P. S., P. B. NUNN, J. HUGON, A. C. LUDOPH, S. M. ROSS, D. N. ROY and R. C. ROBERTSON: Guam amyotrophic lateral sclerosis parkinsonism — dementia linked to a plant excitant neurotoxin. Science 237 (1987) 517–522.

STEELE, J. C., R. L. DOTY, D. P. PERL, K. M. CHEN and L. T. KURLAND: Olfactory dysfunction in patients with amyotrophic lateral sclerosis/parkinsonism-dementia complex of Guam. Neurology 40(1) (1990) 453.

STUCHELL, R. N. and I. D. MANDEL: Salivary gland dysfunction and swallowing disorders. Otolaryngol. Clin. N. Am. 21(4) (1988) 649–661.

TALMI, Y. P., Y. ZOHAR, Y. FINKELSTEIN and N. LAURIAN: Reduction of salivary flow with scopoderm TTS. Ann. Otolol., Rhinol. Laryngol. 97 (1988) 128–130.

UDAKA, T., S. YAMAO, H. NAGATA, S. NAKAMURA and M. KAMEYAMA: Pathologic laughing and crying treated with levodopa. Arch. Neurol. 41 (1984) 1095–1096.

WANZER, S. H., D. FEDERMAN, J. ADELSTEIN, C. CASSEL, E. CASSEM, R. CRANFORD, E. HOOK, B. LO, C. MOERTEL, P. SAFAR, A. STONE and J. VAN EYS: The physicians responsibility toward hopelessly ill patients. N. Engl. J. Med. 320(13) (1989) 844–849.

WARBURTON, A., J. P. ROYSTON, C. J. A. O'NEILL, W. NICHOLSON, R. D. JEE, M. J. DENHAM, S. M. DOBBS and R. J. DOBBS: A quinine a day keeps the leg cramps away? Br. J. Clin. Pharmacol. 23 (1987) 459–465.

WOHLFART, G. and R. L. SWANK: Pathology of amyotrophic lateral sclerosis. Arch. Neurol. Psychiatry 46 (1941) 783–799.

YOUNG, R. and A. W. WEIGHNER: Spasticity. Clin. Orthop. Relat. Res. 219 (1987) 50–62.

Handbook of Clinical Neurology, Vol. 15 (59): Diseases of the Motor System
J.M.B.V. de Jong, editor

CHAPTER 27

Hemiatrophies and hemihypertrophies

G. W. BRUYN[1] and R. P. M. BRUYN[2]

[1]*Department of Neurology, Academic Hospital, State University Leiden, and* [2]*Department of Neurology, Hospital Oudenrijn, Utrecht, The Netherlands*

As every tailor knows, the tacit assumption that the human body develops harmoniously along a longitudinal axis of symmetry is more often than not belied by empiricism. In right-handed normal persons, the left cerebral hemisphere, the left cranial half and the left side of the face are larger, the ribs on the right side and the right arm are longer (and thicker), while the left leg is usually longer and stronger (Ringrose et al. 1965). Within the biologist's and physician's provinces, a certain (or, rather, uncertain) latitude of deviation during the embryological, fetal, and infantile stages of development is allowed beyond which 'normal' variation in symmetry is considered to have passed into the realm of pathology. No consensus, based on physical magnitudes, has been reached as to the borderlines separating the normal from the pathological. As a consequence, one may expect definitions to vary or to be rather blurred, and pathological asymmetries to remain frequently unrecognised. This obtains all the more, because hemiatrophies/-hypertrophies are uncommon, variable in site, surface area and volume, and often associated with abnormalities in other systems that rather captivate the attention. A fairly simple but revealing method to ascertain (and convey to others) an unequivocal measure of facial asymmetry, is to make 2 prints (one inverted left to right, by turning over the negative before printing), cut each of these points into 2 halves along the vertical midline, and join the right + 'inverted right' together, as well as the left and inverted left, in this way obtaining 2 faces (Fig. 1).

DEFINITION AND CLASSIFICATION

Because a fully satisfactory classification is still absent notwithstanding the wealth of material (Rogers, as early as 1964, reported on 772 cases of Parry-Romberg disease), one should at least attempt to provide a non-equivocal definition of concepts to commence with.

In the following pages, the concept of hemiatrophy/-hypertrophy is understood as the visible, observable, measurable reduction in size or volume of one or more parts or even the totality of one half of the body as compared with the size or volume of the homologous part(s) or totality of the other half of the body. It goes without saying that in case of hemiatrophy or -hypertrophy of an entire body-half the observer must refer to measurement tables pertinent to the patient's population (age, sex, race, etc.) in order to decide which one of the 2 halves of the patient's body is hemiatrophic or hemihypertrophic.

Proceeding from this definition, the hemiatro-

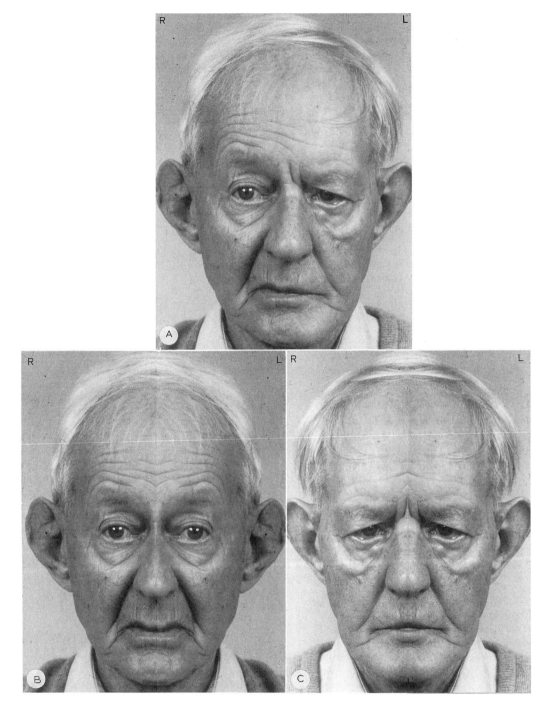

Fig. 1. (A) Leftsided moderate hemiatrophia faciei with slight peripheral n. VII deficit. (B) Face, composed of 2 right halves. (C) Face, composed of 2 left halves.

phies and hemihypertrophies can be subdivided as follows.

A. *True, complete,* or *genuine*: both the skeletal and soft tissue components such as muscles, nerves, vessels and ligaments are either too small or too great.
B. *False, incomplete,* or *quasi*: only the soft parts are involved.

The categories A and B can each be subdivided into

(a) *total*: one body-half in its entirety is involved.
(b) *partial*: one part or some parts of a body-half are involved.

This sub-category may be further subdivided and specified as 'segmental', 'distal', 'proximal', etc.

From a practical point of view, it seems appealing to classify further into 'congenital' and 'acquired' forms. This would be deceptive, however, inasmuch as most if not all of the congenital forms by necessity must be acquired, whether from toxic, metabolic or vascular-occlusive causes. Only the genetic (chromosomal) causation might be considered as 'non-acquired'.

AETIOLOGY

A comprehensive and consistent schema of causes for pathological corporeal asymmetry is lacking. A number of morphogenetic theories have been formulated that cannot claim to be comprehensive. The 'twin-theory' of Gesell (1927) and Sachs (1949) positing a sliding scale from bilateral symmetry to monozygotic twins via in-between-lying gradations of incomplete twins on the basis of unequal distribution of cytoplasm or genetic material in morula- and blastula-cells might account for certain instances of hemihypertrophy but scarcely for hemi-atrophies.

Of the other theories, such as the embryonal, the hypermature ovum, the mechanistic, the vascular, and the neurotrophic, the last-named enjoys most adherence, whether purely neurotrophic or neurotrophic-vascular. Central (autonomous) nervous system anomalies may entail secondary abnormal blood circulation in certain part(s) of the body, such as dermographism, hyperhidrosis and raised skin temperature. Klip-

pel-Trénaunay syndrome seems to be an eloquent example of this kind of pathogenesis. Beyond the satisfaction of mere clinical astuteness in recognition, the fascination of this kind of deranged morphogenesis for the neurologist derives from the intriguing part the (central) nervous system may play in producing them. The relationship between a (causative) CNS lesion and hemiatrophy seems quite natural in view of reports by e.g. Gowers (1879), Bastian and Horsley (1881) and Penfield and Robertson (1943). The observation that mental retardation is a frequent accompaniment of these 'dysmorphogenoses' supports their nervous anchorage, but the frequent association with Wilms-tumor (except in the categories Klippel-Trénaunay, and Silver-Russell syndrome) seems strongly indicative of embryonal (neural crest?) origin.

The toxic effects of thalidomide (polyneuropathy in the adult, phocomelia in the embryo) reported with alarming frequency in the early sixties, constitute an additional potent argument for the neurotrophic theory. The prior difficulties in conceiving of a generalised toxic-metabolic derangement producing a localised defect in one body-half were at least discarded following the thalidomide-story.

Recent developments in the field of growth hormone (HGH), a 190-aminoacid-containing peptide of 21.5 kDa of which the aminoacid sequence is known, and of nerve growth factor (NGF), a 13.2-kDa protein essential for developing sympathetic and sensory neurons, are of potential promise in the explanation of the various hemi-atrophies and -hypertrophies.

By far the greater majority of instances is congenital and non-progressive; only Parry-Romberg's hemifacial atrophy appears to be progressive and we will begin this exposé with it.

HEMIATROPHIES

Progressive hemifacial atrophy (Parry-Romberg's disease, Romberg's disease) (141300)

This condition is one of those hemiatrophies most commonly seen by the practising neurologist. Originally reported by Parry (1825) and described in more detail by Romberg (1846), it

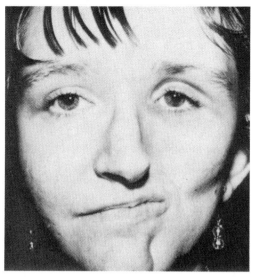

Fig. 2. (A and B) Hemifacial atrophy. Twenty-three-year-old woman with progressive atrophy of the left side of the face, beginning at age 10 and relatively stable for several years. Reproduced from Poskanzer 1975 by courtesy of Dr. Bradford Cannon.

received its technical denomination by von Eulenburg (1871). Exhaustive reviews were presented by Marburg (1912), Archambault and Fromm (1932), Wartenberg (1945), and Rogers (1964), the latter basing his discussion on nearly 800 cases culled from the world literature.

Parry-Romberg's disease may be either true or false, and is not congenital, i.e. not present from birth. It is not a remnant of early Bell's palsy, nor of lipodystrophy, neither is it a part of hemifacial microsomia, although the latter comes up first for differential diagnostic consideration. Progressive hemifacial atrophy usually begins between the 10th and 30th year of life, and becomes stationary after a progressive course lasting 2–10 years. The terminal plateau-stage may be one of minimal deformity or of conspicuous asymmetry (Figs. 2 and 3). From a cosmetic point of view, one has to advise the patient to wait until the end-stage is reached, because premature plastic surgical intervention is bound to be annulled by further atrophy. In about 5% of instances, the 'hemi'-atrophy is bilateral.

Although the atrophy predominantly appears to involve the maxillar and mandibular parts of the face, the fronto-temporal area need not escape. The process usually starts focally with a patch of hyper- or hypopigmentation, above or

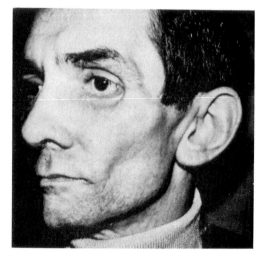

Fig. 3. Hemifacial atrophy. Forty-one-year-old man who noted gradual onset of loss of soft tissue on the left side of the face, about the eye and in the temporal region at age 19. At age 25, he underwent fat transplantation but hemifacial atrophy returned to its previous state within 6 months. No evidence of progression since failure of the surgery. Reproduced from Poskanzer 1975 by courtesy of Dr. Bradford Cannon.

below the eye, subsequently spreading while becoming atrophic. The lower eyelashes or the center of the eyebrow may initially change in colour. In the course of the disorder the scalp-hair and facial hair become involved with discolouration or depigmentation, and occasionally leading to defluvium capillorum even to the point of alopecia areata. The subcutaneous atrophy includes the sebaceous and sweat glands.

The atrophy may spread to involve the soft tissues (panniculus adiposus, muscles, ligaments) and occasionally cartilage and bone, never touching the teeth however. It may spread to involve the ear, tongue, and pharynx.

The observation that the process is' of paramedian origin and that the midline of the face, receiving bilateral innervation, escapes damage, made Wartenburg (1945) stress that the causation is unlikely to issue from or be mediated by trigeminal nerve fibres, an 'anti-neurotrophic' point of view opposed by Rogers (1964). Neither does the process follow the onion-peel-like von Sölden's lines as are observed in hypaesthesia due to lesion of the descending spinal tract and/or nucleus. Nor does it conform to the embryonic closure-lines of the face (Fischer, 1903). Another argument against simple (autonomic) neurotrophism is formed by the fact that the iris ipsilaterally is neither depigmented nor assumes a different colour such as is known in Passow's syndrome (Gibbons and Brooker 1967). Finally, Parry-Romberg's disease is rarely associated, if at all, with Horner-Bernard's syndrome. All this militates against the assumption of involvement of the visceral or somatic nervous system subserving pathogenetic mechanisms. The present author is, however, unaware of modern clinical neurophysiological and neuroradiological procedures to have been applied in this condition.

The differential diagnosis, especially in instances of focal depigmentation and facial disfiguration, in first instance should include scleroderma, inasmuch as the lesion may confusingly closely resemble 'coup de sabre', pointed out as early as 1933 by Pick. Secondly, residual atrophy following bulbar poliomyelitis has to be considered. Hemifacial atrophy has been noticed in instances of Sturge-Weber disease (Louis-Bar 1947; Myle 1950; Kammer 1955): this may, however, have been a case of misrecognition of hemifacial hypertrophy. Finally, the congenital syndrome of *hemifacial microsomia* may deceive the unwary or uninitiated observer. This syndrome is congenital however, rather frequent (1:6000 births) and probably intrauterine-traumatic of origin. Also called the 'first and second branchial arch syndrome', it includes *true* atrophy of the hard and soft tissues deriving from these arches (Coccaro et al. 1975). Hemifacial microsomia may be familial and associated with radial limb defects (Moeschler and Clarren 1982), defects that are also seen in Nager's acromandibulo-facial dysostosis syndrome and in the REAR-syndrome (Kurnit et al. 1978). Microtia may be a forme fruste of it (Bennum et al. 1985), in view of the high incidence of Bell's palsy, labiopalatoschizis, and male preponderance.

One may confuse hemifacial microsomia with another congenital maldevelopmental syndrome, viz. *Goldenhar's* oculo-auriculo-vertebral dysplasia; the latter includes, in addition, vertebral anomalies and an epibulbar dermoid (Gorlin et al. 1963; Rolinick et al. 1987).

The prevalence of the disease is unknown. With respect to the sex ratio, the material reviewed by Rogers (1964) yielded an unequivocal 5:3 predominance in women.

Poskanzer (1975) posited, without providing references, that frequent mention was made of cerebral manifestations in cases of progressive hemifacial atrophy, such as mental retardation, epilepsy and cerebral hemiatrophy. The present author has been able to find some documentation for this statement (Wartenberg 1925; Archambault and Fromm 1932; Merritt et al. 1937; Walsh 1939; Eadie et al. 1963; Moura 1963; Banks and Sugar 1963; Dawson and Beare 1966; Johnson and Kennedy 1969; Kumar et al. 1971). Wolf and Verity (1974) presented an adult male case with cerebellar hemiatrophy, hemiparesis, epilepsy and ophthalmoplegia; in this case the hemiatrophia faciei spreads to involve the ipsilateral shoulder girdle and arm. A link to the phakomatoses is formed by leptomeningeal angioma (Merritt et al. 1937; Eadie et al. 1963; Wolf and Verity 1974), and by the occasional association with buphthalmos (Schneider, 1949).

Cerebral hemiatrophy usually is acquired during fetal or obstetric events, and infant years; it is secondary mainly to traumatic (dystoxic), vascular (Afifi et al. 1987), or infectious lesions (Bruyn 1959). The cranial sequelae properly recognised on roentgenograms (Fig. 4) allow the physician to infer the prae-, peri- or early postnatal cause, because 'es ist das Gehirn dass sich sein Gehäuse formt'.

Other forms of hemiatrophy

The great majority of instances of total or partial hemiatrophy of the body, a limb or part of it, is acquired in utero or later. Causes that readily come to mind are obstetric brachial plexus palsy (Klumpke type, Erb type), infantile acute anterior poliomyelitis, toxic phocomelias already referred to above with respect to thalidomide, and perhaps hemiplegia spastica infantilis. Syringobulbia and lipodystrophy are other established causes.

Linear scleroderma, often beginning before the twentieth year of life and commonly associated with spina bifida occulta, is a lesser known (though well documented) cause of hemifacial, hemimelic or hemicorporeal atrophy (Tuffanelli et al. 1966; Rees 1976; Rosenberg and Greenberg 1979). It has been reported in combination with Fuchs' heterochromic iridocyclitis (Sugar and Banks 1964), with atrophy of the tongue (Schwarz

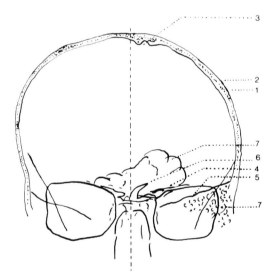

Fig. 4. 1. Unilateral calvarial flattening. 2. Thickening of diploë. 3. Lateral displacement of sagittal sinus, metopic suture and anterior fontanelle. 4. Elevation of sphenoid jugulum with elevation and enlargement of ala parva. 5. Elevation of superior planus of petrosal bone. 6. Inclination of crista galli toward atrophic side. 7. Increased pneumatisation.

et al. 1981), with generalised myopathy (Runne and Fasshauer 1977), and with Schönlein-Henoch nephritis and nocturnal hemoglobinuria (Kuto et al. 1985). The congenital nonprogressive form has been reported in association with glaucoma (Cohen 1979).

Klawans (1981) and Buchman et al. (1988) presented a series of cases of corporeal hemiatrophy with ipsilateral, early onset, benign hemiparkinsonism. This syndrome is distinguished from idiopathic Parkinson's disease by early age of onset, premedication dystonic foot spasms, a limited efficacy of L-dopa and a slowly-progressive course. Neuroradiologic evaluation of their patients in only 1 patient revealed asymmetry of the sylvian fissures; all other patients had normal CT or MRI of the brain. On the other hand, Giladi et al. (1990) reported 11 patients with hemiparkinsonism-hemiatrophy syndrome, 6 of which had brain asymmetry on CT or MRI, with the enlarged ventricle or atrophy contralateral to the side of body hemiatrophy. The infrequent neurological derangements, referred to in the section on hemifacial atrophy, may also be present in hemicorporeal atrophy in the form of ipsilateral porencephaly, cerebral hemiatrophy and contralateral hemiparesis, and cutis marmorata telangiectatica due to congenital generalised fibromatosis (Spraker et al. 1984).

THE HEMIHYPERTROPHIES

An almost limitless variety of hypertrophies have been recorded from a single digit to a body-half, involving both limbs on one side or crossed arm and leg, involving a single (skeletal, integumental) system or more systems, combined or not with hypertrophy of one or more internal organs, including the brain. In many instances of these hemihypertrophies, evidence is present of associated phakomatosis or neurocutaneous syndromes; they are familial, hereditary developmental anomalies due to malformation of the ecto- and/or mesoblast.

In some no such association is found (Fig. 5). The case illustrated is that of hemihypertrophy of both arm and leg including soft tissues and bone on the left side. The patient had adrenocortical carcinoma with a left adrenal tumor and

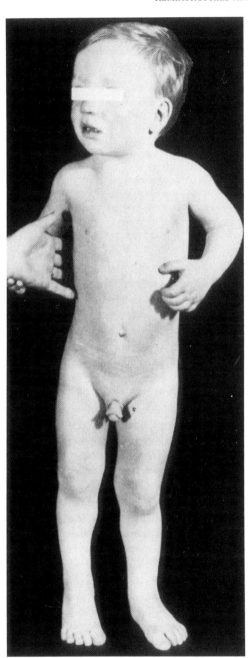

Fig. 5. Seventeen-month-old patient with hypertrophy of the left arm and leg. No other associated organomegaly. Reproduced from Poskanzer 1975 by courtesy of Dr Richard C. Pfister.

Cushing's syndrome and left nephromegaly (Weinstein et al. 1970). This case with, presumably polyploid, adrenocortical carcinoma, an extremely rare lesion in early infancy, underlines the well-known association between Wilms tumor (nephroblastoma) and hemihypertrophy (average age 3 years; as opposed to cases of Wilms tumor with aniridia plus genitourinary tract malformation averaging the age of 1.5 years, Breslow et al. 1988). Another example of this condition was reported by Ogita et al. (1989), in which, again, as review of the literature revealed, the hemihypertrophy proved to be the 'Leitsymptom' rather than the Cushing virilisation. Adrenal adenoma and sponge kidneys in congenital hemihypertrophy (Thompson et al. 1987; Schneider and Fanconi 1987; Tomooka et al. 1988) indicate a link with the phakomatoses; the relationship between congenital hemihypertrophy and renal, adrenal or liver malignancy had already been established by Parker and Skalko (1969) and Saypol and Laudone (1983).

The established association between hemihypertrophies and renal/adrenal malignancies extends to that with cancer of other visceral organs, such as hepatocellular carcinoma in Russell-Silver syndrome (Chitayat et al. 1988), nesidioblastosis and mammary fibroadenoma (Labrune et al. 1988) and of pancreatoblastoma (Drut and Jones 1988) in Beckwith-Wiedemann syndrome. The tissue cells of the hemihypertrophic side have been shown to possess increased growth rate, if not oncogenic potential (Furukawa and Shinohara, 1981).

Partial hemihypertrophy, such as e.g. macromelia, may secondarily and even very late produce nerve deficit, such as happened in the case of posterior interosseous nerve entrapment by hypertrophic supinator and extensor carpi muscles in a case of Dumitru et al. (1988).

HEMIFACIAL HYPERTROPHY (141350)

This condition was described by Meckel (1822), further characterised by Gesell (1921, 1927), Wakefield and Hines (1933), and reviewed by Gruber and Kuss (1937), Ward and Lerner (1947), Gorlin et al. (135 cases, 1976), Sinkovits (1959), Rowe (1962), and Ringrose et al. (105 cases, 1965). It may vary from barely noticeable asymmetry to obvious distortion, with larger and more widely spaced teeth, macroglossia (Gordeeff et al. 1986), and hypertrophy of lips, tonsil and ear. It may be part or start of hemicorporeal hypertrophy. A well-illustrated review, including the histology and radiology of the disorder, was provided by Rowe (1962) who collected an impressive set of illustrative material. The condition has been noted to occur in naevus unius lateris (Muller et al. 1980), a neurocutaneous syndrome, which today is considered part of the 'epidermal naevi syndrome', and associated with cerebral tumor (Andriola 1976). Cerebellar haemangioblastoma has been reported in hemifacial hypertrophy (Furukawa et al. 1973). According to Bergman (1973), the condition is accompanied with ipsilateral hemimegalencephaly and mental retardation in 15–20% of the cases. In the congenital forms, lingual and dental hyperplasia is present. An autosomal dominant form associated with amblyopia, strabismus and cleft palate is known as Bencze syndrome (Kurnit et al. 1979).

Other forms of hemihypertrophy

These show not only a wide variety of associated abnormalities, but are also often associated with enlarged viscera and central nervous system lesions.

The hemihypertrophies occur in about 15% of all cases of Jadassohns's linear nevus sebaceous syndrome (163200), today called the *epidermal nevus syndrome* (Solomon and Esterly 1975) which is well-known for its association with malignant growth, and strongly resembles the *Proteus syndrome* also known for its cranial anomalies (Gorlin 1984; Trappe et al. 1989; Mayatepek et al. 1989) and which should be classified with the phakomatoses and lipomatoses. Cerebral involvement was pointed out by Malamitsi-Pucher et al. (1987). Secondary involvement may also occur of the nervous system, such as spinal cord compression due to vertebral overgrowth (Hornstein et al. 1987). Total corporeal hemihypertrophy and *hemimegalencephaly* (Robain et al. 1988) in the epidermal naevus syndrome was noted by Sakuta et al. (1989), who stressed the importance of differentiation with Bourneville-Pringle's tuberous sclerosis (191100), another one of the phakomatoses, in which oligophrenia, epilepsy and hemimega-

lencephaly occur. Gliomatosis cerebri in the epidermal nevus syndrome was reported by Choi and Kudo (1981). A subgroup was identified by Pavone et al. (1991), who described 4 patients from their own practices and 13 from the literature presenting with facial epidermal nevus, ipsilateral hemimegalencephaly and gyral malformation, mental retardation, seizures and often facial hemihypertrophy. According to Dobyns and Garg (1991) over 70 cases with epidermal nevus syndrome with neurologic presentation are described, 50% of which had the hemimegalencephaly variant.

The hamartomatous common root of the Proteus syndrome (Gorlin 1984) and the *Bannayan-Zonana syndrome* (154380) may pose the physician a diagnostic dilemma; points of distinction are that the congenital Bannayan syndrome is autosomal-dominant and comprises lipomatosis, angiomatosis and macrocephaly with craniofacial anomalies, whereas the non-hereditary Proteus syndrome rather shows hemihypertrophy, macrodactyly, scoliosis and cutaneous lesions (Bialer et al. 1988).

Partial (monomelic) hemihypertrophy, e.g. of a leg, is not unfamiliar to the neurologist whose experience includes patients with the angioosteohypertrophic Klippel-Trénaunay-Weber syndrome (149000) and Cobb syndrome of cutaneous hemangioma with hypertrophy of somatomeric tissues (Devic and Tolot 1906; van Bogaert 1950; Kramer 1972; Kissel and Dureux 1972; Zala and Mumenthaler 1981). The association of hemifacial hypertrophy and Klippel-Trénaunay syndrome was mentioned by Schoch (1956). The Klippel-Trénaunay syndrome is often associated with angiomas or a.v. shunts in the brainstem or brain (Hidano and Arai 1987; case and review by Oyesiku et al. 1988). Cerebrocerebellar hemihypertrophy in Klippel-Trénaunay syndrome has been signalled in a female infant by Anlar et al. (1988). Parkes-Weber syndrome is another example, and again, the link with the phakomatoses is obvious.

The (cerebral) vascular anomalies such as AVM, aneurysm or angioma in the hemihypertrophies as well as the neurological symptoms such as indifference to pain, mental defect, hydrocephalus, epilepsy, and even the probably coincidental metachromatic leukodystrophy one may encounter, were briefly reviewed by Fischer et al. (1984).

Other diseases in which total or partial hemihypertrophy is seen include neurofibromatosis, Kast-Maffucci syndrome (16600; Sun et al. 1985), Russell-Silver syndrome (270050) (Tanner et al. 1975; Duncan et al. 1990), and Ito's hypomelanosis (146150) (Pascual-Castroviejo et al. 1988). In the last-named condition that might be called a melanophacomatosis and may be attributed to microdeletion of 15ql (Turleau et al. 1986), mental retardation is noted in nearly three-quarters of the patients and epilepsy in about half of them. CT-scanning is abnormal in about 50% of patients with Ito's hypomelanosis and appears to indicate heterotopism (Glover et al. 1989).

Rare disorders showing (partial) hemihypertrophy are (partial) *lipodystrophy* which manifests in the first decade of life, and is associated with disturbances of the autonomous nervous system; dysplasia epiphysealis hemimelica (Hinkel and Rupprecht 1989) and Milroy's disease (congenital lymphedema).

Finally, an interesting 'Hemi 3 syndrome' (hypertrophy, hypaesthesia, areflexia) (235000) with scoliosis, all of early developmental origin was reported by Nudleman et al. (1984), who, too, surmise neural crest dysgenesis combined with nerve growth factor dysregulation to be pathogenetically decisive mechanisms. The neural crest theory of origin was propounded strongly by Pollock et al. (1985).

Acknowledgement
The care and professional creativity of Mr G. J. van der Giessen, medical photographer in our department, who, for years made the majority of the illustrations in the present author's papers, is hereby gratefully acknowledged.

REFERENCES

AFIFI, A. K., J. C. GODERSKY and A. MENEZES: Cerebral hematrophy, hypoplasia of internal carotid artery, and intracranial aneurysm. A rare association occurring in an infant. Arch. Neurol. 44 (1987) 232–235.

ANDRIOLA, M.: Nevus Unis Lateralis and Brain Tumor. Am. J. Dis. Child. 130 (1976) 1259–1261.

ANLAR, B., K. YALAZ and C. ERZEN: Klippel-Trénaunay-Weber syndrome; a case with cerebral and cerebellar hemihypertrophy. Neuroradiology 30 (1988) 360.

ARCHAMBAULT, L. and N. K. FROMM: Progressive facial hematrophy. Arch. Neurol. Psychiatry (Chic.) 27 (1932) 529–584.

BANKS, T. L. and H. S. SUGAR: Ocular manifestations of facial hematrophy. Sinai Hosp. Detroit Bull. 11 (1963) 83–88.

BASTIAN, H. C. and V. HORSLEY: Arrest of development in the left arm in association with an extremely small right ascending parietal convolution. Brain 3 (1881) 113–127.

BENCZE, J., A. SCHNITZLER and J. WALAWSKA: Dominant inheritance of hemifacial hyperplasia associated with strabismus. Oral Surg. 35 (1973) 489–500.

BENNUM, R. D., J. B. MULLIKEN and L. B. KABAN: Microtia: A microform of hemifacial microsomia. Plast. Reconst. Surg. 76 (1985) 859–863.

BERGMAN, J. A.: Primary hemifacial hypertrophy. Arch. Otolaryngol. 97 (1973) 490–494.

BIALER, M. G., M. J. RIEDY and W. G. WILSON: Proteus syndrome versus Bannayan-Zonana syndrome: a problem in differential diagnosis. Eur. J. Pediatr. 148 (1988) 122–125.

BRESLOW, N., J. B. BECKWITH, M. CIOL and K. SHARPLES: Age distribution of Wilms' tumor. Cancer Res. 48 (1988) 1653–1657.

BRUYN, G. W.: Pneumoencephalography in the diagnosis of cerebral atrophy. A quantitative study. Thesis, Fac. Med. State Univ. Utrecht. Utrecht, Smits (1959).

BUCHMAN, A. S., C. G. GOETZ and H. L. KLAWANS: Hemiparkinsonism with hematrophy. Neurology 38 (1988) 527–530.

CHITAYAT, D., J. M. FRIEDMAN, L. ANDERSON and J. E. DIMMICK: Hepatocellular carcinoma in a child with familial Russell-Silver syndrome. Am. J. Med. Genet. 31 (1988) 909–914.

CHOI, B. H. and M. KUDO: Abnormal neuronal migration and gliomatosis cerebri in epidermal nevus syndrome. Acta Neuropathol. (Berl.) 53 (1981) 319–325.

COCCARO, P. J., M. H. BECKER and J. M. CONVERSE: Clinical and radiological variation in hemifacial microsomia. In: D. Bergsma (Ed.), Malformation Syndromes. Amsterdam, Excerpta Medica (1975) 314–324.

COHEN, J. S.: Congenital nonprogressive facial hemiatrophy with ipsilateral eye abnormalities and juvenile glaucoma. Ann. Ophthalmol. 11 (1979) 413–416.

DAWSON, T. A. J. and J. M. BEARE: Facial hematrophy (Parry-Romberg syndrome). Br. J. Dermatol. 78 (1966) 545–546.

DEVIC, E. and G. TOLOT: Un cas d'angiosarcome des méninges de la moëlle. Rev. Med. (Paris) 26 (1906) 255–269.

DOBYNS, W. B. and B. P. GARG: Vascular abnormalities in epidermal nevus syndrome. Neurology 41 (1991) 276–279.

DRUT, R. and M. C. JONES: Congenital pancreatoblastoma in Beckwith-Wiedemann syndrome: an emerging association. Pediatr. Pathol. 8 (1988) 331–339.

DUMITRU, D., WALSH, N. and B. VISSER: Congenital hemihypertrophy with posterior interosseus nerve entrapment. Arch. Phys. Med. Rehab. 69 (1988) 696–698.

DUNCAN, P. A., J. G. HALL, L. R. SHAPIRO and B. K. VIBERT: Three generation dominant transmission of the Silver-Russell syndrome. Am. J. Med. Genet. 35 (1990) 245–250.

EADIE, M. J., J. M. SUTHERLAND and J. H. TYRER: The clinical features of hemifacial atrophy. Med. J. Aust. 50 (1963) 177–180.

FISCHER, O.: Ein Beitrag zur Lehre von der Hemiatrophia facialis progressiva. Mschr. Psychiatr. Neurol. 14 (1903) 366.

FISCHER, E. G., R. D. STRAND and F. SHAPIRO: Congenital hemihypertrophy and abnormalities of the cerebral vasculature. J. Neurosurg. 61 (1984) 163–168.

FURUKAWA, T. and T. SHINOHARA: Congenital hemihypertrophy: oncogenic potential of the hypertrophic side. Ann. Neurol. 10 (1981) 199–201.

FURUKAWA, T., T. SHIMIZU and Y. TOYOKURA: Facial hemihypertrophy and cerebellar hemangioblastoma. Neurology (Minneap.) 23 (1973) 1324–1328.

GESELL, A.: Hemihypertrophy and mental defect. Arch. Neurol. Psychiatry (Chic.) 6 (1921) 400–425.

GESELL, A.: Hemihypertrophy and twinning: further study of the nature of hemihypertrophy with report of a new case. Am. J. Med. Sci. 173 (1927) 542–555.

GIBBONS, J. R. and J. P. C. M. BROOKER: Phaeochromocytoma associated with multiple neurofibromatosis and aneurysm of Willis' circle. Br. J. Clin. Pract. 21 (1967) 360–362.

GILADI, N., R. E. BURKE, V. KOSTIC, S. PRZEDBORSKI, M. GORDON, A. HUNT and S. FAHN: Hemiparkinsonism-hemiatrophy syndrome: clinical and neuroradiologic features. Neurology 40 (1990) 1731–1734.

GLOVER, M. T., E. M. BRETT and D. J. ATHERTON: Hypomelanosis of Ito: spectrum of the disease. J. Pediatr. 115 (1989) 75–88.

GORDEEFF, A., J. MERCIER and N. BEDHET: The tongue in facial hemihypertrophy. Rev. Stomatol. Chir. Maxillofac. 87 (1986) 320–326.

GORLIN, R. J.: Proteus syndrome. J. Clin. Dysmorphol. 2 (1984) 8–9.

GORLIN, R. J., K. L. JUE, U. JACOBSEN and E. GOLDSCHMIDT: Oculoauriculovertebral syndrome. J. Pediatr. 63 (1963) 991–999.

GORLIN, R. J., J. J. PINDBORG and M. M. COHEN JR.: Syndromes of the Head and Neck, 2nd ed. New York, McGraw Hill Book Co. (1976) 345–348.

GOWERS, W. R.: The brain in congenital absence of one hand. Brain 1 (1879) 388–397.

GRUBER, G. B. and O. E. KUSS: Der angeborene ortliche Riesenwuchs. In: C. Schwalbe, and G. B. Gruber (Eds.), Die Morphologie der Missbildungen des Menschen und der Tiere, Vol. 3. Jena, Fischer (1937) 423–545.

HIDANO, A. and Y. ARAI: Congenital hemihypertrophy associated with cutaneous pigmento-vascular, cerebral, visceral and bone abnormalities. Ann. Dermatol. Venereol. 114 (1987) 665–669.

HINKEL, G. K. and E. RUPPRECHT: Hemihypertrophy as the main symptom of dysplasia epiphysealis hemimelica. Klin. Paediatr. 201 (1989) 58–62.

HORNSTEIN, L., K. E. BOVE and R. B. TOWBIN: Linear nevi, hemihypertrophy, connective tissue hamartomas, and unusual neoplasms in children. J. Pediatr. 110 (1987) 404–408.

JOHNSON, R. V. and W. R. KENNEDY: Progressive facial hemiatrophy (Parry-Romberg syndrome). Am. J. Ophthalmol. 67 (1969) 561–564.

KAMMER, G.: Beitrag zur Erbbiologie und Klinik der Sturge-Weberschen Erkrankung. Ztschr. Menschl. Vererb. Konstit. Lehre 33 (1955) 203–320.

KISSEL, P. and J. B. DUREUX: Cobb syndrome. Cutaneomeningospinal angiomatosis. In: P. J. Vinken and G. W. Bruyn (Eds.), Handbook of Clinical Neurology, Vol. 14, Ch. 15. Amsterdam, North Holland Publ. Co. (1972) 429–445.

KLAWANS, H. L.: Hemiparkinsonism as a late complication of hemiatrophy: A new syndrome. Neurology (NY) 31 (1981) 625–628.

KRAMER, W.: Klippel Trénaunay syndrome. In: P. J. Vinken and G. W. Bruyn (Eds.), Handbook of Clinical Neurology, Vol. 14, Ch. 12. Amsterdam, North Holland Publ. Co. (1972) 390–404.

KUMAR, P., B. V. AGRAWAL, N. P. SINGH, M. MUKERJI and T. N. EDOLIYA: Progressive right hemifacial atrophy with contralateral cerebral hemiatrophy. J. Assoc. Physicians India 19 (1971) 595–597.

KURNIT, D. M., M. W. STEELE, L. PINSKY and A. DIBBINS: Autosomal dominant transmission of a syndrome of anal, ear, renal and radial congenital malformations. J. Pediatr. 93 (1978) 270–273.

KURNIT, D. M., J. G. HALL, D. B. SHURTLEFF and M. M. COHEN JR.: An autosomal dominantly inherited syndrome of facial asymmetry, esotropia, amblyopia, and submucous cleft palate (Bencze syndrome). Clin. Genet. 16 (1979) 301–304.

KUTO, F., T. SAKAGUCHI, Y. HORASAWA, M. HAYASHI, Y. HIRASAWA and H. TOKUHIRO: Total hemiatrophy. Association with localized scleroderma, Schönlein-Henoch nephritis, and paroxysmal nocturnal hemoglobinuria. Arch. Intern. Med. 145 (1985) 731–733.

LABRUNE, B., LATROBE, M. and J. J. BENICHOU: Neonatal hyperinsulinism, congenital corporeal hemihypertrophy, tumor of the breast in adolescence. Arch. Franç. Pediatr. 45 (1988) 413–415.

LOUIS-BAR, D.: Sur l'hérédité de la maladie de Sturge-Weber-Krabbe. Confin. Neurol. (Basel) 7 (1947) 238–245.

MALAMITSI-PUCHER, A., S. KITSIOU and C. S. BARTSOCAS:

Severe Proteus syndrome in an 18-month-old boy. Am. J. Med. Genet. 27 (1987) 119–125.

MARBURG, O.: Die Hemiatrophia facialis progressiva. Wien, A. Holder (1912).

MAYATEPEK, E., T. W. KURCZYNSKI and E. S. RUPPERT: Expanding the phenotype of the Proteus syndrome: a severely affected patient with new findings. Am. J. Med. Genet. 32 (1989) 402–406.

MECKEL, J. F.: Uber die seitliche Asymmetrie im tierischen Korper, Anatomische physiologische Beobachtungen und Untersuchungen. Halle, Renger (1822) 147.

MERRITT, K. K., H. K. FABER and H. BRUCH: Progressive facial hemiatrophy. J. Pediatr. 10 (1937) 374–395.

MOESCHLER, J. and S. K. CLARREN: Familial occurrence of hemifacial microsomia with radial limb defects. Am. J. Med. Genet. 12 (1982) 371–375.

MOURA, R. A.: Progressive facial hemiatrophia. Am. J. Ophthalmol. 55 (1963) 635–639.

MULLER, J. T., A. B. PICKETT and F. D. FREDERICK: Facial hemihypertrophy associated with nevus unius lateris syndrome. Oral Surg. Oral Med. Oral Pathol. 50 (1980) 226–228.

MYLE, G.: Sémiologie de l'angiomatose encéphalotrigéminé. Acta Neurol. Belg. 50 (1950) 713–736.

NUDLEMAN, K., E. ANDERMANN, F. ANDERMANN, G. BERTRAND and THE LATE E. ROGALA: The Hemi 3 Syndrome. Hemihypertrophy, hemihypaesthesia, hemiareflexia and scoliosis. Brain 107 (1984) 533–546.

OGITA, S., TOKIWA, K. and T. TAKAHASHI: Adrenocortical carcinoma in a child with congenital hemihypertrophy. Z. Kinderchir. 44 (1989) 166–168.

OYESIKU, N. M., N. H. GAHM and R. L. GOLDMAN: Cerebral arteriovenous fistula in the Klippel-Trénaunay-Weber syndrome. Dev. Med. Child Neurol. 30 (1988) 245–248.

PARKER, D. A. and R. G. SKALKO: Congenital asymmetry: report of 10 cases with associated developmental abnormalities. Pediatrics 127 (1969) 584–589.

PARRY, C. H.: Collections from the unpublished medical writings of the late Caleb H. Parry. London, Underwoods (1825) 478.

PASCUAL-CASTROVIEJO, I., L. LÓPEZ-RODRIGUEZ, M. DE LA CRUZ MEDINA, C. SALAMANCA-MAESSO and C. ROCHE HERRERO: Hypomelanosis of Ito. Neurological complications in 34 cases. Can. J. Neurol. Sci. 15 (1988) 124–129.

PAVONE, L., P. CURATOLO, R. RIZZO, G. MICALI, G. INCORPORA, B. P. GARG, D. W. DUNN and W. B. DOBYNS: Epidermal nevus syndrome: A neurologic variant with hemimegalencephaly, gyral malformation, mental retardation, seizures, and facial hemihypertrophy. Neurology 41 (1991) 266–272.

PENFIELD, W. and J. S. M. ROBERTSON: Growth asymmetry due to lesions of the postcentral cortex. Arch. Neurol. Psychiatry 50 (1943) 405–430.

PICK, W.: Sclerodermie en Coup de Sabre mit osteoporotischer Zone im Stirnbein oder Hemiatrophia faciei? Arch. Dermatol. Syph. (Chic.) 167 (1933) 543.

POLLOCK, R. A., M. HASKELL-NEWMAN, A. R. BURDI and D. P. CONDIT: Congenital hemifacial hyperplasia: an embryologic hypothesis and case report. Cleft Palate J. 22 (1985) 173–184.

POSKANZER, D. C.: Hemiatrophies and hemihypertrophies. In: P. J. Vinker and G. W. Bruyn (Eds.), Handbook of Clinical Neurology, Vol. 22. Amsterdam, North-Holland Publ. Co. (1975) 545–554.

REES, T. D.: Facial atrophy. Clin. Plast. Surg. 3 (1976) 637–646.

RINGROSE, R. E., J. T. JABBOUR and D. K. KEELE: Hemihypertrophy. Pediatrics 36 (1965) 434–448.

ROBAIN, O., C. H. FLOQUET, N. HELDT and F. ROZENBERG: Hemimegalencephaly: a clinicopathological study of four cases. Neuropathol. Appl. Neurobiol. 14 (1988) 125–135.

ROGERS, B. O.: Progressive facial hemiatrophy: Romberg's disease. A review of 772 cases. Proc. 3rd Internat. Congr. Plastic Surgery, 1963, Washington D.C., U.S.A. Amsterdam, Excerpta Medica ICS No. 66 (1964) 681–689.

ROLINICK, B. R., C. I. KAYE, K. NAGATOSHI, W. HAUCK and A. O. MARTIN: Oculoauriculovertebral dysplasia and variants: phenotypic characteristics of 294 patients. Am. J. Med. Genet. 26 (1987) 361–375.

ROMBERG, M. H.: Klinische Ergebnisse. Berlin, A. Forstner (1846) 75–81.

ROSENBERG, R. and J. GREENBERG: Linear scleroderma as a cause for hemiatrophy. Ann. Neurol. 16 (1979) 307.

ROWE, N. H.: Hemifacial hypertrophy. Review of the literature and addition of four cases. Oral Surg. 15 (1962) 572–587.

RUNNE, U. and K. FASSHAUER: Idiopathische und sklerodermische Hemiatrophia faciei mit generalisierter Myopathie. Hautarzt 28 (1977) 10–17.

SACHS, B.: Ueber die Genese des angeborenen partiellen Riesenwuchses und ihre Beziehung zur Zwillings- und Geschwülstentstehung. Arch. Kinderheilkd. 137 (1949) 23.

SAKUTA, R., AIKAWA, H. and S. TAKASHIMA: Epidermal nevus syndrome with hemimegalencephaly: a clinical report of a case with acanthosis nigricans-like nevi on the face and neck, hemimegalencephaly, and hemihypertrophy of the body. Brain Dev. 11 (1989) 191–194.

SAYPOL, D. C. and V. P. LAUDONE: Congenital hemihypertrophy with adrenal carcinoma and medullary sponge kidney. Urology 21 (1983) 510–511.

SCHNEIDER, A. and A. FAUCONI: Hemihypertrophy and polycystic kidney disease. Helvet. Pediatr. Acta 42 (1987) 305–307.

SCHNEIDER, G.: Ein Beitrag zur Hemihypertrophie und Hemiatrophia faciei. Dtsch. Ztschr. Zahn-Mund-u. Kieferhlk. 12 (1949) 43–63, 161–168.

SCHOCH, J.: Mitteilung eines Falles von partiellen gekreuzten Riesenwuchs. Ztschr. Orthop. 87 (1956) 286–298.

SCHWARTZ, R. A., A. S. TEDESCO, L. Z. STERN, A. M. KAMINSKA, J. M. HARALDSEN and D. A. GREKIN: Myopathy associated with sclerodermal facial hemiatrophy. Arch. Neurol. 38 (1981) 592–594.

SINKOVITS, V.: Hemihypertrophia faciei. Stoma 12 (1959) 188–199.

SOLOMON, L. and N. ESTERLY: Epidermal and other congenital organoid nevi. Curr. Probl. Pediatr. 6 (1975) 1–56.

SPRAKER, M. K., C. STACK and N. B. ESTERLY: Congenital generalized fibromatosis: A review of the literature and report of a case associated with porencephaly, hemiatrophy, and cutis marmorata telangiectatica congenita. J. Am. Acad. Dermatol. 10 (1984) 365–371.

SUGAR, H. S. and T. L. BANKS: Fuchs heterochromic cyclitis with facial hemiatrophy. Am. J. Ophthalmol. 56 (1964) 627–628.

SUN, T. G., R. G. SWEE and T. C. SHIVES: Chondrosarcoma in Maffucci's syndrome. J. Bone Jt. Dis. 67 A (1985) 1214–1219.

TANNER, J. M., H. LEJARRAGA and N. CAMERON: The natural history of the Silver-Russell syndrome: a longitudinal study of thirty-nine cases. Pediatr. Res. 9 (1975) 611–623.

THOMPSON, I. M., F. R. RODRIGUEZ and C. R. SPENCE: Medullary sponge kidney and congenital hemihypertrophy. South. Med. J. 80 (1987) 1455–1456.

TOMOOKA, Y., ONITSUKA, H. and T. GOYA: Congenital hemihypertrophies with adrenal adenoma and sponge kidneys. Br. J. Radiol. 61 (1988) 851–853.

TRAPPE, B. O., E. RUPPRECHT, E. BOSSELMANN, M. SCHAEPER, K. MOHNIKE and J. GEDSCHOLD: Proteus syndrome. A contribution to the further differential diagnosis of congenital local gigantism. Helv. Paediatr. Acta 43 (1989) 473–482.

TUFFANELLI, D., W. MARMELZAT and C. DORSEY: Linear scleroderma with hemiatrophy: report of three cases associated with collagen-vascular disease. Dermatologica 132 (1966) 51–58.

TURLEAU, C., F. TAILLARD and M. DOUSSAU-DE-BAZIGNAN: Hypomelanosis of Ito (incontinentia pigmenti achromians) and mosaicism for a microdeletion of 15ql. Hum. Genet. 74 (1986) 185–187.

VAN BOGAERT, L.: Pathologie des angiomatoses. Acta Neurol. Psychiatr. Belg. 50 (1950) 525–616.

VON EULENBURG, A.: Lehrbuch der functionellen Nervenkrankheiten. Berlin, Hirschwald (1871).

WAKEFIELD, E. G. and E. A. HINES, JR.: Congenital hemihypertrophy: a report of eight cases. Am. J. Med. Sci. 185 (1933) 493–500.

WALSH, F. B.: Facial hemiatrophy. Am. J. Ophthalmol. 22 (1939) 1–10.

WARD, J. and H. H. LERNER: A review of the subject of congenital hemihypertrophy and a complete case report. J. Pediatr. 31 (1947) 403–414.

WARTENBERG, R.: Zur Klinik und Pathogenese der Hemitrophia faciei progressiva. Arch. Psychiatr. Nervenkrankh. 74 (1925) 602–630.

WARTENBERG, R.: Progressive facial hemiatrophy. Arch. Neurol. Psychiatr. (Chic.) 54 (1945) 75–96.

WEINSTEIN, R. L., B. KLIMAN, J. NEEMAN and R. B. COHEN: Deficient 17-hydroxylation in a corticosterone producing adrenal tumor from an infant with hemihypertrophy and visceromegaly. J. Clin. Endocrinol. 30 (1970) 457–468.

WOLF, S. M. and M. A. VERITY: Neurological complications of progressive facial hemiatrophy. J. Neurol. Neurosurg. Psychiatry 37 (1974) 997–1004.

ZALA, L. and M. MUMENTHALER: Cobb-Syndrom: Assoziation mit verrukosem Angiom, ipsilateraler Hypertrophie der Extremitäten und Café-au-lait-Flecken. Dermatologica 163 (1981) 417–425.

Index

Prepared by W. van Ockenburg

Entries in this index refer both to the present Volume and to Volumes 1–43 in the preceding series of the Handbook of Clinical Neurology